AF323304

LIMIT THEOREMS FOR ASSOCIATED RANDOM FIELDS AND RELATED SYSTEMS

ADVANCED SERIES ON STATISTICAL SCIENCE & APPLIED PROBABILITY

Editor: Ole E. Barndorff-Nielsen

Published

Advanced Series on
Statistical Science &
Applied Probability

Vol. 10

LIMIT THEOREMS FOR ASSOCIATED RANDOM FIELDS AND RELATED SYSTEMS

Alexander Bulinski & **Alexey Shashkin**

Moscow State University, Russia

World Scientific

NEW JERSEY · LONDON · SINGAPORE · BEIJING · SHANGHAI · HONG KONG · TAIPEI · CHENNAI

Published by

World Scientific Publishing Co. Pte. Ltd.

5 Toh Tuck Link, Singapore 596224

USA office: 27 Warren Street, Suite 401-402, Hackensack, NJ 07601

UK office: 57 Shelton Street, Covent Garden, London WC2H 9HE

Library of Congress Cataloging-in-Publication Data
Bulinski, A. V. (Aleksandr Vadimovich)
   Limit theorems for associated random fields and related systems / by Alexander Bulinski & Alexey
Shashkin.
      p. cm. -- (Advanced series on statistical science and applied probability ; v. 10)
   Includes bibliographical references and index.
   ISBN-13: 978-981-270-940-0 (hardcover : alk. paper)
   ISBN-10: 981-270-940-1 (hardcover : alk. paper)
   1. Random fields. 2. Limit theorems (Probability theory) I. Shashkin, Alexey. II. Title.
QA274.45.B85 2007
519.2'3--dc22

                                                                                   2007023102

British Library Cataloguing-in-Publication Data
A catalogue record for this book is available from the British Library.

Printed in Singapore by B & JO Enterprise

To our parents

Preface

The concepts of independence for systems of events or for collections of random variables belong to the principal ones in the Probability Theory. There are numerous beautiful results established for families of independent random variables. One can say that such achievements form a core of the Modern Probability Theory. However in XIX and especially XX centuries interesting stochastic models arose involving dependent random variables. Phenomena studied in physics, chemistry, biology, economics and technics were main sources for these models. Thus the theory of stochastic processes and random fields has emerged and evolved intensively. Of course there were also intrinsic factors in mathematics leading to deep generalizations and new constructions.

Now there are important classes of stochastic processes and random fields, for example, Gaussian, Markov, martingales, mixing ones etc. For each class the appropriate methods of investigation were developed, so we have different complementary tools to describe and analyze various stochastic models.

As far back as the 60 s of the last century the new important classes of positively (and later negatively) dependent random variables were introduced in the pioneering papers by Harris, Lehmann, Esary, Proschan, Walkup, Fortuin, Kasteleyn, Ginibre, Alam, Saxena and Joag-Dev. The interest in such models is to a large extent connected with applications in mathematical statistics, reliability theory, percolation and statistical physics. The concept of association is the basic one here. Note that any family of independent real-valued random variables is automatically associated.

Starting from the seminal paper by Newman (1980), during the last 25 years quite a number of classical limit theorems of Probability Theory, such as central limit theorem (CLT), strong law of large numbers (SLLN), law of the iterated logarithm (LIL), functional LIL, weak and strong invariance principles (IP) etc. were established for these new models and their modifications. The main advantage of dealing with sums of positively or negatively associated random variables is the simplicity of the conditions which ensure the limit theorems. Namely, one can assume for summands the existence of absolute moment of order $s \in (2,3]$ and specify the behavior of the covariance function, e.g., its rate of decrease.

The goal of this book is to introduce the reader to the vast area of recent progress in that research domain. As far as we know, this is the first self-contained exposition of the results obtained during the whole period of development till nowadays. The authors intended to provide detailed proofs rather than reproductions of original journal papers. The bibliography consists of more than 350 items. The word "related" in the title of the book means "related to associated random fields", as there are various modifications of the association concept and a number of such extensions are used below in essential way. In each Chapter we also give some references for further reading as the volume of the book does not permit to include all interesting results.

The book is supplemented with six Appendices. Here one can find an extension of the classical Hoeffding formula given by Khoshnevisan and Lewis, the general information on Markov processes and Poisson spatial process needed to build various examples of dependent random fields, the proofs of auxiliary results from linear algebra and graph theory, the proof of the Móricz inequality, the version of the Berkes–Philipp theorem on normal approximation and some results used to construct the strong approximation of random fields.

Inside any Section the theorems, lemmas, definitions, remarks and examples are numerated in succession. The numeration of formulas takes the form (Section.formula) if the cited formula belongs to current Chapter, and (Chapter.Section.formula) otherwise. Theorems and other statements are numerated analogously. For example, Theorem 1.3.2 belongs to Chapter 1 and Section 3, having the number 2 in this Section. In Appendices instead of the Section number one writes A.1—A.6. The sign $\square$ indicates the end of the proof.

Parts of the book are based on the lectures delivered by the authors at the Moscow State University. The book is addressed to wide audience of probabilists and statisticians who are interested in analysis of various stochastic models. It can be useful for researchers, graduate students and also for academic staff of the Universities.

It is a pleasure to thank our friends and colleagues with whom the authors discussed various problems concerning the theory of limit theorems, especially Professors V.I.Bogachev, A.A.Borovkov, R.C.Bradley, F.Comets, M.Csörgő, Yu.A.Davydov, M.Dekking, J.Dedecker, P.Doukhan, I.A.Ibragimov, M.Iosifescu, J.Jacod, A.Jakubowski, M.S.Keane, K.M.Khanin, A.Yu.Khrennikov, V.Yu.Korolev, V.S.Korolyuk, S.B.Kuksin, I.A.Kurkova, N.N.Leonenko, M.A.Lifshits, S.Louhichi, P.Matula, P.Mladenovic, I.S.Molchanov, S.A.Molchanov, Ya.Yu.Nikitin, O.Penrose, V.V.Petrov, V.I.Piterbarg, Yu.V.Prohorov, G.G.Roussas, G.Samorodnitsky, A.N.Shiryaev, M.Sorensen, Ch.Suquet, M.-C.Viano, V.M.Zolotarev.

The work is partially supported by the INTAS grant 03-51-5018, RFBR grants 05-01-00944-a and 07-01-00373-a.

Alexander Bulinski, Alexey Shashkin

Department of Mathematics and Mechanics of the Moscow State University

Contents

LIMIT THEOREMS FOR ASSOCIATED RANDOM FIELDS AND RELATED SYSTEMS

Chapter 1

Random Systems with Covariance Inequalities

Chapter 1 treats basic concepts and results concerning the properties of positive or negative dependence and their generalizations. Besides, it provides diverse examples to motivate the study of these models. Among remarkable results we mention here the theorem by Pitt showing that a Gaussian system is associated if and only if the covariance function is nonnegative and the theorem by Lee, Rachev and Samorodnitsky establishing the necessary and sufficient conditions for association of a stable vector in terms of its spectral measure. Special attention is paid to Markov processes, random measures (in particular, the so-called cluster and shot-noise fields). For instance, we give a proof of the theorem by Burton, Waymire and Evans claiming that any infinitely divisible random measure on a Polish space is associated. The vector-valued random fields and random variables with values in partially ordered spaces are studied as well. We consider in detail the famous FKG inequalities due to Fortuin, Kasteleyn and Ginibre and theorems by Holley and Preston which are important in statistical physics, e.g., in the study of ferromagnets and in the Percolation Theory. New examples are given for interacting particle systems. It is also demonstrated that negative association, as known, inherent to such popular laws as polynomial and hypergeometric or to distributions generated by order statistics, obtains interesting interpretations in modeling electric currents in spatial nets. To conclude the largest Chapter we discuss various extensions of the principal definitions and present the approach, developed in the last decade, based on the description of the dependence structure of a random field involving upper bounds of covariances for specified "test functions".

1 Basic definitions and simple examples

In this Section the key definitions are introduced and elementary properties of the collections of dependent random variables under consideration are studied. The relations between the concepts of independence, positive and negative dependence, martingales and demimartingales are discussed. Several illustrative examples of random systems obeying the mentioned definitions are included.

1°. Association, positive and negative association. It is instructive to compare the dependence notions discussed in this Section with the basic concept of independence for real-valued random variables X and Y defined on a probability space $(\Omega, \mathcal{F}, \mathsf{P})$. Such X and Y are called *independent* if

$$\mathsf{P}(X \in B, Y \in C) = \mathsf{P}(X \in B)\mathsf{P}(Y \in C) \tag{1.1}$$

for all sets $B, C \in \mathcal{B}(\mathbb{R})$. Recall that the Borel σ-algebra $\mathcal{B}(S)$ for a topological (in particular, metric) space S is the smallest σ-algebra containing all open sets. A mapping $f : S \to V$, for topological spaces S and V, is called *Borel* if $f^{-1}(B) := \{x \in S : f(x) \in B\} \in \mathcal{B}(S)$ for any $B \in \mathcal{B}(V)$. It is a standard exercise to check, using the step functions, that (1.1) is equivalent to the following relation

$$\mathsf{E}f(X)g(Y) = \mathsf{E}f(X)\mathsf{E}g(Y) \tag{1.2}$$

for any bounded Borel functions $f, g : \mathbb{R} \to \mathbb{R}$. As usual, E stands for expectation with respect to the probability measure P.

Instead of (1.2) one can write a formula involving *covariance*, namely,

$$cov(f(X), g(Y)) = 0$$

where $cov(W, Z) := \mathsf{E}WZ - \mathsf{E}W\mathsf{E}Z$ for real-valued random variables W and Z such that W, Z and WZ are integrable with respect to P.

It is well-known that $cov(X, Y) = 0$ if X and Y are independent and integrable, moreover, for any (in general, unbounded) Borel functions $f, g : \mathbb{R} \to \mathbb{R}$, the random variables $f(X)$ and $g(Y)$ are also independent and $cov(f(X), g(Y)) = 0$, provided that $\mathsf{E}|f(X)| < \infty$ and $\mathsf{E}|g(Y)| < \infty$. However, it is equally well-known that there exist random variables X and Y with $cov(X, Y) = 0$ which are *dependent* (that is (1.1) does not hold). For example we can take $Y = X^2$ and set $\mathsf{P}(X = -1) = \mathsf{P}(X = 0) = \mathsf{P}(X = 1) = 1/3$.

In various applications one rather considers $cov(f(X), g(Y))$ for certain classes of *test functions* f and g and assumes that the values of $cov(f(X), g(Y))$ belong to some specified subset of $\mathbb{R}$ (say, $[0, +\infty)$) instead of a single point 0. A further extension is to use random vectors X, Y, taking values in $\mathbb{R}^n$ and $\mathbb{R}^m$ respectively, and Borel functions $f : \mathbb{R}^n \to \mathbb{R}$, $g : \mathbb{R}^m \to \mathbb{R}$. Such natural ideas can be developed in several directions.

With these preliminary remarks we can pass to important definitions leading to construction of interesting stochastic models.

Let $\mathcal{M}(n)$ denote the class of real-valued bounded coordinate-wise nondecreasing Borel functions on $\mathbb{R}^n$, $n \in \mathbb{N}$. For a finite set U its cardinality is denoted by $|U|$ and sometimes by $\sharp U$. Consider a family $\mathbf{X} = \{X_t, t \in T\}$ of real-valued random variables X_t defined on $(\Omega, \mathcal{F}, \mathsf{P})$. For $I \subset T$ set $X_I = \{X_t, t \in I\}$.

The following definitions (and related ones) concerning the concepts of *positive and negative dependence* were introduced in the papers by Harris, Lehmann, Esary, Proschan, Walkup, Joag-Dev, Newman, Alam, Saxena, Burton, Dabrowski and Dehling.

Definition 1.1. ([160]) A family $\mathbf{X}$ is *associated* (abbreviated to $\mathbf{A}$), if for every finite set $I \subset T$ and any functions $f, g \in \mathcal{M}(|I|)$, one has

$$cov(f(X_I), g(X_I)) \geq 0. \tag{1.3}$$

The notation $f(X_I)$ in (1.3) means that one considers any vector $\overrightarrow{X_I}$ in $\mathbb{R}^{|I|}$ constructed by ordering a collection $\{X_t, t \in I\}$. If in a formula there are several functions $f(X_I), g(X_I), f_1(X_I)$ etc., one uses the same vector $\overrightarrow{X_I}$ as an argument of each function. When T is finite it suffices to verify (1.3) only for $I = T$ (e.g., a nondecreasing function $f = f(x_1, x_3)$ is automatically nondecreasing in x_1, x_2, x_3). It is convenient to allow the case $I = \varnothing$ setting $f(X_\varnothing) := 0$.

The next definition introduces a larger class of random variables.

Definition 1.2. ([85, 308]) A family $\mathbf{X}$ is *weakly associated*, or *positively associated* (**PA**) if

$$cov(f(X_I), g(X_J)) \geq 0 \tag{1.4}$$

for any disjoint finite sets $I, J \subset T$ and all functions $f \in \mathcal{M}(|I|)$, $g \in \mathcal{M}(|J|)$.

It will be useful to consider also the following analog of (1.4).

Definition 1.3. ([3, 219]) A family $\mathbf{X}$ is *negatively associated* (**NA**) if, for any disjoint finite sets $I, J \subset T$ and all functions $f \in \mathcal{M}(|I|)$, $g \in \mathcal{M}(|J|)$, one has

$$cov(f(X_I), g(X_J)) \leq 0. \tag{1.5}$$

Note that, in the last two definitions, if $|T| = 1$ then there is no partition of T into two nonempty sets and, by the remark above concerning the case $I = \varnothing$, we stipulate that X is **PA** and X is **NA**.

All these concepts extend the clear idea of positive (negative) correlation. Sometimes we will call a probability measure on the Borel σ-algebra $\mathcal{B}(\mathbb{R}^n)$ *associated* (resp. *positively, negatively associated*) if it is a distribution of a random vector which satisfies the corresponding Definition. In other words, a measure Q is associated ([191]) if and only if

$$\int_{\mathbb{R}^n} f(x)g(x)Q(dx) \geq \int_{\mathbb{R}^n} f(x)Q(dx) \int_{\mathbb{R}^n} g(x)Q(dx), \tag{1.6}$$

for all $f, g \in \mathcal{M}(n)$. In some situations it will be convenient to say that random variables $X_t, t \in T$, are associated (or positively, negatively associated) instead of saying the same about the system $\{X_t, t \in T\}$. In what follows, for a random vector $X = (X_{t_1}, \ldots, X_{t_n})$ we also write $X \in \mathbf{A}$ (resp. $X \in \mathbf{PA}, X \in \mathbf{NA}$) instead of "$X_{t_1}, \ldots, X_{t_n}$ are associated (positively, negatively associated)".

In subsection 5° we consider the functions in infinite number of random variables and in Section 2 we shall provide some generalizations concerning more general partially ordered spaces than $\mathbb{R}^n$. Further we employ also the simple

Remark 1.4. (a) If $Law\{X_t, t \in T\} = Law\{Y_t, t \in T\}$, i.e. the finite-dimensional distributions of these systems are the same, and $\{X_t, t \in T\}$ is associated (resp. **PA**, **NA**) then the family $\{Y_t, t \in T\}$ possesses the same property.

(b) If $\{X_t, t \in T\}$ is associated (**PA** or **NA**) and $L \subset T$, then $\{X_t, t \in L\}$ is associated (resp. **PA** or **NA**).

(c) A random vector $X \in \mathbf{A}$ (**PA** or **NA**) if and only if any of its permutations has the same property.

(d) Using dominated convergence theorem one can replace the boundedness assumption in Definitions 1.1—1.3 with the requirement that all expectations forming the covariance exist.

(e) In Definitions 1.1—1.3 one can use the coordinate-wise nonincreasing functions, since $cov(-f(X), -g(Y)) = cov(f(X), g(Y))$.

2°. Criteria for positive and negative association. Before the discussion, we give a result providing sufficient conditions for validity of Definitions 1.1—1.3. Namely, it is possible to check the conditions of association (positive/negative association) for more narrow classes of functions belonging to $\mathcal{M} = \cup_{n=1}^{\infty} \mathcal{M}(n)$.

As usual, for a set $A \subset S$ the *indicator* of A is a function $\mathbb{I}_A$ such that $\mathbb{I}_A(x) = 1$ for $x \in A$ and 0 otherwise. Sometimes we will write $\mathbb{I}\{A\}$.

A *binary function* is an indicator of some measurable set. A Borel set B in $\mathbb{R}^n$ is called *increasing* if $\mathbb{I}_B \in \mathcal{M}(n)$.

Theorem 1.5. ([160, 330]) *A random vector* $X = (X_1, \ldots, X_n)$ *is associated (resp. positively, negatively associated) if condition* (1.3) *(resp.* (1.4)*,* (1.5)*) holds whenever both f and g belong to $\mathcal{M}$ and to either of the following classes of functions:*
   (a) *binary;*
   (b) *continuous;*
   (c) *having bounded (partial) derivatives of any order.*

Proof. We only deal with association, omitting similar proofs for **PA** and **NA**. One infers in cases (a) – (c) that $cov(f(X), g(X)) \geq 0$ for any $f, g \in \mathcal{M}(n)$.

(a) If Y and Z are square-integrable random variables then the Hoeffding formula[1] ([200]) states that

$$cov(Y, Z) = \int_{\mathbb{R}^2} (\mathsf{P}(Y \geq y, Z \geq z) - \mathsf{P}(Y \geq y)\mathsf{P}(Z \geq z))dydz.$$

Therefore,

$$cov(f(X), g(X)) = \int_{\mathbb{R}^2} cov(\mathbb{I}\{f(X) \geq y\}, \mathbb{I}\{g(X) \geq z\})dydz. \qquad (1.7)$$

The integrand is nonnegative by assumption, since the functions $\mathbb{I}\{f(x) \geq y\}$, $\mathbb{I}\{g(x) \geq z\}$ are binary and nondecreasing in $x \in \mathbb{R}^n$ for any fixed $y, z \in \mathbb{R}$.

(b) Due to the previous fact, it is enough to prove that $cov(f_1(X), f_2(X)) \geq 0$ for $f_i = \mathbb{I}_{B_i}$, $i = 1, 2$, where B_1, B_2 are arbitrary increasing Borel sets in $\mathbb{R}^n$. Suppose

---

[1]See Appendix A.1.

at first that B_1 and B_2 are closed sets. For $x \in \mathbb{R}^n$, $i = 1, 2$ and $k \in \mathbb{N}$ define the functions

$$f_i^k(x) = \begin{cases} 0, & dist(x, B_i) \geq k^{-1}, \\ 1 - k\, dist(x, B_i), & \text{otherwise.} \end{cases}$$

Here *dist* is the distance corresponding to the norm $|\cdot|$ in $\mathbb{R}^n$ where $|x| = \max_{1 \leq r \leq n} |x_r|$, i.e. $dist(x, B) := \inf\{|x - y| : y \in B\}$ for $x \in \mathbb{R}^n$ and $B \subset \mathbb{R}^n$. Each of the functions f_i^k is continuous. One can verify that if $x \leq y\ (x, y \in \mathbb{R}^n)$ then $dist(x, B_i) \geq dist(y, B_i)$, $i = 1, 2$; as usual

$$x = (x_1, \ldots, x_n) \leq y = (y_1, \ldots, y_n) \iff x_i \leq y_i, \ i = 1, \ldots, n. \tag{1.8}$$

Thus all the functions $f_i^k \in \mathcal{M}(n)$ and $f_i^k(x) \to f_i(x)$ as $k \to \infty$, $x \in \mathbb{R}^n$, $i = 1, 2$. Consequently, $cov(f_1(X), f_2(X)) \geq 0$ by the dominated convergence.

Now write Q for the distribution of X in $\mathbb{R}^n$ and let B_1, B_2 be arbitrary increasing sets. Take some $\varepsilon > 0$ and compact sets $C_i \subset B_i$ such that $Q(B_i \setminus C_i) < \varepsilon$, $i = 1, 2$ (this is possible since probability measures are regular, see, e.g., [39, Ch.1, §1, Th. 1.4]). Let $F_i = C_i + \mathbb{R}_+^n = \{x + t : x \in C_i, t = (t_1, \ldots, t_n) \in \mathbb{R}_+^n\}$, $i = 1, 2$.

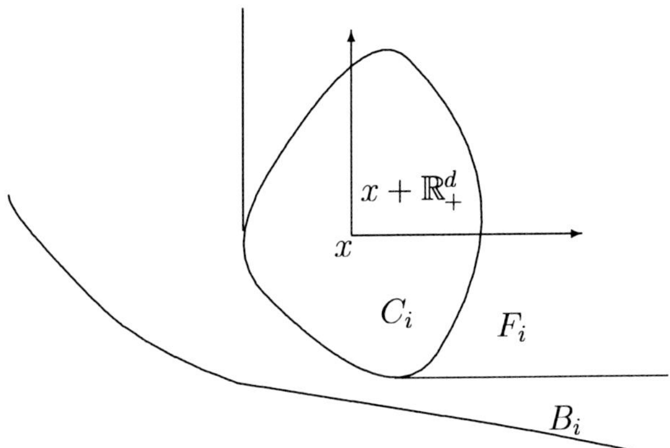

Fig. 1.1

It is easily seen that $C_i \subset F_i \subset B_i$ and F_i is an increasing closed set. Thus, by the already proved part of assertion,

$$cov(\mathbb{I}_{B_1}(X), \mathbb{I}_{B_2}(X)) = Q(B_1 B_2) - Q(B_1)Q(B_2) = Q(F_1 F_2) - Q(F_1)Q(F_2)$$
$$+ Q((B_1 \setminus F_1)F_2) + Q(F_1(B_2 \setminus F_2)) + Q((B_1 \setminus F_1)(B_2 \setminus F_2)) - Q(B_1 \setminus F_1)Q(F_2)$$
$$- Q(F_1)Q(B_2 \setminus F_2) - Q(B_1 \setminus F_1)Q(B_2 \setminus F_2) \geq cov(\mathbb{I}_{F_1}(X), \mathbb{I}_{F_2}(X)) - 3\varepsilon.$$

By the previous argument the last covariance is nonnegative and, since positive ε is arbitrary, the desired result follows.

(c) In view of (b) and by the dominated convergence theorem it suffices to show that, for a continuous function $f \in \mathcal{M}(n)$, there exists a sequence of bounded nondecreasing functions f_k $(k \in \mathbb{N})$ having bounded derivatives of all orders such that $f_k(x) \to f(x)$ as $k \to \infty$ for all $x \in \mathbb{R}^n$. If $\chi_k(x) = (k/2\pi)^{n/2} \exp(-\|x\|^2 k/2)$ where $k \in \mathbb{N}$ and $\|\cdot\|$ stands for Euclidean norm in $\mathbb{R}^n$, then the functions

$$f_k(x) := (f * \chi_k)(x) = \int_{\mathbb{R}^n} f(x - y)\chi_k(y)dy, \quad x \in \mathbb{R}^n,$$

satisfy all the required conditions. Hence $f_k(X(\omega)) \to f(X(\omega))$ for all $\omega \in \Omega$ as $k \to \infty$. $\square$

Remark 1.6. If $f : \mathbb{R} \to \mathbb{R}$ is a nondecreasing function then, for any $c \in \mathbb{R}$, the set $\{x : f(x) \leq c\}$ is an interval of the form $(-\infty, a)$ or $(-\infty, a]$. Thus f is automatically a Borel function. However, it is not true that every coordinate-wise nondecreasing function on $\mathbb{R}^n$ is Borel when $n > 1$. For example, let $B \subset \{(x, y) \in \mathbb{R}^2 : x = -y\}$ be a non-Borel subset of $\mathbb{R}^2$. Consider the function $f : \mathbb{R}^2 \to \mathbb{R}$ such that $f(x, y) = 0$ if $x + y < 0$, $f(x, y) = 1$ if $x = -y$, $(x, y) \in B$, and $f(x, y) = 2$ otherwise. Then $\{(x, y) : f(x, y) \leq 1\} \notin \mathcal{B}(\mathbb{R}^2)$.

3°. Elementary examples. Recall the following basic

Definition 1.7. Real-valued random variables $X_t, t \in T$, are called *(mutually) in-dependent* if, for any finite $I \subset T$ and all $B_t \in \mathcal{B}(\mathbb{R})$, $t \in I$, one has

$$\mathsf{P}\Big(\bigcap_{t \in I}\{X_t \in B_t\}\Big) = \prod_{t \in I}\mathsf{P}(X_t \in B_t).$$

The last relation is equivalent to the independence of σ-algebras $\sigma\{X_t\}, t \in T$, generated by the random variables under consideration.

Let $(S, \mathcal{B})$ be a *measurable space* S endowed with a σ-algebra $\mathcal{B}$. A *random element Z* defined on $(\Omega, \mathcal{F}, \mathsf{P})$ with values in $(S, \mathcal{B})$ is a mapping $Z : \Omega \to S$ which is $\mathcal{F}|\mathcal{B}$-measurable that is $Z^{-1}(B) \in \mathcal{F}$ for any $B \in \mathcal{B}$. One writes $Z \in \mathcal{F}|\mathcal{B}$.

If $h : S \to \mathbb{R}$ and $h \in \mathcal{B}|\mathcal{B}(\mathbb{R})$ then

$$\mathsf{E}h(Z) = \int_{\Omega} h(Z)dP = \int_{S} h(z)P_Z(dz) \tag{1.9}$$

where P_Z is the law of Z on $(S, \mathcal{B})$, i.e. $P_Z(B) := \mathsf{P}(Z^{-1}(B))$, $B \in \mathcal{B}$. More exactly, both integrals (over Ω and S) in (1.9) exist simultaneously and then coincide.

In the next theorem we gather elementary but very useful results providing numerous simple examples of positive and negative association.

Theorem 1.8. ([160]) *The following statements are valid.*

    (a) *A family consisting of a single random variable is associated.*[2]

    (b) *A union of mutually independent collections of associated (resp. **PA, NA**) random variables is associated (resp. **PA, NA**).*

    (c) *A family consisting of independent random variables is **A** and **NA**.*

---

[2]The property (a) is known as one of the Chebyshev inequalities (see, e.g., [189]).

(d) *A family of functions belonging to $\mathcal{M}$, which are taken from finite subsets of associated random variables, is* **A**. *If the arguments belong to finite disjoint subsets of* **PA** (**NA**) *random variables then these functions are* **PA** (**NA**).

(e) *If a sequence $(X_k)_{k\in\mathbb{N}}$ of associated (resp.* **PA**, **NA**) *n-dimensional random vectors converges in law to a random vector X, then $X \in$ **A** (resp.* **PA**, **NA**).

Proof. (a) Let X be an arbitrary random variable and Y its independent copy. Then, for nondecreasing bounded $f, g : \mathbb{R} \to \mathbb{R}$,

$$cov(f(X), g(X)) = \mathsf{E}f(X)g(X) - \mathsf{E}f(X)g(Y) = \frac{1}{2}\mathsf{E}(f(X) - f(Y))(g(X) - g(Y)) \geq 0$$

since the expression under the expectation is nonnegative.

(b) Here we need only consider finite sets of random variables. Suppose that $X^1, \ldots, X^m$ are independent random vectors, each of them associated (number of components may differ). If $m = 1$, the assertion is true, so, inductively, suppose that it holds for $X^1, \ldots, X^{m-1}$. Let r be the dimension of X^m and write Q for the distribution of X^m in $\mathbb{R}^r$. With the notation $X = (X^1, \ldots, X^{m-1})$, one has for $f, g \in \mathcal{M}$ by independence and the Fubini theorem (using also (1.9))

$$cov(f(X), g(X)) = \int_{\mathbb{R}^r} (\mathsf{E}f(X, x)g(X, x) - \mathsf{E}f(X, x)\mathsf{E}g(X, x))Q(dx)$$

$$+ \int_{\mathbb{R}^r} \mathsf{E}f(X, x)\mathsf{E}g(X, x)Q(dx) - \int_{\mathbb{R}^r} \mathsf{E}f(X, x)Q(dx) \int_{\mathbb{R}^r} \mathsf{E}g(X, x)Q(dx).$$

The integrand in the first term of the right-hand side is nonnegative by induction hypothesis. Obviously $\mathsf{E}f(X, x)$ and $\mathsf{E}g(X, x)$, $x \in \mathbb{R}^r$, belong to $\mathcal{M}(r)$. Therefore the difference between the second and third terms is nonnegative since X^m is an associated random vector. The proofs in **PA** and **NA** cases are analogous.

(c) Follows from (a) and (b) already proved.

(d) The functional class $\mathcal{M}$ is closed under compositions.

(e) Follows from the definition of convergence in law and Theorem 1.5, (b). $\square$

Corollary 1.9. *If $X_1, \ldots, X_n$ is a family of associated (in particular, independent) random variables, then the following random systems are associated: a) weighted partial sums $c_1 X_1, c_1 X_1 + c_2 X_2, \ldots, c_1 X_1 + \ldots + c_n X_n$ where $c_1, c_2, \ldots$ are nonnegative numbers; b) partial maxima $\{\max_{j \leq i}(X_1 + \ldots + X_j), i = 1, \ldots, n\}$ or weighted partial maxima with nonnegative weights; c) order statistics[3] $X_{(1)}, \ldots, X_{(n)}$.*

Proof. Application of Theorem 1.8, (d), implies the result. $\square$

Corollary 1.10. *Let $\varphi_k : \mathbb{R} \to \mathbb{R}_+$ be bounded nondecreasing functions where $k = 1, \ldots, n$. If $X_1, \ldots, X_n$ are positively associated random variables then*

$$\mathsf{E} \prod_{k=1}^{n} \varphi_k(X_k) \geq \prod_{k=1}^{n} \mathsf{E}\varphi_k(X_k), \tag{1.10}$$

whereas if $X_1, \ldots, X_n$ are negatively associated, the counterpart of (1.10) holds with the opposite sign of the inequality.

---

[3]For each $\omega \in \Omega$ one rearranges $X_1(\omega), \ldots, X_n(\omega)$ to get $X_{(1)}(\omega) \leq \ldots \leq X_{(n)}(\omega)$.

Proof. The result follows from Theorem 1.8, (d), by induction on n. $\square$

Corollary 1.11. *A family* $\mathbf{X} = \{X_t, t \in T\}$ *consists of independent random variables if and only if simultaneously* $\mathbf{X} \in \mathbf{PA}$ *and* $\mathbf{X} \in \mathbf{NA}$.

Proof. Let $I = \{t_1, \ldots, t_n\} \subset T$ $(n \in \mathbb{N})$. If the components of a vector $X = (X_{t_1}, \ldots, X_{t_n})$ are independent then Theorem 1.8, (b), implies that $X \in \mathbf{PA}$ and $X \in \mathbf{NA}$. The converse statement is an immediate corollary of the independence condition in terms of the distribution functions

$$F_{X_{t_1}, \ldots, X_{t_n}}(x_1, \ldots, x_n) := \mathsf{P}(X_{t_1} \leq x_1, \ldots, X_{t_n} \leq x_n) = \mathsf{E} \prod_{k=1}^{n} \mathbb{I}_{(-\infty, x_k]}(X_{t_k})$$

where $x_1, \ldots, x_n \in \mathbb{R}$. Namely (see, e.g., [383, Ch. II § 2.5]), $X_{t_1}, \ldots, X_{t_n}$ are independent if and only if, for all $x_1, \ldots, x_n \in \mathbb{R}$,

$$F_{X_{t_1}, \ldots, X_{t_n}}(x_1, \ldots, x_n) = \prod_{k=1}^{n} F_{X_{t_k}}(x_k).$$

Now one can use Corollary 1.10 and Remark 1.4, (e). $\square$

Let us provide a simple application of Theorem 1.8. Recall that, for a bounded function $f : [0, 1] \to \mathbb{R}$, the *Bernstein polynomial* of degree n is defined by

$$(B_n f)(x) = \sum_{k=0}^{n} \binom{n}{k} f\,(k/n)\, x^k (1 - x)^{n-k}, \quad x \in [0, 1].$$

As usual, $\binom{n}{k} = n!/(k!(n - k)!)$, $0! := 1$ and $0^0 := 1$.

Example 1.12. ([367]) If $f, g : [0, 1] \to \mathbb{R}$ are bounded nondecreasing functions, then one has everywhere on [0,1]

$$B_n(fg) \geq (B_n f)(B_n g). \tag{1.11}$$

To prove this claim consider a binomial random variable $S_{n,x}$ with parameters $n \in \mathbb{N}$ and $x \in [0, 1]$, that is

$$\mathsf{P}(S_{n,x} = k) = \binom{n}{k} x^k (1 - x)^{n-k}, \quad k = 0, \ldots, n.$$

Then $(B_n f)(x) = \mathsf{E} f(S_{n,x}/n)$ for all $n \in \mathbb{N}$ and $x \in [0, 1]$. Therefore (1.11) is a corollary of Theorem 1.8, (a).

Various notions of dependence introduced above have a natural description in the simplest case of random variables taking only two different values. Without loss of generality we can prove this considering the *binary random variables* X and Y, i.e. with values 0 and 1 (one can then use the transformations $aX + b$, $cY + d$ where $a, c > 0$ and $b, d \in \mathbb{R}$).

Theorem 1.13. ([160]) *Binary random variables* X *and* Y *are associated (resp.* **PA**, **NA**, *independent) if and only if* $cov(X, Y) \geq 0$ *(resp.* $cov(X, Y) \geq 0$, $cov(X, Y) \leq 0$, $cov(X, Y) = 0$*).*

Proof. In view of Theorem 1.5, (a), we establish that $cov(f(X,Y), g(X,Y))$ is nonnegative (resp. nonpositive, zero) for all binary nondecreasing functions f, g. But the only non-constant binary nondecreasing functions from $\{0,1\}^2$ to $\mathbb{R}$ are $X, Y, XY, X + Y - XY$. The direct test of all possible combinations leads to the assertion of the Theorem. $\square$

Theorem 1.13 cannot be carried over to higher dimensions even for binary vectors, as shows the next

Example 1.14. Let $X = (X_1, X_2, X_3)$ be a 3-dimensional random vector with binary components. Consider the functions $f(X_1, X_2) = X_1^+ X_2^+$ and $g(X_3) = X_3$. Set $p_{ijk} := \mathsf{P}(X_1 = i, X_2 = j, X_3 = k)$. If $p_{100} = p_{010} = p_{001} = 1/12$ and $p_{110} = p_{101} = p_{011} = 1/4$, then the components of X are positively correlated but $cov(f(X_1, X_2), g(X_3)) < 0$, hence X is not **PA**. On the other hand, taking $p_{100} = p_{010} = p_{001} = 3/10$, $p_{111} = 1/10$, one obtains a vector with negatively correlated components whereas $cov(f(X_1, X_2), g(X_3)) > 0$, thus $X \notin$ **NA**.

Positive and negative association are stronger conditions than positive and negative correlatedness. To illustrate this consider

Example 1.15. Let X, Y be random variables such that
$$\mathsf{P}(X = -1, Y = 1) = \mathsf{P}(X = -1, Y = -1) = 1/4, \quad \mathsf{P}(X = 1, Y = 0) = 1/2. \quad (1.12)$$
Then X and Y are dependent, $cov(X, Y) = 0$. For any $f, g : \mathbb{R} \to \mathbb{R}$ one has
$$cov(f(X), g(Y)) = f(-1)g(-1)/4 + f(1)g(0)/2 + f(-1)g(1)/4$$
$$-(f(1) + f(-1))(g(-1) + 2g(0) + g(1))/8.$$
Let $f(-1) = 0$, $f(1) = 1$, $g(-1) = g(0) = 0$, $g(1) = 1$. Obviously we can take appropriate functions f, g from $\mathcal{M}(1)$. Then $cov(f(X), g(Y)) = -1/8$. If $f(-1) = 0$, $f(1) = 1$, $g(-1) = 0$, $g(0) = 1$, $g(1) = 1$ then $cov(f(X), g(Y)) = 1/8$. So (X, Y) is neither **PA** nor **NA**.

While positive association is very similar to association, they are also nonequivalent, as the following example of Esary, Proschan and Walkup demonstrates.

Example 1.16. ([160]) Let (X, Y) be a random vector such that, for $i, j = 0, 1, 2$, one has $\mathsf{P}(X = i, Y = j) = p_{ij}$ and p_{ij} are given by the matrix
$$\begin{pmatrix} p_{00} & p_{01} & p_{02} \\ p_{10} & p_{11} & p_{12} \\ p_{20} & p_{21} & p_{22} \end{pmatrix} = \begin{pmatrix} 15/64 & 0 & 1/8 \\ 0 & 9/32 & 0 \\ 1/8 & 0 & 15/64 \end{pmatrix}.$$
Then (X, Y) is **PA** as, according to Theorem 1.5, (a), it is sufficient to verify that $cov(\mathbb{I}\{X \geq x\}, \mathbb{I}\{Y \geq y\}) \geq 0$ for all $x, y \in \mathbb{R}$. Checking this inequality is reduced to the case $x, y \in \{1, 2\}$. Furthermore, take $f(X, Y) = \mathbb{I}\{X \vee Y > 1\}$ and $g(X, Y) = \mathbb{I}\{X \wedge Y > 0\}$, $(x, y) \in \mathbb{R}^2$. Clearly $f, g \in \mathcal{M}(2)$. One has $\mathsf{E}f(X, Y) = 31/64$, $\mathsf{E}g(X, Y) = 33/64$ and $\mathsf{E}f(X, Y)g(X, Y) = 15/64$, hence $cov(f(X, Y), g(X, Y)) < 0$. Thus X and Y are not associated.

4°. Demimartingales. Definitions 1.1—1.3 can be modified, for example, by imposing restrictions on the sets I and J and on the functions f and g.

Definition 1.17. ([310]) The integrable random variables $(S_n)_{n \in T}$ where $T = \mathbb{N}$ or $T = \{1, \ldots, N\} \subset \mathbb{N}$ form a *demimartingale* if, for any $n, n + 1 \in T$ and all $f \in \mathcal{M}(n)$, one has

$$\mathsf{E}\left((S_{n+1} - S_n)f(S_1, \ldots, S_n)\right) \geq 0. \tag{1.13}$$

They form a *demisubmartingale* if the same inequality holds under additional assumption that f is nonnegative.

For example, if $X_1, \ldots, X_n, \ldots$ is a sequence of **PA** centered random variables, then its partial sums $S_n = \sum_{i=1}^{n} X_i, n \in \mathbb{N}$, form a demimartingale. Recall that a sequence $(S_n)_{n \in \mathbb{N}}$ is a *martingale* (with respect to the natural filtration $\mathcal{F}_n = \sigma\{S_1, \ldots, S_n\}$; as usual, $\sigma\{Z_t, t \in T\}$ stands for the smallest σ-algebra with respect to which all the random variables $Z_t, t \in T$, are measurable) if, for all n, one has $\mathsf{E}|S_n| < \infty$ and $\mathsf{E}(S_{n+1}|\mathcal{F}_n) = S_n$ (P-a.s.). One can see that the last relation for integrable random variables S_n is equivalent to the condition that the expectation in (1.13) is zero for any bounded Borel function f. Thus any martingale is automatically a demimartingale. The converse statement is false.

Example 1.18. Let X and Y be random variables such that (1.12) holds. Then (S_1, S_2) with $S_1 = Y$ and $S_2 = X$ is a demimartingale as for any $f \in \mathcal{M}(1)$

$$\mathsf{E}(X - Y)f(Y) = (f(0) - f(-1))/2 \geq 0.$$

However (S_1, S_2) is not a martingale because

$$\mathsf{E}(X|Y = 1) = \frac{\mathsf{E}X\mathbb{I}\{Y = 1\}}{\mathsf{P}(Y = 1)} = -1 \neq 1.$$

Note in passing that binary random variables X and Y can not be used for the above counterexample. Indeed, if $p_{ij} = \mathsf{P}(X = i, X = j)$ where $i, j \in \{0, 1\}$, then (X, Y) is a martingale if and only if $p_{01} = p_{10} = 0$. The same relation provides necessary and sufficient conditions for (X, Y) being a demimartingale.

Let us show that, in general, neither association implies the demimartingale property nor vice versa.

Example 1.19. In view of Theorem 1.13 the condition $p_{01} = p_{10} = 0$ implies that $(X, Y) \in \mathbf{A}$, as in this case $cov(X, Y) = p_{11}(1 - p_{11}) \geq 0$. Now if $p_{11} = p_{00} = p \in (1/4, 1/2)$ and $p_{01} = p_{10} = 1/2 - p$, then $cov(X, Y) = p - 1/4 > 0$. Hence X and Y are associated but do not form a demimartingale as $p_{01} = p_{10} > 0$. Also it is easy to construct a demimartingale which does not possess the association property. Let X take values $-1, 1$ and Y take values $-a, 0, a$, where $a > 0$. Assume that

$$\mathsf{P}(X = 1, Y = -a) = \mathsf{P}(X = -1, Y = a) = p,$$

$$\mathsf{P}(X = 1, Y = 0) = \mathsf{P}(X = -1, Y = 0) = 1/2 - p, \quad 0 < p < 1/2.$$

Set $S_1 = X$, $S_2 = X + Y$. For $f \in \mathcal{M}(1)$ one has

$$\mathsf{E} Y f(X) = ap(f(1) - f(-1)) \geq 0.$$

Thus (S_1, S_2) is a demimartingale. At the same time $cov(S_1, S_2) = \mathsf{Var} X + \mathsf{E} XY = 1 - 2ap < 0$ if $a > 1/(2p)$. Therefore $(S_1, S_2) \notin \mathbf{A}$.

Note also that some elements of martingale technique can be applied to demimartingales. We will give an example of such approach (Doob's inequality for demimartingales) in the next Chapter.

5°. Random functions in infinite number of random variables. Now we consider a simple case (to employ in the sequel for percolation model) of associated random functions in infinite number of random variables. For $x = (x_k)_{k \in \mathbb{N}} \in \mathbb{R}^\infty$ and $y = (y_k)_{k \in \mathbb{N}} \in \mathbb{R}^\infty$ one writes $x \leq y$ if and only if $x_k \leq y_k$ for all $k \in \mathbb{N}$. By $\mathcal{B}(\mathbb{R}^\infty)$ we denote the cylindric σ-algebra in $\mathbb{R}^\infty$, i.e. the smallest σ-algebra of subsets of $\mathbb{R}^\infty$ containing all sets of the form $\{x_{i_1} \leq a_1, \ldots, x_{i_k} \leq a_k\}$ where $(a_1, \ldots, a_k) \in \mathbb{R}^k$, $\{i_1, \ldots, i_k\} \subset \mathbb{N}$, $k \in \mathbb{N}$.

Definition 1.20. A function $f : \mathbb{R}^\infty \to \mathbb{R}$ is called *increasing*[4] if $x \leq y$ $(x, y \in \mathbb{R}^\infty)$ implies $f(x) \leq f(y)$. A set $B \in \mathcal{B}(\mathbb{R}^\infty)$ is called *increasing* if the function $\mathbb{I}_B$ is increasing.

Theorem 1.21. *Let T be a countable set and $\mathbf{X} = \{X_t, t \in T\}$ be a family of independent random variables. Suppose that $f_1, \ldots, f_m$ are $\mathcal{B}(\mathbb{R}^\infty)$-measurable, increasing real-valued functions on $\mathbb{R}^\infty$. Then[5] $f_1(\mathbf{X}), \ldots, f_m(\mathbf{X})$ are associated.*

Proof. Let $T = \{t_1, t_2, \ldots\}$. For any functions $F, G \in \mathcal{M}(m)$ put

$$Y = F(f_1(\mathbf{X}), \ldots, f_m(\mathbf{X})), \quad Z = G(f_1(\mathbf{X}), \ldots, f_m(\mathbf{X})).$$

We will show that $cov(Y, Z) \geq 0$. Introduce $Y_n = \mathsf{E}(Y | X_{t_1}, \ldots, X_{t_n})$, $n \in \mathbb{N}$. The Lévy theorem (see, e.g., [383, Ch. VII, § 4.3]) implies that $Y_n \to \mathsf{E}(Y | X_{t_1}, X_{t_2}, \ldots)$ a.s. and in $L^1(\Omega, \mathcal{F}, \mathsf{P})$ as $n \to \infty$. Obviously $\mathsf{E}(Y | X_{t_1}, X_{t_2}, \ldots) = Y$ a.s. The same conclusion can be drawn for $Z_n = \mathsf{E}(Z | X_{t_1}, \ldots, X_{t_n})$. Thus $\mathsf{E} Y_n \to \mathsf{E} Y$, $\mathsf{E} Z_n \to \mathsf{E} Z$, $\mathsf{E} Y_n Z_n \to \mathsf{E} Y Z$ (Y, Z, Y_n, Z_n are bounded random variables, $n \in \mathbb{N}$) and

$$cov(Y_n, Z_n) \to cov(Y, Z) \text{ as } n \to \infty.$$

We see that it is sufficient to verify that $cov(Y_n, Z_n) \geq 0$ for all $n \in \mathbb{N}$. One may apply the following simple but useful statement (see, e.g., [81]).

Lemma 1.22. *Let $(V, \mathcal{A})$ and $(S, \mathcal{B})$ be some measurable spaces. Assume that ξ and η are independent random elements defined on a probability space $(\Omega, \mathcal{F}, \mathsf{P})$ and taking values in V and S respectively. If $H : V \times S \to \mathbb{R}$ is bounded and $\mathcal{A} \otimes \mathcal{B} | \mathcal{B}(\mathbb{R})$-measurable, one has, for each $v \in V$,*

$$\mathsf{E}(H(\xi, \eta) | \xi = v) = \mathsf{E} H(v, \eta) \text{a.s.} \tag{1.14}$$

---

[4] For $f \in \mathcal{M}(n)$ one rather says that f is coordinate-wise nondecreasing.

[5] We use here the same convention as indicated after Definitions 1.1—1.3.

Continuing the proof of Theorem, we set $v_n = (x_1, \ldots, x_n) \in \mathbb{R}^n$, $h_n(v_n) = \mathsf{E}(Y|X_{t_1} = x_1, \ldots, X_{t_n} = x_n)$ and $W_n = (X_{t_{n+1}}, X_{t_{n+2}}, \ldots)$ where $n \in \mathbb{N}$. Then the independence of random variables X_{t_k} ($k \in \mathbb{N}$) and (1.14) combined imply that

$$h_n(v_n) = \mathsf{E}F(f_1(v_n, W_n), \ldots, f_m(v_n, W_n)). \tag{1.15}$$

Clearly h_n appearing in (1.15) is a bounded and coordinate-wise nondecreasing function. We conclude that $Y_n = h_n(X_{t_1}, \ldots, X_{t_n})$. In the same way $Z_n = g_n(X_{t_1}, \ldots, X_{t_n})$ where g_n also belongs to $\mathcal{M}(n)$. From Theorem 1.8, (d), we deduce that Y_n and Z_n are associated. They are also square-integrable, thus $cov(Y_n, Z_n) \geq 0$, $n \in \mathbb{N}$. $\square$

6°. Conditional distributions and permutations. Next we will be concerned with two theorems providing a large amount of examples of negative association. They exploit an intuitively clear idea that negative association should arise in statistical models which resemble "sampling without replacement". In other words, negative dependence manifests when random variables prevent each other from growing, that is increases in some variables make the other decrease.

Theorem 1.23. ([219]) *Suppose that $X_1, \ldots, X_n$ are independent random variables such that, for any $I \subset \{1, \ldots, n\}$ and arbitrary $f \in \mathcal{M}(|I|)$, the function*

$$\mathsf{E}\left(f(X_i, i \in I) \Big| \sum_{i \in I} X_i = t\right) \tag{1.16}$$

*is nondecreasing in t. Then the conditional distribution of $(X_1, \ldots, X_n)$ given $\sum_{i=1}^n X_i$ is **NA** almost surely. In more detail, there exists a regular [6] conditional distribution $Q(\cdot, \omega)$ of $(X_1, \ldots, X_n)$ given $\sum_{i=1}^n X_i$ which is **NA** for almost all ω.*

Proof. Consider nonempty disjoint subsets $I, J \subset \{1, \ldots, n\}$, $I \cup J = \{1, \ldots, n\}$, and functions $f \in \mathcal{M}(|I|), g \in \mathcal{M}(|J|)$. Let $S_1 = \sum_{i \in I} X_i$, $S_2 = \sum_{j \in J} X_j$, $S = S_1 + S_2$. Then, by the telescopic property[7] of conditional expectation,

$$cov(f(X_I), g(X_J)|S) = \mathsf{E}(cov(f(X_I), g(X_J)|S_1, S_2)|S)$$
$$+ cov(\mathsf{E}(f(X_I)|S_1, S_2), \mathsf{E}(g(X_J)|S_1, S_2)|S) \tag{1.17}$$

where

$$cov(W, Y|Z) := \mathsf{E}(WY|Z) - \mathsf{E}(W|Z)\mathsf{E}(Y|Z), \tag{1.18}$$

provided that $\mathsf{E}WY$, $\mathsf{E}W$ and $\mathsf{E}Y$ exist. We also need a simple

Lemma 1.24. *Let ξ_1 and ξ_2 be independent random vectors with values in $\mathbb{R}^{k_1}$ and $\mathbb{R}^{k_2}$ respectively. Then for arbitrary bounded Borel functions $F_i : \mathbb{R}^{k_i} \to \mathbb{R}$ and any Borel functions $h_i : \mathbb{R}^{k_i} \to \mathbb{R}^{m_i}$ ($m_i, k_i \in \mathbb{N}$, $i = 1, 2$) one has*

$$\mathsf{E}(F_1(\xi_1)F_2(\xi_2)|h_1(\xi_1), h_2(\xi_2)) = \mathsf{E}(F_1(\xi_1)|h_1(\xi_1))\mathsf{E}(F_2(\xi_2)|h_2(\xi_2)). \tag{1.19}$$

In particular,

$$\mathsf{E}(F_i(\xi_i)|h_1(\xi_1), h_2(\xi_2)) = \mathsf{E}(F_i(\xi_i)|h_i(\xi_i)), \; i = 1, 2. \tag{1.20}$$

---

[6]See, e.g., [383, Ch. II, § 7.7].
[7]$\mathsf{E}(\mathsf{E}(\xi|\mathcal{B})|\mathcal{A}) = \mathsf{E}(\xi|\mathcal{A})$ if $\mathsf{E}|\xi| < \infty$ and σ-algebras $\mathcal{A} \subset \mathcal{B} \subset \mathcal{F}$; one writes $\mathsf{E}(\xi|\eta_1, \ldots, \eta_n)$ for $\mathsf{E}(\xi|\sigma(\eta_1, \ldots, \eta_n))$ where $\xi, \eta_1, \ldots, \eta_n$ are random variables defined on $(\Omega, \mathcal{F}, \mathsf{P})$.

Proof. Set $\mathcal{A} = \sigma\{h_1(\xi_1), h_2(\xi_2)\}$. Clearly the variable on the right-hand side of (1.19) is $\mathcal{A}$-measurable. Introduce $A \in \mathcal{A}$, $A = A_1 A_2$ with $A_i = \{h_i(\xi_i) \in B_i\}$, $B_i \in \mathcal{B}(\mathbb{R}^{m_i})$, $i = 1, 2$. Then we see that

$$\mathsf{E} F_1(\xi_1) F_2(\xi_2) \mathbb{I}_A = \mathsf{E}\left(\mathsf{E}(F_1(\xi_1)|h_1(\xi_1))\mathsf{E}(F_2(\xi_2)|h_2(\xi_2))\mathbb{I}_A\right), \tag{1.21}$$

since

$$\mathsf{E}\mathbb{I}_A \mathsf{E}(F_1(\xi_1)|h_1(\xi_1))\mathsf{E}(F_2(\xi_2)|h_2(\xi_2)) = \mathsf{E}\mathsf{E}\left(F_1(\xi_1)\mathbb{I}_{A_1}|h_1(\xi_1)\right)\mathsf{E}\left(F_2(\xi_2)\mathbb{I}_{A_2}|h_2(\xi_2)\right)$$

$$= \mathsf{E} F_1(\xi_1)\mathbb{I}_{A_1}\mathsf{E} F_2(\xi_2)\mathbb{I}_{A_2} = \mathsf{E} F_1(\xi_1)\mathbb{I}_{A_1} F_2(\xi_2)\mathbb{I}_{A_2} = \mathsf{E} F_1(\xi_1) F_2(\xi_2)\mathbb{I}_A.$$

Here we used the following properties of conditional expectation. If a random variable $\zeta \in \mathcal{G}|\mathcal{B}(\mathbb{R})$ and a σ-algebra $\mathcal{G} \subset \mathcal{F}$, then

$$\mathsf{E}(\zeta\nu|\mathcal{G}) = \zeta\mathsf{E}(\nu|\mathcal{G}) \quad \text{whenever} \quad \mathsf{E}|\zeta\nu| < \infty \text{ and } \mathsf{E}|\nu| < \infty.$$

If ξ is an integrable random variable and η is a random vector taking values in $\mathbb{R}^m$, then $\mathsf{E}(\xi|\eta) = \Phi(\eta)$ where $\Phi : \mathbb{R}^m \to \mathbb{R}$ is a Borel function (see, e.g., [383, Ch.II, §7.7]). Thus $\mathsf{E}(F_i(\xi_i)\mathbb{I}_{A_i}|h_i(\xi_i)) = \Phi_i(h_i(\xi_i))$ where Φ_i is a Borel function ($i = 1, 2$). Therefore $\Phi_1(h_1(\xi_1))$ and $\Phi_2(h_2(\xi_2))$ are independent as ξ_1 and ξ_2 are independent. Note also that $\mathsf{E}\mathsf{E}(\zeta|\mathcal{A}) = \mathsf{E}\zeta$ for an integrable random variable ζ.

Obviously (1.21) holds for an algebra $\mathcal{H}$ of finite unions of events having the form $A = A_1 A_2$. This algebra generates $\mathcal{A}$. Now one can obtain (1.21) for all $A \in \mathcal{A}$ using the fact that, for every $\varepsilon > 0$ and any $A \in \mathcal{A}$, there exists some $A_\varepsilon \in \mathcal{H}$ such that $\mathsf{P}(A\triangle A_\varepsilon) < \varepsilon$ where $A\triangle A_\varepsilon = (A \setminus A_\varepsilon) \cup (A_\varepsilon \setminus A)$. $\square$

The first summand on the right-hand side of (1.17) is zero, since by Lemma 1.24

$$\mathsf{E}(f(X_I)g(X_J)|S_1, S_2) = \mathsf{E}(f(X_I)|S_1, S_2)\mathsf{E}(g(X_J)|S_1, S_2).$$

By (1.20) we see that $\mathsf{E}(f(X_I)|S_1, S_2)$ and $\mathsf{E}(g(X_J)|S_1, S_2)$ are Borel functions in S_1 and S_2 respectively. Denote these (bounded) functions by φ_1 and φ_2. In view of (1.16) we conclude that φ_1 and φ_2 are nondecreasing. Then the second summand in (1.17) can be written as $cov(\varphi_1(S_1), \varphi_2(S_2)|S)$.

Let $\mu_z(\cdot)$ be the regular conditional distribution of the vector (S_1, S_2), given $S = z$ ($z \in \mathbb{R}$). Since $S_1 + S_2 = S$, this measure is concentrated on a line $x + y = z$. One has

$$\mathsf{E}(\varphi_1(S_1)\varphi_2(S_2)|S = z) = \int_{\mathbb{R}^2} \varphi_1(x)\varphi_2(y)d\mu_z(x, y)$$

$$= \int_{\{x+y=z\}} \varphi_1(x)\varphi_2(z - x)d\mu_z(x, y) = \int_{\mathbb{R}} \varphi_1(x)\varphi_2(z - x)d\nu_z(x)$$

where ν_z is the projection of μ_z onto x-axis. Note that $\varphi_2(z - x)$ is nonincreasing in x. Thus Theorem 1.8, (a), yields that the last integral is no greater than

$$\int_{\mathbb{R}} \varphi_1(x)d\nu_z(x) \int_{\mathbb{R}} \varphi_2(z - x)d\nu_z(x) = \mathsf{E}(f(S_1)|S = z)\mathsf{E}(g(S_2)|S = z).$$

Hence the conditional covariance in the right-hand side of (1.17) is nonpositive. $\square$

The proof of the next result is omitted.

Theorem 1.25. ([155]) *Suppose that $X_1, \ldots, X_n$ are independent random variables such that each X_i has a positive density $p_i(x)$ and the function $\log p_i(x)$ is concave. Then the function (1.16) is nondecreasing in t.*

Following Joag-Dev and Proschan [219], we say that a random vector $X = (X_1, \ldots, X_n)$ has *permutation distribution* if its distribution is uniform on the set of permutations of some real numbers $a_1, \ldots, a_n$ $(a_1 \leq \ldots \leq a_n$, coincidences are possible), that is, each permutation occurs with probability $1/n!$ Note that this notion is a special case of the so-called *permutational distribution of a random vector* $Y = (Y_1, \ldots, Y_n)$, which means that for any permutation $\{i_1, \ldots, i_n\}$ of $\{1, \ldots, n\}$ the random vector $(Y_{i_1}, \ldots, Y_{i_n})$ is distributed as Y.

Theorem 1.26. ([219]) *A permutation distribution is* **NA**.

Proof. We proceed by induction on n. The case $n = 1$ is trivial; suppose that the assertion is true for $(n-1)$-dimensional vectors. Divide $\{1, \ldots, n\}$ into two nonempty disjoint subsets I, J, such that $1 \in I$, and consider arbitrary functions $f \in \mathcal{M}(|I|), g \in \mathcal{M}(|J|)$. Observe that we can replace f and g with symmetric nondecreasing functions f_1, g_1 such that

$$\mathsf{E}f(X_I) = \mathsf{E}f_1(X_I), \quad \mathsf{E}g(X_J) = \mathsf{E}g_1(X_J), \quad \mathsf{E}f(X_I)g(X_J) = \mathsf{E}f_1(X_I)g_1(X_J).$$

Namely, for $I = \{i_1, \ldots, i_k\}$ and $\vec{a}_I = (a_{i_1}, \ldots, a_{i_k})$, introduce

$$f_1(\vec{a}_I) := \frac{1}{k!} \sum_{\tau \in \mathcal{P}_k} f(\tau(\vec{a}_I))$$

where $\mathcal{P}_k$ is a set of all permutations τ of coordinates of $\vec{a}_I$. Defining g_1 analogously we see that f_1 and g_1 possess the desired properties, e.g., $\mathsf{E}f(X_I) = \mathsf{E}f_1(X_I)$ as $\mathsf{E}f(\tau(\vec{X}_I)) = \mathsf{E}f(\vec{X}_I)$ for any $I \subset \{1, \ldots, n\}$ and all $\tau \in \mathcal{P}_k$. To see this, notice that if $\tau(X_I) = (X_{j_1}, \ldots, X_{j_k})$ then

$$\mathsf{E}f(\tau(\vec{X}_I)) = \sum f(a_{r_1}, \ldots, a_{r_k})\mathsf{P}(X_{j_1} = a_{r_1}, \ldots, X_{j_k} = a_{r_k})$$

$$= \frac{(n-k)!}{n!} \sum f(a_{r_1}, \ldots, a_{r_k}) = \mathsf{E}f(\vec{X}_I)$$

where the sums are taken over all $a_{r_1}, \ldots, a_{r_k} \in \{a_1, \ldots, a_n\}$ and $r_p \neq r_q$ for $p, q \in \{1, \ldots, k\}, p \neq q$.

From (1.18) one derives the well-known *conditional covariance formula*

$$cov(f_1(X_I), g_1(X_J))$$

$$= \mathsf{E}cov(f_1(X_I), g_1(X_J)|X_1) + cov(\mathsf{E}(f_1(X_I)|X_1), \mathsf{E}(g_1(X_J)|X_1)). \qquad (1.22)$$

Note that $\mathsf{P}(X_1 = t) = 1/n$ for any $t \in \{a_1, \ldots, a_n\}$. If $|I| = 1$ then $\mathsf{E}(f(X_I)|X_1 = t) = f(t)$. For $I = \{1, i_2, \ldots, i_k\}, k > 1$, one has

$$\mathsf{E}(f_1(X_I)|X_1 = t) = \frac{\mathsf{E}f_1(X_I)\mathbb{I}\{X_1 = t\}}{\mathsf{P}(X_1 = t)}$$

$$= n \sum {}' f_1(t, a_{i_2}, \ldots, a_{i_k})\mathsf{P}(X_1 = t, X_{i_2} = a_{i_2}, \ldots, X_{i_k} = a_{i_k})$$

$$= n\frac{(n-k)!}{n!} \sum {}' f_1(t, a_{r_2}, \ldots, a_{r_k})$$

where the sum $\sum'$ is taken over all $(a_{r_2}, \ldots, a_{r_k})$ with $a_{r_p} \in A_t := \{a_1, \ldots, a_n\} \setminus \{t\}$ and $r_p \neq r_q$ for $p, q \in \{2, \ldots, k\}$, $p \neq q$. Thus using the symmetry of f_1 we conclude that $\mathsf{E}(f_1(X_I)|X_1 = t)$ is a nondecreasing function in $t \in \{a_1, \ldots, a_n\}$.

Similarly one can verify that if $J = \{j_1, \ldots, j_m\}$, $m \geq 1$, then

$$\mathsf{E}(g_1(X_J)|X_1 = t) = n \frac{(n - m - 1)!}{n!} \sum'' g_1(a_{s_1}, \ldots, a_{s_m})$$

where $\sum''$ is taken over all m-tuples $(a_{s_1}, \ldots, a_{s_m})$ such that $a_{s_u} \in A_t$ and $s_u \neq s_v$ for $u, v \in \{1, \ldots, m\}$, $u \neq v$. Consequently $\mathsf{E}(g_1(X_J)|X_1 = t)$ is nonincreasing in $t \in \{a_1, \ldots, a_n\}$, g_1 being a symmetric function. By Theorem 1.8, (a), the second term on the right-hand side of (1.22) is nonpositive.

Furthermore, for any $t \in \{a_1, \ldots, a_n\}$ and any function $h : \mathbb{R}^n \to \mathbb{R}$, by the same reasons,

$$\mathsf{E}(h(X)|X_1 = t) = \frac{1}{(n-1)!} \sum''' h(t, a_{j_2}, \ldots, a_{j_n}) = \mathsf{E}h(t, Y^t) \tag{1.23}$$

where $\sum'''$ is the sum over all permutations $(a_{j_2}, \ldots, a_{j_n})$ of the set A_t and Y^t is the vector having $(n-1)$-dimensional permutation distribution on A_t. Whence, by the induction hypothesis and (1.23),

$$\mathsf{E}(f_1(X_I)g_1(X_J)|X_1 = t) = \mathsf{E}(f_1(t, Y^t_{I \setminus \{1\}})g_1(Y^t_J))$$

$$\leq \mathsf{E}(f_1(t, Y^t_{I \setminus \{1\}})\mathsf{E}g_1(Y^t_J)) = \mathsf{E}(f_1(X_I)|X_1 = t)\mathsf{E}g_1(X_J|X_1 = t).$$

Therefore, the conditional covariance in the first summand of (1.22) is a.s. nonpositive by the induction hypothesis. $\square$

Theorem 1.27. ([219]) *The following distributions are* **NA**: *a) multinomial, b) that of random sampling without replacement, c) multiparameter hypergeometric.*

Proof. (a) Suppose that $X = (X_1, \ldots, X_n)$ has a multinomial distribution with parameters $m \in \mathbb{N}$ and $p = (p_1, \ldots, p_n) \in \mathbb{R}^n$ where $p_i \geq 0$, $i = 1, \ldots, n$ and $\sum_{i=1}^n p_i = 1$. That is X_i is a number of outcomes "i" in m independent experiments, in each experiment the i-th outcome appears with probability p_i. We may assume that $p_i > 0$ for $i = 1, \ldots, n$, since if $p_i = 0$, then X_i vanishes a.s. and does not affect the required property.

In other words, $X = \sum_{i=1}^m \xi^{(i)}$ where $\xi, \xi^{(1)}, \ldots, \xi^{(m)}$ are i.i.d. random vectors taking values in $\mathbb{R}^n$ such that for $\xi = (\xi_1, \ldots, \xi_n)$ and m_r equal to 0 or 1, $\sum_{r=1}^n m_r = 1$, one has $\mathsf{P}(\xi_1 = m_1, \ldots, \xi_n = m_n) = p_j$ if $m_j = 1$ (and therefore $m_r = 0$ for $r \neq j$).

Due to Theorem 1.8, (d), it is enough to verify that $\xi \in$ **NA**. Let $I = \{i_1, \ldots, i_k\}$, $J = \{j_1, \ldots, j_s\}$ be disjoint subsets of $\{1, \ldots, n\}$. We will show that $cov(f(\xi_I), g(\xi_J)) \leq 0$ for any $f \in \mathcal{M}(k)$ and $g \in \mathcal{M}(s)$. Set

$$Q_1 = f(1, 0, \ldots, 0)p_{i_1} + \ldots + f(0, \ldots, 0, 1)p_{i_k},$$

$$Q_2 = g(1, 0, \ldots, 0)p_{j_1} + \ldots + g(0, \ldots, 0, 1)p_{j_s},$$

$$f(\mathbf{0}) = f(0, \ldots, 0), \quad g(\mathbf{0}) = g(0, \ldots, 0), \quad p_I = \sum_{i \in I} p_i, \quad p_J = \sum_{j \in J} p_j.$$

Then $cov(f(\xi_I), g(\xi_J))$ obtains the following expression:

$$f(0)Q_2 + g(0)Q_1 + f(0)g(0)(1 - p_I - p_J) - (Q_1 + f(0)(1 - p_I))(Q_2 + g(0)(1 - p_J))$$

$$= -Q_1 Q_2 + f(0)Q_2 p_I + g(0)Q_1 p_J - f(0)g(0)p_I p_J = -(Q_1 - f(0)p_I)(Q_2 - g(0)p_J) \leq 0$$

as $Q_1 \geq f(0)p_I$ and $Q_2 \geq g(0)p_J$ due to monotonicity of f and g.

(b) Consider an urn with n distinct balls labelled by numbers $a_1, \ldots, a_n$. Suppose that m balls are taken randomly without replacement. Let $X_1, \ldots, X_m$ be the numbers of drawn balls. These random variables are **NA**. Indeed, if $m = n$ then clearly $(X_1, \ldots, X_n)$ is a vector with permutation distribution and by Theorem 1.26 it is **NA**. If $m < n$ then $(X_1, \ldots, X_m)$ is **NA** due to Remark 1.4, (b).

(c) Assume that there is an urn containing N different balls of k colors. Let $N_0 = 0$ and N_i be the number of balls of i-th color, $i = 1, \ldots, k$, $N_1 + \ldots + N_k = N$. We can enumerate the balls so that the balls of i-th color have the numbers $\sum_{j=1}^{i-1} N_j + 1, \ldots, \sum_{j=1}^{i} N_j$. Assume that m balls are taken randomly without replacement. Let $m = N$ and let Y_j be the random variable equal to 1 if the j-th ball has been drawn and 0 otherwise, $j = 1, \ldots, N$. Obviously $(Y_1, \ldots, Y_N)$ has a permutation distribution and thus is **NA** by Theorem 1.26. Therefore $(Y_1, \ldots, Y_m)$ is **NA** for $m \leq N$. Introduce

$$X_i = \sum_{t: N_1 + \ldots + N_{i-1} < t \leq N_1 + \ldots + N_i} Y_t, \quad i = 1, \ldots, k.$$

Then X_i is the total number of drawn balls of color i, i.e. the random vector $X = (X_1, \ldots, X_k)$ has a multivariate hypergeometric distribution. Since the sum is taken over disjoint sets, by Theorem 1.8, (d), X is **NA**. $\square$

7°. The Kimball theorem. We finish the survey of elementary examples with an application to statistics. Suppose that $\{X_{ij}, i = 1, \ldots, r, j = 1, \ldots, s\}$ are independent observations, $X_{ij} \sim N(m_{ij}, \sigma^2)$ where $\sigma > 0$. In analysis of variance, i and j are interpreted as levels of factors A and B (the detailed description of the model is given , e.g., in [416, p. 177]). Set

$$\overline{X} = \frac{1}{rs} \sum_{i=1}^{r} \sum_{j=1}^{s} X_{ij}, \quad \overline{X_{i\cdot}} = \frac{1}{s} \sum_{j=1}^{s} X_{ij}, \quad \overline{X_{\cdot j}} = \frac{1}{r} \sum_{i=1}^{r} X_{ij},$$

$$m = \frac{1}{rs} \sum_{i=1}^{r} \sum_{j=1}^{s} m_{ij}, \quad m_{i\cdot} = \frac{1}{s} \sum_{j=1}^{s} m_{ij}, \quad m_{\cdot j} = \frac{1}{r} \sum_{i=1}^{r} m_{ij}$$

where $i = 1, \ldots, r$, $j = 1, \ldots, s$. Consider the hypotheses

$$H_A = \{m_{1\cdot} = \ldots = m_{r\cdot}\}, \quad H_B = \{m_{\cdot 1} = \ldots = m_{\cdot s}\}.$$

The usual way of testing these hypotheses leads to the construction of three quadratic forms

$$q_1 = s \sum_{i=1}^{r} (\overline{X_{i\cdot}} - \overline{X})^2, \quad q_2 = r \sum_{j=1}^{s} (\overline{X_{\cdot j}} - \overline{X})^2, \quad q_3 = \sum_{i=1}^{r} \sum_{j=1}^{s} (X_{ij} - \overline{X_{i\cdot}})(X_{ij} - \overline{X_{\cdot j}}).$$

If the hypotheses H_A and H_B hold, then the quadratic forms q_1, q_2, q_3 are independent and have the chi-square distribution with degrees of freedom $r - 1, s - 1, m = (r - 1)(s - 1)$ respectively ([111]). The Fisher test statistics are

$$F_A = \frac{(s - 1)q_1}{q_3}, \quad F_B = \frac{(r - 1)q_2}{q_3},$$

here $F_A := 0$ and $F_B := 0$ on the set of zero probability where $q_3 = 0$. The following Kimball's theorem ([235]) can be easily obtained using association.

Theorem 1.28. ([160, 235]) *The probability that no first kind error occurs during the simultaneous test of H_A and H_B is no less than the product of the same probabilities for H_A and H_B tested independently.*

Proof. Note that F_A and F_B are nondecreasing functions in independent random variables q_1, q_2, q_3^{-1}. By Theorem 1.8 (c), (d), they are associated, therefore, for any $a > 0$ one has $\mathsf{P}(F_A \le a, F_B \le a) \ge \mathsf{P}(F_A \le a)\mathsf{P}(F_B \le a)$. $\square$

2 Classes of associated and related systems

In this Section we provide further examples of various random systems for which association (**PA** or **NA**) can be proved by direct checking the corresponding definition. The required inequalities for covariances are established by methods which are determined by intrinsic properties of the process or field under consideration. In Section 4 we will discuss a more complicated way leading to sufficient conditions of association (**NA**) which can be applied to rather general random systems without identifying their exact finite-dimensional distributions.

$1°$. **Normal random systems.** Recall that a family $\mathbf{X} = \{X_t, t \in T\}$ of real-valued random variables is called *normal* (or *Gaussian*) *system* if all its finite-dimensional distributions are Gaussian (see, e.g., [383, Ch. II, § 13.6]). An arbitrary normal system $\mathbf{X}$ consists of independent random variables whenever $cov(X_t, X_s) = 0$ for all $s, t \in T$, $s \ne t$ (see, e.g., [383, Ch. II, § 13.4]). This property has a beautiful modification due to Pitt ([330]) and Joag-Dev and Proschan ([219]), answering the question when normal system is associated or **NA**. The proof is, nevertheless, notably more difficult.

Theorem 2.1. ([330]) *Let $X = (X_1, \ldots, X_n)$ be a normal random vector. Then it is associated if and only if its components are nonnegatively correlated.*

Proof. Necessity. If $X \in \mathbf{A}$, then $\sigma_{ij} = cov(X_i, X_j) \ge 0$, $i, j = 1, \ldots, n$.

Sufficiency. If X is associated and $a \in \mathbb{R}^n$ then $X + a$ is associated in view of Theorem 1.8, (d). Thus without loss of generality we may and will assume that X has mean zero. Let $\Sigma = (\sigma_{ij})$ be the covariance matrix of X. Suppose at first that $\det \Sigma > 0$. Then there exists a density of X

$$\phi(x) = (2\pi)^{-n/2}(\det \Sigma)^{-1/2}e^{-(\Sigma^{-1}x, x)/2}$$

where $(\cdot, \cdot)$ denotes the inner product in $\mathbb{R}^n$. Due to Theorem 1.5, (c), it suffices to show that for any pair of functions $f, g \in C_b^1(\mathbb{R}^n) \cap \mathcal{M}(n)$ one has $cov(f(X), g(X)) \geq 0$. Take a random vector $Z \sim N(0, \Sigma)$ independent of X and for $\lambda \in [0, 1]$ set $Y(\lambda) = \lambda X + (1 - \lambda^2)^{1/2} Z$. Clearly, $Y(\lambda) \sim N(0, \Sigma)$ and $cov(X_i, Y_j(\lambda)) = \lambda \sigma_{ij}$, $i, j = 1, \ldots, n$. Introduce a function

$$F(\lambda) = \mathsf{E} f(X) g(Y(\lambda)), \ \lambda \in [0, 1].$$

One can easily see that F is continuous on $[0, 1]$ and $F(1) - F(0) = cov(f(X), g(X))$. Therefore, it is enough to prove that there exists a derivative $F'(\lambda) \geq 0$ when $\lambda \in (0, 1)$. For such λ introduce a conditional probability density function[8]

$$p_{Y(\lambda)|X=x}(y) = \frac{p_{Y(\lambda), X}(y, x)}{p_X(x)} = \frac{\partial^n}{\partial y_1 \ldots \partial y_n} P(Y_1(\lambda) \leq y_1, \ldots, Y_n(\lambda) \leq y_n | X = x)$$

where $p_{Y(\lambda), X}$ and p_X are the densities of $(Y(\lambda), X)$ and X respectively. Then by Lemma 1.22 one has

$$p(\lambda, x, y) := p_{Y(\lambda)|X=x}(y) = \frac{\partial^n}{\partial y_1 \ldots \partial y_n} P\left(Z_1 \leq \frac{y_1 - \lambda x_1}{\sqrt{1 - \lambda^2}}, \ldots, Z_n \leq \frac{y_n - \lambda x_n}{\sqrt{1 - \lambda^2}}\right)$$

$$= (1 - \lambda^2)^{-n/2} \phi((1 - \lambda^2)^{-1/2}(\lambda x - y)), \quad x, y \in \mathbb{R}^n.$$

Thus, for $\lambda \in (0, 1)$,

$$F(\lambda) = \int_{\mathbb{R}^n} \int_{\mathbb{R}^n} f(x) g(y) p_{Y(\lambda), X}(y, x) dy dx = \int_{\mathbb{R}^n} f(x) p_X(x) \int_{\mathbb{R}^n} p(\lambda, x, y) g(y) dy dx$$

$$= \int_{\mathbb{R}^n} \phi(x) f(x) \int_{\mathbb{R}^n} p(\lambda, x, y) g(y) dy dx = \int_{\mathbb{R}^n} \phi(x) f(x) h(\lambda, x) dx$$

where $h(\lambda, x) = \int_{\mathbb{R}^n} p(\lambda, x, y) g(y) dy$. Set

$$\phi_\lambda(x) = (1 - \lambda^2)^{-n/2} \phi((1 - \lambda^2)^{-1/2} x),$$

whence

$$h(\lambda, x) = (\phi_\lambda * g)(\lambda x) = \int_{\mathbb{R}^n} \phi_\lambda(\lambda x - y) g(y) dy = \int_{\mathbb{R}^n} \phi_\lambda(y) g(\lambda x - y) dy.$$

So $h(\lambda, x)$ has bounded, continuous and nonnegative derivatives $\partial h / \partial x_k$, $k = 1, \ldots, n$.

Lemma 2.2. *For any* $y, x \in \mathbb{R}^n$ *and* $\lambda \in (0, 1)$ *the function* $p(\lambda, x, y)$ *satisfies the partial differential equation*

$$\frac{\partial p}{\partial \lambda} = -\frac{1}{\lambda} \left(\sum_{j,k=1}^{n} \sigma_{jk} \frac{\partial^2 p}{\partial x_j \partial x_k} - \sum_{j=1}^{n} x_j \frac{\partial p}{\partial x_j}\right). \tag{2.1}$$

---

[8]See, e.g., [383, Ch. II, § 7.6].

Proof. For any fixed y and λ, all the functions entering (2.1) surely belong to $L^1(\mathbb{R}^n) \cap L^2(\mathbb{R}^n)$. Therefore it suffices to prove the equality of the Fourier transforms on both sides of (2.1). For simplicity we define the Fourier transform of an integrable function $h : \mathbb{R}^n \to \mathbb{R}$ as

$$F[h](t) = \int_{\mathbb{R}^n} e^{i(t,x)} h(x)\,dx, \quad i^2 = -1.$$

Then the desired equality is

$$\frac{\partial F[p](t)}{\partial \lambda} = \frac{1}{\lambda}\left(\sum_{j,k=1}^n \sigma_{jk} t_j t_k F[p](t) - \sum_{j=1}^n \frac{\partial}{\partial t_j}(t_j F[p](t)) \right). \tag{2.2}$$

Since for any fixed $\lambda \in (0,1)$ and $y \in \mathbb{R}^n$ the function $q = q(x) = \lambda^n p(\lambda, x, y)$ is the density of normal distribution with mean $\lambda^{-1}y$ and covariance matrix $\lambda^{-2}(1-\lambda^2)\Sigma$, its Fourier transform is just the corresponding characteristic function, i.e.

$$\lambda^n F[p](t) = \exp\left\{ i(t, \lambda^{-1}y) - \frac{1}{2}(\lambda^{-2}-1)(\Sigma t, t) \right\}, \quad t \in \mathbb{R}^n.$$

It is not difficult to verify that this function satisfies (2.2). $\square$

Note that since p exponentially decreases at infinity, one has $\partial h(\lambda, x)/\partial \lambda = \int_{\mathbb{R}^n} g(y)(\partial p(\lambda, x, y)/\partial \lambda)\,dy$. The same rule is true for taking derivatives in x_j, with $j = 1, \ldots, n$. Thus from the definition of $F(\lambda)$ and Lemma 2.2 it follows that

$$F'(\lambda) = -\frac{1}{\lambda}\int_{\mathbb{R}^n} \phi(x) f(x) \left(\sum_{j,k=1}^n \sigma_{jk} \frac{\partial^2 h(\lambda, x)}{\partial x_j \partial x_k} - \sum_{j=1}^n x_j \frac{\partial h(\lambda, x)}{\partial x_j} \right) dx. \tag{2.3}$$

Lemma 2.3. *For any* $\lambda \in (0,1)$

$$F'(\lambda) = \frac{1}{\lambda}\int_{\mathbb{R}^n} \phi(x) \left(\sum_{j,k=1}^n \sigma_{jk} \frac{\partial f(x)}{\partial x_j} \frac{\partial h(\lambda, x)}{\partial x_k} \right) dx.$$

Proof. Let $D = (d_{jk}) = \Sigma^{-1}$. Note that D is symmetric. Then we get

$$\int_{\mathbb{R}^n} \phi(x) \left(\sum_{j,k=1}^n \sigma_{jk} f(x) \frac{\partial^2 h(\lambda, x)}{\partial x_j \partial x_k} \right) dx + \int_{\mathbb{R}^n} \phi(x) \left(\sum_{j,k=1}^n \sigma_{jk} \frac{\partial f(x)}{\partial x_j} \frac{\partial h(\lambda, x)}{\partial x_k} \right) dx$$

$$= (2\pi)^{-n/2}(\det \Sigma)^{-1/2} \int_{\mathbb{R}^n} e^{-(Dx,x)/2} \sum_{j,k=1}^n \sigma_{jk} \frac{\partial}{\partial x_j}\left(f(x) \frac{\partial h(\lambda, x)}{\partial x_k} \right) dx$$

$$= ((2\pi)^n \det \Sigma)^{-1/2} \int_{\mathbb{R}^n} \sum_{j,k=1}^n \sigma_{jk} f(x) \frac{\partial h(\lambda, x)}{\partial x_k} e^{-(Dx,x)/2} \frac{\partial}{\partial x_j} \frac{1}{2}\left(\sum_{l,m=1}^n d_{lm} x_l x_m \right) dx$$

using integration by parts in x_j. Introducing the Kronecker symbol δ_{mk} we may rewrite the last expression as

$$\int_{\mathbb{R}^n} \phi(x)f(x) \sum_{j,k=1}^{n} \sigma_{jk} \frac{\partial h(\lambda, x)}{\partial x_k} \sum_{m=1}^{n} d_{mj}x_m dx$$

$$= \int_{\mathbb{R}^n} \phi(x)f(x) \sum_{k=1}^{n} \frac{\partial h(\lambda, x)}{\partial x_k} \sum_{m=1}^{n} x_m \sum_{j=1}^{n} d_{mj}\sigma_{jk} dx$$

$$= \int_{\mathbb{R}^n} \phi(x)f(x) \sum_{k=1}^{n} \frac{\partial h(\lambda, x)}{\partial x_k} \sum_{m=1}^{n} x_m \delta_{mk} dx = \int_{\mathbb{R}^n} \phi(x)f(x) \sum_{k=1}^{n} \frac{\partial h(\lambda, x)}{\partial x_k} x_k dx,$$

which together with (2.3) provides the desired equality. $\square$

Continuing the proof of Theorem 2.1, observe that $F'(\lambda) \geq 0$ if all $\sigma_{jk} \geq 0$. Consequently $F(1) - F(0) = cov(f(X), g(X)) \geq 0$.

Suppose now that $\det \Sigma = 0$. Then a vector $X_k = X + k^{-1}W$, where $W \sim N(0, I_n)$ is independent of X ($k \in \mathbb{N}$ and I_n is the unit $n \times n$ matrix), is associated by the already proved part of the Theorem. The result follows by letting $k \to \infty$ and invoking Theorem 1.8, (e). The proof is complete. $\square$

Remark 2.4. Let $X = \{X_t, t \in T\}$ be a Gaussian system. Then $X \in \mathbf{A}$ if and only if $X \in \mathbf{PA}$ (if $X \in \mathbf{PA}$ then $cov(X_t, X_s) \geq 0$ for any $t, s \in T$ and one can apply Theorem 2.1).

Corollary 2.5. ([152]) *Let* $W = (W_1(t), \ldots, W_m(t))_{t \geq 0}$ *be a Brownian motion in* $\mathbb{R}^m$. *Suppose that* $u_i, V_{ij} : \mathbb{R} \to \mathbb{R}$ ($i = 1, \ldots, k$, $j = 1, \ldots, m$) *are measurable functions such that* $\int_0^t (|u_i(s)| + V_{ij}^2(s))ds < \infty$ *for any* $t \geq 0$ *and all* i, j. *Set*

$$X_i(t) = x_{0,i} + \int_0^t u_i(s)ds + \sum_{j=1}^{m} \int_0^t V_{ij}(s)dW_j(s)$$

where $x_{0,i} \in \mathbb{R}$, $i = 1, \ldots, k$. *If, moreover,* $V_{ij}(t)V_{lj}(t) \geq 0$ *for all* $t \geq 0$, *and* $i, l \in \{1, \ldots, k\}$, $j \in \{1, \ldots, m\}$, *then the family* $\{X_i(t), i = 1, \ldots, k, t \geq 0\} \in \mathbf{A}$.

Proof. The random system under consideration is Gaussian, so on account of Theorem 2.1 it suffices to prove that it is nonnegatively correlated. Note that the components of Brownian motion are independent. Thus, using the well-known properties of Itô integrals (see, e.g., [412, Ch. 12]), for any $t, s \geq 0$ one has

$$cov(X_i(t), X_l(s)) = \sum_{j=1}^{m} cov\left(\int_0^t V_{ij}(u)dW_j(u), \int_0^s V_{lj}(z)dW_j(z)\right)$$

$$= \sum_{j=1}^{m} \mathsf{E} \int_0^t V_{ij}(u)dW_j(u) \int_0^s V_{lj}(z)dW_j(z) = \sum_{j=1}^{m} \int_0^{s \wedge t} V_{ij}(z)V_{lj}(z)dz \geq 0. \ \square$$

Theorem 2.6. ([219]) *Let* $X = (X_1, \ldots, X_n)$ *be a normal random vector,* $n \geq 2$. *Then it is* $\mathbf{NA}$ *if and only if its components are non-positively correlated.*

Proof is analogous to that of the previous theorem. The only distinction is that if all the σ_{jk} with $j \neq k$ are non-positive, and the arguments of functions f and g belong to disjoint sets of variables, then Lemma 2.3 implies that $F'(\lambda) \leq 0$, because for any $i = 1, \ldots, n$ at least one of the derivatives $\partial f(x)/\partial x_i$, $\partial g(\lambda, x)/\partial x_i$ is identically zero. $\square$

In [219] a more simple proof of Theorem 2.6 was given using the result of [218] (which is close to Theorem 2.1 but less general).

Definition 2.7. A probability measure μ on $\mathcal{B}(\mathbb{R}^n)$ is called a *correlation measure* if, for all closed, symmetric, convex subsets of $\mathbb{R}^n$, one has

$$\mu(A \cap B) \geq \mu(A)\mu(B).$$

Let $\mu_n = N(0, I_n)$ be a standard Gaussian measure on $\mathcal{B}(\mathbb{R}^n)$, $n \geq 1$. A famous conjecture asserts that μ_n is a correlation measure for each $n \geq 1$. This conjecture is obviously true if $n = 1$. Pitt [329] proved it for $n = 2$. For $n \geq 3$ the conjecture remains unsettled.

Lewis and Pritchard [260] showed that, for each $n \geq 3$ and $0 < p < 2/(n-1)$, there exists a spherically symmetric correlation measure μ on $\mathbb{R}^n$ with

$$\lim_{r \to \infty} r^{-p} \log \mu(B_r(0)^c) = -1$$

where $B_r(0) = \{x \in \mathbb{R}^n : \|x\| \leq r\}$, $r > 0$. Moreover, if μ is a symmetric correlation measure on $\mathcal{B}(\mathbb{R}^n)$, $n \geq 2$, and if its support is not contained in a one-dimensional subspace of $\mathbb{R}^n$, then, for some $a > 0$,

$$\int_{\mathbb{R}^n} e^{a\|x\|^2} d\mu(x) = \infty.$$

Consequently we have examples of positively correlated indicators of the Borel sets A and B which are not increasing and at the same time $cov(\mathbb{I}_A, \mathbb{I}_B) \geq 0$.

Now we formulate the inequality by Hargé, based on a result of Caffarelli [89] concerning optimal transport of measure. The omitted proof uses also the properties of Ornstein–Uhlenbeck semigroups. In the following theorem it is assumed that all the integrals involved are taken over $\mathbb{R}^n$ and well-defined.

Theorem 2.8. ([190]) *Let f and g be convex functions defined on $\mathbb{R}^n$ and $\mu = N(0, I_n)$. Then*

$$\int fg\,d\mu \geq \Big(1 + (m(f), m(g))\Big) \int f\,d\mu \int g\,d\mu$$

where

$$m(f) = \int x \frac{f(x)}{\int f\,d\mu} d\mu(x), \quad m(g) = \int x \frac{g(x)}{\int g\,d\mu} d\mu(x)$$

and $(\cdot, \cdot)$ stands for the inner product in $\mathbb{R}^n$.

In particular, if $m(f) = 0$ or $m(g) = 0$ then

$$\int fg\,d\mu \geq \int f\,d\mu \int g\,d\mu.$$

In most cases $\int xf\,d\mu = \int \nabla f\,d\mu$ and then for $g = f$ we have

$$\left(\int \nabla f\,d\mu\right)^2 \leq \int f^2\,d\mu - \left(\int f\,d\mu\right)^2.$$

It is interesting to compare the last inequality with the well-known one by Poincaré

$$\int f^2\,d\mu - \left(\int f\,d\mu\right)^2 \leq \int \|\nabla f\|^2\,d\mu.$$

We have no possibility to tackle here the relations with development of such important research field as the Sobolev logarithmic inequalities , see, e.g., [358].

2°. Associated measures on partially ordered spaces. Since the only notion needed essentially for Definitions 1.1—1.3 is that of nondecreasing function, they admit natural extensions. Let $(S, \mathcal{B})$ be a partially ordered measurable space (see, e.g., [239, Ch. I, §4, 1]) with partial order $\leq_S$. We write now $\leq_S$ to avoid misunderstanding whereas $\leq$ denotes the usual total order in $\mathbb{R}$.

Definition 2.9. A function $f : S \to \mathbb{R}$ is $\leq_S$ *increasing* if, for any $x, y \in S$ such that $x \leq_S y$, one has $f(x) \leq f(y)$.

Definition 2.10. A probability measure μ on $(S, \mathcal{B})$ is called *positively correlated*, or *associated*, if

$$\int_S fg\,d\mu \geq \int_S f\,d\mu \int_S g\,d\mu \qquad (2.4)$$

for any bounded $\leq_S$ increasing $\mathcal{B}|\mathcal{B}(\mathbb{R})$-measurable functions $f, g : S \to \mathbb{R}$. A random element X defined on a probability space $(\Omega, \mathcal{F}, \mathsf{P})$ and taking values in $(S, \mathcal{B})$ is called associated if $Law(X)$ possesses this property.

To emphasize the role of the partial order in this definition it is worthwhile to write $(S, \mathcal{B}, \mu, \leq_S) \in \mathbf{A}$.

Obviously, if $\mu = \delta_x$ (the *Dirac measure* concentrated at a point $x \in S$), then μ is associated for any partial order defined for elements of S. Condition (1.6) is a particular case of (2.4) when one uses in $S = \mathbb{R}^n$ the partial order (1.8). Note that, in contrast to Theorem 1.8, (a), a single random element with values in S need not be associated (see Theorem 2.17 below). At the same time we have the following analog of Theorem 1.5.

Lemma 2.11. $(S, \mathcal{B}, \mu, \leq_S) \in \mathbf{A}$ *if and only if*

$$\mu(A \cap B) \geq \mu(A)\mu(B) \qquad (2.5)$$

for any $\leq_S$ increasing sets $A, B \in \mathcal{B}$ (i.e. with $\mathbb{I}_A$ and $\mathbb{I}_B$ increasing on $(S, \leq_S)$).

Proof. Necessity is clear.

Sufficiency. For $\leq_S$ increasing $f : S \to \mathbb{R}$, $f \in \mathcal{B}|\mathcal{B}(\mathbb{R})$, $0 \leq f < 1$ and $n \in \mathbb{N}$, define

$$f_n(x) := \sum_{k=1}^{n} \frac{k}{n} \mathbb{I}\left\{ \frac{k-1}{n} \leq f(x) < \frac{k}{n} \right\} \equiv \frac{1}{n} \sum_{k=1}^{n} \mathbb{I}\left\{ f(x) \geq \frac{k-1}{n} \right\}, \quad x \in S. \quad (2.6)$$

Obviously $f_n(x) \to f(x)$ for all $x \in S$ as $n \to \infty$ and $\mathbb{I}\{x : f(x) \geq v\}$ is $\leq_S$ increasing for each $v \in [0,1)$. The general case is reduced to considered one. $\square$

To provide important examples of associated measures we need some auxiliary results and new notation. The next result is similar to Theorem 1.8, (d).

Theorem 2.12. *Let $(S_i, \mathcal{B}_i, \mathsf{P}_i, \leq_i)$ be partially ordered probability spaces, $i = 1, 2$. Assume that $h : S_1 \to S_2$ is $\mathcal{B}_1|\mathcal{B}_2$-measurable and increasing (i.e. $x \leq_1 y$ implies $h(x) \leq_2 h(y)$). If, moreover, $\mathsf{P}_2 = Law(h)$ and $(S_1, \mathcal{B}_1, \mathsf{P}_1, \leq_1) \in \mathbf{A}$ then $(S_2, \mathcal{B}_2, \mathsf{P}_2, \leq_2) \in \mathbf{A}$.*

Proof. Take any increasing $A_2, B_2 \in \mathcal{B}_2$. Then $A_1 := h^{-1}(A_2)$ and $B_1 := h^{-1}(B_2)$ are increasing in $(S_1, \leq_1)$. Indeed, if for example $x_1 \in A_1$ and $y_1 \geq_1 x_1$, then $h(x_1) \in A_2$ and $h(y_1) \geq_2 h(x_1)$. Therefore, $h(y_1) \in A_2$ as A_2 is $\leq_2$ increasing. Consequently, $y_1 \in h^{-1}(A_2) = A_1$. Taking into account that $\mathsf{P}_2 = \mathsf{P}_1 h^{-1}$ one has

$$\mathsf{P}_2(A_2 \cap B_2) - \mathsf{P}_2(A_2)\mathsf{P}_2(B_2) = \mathsf{P}_1(A_1 \cap B_1) - \mathsf{P}(A_1)\mathsf{P}_1(B_1) \geq 0$$

where the last inequality is due to Lemma 2.11. $\square$

Corollary 2.13. *Let $(S_t, \mathcal{B}_t, \preceq_t)_{t \in \mathbb{T}}$ be a family of partially ordered spaces, $S_{\mathbb{T}} = \prod_{t \in \mathbb{T}} S_t$, $\mathcal{B}_{\mathbb{T}}$ be a cylindric σ-algebra in $S_{\mathbb{T}}$ and*

$$x \preceq_{\mathbb{T}} y \text{ for } x = \{x_t, t \in \mathbb{T}\}, \ y = \{y_t, t \in \mathbb{T}\} \in S_{\mathbb{T}} \iff x_t \preceq_t y_t, \ t \in \mathbb{T}. \quad (2.7)$$

If $\mu_{\mathbb{T}}$ is an associated measure on $(S_{\mathbb{T}}, \mathcal{B}_{\mathbb{T}}, \preceq_{\mathbb{T}})$ then for any $U \subset \mathbb{T}$ its projection $\mu_{\mathbb{T},U}$ on $(S_U, \mathcal{B}_U, \preceq_U)$ is associated (for $(S_U, \mathcal{B}_U, \preceq_U)$ one uses the similar definition as for $(S_{\mathbb{T}}, \mathcal{B}_{\mathbb{T}}, \preceq_{\mathbb{T}})$) and $\mu_{\mathbb{T},U} = \pi_{\mathbb{T},U}^{-1} \mu_{\mathbb{T}}$ where $\pi_{\mathbb{T},U} x = x|_U$ is a restriction of a function $x \in S_{\mathbb{T}}$ producing a function defined on U.

Let $\leq_1$ and $\leq_2$ be partial orders on S. One says that $\leq_2$ is finer than $\leq_1$ if for any $x \leq_1 y$, $x, y \in S$ one has $x \leq_2 y$. Clearly, in such case if $f : S \to \mathbb{R}$ is increasing on $(S, \leq_2)$ then f is also increasing on $(S, \leq_1)$. Therefore, if $(S, \mathcal{B}, \mu, \leq_1) \in \mathbf{A}$ then $(S, \mathcal{B}, \mu, \leq_2) \in \mathbf{A}$.

Introduce the following partial order "inc" (based on increments of the functions) in the space $S = C_0[0, T] = \{f \in C[0, T] : f(0) = 0\}$. Let

$$x \leq_{\text{inc}} y \text{ if } x(t) - x(s) \leq y(t) - y(s) \text{ for any } 0 \leq s \leq t \leq T. \quad (2.8)$$

We can endow S with usual partial order $\preceq$ of the type (2.7) defined by way of

$$x \preceq y \text{ for } x, y \in S \text{ if } x(t) \leq y(t) \text{ for all } t \in [0, T]. \quad (2.9)$$

Note that $\preceq$ is finer than $\leq_{\text{inc}}$.

Let $\mathbb{W}$ be a Wiener measure on $(S, \mathcal{B}(S))$, i.e. a law of a standard (continuous) Wiener process W defined on $[0, T]$. The following result is due to Barbato.

Theorem 2.14. ([18]) *Let* $S = C_0[0, T]$, $\mathcal{B} = \mathcal{B}(C_0[0, T])$ *and* $\mu = \mathbb{W}$. *Then* μ *is associated in the sense* $(S, \mathcal{B}, \mu, \leq_{\text{inc}}) \in \mathbf{A}$.

Proof. It is convenient to employ the canonical version of the Wiener process, that is $(\Omega, \mathcal{F}, \mathsf{P}) = (S, \mathcal{B}, \mu)$ and $W(\omega) = \omega$ for $\omega \in S$.

At first we shall verify (2.5) for an auxiliary system of events A, B. Let $H = \{t_1, \ldots, t_n\}$ be a partition of $[0, T]$ with $0 = t_0 < \ldots < t_n = T$ $(n \in \mathbb{N})$. Set $X_i(\omega) := W_{t_i}(\omega) - W_{t_{i-1}}(\omega)$, $\omega \in \Omega$, $i = 1, \ldots, n$. Then $X_H = (X_1, \ldots, X_n) \in \mathbf{A}$ in view of Theorem 1.8, (c). Let $\mathcal{A}_H := \sigma\{X_H\}$ and

$$x \leq_H y \text{ for } x, y \in S \iff x(t_i) - x(t_{i-1}) \leq y(t_i) - y(t_{i-1}), \quad i = 1, \ldots, n. \quad (2.10)$$

Thus $(S, \mathcal{A}_H, \mu, \leq_H) \in \mathbf{A}$. Note that $\leq_H$ is finer than $\leq_{\text{inc}}$. Nevertheless, one has

Lemma 2.15. *A set* $A \in \mathcal{A}_H$ *is* $\leq_{\text{inc}}$ *increasing if and only if it is* $\leq_H$ *increasing.*

Proof. Necessity. Assume that A is $\leq_{\text{inc}}$ increasing. Let $x \in A$ and $y \geq_H x$ (i.e. $x \leq_H y$). We have to show that $y \in A$. Set $p_i = y(t_i) - x(t_i)$, $i = 0, \ldots, n$. Then according to (2.10) the relation $x \leq_H y$ can be written as follows

$$0 = p_0 \leq p_1 \leq \ldots \leq p_n.$$

Take an $\leq_{\text{inc}}$ increasing function $f \in S$ such that $f(t_i) = p_i$, $i = 0, \ldots, n$. Put $z(t) := x(t) + f(t)$, $t \in [0, T]$. Then obviously $z \geq_{\text{inc}} x$ (i.e. $x \leq_{\text{inc}} z$) and consequently $z \in A$ because A is $\leq_{\text{inc}}$ increasing. At the same time $y(t_i) = z(t_i)$, $i = 0, \ldots, n$. Thus $y \in A$ as $A \in \mathcal{A}_H$ (indeed, any A from $\mathcal{A}_H$ is determined only by the values of its elements at the points $t_i \in H$, $i = 0, \ldots, n$).

Sufficiency. Suppose that A is $\leq_H$ increasing. Let $x \in A$ and $y \geq_{\text{inc}} x$. Then $x \leq_H y$ as the order $\leq_H$ is finer than $\leq_{\text{inc}}$. Thus $y \in A$ and A is $\leq_{\text{inc}}$ increasing. $\square$.

Introduce an algebra $\mathcal{A} := \cup_H \mathcal{A}_H$ where the union is taken over all finite partitions of $[0, T]$. Take any $\leq_{\text{inc}}$ increasing $A, B \in \mathcal{A}$. Then there exists a partition H_0 such that $A, B \in \mathcal{A}_{H_0}$, since $\mathcal{A}_{H'} \subset \mathcal{A}_{H''}$ if $H' \subset H''$. Due to Lemma 2.15, A and B are $\leq_{H_0}$ increasing. We have seen that $(S, \mathcal{A}_{H_0}, \mu, \leq_{H_0}) \in \mathbf{A}$. Therefore, $cov(\mathbb{I}_A, \mathbb{I}_B) \geq 0$ for any $\leq_{\text{inc}}$ increasing $A, B \in \mathcal{A}$. Thus to complete the proof of the Theorem we only need the following

Lemma 2.16. *For each* $\varepsilon > 0$ *and any* $\leq_{\text{inc}}$ *increasing* $B \in \mathcal{B}$ *there exists* $\leq_{\text{inc}}$ *increasing* $A \in \mathcal{A}$ *such that* $\mu(A \triangle B) < \varepsilon$.

Proof. Note that $\mathcal{B} = \sigma\{\mathcal{A}\}$. Therefore one can find $C \in \mathcal{A}$, i.e. $C \in \mathcal{A}_H$ for some partition H, such that $\mu(B \triangle C) < \varepsilon$. The problem is to find a $\leq_{\text{inc}}$ increasing approximating set $A \in \mathcal{A}_H$.

Let $E \subset S$ consist of f which are linear at every interval $[t_{i-1}, t_i]$, $i = 1, \ldots, n$, and $F = \{f \in S : f(t) = 0, t \in H\}$. For $f \in S$ let $\Pi_H f$ be obtained by linear interpolation, connecting the points $(t_i, f(t_i))$, $i = 1, \ldots, n$. Clearly, E and F are linear subspaces of S and $S = E \oplus F$. For any $f \in S$ one has $f = \Pi_H f + (f - \Pi_H f)$ where $\Pi_H f \in E$ and $g = f - \Pi_H f \in F$. Let $\mathcal{E}$ and $\mathcal{G}$ be the traces of $\mathcal{B}$ on E and F respectively. We can write

$$W(\omega) = \Pi_H W(\omega) + (W(\omega) - \Pi_H W(\omega)) =: Y(\omega) + Z(\omega).$$

It is easily seen that Y and Z are random elements defined on $(S, \mathcal{B}, \mu)$ and taking values in $(E, \mathcal{E})$ and $(F, \mathcal{G})$ respectively. Moreover, $\mathcal{A}_H = \sigma\{Y\}$ and Y, Z are independent. To verify the last assertion it suffices to take into account that Y and Z are Gaussian processes with $cov(Y(s), Z(t)) = 0$ for any $s, t \in [0, t]$. Let us write $\mathbb{P}_1 := Law(Y)$ and $\mathbb{P}_2 := Law(Z)$.

For $\omega = f + g$ where $f \in E$ and $g \in F$ introduce

$$\varphi(\omega) := \mathsf{E}(\mathbb{I}\{W \in B\}|Y = f) = \mathsf{E}(\mathbb{I}_B(Y + Z)|Y = f),$$

then Lemma 1.22 and (1.9) yield

$$\varphi(\omega) = \int_F \mathbb{I}_B(f + g)d\mathbb{P}_2(g) \tag{2.11}$$

where E stands for expectation with respect to μ. Thus, for any $D \in \sigma\{Y\}$, using the elementary properties of the conditional expectation we can write

$$\mu(B \triangle D) = \mathsf{E}|\mathbb{I}_B - \mathbb{I}_D| = \mathsf{E}(1 - \mathbb{I}\{W \in B\})\mathbb{I}_D + \mathsf{E}\mathbb{I}\{W \in B\}\mathbb{I}_{D^c}$$

$$= \mathsf{E}(\mathsf{E}(1 - \mathbb{I}\{W \in B\})\mathbb{I}_D)|Y) + \mathsf{E}(\mathsf{E}\mathbb{I}\{W \in B\}\mathbb{I}_{D^c}|Y)$$

$$= \mathsf{E}((1 - \varphi)\mathbb{I}_D) + \mathsf{E}(\varphi\mathbb{I}_{D^c}) = \mathsf{E}|\varphi - \mathbb{I}_D|.$$

Note that $0 \leq \varphi \leq 1$. Therefore,

$$|\varphi - \mathbb{I}_D| = (1 - \varphi)\mathbb{I}_D + \varphi\mathbb{I}_{D^c}$$

$$= (1 - \varphi)\mathbb{I}_{D \cap \{\varphi \geq 1/2\}} + (1 - \varphi)\mathbb{I}_{D \cap \{\varphi < 1/2\}} + \varphi\mathbb{I}_{D^c \cap \{\varphi \geq 1/2\}} + \varphi\mathbb{I}_{D^c \cap \{\varphi < 1/2\}}$$

$$\geq (1 - \varphi)\mathbb{I}_{D \cap \{\varphi \geq 1/2\}} + \varphi\mathbb{I}_{D \cap \{\varphi < 1/2\}} + (1 - \varphi)\mathbb{I}_{D^c \cap \{\varphi \geq 1/2\}} + \varphi\mathbb{I}_{D^c \cap \{\varphi < 1/2\}}$$

$$= (1 - \varphi)\mathbb{I}_{\{\varphi \geq 1/2\}} + \varphi\mathbb{I}_{\{\varphi < 1/2\}} = |\varphi - \mathbb{I}_A|$$

where $A = \{\varphi \geq 1 - \varphi\} = \{\varphi \geq 1/2\} \in \mathcal{A}_H$ as $\sigma\{Y\} = \mathcal{A}_H$. Consequently

$$\mu(A \triangle B) = \mathsf{E}|\varphi - \mathbb{I}_A| \leq \mathsf{E}|\varphi - \mathbb{I}_C| = \mu(B \triangle C) < \varepsilon.$$

Now we show that φ is $\leq_{\text{inc}}$ increasing if B is $\leq_{\text{inc}}$ increasing. Let $\omega_1, \omega_2 \in S$ and $\omega_1 \leq_{\text{inc}} \omega_2$. Write $\omega_i = f_i + g_i$, $f_i \in E$, $g_i \in F$, $i = 1, 2$. Then in view of (2.11) we have

$$\varphi(\omega_i) = \int_F \mathbb{I}_B(f_i + g)d\mathbb{P}_2(g), \quad i = 1, 2.$$

As $\omega_1 \leq_{\text{inc}} \omega_2$ we conclude that $f_1 \leq_{\text{inc}} f_2$ and $f_1 + g \leq_{\text{inc}} f_2 + g$ for any $g \in F$. Therefore, $\mathbb{I}_B(f_1 + g) \leq \mathbb{I}_B(f_2 + g)$ as B is $\leq_{\text{inc}}$ increasing. Thus $\varphi(\omega_1) \leq \varphi(\omega_2)$ in view of (2.11). Consequently A is $\leq_{\text{inc}}$ increasing which completes the proof of the Theorem. $\square$

Let $(S, \mathcal{B} \leq)$ be a partially ordered measurable space. For a subset $B \subset S$, denote by C_B the *upper closure* of B, that is,

$$C_B = \{y \in S : y \geq x \text{ for some } x \in B\}. \tag{2.12}$$

We will assume further on that *partial order is measurable*, that is, for any $x \in S$, the sets $\{y : y \geq x\}$ and $\{y : y \leq x\}$ belong to $\mathcal{B}$. Thus, for any $x \in S$,

$$\{x\} = \{y : y \leq x\} \cap \{y : y \geq x\} \in \mathcal{B} \quad \text{and} \quad C_{\{x\}} \in \mathcal{B}.$$

Now we provide the result by Lindqvist, which can be regarded as "generalized Chebyshev inequality".

Theorem 2.17. ([269]) *A partially ordered space $(S, \leq)$ endowed with a σ-algebra $\mathcal{B}$ is totally ordered if and only if every probability measure on $\mathcal{B}$ is associated.*

Proof. Necessity. Suppose that S is not totally ordered. Then there are two non-comparable elements $x, y \in S$. Take a probability measure P such that

$$\mathsf{P}(\{x\}) = \mathsf{P}(\{y\}) = 1/2.$$

Assume that P is associated. Introduce increasing sets $C_{\{x\}}$ and $C_{\{y\}}$ according to (2.12). Then $\mathsf{P}(C_{\{x\}})\mathsf{P}(C_{\{y\}}) > 0$. However, since $y \notin C_{\{x\}}$ and $x \notin C_{\{y\}}$, one has $\mathsf{P}(C_{\{x\}} \setminus \{x\}) = \mathsf{P}(C_{\{y\}} \setminus \{y\}) = 0$. But then $\mathsf{P}(C_{\{x\}}C_{\{y\}}) = 0$. We come to a contradiction.

Sufficiency. Now let S be totally ordered. As everywhere, for a set $B \subset S$ we write $B^c = S \setminus B$. Take a probability measure P on S and any increasing sets $C_1, C_2 \in \mathcal{B}$. Note that either $C_1 C_2^c = \varnothing$ or $C_2 C_1^c = \varnothing$. If that were not true, there would exist different points $x \in C_1 C_2^c$, $y \in C_2 C_1^c$. But then neither $y \geq x$ nor $x \geq y$, which is impossible as S is totally ordered. Suppose at first that $C_1 C_2^c = \varnothing$. Then

$$\mathsf{P}(C_1 C_2) = \mathsf{P}(C_1) - \mathsf{P}(C_1 C_2^c) = \mathsf{P}(C_1) \geq \mathsf{P}(C_1)\mathsf{P}(C_2).$$

The same situation occurs for the case $C_2 C_1^c = \varnothing$. Now the proof of the claim follows by Lemma 2.11. $\square$

Example 2.18. Introduce in $\mathbb{R}^2$ the usual partial order $\leq$ (see (1.8)) and lexicographical order $\leq_{\mathrm{lex}}$. Take a binary random vector $X = (X_1, X_2)$ with values in $\mathbb{R}^2$ such that $cov(X_1, X_2) < 0$. Then obviously $(\mathbb{R}^2, \mathcal{B}(\mathbb{R}^2), \mathsf{P}_X, \leq) \notin \mathbf{A}$. However, due to Theorem 2.17, $(\mathbb{R}^2, \mathcal{B}(\mathbb{R}^2), \mathsf{P}_X, \leq_{\mathrm{lex}}) \in \mathbf{A}$. Thus we have an example of probability space endowed with two different partial orders such that the first one does not provide an association whereas the second one does.

To finish this subsection note that first correlation inequalities in general partially ordered spaces were obtained in the papers by Kamae, Krengel, O'Brien [223] and Ahlswede, Daykin [1]. Associated measures on partially ordered spaces were considered by Lindqvist ([269]) with application to the Bayesian risk theory. In subsection 4° we use Theorem 2.14 to study an association property for the solution of a stochastic differential equation.

3°. Markov processes. Monotonicity and association. Now we intend to describe a method of constructing new associated random vectors from a given one, using here the notation and basic properties of the Markov processes given in Appendix A.2. To this end, introduce

Definition 2.19. A homogeneous Markov process $\mathbf{X} = \{X_t, t \geq 0\}$ with values in S (endowed with partial order "$\leq$") is called *monotone* if, for all $t > 0$ and any nondecreasing function $f : S \to \mathbb{R}$, the image $T_t f(\cdot)$ is nondecreasing, and $\mathbf{X}$ is called *preserving positive correlations*, if, for any $t > 0$ and every positively correlated probability measure μ, the measure μT_t is positively correlated.

In other words $\mathbf{X} = \{X_t, t \geq 0\}$ preserves positive correlations if, for any $t > 0$ and any associated probability measure μ, the law of X_t is associated whenever $Law(X_0) = \mu$. Obviously, if $\mathbf{X}$ is monotone and $f : S \to \mathbb{R}$ is a nonincreasing function (i.e. if $(-f)$ is nondecreasing) then $(T_t f)(\cdot)$ is also nonincreasing.

Theorem 2.20. ([109, 192]) *Suppose that a homogeneous Markov chain $\mathbf{X}$ with values in a finite space S is monotone. Assume that the infinitesimal matrix A of $\mathbf{X}$ exists[9]. Then $\mathbf{X}$ preserves positive correlations if and only if*

$$A(x, y) = 0 \text{ for incomparable } x, y \in S. \tag{2.13}$$

In particular $\mathbf{X}$ has the desired property if S is totally ordered (e.g., $S \subset \mathbb{R}$).

Proof. Necessity. Let $\mathbf{X}$ preserve positive correlations, i.e. $Law(X_t)$ be associated for $t \geq 0$. Suppose that $A(x, y) > 0$ for some incomparable $x, y \in S$ (if x and y are incomparable then $x \neq y$ and therefore $A(x, y) \geq 0$). Let $Law(X_0) = \delta_x$ which is associated. As usual the symbol P_x will remind that the process $\mathbf{X}$ starts at the point x at the moment $t = 0$. Note that by the total probability formula

$$\mathsf{P}_x(X_t = y) = \sum_{z \in S} \mathsf{P}(X_0 = z)\mathsf{P}(X_t = y | X_0 = z) = \mathsf{P}(X_t = y | X_0 = x) = \mathsf{P}(t, x, y).$$

In view of (A.2.10) one has, as $t \to 0+$,

$$\mathsf{P}_x(X_t = y) = \begin{cases} 1 + A(x, x)t + o(t), & y = x, \\ A(x, y)t + o(t), & \text{otherwise.} \end{cases}$$

Consequently,

$$\mathsf{P}_x(X_t \geq x) \geq \mathsf{P}_x(X_t = x) \to 1, \quad t \to 0+.$$

If $z \geq y$ then $z \neq x$ as x and y are incomparable. Therefore, taking into account that S is finite, one has

$$\mathsf{P}_x(X_t \geq y) = \sum_{z \geq y} \mathsf{P}_x(X_t = z) = t\sum_{z \geq y} A(x, z) + o(t) = tA(x, y) + t\sum_{z > y} A(x, z) + o(t)$$

---

[9]It means existence of the infinitesimal matrix (A.2.10) for $(p(t, x, y))$ corresponding to $\mathbf{X}$.

as $t \to 0+$ where $z > y$ means that $z \geq y$ and $z \neq y$. Thus

$$P_x(X_t \geq x)P_x(X_t \geq y) = tA(x,y) + t\sum_{z>y} A(x,z) + o(t), \quad t \to 0+. \tag{2.14}$$

Furthermore,

$$P_x(X_t \geq x, X_t \geq y) = \sum_{z \geq x, z \geq y} P_x(X_t = z)$$

$$= t\sum_{z \geq x, z \geq y} A(x,z) + o(t) \leq t\sum_{z>y} A(x,z) + o(t), \quad t \to 0+, \tag{2.15}$$

as $z \geq x$ and $z \geq y$ imply that $z \neq x$ and $z \neq y$ (thus $\{z \geq x, z \geq y\} \subset \{z > y\}$). Relations (2.14) and (2.15) yield

$$q_t(x,y) := P_x(X_t \geq x, X_t \geq y) - P_x(X_t \geq x)P_x(X_t \geq y) \leq -tA(x,y) + o(t), \quad t \to 0+.$$

So there exists $t_0 > 0$ such that $q_t(x,y) < 0$ for $0 < t < t_0$.

On the other hand, for any $w \in S$, the function $\mathbb{I}\{z : z \geq w\}$ is nondecreasing (if z and w are incomparable then $\mathbb{I}\{z : z \geq w\}(z) = 0$). For nondecreasing functions $f(z) = \mathbb{I}\{z \geq x\}$ and $g(z) = \mathbb{I}\{z \geq y\}$ and any $t > 0$ the association of $Law(X_t)$ implies that

$$P_x(X_t \geq x, X_t \geq y) = E_x f(X_t)g(X_t) \geq E_x f(X_t)E_x g(X_t) \geq P_x(X_t \geq x)P_x(X_t \geq y),$$

here the symbol E_x reminds that expectations $Ef(X_t)$ and $Eg(X_t)$ are taken for X_t such that $X_0 = x$. However the last inequality contradicts the bound obtained for $q_t(x,y)$, hence we have established the necessity of condition (2.13).

Sufficiency. For any $h : S \to \mathbb{R}$, every $x \in S$ and all $u \geq 0$ in view of (A.2.12), (A.2.13) and (A.2.15) one has

$$(T_u h)(x) = h(x) + u\sum_{y \in S} A(x,y)h(y) + \theta_1 c_1 u^2|h| \tag{2.16}$$

where $\theta_1 = \theta_1(h,x,u)$, $|\theta_1| \leq 1$ and $c_1 = \|A\|^2 e^{\|A\|}/2$. Using (2.16) and the equality $\sum_{y \in S} A(x,y) = 0$, from (A.2.11) we have, for any functions $f, g : S \to \mathbb{R}$, $u \in [0,1]$,

$$T_u(fg)(x) - T_u f(x)T_u g(x)$$

$$= u\sum_{y \in S} A(x,y)(f(x) - f(y))(g(x) - g(y)) + \theta_2 c_2 u^2|f||g|, \tag{2.17}$$

here $\theta_2 = \theta_2(f,g,x,u)$ and $c_2 = c_2(\|A\|) > 0$.

Let f and g be nondecreasing functions. Then for comparable x and y one has

$$(f(x) - f(y))(g(x) - g(y)) \geq 0$$

and $A(x,y) = 0$ for incomparable x and y due to (2.13). Therefore for such f, g and $u \in [0,1]$ from (2.17) one obtains

$$T_u(fg)(x) \geq T_u f(x)T_u g(x) - c_2 u^2|f||g|. \tag{2.18}$$

Note (see (A.2.7)) that T_t is a linear operator such that

$$T_t h = h \text{ if } h(z) = c \text{ for all } z \in S \text{ and some } c \in \mathbb{R}, \qquad (2.19)$$

$$T_t h \geq 0 \text{ if } h(z) \geq 0 \text{ for all } z \in S. \qquad (2.20)$$

Moreover, for any $u \geq 0$

$$|T_u f| \leq \|T_u\| |f| \leq |f|. \qquad (2.21)$$

For $k \in \mathbb{N}$ and $u_1, \ldots, u_k \in [0,1]$, set $v = u_2 + \ldots + u_k$ and $w = u_2^2 + \ldots + u_k^2$. The process $\mathbf{X}$ is monotone, so $T_u f$ and $T_u g$ are nondecreasing functions. Thus (2.18)—(2.21) and induction on k guarantee the relation

$$T_{u_1+u_2+\ldots+u_k}(fg)(x) = (T_{u_1}(T_v(fg)))(x) \geq T_{u_1}\Big((T_v f)(T_v g) - c_2 w |f||g|\Big)(x)$$

$$\geq (T_{u_1}(T_v f))(x)(T_{u_1}(T_v g))(x) - c_2 u_1^2 |T_v f||T_v g| - c_2 w |f||g|$$

$$\geq (T_{u_1+u_2+\ldots+u_k} f)(x)(T_{u_1+u_2+\ldots+u_k} g)(x) - c_2(u_1^2 + u_2^2 + \ldots + u_k^2)|f||g|. \qquad (2.22)$$

Taking $t > 0$, $n > t$ $(n \in \mathbb{N})$ and applying (2.22) one may conclude that

$$(T_t fg)(x) = (T_{n(t/n)} fg)(x) \geq (T_t f)(x)(T_t g)(x) - c_2 n \left(\frac{t}{n}\right)^2 |f||g|.$$

Letting $n \to \infty$ we see that, for any $t > 0$, nondecreasing $f, g : S \to \mathbb{R}$ and $x \in S$,

$$(T_t fg)(x) \geq (T_t f)(x)(T_t g)(x). \qquad (2.23)$$

Now, if μ is an associated probability measure, then for such f and g we infer from (A.2.7) and (2.23) that, for $t > 0$,

$$\int_S fg \, d(\mu T_t) = \sum_{y \in S} f(y)g(y) \sum_{x \in S} \mu(x) p(t, x, y) = \int_S (T_t fg) d\mu$$

$$\geq \int_S T_t f T_t g \, d\mu \geq \int_S T_t f \, d\mu \int_S T_t g \, d\mu = \int_S f \, d(\mu T_t) \int_S g \, d(\mu T_t),$$

so μT_t is associated. Therefore $\mathbf{X}$ preserves positive correlations. $\square$

Remark 2.21. One can interpret condition (2.13) for a Markov chain $\mathbf{X}$ in terms of trajectories' jumps. If A is the infinitesimal matrix of $\mathbf{X}$ the trajectories can be constructed as follows (see, e.g., [272, p. 599]). Let $X_0 = x$ with probability $\mu_0(\{x\})$, $x \in S$. Then $\mathbf{X}$ stays at a point x during the random time τ_0. If $A(x, x) = 0$ then $\tau_0 = +\infty$ a.s., that is $X(t) = x$ for all $t \geq 0$. If $A(x, x) < 0$ then τ_0 has exponential distribution with parameter $-A(x, x)$. At the moment τ_0 the trajectory makes a jump to a point $y \neq x$ with probability $-A(x, y)/A(x, x)$. Let τ_1 denote the sojourn time at the point y. Again there is an alternative. If $A(y, y) = 0$ then $\tau_1 = +\infty$ a.s. Otherwise τ_1 has exponential distribution with parameter $-A(y, y)$. Now (if $\tau_0 \vee \tau_1 < \infty$) the trajectory makes a jump to the point $z \neq y$ with probability $-A(y, z)/A(y, y)$, etc. Therefore condition (2.13) means that for incomparable x and y the jumps from x to y or from y to x are possible with probability zero.

A simple condition on infinitesimal matrix A ensuring monotonicity of $\mathbf{X}$ is provided by

Theorem 2.22. *([268]) A homogeneous Markov chain* $\mathbf{X} = \{X_t, t \geq 0\}$ *with values in a finite space* S *(endowed with* σ-*algebra* $\mathcal{B} = 2^S$ *and measurable partial order "$\leq$") is monotone if*

$$\sum_{z \in U} A(x, z) \leq \sum_{z \in U} A(y, z) \tag{2.24}$$

for any increasing set $U \subset S$ *and all* $x, y \in S$ *such that* $x \leq y$ *and* x, y *belong simultaneously either to* U *or to* $S \setminus U$.

Proof. Let $f : S \to \mathbb{R}$ be a nondecreasing function. We claim that, for any $x, y \in S$, $x \leq y$,

$$(T_t f)(x) \leq (T_t f)(y), \quad t > 0. \tag{2.25}$$

In view of linearity of T_t (see (A.2.7)) one can assume that $0 \leq f < 1$.

At first we prove that there exists some $h > 0$ determined by the matrix A such that the function $(I + tA)f$ is nondecreasing for any $t \in (0, h)$, that is, for $x, y \in S$, $x \leq y$, one has

$$((I + tA)f)(x) - ((I + tA)f)(y) \leq 0. \tag{2.26}$$

We employ the fact that f is a pointwise limit of linear combinations of increasing sets indicators taken with nonnegative coefficients, see (2.6). Consequently it suffices to check (2.26) for $f = \mathbb{I}_U$, U being an increasing set. Let us show that, for $x, y \in S$, $x \leq y$, and any $t \in (0, h)$, the left-hand side of (2.26) can be written as follows:

$$f(x) + t \sum_{z \in S} A(x, z) f(z) - f(y) - t \sum_{z \in S} A(y, z) f(z)$$

$$= f(x) - f(y) + t \left(\sum_{z \in U} A(x, z) - \sum_{z \in U} A(y, z) \right) \leq 0, \tag{2.27}$$

if $h = h(A)$ is defined by

$$h = \left(2 \max_{x \in S} \sum_{z \in S} |A(x, z)| \right)^{-1} \tag{2.28}$$

where $0^{-1} := +\infty$. Really, if x and y both belong simultaneously either to U or to $S \setminus U$, then inequality (2.27) follows by Theorem's assumption (without additional requirements on $t > 0$). The case $x \in U, y \notin U$ is impossible since U is increasing and $y \geq x$. Finally, if $x \notin U$ but $y \in U$, then $f(x) = 0$ and $f(y) = 1$, so

$$f(x) - f(y) + t \left(\sum_{z \in U} A(x, z) - \sum_{z \in U} A(y, z) \right)$$

$$\leq -1 + 2t \max_{x \in S} \sum_{z \in S} |A(x, z)| \leq 0$$

because of the choice of h in (2.28).

Consequently, if $f : S \to \mathbb{R}$ is nondecreasing, then for any $n \in \mathbb{N}$ the function $(I + sA)^n f$ is also nondecreasing, for $s \in (0, h(f, A))$. By (A.2.13) and (A.2.14) for any $t > 0$ we have the pointwise equality

$$T_t f = \lim_{n \to \infty} \left(I + \frac{t}{n} A \right)^n f,$$

and by the previous argument the function inside the limit is nondecreasing for all n so large that $t/n < h(f, A)$. Thus $T_t f$ is nondecreasing which proves the Theorem.

$\square$

If the phase space S is a part of real line (with a partial order inherited from the usual one on $\mathbb{R}$), e.g., the set $\{1, \ldots, n\}$, then each distribution on S is automatically associated by Theorem 1.8, (a). Thus in this case if $\mathbf{X}$ is monotone and its infinitesimal matrix A exists, then, for any initial distribution $\mu = Law(X_0)$, the measure $Law(X_t)$ is associated for all $t \geq 0$.

Example 2.23. Let $\mathbf{X}$ be a Markov chain with values in a space $S = \{s_1, s_2\}$ where $s_1 \leq s_2$. Assume that there exists the corresponding infinitesimal matrix A. Then obviously condition (2.24) holds. Therefore such Markov chain is monotone.

We have discussed the properties of $Law(X_t)$ for each $t \geq 0$. Now a non-trivial question is what happens for finite-dimensional distributions of $\mathbf{X}$ of order greater than one.

Definition 2.24. A process[10] $\{X(t), t \geq 0\}$ is called *time-associated* if it is an associated random process as a system of random variables.

The time-association of Markov chains and its applications were studied, e.g., by Hjort, Natvig and Funnemark.

Theorem 2.25. ([199]) *A Markov process* $\mathbf{X} = \{X(t), t \geq 0\}$ *with values in a finite subset of* $\mathbb{R}$ *is time-associated if it is standard and monotone.*

Proof will be given in the first subsection of Section 3. However, in view of Example 2.23 it is interesting to mention now a simple

Corollary 2.26. *If* $\mathbf{X}$ *is a Markov chain with state space* $S = \{s_1, s_2\} \subset \mathbb{R}$ *and the infinitesimal matrix* A *exists then* $\mathbf{X}$ *is automatically associated in time. Moreover, the covariance function of* X *decreases exponentially at infinity, i.e.*

$$cov(X_s, X_{s+t}) \leq Ce^{-\lambda t}, \quad s, t \geq 0,$$

where $C, \lambda > 0$ *are independent of* s *and* t.

---

[10]It is not required in general that the state space of the process is finite.

Proof. Evidently one may assume $s_1 = 0$, $s_2 = 1$. Let $X = \{X_t, t \geq 0\}$ be a standard (homogeneous) Markov chain taking values in $S = \{0, 1\} \subset \mathbb{R}$. Set $P(t) = (p_{ij}(t))_{i,j=0}^1$ where $p_{ij}(t) = P(X(t) = j | X(0) = i)$, $i, j \in \{0, 1\}$, $t \geq 0$. For some $\alpha, \beta \geq 0$, in view of (A.2.13) one has

$$A = \begin{pmatrix} -\alpha & \alpha \\ \beta & -\beta \end{pmatrix} \quad \text{and} \quad P(t) = e^{tA}, \quad t \geq 0, \tag{2.29}$$

where A is the infinitesimal matrix of X. The eigenvalues of A are $\lambda_0 = 0$ and $\lambda_1 = -\alpha - \beta$. If $\alpha = \beta = 0$ then $P(t) = I$ for all $t \geq 0$, here I is the identity matrix. Now let $\alpha + \beta \neq 0$. We can find a matrix B such that $BAB^{-1} = D$ where $D = diag(\lambda_0, \lambda_1)$ is a diagonal matrix. Therefore

$$P(t) = Be^{tD}B^{-1} = B \begin{pmatrix} 1 & 0 \\ 0 & e^{-(\alpha+\beta)t} \end{pmatrix} B^{-1}, \quad t \geq 0.$$

Furthermore, for $t \geq 0$, one obtains

$$p_{00}(t) = \frac{\beta}{\alpha+\beta} + \frac{\alpha}{\alpha+\beta}e^{-(\alpha+\beta)t}, \quad p_{01}(t) = \frac{\alpha}{\alpha+\beta} - \frac{\alpha}{\alpha+\beta}e^{-(\alpha+\beta)t},$$

$$p_{10}(t) = \frac{\beta}{\alpha+\beta} - \frac{\beta}{\alpha+\beta}e^{-(\alpha+\beta)t}, \quad p_{11}(t) = \frac{\alpha}{\alpha+\beta} + \frac{\beta}{\alpha+\beta}e^{-(\alpha+\beta)t}.$$

Of course we can also use the Kolmogorov backward or forward equations (i.e. $p'_{ij}(t) = \sum_{k=0}^1 a_{ik}p_{kj}(t)$ or $p'_{ij}(t) = \sum_{k=0}^1 p_{ik}(t)a_{kj}$ respectively, with $(a_{ij})_{i,j=0}^1 = A$) to find $p_{ij}(t)$.

Set $p_i = P(X(0) = i)$, $i = 0, 1$. Then

$$cov(X(t), X(0)) = P(X(t) = X(0) = 1) - P(X(t) = 1)P(X(0) = 1)$$

$$= p_1 p_{11}(t) - (p_0 p_{01}(t) + p_1 p_{11}(t)) = p_0 p_1 e^{-(\alpha+\beta)t}, \quad t \geq 0. \quad \square$$

Note also that the same claim is valid for a discrete time (homogeneous) Markov chain $Y = \{Y(n), n \in \mathbb{Z}_+\}$ taking values in $S = \{0, 1\}$. In fact, if the transition matrix $P = (p_{ij})_{i,j=0}^1$ with

$$p_{ij} = P(Y(n+1) = j | Y(n) = i), \quad i, j \in \{0, 1\}, \quad n \in \mathbb{Z}_+,$$

is degenerate, then $\{Y(n), n \in \mathbb{Z}_+\}$ consists of independent random variables, hence associated. If $\det P \neq 0$, there exists a matrix A of the form (2.29) such that $e^A = P$, and therefore $P(n) = e^{nA} = P^n$, here $P(n)$ is the matrix of transition in n steps. Thus we can embed Y to some standard homogeneous Markov chain $X = \{X(t), t \geq 0\}$, that is the finite-dimensional distributions of $Y = \{Y(n), n \in \mathbb{Z}_+\}$ and $X = \{X(n), n \in \mathbb{Z}_+\}$ coincide. Then the time-association of X implies that Y possesses the same property.

The exponential decrease of covariances, together with boundedness of random variables, is a very strong condition permitting to establish many limit theorems (see Chapters 3—8).

4°. Markov processes and diffusion. We shall consider $S = \mathbb{R}^n$ (with the usual partial order, see (1.8)) and $\mathcal{B} = \mathcal{B}(\mathbb{R}^n)$.

The *diffusion processes* constitute an important class of Markov processes. One often introduces such processes by means of a generator having the form

$$(Gf)(x) = \frac{1}{2} \sum_{i,j=1}^{n} a^{i,j}(x) \frac{\partial^2 f(x)}{\partial x_i \partial x_j} + \sum_{i=1}^{n} b^i(x) \frac{\partial f(x)}{\partial x_i}, \ x \in \mathbb{R}^n. \tag{2.30}$$

Here $a(x) = (a^{ij}(x))_{i,j=1}^{n}$ is a symmetric matrix, positively definite for each x, and $b(x) = (b^j(x))_{j=1}^{n}$ is a vector function. More exactly one supposes that G has such form for all f which are "smooth enough" (e.g., $f \in C_0^2(\mathbb{R}^n)$, that is f has a compact support and possesses continuous partial derivatives of the second order). Moreover, one assumes that $a(x)$ and $b(x)$ have some "nice properties". After that one considers an appropriate extension of G as a linear operator in a due Banach space. In general, one also assumes that the "corresponding" Markov process has continuous trajectories. Note in passing that there arise many problems on this way. For instance, it would be desirable to construct a Markov semigroup $(T_t)_{t \geq 0}$ having the generator G (that is to construct a *contraction*[11] semigroup $(T_t)_{t \geq 0}$ such that T_t is given by formula (A.2.16) with a Markov transition function $P(t, x, B)$ satisfying conditions $(1°) - (4°)$. If we can obtain a Markov transition function then it is not difficult to introduce the consistent finite dimensional distributions and construct the related Markov process via the Kolmogorov fundamental theorem (for the Borel space $S = \mathbb{R}^n$). Of course there is the classical Hille – Iosida theorem (see, e.g., [151, Ch.1, §4]) providing the necessary and sufficient conditions to have G as generator of a contraction semigroup in Banach space. However this approach is not easy when we want to realize the construction of a desired diffusion process.

There is another approach based on the Itô stochastic differential equations (see, e.g., [225, 311]). Namely, it is possible to define $\mathbf{X} = \{X_t, t \geq 0\}$ as a solution of the equation

$$dX_t = b(X_t)dt + \sigma(X_t)dW_t, \ t \geq 0, \ X_0 = x, \tag{2.31}$$

where $\{W_t, t \geq 0\}$ is a Brownian motion in $\mathbb{R}^n$, $b : \mathbb{R}^n \to \mathbb{R}$, $\sigma : \mathbb{R}^n \to \mathbb{R}^{n \times n}$ and

$$\|b(x) - b(y)\| + \|\sigma(x) - \sigma(y)\|_2 \leq D|x - y|, \ x, y \in \mathbb{R}^n. \tag{2.32}$$

Here $\| \cdot \|$ is Euclidean norm in $\mathbb{R}^n$ and $\|\sigma(x)\|_2 = (\sum_{i,j=1}^{n} (\sigma^{i,j}(x))^2)^{1/2}$. In fact one uses in (2.31) $\sigma(x) = (a(x))^{1/2}$, that is the square root of the matrix $a(x)$ occurring in (2.30), and $b(\cdot)$ is the same as in (2.30). Fortunately one can provide sufficient conditions in terms of $a(x)$ instead of $\sigma(x)$ in (2.32) to guarantee the existence of the solution of (2.31) which is a Markov process with corresponding generator (2.30).

Now we assume that an operator G in (2.30) is such that coefficients $a^{i,j}(x), b^i(x)$ for $i, j \in \{1, \ldots, n\}$ have bounded partial derivatives of all orders and the inverse matrix $a(x)^{-1}$ exists and its entries are bounded on $\mathbb{R}^n$. These conditions ensure

---

[11]This means $\|T_t\| \leq 1$ for all $t \geq 0$.

that there exists a diffusion process $\mathbf{X}$ with generator G. For many important details (which we have to omit here) concerning the theory of diffusion processes and stochastic differential equations see, e.g., [151], [412], [394], [347].

To formulate the next result we define the class $\mathbb{L}_1 = \{f : \mathbb{R}^n \to \mathbb{R}, Lip(f) < \infty\}$ of Lipschitz functions where the *Lipschitz constant*

$$Lip(f) := \sup_{x,y \in \mathbb{R}^n,\, x \neq y} \frac{|f(x) - f(y)|}{\|x - y\|_1}$$

and $\|x\|_1 = \sum_{i=1}^n |x_i|$ for $x = (x_1, \ldots, x_n) \in \mathbb{R}^n$.

Theorem 2.27. ([196]) *Let* $\mathbf{X} = \{X_t, t \geq 0\}$ *be a Markov (diffusion) process introduced above. Suppose that for all* i, j, k *with* $k \notin \{i, j\}$ *one has*

$$\frac{\partial a^{ij}}{\partial x_k} = 0, \quad \frac{\partial b^j}{\partial x_k} \geq 0.$$

Then $(T_t)_{t \geq 0}$ *is monotone and* $Lip(T_t f) \leq e^{tK} Lip(f)$ *for* $f \in \mathbb{L}_1$ *and* $t \geq 0$ *where*

$$K = \sup_x \max_i \Big(\partial b^i(x)/\partial x_i + \sum_{j \neq i} |\partial b^j(x)/\partial x_i| \Big).$$

Let, moreover, $a^{ij}(x) \geq 0$ *for all* i, j, x. *Then* $(T_t)_{t \geq 0}$ *preserves positive correlations.*

Because of its length, we do not reproduce the proof of this result.

Corollary 2.28. ([196, 363]) *Let* $p(x)$ *be a strictly positive infinitely differentiable probability density in* $\mathbb{R}^n$ *such that, for all* $i \neq j$ *and all* $x \in \mathbb{R}^n$,

$$\frac{\partial^2 \log p(x)}{\partial x_i \partial x_j} \geq 0.$$

Then the distribution in $\mathbb{R}^n$ *having density* $p(x)$ *is associated.*

Proof. Set $h(x) = \log p(x)$ and consider the differential operator

$$G = \Delta + \sum_{j=1}^n \frac{\partial h}{\partial x_j} \frac{\partial}{\partial x_j}$$

where the Laplace operator $\Delta = \sum_{j=1}^n \partial^2/\partial x_j^2$. By Theorem 2.27 the Markov semigroup with infinitesimal operator[12] G is monotone and preserves positive correlations. Furthermore, by Khasminskii's theorem (Theorem 5 in [229]) and Echeverria's theorem ([154]) this semigroup is *ergodic*, that is, for any initial distribution ν, the measures νP^t converge weakly (as $t \to \infty$) to the invariant measure μ, which is the unique measure such that

$$\int_{\mathbb{R}^n} G f(x) d\mu(x) = 0 \quad \text{for all } f \in C_0^\infty.$$

The measure $\mu(dx) = p(x)dx$ satisfies the last relation, hence it is the invariant measure. If we take ν to be associated (for example, concentrated at a single point), then all the νP^t are also associated. By Theorem 1.8, (e), μ is also associated. $\square$

---

[12]More precisely, the infinitesimal operator is an extension of G.

Applying the same ergodic argument, one deduces the part of Theorem 2.1. Namely, it is enough to consider the Ornstein–Uhlenbeck differential operator

$$G = \sum_{i,j=1}^{n} \sigma_{ij} \frac{\partial^2}{\partial x_i \partial x_j} - \sum_{j=1}^{n} x_j \frac{\partial}{\partial x_j},$$

since the corresponding invariant distribution is the $(0, \Sigma)$-normal law.

One can extend Corollary 2.28, using approximation by smooth functions, to comprise $h \in C^2(\mathbb{R}^n)$, see [196]. A similar statement proved by Thomas ([400]) uses a functional-analytic representation of the semigroup $(T_t)_{t \geq 0}$.

Now we provide the result by Barbato concerning the association property of the solution of a stochastic differential equation

$$\begin{cases} dX_t = b(X_t)dt + \sigma(X_t)dW_t, \\ X_0 = x_0, \end{cases} \tag{2.33}$$

studied on a Wiener space $(\Omega, \mathcal{F}, \mathsf{P}) = (C_0[0, T], \mathcal{B}(C_0[0, T]), \mathbb{W})$ endowed with a partial order $\leq_{\text{inc}}$ (see subsection 3°).

Theorem 2.29. ([18]) *Let $b : \mathbb{R} \to \mathbb{R}$ be a Lipschitz function and the same be true of the derivative σ' of $\sigma : \mathbb{R} \to [\varepsilon, M]$, for some $\varepsilon, M > 0$.*

Then there exists a stochastic process $X = \{X_t, t \in [0, T]\}$ (defined for all $\omega \in \Omega$) such that

(i) X is a strong solution of (2.33);

(ii) for each $t \in [0, T]$, $X_t(\cdot)$ is an increasing function on $(\Omega, \mathcal{F}, \mathsf{P}, \leq_{\text{inc}})$.

Proof. (i) Introduce

$$F(x) := \int_0^x \frac{1}{\sigma(s)} ds, \quad x \in \mathbb{R},$$

and the inverse function $F^{-1}(x) =: G(x)$. By the conditions imposed, F and G are strictly increasing and, for $x \in \mathbb{R}$,

$$G'(x) = \sigma(G(x)), \quad G''(x) = \sigma(G(x))\sigma'(G(x)). \tag{2.34}$$

Set

$$\widehat{b}(x) := \frac{b(G(x))}{\sigma(G(x))} - \frac{1}{2}\sigma'((G(x)), \quad x \in \mathbb{R}.$$

Note that $\widehat{b}$ is a locally Lipschitz function as in view of (2.34), for any $[u, v] \subset \mathbb{R}$,

$$\sup_{x,y \in [u,v]} \frac{|\widehat{b}(x) - \widehat{b}(y)|}{|x - y|} \leq M \left(\frac{1}{\varepsilon} Lip(b) + \frac{1}{\varepsilon^2} \sup_{z \in \Delta} |b(z)| \sup_{z \in \Delta} |\sigma'(z)| + \frac{1}{2} Lip(\sigma') \right)$$

where $\Delta = [G(u), G(v)]$. Moreover, taking into account that $x/M \leq F(x) \leq x/\varepsilon$ for $x \geq 0$ and $x/\varepsilon \leq F(x) \leq x/M$ for $x < 0$ (and consequently $\varepsilon y \leq G(y) \leq My$ for all $y \geq 0$ and $My \leq G(y) \leq \varepsilon y$ for $y < 0$) we may write

$$|\widehat{b}(x)| \leq c_1 + c_2|x|, \quad x \in \mathbb{R}$$

where

$$c_1 = \frac{1}{\varepsilon}|b(0)| + \frac{1}{2}|\sigma'(0)|, \quad c_2 = \frac{M}{\varepsilon}Lip(b) + \frac{M}{2}Lip(\sigma').$$

Therefore, a theorem by Engelbert–Schmidt (see, e.g., [100], p. 10) guarantees the existence of a strong solution $Z = \{Z_t, t \geq 0\}$ to a stochastic differential equation

$$\begin{cases} dZ_t = \widehat{b}(Z_t)dt + dW_t, \\ Z_0 = F(x_0), \end{cases} \tag{2.35}$$

which is unique (i.e. if $\widetilde{Z} = \{\widetilde{Z}_t, t \geq 0\}$ is a solution of (2.35) then the trajectories of Z and $\widetilde{Z}$ coincide a.s.)

Now we show that $X(t) = G(Z_t)$, $t \in [0, T]$, is a (unique strong) solution of (2.33). Indeed, employing the Itô formula one has

$$dG(Z_t) = G'(Z_t)dZ_t + \frac{1}{2}G''(Z_t)dt.$$

Thus (2.34) and (2.35) yield

$$dX_t = \sigma(X_t)dZ_t + \frac{1}{2}\sigma(X_t)\sigma'(X_t)dt,$$

$$dX_t = b(X_t)dt - \frac{1}{2}\sigma(X_t)\sigma'(X_t)dt + \sigma(X_t)dW_t + \frac{1}{2}\sigma(X_t)\sigma'(X_t)dt,$$

$$dX_t = b(X_t)dt + \sigma(X_t)dW_t,$$

which completes the proof of (i).

(ii) As $X_t = G(Z_t)$ with increasing function $G : \mathbb{R} \to \mathbb{R}$, it is sufficient to check that $\omega_1 \leq_{\mathrm{inc}} \omega_2$ implies $Z_t(\omega_1) \leq Z_t(\omega_2)$ for all $t \geq 0$. Assume that the last statement is not true. Then there exists $t^* \in (0, T]$ such that $Z_{t^*}(\omega_1) > Z_{t^*}(\omega_2)$. Set $t_0 := \sup\{t \in [0, T] : Z_t(\omega_1) = Z_t(\omega_2)$. Then due to the continuity of the trajectories of Z we have $t_0 < t^*$ and $Z_t(\omega_1) > Z_t(\omega_2)$ for each $t \in (t_0, t^*]$. Take $L > Lip(\widehat{b})$ and choose $t_* \in (t_0, t^*]$ in such a way that $0 < t_* - t_0 < 1/L$. Note that in view of (2.35) for $t \in [t_0, T]$

$$Z_t = Z_{t_0} + \int_{t_0}^{t} \widehat{b}(Z_s)ds + W_t - W_{t_0}.$$

Taking into account that $Z_{t_0}(\omega_1) = Z_{t_0}(\omega_2)$ we come, for any $t \in [t_0, t_*]$, to the following relation

$$Z_t(\omega_1) - Z_t(\omega_2) = \int_{t_o}^{t} (\widehat{b}(Z_s(\omega_1)) - \widehat{b}(Z_s(\omega_2)))ds + (\omega_1(t) - \omega_1(t_0)) - (\omega_2(t) - \omega_2(t_0))$$

(we use the canonical representation for W), and due to the inequality $\omega_1 \leq_{\mathrm{inc}} \omega_2$ conclude that

$$\sup_{t \in [t_0, t_*]} |Z_t(\omega_1) - Z_t(\omega_2)| \leq L(t_* - t_0) \sup_{t \in [t_0, t_*]} |Z_t(\omega_1) - Z_t(\omega_2)|.$$

However, $0 < L(t_* - t_0) < 1$ implies $\sup_{t \in [t_0, t_*]} |Z_t(\omega_1) - Z_t(\omega_2)| = 0$. The last equality cannot be satisfied since $Z_t(\omega_1) - Z_t(\omega_2) > 0$ for all $t \in (t_0, t_*]$. $\square$

Corollary 2.30. ([18]) *Let the conditions of Theorem 2.29 be satisfied and $X = \{X_t, t \in [0, T]\}$ be the strong solution of (2.33). Then a family X is associated.*

The proof follows immediately from Theorems 2.14, 2.29 and Corollary 2.13.

3 Random measures

This Section is in fact a continuation of the previous one as we study here an important class of random systems described by means of random measures. On this way, for example, the full description of positively and negatively associated stable random vectors is provided.

1°. Poisson spatial stochastic process. We start with a definition of kernel which plays an important role in various applications.

Definition 3.1. Given two measurable spaces $(V, \mathcal{A})$ and $(S, \mathcal{B})$, a mapping $\mu : V \times \mathcal{B} \to \overline{\mathbb{R}}_+$ is called a (*probability*) *kernel* from V to S (more exactly from $(V, \mathcal{A})$ to $(S, \mathcal{B})$) if the function $\mu(s, B)$ is $\mathcal{A}$-measurable for any fixed $B \in \mathcal{B}$ and a (probability) measure in $B \in \mathcal{B}$ for every fixed $s \in V$. Kernels on a basic probability space $(\Omega, \mathcal{F}, \mathsf{P})$, that is for $(V, \mathcal{A}) = (\Omega, \mathcal{F})$, are called *random measures*. In the last case we often write M instead of μ and omit $\omega \in \Omega$, just saying M is given on $\mathcal{B}(S)$.

Recall that a probability space $(\Omega, \mathcal{F}, \mathsf{P})$ is *complete* if the relations $D \subset A$, $A \in \mathcal{F}$ and $\mathsf{P}(A) = 0$ imply that $D \in \mathcal{F}$ (and consequently $\mathsf{P}(D) = 0$). It is well-known that without loss of generality one may assume that the probability space is complete (see, e.g., [383, Ch. II, §3.1]). Then, if ξ is a random vector with values in $\mathbb{R}^n$ and $\eta = \eta(\omega)$ is a function defined on Ω with values in $\mathbb{R}^n$ such that $\{\omega : \eta(\omega) \neq \xi(\omega)\} \subset A$ where $\mathsf{P}(A) = 0$, we conclude that η is also a random vector (with the same law as ξ). If $(\zeta_k)_{k \in \mathbb{N}}$ is a sequence of random vectors in $\mathbb{R}^n$ and $\zeta_k \to \zeta$ P-a.s., $k \to \infty$, then ζ is a random vector.

Recall also (see, e.g., [239, Ch. 1, §5 2,3]) that a nonempty family $\mathcal{R}$ consisting of subsets of S is a *semiring* if $\varnothing \in \mathcal{R}$, $\mathcal{R}$ is closed under taking finite intersections, and for any sets $B, C \in \mathcal{R}$ with $C \subset B$ there exists a finite collection of sets $A_1, \ldots, A_n \in \mathcal{R}$ such that $B \setminus C = \cup_{k=1}^n A_k$. The family of blocks $(a, b] \subset \mathbb{R}^n$ is a semiring $((a, a] = \varnothing)$. A family $\mathcal{R}$ is a *ring* if $A, B \in \mathcal{R}$ implies $A \cap B \in \mathcal{R}$, $A \triangle B \in \mathcal{R}$ (then $\mathcal{R}$ is closed under taking differences, intersections and unions of finite number of sets from $\mathcal{R}$). For example, finite unions of blocks $(a, b] \subset \mathbb{R}^n$ form a ring. A system $\mathcal{B}_0(S)$ of all bounded Borel sets in a metric space (S, ρ) is a ring. For a family $\mathcal{R}$ of subsets of a set S there is a smallest ring containing $\mathcal{R}$. If S belongs to a ring one says that this ring is an *algebra*.

Let $S = \mathbb{R}^n$ (endowed with Euclidean distance) and Λ be a *locally finite* (non-random) measure[13], i.e. $\Lambda(B) < \infty$ for any $B \in \mathcal{B}_0(\mathbb{R}^n)$. As usual, the space $(\mathbb{R}^n, \mathcal{B}(\mathbb{R}^n), \Lambda)$ is assumed complete. We exclude the trivial case $\Lambda \equiv 0$.

Definition 3.2. A *Poisson spatial stochastic process* in $\mathbb{R}^n$ with (locally finite) *intensity measure* Λ is a random process $Z = \{Z(B), B \in \mathcal{B}_0(\mathbb{R}^n)\}$, defined on a probability space $(\Omega, \mathcal{F}, \mathsf{P})$, such that

---

[13]In general $\Lambda(B) \in \mathbb{R}_+ \cup \{\infty\}$ for $B \in \mathcal{B}(\mathbb{R}^n)$.

1) for any $B \in \mathcal{B}_0(\mathbb{R}^n)$ one has $Z(B) \sim Pois(\Lambda(B))$, i.e.

$$P(Z(B) = N) = \frac{(\Lambda(B))^N}{N!} \, e^{-\Lambda(B)}, \quad N = 0, 1, \ldots; \tag{3.1}$$

2) for every integer $k > 1$ and any pairwise disjoint sets $B_1, \ldots, B_k \in \mathcal{B}_0(\mathbb{R}^n)$, the random variables $Z(B_1), \ldots, Z(B_k)$ are independent.

One says that Z has *intensity function* $\lambda(x)$, $x \in \mathbb{R}^n$, if

$$\Lambda(B) = \int_B \lambda(x) dx, \quad B \in \mathcal{B}_0(\mathbb{R}^n), \tag{3.2}$$

that is Λ is absolutely continuous with respect to the Lebesgue measure in $\mathbb{R}^n$.

Recall that if $\xi \sim Pois(a)$ for some $a \geq 0$ then its characteristic function

$$\varphi_\xi(t) = \exp\{a(e^{it} - 1)\}, \quad t \in \mathbb{R}, \quad i^2 = -1 \tag{3.3}$$

($\xi = 0$ a.s. if $a = 0$). Note that $\mathsf{E}\xi = a$. Thus $\Lambda(B) = \mathsf{E}Z(B)$ for $B \in \mathcal{B}_0(\mathbb{R}^n)$.

Remark 3.3. The above conditions 1) and 2) are equivalent to the following one: for every $k \in \mathbb{N}$, any pairwise disjoint $B_1, \ldots, B_k \in \mathcal{B}_0(\mathbb{R}^n)$ and all k-tuples $t_1, \ldots, t_k \in \mathbb{R}$,

$$\varphi_{Z(B_1), \ldots, Z(B_k)}(t_1, \ldots, t_k) = \prod_{r=1}^{k} \exp\{\Lambda(B_r)(e^{it_r} - 1)\}, \tag{3.4}$$

here φ stands for the characteristic function of a random vector $(Z(B_1), \ldots, Z(B_k))$.

It is convenient (see, e.g., [119]) to view Z as a collection of "random points" x_i in $\mathbb{R}^n$ such that probability to have N such points in a set $B \in \mathcal{B}_0(\mathbb{R}^n)$ is given by (3.1). One writes $Z = \{x_i\}$. However this with values in $\mathbb{R}^n$ which are used to form certain sums with random number of random summands. We give the details of this well-known construction in Appendix A.3 because we will use it.

A Poisson spatial stochastic process is an example of random measure obeying the following

Definition 3.4. Let M be a random measure defined on $\mathcal{B}(S)$ such that $M(B) < \infty$ a.s. for any $B \in \mathcal{B}_0(S)$. This random measure is called *independently scattered* if, for each $k \in \mathbb{N}, k > 1$, and any collection of pairwise disjoint sets $B_1, \ldots, B_k \in \mathcal{B}_0(S)$, the random variables $M(B_1), \ldots, M(B_k)$ are independent.

Theorem 3.5. ([87]) *A family $\{M(B), B \in \mathcal{B}_0(S)\}$ is associated whenever M is an independently scattered random measure.*

Proof. Let $B_1, \ldots, B_n \in \mathcal{B}_0(S)$. Then there are a finite number of pairwise disjoint sets $C_1, \ldots, C_m \in \mathcal{B}_0(S)$ such that every B_r is a union of some of the C_j. The random variables $M(C_1), \ldots, M(C_m)$ are independent, hence Theorem 1.8, (d), implies that $M(B_1), \ldots, M(B_n)$ are also associated. $\square$

2°. Shot-noise random fields. Now we will discuss the celebrated examples of random measures provided by the Poisson spatial stochastic process.

One often employs the notation

$$Y(u) = \sum_i F(u, x_i, \xi_i) \qquad (3.5)$$

where $u \in T$ (T is some set), $\{x_i\}$ is a Poisson spatial process in $\mathbb{R}^n$, $\{\xi_i\}$ is a countable family of i.i.d. random variables (or vectors) which is independent of $\{x_i\}$, and F is a given deterministic function. In fact, we consider an array $\{\xi, \xi_{mj}\}_{m,j \in \mathbb{N}}$ of i.i.d. random variables (or vectors) which is independent of $(\tau_m, X_{mj})_{m,j \in \mathbb{N}}$, introduced to construct a Poisson random field in (A.3.1), and use instead of (3.5) the following formula

$$Y(u) = \sum_{m=1}^{\infty} \sum_{j=1}^{\tau_m} F(u, X_{mj}, \xi_{mj}), \quad u \in T. \qquad (3.6)$$

We stipulate that formula (3.5) is meaningful if (3.6) is well defined, in particular, (the domain of definition of) F admits the substitution of u, X_{mj}, ξ_{mj} as its arguments and the result is a random variable for each $u \in T$. It is natural to occur in the situation when the sum does not depend on the order of summation. To this end one supposes that a.s.

$$\sum_i |F(u, x_i, \xi_i)| < \infty, \quad \text{i.e.} \quad \sum_{m=1}^{\infty} \sum_{j=1}^{\tau_m} |F(u, X_{mj}, \xi_{mj})| < \infty, \quad u \in T. \qquad (3.7)$$

Note also that it is necessary to verify that the law of a random function introduced in (3.5), i.e. in (3.6), (as well as the expression in (3.7)) is independent of the choice of a partition $K_1, K_2, \ldots$ and the random elements used to construct $\{x_i\}$. This property is a corollary of the following two simple lemmas.

Lemma 3.6. *Let $F : T \times \mathbb{R}^n \times W \to \mathbb{R}$ where (W, ρ) is a separable metric space. Assume that $\xi : \Omega \to W$, $\xi \in \mathcal{F}|\mathcal{B}(W)$ and $F(u, \cdot, \cdot)$ is $\mathcal{B}(R^n) \otimes \mathcal{B}(W)$-measurable for each $u \in T$. Then (3.7) is satisfied if*

$$\int_{\mathbb{R}^n} \mathsf{E}|F(u, x, \xi)|\Lambda(dx) < \infty, \quad u \in T.$$

Proof. Due to the assumptions $F(u, X_{mj}, \xi_{mj})$ is a random variable for each $u \in T$ and any $m, j \in \mathbb{N}$. One has for the expectation of a series of nonnegative random variables

$$\mathsf{E} \sum_{m=1}^{\infty} \sum_{j=1}^{\tau_m} |F(u, X_{mj}, \xi_{mj})| = \sum_{m=1}^{\infty} \sum_{N=0}^{\infty} \frac{\Lambda(K_m)^N}{N!} e^{-\Lambda(K_m)} N \mathsf{E}|F(u, X_{m1}, \xi)|$$

$$= \sum_{m=1}^{\infty} \Lambda(K_m) \int_{K_m \times W} |F(u, x, z)| P_{X_{m1}}(dx) P_\xi(dz) = \int_{\mathbb{R}^n} \mathsf{E}|F(u, x, \xi)|\Lambda(dx) < \infty,$$

$P_{X_{mj}}$ and P_ξ being the laws of X_{m1} and ξ respectively. Here we used the Fubini theorem and formula (1.9). The proof is complete. $\square$

Lemma 3.7. *Assume that conditions of Lemma 3.6 hold. Then, for any $k \in \mathbb{N}$, all $t_1, \ldots, t_k \in \mathbb{R}$ and arbitrary $u_1, \ldots, u_k \in T$, the characteristic function of the vector $(Y(u_1), \ldots, Y(u_k))$ is equal to*

$$\exp\left\{ \int_{\mathbb{R}^n} \mathsf{E}\left(\exp\left\{ i \sum_{r=1}^{k} t_r F(u_r, x, \xi) \right\} - 1 \right) \Lambda(dx) \right\} \tag{3.8}$$

where $i^2 = -1$; thus the law of a random function Y introduced in (3.6) does not depend on the choice of explicit construction for a Poisson spatial process $\{x_i\}$.

Proof. The expression $\mathsf{E} \exp\left\{ i \sum_{r=1}^{k} t_r Y(u_r) \right\}$ equals

$$\prod_{m=1}^{\infty} \sum_{N=0}^{\infty} \left(\mathsf{E} \exp\left\{ i \sum_{r=1}^{k} t_r F(u_r, X_{m1}, \xi) \right\} \right)^N \frac{\Lambda(K_m)^N}{N!} e^{-\Lambda(K_m)}$$

$$= \prod_{m=1}^{\infty} \exp\left\{ \Lambda(K_m) \left(\mathsf{E} \exp\left\{ i \sum_{r=1}^{k} t_r F(u_r, X_{m1}, \xi) \right\} - 1 \right) \right\},$$

which coincides with (3.8). The proof is complete. $\square$

Let $Z = \{x_i\}$ be a Poisson spatial process in $\mathbb{R}^n$ with intensity measure Λ. Suppose that $\{\xi, \xi_i\}$ is a countable family of i.i.d. random variables which are independent of Z. Assume that $\psi : \mathbb{R}^n \to \mathbb{R}$ is a Borel function such that for almost every $\omega \in \Omega$ and any $u \in \mathbb{R}^n$ the following expression is well defined:

$$Y(u) = \sum_i \xi_i \psi(u - x_i). \tag{3.9}$$

In view of Lemma 3.6 one can assume that

$$\mathsf{E}|\xi| < \infty \quad \text{and} \quad \int_{\mathbb{R}^n} |\psi(u - x)| \Lambda(dx) < \infty, \quad u \in \mathbb{R}^n.$$

Note that Y provides an example of *shot-noise* random fields which are widely used in the theory of disordered structures (see, e.g., [271]).

Theorem 3.8. *Let $Y = \{Y(u), u \in \mathbb{R}^n\}$ be a shot-noise random field defined in (3.9) such that $\xi \geq 0$ a.s. and a measurable function ψ is nonnegative. Assume that*

$$Y(u) < \infty \quad \text{a.s. for any } u \in \mathbb{R}^n. \tag{3.10}$$

Then Y is an associated random field.

Proof. As in (3.5) we shall consider Y as a formal notation for a field

$$Y(u) = \sum_{m=1}^{\infty} Y_m(u) \quad \text{where} \quad Y_m(u) = \sum_{j=1}^{\tau_m} \xi_{mj} \psi(u - X_{mj}), \quad m \in \mathbb{N}, \ u \in \mathbb{R}^n.$$

For any $u \in \mathbb{R}^n$ one has

$$Y(u) = \lim_{N \to \infty} \sum_{m=1}^{N} Y_m(u) \quad \text{a.s.}$$

Clearly $Y_1, Y_2, \ldots$ are independent random fields. Thus, due to Theorem 1.8, (e), and Corollary 1.9, (a), it suffices to verify that $Y_m(\cdot)$ is associated for any $m \in \mathbb{N}$. Having fixed arbitrary $m \in \mathbb{N}$, we write instead of $K_m, \tau_m, \xi_{mj}, X_{mj}$ and Y_m simply K, τ, ξ_j, X_j and Y_0 respectively. Thus

$$Y_0(u) = \sum_{j=1}^{\tau} \xi_j \psi(u - X_j), \quad u \in \mathbb{R}^n.$$

Consider at first the case $\xi = a$ for some $a \geq 0$ (that is $\xi_j = a$ a.s. for all $j \in \mathbb{N}$) and $\psi = \sum_{l=1}^{N} c_l \mathbb{I}_{B_l}$, here $c_l \in \mathbb{R}_+$, $B_l \in \mathcal{B}_0(\mathbb{R}^n)$, $l = 1, \ldots, N$ ($N \in \mathbb{N}$). Then we have

$$Y_0(u) = a \sum_{j=1}^{\tau} \sum_{l=1}^{N} c_l \mathbb{I}_{B_l}(u - X_j) = a \sum_{l=1}^{N} c_l \sum_{j=1}^{\tau} \mathbb{I}\{X_j \in u - B_l\} = a \sum_{l=1}^{N} c_l U_0(u - B_l)$$

where

$$U_0(B) = \sum_{j=1}^{\tau} \mathbb{I}\{X_j \in B\}, \quad B \in \mathcal{B}_0(\mathbb{R}^n).$$

Note that U_0 is a Poisson spatial process in $\mathbb{R}^n$ with intensity measure $\Lambda_0(B) = \Lambda(B \cap K)$, $B \in \mathcal{B}(\mathbb{R}^n)$. Theorems 3.5 and 1.8, (e), imply that Y_0 is associated.

Let now ξ be a simple random variable taking values $a_1, \ldots, a_s$ with probabilities $p_1, \ldots, p_s$ respectively and ψ be the same simple function as above. Consider s independent auxiliary Poisson spatial processes $U^k = \{x_i^k\}$, in other words,

$$U^k(B) = \sum_{j=1}^{\tau^k} \mathbb{I}(X_j^k \in B), \quad B \in \mathcal{B}_0(\mathbb{R}^n), \quad k = 1, \ldots, s,$$

here U^k has intensity measure $p_k \Lambda_0$, $(\tau^k, X_j^k)_{j \in \mathbb{N}}$ are s independent arrays used to construct U^k in the same way as it was described for U_0 ($k = 1, \ldots, s$). Thus

$$\tau^k \sim Pois(p_k \Lambda(K)), \quad \mathsf{P}(X_j^k \in B) = \Lambda(B \cap K)/\Lambda(K), \quad B \in \mathcal{B}_0(\mathbb{R}^n),$$

where $0/0$ means 0. Introduce, for $a_1, \ldots, a_s \in \mathbb{R}_+$, the fields

$$Y^k(x) = a_k \sum_{j=1}^{\tau^k} \psi(x - X_j^k), \quad x \in \mathbb{R}^n, \quad k = 1, \ldots, s.$$

Lemma 3.9. *One has $Law\{Y_0(u), u \in \mathbb{R}^n\} = Law\{Y^1(u) + \ldots + Y^s(u), u \in \mathbb{R}^n\}$.*

Proof. Take any $v \in \mathbb{N}$, arbitrary $u_1, \ldots, u_v \in \mathbb{R}^n$ and $t_1, \ldots, t_v \in \mathbb{R}$. Then (3.8) implies that

$$\mathsf{E}\exp\Big\{i\sum_{r=1}^{v} t_r Y_0(u_r)\Big\} = \exp\Big\{\Lambda(K)\Big(\sum_{k=1}^{s} p_k \mathsf{E}\exp\Big\{i a_k \sum_{r=1}^{v} t_r \psi(u_r - X_1)\Big\} - 1\Big)\Big\}.$$

Likewise, we obtain the same expression for $\mathsf{E}\exp\big\{i\sum_{r=1}^{v} t_r \sum_{k=1}^{s} Y^k(u_r)\big\}$. The proof is complete. $\square$

Due to Lemma 3.9 we conclude that the assertion of Theorem is true if ξ takes finite number of nonnegative values and ψ is a simple function.

Let now ξ take finite number of nonnegative values and $\psi : \mathbb{R}^n \to \mathbb{R}_+$. We can easily construct a sequence of simple functions $\psi_q \geq 0$ such that $\psi_q \nearrow \psi$ for all $u \in \mathbb{R}^n$ as $q \to \infty$. Therefore (in view of (3.10)), for any $u \in \mathbb{R}^n$,

$$Y_{0q}(u) := \sum_{j=1}^{\tau} \xi_j \psi_q(u - X_j) \to \sum_{j=1}^{\tau} \xi_j \psi(u - X_j) = Y_0(u) \ \text{ a.s.,} \ q \to \infty.$$

By virtue of already proved cases and Theorem 1.8,(e), we infer that, for each $k \in \mathbb{R}$ and all $u_1, \ldots, u_k \in \mathbb{R}^n$, the random variables $Y_0(u_1), \ldots, Y_0(u_k)$ are associated.

To conclude the proof consider now any nonnegative random variable ξ and arbitrary measurable $\psi : \mathbb{R}^n \to \mathbb{R}_+$ (under condition (3.10)). For a sequence of i.i.d random variables $\xi, \xi_1, \ldots$ take the sequences of simple nonnegative random variables $\xi_j^{(q)}$, $q, j \in \mathbb{N}$, such that $\xi_j^{(q)} \nearrow \xi_j$ for all $\omega \in \Omega$ and $j \in \mathbb{N}$, as $q \to \infty$, e.g., $\xi_j^{(q)} := \sum_{r=0}^{m(q)} r2^{-q}\mathbb{I}\{r2^{-q} \leq \xi_j < (r+1)2^{-q}\}$, $m(q) = 2^{2q} - 1$, $q \in \mathbb{N}$. Obviously, $Y_0^{(q)}(u) := \sum_{j=1}^{\tau} \xi_j^{(q)} \psi(u - X_j) \to Y_0(u)$ a.s., $q \to \infty$, for any $u \in \mathbb{R}^n$, and application of Theorem 1.8,(e), once again leads to the desired result. $\square$

3°. Cluster random fields. Let $\mathcal{R}$ be a ring of subsets of $\mathbb{R}^n$ consisting of finite unions of the parallelepipeds (i.e. *rectangles* or *blocks*) of the form

$$C = (a_1, b_1] \times \ldots \times (a_n, b_n], \quad a_i \leq b_i, \ i = 1, \ldots, n.$$

Take a random measure M given on $\mathcal{B}(\mathbb{R}^n)$. Then clearly $\mu(B) := \mathsf{E}M(B)$ is a nonrandom measure on $(\mathbb{R}^n, \mathcal{B}(\mathbb{R}^n))$. Suppose that μ is σ-finite, so there exist at most countably many hyperplanes $H_1, H_2, \ldots$ orthogonal to a coordinate axis of $\mathbb{R}^n$ and such that $\mu(H_m) > 0$.

For $C \in \mathcal{R}$, set $H(C) := \{x \in \mathbb{R}^n : \mu(\partial(C + x)) > 0\}$. Thus evidently $\mathrm{mes}(H(C)) = 0$ for each $C \in \mathcal{R}$. Let Z be a Poisson spatial process with intensity measure Λ possessing intensity function λ (see (3.2)). Whence $\Lambda(H(C)) = 0$ for any $C \in \mathcal{R}$ as well. Introduce $G(C) := \mathbb{R}^n \setminus H(C)$, $C \in \mathcal{R}$.

Remark 3.10. If μ is a finite nonrandom measure on $(\mathbb{R}^n, \mathcal{B}(\mathbb{R}^n))$ and B is a Borel set then $f_B : \mathbb{R}^n \to \mathbb{R}$, where $f_B(x) = \mu(x + B)$, $x \in \mathbb{R}^n$, is a Borel function. Indeed, for any rectangle B this is true since the distribution function of μ is upper-continuous, hence Borel. Besides, the class $\{B \subset \mathbb{R}^n : f_B$ is a Borel function$\}$ is monotone (see, e.g., [81, Ch. 1, §3]), as one can verify directly, therefore, it contains $\mathcal{B}(\mathbb{R}^n)$. Consequently $H(C)$ and $G(C)$ are Borel sets for any $C \in \mathcal{R}$.

Now take a sequence $\{M, M_i, i \in \mathbb{N}\}$ of i.i.d. random measures independent of the above mentioned Poisson spatial process $Z = \{x_i\}$. Consider

$$X(B) = \sum_i M_i(B + x_i) \tag{3.11}$$

defined on $\Omega \times \mathcal{B}(\mathbb{R}^n)$ and assuming values in $\overline{\mathbb{R}}_+$.

If $X = \{X(B), B \in \mathcal{B}(\mathbb{R}^n)\}$ is a random function (i.e. a collection of random variables with values in $\overline{\mathbb{R}}_+$) then one says that X is a *cluster point random field*[14]. In the same way as we interpreted formula (3.5), we shall consider instead of (3.11) a random function

$$X(B) = \sum_{m=1}^{\infty} \sum_{j=1}^{\tau_m} M_{mj}(B + X_{mj}), \quad B \in \mathcal{B}(\mathbb{R}^n), \tag{3.12}$$

where M_{mj} are independent copies of M which are independent of the array $(\tau_m, X_{mj})_{m,j \in \mathbb{N}}$ used to construct Z. Thus a family $\{M(B), M_{mj}(B), B \in \mathcal{B}(\mathbb{R}^n)\}$ is independent of $(\tau_m, X_{mj})_{m,j \in \mathbb{N}}$. The expression (3.12) is more involved than (3.6) as $M_{mj} = M_{mj}(\omega, B)$, $\omega \in \Omega$, $B \in \mathcal{B}(\mathbb{R}^n)$. To discuss the convergence of series (3.12) we need some auxiliary results.

Lemma 3.11. *Let M be a random measure on $(\mathbb{R}^n, \mathcal{B}(\mathbb{R}^n))$ with σ-finite intensity measure μ. Assume that $M(C) < \infty$ a.s. for any $C \in \mathcal{R}$. Then a function $M(\omega, C + x)$, defined for $\omega \in \Omega$ and $x \in G(C)$, is $\mathcal{F} \otimes \mathcal{B}(G(C))$-measurable for every $C \in \mathcal{R}$. If $\mu(C) < \infty$ (consequently $M(C) < \infty$ a.s.) for any $C \in \mathcal{R}$, then a function $\mu(C + x)$ is $\mathcal{B}(G(C))$-measurable for every $C \in \mathcal{R}$.*

Proof. Let us show that a function $x \mapsto M(\omega, C + x)$ is a.s. continuous on $G(C)$. For any $x \in G(C)$ and $h \in \mathbb{R}^n$, one has

$$|M(\omega, C + x + h) - M(\omega, C + x)| \le M((\partial(C+x))^{2|h|}) \to M(\partial(C+x)) \text{ a.s.,} \quad h \to 0,$$

where, as always, $B^\varepsilon = \{x \in \mathbb{R}^n : dist(x, B) < \varepsilon\}$, $\varepsilon > 0$. Here we have used the condition $M(C) < \infty$ a.s. for any $C \in \mathcal{R}$ and the fact that ∂D is a closed set for any $D \subset \mathbb{R}^n$. Note that $\mu(\partial(C + x)) = 0$ for $x \in G(C)$. Hence for such x we conclude that $M(\partial(C + x)) = 0$ a.s. because $M(\partial(C + x))$ is a nonnegative random variable. In other words the random field $\{M(C + x), x \in G(C)\}$ has a.s. continuous trajectories. Therefore it is $(\mathcal{F} \otimes \mathcal{B}(G(C)))$-measurable (see, e.g., [412, Ch. 1]). The second statement is an immediate corollary of the Fubini theorem. $\square$

Lemma 3.12. *Let M be a random measure on $(\mathbb{R}^n, \mathcal{B}(\mathbb{R}^n))$ with σ-finite intensity measure μ. Suppose that, for any $B \in \mathcal{R}$, $\mu(B) < \infty$ and*

$$\int_{\mathbb{R}^n} \mu(B + x)\Lambda(dx) < \infty \tag{3.13}$$

where Λ satisfies[15] (3.2). Then, for every $B \in \mathcal{R}$,

$$\sum_{m=1}^{\infty} \mathsf{E} \sum_{j=1}^{\tau_m} M_{mj}(B + X_{mj}) < \infty, \tag{3.14}$$

whereas the series (3.12) converges in $L^1(\Omega, \mathcal{F}, \mathsf{P})$ and its sum is a.s. additive on $\mathcal{R}$. Moreover, the law of $\{X(B), B \in \mathcal{R}\}$ does not depend on a choice of explicit construction for Z.

---

[14]See, e.g., [87, 119].
[15]Instead of $\Lambda(dx)$ one can write $\lambda(x)dx$ where λ is the intensity function.

Proof. Note that $\Lambda(H(B)) = 0$. Therefore Lemma 3.11 guarantees that the expression in the left-hand side of (3.13) is meaningful.

It is convenient to suppose that all M_{mj} and τ_m, X_{mj} are defined at first on some probability spaces $(\Omega_1, \mathcal{F}_1, \mathsf{P}_1)$ and $(\Omega_2, \mathcal{F}_2, \mathsf{P}_2)$ respectively. After that we can extend them to $(\Omega, \mathcal{F}, \mathsf{P}) = (\Omega_1, \mathcal{F}_1, \mathsf{P}_1) \otimes (\Omega_2, \mathcal{F}_2, \mathsf{P}_2)$. Then $(M_{mj})_{m,j\in\mathbb{N}}$ and $(\tau_m, X_{mj})_{m,j\in\mathbb{N}}$ are independent automatically (more exactly we consider the extensions of these random elements on $(\Omega, \mathcal{F}, \mathsf{P})$).

Let us fix arbitrary $B \in \mathcal{R}$ and take any $z_0 \in G(B)$. Introduce $X^*_{mj}(\omega) = X_{mj}(\omega)$ if $X_{mj}(\omega) \in G(B)$ and $X^*_{mj}(\omega) = z_0$ otherwise, then $X_{mj} = X^*_{mj}$ P-a.s. for all $m, j \in \mathbb{N}$. Therefore $M_{mj}(B + X_{mj}) = M_{mj}(B + X^*_{mj})$ P-a.s. on the complete probability space $(\Omega, \mathcal{F}, \mathsf{P})$. A composition of measurable functions is measurable. So $M_{mj}(\omega_1, B + X^*_{mj}(\omega_2)) \in (\mathcal{F}_1 \otimes \mathcal{F}_2)|\mathcal{B}(\mathbb{R})$ for any $m, j \in \mathbb{N}$ by Lemma 3.11.

The Fubini theorem and formula (1.9) imply that

$$\mathsf{E}M_{mj}(B + X^*_{mj}) = \int_{\Omega_1 \times \Omega_2} M_{mj}(\omega_1, B + X^*_{mj}(\omega_2))\, d\mathsf{P}_1\, d\mathsf{P}_2$$

$$= \int_{\Omega_2} \mu(B + X^*_{mj}(\omega_2))d\mathsf{P}_2 = \int_{K_m} \mu(B + x)P_{X_{mj}}(dx).$$

Thus the left-hand side of (3.14) can be written as follows:

$$\sum_{m=1}^{\infty} \sum_{N=0}^{\infty} N\mathsf{E}M_{m1}(B + X^*_{m1})\frac{\Lambda(K_m)^N}{N!} e^{-\Lambda(K_m)}$$

$$= \sum_{m=1}^{\infty} \int_{K_m} \mu(B + x)\Lambda(dx) = \int_{\mathbb{R}^n} \mu(B + x)\Lambda(dx) < \infty. \tag{3.15}$$

The second assertion is trivial due to (3.14). The third one is established in a way similar to the proof of Lemma 3.7. $\square$

Remark 3.13. As the law of $\{X(B), B \in \mathcal{R}\}$ does not depend on a partition of $\mathbb{R}^n$ into bounded Borel sets $K_1, K_2, \ldots$, we shall further consider a partition formed by unit cubes $Q_j = [j, j + 1)$, $j \in \mathbb{Z}^n$, $\mathbf{1} = (1, \ldots, 1) \in \mathbb{N}^n$. Note that $B + x \subset Q_j^r$ if $x \in Q_j$ and $r = [diam(B)] + 1$. Here, as usual, $Q_j^r = \{x \in \mathbb{R}^n : dist(x, Q_j) < r\}$ and $[\cdot]$ stands for the integer part of a number. Therefore (3.13) is satisfied if

$$\sum_{j\in\mathbb{Z}^n} \mu(Q_j^r)\Lambda(Q_j) < \infty, \quad r \in \mathbb{N}. \tag{3.16}$$

For example, (3.16) holds whenever Λ has bounded support. If $\sup_j \Lambda(Q_j) < \infty$ and $\mu(\mathbb{R}^n) < \infty$, then (3.16) is valid.

To study the association property of X we invoke other auxiliary statements.

Lemma 3.14. *Let $\{M^k, k \in \mathbb{N}\}$ be a family of random measures on $(\mathbb{R}^n, \mathcal{B}(\mathbb{R}^n))$ such that $\mu^k(B) := \mathsf{E}M^k(B) < \infty$ for each $B \in \mathcal{R}$ (consequently μ^k is σ-finite, $k \in \mathbb{N}$). Assume that Λ satisfies (3.2) and, for any $B \in \mathcal{R}$, the integrals*

$$\int_{\mathbb{R}^n} \mu^k(B + x)\Lambda(dx), \quad k \in \mathbb{N}, \tag{3.17}$$

converge uniformly in k. Then uniformly in k converges the series [16]

$$\sum_{m=1}^{\infty} \mathsf{E} \sum_{j=1}^{\tau_m} M_{mj}^k(B + X_{mj}), \tag{3.18}$$

for each $B \in \mathcal{R}$ and any enumeration of $Q_j, j \in \mathbb{Z}^n$, into a sequence $K_1, K_2, \ldots$.

Proof. For each $B \in \mathcal{R}$ we modify X_{mj} as in the proof of Lemma 3.12, so $M_{mj}^k(B + X_{mj})$ are random variables $(m, j, k \in \mathbb{N})$. Set

$$T(R) = \{j \in \mathbb{Z}^n : Q_j \subset \{x \in \mathbb{R}^n : |x| \leq R\}\}, \quad m_0(R) = \max\{m \in \mathbb{N} : K_m \in T(R)\}.$$

Uniform convergence of integrals (3.17) implies that, for any $\varepsilon > 0$, there exists $R = R(\varepsilon) \in \mathbb{N}$ such that

$$\sup_{k \in \mathbb{N}} \int_{|x| \geq R} \mu^k(B + x)\, \Lambda(dx) < \varepsilon.$$

By virtue of Lemma 3.12 (see (3.15)) we know that, for any $I \subset \mathbb{Z}^n$,

$$\sum_{m \in I} \mathsf{E} \sum_{j=1}^{\tau_m} M_{mj}^k(B + X_{mj}) = \sum_{m \in I} \int_{K_m} \mu^k(B + x)\Lambda(dx).$$

Thus, for any $k \in \mathbb{N}$, one has

$$\sum_{m \geq m_0(R)} \mathsf{E} \sum_{j=1}^{\tau_m} M_{mj}^k(B + X_{mj}) \leq \int_{|x| \geq R} \mu^k(B + x)\Lambda(dx) < \varepsilon. \quad \square$$

Remark 3.15. Using Remark 3.13 one can verify the following. If $\{M^k, k \in \mathbb{N}\}$ is a family of random measures on $(\mathbb{R}^n, \mathcal{B}(\mathbb{R}^n))$ such that $\mu^k(B) := \mathsf{E}M^k(B) < \infty$ for each $B \in \mathcal{R}$, then from the uniform in k convergence of the series

$$\sum_{j \in \mathbb{Z}^n} \mu^k(Q_j^r)\Lambda(Q_j) < \infty,$$

for any $r \in \mathbb{N}$, ensues uniform in k convergence of the integrals (3.17).

Lemma 3.16. *Let $Q_j, j \in \mathbb{Z}^n$, be unit cubes used in the proof of Lemma 3.14. Assume that a σ-finite measure Λ satisfies (3.2),*

$$\sup_{j \in \mathbb{Z}^n} \Lambda(Q_j) < \infty \quad \text{and} \quad \limsup_{k \to \infty} \mu^k(\{x : |x| \geq R\}) \to 0, \quad R \to \infty,$$

then the integrals (3.17) converge uniformly in k.

Proof. Let $r = [diam(B)] + 1$ where $B \in \mathcal{R}$. For $R \in \mathbb{N}$ introduce a set

$$J(R) = \{j \in \mathbb{Z}^n : Q_j \subset \{x \in \mathbb{R}^n : |x| \geq R\}\}.$$

Then one easily obtains the bounds

$$\int_{|x| > R} \mu^k(B + x)\Lambda(dx) \leq \sum_{j \in J(R)} \int_{Q_j} \mu^k(B + x)\Lambda(dx)$$

---

[16] We use the partition of $\mathbb{R}^n$ mentioned in Remark 3.13.

$$\leq \sum_{j\in J(R)} \int_{Q_j} \mu^k(Q_j^r)\Lambda(dx) = \sum_{j\in J(R)} \mu^k(Q_j^r)\Lambda(Q_j) \leq \sup_{t\in\mathbb{Z}^n} \Lambda(Q_t) \sum_{j\in J(R)} \mu^k(Q_j^r)$$

$$\leq (2r+1)^n \sup_{t\in\mathbb{Z}^n} \Lambda(Q_t)\mu^k(\{x\in\mathbb{R}^n : |x| \geq R-r\}) \to 0, \ \ R\to\infty. \ \ \square$$

Definition 3.17. We say that M is a *simple random measure* on $\mathcal{B}(\mathbb{R}^n)$, i.e. on $(\mathbb{R}^n, \mathcal{B}(\mathbb{R}^n))$, if there are points $a_1, \ldots, a_s \in \mathbb{R}^n$ and random variables $\xi_1, \ldots, \xi_s$ taking values in a set $\{b_0, \ldots, b_q\} \subset \mathbb{R}_+$ such that

$$M(B) = \sum_{v=1}^{s} \xi_v \delta_{a_v}(B), \ \ B \in \mathcal{B}(\mathbb{R}^n),$$

where δ_a denotes a Dirac measure concentrated at a point $a \in \mathbb{R}^n$.

Lemma 3.18. *Let M be a random measure on $\mathcal{B}(\mathbb{R}^n)$ with intensity measure μ such that $\mu(B) < \infty$ for any $B \in \mathcal{R}$. Then there exists a sequence of simple random measures $M^k, k \in \mathbb{N}$, on $\mathcal{B}(\mathbb{R}^n)$ such that*
  1) *if $B \in \mathcal{R}$ then*

$$\mathsf{E}|M(B) - M^k(B)| \leq 2k^{-1} + \mathsf{E}M((\partial B)^{2/k}), \ \ k \in \mathbb{N}; \tag{3.19}$$

  2) *if, moreover, $\mu(\mathbb{R}^n) < \infty$ then*

$$\lim_{R\to\infty} \limsup_{k\to\infty} \mu^k(\mathbb{R}^n \setminus [-R,R]^n) = 0$$

where μ^k is the intensity measure of M^k, $k \in \mathbb{N}$.

Proof. For any $k \in \mathbb{N}$ divide the cube $[-k,k)^n$ into $s = (2k)^{2n}$ cubes with edge length $1/(2k)$. This can be done by marking the points $0, \pm 1/(2k), \pm 2/(2k), \ldots$ at the coordinate axes and drawing through them hyperplanes orthogonal to the corresponding axis. Let $D_1, \ldots, D_s$ be the obtained cubes and $a_1, \ldots, a_s$ be their minimal points (with respect to the usual partial order in $\mathbb{R}^n$). Clearly $s = s(k)$, $a_v = a_v(k)$ and $D_v = D_v(k)$, $v = 1, \ldots, s$. To simplify the notation we write s, a_v and D_v. Define a random measure $M^{[k]}$ ($k \in \mathbb{N}$) such that

$$M^{[k]}(\{a_v\}) = M(D_v), v = 1, \ldots, s,$$

$$M^{[k]}(B) = \sum_{v:a_v\in B} M^{[k]}(\{a_v\}), \ \ B \in \mathcal{B}(\mathbb{R}^n).$$

For every $B \in \mathcal{B}(\mathbb{R}^n)$,

$$M^{[k]}(B) = \sum_{v:D_v\subset B} M(D_v) + \sum_{v\in I} M(D_v)$$

where $I = \{v : a_v \in B, D_v \cap B \neq D_v\}, I = I(B,k)$. Then, for any $B \in \mathcal{B}(\mathbb{R}^n)$,

$$M(B) \geq \sum_{v:D_v\subset B} M(D_v) = M^{[k]}(B) - \sum_{v\in I} M(D_v) \geq M^{[k]}(B) - M((\partial B)^{2/k}),$$

$$M(B) \leq \sum_{v:D_v \subset B} M(D_v) + \sum_{v:D_v \cap \partial B \neq \varnothing} M(D_v) \leq M^{[k]}(B) + M((\partial B)^{2/k})$$

where for $D \subset \mathbb{R}^n$, as usual, $D^\varepsilon = \{x \in \mathbb{R}^n : dist(x, D) < \varepsilon\}$, $\varepsilon > 0$. Therefore

$$\mathsf{E}|M(B) - M^{[k]}(B)| \leq \mathsf{E}M((\partial B)^{2/k}), \quad B \in \mathcal{B}(\mathbb{R}^n), \quad k \in \mathbb{N}. \tag{3.20}$$

For any $L > 0$ and all $k \in \mathbb{N}$, one has

$$\mathsf{E}M^{[k]}(B)\mathbb{I}\left\{\bigcup_{v=1}^{s}\{M^{[k]}(\{a_v\}) \geq L\}\right\} \leq \mathsf{E}M([-k,-k)^n)\mathbb{I}\{M([-k,k)^n) \geq L\}.$$

The last expression can be made less than k^{-1} if $L = L(k) > 0$ is large enough as $\mathsf{E}M([-k,k)^n) < \infty$ ($\mu(B) < \infty$ for any $B \in \mathcal{R}$).

Having taken $q = q(k) \in \mathbb{N}$ ($k \in \mathbb{N}$), divide the interval $[0, L)$ into equal intervals of length Lq^{-1} with boundary points $0 = b_0 < b_1 < \ldots < b_q = L$. Set

$$M^k(B) = \sum_{v:a_v \in B} M^k(\{a_v\}), \quad B \in \mathcal{B}(\mathbb{R}^n),$$

where $M^k(\{a_v\}) = b_{j-1}$ if $M^{[k]}(\{a_v\}) \in [b_{j-1}, b_v)$, $1 \leq j \leq q$, and $M^k(\{a_v\}) = 0$ otherwise. Note that $M^k(\{a_v\}) \leq M^{[k]}(\{a_v\})$, $v = 1, \ldots, s$. Consequently, for any $B \in \mathcal{B}(\mathbb{R}^n)$,

$$0 \leq M^{[k]}(B) - M^k(B) \leq M^{[k]}(B)\mathbb{I}\left\{\bigcup_{v=1}^{s}\{M^{[k]}(\{a_v\}) \geq L\}\right\}$$

$$+ \sum_{v:a_v \in B} (M^{[k]}(\{a_v\}) - M^k(\{a_v\}))\mathbb{I}\left\{\bigcap_{v=1}^{s}\{M^{[k]}(\{a_v\}) < L\}\right\}$$

$$\leq M^{[k]}(B)\mathbb{I}\left\{\bigcup_{v=1}^{s}\{M^{[k]}(\{a_v\}) \geq L\}\right\} + sLq^{-1}.$$

Thus

$$\mathsf{E}|M^{[k]}(B) - M^k(B)| \leq k^{-1} + sLq^{-1} \leq 2k^{-1}, \quad k \in \mathbb{N}, \tag{3.21}$$

if $L = L(k)$ is chosen large enough and $q = q(k)$ is taken in such a way that $(2k)^{2n}L(k)q(k)^{-1} \leq k^{-1}$, $k \in \mathbb{N}$.

In view of (3.20) and (3.21) we come to (3.19), so $(M^k)_{k \in \mathbb{N}}$ is the desired sequence of simple random measures. Now the second assertion is obvious because

$$\mathsf{E}M^k(\mathbb{R}^n \setminus [-R, R]^n) \leq 2k^{-1} + \mathsf{E}M((\partial([-R, R]^n))^{2/k}) + \mathsf{E}M(\mathbb{R}^n \setminus [-R, R]^n)$$

$$\leq 2k^{-1} + 2\mathsf{E}M(\mathbb{R}^n \setminus [-R+2, R-2]^n),$$

for $k \in \mathbb{N}$ and $R > 2$. Since $\mu(\mathbb{R}^n) < \infty$, we see that

$$\lim_{R \to \infty} \limsup_{k \to \infty} \mu^k(\mathbb{R}^n \setminus [-R, R]^n) \leq 2 \lim_{R \to \infty} \mu(\mathbb{R}^n \setminus [-R+2, R-2]^n) = 0. \quad \square$$

4°. Association of a cluster random field. The following theorem extends a result by Burton and Waymire ([87]).

Theorem 3.19. *Let $Z = \{x_i\}$ be a Poisson spatial process with intensity measure Λ satisfying (3.2) and such that*

$$\sup_{j \in \mathbb{Z}^n} \Lambda(Q_j) < \infty$$

where $Q_j = [j, j+1)$, $j \in \mathbb{Z}^n$, are unit cubes in $\mathbb{R}^n$. Suppose that M is a random measure on $(\mathbb{R}^n, \mathcal{B}(\mathbb{R}^n))$ with intensity measure μ such that $\mu(\mathbb{R}^n) < \infty$. Then a family $X = \{X(B), B \in \mathcal{R}\}$ introduced by (3.12) is associated.

Proof. Suppose at first that M is *deterministic and atomic* with finite number of atoms, i.e. $M = \sum_{v=1}^{s} d_v \delta_{a_v}$ where $a_1, \ldots, a_s \in \mathbb{R}^n$ and $d_1, \ldots, d_s \geq 0$. Then, for any bounded Borel set $B \subset \mathbb{R}^n$, one has

$$X(B) = \sum_i M_i(B + x_i) = \sum_{v=1}^{s} d_v \sum_i \mathbb{I}\{x_i \in B - a_v\} = \sum_{v=1}^{s} d_v Z(B - a_v).$$

Whence a family $\{X(B), B \in \mathcal{B}_0(R^n)\}$ is associated by Theorems 1.8, (d), and 3.5.

Now let M be a simple random measure, that is $M = \sum_{v=1}^{s} \xi_v \delta_{a_v}$ where $\xi_1, \ldots, \xi_s$ are (possibly dependent) nonnegative random variables taking finite number of values. Let $\xi = (\xi_1, \ldots, \xi_s)$ take values in a set $\{b^t = (b_1^t, \ldots, b_s^t), t \in T\}$, $b^t \in \mathbb{R}_+^s$ for all $t \in T$, T being a finite subset of $\mathbb{N}$. Thus for $B \in \mathcal{B}_0(\mathbb{R}^n)$

$$X(B) = \sum_i \sum_{v=1}^{s} \xi_v^i \mathbb{I}\{a_v \in B + x_i\} = \sum_i \sum_{t \in T} \mathbb{I}\{\xi^i = b^t\} \sum_{v=1}^{s} b_v^t \mathbb{I}\{a_v \in B + x_i\}.$$

Here $\{\xi^i\}_{i \in \mathbb{N}}$ are i.i.d. random vectors distributed as ξ. One assumes that $\{\xi^i\}_{i \in \mathbb{N}}$ and $\{x_i\}$ are independent.

Let $Z^{(t)} = \{x_i^t\}$ ($t \in T$) be mutually independent Poisson random fields in $\mathbb{R}^n$ such that $Z^{(t)}$ has intensity measure $p_t \Lambda$ where $p_t = \mathsf{P}(\xi = b^t)$, $t \in T$. Define the random measures $X^{(t)}$ by the relation

$$X^{(t)}(B) = \sum_i \sum_{v=1}^{s} b_v^t \mathbb{I}\{x_i^t \in a_v - B\} = \sum_{v=1}^{s} b_v^t Z^{(t)}(a_v - B), \qquad (3.22)$$

here $B \in \mathcal{B}_0(\mathbb{R}^n)$, $t \in T$. Then the measures $X^{(t)}, t \in T$, are mutually independent and associated (this case has already been studied), hence their sum is associated. In the next lemma we show that the above mentioned sum is distributed as X and, consequently, X is associated by Remark 1.4, (a).

Lemma 3.20. *For any $r \in \mathbb{N}$ and any collection of sets $B_1, \ldots, B_r \in \mathcal{B}_0(\mathbb{R}^n)$, one has*

$$Law(X(B_1), \ldots, X(B_r)) = Law\left(\sum_{t \in T} X^{(t)}(B_1), \ldots, \sum_{t \in T} X^{(t)}(B_r)\right). \qquad (3.23)$$

Proof. Clearly it suffices to consider pairwise disjoint sets $B_1, \ldots, B_r$. Introduce $C_{11} := a_1 - B_1, \ldots, C_{1r} := a_1 - B_r$ and

$$C_{vl} := (a_v - B_l) \setminus \left(\left(\bigcup_{z=1}^{v-1} \bigcup_{q=1}^{r} C_{zq} \right) \cup \left(\bigcup_{m=1}^{l-1} C_{vm} \right) \right) \quad v = 2, \ldots, s, \quad l = 1, \ldots, r.$$

As usual, the union over an empty set is agreed to be empty. For any $v = 1, \ldots, s$, $l = 1, \ldots, r$, all these sets C_{vl} are pairwise disjoint and

$$a_v - B_l = \left(\bigcup_{z=1}^{v-1} \bigcup_{q=1}^{r} C_{zq} \right) \cup \left(\bigcup_{m=1}^{l} C_{vm} \right). \tag{3.24}$$

We check that the multidimensional Laplace transforms[17] of both random vectors in (3.23) coincide which implies the coincidence of their distributions. Let $u = (u_1, \ldots, u_r) \in \mathbb{R}_+^r$. Then by (3.22) and (3.24)

$$\mathsf{E} \exp \left\{ -\sum_{l=1}^{r} u_l \sum_{t \in T} X^{(t)}(B_l) \right\} = \prod_{t \in T} \mathsf{E} \exp \left\{ -\sum_{l=1}^{r} u_l \sum_{v=1}^{s} b_v^t Z^{(t)}(a_v - B_l) \right\}$$

$$= \prod_{t \in T} \mathsf{E} \exp \left\{ -\sum_{l=1}^{r} u_l \sum_{v=1}^{s} b_v^t \left(\sum_{z=1}^{v-1} \sum_{q=1}^{r} Z^{(t)}(C_{zq}) + \sum_{q=1}^{l} Z^{(t)}(C_{vq}) \right) \right\}.$$

By changing the order of summation we establish that the last expression can be rewritten as

$$\prod_{t \in T} \mathsf{E} \exp \left\{ -\sum_{q=1}^{r} \sum_{l=1}^{r} u_l \sum_{z=1}^{s-1} \sum_{v=z+1}^{s} b_v^t Z^{(t)}(C_{zq}) - \sum_{q=1}^{r} \sum_{l=q}^{r} u_l \sum_{v=1}^{s} b_v^t Z^{(t)}(C_{vq}) \right\}$$

$$= \prod_{t \in T} \mathsf{E} \exp \left\{ -\sum_{z=1}^{s} \sum_{q=1}^{r} \alpha^{(t)}(z, q) Z^{(t)}(C_{zq}) \right\} := \Delta, \tag{3.25}$$

after we also change in one of the sums the summation index from v to z and introduce

$$\alpha^{(t)}(z, q) = \sum_{l=1}^{r} u_l \sum_{v=z+1}^{s} b_v^t + \sum_{l=q}^{r} u_l \sum_{z=1}^{s} b_z^t, \quad b_{s+1}^t := 0, \quad q = 1, \ldots, r, \ z = 1, \ldots, s.$$

These $\alpha^{(t)}(z, q)$ depend on r, s, b^t ($t \in T$) and u_l ($l = 1, \ldots, r$). If, for $\lambda \geq 0$, $\zeta \sim Pois(\lambda)$ and $h \geq 0$, then $\mathsf{E} e^{-h\zeta} = \exp\{\lambda(e^{-h} - 1)\}$. Thus, remembering that $Z^{(t)}(C_{vl}) \sim Pois(p_t \Lambda(C_{vl}))$, and as $\{Z^{(t)}(C_{vl}); t \in T, v = 1, \ldots, s, l = 1, \ldots, r\}$ are independent, we conclude that

$$\Delta = \prod_{t \in T} \prod_{z=1}^{s} \prod_{q=1}^{r} \exp \left\{ p_t \Lambda(C_{zq}) \left(e^{-\alpha^{(t)}(z,q)} - 1 \right) \right\} \tag{3.26}$$

---

[17] We do not calculate here the characteristic functions to avoid mixing of the appearing imaginary unit i with the index used for Poisson random fields $\{x_i\}$ and $\{x_i^t\}$.

where Δ was defined in (3.25).

Now we deal with the random vector at the left-hand side of (3.23). Again by (3.24)

$$\mathsf{E}\exp\left\{-\sum_{l=1}^{r}u_l X(B_l)\right\} = \mathsf{E}\exp\left\{-\sum_{l=1}^{r}u_l\sum_{i}\sum_{t\in T}\mathbb{I}\{\xi^i = b^t\}\sum_{v=1}^{s}b_v^t\mathbb{I}_i(a_v - B_l)\right\}$$

$$= \mathsf{E}\exp\left\{-\sum_{l=1}^{r}u_l\sum_{i}\sum_{t\in T}\mathbb{I}\{\xi^i = b^t\}\sum_{v=1}^{s}b_v^t\left(\sum_{z=1}^{v-1}\sum_{q=1}^{r}\mathbb{I}_i(C_{zq}) + \sum_{q=1}^{l}\mathbb{I}_i(C_{vq})\right)\right\}$$

$$= \mathsf{E}\exp\left\{-\sum_{q=1}^{r}\sum_{z=1}^{s}\sum_{i}\sum_{t\in T}\alpha^{(t)}(z,q)\mathbb{I}\{\xi^i = b^t\}\mathbb{I}_i(C_{zq})\right\} := \Delta_1,$$

analogously to (3.26), here $\mathbb{I}_i(C) := \mathbb{I}\{x_i \in C\}$.

Next take a partition of $\mathbb{R}^n$ by means of C_{zq}, $z \in I = \{1,\ldots,s\}, q \in J = \{1,\ldots,r\}$ and some other sets C_{zq} in $\mathcal{B}_0(\mathbb{R}^n)$, with $(z,q) \in \mathbb{N}^2 \setminus (I \times J)$. Consider a version of Poisson random field Z with intensity measure Λ constructed in Theorem A.5. Let $\{\tau_{zq}, X_{zqj}, z,q,j \in \mathbb{N}\}$ be a family of independent random variables such that for $z,q,j \in \mathbb{N}$

$$\tau_{zq} \sim Pois(\Lambda(C_{zq})) \text{ and } \mathsf{P}(X_{zqj} \in B) = \Lambda(B \cap C_{zq})/\Lambda(C_{zq}), \ \ B \in \mathcal{B}(\mathbb{R}^n),$$

if $\Lambda(C_{zq}) > 0$ (if $\Lambda(C_{zq}) = 0$ then $\tau_{zq} = 0$ a.s. and the distribution of X_{zqj} can be taken arbitrarily). Let $(\xi^{zqj})_{z,q,j\in\mathbb{N}}$ be a family of independent copies of a random vector ξ. Assume that this family is also independent of $\{\tau_{zq}, X_{zqj}, z,q,j \in \mathbb{N}\}$. Then we see that

$$\Delta_1 = \mathsf{E}\exp\left\{-\sum_{z=1}^{s}\sum_{q=1}^{r}\sum_{j=1}^{\tau_{zq}}\sum_{t\in T}\alpha^{(t)}(z,q)\mathbb{I}\{\xi^{zqj} = b^t\}\mathbb{I}\{X_{zqj} \in C_{zq}\}\right\}$$

$$= \prod_{z=1}^{s}\prod_{q=1}^{r}\sum_{N=0}^{\infty}\frac{\Lambda(C_{zq})^N}{N!}e^{-\Lambda(C_{zq})}\left(\mathsf{E}\exp\left\{-\sum_{t\in T}\alpha^{(t)}(z,q)\mathbb{I}\{\xi^{zq1} = b^t)\}\right\}\right)^N$$

$$= \prod_{z=1}^{s}\prod_{q=1}^{r}\sum_{N=0}^{\infty}\frac{\Lambda(C_{zq})^N}{N!}e^{-\Lambda(C_{zq})}\left(\sum_{t\in T}p_t\exp\{-\alpha^{(t)}(z,q)\}\right)^N$$

$$= \prod_{z=1}^{s}\prod_{q=1}^{r}\exp\left\{-\Lambda(C_{zq})\left(1 - \sum_{t\in T}p_t\exp\left\{-\alpha^{(t)}(z,q)\right\}\right)\right\}.$$

Consequently $\Delta = \Delta_1$, which completes the proof of Lemma. $\square$

Now suppose that M is a general measure satisfying Theorem's conditions. Consider $(\tau_{mj}, X_{mj})_{mj\in\mathbb{N}}$ and $(M, M_{mj})_{m,j\in\mathbb{N}}$ used to construct a process X introduced by (3.12). According to Remark 3.13 we use the partition of $\mathbb{R}^n$ into cubes $Q_j, j \in \mathbb{Z}^n$. For $k \in \mathbb{N}$ take a random measure M^k described in Lemma 3.18. Let, for each $k \in \mathbb{N}$, $(M_{mj}^k)_{m,j \in \mathbb{N}}$ be independent copies of M^k such that the family $\{M_{mj}^k(B), B \in \mathcal{R}, m, j, k \in \mathbb{N}\}$ be independent of $(\tau_{mj}, X_{mj})_{m,j \in \mathbb{N}}$. Set

$$X^k(B) = \sum_{m=1}^{\infty} \sum_{j=1}^{\tau_m} M_{mj}^k(B + X_{mj}), \quad B \in \mathcal{B}(\mathbb{R}^n), \quad k \in \mathbb{N}. \tag{3.27}$$

Lemma 3.21. *Assume that conditions of Theorem 3.19 hold. Then, for any $B \in \mathcal{R}$,*

$$X^k(B) \to X(B) \text{ in } L^1(\Omega, \mathcal{F}, \mathsf{P}) \text{ as } k \to \infty.$$

Proof. Lemmas 3.12 and 3.14 demonstrate that, for any $B \in \mathcal{R}$ and all $k \in \mathbb{N}$, the series (3.12) and (3.27) converge in $L^1(\Omega, \mathcal{F}, \mathsf{P})$. Lemmas 3.14, 3.16, 3.18, 2), and Remark 3.13 imply that the series (3.14) converges and the series (3.18) converge uniformly in k. Thus, for any $L \in \mathbb{N}$,

$$\mathsf{E}|X(B) - X^k(B)| \le \sum_{m=1}^{L} \mathsf{E} \sum_{m=1}^{\tau_m} \left| M_{mj}(B + X_{mj}) - M_{mj}^k(B + X_{mj}) \right|$$

$$+ \sum_{m=L+1}^{\infty} \mathsf{E} \sum_{j=1}^{\tau_m} M_{mj}(B + X_{mj}) + \sum_{m=L+1}^{\infty} \mathsf{E} \sum_{j=1}^{\tau_m} M_{mj}^k(B + X_{mj}).$$

Here and subsequently we use instead of X_{mj} their modifications X_{mj}^* (omitting the star) introduced in the proof of Lemma 3.13. Moreover, for every $B \in \mathcal{R}$ and any $\varepsilon > 0$, there exists $L_0 = L_0(B, \varepsilon)$ such that

$$\sum_{m=L_0+1}^{\infty} \mathsf{E} \sum_{j=1}^{\tau_m} M_{mj}(B + X_{mj}) + \sum_{m=L_0+1}^{\infty} \mathsf{E} \sum_{j=1}^{\tau_m} M_{mj}^k(B + X_{mj}) < \varepsilon, \quad k \in \mathbb{N}. \tag{3.28}$$

It is convenient to make assumption that initially all M_{mj} (and therefore all M_{mj}^k) be defined on $(\Omega_1, \mathcal{F}_1, \mathsf{P}_1)$ and $(\tau_m, X_{mj})_{m,j \in \mathbb{N}}$ be defined on $(\Omega_2, \mathcal{F}_2, \mathsf{P}_2)$. After that take $(\Omega, \mathcal{F}, \mathsf{P}) = (\Omega_1, \mathcal{F}_1, \mathsf{P}_1) \otimes (\Omega_2, \mathcal{F}_2, \mathsf{P}_2)$ and consider extensions of M_{mj} and τ_m, X_{mj} $(m, j \in \mathbb{N})$ on $(\Omega, \mathcal{F}, \mathsf{P})$. Denote by E_1 and E_2 the integration over $(\Omega_1, \mathcal{F}_1, \mathsf{P}_1)$ and $(\Omega_2, \mathcal{F}_2, \mathsf{P}_2)$ respectively.

One has

$$\sum_{m=1}^{L_0} \mathsf{E} \sum_{m=1}^{\tau_m} \left| M_{mj}(B + X_{mj}) - M_{mj}^k(B + X_{mj}) \right|$$

$$\le \sum_{m=1}^{L_0} \sum_{N=0}^{\infty} \sum_{j=1}^{N} \mathsf{E} \left| M_{mj}(B + X_{mj}) - M_{mj}^k(B + X_{mj}) \right| \frac{\Lambda(K_m)^N}{N!} e^{-\Lambda(K_m)}$$

$$= \sum_{m=1}^{L_0} \Lambda(K_m) \mathsf{E} |M(B + X_{m1}) - M^k(B + X_{m1})| =: \Delta(L_0, k).$$

For each $\omega_2 \in \Omega_2$ and all $k \in \mathbb{N}$ one infers, on account of Lemma 3.18, 1), that

$$\mathsf{E}_1 |M(B + X_{m1}(\omega_2)) - M^k(B + X_{m1}(\omega_2))| \le 2k^{-1} + \mathsf{E}_1 M(\partial(B + X_{m1}(\omega_2))^{2/k}).$$

Consequently, taking E_2 of both sides of this inequality we observe (using the Fubini theorem and the condition $\mathsf{E}M(\mathbb{R}^n) < \infty$) that

$$\mathsf{E}|M(B + X_{m1}(\omega_2)) - M^k(B + X_{m1}(\omega_2))| \leq 2k^{-1} + \mathsf{E}M(\partial(B + X_{m1})^{2/k}). \quad (3.29)$$

The monotone convergence theorem yields that, for any $B \in \mathcal{R}$,

$$\mathsf{E}M(\partial(B + X_{m1})^{2/k}) \to \mathsf{E}M(\partial(B + X_{m1})) \text{ as } k \to \infty.$$

However,

$$\mathsf{E}M(\partial(B + X_{m1})) = \mathsf{E}_2\mathsf{E}_1 M(\partial(B + X_{m1})) = \int_{\Omega_2} \mu(\partial B + X_{m1}(\omega_2))d\mathsf{P}_2$$

$$= \int_{K_m} \mu(\partial(B + x))P_{X_{m1}}(dx) = 0 \quad (3.30)$$

since $\Lambda(H(B)) = 0$. In view of (3.29) and (3.30) we conclude that, for any $B \in \mathcal{R}$,

$$\Delta(L_0, k) \leq L_0 \mathsf{E}\left|M(B + X_{m1}) - M^k(B + X_{m1})\right| \sup_{m \in \mathbb{N}} \Lambda(K_m) \to 0 \quad (3.31)$$

as $k \to \infty$. Therefore, as $\varepsilon > 0$ could be taken arbitrarily small, (3.28) and (3.31) imply the result. The proof is complete. $\square$

Corollary 3.22. *Assume that conditions of Theorem 3.19 are satisfied. Then a family[18] $\{X(B), B \in \mathcal{S}\}$ is associated where $\mathcal{S}$ is a system of open and closed bounded sets in $\mathbb{R}^n$. If $\mathcal{G} = \{B \in \mathcal{B}_0(R^n) : \Lambda(H(B)) = 0\}$ where $H(B) = \{x \in \mathbb{R}^n : \mu(\partial(B + x)) > 0\}$, then a family $\{X(B), B \in \mathcal{G}\}$ is associated.*

Proof. Let $B_1, \ldots, B_r$ be pairwise disjoint open subsets of $\mathbb{R}^n$. Each open set in $\mathbb{R}^n$ is a countable union of open balls. Thus $B_v = \cup_{j=1}^{\infty} C_{vj}$ for $v = 1, \ldots, r$ where $C_{vj} \in \mathcal{R}$, $C_{vj} \subset C_{v,j+1}$, $j \in \mathbb{N}$. Therefore $(X(C_{1j}), \ldots, X(C_{rj})) \to (X(B_1), \ldots, X(B_r))$ a.s., $j \to \infty$. In view of Theorems 1.8, (d), and 3.19 we conclude that $X(B_1), \ldots, X(B_r)$ are associated. If B is a bounded closed set in $\mathbb{R}^n$ then one can take bounded open sets $B^{\varepsilon_k} = \{x \in \mathbb{R}^n : dist(x, B) < \varepsilon_k\}$, $k \in \mathbb{N}$, such that $B^{\varepsilon_k} \downarrow B$ as $k \to \infty$. Using the already proved part for open bounded sets and once again Theorem 1.8, (d), we come to the first statement of the Corollary.

Now consider an arbitrary $B \in \mathcal{G}$. Let us take open sets $O_k = B^{\varepsilon_k}$, $k \in \mathbb{N}$. Set $D = \cap_{k=1}^{\infty} O_k$. Note that D is the closure of B, hence $D \setminus B \subset \partial B$. Therefore,

$$X(D) - X(B) = \sum_{m=1}^{\infty} \sum_{j=1}^{\tau_m} (M_{mj}(D + X_{mj}) - M_{mj}(B + X_{mj}))$$

$$\leq \sum_{m=1}^{\infty} \sum_{j=1}^{\tau_m} M_{mj}((D \setminus B) + X_{mj}) \leq \sum_{m=1}^{\infty} \sum_{j=1}^{\tau_m} M_{mj}(\partial(B + X_{mj})).$$

---

[18]Note that a ring $\mathcal{R}$ consists of subsets of $\mathbb{R}^n$ which are neither open nor closed.

By the same reasoning as in the proof of Lemmas 3.11 and 3.12 we come, for every $m \in \mathbb{N}$, to the equality

$$\mathsf{E}M_{mj}(\partial(B + X_{mj})) = \int_{K_m} \mu(\partial(B + x))\Lambda(dx) = 0$$

as $\Lambda(H(B)) = 0$. Consequently $\mathsf{E}|X(D) - X(B)| \leq 0$ and $X(B) = X(D)$ a.s. Now in view of the case of bounded open sets considered above, Theorem 1.8, (d) and Remark 1.4, (a), the last statement of the Corollary is valid. $\square$

Note in passing that cluster random measures were used by Neumann and Scott to model the distribution of galaxies, see, e.g., [119]. Some corollaries of this result and other examples of associated random measures can be found in papers [87, 164].

$5°$. **Point random fields.** A Poisson random field provides a simple example of a point random field. A random measure X in $\mathbb{R}^n$ taking nonnegative integer values is called a *point field* if, for any $m \in \mathbb{N}$ and $D \in \mathcal{B}_0(\mathbb{R}^n)$, there is a function $r_{D,m} : D^m \to \mathbb{R}_+$ such that

$$P(X(B_1) = k_1, \ldots, X(B_v) = k_v) =$$

$$\sum_{j=0}^{\infty} \binom{j+k}{k_1, \ldots, k_v, j} \int_{B_1^{k_1} \times \ldots \times B_v^{k_v} \times C^j} \cdots \int \frac{1}{(j+k)!} r_{D,j+k}(x_1, \ldots, x_{j+k}) dx \qquad (3.32)$$

for all disjoint measurable $B_1, \ldots, B_v \subset D$ and $k_1, \ldots, k_v \in \mathbb{Z}_+$. Here $k = \sum_{i=1}^{v} k_i$, $C = D \setminus (\cup_{i=1}^{v} B_i)$,

$$\binom{N}{q_1, \ldots, q_s} = \frac{N!}{q_1! \ldots q_s!}, \quad N \in \mathbb{N}, \quad q_i \geq 0, \quad i = 1, \ldots, s, \quad \sum_{i=1}^{s} q_i = N,$$

and integration is with respect to the Lebesgue measure in $\mathbb{R}^{j+k}$.

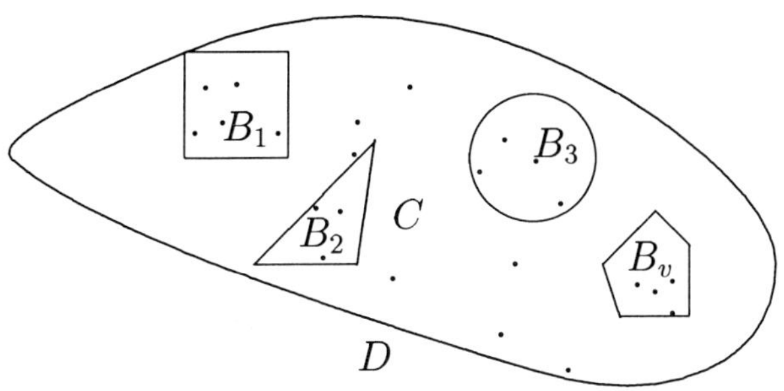

Fig. 1.2 $(v = 4, k_1 = 5, k_2 = 3, k_3 = 4, k_4 = 4, j = 7)$

The functions $r_{D,m}$ are referred to as the *absolute product densities*. Their informal sense is that the probability to find exactly m points in D, each of them inside a small volume Δx_i near the point x_i, $i = 1, \ldots, m$, amounts to $r_{D,m}(x_1, \ldots, x_m)\Delta x_1 \ldots \Delta x_m / m!$.

Let $\lambda : \mathbb{R}^n \to \mathbb{R}$ be a nonnegative function, locally integrable with respect to the Lebesgue measure. If one takes

$$r_{D,m}(x_1,\ldots,x_m) = \exp\left\{-\int_D \lambda(x)dx\right\}\lambda(x_1)\ldots\lambda(x_m)$$

for any $m \in \mathbb{N}$ and any $D \in \mathcal{B}_0(\mathbb{R}^n)$, then an easy calculation shows that the corresponding point field is a Poisson one with intensity function λ. The following theorem belongs to Burton and Waymire; for some further results on association of point processes see, e.g., [245, 246].

Theorem 3.23. ([87]) *Suppose that a point random field X in $\mathbb{R}^n$ has continuous absolute product densities such that, for any cube $D \subset \mathbb{R}^n$, any $m \in \mathbb{N}$ and all points $x_1,\ldots,x_i,\ldots,x_j,\ldots,x_m$ $(1 \leq i \leq j \leq m)$, the inequality*

$$r_{D,m}(x_1,\ldots,x_m)r_{D,j-i+1}(x_i,\ldots,x_j) \geq r_{D,j}(x_1,\ldots,x_j)r_{D,m-i+1}(x_i,\ldots,x_m)$$

holds. Then X is associated.

The proof is postponed until next Section, as it relies on the lattice correlation inequalities techniques.

6°. Stochastic integral for nonrandom functions. To construct further examples of associated random measures we employ the concept of stochastic integral for nonrandom functions. Again we assume that all probability spaces and measure spaces are complete.

Definition 3.24. Let $(S,\mathcal{B})$ be a measurable space and M be a random measure[19] on $(S,\mathcal{B})$. We say that f is *integrable* with respect to M if there exists an event $A = A(f)$ such that $\mathsf{P}(A) = 1$ and, for any $\omega \in A$, the function f is integrable with respect to $M(\omega)$. The *integral* (or *stochastic integral*) of f with respect to M is the function $\mathbf{I}(f) : \Omega \to \mathbb{R}$ which is defined as follows:

$$\mathbf{I}(f) = \begin{cases} \int_S f(x)M(\omega,dx) & \text{if } \omega \in A, \\ 0 & \text{otherwise.} \end{cases} \tag{3.33}$$

For integrable f we write instead of (3.33)

$$\mathbf{I}(f) = \int_S f(x)M(dx).$$

Note that $f : S \to \mathbb{R}$ is integrable (with respect to M) if and only if $|f|$ is integrable. It is easy to see that integrable (with respect to M) functions form a linear space.

It is natural to extend this Definition and introduce, for measurable $f : S \to \mathbb{R}$,

$$\mathbf{I}(f) := \mathbf{I}(f_+) - \mathbf{I}(f_-) \tag{3.34}$$

if there exists an event $A = A(f)$ with $\mathsf{P}(A) = 1$ such that the right-hand side of (3.34) is well defined for $\omega \in A$, i.e. the case $\infty - \infty$ is excluded. Here, as usual,

---

[19]We consider only σ-finite measure $M(\omega,\cdot)$ for each $\omega \in \Omega$.

$f_+ = f\mathbb{I}\{f \geq 0\}$ and $f_- = -f\mathbb{I}\{f < 0\}$. Then $\mathbf{I}(f) \in [-\infty, \infty]$ and f is called integrable whenever $\mathbf{I}(f) \in \mathbb{R}$. If $f \leq g$ for f and g integrable with respect to M, then $\mathbf{I}(f) \leq \mathbf{I}(g)$ a.s.

Furthermore, let (S, ρ) be a metric space. Till the end of Section 3 we will consider a random measure M on $\mathcal{B}(S)$ such that $M(B) < \infty$ a.s. if $B \in \mathcal{B}_0(S)$, i.e. if B belongs to the ring of bounded Borel subsets of S.

We say that (nonrandom) $f : S \to \mathbb{R}$ is a *simple function* if

$$f = \sum_{k=1}^{L} c_k \mathbb{I}\{B_k\}, \quad c_k \in \mathbb{R}, \quad B_k \in \mathcal{B}_0(S), \quad k = 1, \ldots, L, \quad L \in \mathbb{N}.$$

Lemma 3.25. *Suppose that $f : S \to \mathbb{R}$ is integrable with respect to M, then there exists a sequence of simple functions $f_N : S \to \mathbb{R}$ such that $\mathbf{I}(f_N) \to \mathbf{I}(f)$ a.s., $N \to \infty$ (thus $\mathbf{I}(f)$ is a random variable). If $f : S \to \mathbb{R}_+$ is a measurable function, then there are simple functions f_N such that $0 \leq f_N \nearrow f$ on S as $N \to \infty$. Moreover, for any such sequence of simple functions f_N one has $0 \leq \mathbf{I}(f_N) \nearrow \mathbf{I}(f)$ a.s., $M \to \infty$, where $\mathbf{I}(f) \in [0, \infty]$ a.s. In this case*

$$\mathsf{EI}(f) = \lim_{N \to \infty} \mathsf{EI}(f_N) = \int_S f(x)\mu(dx) \in [0, +\infty],$$

here μ is the intensity measure for M. If $f : S \to \mathbb{R}$ then f is integrable with respect to M when $\int_S |f(x)|\mu(dx) < \infty$.

This statement follows by the Lebesgue integral construction with nonrandom measures, thus the proof is omitted.

Lemma 3.26. *Let $(f_N)_{N \in \mathbb{N}}$ be a sequence of integrable (with respect to M) functions such that $f_N \to f$ on S as $N \to \infty$ and $|f| \leq g$ where g is a nonnegative integrable (w.r.t. M) function. Then f is integrable (w.r.t. M) and $\mathbf{I}(f_N) \to \mathbf{I}(f)$ a.s., $N \to \infty$.*

Proof. The assertion is due to a standard construction of integral with respect to $M(\omega, \cdot)$ for each $\omega \in \Omega$ and the dominated convergence theorem. $\square$

Theorem 3.27. *Let M be an associated random measure on a measurable space $(S, \mathcal{B})$, i.e. a family $\{M(B), B \in \mathcal{B}_0(S)\}$ is associated. If $f_r : S \to \mathbb{R}_+$ is a (nonrandom) function integrable with respect to M ($r = 1, \ldots, m$), then the random vector $(\mathbf{I}(f_1), \ldots, \mathbf{I}(f_m))$ is associated.*

Proof. If all f_r are simple nonnegative functions, then

$$\mathbf{I}(f_r) = \sum_{k=1}^{N_r} c_{r,k} M(B_{r,k}), \quad c_{r,k} \in \mathbb{R}_+, \quad B_{r,k} \in \mathcal{B}_0(S), \quad r = 1, \ldots, m.$$

Therefore Corollary 1.9 applies. Otherwise on account of Lemma 3.25 we can take a sequence f_{rN} of nonnegative simple functions such that $\mathbf{I}(f_{rN}) \to \mathbf{I}(f_r)$ a.s. when $N \to \infty$, $r = 1, \ldots, m$. Theorem 1.8, (e) yields the result. $\square$

7°. Infinitely divisible random measures.

Definition 3.28. A random measure M on a space $(S, \mathcal{B}(S))$ is called *infinitely divisible* if, for any $m \in \mathbb{N}$, there exist independent identically distributed random measures $M_1, \ldots, M_m$ on $(S, \mathcal{B}(S))$ such that $Law(M) = Law(M_1 + \ldots + M_m)$.

Actually, in the above Definition one considers $(S, \mathcal{B}_0(S))$ because we assume that $M(B) < \infty$ a.s. for $B \in \mathcal{B}_0(S)$. Thus, for any $n \in \mathbb{N}$ and $B_1, \ldots, B_n \in \mathcal{B}_0(S)$,

$$Law(M(B_1), \ldots, M(B_n)) = Law\left(\sum_{j=1}^m M_j(B_1), \ldots, \sum_{j=1}^m M_j(B_n)\right). \qquad (3.35)$$

Recall that a complete separable metric space (S, ρ) is called *Polish*. The following general result is due to Burton, Waymire and (independently) Evans.

Theorem 3.29. ([88, 164]) *An infinitely divisible random measure M on the Borel σ-algebra of a Polish space (S, ρ) is associated.*

Proof. Let at first $S = \{s_1, \ldots, s_k\}$ for some $k \in \mathbb{N}$ and $\mathcal{B}(S) = 2^S$. Then

$$M(B) = \sum_{r:s_r \in B} M(\{s_r\}), \quad B \in \mathcal{B}(S).$$

Therefore, in view of Theorem 1.8, (d), it suffices to verify that the vector $X = (M(\{s_1\}), \ldots, M(\{s_k\})) \in \mathbf{A}$. Due to (3.35) (with $n = k$, $B_r = \{s_r\}$, $r = 1, \ldots, k$) we conclude that X is an infinitely divisible vector with values in $\mathbb{R}_+^k$.

Now recall the Lévy–Khintchine formula written as

$$\mathsf{E}e^{i(t,X)} = \exp\left\{i(t,b) - \frac{(\Sigma t, t)}{2} + \int_{\mathbb{R}^k} (e^{i(t,x)} - 1 - i(t,x)\mathbb{I}\{\|x\| \le 1\})L(dx)\right\}.$$

Here $i^2 = -1$, $t \in \mathbb{R}^k$, $b \in \mathbb{R}^k$, Σ is a symmetric nonnegatively definite matrix and the *Lévy measure* L is a σ-finite measure on $\mathbb{R}^k$ such that $L(\{0\}) = 0$ and

$$\int_{\mathbb{R}^k} (\|x\|^2 \wedge 1)L(dx) < \infty. \qquad (3.36)$$

There is a one-to-one correspondence between the distributions of infinitely divisible random vectors with values in $\mathbb{R}^k$ and the triples (b, Σ, L).

Assume at first that

$$\int_{\mathbb{R}^k} (\|x\| \vee 1)L(dx) < \infty. \qquad (3.37)$$

Then the integrals

$$\int_{\{x:\|x\| \le 1\}} (t,x)L(dx) \quad \text{and} \quad \int_{\mathbb{R}^k} (e^{i(t,x)} - 1)L(dx)$$

converge for any $t \in \mathbb{R}^k$ because $|(t,x)| \le \|t\|\|x\|$ and $|e^{i(t,x)} - 1| \le |(t,x)| \wedge 2$ for all $t, x \in \mathbb{R}^k$. So one can rewrite $\mathsf{E}\exp\{i(t, X)\}$ as

$$\exp\left\{i(t,b') - \frac{(\Sigma t, t)}{2} + \int_{\mathbb{R}^k} (e^{i(t,x)} - 1)L(dx)\right\}, \quad t \in \mathbb{R}^k,$$

where

$$b' = b - \int_{\{x:\|x\|\le 1\}} x L(dx).$$

For a vector-valued function an integral is taken, as usual, componentwise.

We shall prove that

$$\varphi(t) = \exp\left\{ \int_{\mathbb{R}^k} (e^{i(t,x)} - 1) L(dx) \right\}, \quad t \in \mathbb{R}^k,$$

is a characteristic function with the help of an auxiliary field Z.

Let Z be a Poisson spatial process in on $\mathbb{R}^k$ with intensity measure L. Introduce a random vector $Y = (Y_1, \ldots, Y_k)$ letting

$$Y_j = \int_{\mathbb{R}^k} x_j Z(dx), \quad j = 1, \ldots, k.$$

Due to (3.37) one has

$$\int_{\mathbb{R}^n} |x_j| L(dx) < \infty, \quad j = 1, \ldots, k. \tag{3.38}$$

Therefore, according to Lemma 3.25, all Y_j are real-valued random variables.

Lemma 3.30. *Under condition* (3.37) *the function* $\varphi(t)$, $t \in \mathbb{R}^k$, *is a characteristic function of a vector* Y.

Proof. Consider the sets $B_{m,l} = \left(\frac{l_1-1}{m}, \frac{l_1}{m}\right] \times \ldots \times \left(\frac{l_k-1}{m}, \frac{l_k}{m}\right] \subset \mathbb{R}^k$, $m \in \mathbb{N}$, $l \in \mathbb{Z}^k$. Recall that Z is an independently scattered random measure. Hence for any $t \in \mathbb{R}^k$ we have, by Lemma 3.26 and formula (3.3),

$$\mathsf{E} \exp\left\{ i \sum_{j=1}^{k} t_j \int_{\mathbb{R}^k} x_j Z(dx) \right\} = \lim_{m\to\infty} \mathsf{E} \exp\left\{ i \sum_{l\in\mathbb{Z}^k,\, |l|\le m^2} \sum_{j=1}^{k} t_j \frac{l_j}{m} Z(B_{m,l}) \right\}$$

$$= \lim_{m\to\infty} \prod_{l\in\mathbb{Z}^k,\, |l|\le m^2} \mathsf{E} \exp\left\{ i \sum_{j=1}^{k} t_j \frac{l_j}{m} Z(B_{m,l}) \right\}$$

$$= \lim_{m\to\infty} \exp\left\{ \sum_{l\in\mathbb{Z}^k,\, |l|\le m^2} L(B_{m,l}) \left(\exp\left\{ i \sum_{j=1}^{k} t_j \frac{l_j}{m} \right\} - 1 \right) \right\}$$

$$= \exp\left\{ \int_{\mathbb{R}^k} \left(e^{i(t,x)} - 1 \right) L(dx) \right\}. \quad \square$$

Thus $Law(X) = Law(Y+W)$ where Y and W are independent random vectors, $W \sim N(b', \Sigma)$. We prove that $\Sigma = 0$. Assume that $\Sigma \ne 0$. Then there exists a linear manifold $\mathcal{L} = b' + \mathbb{R}^q$ in $\mathbb{R}^k$ such that $1 \le q < k$, $\mathsf{P}(W \in \mathcal{L}) = 1$ and the restriction of W to $\mathcal{L}$ has strictly positive density on $\mathcal{L}$. The convolution formula yields that the vector $Y + W$ has strictly positive density on $\mathcal{L}$ as well. Therefore,

$P(X \in \mathbb{R}^k \setminus \mathbb{R}_+^k) \geq P(Y + W \in \mathcal{L} \setminus \mathbb{R}_+^k) > 0$. Consequently $P(X \in \mathbb{R}_+^k) < 1$ which is not true. Hence $\Sigma = 0$.

We have established that $Law(X) = Law(b' + Y)$. We claim now that

$$L(\mathbb{R}^k \setminus \mathbb{R}_+^k) = 0. \tag{3.39}$$

If that were not the case, we could choose $j \in \{1, \dots, k\}$ and a compact set $K \subset \mathbb{R}^k$ such that $L(K) > 0$ and $a = \sup_{x \in K} x_j < 0$. Due to (3.38) $U := \int_{x : x_j \geq 0} x_j Z(dx)$ is a random variable. Therefore there exists $x_0 > 0$ such that $P(U \leq x_0) > 0$. Note that

$$Y_j \leq \int_K x_j Z(dx) + U \leq aZ(K) + U$$

where $Z(K)$ and U are independent. So, for any $y < 0$,

$$P(Y_j \leq y) \geq P(aZ(K) + U \leq y) \geq P(aZ(K) + U \leq y, U \leq x_0)$$

$$\geq P(aZ(K) + x_0 \leq y, U \leq x_0) \geq P(Z(K) \geq a^{-1}(y - x_0))P(U \leq x_0) > 0$$

since $Z(K) \sim Pois(L(K))$ with $L(K) > 0$. However, $Law(X_j) = Law(b'_j + Y_j)$. Therefore $Y_j \geq -b'_j$ a.s. because $X_j = M(\{s_j\}) \geq 0$ a.s. The contradiction shows that (3.39) holds and

$$Y_j = \int_{\mathbb{R}_+^k} x_j Z(dx), \quad j = 1, \dots, k.$$

By Theorems 3.5 and 3.27 the vector Y is associated, hence X is also associated. If (3.37) does not hold, consider the random vectors $(X^n)_{n \in \mathbb{N}}$ such that

$$\mathsf{E}e^{i(t, X^n)} = \exp\left\{i(t, b) + \int_{\mathbb{R}^k} (e^{i(t, x)} - 1 - i(t, x)\mathbb{I}\{\|x\| \leq 1\})L_n(dx)\right\}$$

where $t \in \mathbb{R}^k$, $n \in \mathbb{N}$ and $L_n(B) := L\left(B \cap \{x \in \mathbb{R}^k : n^{-1} \leq \|x\| \leq n\}\right)$, $n \in \mathbb{N}$. Then each X^n is associated by the proved part of Theorem, and condition (3.36) implies that $X_n \to X$ in law as $n \to \infty$. So X is associated by Theorem 1.8, (e).

Now suppose that S is infinite. As usual, we will denote by $B_r(s)$ the open ball of radius r with center point s, that is $B_r(s) = \{x \in S : \rho(s, x) < r\}$.

Take any pairwise disjoint compacts $C_1, \dots, C_m$ in S and $C = \cup_{j=1}^m C_j$. Then

$$\rho_0 := \inf_{1 \leq k, j \leq m, k \neq j} \inf_{x \in C_k, y \in C_j} \rho(x, y) > 0.$$

We will prove that $(M(C_1), \dots, M(C_m))$ is an associated vector. For any $n \in \mathbb{N}$ such that $2n^{-1} < \rho_0$ let $J_n = \{s_1^n, \dots, s_{q(n)}^n\} \subset C$ be a $(1/n)$-net in C. Define the random measure M^n concentrated on J_n $(n \in \mathbb{N})$ by way of

$$M^n(B) = \sum_{k : s_k^n \in B} M^n(\{s_k^n\}), \quad B \in \mathcal{B}(S),$$

where $M^n(\{s_1^n\}) := M(B_{1/n}(s_1^n))$ and

$$M^n(\{s_k^n\}) := M\left(B_{1/n}(s_k^n) \setminus \left(\bigcup_{j=1}^{k-1} B_{1/n}(s_j^n)\right)\right), \quad k = 2, \dots, q(n).$$

Thus $Y_n = (M^n(\{s_1^n\}), \ldots, M^n(\{s_{q(n)}^n\}))$ is infinitely divisible, hence associated, by the already proved part of the Theorem. Therefore M^n is an associated random measure. Moreover, for $j = 1, \ldots, m$,

$$M(C_j) \leq M \left(\bigcup_{k:s_k^n \in C_j} B_{1/n}(s_k^n) \right) = M^n(C_j) \leq M(C_j^{(1/n)}), \qquad (3.40)$$

here $C_j^{(1/n)} = \{x \in S : \rho(x, C_j) := \inf_{y \in C_j} \rho(x, y) < 1/n\}$. Note that $C_j^{(1/n)} \downarrow C_j$ as $n \to \infty$. Accordingly, $M(C_j^{(1/n)}) \to M(C_j)$ a.s. for each $j = 1, \ldots, m$ as $n \to \infty$. In view of (3.40) we conclude that $M^n(C_j) \to M(C_j)$ almost surely as $n \to \infty$, $j = 1, \ldots, m$. By Theorem 1.8, (e), the vector $(M(C_1), \ldots, M(C_m))$ is associated.

Now we use the following result being of independent interest.

Lemma 3.31. *Let M be a random measure on a Polish space (S, ρ) such that $M(B) < \infty$ a.s. if $B \in \mathcal{B}_0(S)$. Then, for any $B \in \mathcal{B}_0(S)$, there exists a sequence of compact sets $C_n, n \in \mathbb{N}$, such that $C_n \subset B$ for any $n \in \mathbb{N}$ and $M(C_n) \to M(B)$ in probability as $n \to \infty$.*

Proof. For $B \in \mathcal{B}_0(S)$ and $n \in \mathbb{N}$ introduce an event $A_n = \{M(B) \leq n\}$. Then $\mathsf{P}(A_n^c) \to 0$ as $n \to \infty$ because $M(B)$ is a random variable. The following set functions

$$\mu_n(D) = \mathsf{E}M(D \cap B)\mathbb{I}\{A_n\}, \quad D \in \mathcal{B}(S), \quad n \in \mathbb{N},$$

are finite measures on $\mathcal{B}(S)$. Consequently, as (S, ρ) is a Polish space, for any $n \in \mathbb{N}$ we can find a compact set $C_n \subset B$ such that $\mu_n(B \setminus C_n) < 1/n$ (see, e.g., [39, §1, Th. 1.4]). Then, for any $\varepsilon > 0$,

$$\mathsf{P}(|M(B) - M(C_n)| > \varepsilon) = \mathsf{P}(M(B \setminus C_n) > \varepsilon) \leq \mathsf{P}(A_n \cap \{M(B \setminus C_n) > \varepsilon\}) + \mathsf{P}(A_n^c)$$

$$\leq \varepsilon^{-1}\mathsf{E}M(B \setminus C_n)\mathbb{I}\{A_n\} + \mathsf{P}(A_n^c) = \varepsilon^{-1}\mu_n(B \setminus C_n) + \mathsf{P}(A_n^c) < (\varepsilon n)^{-1} + \mathsf{P}(A_n^c) \to 0$$

as $n \to \infty$, which is our claim. $\square$

To finish the proof of Theorem let $B_1, \ldots, B_m$ be arbitrary sets in $\mathcal{B}_0(S)$. Then we have to show the association of $Z = (M(B_1), \ldots, M(B_m))$. Without loss of generality we may assume that $B_1, \ldots, B_m$ are pairwise disjoint (see Corollary 1.9).

For any $j = 1, \ldots, m$, in view of Lemma 3.31 we can find a sequence of compacts $(C_{jn})_{n \in \mathbb{N}}$ such that $M(C_{jn}) \to M(B_j)$ in probability as $n \to \infty$. Clearly, for each $n \in \mathbb{N}$, the compacts $C_{1n}, \ldots, C_{mn}$ are disjoint. Therefore the vector $Z_n = (M(C_{1n}), \ldots, M(C_{mn}))$ is associated. Since $Z_n \to Z$ in probability as $n \to \infty$, the required statement ensues by Theorem 1.8, (e). $\square$

Remark 3.32. In particular, it was proved that an infinitely divisible random vector in $\mathbb{R}^m$ is associated, provided that all its components are a.s. positive. See [362] for related results.

8°. Stable random systems. A non-degenerate random vector $X = (X_1, \ldots, X_n)$ is *stable* if, for any $k \in \mathbb{N}$, there are some $c(k) \geq 0$ and $d(k) \in \mathbb{R}^n$ such that the laws of vectors $X^1 + \ldots + X^k$ and $c(k)X + d(k)$ coincide, if $X^1, \ldots, X^k$ are mutually independent and distributed as X. In this case $c(k) = k^{1/\alpha}$ for some $\alpha \in (0, 2]$, see [167, Ch. VI, § 1]. Recall that a random vector $X = (X_1, \ldots, X_n)$ has *α-stable distribution*, $0 < \alpha < 2$, if and only if

$$\mathsf{E}\exp\{i(X, t)\} = \exp\left\{ -\int_{S^{n-1}} |(s, t)|^\alpha (1 - i\, sgn((s, t))\phi(\alpha, s, t))\Gamma(ds) + i(b, t) \right\}$$

where S^{n-1} is the unit sphere in $\mathbb{R}^n$, $t, b \in \mathbb{R}^n$, $s \in S^{n-1}$, Γ is a finite measure on $\mathcal{B}(S^{n-1})$ (the so-called *spectral measure* of X) and

$$\phi(\alpha, s, t) = \begin{cases} \tan \frac{\pi\alpha}{2}, & \alpha \neq 1, \\ -\frac{2}{\pi} \log |(s, t)|, & \alpha = 1. \end{cases} \tag{3.41}$$

The question when the association of X holds arises, for example, in the study of multidimensional domains of attraction of stable laws and in analysis of assets portfolio ([339]). The following result by Lee, Rachev and Samorodnitsky describes the association of stable vectors.

Theorem 3.33. ([254]). *An α-stable vector X $(0 < \alpha < 2)$ is associated if and only if its spectral measure Γ satisfies the condition*

$$\Gamma(S_-) = 0 \tag{3.42}$$

where $S_- = \{(s_1, \ldots, s_n) \in S^{n-1} : s_i s_j < 0 \text{ for some } i, j \in \{1, \ldots, n\}\}$. The vector $X \in \mathbf{NA}$ if and only if

$$\Gamma(S_+) = 0 \tag{3.43}$$

where $S_+ = \{(s_1, \ldots, s_n) \in S^{n-1} : s_i s_j > 0 \text{ for some } i, j \in \{1, \ldots, n\}, i \neq j\}$.

Proof. Sufficiency. Assume that (3.42) holds. The measure Γ is concentrated on the set $(S^{n-1} \cap \mathbb{R}^n_+) \cup (S^{n-1} \cap \mathbb{R}^n_-)$ where $\mathbb{R}^n_- := \{x \in \mathbb{R}^n : x_j \leq 0, j = 1, \ldots, n\}$. It is enough to assume that Γ is supported by $S^{n-1} \cap \mathbb{R}^n_+$. Indeed, then for Γ supported by $S^{n-1} \cap \mathbb{R}^n_-$ the proof is done by considering $(-X_1, \ldots, -X_n)$. Since the sum of two independent stable vectors has spectral measure equal to the sum of spectral measures of added vectors, the general case follows due to Theorem 1.8, (c), (d).

Case 1: $\alpha \in (1, 2)$. Define a σ-finite measure L on $\mathbb{R}^n$ letting

$$L(\{0\}) = 0, \; L(\mathbb{R}^n \setminus \mathbb{R}^n_+) = 0,$$

$$L(\{(xs_1, \ldots, xs_{n-1}) : (s_1, \ldots, s_{n-1}) \in B, x \in C\}) = \frac{1}{c(\alpha)}\Gamma(B)\int_C y^{-\alpha-1}dy$$

where B is a Borel subset of $S^{n-1} \cap \mathbb{R}^n_+$, a set $C \in \mathcal{B}_0(\mathbb{R})$ is such that $\inf_{x \in C} x > 0$ and $c(\alpha) > 0$ will be chosen later.

Then L is the Lévy measure of X (considered as an infinitely divisible random vector) multiplied by a positive constant. Indeed, we have

$$\exp\left\{i(t,a) + \int_{\mathbb{R}^n}\left(e^{i(t,y)} - 1 - i(t,y)\mathbb{I}\{\|y\| \le 1\}\right)L(dy)\right\}$$

$$= \exp\left\{\int_{\mathbb{R}^n}\left(e^{i(t,y)} - 1 - i(t,y)\right)L(dy)\right\}$$

$$= \exp\left\{\frac{1}{c(\alpha)}\int_{S^{n-1}\cap\mathbb{R}^n_+}\int_0^\infty\left(e^{ir(s,t)} - 1 - ir(s,t)\right)r^{-\alpha-1}dr\Gamma(ds)\right\} \tag{3.44}$$

where $a = -\int_{y:\|y\|>1} yL(dy)$, this integral clearly converges as $\alpha > 1$. In order to show that the right-hand side of (3.44) is the characteristic function of the vector $X - b$, with some nonrandom $b \in \mathbb{R}^n$, we evaluate the following integral passing to the complex variable z. If $(s,t) > 0$, then

$$\int_0^\infty\left(e^{i(s,t)r} - 1 - ir(s,t)\right)r^{-\alpha-1}dr$$

$$= (s,t)^\alpha\int_0^\infty\frac{e^{iz} - 1 - iz}{z^{\alpha+1}}dz = (s,t)^\alpha c_0(\alpha)\left(1 - i\tan\left(\frac{\pi\alpha}{2}\right)\right) \tag{3.45}$$

where $c_0(\alpha) > 0$. In more detail, consider $I = \int_0^\infty(e^{iz} - 1 - iz)z^{-\alpha-1}dz$ where $z^{\alpha+1}$ for $z \in \{\operatorname{Im} z \ge 0\}$ denotes the principal branch of the analytic function taking real values on the positive half-line. Then, by the Cauchy theorem,

$$0 = \int_{-\infty}^0\frac{e^{iz} - 1 - iz}{z^{\alpha+1}}dz + \int_0^\infty\frac{e^{iz} - 1 - iz}{z^{\alpha+1}}dz$$

$$= I + \int_0^\infty\frac{e^{-i\zeta} - 1 + i\zeta}{(-1)^{\alpha+1}\zeta^{\alpha+1}}d\zeta = I + (-1)^{-\alpha-1}\overline{I} = I - \overline{I}e^{-\alpha\pi i}$$

where $\overline{I}$ stands for the complex conjugate of I. Since $\operatorname{Re} I < 0$, the last relation implies that $I = \rho e^{-i\pi\alpha/2}$ for some $\rho > 0$, or equivalently that

$$I = c_1(\alpha)\cos(\pi\alpha/2)(1 - i\tan\pi\alpha/2), \quad c_1(\alpha) > 0.$$

This entails (3.45). The same reasoning applies to $(s,t) \le 0$. Thus the characteristic function of X plus some constant vector coincides with the left-hand side of (3.44) if $c(\alpha) = c_0(\alpha)$.

The desired result follows by Theorem 3.29 since L is supported by $\mathbb{R}^n_+$.

Case 2: $\alpha \in (0,1)$. The argument in this case is similar, with the only exception that one writes the characteristic function of X in the form

$$\exp\left\{i(t,v) + \int_{\mathbb{R}^n}\left(e^{i(t,y)} - 1\right)L(dy)\right\}$$

with $v \in \mathbb{R}^n$.

Case 3: $\alpha = 1$. Define a σ-finite measure L on $\mathbb{R}^n$ as it was made in Case 1. To see that L is a (normalized) Lévy measure of X, write

$$\exp\left\{\int_{\mathbb{R}^n}\left(e^{i(t,y)} - 1 - i(t,y)\mathbb{I}\{\|y\| \leq 1\}\right)L(dy)\right\}$$

$$= \exp\left\{\frac{1}{c(1)}\int_{S^{n-1}\cap\mathbb{R}_+^n}\int_{\mathbb{R}_+}\left(e^{ir(s,t)} - 1 - ir(s,t)\mathbb{I}\{|r| \leq 1\}\right)r^{-2}dr\Gamma(ds)\right\}.$$

If $(s,t) > 0$, then, passing to complex variable, we observe that

$$\int_{\mathbb{R}_+}\left(e^{ir(s,t)} - 1 - ir(s,t)\mathbb{I}\{|r| \leq 1\}\right)r^{-2}dr = (s,t)\int_{\mathbb{R}_+}\frac{e^{iz} - 1 - iz\mathbb{I}\{|z| \leq (s,t)\}}{z^2}dz$$

$$= (s,t)\int_{\mathbb{R}_+}\frac{e^{iz} - 1 - iz\mathbb{I}\{|z| \leq 1\}}{z^2}dz - 2i(s,t)\int_1^{(s,t)}\frac{dz}{z} = (s,t)\int_{\mathbb{R}_+}\frac{\cos z - 1}{z^2}dz$$

$$-i(s,t)\log(s,t) + i(s,t)\int_0^\infty\frac{z - z\mathbb{I}\{z \leq 1\}}{z^2}dz$$

$$= (s,t)\left(-\pi/2 - i\log(s,t) + i\int_0^\infty\frac{z - z\mathbb{I}\{z \leq 1\}}{z^2}dz\right),$$

since by standard complex analysis $\int_{\mathbb{R}}(1-\cos z)z^{-2}dz = \pi$. Repeating the argument for $(s,t) < 0$ with appropriate changes and taking into account that $|x|sgn(x) = x$ $(x \in \mathbb{R})$ one sees that

$$\int_{\mathbb{R}_+}\left(e^{ir(s,t)} - 1 - ir(s,t)\mathbb{I}\{|r| \leq 1\}\right)r^{-2}dr$$

$$= -|(s,t)|(\pi/2 + i\log(s,t)) + i(s,t)\int_0^\infty\frac{z - z\mathbb{I}\{z \leq 1\}}{z^2}dz.$$

Thus L is the Lévy measure of X (plus some constant) if $c(1) = \pi/2$. The proof is finished as in Case 1.

Now assume that (3.43) holds. Evidently, the condition on Γ implies that for any two different pairs of indices $\{m,j\}, \{p,q\}$ one has $\Gamma(E_{mj} \cap E_{pq}) = 0$ where $E_{kl} = \{s \in S^{n-1} : s_k s_l \neq 0\}$. We also use the notation

$$D_m = \{(s_1,\ldots,s_n) \in S^{n-1} : s_j = 0, j \neq m, \text{ and } s_m \neq 0\}.$$

Then X is equal in distribution to the random vector

$$\sum_{m=1}^{n-1}\sum_{j=m+1}^n (X_1(m,j),\ldots,X_n(m,j)) + (W_1,\ldots,W_n)$$

where all the random vectors entering the double sum are mutually independent, the characteristic function of $X(m,j)$ is

$$\exp\left\{-\int_{E_{mj}}|(s,t)|^\alpha(1 - isgn((s,t))\phi(\alpha,s,t))\Gamma(ds) + i(b^{m,j},t)\right\}$$

for $m = 1, \ldots, n-1$, $j = m+1, \ldots, n$, and

$$\mathsf{E}\exp\{i(W_k, t)\} = \exp\left\{-\int_{D_k} |(s,t)|^\alpha (1 - i\,sgn((s,t))\phi(\alpha, s, t))\Gamma(ds) + i(b^k, t)\right\},$$

$k = 1, \ldots, n$. By Theorem 1.8, (c), we only have to prove that each of these vectors is **NA**. The vector $(W_1, \ldots, W_n)$ has independent components, hence is **NA**. Moreover, $X_k(m, j) = 0$ a.s. if $k \notin \{m, j\}$. Thus, it remains to show that $X_m(m, j)$ and $X_j(m, j)$ are negatively associated. But since $s_m s_j \leq 0$ Γ-a.e. on the set E_{mj}, the random variables $X_m(m, j)$ and $-X_j(m, j)$ are associated by the already proved part of Theorem.

Necessity. Suppose that X is associated and $\Gamma(S_-) > 0$. Let $Y = (Y_1, \ldots, Y_n)$ be an independent copy of X and let $U_i = X_i - Y_i$. If Y is associated, the same is true of $(-Y)$. Then the random vector $U = (U_1, \ldots, U_n)$ is associated by Theorem 1.8, (d), and is symmetric α-stable, that is, it has shift $\beta = 0$ and spectral measure Π defined by the equality $\Pi(B) = \Gamma(B) + \Gamma(-B)$, $B \in \mathcal{B}(S^{n-1})$. Clearly, $\Pi(S_-) = \Gamma(S_-) + \Gamma(-S_-) > 0$, therefore, for some $i, j \in \{1, \ldots, n\}$ one has

$$\Pi(\{s \in S^{n-1}; s_i > 0, s_j < 0\}) > 0.$$

Without loss of generality we assume that $i = 1, j = 2$. By the change of variables of integration, one can see that the spectral measure $\Gamma_{1,2}$ of (U_1, U_2) has the property

$$\Gamma_{1,2}(\{s \in S^1 : s_1 > 0, s_2 < 0\}) > 0. \tag{3.46}$$

For any $x > 0$, the association of U_1 and U_2 implies

$$\mathsf{P}(U_2 \leq -x | U_1 > x) \leq \mathsf{P}(U_2 \leq -x). \tag{3.47}$$

Note that $(U_1, -U_2)$ is a symmetric α-stable vector. By (3.46) and the theorem of Samorodnitsky ([361]) the following limit is strictly positive

$$\lim_{x \to \infty} \mathsf{P}(U_2 \leq -x | U_1 > x) = \left(\int_{S^1} (s_1^+)^\alpha \Gamma_{1,2}(ds)\right)^{-1} \int_{s_1 > 0, s_2 < 0} (s_1 \wedge (-s_2))^\alpha \Gamma_{1,2}(ds).$$

However this contradicts an obvious fact that the right-hand side of (3.47) tends to zero as $t \to \infty$.

The argument in the **NA** case is similar. $\square$

An analog of this theorem for infinitely divisible random vector is not valid. As shown by Samorodnitsky ([362]), there exists a 2-dimensional associated random vector having characteristic function

$$\mathsf{E}\{\exp i(X, t)\} = \exp\left\{\int_{\mathbb{R}^2} \left(e^{i(x,t)} - 1 - i\mathbb{I}\{\|x\| \leq 1\}(x, t)\right) L(dx) + i(\beta, t)\right\},$$

with the Lévy measure L such that $L(\{x : x_1 > 0, x_2 < 0\}) > 0$. In fact, the condition

$$L(\mathbb{R}^n_+ \cup \mathbb{R}^n_-) = 1$$

is equivalent to the statement that, for every $\gamma > 0$, the random vector with characteristic function $(\mathsf{E}\exp i(X, t))^\gamma$ is infinitely divisible ([341]). There are also close results on association of random vectors with extreme value distribution ([282, 342]).

4   Association and probability measures on lattices

As was mentioned in the previous Section, there are not too many important examples of associated, **PA** or **NA** random vectors which admit the direct verifying of Definitions 1.1—1.3. Thus, one needs simple enough sufficient conditions for a random system to be associated (**PA** or **NA**, or have similar properties). This approach can be realized by considering more complicated structure of the distribution of a vector X. In view of great number of formalizations of *positive dependence* in different sources (see, e.g., [103–105, 203, 206, 226, 261, 368] and references therein), we will not list all of them, only the most important ones.

    1°. Totally positive functions and stochastic orders.

Definition 4.1. ([150]) Real-valued random variables $X_1, X_2, \ldots$ are *stochastically increasing in sequence (SIS)* if, for each $n > 1$ and any $x \in \mathbb{R}$, the regular conditional probability distribution function[20] $\mathsf{P}(X_n \leq x | X_i = x_i, i = 1, \ldots, n - 1)$ is nonincreasing in $x_1, \ldots, x_{n-1}$.

Theorem 4.2. ([24]) *A SIS random sequence is associated.*

    Proof. We will show that, for any $n > 1$, there exists a coordinate-wise nondecreasing function $h : \mathbb{R}^n \to \mathbb{R}$ and a random variable U independent of $Y = (X_1, \ldots, X_{n-1})$ such that $X = (Y, X_n)$ equals in distribution to the random vector $(Y, h(U, Y))$. Then the desired conclusion will follow by induction on n and Theorem 1.8, (c) and (d). More exactly, to construct such U, in general, an extension of the initial probability space $(\Omega, \mathcal{F}, \mathsf{P})$ is required. That is we can consider U initially defined on $(\Omega', \mathcal{F}', \mathsf{P}')$ and then introduce the product space $(\Omega \times \Omega', \mathcal{F} \otimes \mathcal{F}', \mathsf{P} \otimes \mathsf{P}')$. After that we set $X(\omega, \omega') := X(\omega)$, $U(\omega, \omega') := U(\omega')$, see, e.g., [383, Ch. II, § 2.8] for details. Further we use U uniformly distributed on $[0, 1]$. In this case one takes $(\Omega', \mathcal{F}', \mathsf{P}')$ as $[0, 1]$ with a class of Borel sets and P' as the Lebesgue measure μ ($\mathcal{F}'$ being $\mathcal{B}([0, 1])$ completed with respect to μ). The mentioned procedure with $\Omega' = [0, 1]$ is called a *standard extension* of initial probability space. To simplify the notation we will write $(\Omega, \mathcal{F}, \mathsf{P})$ for such an extension.

    Fix $n \geq 2$. For any $t \in \mathbb{R}^{n-1}$, $s \in \mathbb{R}$ and $u \in (0, 1)$ set

$$F_t(s) = \mathsf{P}(X_n \leq s | Y = t), \quad h(u, t) = \inf\{s : u \leq F_t(s)\}$$

where F_t is the so called *regular conditional distribution function*. It means that there exist Ω_0 having probability one and a version of $G(s, \omega) = \mathsf{P}(X_n \leq s | Y)$, with $s \in \mathbb{R}$, such that $G(\cdot, \omega)$ is a distribution function for every $\omega \in \Omega_0$. This version exists as X_n takes values in the Polish space $\mathbb{R}$. Exclude the event $\Omega \setminus \Omega_0$ from the probability space. Then for any $s \in \mathbb{Q}$ we can take $\mathsf{P}(X_n \leq s | Y) = \varphi(s, Y)$ where for every $s \in \mathbb{R}$ the function $\varphi(s, \cdot)$ is a Borel map from $\mathbb{R}^{n-1}$ to $\mathbb{R}$. Indeed, that function can be taken to be a Borel one for rational s. Then for general s the corresponding function is Borel as a monotone limit of Borel functions.

---

[20]We recall this notion in the proof below.

Thus h is nondecreasing in u and t as we consider a SIS random sequence. Besides that, for any $u \in (0,1)$, $t \in \mathbb{R}^{n-1}$ and $z \in \mathbb{R}$,

$$h(u,t) \le z \Leftrightarrow u \le F_t(z) = \varphi(z,t) \qquad (4.1)$$

where we took into account that $F_t(\cdot)$ is right-continuous for every $t \in \mathbb{R}^{n-1}$.

Let $k \in \mathbb{N}$ and suppose that $g : \mathbb{R}^k \to \mathbb{R}$ is a Borel function. Then the set

$$B(g) := \{(x_1,\ldots,x_k,x_{k+1}) \in \mathbb{R}^{k+1} : x_{k+1} \le g(x_1,\ldots,x_k)\} \in \mathcal{B}(\mathbb{R}^{k+1}).$$

Indeed, g can be viewed as a Borel mapping from $\mathbb{R}^{k+1}$ to $\mathbb{R}$ ($g(x_1,\ldots,x_k,x_{k+1}) = g(x_1,\ldots,x_k)$), as well as x_{k+1}. Therefore $v(x_1,\ldots,x_k,x_{k+1}) := g(x_1,\ldots,x_k) - x_{k+1}$ is a Borel function. Note that $B(g) = \{x \in \mathbb{R}^{k+1} : v(x) \in [0,+\infty)\}$.

Consequently, in view of (4.1) we conclude that the function h is Borel. Hence for any random variable U and every $t \in \mathbb{R}^{n-1}$ the function $h(U(\omega),t)$ is a random variable. Now take U uniformly distributed on $[0,1]$ and independent of Y. Then

$$P(h(U,t) \le s) = P(U \le F_t(s)) = F_t(s) = P(X_n \le s | Y = t).$$

Since U and Y are independent, we have on account of Lemma 1.22

$$P(h(U,Y) \le s | Y = t) = E\left(\mathbb{I}\{h(U,Y) \le s\} | Y = t\right)$$

$$= E\left(\mathbb{I}\{h(U,t) \le s\}\right) = P(h(U,t) \le s).$$

Hence, conditionally on $Y = t$, $h(U,Y)$ is distributed as X_n (given Y). Therefore $Law(Y,h(U,Y)) = Law(Y,X_n)$. In fact one can easily verify that the characteristic functions of $(Y,h(U,Y))$ and (Y,X_n) coincide. To this end one can use the well-known properties of conditional expectation, in particular, the formula (see, e.g., [383, Ch. II, §7.7, Th. 3])

$$E(f(Z)|Y = t) = \int_{\mathbb{R}} f(s)\, dP(Z \le s | Y = t)$$

for a random variable Z and a bounded measurable function f, $P(Z \le s | Y)$ being a regular conditional distribution function with respect to a random vector Y. $\square$

As a corollary to Theorem 4.2 we can establish the postponed Theorem 2.25.

Proof of Theorem 2.25. For $n \in \mathbb{N}$, take any $0 \le t_1 < t_2 < \ldots < t_n$ and arbitrary $x_1,\ldots,x_n \in S$. For each $t \in \mathbb{R}$ a function $f_t(x) = \mathbb{I}\{x \ge t\}$, $x \in \mathbb{R}$, is nondecreasing. Then, by Markov property (A.2.1) and the definition of transition operator (A.2.7),

$$P(X(t_n) \ge x_n | X(t_{n-1}) = x_{n-1}, \ldots, X(t_1) = x_1) = P(X(t_n) \ge x_n | X(t_{n-1}) = x_{n-1})$$

$$= E(f_{x_n}(X(t_n)) | X(t_{n-1}) = x_{n-1}) = T_{t_n - t_{n-1}} f_{x_n}(x_{n-1})$$

which is, by monotonicity, a nondecreasing function in x_{n-1} (and consequently in $x_1,\ldots,x_{n-1}$). Therefore, the random sequence $(X(t_1),\ldots,X(t_n))$ is SIS, hence associated. $\square$

The converse to Theorem 4.2 is not true even when $n = 2$, as shown by Esary, Proschan and Walkup.

Example 4.3. ([160]). Let (X, Y) be a random vector and $\mathsf{P}(X = i, Y = j) = p_{ij}$ for $i, j \in \{0, 1, 2\}$ with p_{ij} given by a matrix

$$\begin{pmatrix} p_{00} & p_{01} & p_{02} \\ p_{10} & p_{11} & p_{12} \\ p_{20} & p_{21} & p_{22} \end{pmatrix} = \begin{pmatrix} 1/4 & 0 & 1/8 \\ 0 & 1/4 & 0 \\ 1/8 & 0 & 1/4 \end{pmatrix}.$$

Then $(X, Y) \in \mathbf{A}$ but is not SIS, since $\mathsf{P}(Y \geq 2 | X = 0) > 0 = \mathsf{P}(Y \geq 2 | X = 1)$.

Theorem 4.2 and analogous results on stochastic orders are especially useful in reliability theory (see [24, 161]), when the random variables involved are binary. Then the condition of increasing takes discrete form. We provide one application of association property in that domain.

A *structure* X is a finite collection of binary random variables $\{X_t, t \in T\}$, possibly dependent, and a binary function $\Phi : \{0, 1\}^T \to \{0, 1\}$. As usual, for $x = (x_t)_{t \in T}$, $y = (y_t)_{t \in T} \in S = \{0, 1\}^T$ the relation $x \leq y$ means that $x_t \leq y_t$ for all $t \in T$ and $x < y \Leftrightarrow (x \leq y$ and $x \neq y)$. The structure is called *monotone* if Φ is a *monotone function*, i.e. if either Φ or $(-\Phi)$ is nondecreasing. The vector (function) $x \in S$ is a *minimal way* of Φ (resp. *minimal cut* of Φ) if $\Phi(x) = 1$ and $y < x \Rightarrow \Phi(y) = 0$ (resp. $\Phi(x) = 0$ and $y > x \Rightarrow \Phi(y) = 1$). For example if $T = \{1, \ldots, n\}$ and $\Phi(x) = \min_{t \in T} x_t$, $x \in S = \{0, 1\}^T$, then x with $x_t = 1$ for all $t \in T$ is a minimal way, whereas if $x_{t_0} = 0$ and $x_t = 1$ for $t \in T \setminus \{t_0\}$ then such x is a minimal cut. Note that this Φ has a unique minimal way.

For a minimal way P_j, let $\rho_j(x) = \prod_{i \in P_j} x_i$. For a minimal cut K_j, introduce $\kappa_j(x) = 1 - \prod_{i \in K_j}(1 - x_i)$. A collection $\{\rho_1(x), \ldots, \rho_p(x)\}$ (resp. $\kappa_1(x), \ldots, \kappa_k(x)$) is called a *set of successive minimal ways* (resp. *parallel minimal cuts*) if $\Phi(x) = 1 - \prod_{j=1}^{p}(1 - \rho_j(x))$ (resp. $\Phi(x) = \prod_{j=1}^{k} \kappa_j(x)$), for all $x \in S$.

Theorem 4.4. ([24]) *Let Φ be a monotone structure of associated elements. Let $\rho_1(x), \ldots, \rho_p(x)$ be all the successive minimal ways of Φ, and $\kappa_1(x), \ldots, \kappa_k(x)$ be all its parallel minimal cuts. Then*

$$\prod_{j=1}^{k} \mathsf{P}(\kappa_j(X) = 1) \leq \mathsf{P}(\Phi(X) = 1) \leq 1 - \prod_{j=1}^{p} \mathsf{P}(\rho_j(X) = 0).$$

Proof. Note that the set $\{X_i, \kappa_j(X), \rho_v(X); i \in T, j = 1, \ldots, k, v = 1, \ldots, p\}$ is associated by Theorem 1.8, (d). Hence, by Corollary 1.10 one has

$$\mathsf{P}(\kappa_1(X) = \ldots = \kappa_k(X) = 1) \geq \prod_{j=1}^{k} \mathsf{P}(\kappa_j(X) = 1).$$

However the inequality $\mathsf{P}(\kappa_1(X) = \cdots = \kappa_k(X) = 1) \leq \mathsf{P}(\Phi(X) = 1)$ is trivial. Similar argument yields the second inequality. $\square$

Recall that, for a 2-dimensional binary random vector X, a necessary and sufficient condition of association is, by Theorem 1.13,

$$p_{11} \geq (p_{11} + p_{10})(p_{11} + p_{01}) \tag{4.2}$$

where $p_{ij} = \mathsf{P}(X = (i,j))$, $i, j \in \{0,1\}$. Rewrite this condition as

$$p_{11}(1 - p_{11}) \geq p_{10}p_{01} + p_{11}p_{01} + p_{11}p_{10}.$$

Taking into account that $1 - p_{11} = p_{00} + p_{01} + p_{10}$ we see that (4.2) is equivalent to

$$p_{00}p_{11} \geq p_{10}p_{01}.$$

It is desirable to obtain an extension of this fact for application to much more general random systems. To accomplish this task we need some further definitions.

Definition 4.5. A function $f : \mathbb{R}^2 \to \mathbb{R}$ is called *totally positive of order* 2 (abbreviated notation: f is TP_2) if

$$\det \| f(x_i, y_j) \|^2_{i,j=1} \geq 0 \text{ for all } x_1, x_2, y_1, y_2 \in \mathbb{R} \text{ with } x_1 < x_2, y_1 < y_2.$$

A function $f : \mathbb{R}^n \to \mathbb{R}$ $(n \geq 2)$ is called TP_2 *in pairs* if it is TP_2 in any pair of arguments, other $n - 2$ arguments being arbitrary but fixed. It is called *multiparameter totally positive of order* 2 (one writes MTP_2) if

$$f(x \vee y)f(x \wedge y) \geq f(x)f(y) \text{ for any } x, y \in \mathbb{R}^n$$

where $x \vee y = (x_1 \vee y_1, \ldots, x_n \vee y_n)$, $x \wedge y = (x_1 \wedge y_1, \ldots, x_n \wedge y_n)$.

Lemma 4.6. *A twice continuously differentiable positive function $f : \mathbb{R}^n \to \mathbb{R}$ is MTP_2 whenever $\partial^2 (\log f)/\partial x_i \partial x_j \geq 0$, for all $i, j = 1, \ldots, n$, $i \neq j$, and all $x \in \mathbb{R}^n$.*

Proof. The necessity part is simple. In fact, if $f : \mathbb{R}^n \to \mathbb{R}$ is MTP_2, then the derivative

$$\frac{\partial^2 \log f(x)}{\partial x_i \partial x_j} = \lim_{\delta \searrow 0} \frac{\log f(x + \delta(e_i + e_j)) - \log f(x + \delta e_i) - \log f(x + \delta e_j) + \log f(x)}{\delta^2}$$

is nonnegative for $x \in \mathbb{R}^n, 1 \leq i < j \leq n$, here $e_k = (0, \ldots, 1, \ldots, 0) \in \mathbb{R}^n$ is the k-th unit vector of standard basis, $k = 1, \ldots, n$.

The proof of sufficiency uses induction on n. Assume that $n > 1$ and the sufficiency part of the assertion is true for dimension $n - 1$ (the induction base will automatically follow from the case $n = 2$). Let $f : \mathbb{R}^n \to \mathbb{R}$ be a twice continuously differentiable positive function, $h = \log f$, and assume that $\partial^2 h(x)/\partial x_i \partial x_j \geq 0$ for all $i, j = 1, \ldots, n$, $i \neq j$, and any $x \in \mathbb{R}^n$. We intend to show that

$$h(x \vee y) + h(x \wedge y) \geq h(x) + h(y) \tag{4.3}$$

for any $x, y \in \mathbb{R}^n$. Renumeration of the coordinates permits to reckon that there exists $m \leq n$ such that $x_i \leq y_i$ for $i = 1, \ldots, m$ and $x_i \geq y_i$ for $i = m + 1, \ldots, n$ (if $m = 0$ or $m = n$ then (4.3) is obviously true). To shorten the formulas we write

$\vec{x}_m = (x_1, \ldots, x_m) \in \mathbb{R}^m$, $\vec{y}_m = (y_1, \ldots, y_m) \in \mathbb{R}^m$ and if $m \leq n - 2$ let us introduce $\vec{x}'_m = (x_{m+1}, \ldots, x_{n-1})$, $\vec{y}'_m = (y_{m+1}, \ldots, y_{n-1})$. So we have

$$h(x \wedge y) - h(x) - h(y) + h(x \vee y)$$

$$= h(\vec{x}_m, \vec{y}'_m, y_n) - h(\vec{x}_m, \vec{x}'_m, x_n) - h(\vec{y}_m, \vec{y}'_m, y_n) + h(\vec{y}_m, \vec{x}'_m, x_n)$$

$$= h(\vec{x}_m, \vec{y}'_m, y_n) - h(\vec{x}_m, \vec{x}'_m, y_n) - h(\vec{y}_m, \vec{y}'_m, y_n) + h(\vec{y}_m, \vec{x}'_m, y_n)$$

$$- h(\vec{x}_m, \vec{x}'_m, x_n) + h(\vec{y}_m, \vec{x}'_m, x_n) + h(\vec{x}_m, \vec{x}'_m, y_n) - h(\vec{y}_m, \vec{x}'_m, y_n)$$

$$\geq h(\vec{y}_m, \vec{x}'_m, x_n) - h(\vec{y}_m, \vec{x}'_m, y_n) + h(\vec{x}_m, \vec{x}'_m, y_n) - h(\vec{x}_m, \vec{x}'_m, x_n)$$

by the induction hypothesis. The last expression can be rewritten as

$$\int_{y_n}^{x_n} \left(\frac{\partial h}{\partial t}(\vec{y}_m, \vec{x}'_m, t) - \frac{\partial h}{\partial t}(\vec{x}_m, \vec{x}'_m, t) \right) dt \geq 0.$$

The arising inequality is due to the fact that $\partial h(u, t) / \partial t$ is, for any fixed t, a coordinatewise nondecreasing function in $u = (u_1, \ldots, u_{n-1}) \in \mathbb{R}^{n-1}$. Thus relation (4.3) follows, and the proof is complete. $\square$

Other examples of MTP_2 functions can be found, e.g., in [226]. The following result originates from the paper by Sarkar. Note that Corollary 2.28 ensues from

Theorem 4.7. ([363]) *Suppose that a random vector $X = (X_1, \ldots, X_n)$ has an everywhere positive density p with respect to Lebesgue measure, and this density is MTP_2. Then X is associated.*

Proof. Let $n \in \mathbb{N}$, $n \geq 2$, $s \in \mathbb{R}$, $x, y \in \mathbb{R}^{n-1}$ and $x_1 \leq y_1, \ldots, x_{n-1} \leq y_{n-1}$. Employing the MTP_2 property, for any $t \geq s$ and $u \leq s$, we have $p(x, t)p(y, u) \leq p(y, t)p(x, u)$. Consequently,

$$\int_s^\infty p(x, t)dt \int_{-\infty}^s p(y, u)du \leq \int_s^\infty p(y, t)dt \int_{-\infty}^s p(x, u)du.$$

Adding to both sides the term $\int_s^\infty p(x, t)dt \int_s^\infty p(y, u)du$ one obtains

$$\int_s^\infty p(x, t)dt \int_{\mathbb{R}} p(y, u)du \leq \int_s^\infty p(y, t)dt \int_{\mathbb{R}} p(x, u)du. \tag{4.4}$$

We can divide (4.4) by $\int_{\mathbb{R}} p(y, u)du \int_{\mathbb{R}} p(x, u)du$, which leads to the inequality

$$\mathsf{P}(X_n \geq s | X_1 = x_1, \ldots, X_{n-1} = x_{n-1}) \leq \mathsf{P}(X_n \geq s | X_1 = y_1, \ldots, X_{n-1} = y_{n-1}).$$

Thus, the sequence $(X_1, \ldots, X_n)$ is SIS, hence associated by Theorem 4.2. $\square$

Remark 4.8. The Lebesgue measure in the conditions of Theorem 4.7 can actually be replaced with some other measure $\mu \otimes \ldots \otimes \mu$ when μ is a finite or σ-finite measure on $(\mathbb{R}, \mathcal{B}(\mathbb{R}))$. The proof remains the same.

Observe that conditions given by the last Theorem (and by the above Remark) are not necessary, as a slight modification of Example 4.3 shows. Namely, to make all the entries positive add a small quantity $\varepsilon > 0$ to zero entries of the matrix and subtract $4\varepsilon/5$ from positive ones; for small enough ε the association survives. Another example is given by a normal random vector with mean 0 and nondegenerate covariance matrix Σ. Its density is MTP_2 if and only if the matrix Σ^{-1} has non-positive off-diagonal elements (by Lemma 4.6). Due to Theorem 2.1 the association of X is equivalent to nonnegativity of all the elements of Σ, which is a weaker condition if dimension is higher than 2 (consider a matrix $\Sigma = (\sigma_{ij})_{i,j=1,2,3}$ with $\sigma_{11} = \sigma_{22} = \sigma_{33} = 1$, $\sigma_{23} = \sigma_{32} = 0$ and other entries equal to $2/3$).

2°. FKG-inequalities and related results. Now we turn to general partially ordered spaces. Again we assume that partial order is measurable (see the definition in subsection 2.2). For finite spaces, we will assume that all their subsets are measurable.

Definition 4.9. A partially ordered set L is called a *lattice*, if any two elements x and y in L have a least upper bound $x \vee y$ and a greatest lower bound $x \wedge y$ (i.e. $x \leq x \vee y$, $y \leq x \vee y$ and $x \leq z, y \leq z$ imply $x \vee y \leq z$, while $x \wedge y$ is defined analogously). A lattice is called *distributive* if these operations satisfy either of the following two equivalent (see [45]) conditions

$$x \wedge (y \vee z) = (x \wedge y) \vee (x \wedge z), \quad x \vee (y \wedge z) = (x \vee y) \wedge (x \vee z),$$

for all $x, y, z \in L$.

A typical example of a finite distributive lattice is a collection W of subsets of $\{1, \ldots, n\}$ such that $A, B \in W \Rightarrow A \cap B, A \cup B \in W$, with partial order

$$A \leq B \Leftrightarrow A \subset B. \tag{4.5}$$

By the Birkhoff theorem [45, p. 59], every finite distributive lattice L is *isomorphic* to some lattice W of subsets of $\{1, \ldots, n\}$ with partial order of inclusion. That is, there exists a bijection F from L to W such that $x \leq y$ if and only if $F(x) \leq F(y)$ for any $x, y \in L$.

The following theorems by Fortuin, Kasteleyn and Ginibre [172], Holley [202] and Preston [334] play a fundamental role in analysis of measures related to various statistical physics models. The condition (4.6) in Theorem 4.10 is usually referred to as *FKG-inequalities*. To simplify the notation, for a finite set T with measure μ on T and $t \in T$, we sometimes write $\mu(t)$ instead of $\mu(\{t\})$.

Theorem 4.10. ([172]) *Let L be a finite distributive lattice. Let μ be a probability measure on $(L, 2^L)$ such that for any $x, y \in L$*

$$\mu(x \vee y)\mu(x \wedge y) \geq \mu(x)\mu(y). \tag{4.6}$$

Then μ is associated, i.e. for any increasing[21] *functions $f, g : L \to \mathbb{R}$, one has*

$$\int_L fg\,d\mu - \int_L f\,d\mu \int_L g\,d\mu \geq 0. \tag{4.7}$$

---

[21]See Definition 2.9.

Without loss of generality we can consider $L = W$ (see (4.5)). In this case instead of inequalities (4.6) one writes

$$\mu(A \cup B)\mu(A \cap B) \geq \mu(A)\mu(B)$$

and, for increasing $f, g : W \to \mathbb{R}$,

$$\sum_{A \in W} f(A)g(A)\mu(A) \geq \sum_{A \in W} f(A)\mu(A) \sum_{A \in W} g(A)\mu(A).$$

Theorem 4.11. ([202]) *Let L be a finite distributive lattice and μ_1, μ_2 be two probability measures on $(L, 2^L)$. Suppose that for any $x, y \in L$*

$$\mu_1(x \vee y)\mu_2(x \wedge y) \geq \mu_1(x)\mu_2(y). \tag{4.8}$$

Then there exists a probability measure ν on the product space $(L, 2^L) \otimes (L, 2^L)$, having marginal projections onto the first and second factors μ_1 and μ_2 respectively, and such that $\nu(\{(x, y) : x \geq y\}) = 1$.

The proofs of these results will be given after the proof of Theorem 4.12.

Now let T be an arbitrary finite set and for any $t \in T$ let $(S_t, \mathcal{B}_t, \lambda_t)$ be a totally ordered measure space with a finite or σ-finite measure λ_t. It is convenient to use $T = \{1, \ldots, n\}$ and consider $S = S_1 \times \ldots \times S_n$, $n \in \mathbb{N}$. For $x, y \in S$, set $x \vee y = (x_t \vee y_t)_{t \in T}, x \wedge y = (x_t \wedge y_t)_{t \in T}$. S is a partially ordered space, if we write $x \leq y$ for x, y such that $x_t \leq y_t, t \in T$. Denote the *product measure space* (see, e.g., [383, Ch. II, § 2.8]) over $t \in T$ by $(S, \mathcal{B}, \lambda)$.

Theorem 4.12. ([334]) *Let μ_1, μ_2 be two probability measures on S having densities f_1, f_2 with respect to λ. Suppose that*

$$f_1(x \vee y)f_2(x \wedge y) \geq f_1(x)f_2(y) \text{ for all } x, y \in S. \tag{4.9}$$

Then there exists a probability measure ν on the product space $S \times S$ having projections μ_1, μ_2 onto the first and second factors S respectively and such that $\nu(\{(x, y) : x \geq y\}) = 1$.

Proof of Theorem 4.12. The proof is by induction on n. Consider the case $n = 1$. Define a probability measure ν on $\mathcal{B} \otimes \mathcal{B}$ by the equality

$$\nu(C) = \int_C (f_1(x) \wedge f_2(y))d\gamma + \int_C \frac{h_1(x)h_2(y)}{\int_S h_2(z)d\lambda(z)} d(\lambda \otimes \lambda), \tag{4.10}$$

here $\gamma(B) = \lambda(\{x \in S : (x, x) \in B\})$ for $B \in \mathcal{B} \otimes \mathcal{B}$, $h_1(x) = (f_1(x) - f_2(x))^+$ and $h_2(y) = (f_2(y) - f_1(y))^+$. If the denominator appearing in the right-hand side of (4.10) is zero, then the whole term containing the fraction is agreed to be 0. To see that the definition of ν is correct note that the class of subsets

$$\mathcal{D} = \{B \subset S \times S : \{x \in S : (x, x) \in B\} \in \mathcal{B}\}$$

is a σ-algebra. It contains all sets of the form $B = B_1 \times B_2$ where $B_1, B_2 \in \mathcal{B}$, since the set $\{x \in S : (x, x) \in B\} = B_1 \cap B_2 \in \mathcal{B}$. Therefore, $\mathcal{B} \otimes \mathcal{B} \subset \mathcal{D}$.

Clearly for $C = B \times S$ where $B \in \mathcal{B}$ one has

$$\nu(C) = \int_B (f_1(x) \wedge f_2(x))\, d\lambda + \int_B (f_1(x) - f_2(x))^+\, d\lambda = \int_B f_1(x) d\lambda = \mu_1(B).$$

In the same manner $\nu(S \times B) = \mu_2(B)$. Evidently $\nu(S \times S) = 1$ and ν has all other properties of a probability measure. Suppose that there exist x, y such that $x < y$ (i.e. $x \leq y$ and $x \neq y$) and $h_1(x)h_2(y) > 0$. Then $f_1(x) > f_2(x), f_2(y) > f_1(y)$, but in this case $f_1(y)f_2(x) < f_1(x)f_2(y)$ which contradicts (4.9). Hence the second term in the right-hand side of (4.10) is equal to 0 for $C = \{(x, y) \in S \times S : x < y\}$. The first term there is also 0 as $\{x \in S : (x, x) \in C\} = \varnothing$. Thus, the third condition for ν is true.

To proceed now with the induction, for $n \geq 2$ take $V = \{1, \ldots, n-1\}$ and introduce

$$S(V) = \prod_{t \in V} S_t, \quad \mathcal{B}(V) = \bigotimes_{t \in V} \mathcal{B}_t, \quad \lambda_V = \bigotimes_{t \in V} \lambda_t.$$

Let $\pi(\mu_i)$ be the projection of μ_i onto $\mathcal{B}(V)$, $i = 1, 2$. Then $\pi(\mu_i)$ has the Radon–Nikodym derivative

$$g_i(z) = \int_{S_n} f_i(z, u) d\lambda_n(u), \quad z \in S(V),$$

with respect to λ_V. Here and in what follows, we sometimes write $f_1(x) = f_1(z, u)$ for $x = (z, u)$, $z \in S(V), u \in S_n$, and use similar notation for f_2.

Lemma 4.13. *If a, b, c, d are nonnegative numbers such that $a \geq (c \vee d)$ and $ab \geq cd$, then $a + b \geq c + d$.*

Proof. If $a = 0$, then $c = d = 0$ and the assertion is obvious; otherwise note that it follows from the inequality $a + (cd)/a \geq c + d$, which is in turn equivalent (for $a > 0$) to $(a - d)(a - c) \geq 0$. $\square$

Lemma 4.14. *For all $z, w \in S(V)$ one has*

$$g_1(z \vee w)g_2(z \wedge w) \geq g_1(z)g_2(w). \tag{4.11}$$

Proof. Let $S(=) = \{(u, v) \in S_n \times S_n : u = v\}$ and

$$S(+) = \{(u, v) \in S_n \times S_n : u > v\}, \quad S(-) = \{(u, v) \in S_n \times S_n : u < v\}.$$

With this notation,

$$g_1(z \vee w)g_2(z \wedge w) = \iint_{S(+)\cup S(-)\cup S(=)} f_1(z \vee w, u) f_2(z \wedge w, v) d\lambda_n(u) d\lambda_n(v)$$

$$= \iint_{S(=)} f_1(z \vee w, u) f_2(z \wedge w, v) d\lambda_n(u) d\lambda_n(v)$$

$$+ \iint_{S(+)} (f_1(z \vee w, u) f_2(z \wedge w, v) + f_1(z \vee w, v) f_2(z \wedge w, u))\, d\lambda_n(u) d\lambda_n(v).$$

Similarly,

$$g_1(z)g_2(w) = \iint_{S(=)} f_1(z,u)f_2(w,v)d\lambda_n(u)d\lambda_n(v)$$

$$+ \iint_{S(+)} (f_1(z,u)f_2(w,v) + f_1(z,v)f_2(w,u))\, d\lambda_n(u)d\lambda_n(v).$$

By assumption, for all $z, w \in S(V)$ and $t \in S_n$,

$$f_1(z \vee w, t)f_2(z \wedge w, t) \geq f_1(z,t)f_2(w,t), \tag{4.12}$$

so one can ignore the terms involving integration over $S(=)$. The proof of Lemma
will be completed as soon as we can show that, for $u > v$,

$$a + b := f_1(z \vee w, u)f_2(z \wedge w, v) + f_1(z \vee w, v)f_2(z \wedge w, u)$$

$$\geq f_1(z,u)f_2(w,v) + f_1(z,v)f_2(w,u) =: c + d. \tag{4.13}$$

By Theorem's hypothesis applied to vectors $x = (z,u)$, $y = (w,v)$ and $x' = (z,v)$,
$y' = (w,u)$ respectively we have (here $u > v$)

$$a = f_1(z \vee w, u)f_2(z \wedge w, v) \geq f_1(z,u)f_2(w,v) = c. \tag{4.14}$$

$$a = f_1(z \vee w, u)f_2(z \wedge w, v) \geq f_1(z,v)f_2(w,u) = d. \tag{4.15}$$

Thus $a \geq c \vee d$. Using (4.12) for $t = u$ and $t = v$ we conclude that

$$ab = f_1(z \vee w, u)f_2(z \wedge w, v)f_1(z \vee w, v)f_2(z \wedge w, u)$$

$$\geq f_1(z,u)f_2(w,v)f_1(z,v)f_2(w,u) = cd.$$

Therefore, relation (4.13) follows from Lemma 4.13, hence (4.11) is established. $\square$

Return now to the proof of Theorem 4.12. Suppose that it is true for all sets with
cardinality less than n. By Lemma 4.14, measures $\pi(\mu_1)$, $\pi(\mu_2)$ on $(S(V), \mathcal{B}(V), \lambda_V)$
satisfy the conditions of Theorem. Thus the induction hypothesis implies that there
exists a probability measure τ on the space $(S(V) \times S(V), \mathcal{B}(V) \otimes \mathcal{B}(V), \lambda_V \otimes \lambda_V)$
such that

$$\tau(A \times S(V)) = \pi(\mu_1)(A) \ \text{ and } \ \tau(S(V) \times B) = \pi(\mu_2)(B) \ \text{ for all } A, B \in \mathcal{B}(V),$$

$$\tau(\{(z,w) \in S(V) \times S(V) : z \geq w\}) = 1.$$

Let $\psi : S_n \to \mathbb{R}$ be some nonnegative measurable function having $\int_{S_n} \psi d\lambda_n = 1$.
For $z \in S(V)$, $u \in S_n$ and $i = 1, 2$ define

$$q_i(z,u) = \begin{cases} \left(\int_{S_n} f_i(z,a)d\lambda_n(a)\right)^{-1} f_i(z,u) & \text{if } \int_{S_n} f_i(z,a)d\lambda_n(a) > 0, \\ \psi(u), & \text{otherwise.} \end{cases}$$

Then q_i is (a version of) the Radon–Nikodym derivative of μ_i with respect to
$\pi(\mu_i) \otimes \lambda_n$. This is easily checked as it suffices to consider the sets of a form $B_V \times B_n$,

$B_V \in \mathcal{B}(V)$, $B_n \in \mathcal{B}_n$. Now define the functions $q, r : S(V) \times S(V) \times S_n \times S_n \to \mathbb{R}$, setting $q(z, w, u, v) = q_1(z, u) \wedge q_2(w, v)$ and

$$r(z, w, u, v) = \frac{(q_1(z, u) - q_2(w, u))^+ (q_2(w, v) - q_1(z, v))^+}{\int_{S_n} (q_2(w, v) - q_1(z, v))^+ d\lambda_n(v)},$$

with the same convention on the denominator as in the paragraph after (4.10). Introduce the measure ν on $(S \times S, \mathcal{B} \otimes \mathcal{B})$ by the formula

$$\nu(C) = \int_C q(z, w, u, v) d\left(\tau(z, w) \otimes \gamma_n(u, v)\right)$$

$$+ \int_C r(z, w, u, v) d\left(\tau(z, w) \otimes (\lambda_n(u) \otimes \lambda_n(v))\right)$$

where γ_n and $\lambda_n \otimes \lambda_n$ are measures on $(S_n \times S_n, \mathcal{B}_n \otimes \mathcal{B}_n)$ and the former is defined analogously to γ in (4.10).

The measure ν possesses the first and second required properties. For example, for $C = B \times S$ it suffices to verify that $\nu(C) = \mu_1(C)$ for "rectangle" sets of the type $B = B_V \times B_n$ where $B_V \in \mathcal{B}(V)$ and $B_n \in \mathcal{B}_n$ (such sets form a semiring generating $\mathcal{B}$). In that case,

$$\nu(C) = \int_{B_V \times B_n \times S} (q_1(z, u) \wedge q_2(w, v)) \, d(\tau \otimes \gamma_n)$$

$$+ \int_{B_V \times B_n \times S} \frac{(q_1(z, u) - q_2(w, u))^+ (q_2(w, v) - q_1(z, v))^+}{\int_{S_n} (q_2(w, v) - q_1(z, v))^+ d\lambda_n(v)} d(\tau \otimes (\lambda_n \otimes \lambda_n))$$

$$= \int_{R(V)} (q_1(z, u) \wedge q_2(w, u)) \, d(\tau \otimes \lambda_n) + \int_{R(V)} (q_1(z, u) - q_2(w, u))^+ d(\tau \otimes \lambda_n)$$

$$= \int_{B_V \times B_n} q_1(z, u) d(\pi(\mu_1)(z) \otimes \lambda_n(u)) = \mu_1(B)$$

with notation $R(V) = B_V \times B_n \times S(V)$. In particular $\nu(S \times S) = \mu_1(S) = 1$. The second condition is verified in the same way. To prove that ν satisfies the third requirement, consider the sets

$$B_i = \left\{ z \in S(V) : \int_{S_n} f_i(z, u) d\lambda_n(u) = 0 \right\}, \quad i = 1, 2.$$

If $z \notin B_1, w \notin B_2$, $u < v$ and $z \geq w$ then $q_1(z, u) q_2(w, v) \leq q_1(z, v) q_2(w, u)$ by (4.9). Analogously to the end of the proof in the $n = 1$ case, we conclude that $r(z, w, u, v) = 0$ unless $u \geq v$. Therefore it remains to show that $\nu(B_1 \times S_n \times S) = \nu(S \times B_2 \times S_n) = 0$. According to the already established property of the measure,

$$\nu(B_1 \times S_n \times S) = \mu_1(B_1 \times S_n) = \int_{B_1} \int_{S_n} f_1(z, u) d\lambda_n(u) d\lambda_n(z) = 0,$$

and similarly $\nu(S \times B_2 \times S_n) = 0$. $\square$

Corollary 4.15. *Suppose that conditions of Theorem 4.12 are fulfilled. Then*

$$\int_S h d\mu_1 \geq \int_S h d\mu_2 \tag{4.16}$$

for any bounded increasing function $h : S \to \mathbb{R}$

Proof. Evidently

$$\int_S h d\mu_1 - \int_S h d\mu_2 = \int_S (h(x) - h(y)) d\nu(x,y).$$

The right-hand side integral is actually taken over the set $\{(x,y) \in S : x \geq y\}$, since it supports ν, and on this set $h(x) \geq h(y)$. $\square$

Remark 4.16. If $|S| < \infty$, the validity of (4.16) for any bounded increasing $h :$ $S \to \mathbb{R}$ is equivalent to the existence of measure ν described in the formulation of Theorem 4.12. See [335] for the proof based on graph theory.

Proof of Theorem 4.11. By Birkhoff's theorem we may consider L which is a lattice of subsets of $\{1, \ldots, n\}$, for some $n \in \mathbb{N}$, with usual partial order. Take $T = \{1, \ldots, n\}$, $S_t = \{0, 1\}$ for any $t \in T$ and let λ_t be the *counting measure* on $\{0, 1\}$, i.e. $\lambda_t(0) = \lambda_t(1) = 1$. Define the functions $q_1, q_2 : 2^T \to \mathbb{R}$ as follows:

$$q_i(A) = \begin{cases} \mu_i(\{A\}), & A \in L, \\ 0, & \text{otherwise,} \end{cases}$$

$i = 1, 2$. Consider the product space $S = \{0, 1\}^n$ with counting measure $\lambda = \lambda_1 \otimes \ldots \otimes \lambda_n$. If $A, B \in L$, then by (4.8) the densities q_1 and q_2 satisfy inequality (4.9). If $A \notin L$ or $B \notin L$, then (4.9) is trivially satisfied as its right-hand side is zero. Therefore, due to Theorem 4.12 there exists a probability measure ν on $2^T \times 2^T$ with marginal densities q_1 and q_2 such that $\nu(A \times B) = 0$ unless $B \subset A$. Besides, this measure ν is concentrated on $L \times L$, since $\nu(L \times L) = \mu_1(L) = 1$. $\square$

Proof of Theorem 4.10. If $x, y \in L$ and $\mu(x) \wedge \mu(y) > 0$, then, because of (4.6), $\mu(x \wedge y) > 0$ and $\mu(x \vee y) > 0$. That is, the elements $x \in L$ for which $\mu(x) > 0$ form a finite distributive lattice. The elements having $\mu(x) = 0$ do not take part in the expectations in (4.7). So we may suppose from the beginning that μ is strictly positive. Also we may assume that g is strictly positive, adding a constant to g when necessary. Take measures $\lambda = \mu_2 = \mu$ and μ_1 so that its Radon-Nikodym derivative with respect to μ equals

$$\frac{d\mu_1}{d\mu}(x) = \frac{g(x)}{\int_L g d\mu}.$$

Then (4.9) holds since, for $x, y \in L$, one has $d\mu_2/d\lambda = 1$ and

$$\frac{d\mu_1}{d\lambda}(x \vee y) \frac{d\mu_2}{d\lambda}(x \wedge y) = \frac{g(x \vee y)}{\int_L g d\mu} \geq \frac{g(x)}{\int_L g d\mu} = \frac{d\mu_1}{d\lambda}(x) \frac{d\mu_2}{d\lambda}(y).$$

Due to monotonicity of f, conditions of Corollary 4.15 are satisfied. Therefore,

$$\int_L f g d\mu \left(\int_L g d\mu \right)^{-1} \geq \int_L f d\mu. \ \square$$

Analogous argument allows to deduce Theorem 4.7 from Theorem 4.12. Note that in the former theorem p need not be a density with respect to Lebesgue measure, but may be the density with respect to a product of an arbitrary finite collection of σ-finite measures $\lambda_1, \ldots, \lambda_n$ on $\mathbb{R}$, and may somewhere equal zero.

Remark 4.17. If a probability measure μ on a lattice $(L, 2^L)$ of subsets of W satisfies the FKG-inequalities (hence is associated), then the random variables $\{X_i, i \in W\} \in \mathbf{A}$, where $X_i = \mathbb{I}\{i \in R\}$ and R is the random subset of W chosen according to μ. In fact, if $T \subset W$ and $f, g \in \mathcal{M}(|T|)$, then $f(X_i, i \in T)$, $g(X_i, i \in T)$ are increasing functions on L, consequently (4.7) holds.

Before going over to more complicated examples, we present a nice combinatorial result by Seymour and Welsh ([367]) and then postponed proof of Theorem 3.23.

Corollary 4.18. ([367]) *Let S be a finite nonempty set and $\mathcal{C}$ be a collection of subsets of S which is increasing (that is, if $B \in \mathcal{C}$ and $B \subset B' \subset S$, then $B' \in \mathcal{C}$). Then the average cardinality[22] of sets in $\mathcal{C}$ is not less than $|S|/2$.*

Proof. Consider on the finite lattice $L = 2^S$ the counting measure μ, divided by $2^{|S|}$. This measure satisfies (4.6) as an equality. Apply Theorem 4.10 for increasing functions $f, g : 2^S \to \mathbb{R}$, $f(A) = \mathbb{I}\{A \in \mathcal{C}\}$, $g(A) = |A|$. We obtain

$$2^{-|S|} \sum_{A \in \mathcal{C}} |A| \geq 2^{-2|S|} |\mathcal{C}| \sum_{A \subset S} |A|,$$

hence

$$|\mathcal{C}|^{-1} \sum_{A \in \mathcal{C}} |A| \geq 2^{-|S|} \sum_{A \subset S} |A|.$$

The right-hand side is the average cardinality of an arbitrary subset of S, which is $|S|/2$, as for $n \in \mathbb{N}$

$$\sum_{k=0}^{n} k \binom{n}{k} (1/2)^k (1 - 1/2)^{n-k} = \frac{n}{2}. \quad \square$$

Now we are able to establish the result concerning the association of a point random field discussed in Section 2.

Proof of Theorem 3.23. Clearly, it suffices to study the case when the value of measure X on the completion of some cube $U_N = [-N, N)^n$ is almost surely zero. In fact, one can construct a sequence of measures converging in law to X and having the above mentioned property with corresponding $N \in \mathbb{N}$. For $k \in \mathbb{N}$ divide U_N into cubes, open "from above" and closed "from below" (i.e. into the cubes of the type $[a, b)$), each having edge length 2^{-k}. The total number of such "small" cubes is $(2N2^k)^n$. Let E_k be the set of all least points of "small" cubes. To each subset $R = \{x_1, \ldots, x_m\}$ of E_k we assign a measure

$$\mathsf{P}_k(R) = Z_k\, r_{U_N, m}(x_1, \ldots, x_m) 2^{-km},$$

---

[22]That is the sum of cardinalities of sets belonging to $\mathcal{C}$ divided by $|\mathcal{C}|$.

using the functions $r_{D,m}$ (with $D = U_N$) defined in (3.32), Z_k being a normalizing constant taken so that $\sum_{R \subseteq E_k} \mathsf{P}_k(R) = 1$. Now introduce a random measure X^k as follows: for any $B \in \mathcal{B}_0(\mathbb{R}^n)$ let $X^k(B) = |B \cap R|$ where the subset $R \subset E_k$ is drawn according to probability measure P_k. In view of the Theorem condition P_k satisfies the FKG inequality and consequently, by Remark 4.17, the random variables

$$\{\mathbb{I}\{x_i \in R\}, \ x_i \text{ is a point in } E_k\}$$

are associated. Therefore, the random measure $X^k \in \mathbf{A}$ by Theorem 1.8, (d).

What is left is to show that finite-dimensional distributions of the random measure X^k converge to those of X as $k \to \infty$. Then, by Theorem 1.8, (e), X will be associated. We will check that $(X^k(A_1), \ldots, X^k(A_v)) \to (X(A_1), \ldots, X(A_v))$ in law for any $r, v \in \mathbb{N}$, where $A_1, \ldots, A_v$ are finite disjoint unions of small cubes defined above with edge length 2^{-r}. Note that, for any $k > r$, each small cube $[a, b)$ with edge length 2^{-k} either belongs to A_l or belongs to its complement, $l = 1, \ldots, v$.

Fix $m_1, \ldots, m_{v+1} \in \mathbb{Z}_+$. Set $m = m_1 + \ldots + m_v$, $q = m + m_{v+1}$, $A_{m+1} = [-N, N)^n \setminus \cup_{j=1}^v A_j$ and $a = (m_1! \ldots m_{v+1}!)^{-1}$. For $k > r$ we have

$$\mathsf{P}\Big(X^k(A_1) = m_1, \ldots, X^k(A_{v+1}) = m_{v+1}\Big)$$

$$= a \sum_{\substack{x_i \in A_1 \cap E_k \\ 1 \le i \le m_1}} \cdots \sum_{\substack{x_i \in A_{v+1} \cap E_k \\ m < i \le q}} r_{U_N, q}(x_1, \ldots, x_q) \frac{Z_k}{2^{kq}} \to a \int_{A_1^{m_1} \times \ldots \times A_{v+1}^{m_{v+1}}} \cdots \int r_{U_N, q}(x) dx$$

as $k \to \infty$, since the multiple sum above is the Riemann sum approximation for the integral which is equal to $\mathsf{P}(X(A_1) = m_1, \ldots, X(A_{v+1}) = m_{v+1})$.

Note that $X(B) = 0$ a.s. for any set B of zero Lebesgue measure. Next, for any collection of rectangles $B_1, \ldots, B_m \subset \mathbb{R}^k$, it is easily seen (when rectangles are approximated from inside by unions of small cubes with dyadic-rational edges) that $(X(B_1), \ldots, X(B_m))$ is associated. The rest of the proof is performed as in the proof of Theorem 3.19. $\square$

3°. The Ising model. Now we pass to anticipated examples of applications. Let each point of the lattice $\mathbb{Z}^d$ be occupied by a particle (of lattice gas). Assign to this particle, located at a point (or site) $i \in \mathbb{Z}^d$, a *spin*, which is a random variable X_i taking values in a finite subset D of $\mathbb{R}$. The simplest situation is when all the spins are independent and $D = \{-1, 1\}$. If the spins of particles can influence each other, it becomes necessary to study not only one-dimensional distributions of $\{X_i\}$ but at least finite-dimensional ones.

Let E be a finite subset of $\mathbb{Z}^d$. Suppose at first that there is no gas at points outside E, and the distribution of $\{X_i, i \in E\}$ is an everywhere positive measure on the finite set $D^{|E|}$. Then each possible combination of values of X_i is called *configuration*. To each configuration $\sigma = \{\sigma_i, i \in E\}$, $\sigma = \sigma(E)$, we assign the *energy* $U(\sigma) > 0$. It is assumed that the probability of a configuration σ is proportional to $\exp(-U(\sigma))$, i.e.

$$\mathsf{P}(X_i = \sigma_i, i \in E) = \frac{e^{-U(\sigma)}}{Z_E}, \quad Z_E = \sum_{\sigma \in D^{|E|}} e^{-U(\sigma)}.$$

One often writes simply $\mathsf{P}(\sigma)$ considering P as the distribution of a random field $\{X_i, i \in E\}$. For example, the classical *Ising ferromagnet* is a (finite) set of particles having spin either 1 or -1 and the corresponding energy of a configuration being

$$U(\sigma) = \sum_{i,j \in E} J_{ij}\sigma_i\sigma_j + \sum_{i \in E} \mu_i\sigma_i$$

where $J_{ij}, \mu_i \in \mathbb{R}$, $i, j \in E$.

Assume, for simplicity, that L is the lattice of all subsets of a finite set E ($|E| > 1$) and μ is a measure on $(L, 2^L)$ such that $\mu(R) > 0$ for all $R \in L$. Set $\lambda(R) = \log \mu(R)$. Then the FKG-inequalities can be rewritten as follows

$$\lambda(R \cup S) + \lambda(RS) \geq \lambda(R) + \lambda(S), \quad R, S \in L. \tag{4.17}$$

Lemma 4.19. *A condition equivalent to (4.17) reads: for all $r, s \in E$, $r \neq s$, and any $R \subset (E \setminus \{s, r\})$,*

$$\lambda(R \cup \{r, s\}) + \lambda(R) \geq \lambda(R \cup \{r\}) + \lambda(R \cup \{s\}). \tag{4.18}$$

Proof. The implication $(4.17) \Rightarrow (4.18)$ is obvious if we take in (4.17) instead of R and S the sets $R \cup \{r\}$ and $R \cup \{s\}$ respectively. To establish the converse, rewrite (4.17) as

$$\lambda(R \cup T_1 \cup T_2) + \lambda(R) \geq \lambda(R \cup T_1) + \lambda(R \cup T_2) \tag{4.19}$$

where R, T_1, T_2 are pairwise disjoint subsets of E, $|T_1| \geq |T_2|$. The proof is by induction on $|T_1| + |T_2|$. If $|T_1| = |T_2| = 1$, evidently (4.19) coincides with (4.18). Let now $|T_1| \geq 2, |T_2| \geq 1$. We take $t \in T_1$, $V = T_1 \setminus \{t\}$ and introduce $R' = R \cup V$, $T_1' = \{t\}$, $T_2' = T_2$. Then $R \cup T_1 \cup T_2 = R' \cup T_1' \cup T_2'$. By the induction hypothesis,

$$\lambda((R \cup V) \cup \{t\} \cup T_2) + \lambda(R \cup V) \geq \lambda((R \cup V) \cup \{t\}) + \lambda((R \cup V) \cup T_2),$$

which is the same as

$$\lambda(R \cup T_1 \cup T_2) + \lambda(R \cup V) \geq \lambda(R \cup T_1) + \lambda(R \cup V \cup T_2).$$

Using the induction hypothesis again we can continue to obtain

$$\lambda(R \cup V \cup T_2) \geq -\lambda(R) + \lambda(R \cup V) + \lambda(R \cup T_2)$$

which leads to (4.19). $\square$

If one defines, for $P, R \subset E$, the functions $n_P(R) = \mathbb{I}\{P \subset R\}$, then such n_P form a basis in the linear space of functions on L (for fixed R). Really, the total number of these functions is $|L|$. Besides that, for any $n = 2, \ldots, |L|$ and any collection of different sets $P_1, \ldots, P_n \in L$ one can find a set $P_i, 1 \leq i \leq n$, such that $P_j \not\subset P_i$ for $j \neq i$. Hence, for all real numbers $a_1, \ldots, a_n$ such that $\sum_{k=1}^{n} a_k \mathbb{I}\{P_k \subset R\} \equiv 0$ we may substitute $R = P_i$ and verify that $a_i = 0$. Repeating this argument $n - 2$ times we see that $a_1 = \ldots = a_n = 0$. Thus, the functions $\{n_P(\cdot), P \in L\}$ are linearly independent. Note that the function $n_\varnothing(\cdot)$ is identically 1.

For $R \neq \varnothing$ let us write the expansion of λ

$$\lambda(R) = \sum_P \varphi(P) n_P(R) = \sum_{P:P \subset R} \varphi(P)$$

where, **by the** *Möbius inversion* (inclusions-exclusions) *formula,*

$$\varphi(R) = \lambda(R) - \sum_{Q:Q \subset R, |Q| \leq |R|-1} \varphi(Q) = \lambda(R) - \sum_{Q:Q \subset R, |Q|=|R|-1} \lambda(Q)$$

$$+ \sum_{T:T \subset R, |T| \leq |R|-2} \varphi(T) = \ldots = \sum_{Q:Q \subset R} (-1)^{|R|-|Q|} \lambda(Q).$$

Condition (4.18) can be rewritten as the following inequality:

$$\sum_{P:P \subset R} \varphi(P \cup \{r,s\}) \geq 0 \qquad\qquad (4.20)$$

for any $r, s \in E$, $r \neq s$, and for all $R \subset E \setminus \{r, s\}$.

In statistical mechanics it is important to know whether a lattice gas distribution satisfies the FKG-inequalities, since these inequalities help to study the phase transition properties, see, e.g., books [335, Ch. 9], [177]. Here E is interpreted as a subset of $\mathbb{Z}^d$, and $n_r = n_{\{r\}}(\cdot), r \in \mathbb{Z}^d$, stands for the state of the site r which can be occupied or empty, i.e. $n_r = 1$ or $n_r = 0$. The probability of a configuration R in E is $Z^{-1} \exp\{-\lambda(R)\}$ where λ satisfies (4.18) and Z is the normalizing constant. The function $H = -\lambda$ is called the *Hamiltonian.* The alternate interpretation as the (general) Ising spin system appears if we define

$$\sigma_r = 2n_r - 1, \quad r \in E, \quad \sigma_R = \prod_{r \in R} \sigma_r, \quad R \subset E.$$

Then, for any $Q \in L$,

$$\sigma_R(Q) = \prod_{r \in R} (2\mathbb{I}\{r \in Q\} - 1) = \sum_{P \subset R} 2^{|P|} (-1)^{|R|-|P|} n_P(Q).$$

Inverting this we can obtain the expressions for n_P in terms of σ_R. Consequently $\{\sigma_R, R \subset E\}$ is also a basis and one can rewrite

$$H = -\sum_P \varphi(P) n_P = -\sum_R J(R) \sigma_R = -\sum_R \sum_{P \subset R} 2^{|P|} J(R)(-1)^{|R|-|P|} n_P$$

where $J(R)$ is the so-called *interaction potential.* Note that

$$\varphi(P) = \sum_{R:R \supset P} 2^{|P|} (-1)^{|R|-|P|} J(R) \qquad\qquad (4.21)$$

and, again by inversion,

$$J(R) = \sum_{P:P \supset R} 2^{-|P|} \varphi(P), \quad R \subset E.$$

The next result is due to Fortuin, Kasteleyn and Ginibre.

Theorem 4.20. ([172]) *Suppose that J meets the inequality*

$$\sum_{S \not\ni r,s} J(S \cup \{r,s\}) \sigma_S(R) \geq 0$$

for any $R \subset E$ and all $r, s \in E$, $r \neq s$. Then the Ising system determined by J satisfies the FKG-inequalities.

Proof. On applying (4.21), for pairwise disjoint sets $R, \{r\}$ and $\{s\}$ we obtain from (4.20) that the required inequality is the following:

$$\sum_{P \subset R} \varphi(P \cup \{r, s\}) = \sum_{P \subset R} 2^{|P|+2} \sum_{\substack{S \supset P \\ S \not\ni r,s}} (-1)^{|S|-|P|} J(S \cup \{r, s\}) \geq 0.$$

The above inequality should hold for all sets R not containing r and s. Changing the order of summation we can simplify this to

$$\sum_{S \not\ni r,s} J(S \cup \{r, s\})(-1)^{|S|+|RS|} = \sum_{S \not\ni r,s} J(S \cup \{r, s\})(-1)^{|S \cap (E \setminus R)|} \geq 0 \qquad (4.22)$$

(the factor $(-1)^{|RS|}$ appears while calculating the sum in P; we have also used that $|S| + |RS| = |S \setminus R| + 2|RS|$). Changing the notation from R to $E \setminus R$ and taking into account that $\sigma_S(Q) = (-1)^{|S \cap Q|}$, one sees that the Theorem condition implies (4.22). $\square$

A spin system is called *two-body interaction* if $J(R) = 0$ for all R with $|R| > 2$.

From Theorem 4.20 one easily deduces the following

Corollary 4.21. *A two-body interacting system such that $J(s, t) \geq 0$ when $s \neq t$ satisfies the FKG-inequalities.*

Remark 4.22. Though these results were proved for a finite lattice, they can be also extended to infinite sets if the potentials are defined in such a way that all the set-indexed series be absolutely convergent. In particular it is true for Ising model when $J_{ij} \neq 0$ only for i and j such that $|i - j| = 1$.

FKG-inequalities and their generalizations remain a subject of constant interest in reliability theory, statistical physics, quantum physics, discrete mathematics (see, e.g., [89, 184, 196, 201, 226, 252, 307, 344]). Interesting modifications appear when index sets are more general than $\mathbb{Z}^d$, then the corresponding algebraic lattice may be non-distributive. For instance, on triangular nets the "right" FKG inequalities differ from the usual [96]; even if we consider a finite graph consisting of one triangle, the simplest example might fail to satisfy classical FKG. The modified FKG were also applied to phase transitions problem for the corresponding Potts model and to analyzing existence of infinite clusters in percolation models.

4°. **Percolation.** In 1957 Broadbent and Hammersley [57] proposed a remarkable stochastic model which gave the birth to Percolation Theory. Suppose that a large porous stone is put in a vessel with water. What is the probability that the water will reach some fixed point of a stone? Of course one has to formalize this problem. Let us imagine that the stone contains a number of "broad" and "narrow" channels. More exactly, assume for simplicity that we have the lattice $\mathbb{Z}^3$ and the channels are the edges connecting the neighbor vertices of $\mathbb{Z}^3$ (x and y are neighbors if $\|x - y\|_1 = \sum_{i=1}^{3} |x_i - y_i| = 1$). Assume that each of these edges can be "open" (broad) or "closed" (narrow) with respective probability p and $q = 1 - p$, $0 < p < 1$. Moreover, we let all the edges be controlled independently.

A *path of length* n from the vertex x to vertex y is a collection of edges $e_0, \ldots, e_{n-1}$ such that e_k connects the neighbor vertices x_k and x_{k+1} where $x_0 = x$, $x_n = y$ and $x_i \neq x_j$ ($i \neq j$). One writes $x \leftrightarrow y$ if there exists a path from x to y.

An *infinite path* starting at $x \in \mathbb{Z}^3$ is an infinite collection of edges $e_0, \ldots, e_{n-1}, \ldots$ such that e_k connects the neighbor vertices x_k and x_{k+1} ($k \in \mathbb{N}$) where $x_0 = x$ and $x_i \neq x_j$ for $i \neq j$. In this case one writes $x \leftrightarrow \infty$.

A path is called *open* if all its edges are open. We can consider a stone as a subset of $\mathbb{Z}^3$. Then the above mentioned problem (whether a fixed point x of a stone will be wet) can be reformulated as follows. Is there an open way from x to the surface of a stone? The last step in formulation of this percolation problem is to imagine, using scaling, that the stone is large enough and instead of a subset of $\mathbb{Z}^3$ we can consider the whole lattice. In this case a precise mathematical problem admits simple formulation: does there exist an *infinite open path* starting at x?

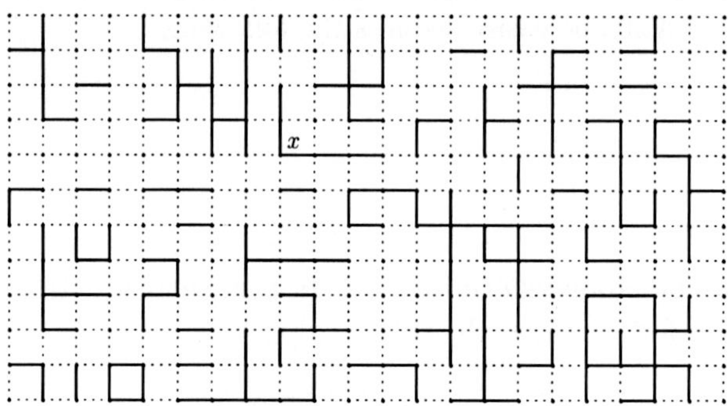

Fig. 1.3   ($d = 2$, solid lines denote open paths, thick lines denote C_x)

Introduce a set C_x as a collection of all vertices of $\mathbb{Z}^3$ which can be reached by means of some open path starting at x. It is not difficult to see that the posed problem on infinite path is equivalent to the statement that $|C_x| = \infty$.

Recall that configurations of open and closed edges are random. Thus we have to specify a stochastic model.

Now we consider $\mathbb{Z}^d$, $d \geq 1$, as a set of vertices and $\mathbb{E}^d$ as a set of edges connecting the neighboring vertices ($x = (x_1, \ldots, x_d), y = (y_1, \ldots, y_d) \in \mathbb{Z}^d$ are neighbors if $\|x - y\|_1 := \sum_{i=1}^{d} |x_i - y_i| = 1$.) As a sample space Ω we take $\{0, 1\}^{\mathbb{E}^d}$. A point $\omega = (\omega(e), e \in \mathbb{E}^d)$ is called a *configuration*. For $e \in \mathbb{E}^d$ the value $\omega(e) = 0$ corresponds to the fact that e is closed and $\omega(e) = 1$ corresponds to e being open.

Let $\mathcal{F}$ be an σ-algebra generated by the finite-dimensional cylinders (here one can say that $\mathcal{F}$ is generated by the "cylinders" of the type

$$C = \{\omega \in \Omega : \omega(e_{j_1}) = i_1, \ldots, \omega(e_{j_k}) = i_k\} \qquad (4.23)$$

where $e_{j_r} \in \mathbb{E}^d$, $i_r \in \{0, 1\}$, $r = 1, \ldots, k$ and $k \in \mathbb{N}$).

Introduce for $0 \leq p \leq 1$ on $(\Omega, \mathcal{F})$ a measure $\mathbb{P}_p$ as a product over $\mathbb{E}^d$ of the Bernoulli measures $\mu_{p,e}$, that is $\mathbb{P}_p = \otimes_{e \in \mathbb{E}^d} \mu_{p,e}$ where $\mu_{p,e}(\{1\}) = p$, $\mu_{p,e}(\{0\}) = q$ $(q = 1 - p)$. Thus for C appearing in (4.23) one has

$$\mathbb{P}_p(C) = p^{\sum_{r=1}^n i_r} q^{n - \sum_{r=1}^n i_r}. \tag{4.24}$$

It is very convenient to construct on $(\Omega, \mathcal{F})$ all measures $\mathbb{P}_p$ for $0 \leq p \leq 1$ in the following way. Consider a family $\{X(e), e \in \mathbb{E}^d\}$ of i.i.d. random variables defined on some probability space and having uniform distribution on $[0, 1]$. Set

$$\eta_p(e) = \mathbb{I}\{X(e) < p\}, \quad 0 \leq p \leq 1, \quad e \in \mathbb{E}^d. \tag{4.25}$$

Then the probability of event $\{(\eta_p(e_{j_1}), \ldots, \eta_p(e_{j_k})) \in C\}$ for cylinders C introduced in (4.23) is given by formula (4.24). In other words, $\mathbb{P}_p$ is the distribution of the field $\{\eta_p(e), e \in \mathbb{E}^d\}$.

Theorem 4.23. *Let Y be a real-valued, increasing[23], measurable function on $(\Omega, \mathcal{F})$. Then $\mathbb{E}_{p_1} Y \leq \mathbb{E}_{p_2} Y$ whenever $p_1 \leq p_2$, provided that both expectations exist. In particular, for any increasing event A one has $\mathbb{P}_{p_1}(A) \leq \mathbb{P}_{p_2}(A)$ if $p_1 \leq p_2$.*

Proof. Recall that $\eta_p(e) = 1$ if $X(e) < p$ and $\eta_p(e) = 0$ otherwise. Thus $p_1 \leq p_2$ implies $\eta_{p_1}(e) \leq \eta_{p_2}(e)$ for any $e \in \mathbb{E}^d$, that is $\eta_{p_1}(\omega) \leq \eta_{p_2}(\omega)$ for any $\omega \in \Omega$. If a family of random variables $X(e), e \in \mathbb{E}^d$, is defined on a probability space endowed with measure P, then due to (4.25) one has

$$\mathbb{E}_{p_1} Y = \mathsf{E} Y(\eta_{p_1}(\cdot)) \leq \mathsf{E} Y(\eta_{p_2}(\cdot)) = \mathbb{E}_{p_2} Y$$

as Y is an increasing function. Here E stands for the integration with respect to measure P. $\square$

It is easily seen that for any $x \in \mathbb{Z}^d$ the set $\{|C_x| = +\infty\} \in \mathcal{F}$. Although we have defined C_x above when $d = 3$, all definitions of this subsection can be extended for any dimension d. Moreover, note that $\{|C_x| = +\infty\}$ and $\{x \leftrightarrow y\}$ are examples of increasing events for all $x, y \in \mathbb{Z}^d$.

Let $\theta(p, d, x) = \mathbb{P}_p(|C_x| = \infty)$ for $x \in \mathbb{Z}^d$, $d \in \mathbb{N}$ and $0 \leq p \leq 1$. Obviously $\theta(0, d, x) = 0$ and $\theta(1, d, x) = 1$ for all $x \in \mathbb{Z}^d$ and $d \geq 1$. Due to Theorem 4.23 we conclude that $\theta(x, d, p)$ is non-decreasing function in $p \in [0, 1]$ for every $x \in \mathbb{Z}^d$, $d \geq 1$.

Now consider the *critical probability*

$$p_c(d, x) := \sup\{p : \theta(p, d, x) = 0\}.$$

Note that $p_c(d, x) = \inf\{p : \theta(p, d, x) > 0\}$. Clearly the Borel–Cantelli lemma implies that, for $d = 1$ and every $x \in \mathbb{Z}$, one has $p_c(1, x) = 1$. One of the beautiful results of the Percolation Theory is the following

Theorem 4.24. *For every $d \geq 2$ and any $x \in \mathbb{Z}^d$ one has $0 < p_c(d, x) < 1$.*

---

[23] We use on Ω the partial order meaning that $\omega \leq \tau$ if $\omega(e) \leq \tau(e)$ for each $e \in \mathbb{E}^d$.

The proof can be found, e.g., in [184, p. 14].

One can say that for dimensions greater than 1 there exists non-trivial critical value of probability separating the phenomena of existence of finite or infinite open clusters containing a point x.

Theorem 4.25. *For any $d \geq 1$ the critical probability $p_c(d, x)$ does not depend on $x \in \mathbb{Z}^d$.*

Proof. The case $d = 1$ is trivial. Let us consider $d \geq 2$. For any $x, y \in \mathbb{Z}^d$

$$\theta(p, d, x) = \mathbb{P}_p(|C_x| = \infty) = \mathbb{P}_p(x \leftrightarrow \infty)$$

$$\geq \mathbb{P}(\{x \leftrightarrow y\} \cap \{y \leftrightarrow \infty\}) \geq \mathbb{P}_p(x \leftrightarrow y)\theta(p, d, y).$$

The last inequality is valid because the events $\{x \leftrightarrow y\}$ and $\{y \leftrightarrow \infty\}$ are increasing in ω and we apply Theorem 1.21, as instead of $\mathbb{P}_p$ we can use the measure P given on a probability space where a family of independent random variables η_p is defined (see (4.25)).

Now we note that $\mathbb{P}_p(x \leftrightarrow y) > 0$ for any $x, y \in \mathbb{Z}^d$. Thus, if $\theta(p, d, y) > 0$ then $\theta(p, d, x) > 0$. We can interchange x and y to conclude that $\theta(p, d, y) > 0$ if $\theta(p, d, x) > 0$. Therefore, $\inf\{p : \theta(p, d, x) > 0\} = \inf\{p : \theta(p, d, y) > 0\}$. The proof is complete. $\square$

Remark 4.26. The ideas similar to the proof of the last theorem allow to estimate the behavior of the largest cycle enclosing the origin (namely, how much will it differ from its convex hull) [6] and to prove that in two-infections interacting model the coexistence appears with probability one [176].

We tackle also the *bond percolation* on graphs which, in general, need not coincide with $(\mathbb{Z}^d, \mathbb{E}^d)$. However, we restrict ourselves to finite graphs.

Let V be the set of vertices of a finite graph[24] $G = (V, E)$, where E is a set of edges. For any $r \in E$, let p_r and q_r be positive numbers, $p_r + q_r = 1$. The measure defined on 2^E by

$$\mu(R) = \prod_{r \in R} p_r \prod_{s \notin R} q_s, \quad R \subset E,$$

satisfies the FKG. Note that in this case one has equality in (4.6). The arising random graph is called a *percolation model*. The first example of association property for this model was given by Harris ([191]).

Now for any $R \subset E$ consider a subgraph $G_R = (V, R)$, obtained by omitting in (V, E) all edges which are not in R. Recall that a connected subgraph H of G is called *maximal*, if there is no other connected subgraph in G having H as a proper subgraph. A *cluster* of G_R is the maximal connected subgraph of G_R.

---

[24]See, e.g., [48, Ch.I, 1].

For any subgraph H of G, let $\gamma_H(R) = \mathbb{I}\{H$ is a cluster of $G_R\}$. Define on 2^E the measure $e^{\lambda(R)}$ setting

$$\lambda(R) = \sum_H \phi(H)\gamma_H(R), \quad R \subset E,$$

where $\phi : H \to \mathbb{R}$ is a function, and the summation is over all subgraphs of G.

To satisfy (4.18) we impose on λ the condition

$$\sum_H \phi(H)\left(\gamma_H(R \cup \{r,s\}) + \gamma_H(R) - \gamma_H(R \cup \{r\}) - \gamma_H(R \cup \{s\})\right) \geq 0, \quad (4.26)$$

for all $R \subset E$ and all $r, s \in E \setminus R$, $r \neq s$.

Theorem 4.27. ([172]) *The following examples of functions ϕ satisfy (4.26):*
(a) $\phi(H) = \sum_e \xi(e)$, *the summation is taken over edges of H and $\xi : E \to \mathbb{R}$.*
(b) $\phi(H) = \sum_v \psi(v)$, *the summation is taken over vertices of G and $\psi : V \to \mathbb{R}$.*
(c) $\phi(H) = c$.
(d) $\phi(H) = \psi(v) \geq 0$ *if H is an isolated vertex v, and 0 otherwise.*

Proof. Straightforward calculations. $\square$

For study of various problems of Percolation Theory one can refer, e.g., to [97], [184].

5°. Negative dependence and networks. It would be nice to obtain general results analogous to FKG-inequalities for negative association. Let λ_1 and λ_2 be some σ-finite measures on $\mathbb{R}$ and λ be their product on $\mathbb{R}^2$. If one considers the density of a 2-dimensional random vector (X, Y) with respect to λ as an infinite matrix of non-countable order (finite-dimensional matrix, if X and Y have finite number of possible values), then by Theorem 4.7 association is ensured by the condition that "second order minors" of the density function are nonnegative[25]. Loosely speaking, this is a general variant of Theorem 1.13 applicable to all possible distributions. Thus, one may think that negative dependence will arise through reversing the sign in all inequalities. Unfortunately, the analog of Theorem 4.7 in the **NA** case is not true, as an example below shows. Call a function $f : \mathbb{R}^2 \to \mathbb{R}$ *reverse regular of second order* (RR_2), if $\det \|f(x_i, y_j)\|_{i,j=1}^2 \leq 0$ for all $x_1, x_2, y_1, y_2 \in \mathbb{R}$ with $x_1 < x_2, y_1 < y_2$.

Example 4.28. Consider a random vector (X, Y) such that $\mathsf{P}(X = i, Y = j) = p_{ij}$, $i \in \{0, 1, 2\}$, $j \in \{0, 1\}$ and p_{ij} are given by a matrix

$$\begin{pmatrix} p_{00} & p_{01} & p_{02} \\ p_{10} & p_{11} & p_{12} \end{pmatrix} = \begin{pmatrix} 1/40 & 1/10 & 3/10 \\ 1/4 & 1/5 & 1/8 \end{pmatrix}.$$

Then the RR_2 condition is satisfied by the density of (X, Y) with respect to counting measure on $\mathbb{Z}^2$, but $\mathsf{P}(X > 1, Y > 0) > \mathsf{P}(X > 1)\mathsf{P}(Y > 0)$. Hence, $(X, Y) \notin \mathbf{NA}$.

---

[25] Provided that the matrix entries are numerated from the bottom left corner, which is an unusual way. In this book we use standard notation, numerating the entries from the top left corner.

It is possible to avoid this asymmetry introducing various forms of negative dependence having less general applicability (see, e.g., [46, 185, 207, 219, 227, 325]). We concentrate on one example which illustrates the interesting phenomenon called "positive influence" ([147, 166]). Its informal description is that negative correlatedness, together with existence of a random variable **positively** depending (in some restricted sense) on others, imply stronger negative dependence property, say, in our situation, negative association.

Let $G = (V, E)$ be a *connected graph* (see, e.g., [48] for basic information on graphs and trees). Recall the following

Definition 4.29. The set $\mathcal{E} = \{e_1, \ldots, e_k\} \subset E$ is called a *cycle* if there are vertices $v_1, \ldots, v_k, v_{k+1} = v_1$ in V such that $e_1 = (v_1, v_2), \ldots, e_{k-1} = (v_{k-1}, v_k), e_k = (v_k, v_{k+1})$. A *spanning tree* of G is a connected graph $T = (V, E_1)$ such that $E_1 \subset E$ and T does not have cycles.

Let $\mathcal{T}$ be a (finite) set of all spanning trees T of G. Choose at random an element in $\mathcal{T}$, assigning these trees equal probability (one says that this is a uniform spanning tree). Introduce a random variable $X_e = \mathbb{I}\{e \in T\}$, $e \in E$. The behavior of the uniform spanning tree is an important field of study in electric and information network theory. Among different results, various forms of dependence are established. Informally, negative dependence arises like in permutation distribution. Since all the spanning trees have $|V| - 1$ edges, if some edges are selected, each of other ones is less likely to be chosen. Formalization of this argument is a difficult problem. We start with several lemmas. The first two of them are of algebraic character and we place their proofs in Appendix A.4.

Lemma 4.30. *Let $\{e^1, \ldots, e^n\}$ be the standard orthonormal basis in the Euclidean space $\mathbb{R}^n$ and let M and L be linear subspaces of $\mathbb{R}^n$ such that $M \subset L$. Then, for any $i = 1, \ldots, n$, one has $0 \leq a_i^i \leq b_i^i$, where $a^i = (a_1^i, \ldots, a_n^i)$ and $b^i = (b_1^i, \ldots, b_n^i)$ are the respective orthogonal projections of e^i onto M and L.*

We will consider the graph $G = (V, E)$ as directed one, i.e. to each edge $e = (x, y)$ we assign the direction in which it is drawn, either from x to y or vice versa. We write respectively $e = (x \rightarrowtail y)$ and $e = (y \rightarrowtail x)$ to distinguish between these cases. For a finite directed graph $G = (V, E)$ let $H(G)$ be the linear space of real functions on the finite set E. Obviously $H(G)$ can be viewed as Euclidean space isomorphic to the space $\mathbb{R}^{|E|}$. The inner product in $H(G)$ is defined in a usual way, i.e.

$$(\varphi, \psi) := \sum_{e \in E} \varphi(e)\phi(e).$$

Note that if $\mathcal{E} = \{e_1, \ldots, e_k\} \subset E$ is a cycle, then, in general, $e_1, \ldots, e_k$ can have various directions. A function $F : E \to \mathbb{R}$ is called *cyclic* over a cycle $\mathcal{E}$ if

$$F(e) = \begin{cases} 1, & e \in \mathcal{E} \text{ and } e = (v_i \rightarrowtail v_{i+1}) \text{ for some } i \in \{1, \ldots, k\}, \\ -1, & e \in \mathcal{E} \text{ and } e = (v_{i+1} \rightarrowtail v_i) \text{ for some } i \in \{1, \ldots, k\}, \\ 0, & e \notin \mathcal{E}. \end{cases}$$

Denote by $\Diamond(G)$ the linear span of all cyclic functions in $H(G)$. For an edge $e = (x, y)$ define the function $\chi^e \in H(G)$ as indicator function of e, that is $\chi^e(h) = \mathbb{I}\{h = e\}$. Denote by $\tau(G)$ (resp. $\tau_e(G)$, $\tau_f(G)$ $\tau_{ef}(G)$) the total number of spanning trees in a graph G (resp. the number of spanning trees which contain the edge e, the edge f, both edges e and f).

Lemma 4.31. *Let G be a finite directed graph. Let $e = (x, y) \in E$ and P be the linear operator projecting $H(G)$ onto the orthogonal complement of $\Diamond(G)$. Then*

$$(P\chi^e)(e) = \frac{\tau_e(G)}{\tau(G)}. \tag{4.27}$$

Definition 4.32. The function $P\chi^e$ is called the *unit electric current* from x to y.

Lemma 4.33. *The random variables $\{X_e, e \in E\}$ introduced before Lemma 4.30 are nonpositively correlated.*

Proof. This fact is a particular case of the so-called Rayleigh monotonicity principle of electrical networks theory (see, e.g., [27] for a more thorough account on the subject). Since the spanning tree T of G is uniform, we have

$$\mathsf{P}(e \in T) = \frac{\tau_e(G)}{\tau(G)} \quad \text{and} \quad \mathsf{P}(e \in T | f \in T) = \frac{\tau_{ef}(G)}{\tau_f(G)}.$$

Let G^f be the graph obtained from G by contraction of the edge f into a single point. That is, if f connects vertices $a, b \in V$, then $G^f = (V \setminus \{b\}, E \setminus \{f\})$; and each edge h which connected some vertex $c \in V$ with b in graph G connects c with a in graph G^f. It is easy to see that there exists a one-to-one correspondence between spanning trees in G containing f and spanning trees in G^f. In fact, having a spanning tree in G containing f, contract the edge f to obtain the spanning tree in G^f. Conversely, having a spanning tree in G^f consisting of edges $t_1, \ldots, t_n$ we obtain a spanning tree in G taking edges $f, t_1, \ldots, t_n$.

So, the Lemma's assertion is equivalent to the following: for any $e, f \in E$, $e \neq f$,

$$\frac{\tau_e(G^f)}{\tau(G^f)} \leq \frac{\tau_e(G)}{\tau(G)}. \tag{4.28}$$

By Lemma 4.31 the right-hand side equals $P\chi^e(e)$, see (4.27). Clearly one has $H(G) = H(G^f) \oplus < \chi^f >$. Here $< \cdot >$ denotes the linear span of a vector and $\oplus$ is a direct sum of linear subspaces of $H(G)$. By the same Lemma 4.31 the left-hand side of (4.28) is $P^f \chi^e(e)$, where P^f is the linear operator in $H(G^f)$ projecting onto the complement of $\Diamond(G^f)$. We may extend P^f to the whole space $H(G)$ if we let $P^f(\chi^f) = 0$. This definition is natural, since one may think that when being contracted, the edge f becomes a loop[26], consequently χ^f becomes a cyclic function. Each cycle in G is, of course, a cycle in G^f. Consequently, $\Diamond(G) \subset \Diamond(G^f)$, therefore $\Diamond(G)^{\perp} \supset \Diamond(G^f)^{\perp}$. Then P^f is a projection onto a smaller space than the range of P is. Thus, $P\chi^e(e) \geq P^f \chi^e(e)$ by Lemma 4.30, hence (4.28) is true. $\square$

---

[26]i.e. an edge connecting some point with itself.

Lemma 4.34. *For any set $U \subset E$, each binary increasing function $f : \mathbb{R}^{|U|} \to \mathbb{R}$ and all $e \in E \setminus U$, one has*

$$cov(X_e, f(X_i, i \in U)) \leq 0.$$

Proof. The Lemma is proved by induction on $|V| + |E|$. The case $|V| = 2$ is handled by direct calculation using the fact that f is increasing. Namely, in this case the spanning tree consists of two vertices and one edge chosen with equal probability n^{-1} from the set of edges $e_1, \ldots, e_n$ connecting two vertices. Thus, the random variables $X_{e_1}, \ldots, X_{e_n}$ are **NA** by Corollary 1.27.

Now let us suppose that the Lemma is already proved for all the graphs having the total number of vertices and edges less than given one. To simplify the notation, denote by A the event $\{f(X_i, i \in U) = 1\}$ and by E_k the event $\{X_k = 1\}$, $k \in E$. We show that $\mathsf{P}(AE_k) \leq \mathsf{P}(A)\mathsf{P}(E_k)$, which is our claim. The case $\mathsf{P}(E_k) = 0$ is trivial. Otherwise we may rewrite the inequality as $\mathsf{P}(A|E_k) \leq \mathsf{P}(A)$. For any edge $l \in E, l \neq e$, such that $\mathsf{P}(E_l E_k) > 0$,

$$\mathsf{P}(A|E_k) = \mathsf{P}(E_l|E_k)\mathsf{P}(A|E_l E_k) + \mathsf{P}(E_l^c|E_k)\mathsf{P}(A|E_l^c E_k),$$

$$\mathsf{P}(A) = \mathsf{P}(E_l)\mathsf{P}(A|E_l) + \mathsf{P}(E_l^c)\mathsf{P}(A|E_l^c),$$

if we agree that the conditional probability given condition having zero probability is zero.

By Lemma 4.33, $\mathsf{P}(E_k E_l) \leq \mathsf{P}(E_k)\mathsf{P}(E_l)$, that is $\mathsf{P}(E_l|E_k) \leq \mathsf{P}(E_l)$. By the induction hypothesis applied to the graph G^l (respectively G with deleted edge l) and the increasing function f restricted to E_l (respectively E_l^c), we have $\mathsf{P}(A|E_l E_k) \leq \mathsf{P}(A|E_l)$, $\mathsf{P}(A|E_l^c E_k) \leq \mathsf{P}(A|E_l^c)$.

If we could take l in such a way that

$$\mathsf{P}(A|E_l E_k) \geq \mathsf{P}(A|E_l^c E_k), \tag{4.29}$$

then we would be able to write

$$\mathsf{P}(A|E_k) = \mathsf{P}(E_l|E_k)\mathsf{P}(A|E_l E_k) + \mathsf{P}(E_l^c|E_k)\mathsf{P}(A|E_l^c E_k)$$

$$\leq \mathsf{P}(E_l)\mathsf{P}(A|E_l E_k) + \mathsf{P}(E_l^c)\mathsf{P}(A|E_l^c E_k) \leq \mathsf{P}(E_l)\mathsf{P}(A|E_l) + \mathsf{P}(E_l^c)\mathsf{P}(A|E_l^c) = \mathsf{P}(A)$$

(we have used here that if a, b, x, y are such numbers that $a \leq b$ and $x \geq y$, then $ax + (1-a)y \leq bx + (1-b)y$). But such l can always be chosen. Indeed, the inequality (4.29) is equivalent to $\mathsf{P}(A|E_l E_k) \geq \mathsf{P}(A|E_k)$, which is the same as $\mathsf{P}(E_l|AE_k) \geq \mathsf{P}(E_l|E_k) > 0$. Note that

$$\sum_{l \neq e} \mathsf{P}(E_l|AE_k) = |V| - 2 = \sum_{l \neq e} \mathsf{P}(E_l|E_k),$$

since any spanning tree in G has $|V| - 1$ edges. Both sums in the last relation actually involve l such that $l \neq e$, $\mathsf{P}(E_l|E_k) > 0$. Thus there exists at least one such l for which $\mathsf{P}(E_l|AE_k) \geq \mathsf{P}(E_l|E_k) > 0$. $\square$

The following result is due to Feder and Mihail.

Theorem 4.35. ([166]) *The random variables $\{X_e, e \in E\}$ are **NA**.*

Proof. Again we use induction, now on $|E|$. The case $|E| = 1$ is trivial. For the next step, take disjoint non-empty sets $I, J \subset E$, such that their union is E, and binary functions $f \in \mathcal{M}(|I|)$, $g \in \mathcal{M}(|J|)$. Since $\sum_{j \in E} X_j = |V| - 1$ a.s., we have

$$0 = cov(f(X_I), |V| - 1) = cov\left(f(X_I), \sum_{j \in E} X_j\right) = \sum_{j \in E} cov(f(X_I), X_j).$$

The last sum is zero, therefore there exists some $k \in E$ such that

$$cov(f(X_I), X_k) \geq 0. \tag{4.30}$$

By the conditional covariance formula (1.22), $cov(f(X_I), g(X_J))$ can be written as

$$\mathsf{E}cov(f(X_I), g(X_J)|X_k) + cov(\mathsf{E}(f(X_I)|X_k), \mathsf{E}(g(X_J)|X_k)). \tag{4.31}$$

The conditional covariance is a.s. nonpositive by induction hypothesis (applied either to graph G^l or to the graph G with deleted edge l).

For a pair of binary random variables, due to Theorem 1.13, independence is the same as uncorrelatedness. If the covariance (4.30) is zero, then X_k and $f(X_I)$ are independent, hence $\mathsf{E}(f(X_I)|X_k)$ is a.s. constant and the second covariance in (4.31) is zero. If the covariance in (4.30) is strictly positive, then, by Lemma 4.34, $k \in I$, and therefore, by the same Lemma

$$cov(g(X_J), X_k) \leq 0. \tag{4.32}$$

Now relations (4.30), (4.32) and Theorem 1.13 imply that $\mathsf{E}(f(X_I)|X_k)$ and $\mathsf{E}(g(X_J)|X_k)$ are almost surely *discordant* functions of the random variable X_k (that is one of them is increasing and another is nonincreasing). By Theorem 1.8, (a), the second summand in (1.22) is nonpositive. $\square$

Uniform spanning tree is an example of a balanced matroid. A *matroid* is a pair of a finite set S and a collection $\mathcal{S}$ of its subsets such that all subsets belonging to $\mathcal{S}$ have the same cardinality and for each pair of $B_1, B_2 \in \mathcal{S}$ the following property holds: for any $e_1 \in B_1$ there exists $e_2 \in B_2$ such that $(B_1 \setminus \{e_1\}) \cup \{e_2\} \in \mathcal{S}$.

A matroid is called *balanced* ([166]) if the presence of an element $e \in S$ in this matroid makes presence of any other element less likely, that is, an analog of Lemma 4.33 is true. Theorem 4.35 can be proved, in fact, for all balanced matroids [166]. Thus an analogous result is true, for example, for uniformly selected basis of a finite linear space.

The uniform spanning tree is a simple example of a so-called determinantal probability measure on the edge set E. The complete account of the connection of determinantal probability measures with representation theory as well as the extension of the Feder–Mihail theorem on more general matroids (in particular the non-equal conductivities case) can be found in [281]. The study of situation when the electric network is infinite is also treated there.

5　Further extensions of dependence notions

1°. Covariances for test functions. Association and related notions can be viewed as descriptions of dependence between the elements of a process or field, like Markov properties, martingale properties, or mixing, which is especially close to association. A counterpart of the mixing coefficient for fields under consideration is the covariance function. An important property of **PA/NA** systems, resembling the behavior of Gaussian systems, is that uncorrelatedness implies (mutual) independence as will be shown below, see Corollary 5.5. More generally, weak correlatedness, or rapid decrease of the covariance function off the diagonal, entails that the random systems exhibit the behavior typical for weakly dependent systems (see, e.g., [110, 251, 307, 385]). To state this rigorously, we need the following definitions.

Definition 5.1. Let (S, ρ) and (V, τ) be metric spaces. A map $F : S \to V$ is *Lipschitz* if

$$Lip(F) := \sup_{\substack{x,y \in S \\ x \neq y}} \frac{\tau(F(x), F(y))}{\rho(x, y)} < \infty.$$

For a map $F = (F_1, \ldots, F_k) : \mathbb{R}^n \to \mathbb{R}^k$ we will always use the metric ρ induced by the l^1-norm $\|x\|_1 = \sum_i |x_i|$ unless otherwise indicated, here $i = 1, \ldots, n$. For such F we introduce also the *partial Lipschitz constants*

$$Lip_i(F) = \sup \frac{\|F(x_1, \ldots, x_{i-1}, x_i, x_{i+1}, \ldots, x_n) - F(x_1, \ldots, x_{i-1}, y_i, x_{i+1}, \ldots, x_n)\|}{|x_i - y_i|}$$

where $\| \cdot \|$ is the Euclidean norm in $\mathbb{R}^k$ and the supremum is taken over $x_1, x_2, \ldots, x_n, y_i \in \mathbb{R}$, $x_i \neq y_i$. Clearly, $Lip_i(F) \leq Lip(F)$, $i = 1, \ldots, n$. If there exists $\partial f / \partial x_i$ for all $x \in \mathbb{R}^n$ $(i = 1, \ldots, n)$, then $L_i(f) \leq \sup_{x \in \mathbb{R}^n} |\partial f / \partial x_i|$.

We denote by $BL(n)$ the class of bounded Lipschitz functions $f : \mathbb{R}^n \to \mathbb{R}, n \in \mathbb{N}$. The following simple facts are noteworthy.

Lemma 5.2. *(a) If (S, ρ), (V, τ), and (K, ν) are metric spaces and there are Lipschitz maps $F : S \to V, G : V \to K$, then their composition $G \circ F$ is Lipschitz and $Lip(G \circ F) \leq Lip(F)Lip(G)$. In particular, if these spaces are Euclidean $\mathbb{R}^n, \mathbb{R}^m$ and $\mathbb{R}^k$ respectively, endowed with the l^1-norm, then, for any $i = 1, \ldots, n$, $v = 1, \ldots, k$, one has*

$$Lip_i(G \circ F)_v \leq \sum_{j=1}^{m} Lip_j G_v Lip_i F_j.$$

(b) For any $f \in BL(n)$ there exists a uniformly bounded sequence of smooth functions $f_k : \mathbb{R}^n \to \mathbb{R}$, converging uniformly to f and such that $Lip_i f_k \to Lip_i f$ as $k \to \infty$, $i = 1, \ldots, n$.

Proof. The assertion (a) is verified directly. The proof of assertion (b) is similar to that of Theorem 1.5, (c). $\square$

The next simple theorem clarifies the above mentioned relation between the covariance function and dependence property. In this regard we refer to the papers by Simon, Newman, Birkel, Roussas, Peligrad and Shao, Lewis, Zhang [44, 259, 307, 320, 352, 429]; related topics are discussed in [51, 116, 134, 278]. Here we use an observation made by Bulinski and Shabanovich.

Theorem 5.3. ([78]) *Let* $X = \{X_j, j \in \mathbb{Z}^d\}$ *be an associated, or positively or negatively associated, random field such that* $\mathsf{E}X_j^2 < \infty$ *for any* $j \in \mathbb{Z}^d$. *Let* I *and* J *be two finite subsets of* $\mathbb{Z}^d$. *In the* **PA** *and* **NA** *cases suppose, in addition, that* I *and* J *are disjoint. Then, for any Lipschitz functions* $f : \mathbb{R}^{|I|} \to \mathbb{R}$, $g : \mathbb{R}^{|J|} \to \mathbb{R}$, *one has*

$$|cov(f(X_I), g(X_J))| \le \sum_{i \in I} \sum_{j \in J} Lip_i(f) Lip_j(g) |cov(X_i, X_j)|. \tag{5.1}$$

Proof. All expressions in (5.1) are meaningful, since any Lipschitz function of a finite number of elements of X is square-integrable (as $\mathsf{E}X_j^2 < \infty$, $j \in \mathbb{Z}^d$, and as $|f|$ and $|g|$ are majorized by linear functions in $|x_i|$, $i \in I$, and $|x_j|$, $j \in J$, respectively). Consider the functions $f_+(X_I), f_-(X_I), g_+(X_J), g_-(X_J)$ defined by way of

$$f_\pm(X_I) = f(X_I) \pm \sum_{i \in I} Lip_i(f) X_i, \ g_\pm(X_J) = g(X_J) \pm \sum_{j \in J} Lip_j(g) X_j,$$

with simultaneous choice of plus or minus at both sides of the formulas. Then one can easily check that f_+ and g_+ are coordinate-wise nondecreasing functions, whereas f_- and g_- are nonincreasing. In the association or **PA** case we have

$$cov(f_+(X_I), g_+(X_J)) \ge 0, \ cov(f_-(X_I), g_-(X_J)) \ge 0,$$
$$cov(-f_+(X_I), g_-(X_J)) \ge 0, \ cov(f_-(X_I), -g_+(X_J)) \ge 0.$$

Addition of the first two inequalities yields the desired upper estimate for $cov(f(X_I), g(X_J))$, while by adding the third and fourth the lower estimate ensues.

The **NA** case is considered in just the same way, with reversing all four inequalities. $\square$

In paper [429] instead of Lipschitz functions the compositions of functions of bounded variation were considered. In that case at the right-hand side of (5.1) appear covariances of monotone functions of X_i (the total variation of f and g).

Remark 5.4. Let X be a random vector with $\mathsf{E}\|X\|^2 < \infty$ and $cov(X_i, X_j) \ge 0$ for all $i, j \in \{1, \ldots, n\}$. One has equality in (5.1) if f and g are linear functions. Thus in general inequality (5.1) is optimal in a sense. Note that this Theorem may be applied to complex-valued Lipschitz functions[27] (i.e. the real and imaginary parts of these functions are Lipschitz) by representing them as algebraic sum of real and imaginary parts, the cost is a factor 4 arising in the right-hand side of (5.1). In particular, we may apply it to characteristic functions of random vectors, i.e. to complex exponential functions.

---

[27] If ξ and η are square-integrable complex-valued random variables then $cov(\xi, \eta) := \mathsf{E}\xi\overline{\eta} - \mathsf{E}\xi\mathsf{E}\overline{\eta}$, the bar denoting the complex conjugation.

Corollary 5.5. *Suppose that* $X = (X_1, \ldots, X_n)$ *is an associated* $(\mathbf{PA}, \mathbf{NA})$ *random vector such that* $\mathsf{E}\|X\|^2 < \infty$. *Then, for any* $t_1, \ldots, t_n \in \mathbb{R}$, *one has*

$$|\mathsf{E} \exp\{it_1 X_1 + \ldots + it_n X_n\} - \mathsf{E} \exp\{it_1 X_1\} \ldots \mathsf{E} \exp\{it_n X_n\}|$$

$$\leq 4 \sum_{\substack{1 \leq j,k \leq n \\ j \neq k}} |t_j t_k| |cov(X_j, X_k)|. \tag{5.2}$$

An associated $(\mathbf{PA}, \mathbf{NA})$ *system having zero correlations consists of independent random variables.*

The celebrated Newman's inequality [307] has the form (5.2) where the factor 4 is replaced with 1.

Proof. The first assertion follows from Theorem 5.3 and Remark 5.4 by writing the complex exponent as the sum of trigonometric functions and using induction on n. The second one is due to description of independence in terms of characteristic functions. $\square$

Moreover, from Theorem 5.3 one deduces an interesting

Corollary 5.6. *For a* $\mathbf{PA}$ *(or* $\mathbf{NA}$) *random vector* (X, Y) *with values in* $\mathbb{R}^2$ *and strictly increasing functions* $f, g : \mathbb{R} \to \mathbb{R}$, *the inequality in Definition 1.2 (resp. 1.3) is strict, unless* X *and* Y *are independent.*

Proof. If $cov(f(X), g(Y)) = 0$, then Corollary 5.5 yields that $f(X)$ and $g(Y)$ are independent. We conclude (see Remark 1.6) that X and Y are independent as f and g are strictly increasing, therefore f^{-1} and g^{-1} have the same property. $\square$

Definition 5.7. ([68, 82]) A random field $X = \{X_j, j \in \mathbb{Z}^d\}$, such that $\mathsf{E} X_j^2 < \infty$ for each $j \in \mathbb{Z}^d$, is called *quasi-associated*[28] if it satisfies condition (5.1).

A situation where Theorem 5.3 is especially useful occurs when the covariance function decreases rapidly at infinity. We say that a wide-sense stationary field X satisfies the *finite susceptibility condition* of Newman ([307]) if

$$\sigma^2 := \sum_{j \in \mathbb{Z}^d} cov(X_0, X_j) < \infty. \tag{5.3}$$

Recall that for an array $\{a_j, j \in \mathbb{Z}^d\}$ of real numbers one writes $a = \sum_{j \in \mathbb{Z}^d} a_j$ whenever, for any $\varepsilon > 0$, there exists $N = N(\varepsilon) \in \mathbb{N}^d$ such that

$$\left| a - \sum_{-m \leq j \leq n} a_j \right| < \varepsilon$$

if $m, n \in \mathbb{N}^d$, $m, n \geq N$, here "$\leq$" means the usual partial order in $\mathbb{R}^d$. The value σ^2 defined in (5.3) is in fact the asymptotic variance for normalized partial sums of a random process ($d = 1$) or a field (for $d > 1$) as will be seen in Chapter 3, and is therefore nonnegative.

---

[28]Other usage of this term is given in [230].

For non-stationary random fields one uses the *Cox–Grimmett coefficient*[29] ([110])

$$u_r = \sup_{k \in \mathbb{Z}^d} \sum_{j \in \mathbb{Z}^d, \, |j-k| \geq r} |cov(X_k, X_j)|, \quad r \in \mathbb{Z}_+, \tag{5.4}$$

assuming that it is finite for any $r \in \mathbb{N}$ and tends to zero as $r \to \infty$.

Remark 5.8. All the norms in a finite-dimensional space are equivalent, thus the choice of the sup-norm $|\cdot|$ in (5.4) and in other similar occasions is not essential. However it is convenient for calculations, since r-neighborhoods $(r > 0)$ of a point in $\mathbb{R}^d$ with respect to this norm are cubes.

It is well-known that a number of classical limit theorems of Probability Theory (see, e.g., [167, 326]) can be established under two natural hypotheses. Namely, the random variables under consideration be independent and possess the certain absolute moment, e.g. the first, second or $2 + \delta$ for some positive δ. The advantage of studying the associated (**PA**, **NA**) systems of random variables is the following. It is possible to prove the analogs of classical limit theorems (e.g., central limit theorem, invariance principles and other) for such dependent systems using simple moment conditions for random variables of these systems and imposing conditions on the asymptotical behavior of the covariance function, e.g., in terms of the Cox–Grimmett coefficient (when the random variables are square-integrable). Thus, the role of the Cox–Grimmett coefficient is similar to that of the mixing coefficients (see, e.g., [54, 55, 138, 210]), while it is much easier to estimate the former one.

2°. Association and mixing. The question when an associated system is mixing has not been completely answered. One can say that association (**PA or NA**) and mixing provide complementary tools for study of stochastic systems. Moreover, for random fields defined on $\mathbb{Z}^d$ (or $\mathbb{R}^d$) there are no rich collections of examples possessing mixing properties if $d > 1$ (see, e.g., [53, 61, 136, 138, 306]). The most important example is provided by *m-dependent random fields* (i.e. σ-algebras $\sigma\{X_t, t \in I\}$ and $\sigma\{X_t, t \in J\}$ are independent when $dist(I, J) \geq m$).

We present two results showing that the relation between mixing and positive (or negative) association is rather complicated. They both are concerned with random processes and Rosenblatt's strong mixing. Recall that a stochastic process $X = \{X_j, j \in \mathbb{Z}\}$ is *strongly mixing* if

$$\alpha_k(X) = \sup_{\substack{n \in \mathbb{Z} \\ A \in \mathcal{F}_{\leq n} \\ B \in \mathcal{F}_{\geq n+k}}} |\mathsf{P}(AB) - \mathsf{P}(A)\mathsf{P}(B)| \to 0 \text{ as } k \to \infty \tag{5.5}$$

where $\mathcal{F}_{\leq n} = \sigma\{X_j, j \leq n\}$, $\mathcal{F}_{\geq n+k} = \sigma\{X_j, j \geq n + k\}$. Clearly the second supremum in (5.5) can be taken over the sets[30] of the type

$$A = \{X_I \in C\}, \quad B = \{X_J \in D\}, \quad C \in \mathcal{B}(\mathbb{R}^l), \quad D \in \mathcal{B}(\mathbb{R}^m), \tag{5.6}$$

---

[29] In [110] the covariances were used without absolute values as the field was associated.
[30] As these sets constitute an algebra generating $\mathcal{F}_{\leq n}$ and $\mathcal{F}_{\geq n+k}$.

where $X_I = (X_{i_1}, \ldots, X_{i_l})$, $X_J = (X_{j_1}, \ldots, X_{j_m})$, $I = \{i_1, \ldots, i_l\} \subset \mathbb{Z} \cap (-\infty, n]$, $J = \{j_1, \ldots, j_m\} \subset \mathbb{Z} \cap [n + k, +\infty)$ and $l, m \in \mathbb{N}$. Also we may assume all the σ-algebras under consideration to be completed by the collection of null sets with respect to measure P.

The next result is due to Matula.

Theorem 5.9. ([292]) *Let* $X = \{X_j, j \in \mathbb{Z}\}$ *be a* $\mathbb{Z}$*-valued quasi-associated random process, such that* $\mathsf{E}X_j^2 < \infty$ *for any* $j \in \mathbb{Z}$*. Then it is strongly mixing with*

$$\alpha_k \leq 4 \sum_{i=0}^{\infty} u_{k+i}, \quad k \in \mathbb{N},$$

provided that the series is convergent (u_r *is introduced in* (5.4)).

By scaling, this Theorem can be applied to X_j's having lattice distribution on $\mathbb{R}$ if that lattice can be chosen the same for all X_j.

Proof. Let $h : \mathbb{R} \to \mathbb{R}$ be a Lipschitz function such that $h(0) = 1$, $0 \leq h(x) \leq 1$ for any $x \in \mathbb{R}$, $h(x) = 0$ whenever $|x| \geq 1/2$ and $Lip(h) = 2$.

For any $d \in \mathbb{N}$ and $k \in \mathbb{Z}^d$, define a function $h_{d,k} : \mathbb{R}^d \to \mathbb{R}$ as follows:

$$h_{d,k}(x) = \prod_{i=1}^{d} h(x_i - k_i), \quad x = (x_1, \ldots, x_d) \in \mathbb{R}^d.$$

Finally, for a Borel set $C \subset \mathbb{R}^d$, take

$$h_C(x) = \sum_{k \in C \cap \mathbb{Z}^d} h_{d,k}(x), \quad x \in \mathbb{R}^d. \tag{5.7}$$

Note that only finite number of summands in (5.7) can be different from 0 since h has compact support.

For all $x \in \mathbb{Z}^d$ one has $\mathbb{I}_C(x) = h_C(x)$. Moreover, h_C takes values in $[0, 1]$, and all its partial Lipschitz constants are no greater than 2.

For arbitrary events A, B appearing in (5.6), by Theorem 5.3,

$$|\mathsf{P}(AB) - \mathsf{P}(A)\mathsf{P}(B)| = |cov(\mathbb{I}\{A\}, \mathbb{I}\{B\})| = |cov(h_C(X_I), h_D(X_J))|$$

$$\leq 4 \sum_{i \in I} \sum_{j \in J} |cov(X_i, X_j)| \leq 4 \sum_{i=0}^{\infty} u_{k+i}. \quad \square$$

In general, "good" associated random system (e.g., bounded and having exponential rate of decrease of the Cox–Grimmett coefficient) need not be strongly mixing, as the following examples show.

Example 5.10. Let $(\varepsilon_n)_{n \in \mathbb{Z}}$ be a sequence of i.i.d. random variables and ξ be a random variable independent of this sequence and taking values ± 1 with probability $1/2$. Set $X_n = \xi$ for $n \in U = \{n_k, k \in \mathbb{Z}_+\}$ where $n_0 = 1$ and $n_k = n_{k+1} + k$, $k \in \mathbb{N}$, and let $X_n = \varepsilon_n$ for $n \in \mathbb{Z} \setminus U$. Clearly $X = (X_n)_{n \in \mathbb{Z}}$ is not strongly mixing as $\alpha(\sigma\{\xi\}, \sigma\{\xi\}) = 1/4$ and therefore $\alpha_k(X) = 1/4$, $k \in \mathbb{N}$. However $X \in \mathbf{A}$. Indeed, in view of Theorem 1.8 the families $\{X_n, n \in U\}$ and $\{X_n, n \notin U\}$ are associated and independent. Moreover, we can modify this example to have $u_r \to 0$ as $r \to \infty$. Namely, set $Y_n = a_n \xi$ for $n \in U$ (with appropriate choice of $a_n \geq 0$) and $Y_n = \varepsilon_n$ for $n \in \mathbb{Z} \setminus U$.

As shown by Doukhan and Louhichi, one can even give an example in which the random process is strictly stationary.

Example 5.11. ([141]) Let $\varepsilon = (\varepsilon_j)_{j \in \mathbb{Z}}$ be a sequence of independent random variables on a complete probability space $(\Omega, \mathcal{F}, \mathsf{P})$, taking values $\pm 1/2$ with equal probability $1/2$. Consider a stationary random sequence $X = (X_n)_{n \in \mathbb{Z}}$ satisfying the autoregression equation

$$X_n = \frac{X_{n-1}}{2} + \varepsilon_n, \quad n \in \mathbb{Z}.$$

Then $X_n = \sum_{j=0}^{\infty} 2^{-j} \varepsilon_{n-j}$ a.s. for any $n \in \mathbb{Z}$ (the series of independent summands converges a.s. and in quadratic mean). Hence, $X \in \mathbf{A}$ due to Theorem 1.8. Moreover, it is easy to calculate that

$$cov(X_0, X_n) = 2^{-|n|} \mathsf{E} X_0^2 = 2^{-|n|}/3, \quad n \in \mathbb{Z}.$$

Consequently $u_r \leq 2^{2-r}/3$, $r \in \mathbb{N}$. To see that X is not strongly mixing, consider an event $E_n = \{X_n > 0\} \in \mathcal{F}_{\leq n}$ where σ-algebras $\mathcal{F}_{\leq n}(X)$ and $\mathcal{F}_{\geq n+k}(X)$ are defined after (5.5).

Since X_n has uniform distribution[31] on $[-1, 1]$, one has $\mathsf{P}(E_n) = 1/2$. Note that $X_{n-1} = 2(X_n - \varepsilon_n)$, $n \in \mathbb{Z}$, and consequently,

$$\frac{X_{n-1} + 1}{2} = \left\{ \frac{X_{n-1} + 1}{2} \right\} = \left\{ \frac{2X_n - 2\varepsilon_n + 1}{2} \right\} = \left\{ \frac{2X_n + 1}{2} \right\},$$

hence

$$X_{n-1} = 2 \left\{ \frac{2X_n + 1}{2} \right\} - 1 \text{ a.s.} \tag{5.8}$$

for any $n \in \mathbb{Z}$, here $\{x\}$ is the fractional part of $x \in \mathbb{R}$. Iterating the relation (5.8), one sees that, for any $k \in \mathbb{N}$, there exists an event A_n such that $\mathsf{P}(A_n) = 0$ and $E_n \triangle A_n \in \mathcal{F}_{\geq n+k} \cap \mathcal{F}_{\leq n}$. Therefore $\alpha_k(X) \geq \mathsf{P}(E_n) - \mathsf{P}(E_n)^2 = 1/4$ for any k.

3°. Generalizations of positive and negative dependence. The definition of quasi-associated random field unifies the positively and negatively associated random systems under the assumption that the covariance function is summable (which is usually required in limit theorems). A natural idea is to replace this definition with a more general one in which the left-hand side is the same as in (5.1) but the estimate is given without the help of the Cox–Grimmett coefficient.

We shall discuss the approach based on description of the dependence structure of a random field by means of the upper bound for covariances involving specified

---

[31] This is seen, e.g., by calculating the characteristic function of $\sum_{j=0}^{\infty} 2^{-j} \varepsilon_{n-j}$.

"test functions". This approach has the origin in the study of mixing processes and random fields (see, e.g., [55, 418]). In [16] it was proposed to use the power-type test functions to obtain the estimates of absolute moments of order $s > 2$ for partial sums. In [78] it was demonstrated that for Lipschitz functions in **PA** or **NA** random fields one can obtain the natural covariance inequality without the hypothesis that such functions are monotone, see (5.1). The definition below, which is principal in this Section, was given for stochastic processes by Doukhan and Louhichi in [141], and for random fields by Bulinski and Suquet in [82], Doukhan and Lang in [139] . Set $BL = \cup_{n \in \mathbb{N}} BL(n)$ ("bounded Lipschitz") and let $\psi : \mathbb{R}^2 \times BL^2 \to \mathbb{R}$ be some nonnegative functional.

Definition 5.12. ([82, 139, 141]) A real-valued random field $X = \{X_j, j \in \mathbb{Z}^d\}$ is called (BL, ψ, θ)-*dependent* if there exists a nonincreasing sequence $\theta = (\theta_r)_{r \in \mathbb{N}}$, $\theta_r \to 0$ as $r \to \infty$, such that for any finite disjoint sets $I, J \subset \mathbb{Z}^d$ with $dist(I, J) = r \in \mathbb{N}$, and any functions $f \in BL(|I|)$, $g \in BL(|J|)$, one has

$$|cov(f(X_I), g(X_J))| \leq \psi(|I|, |J|, f, g)\theta_r. \tag{5.9}$$

Here $dist(I, J) = \min\{|i - j| : i \in I, j \in J\}$, and $|\cdot|$ is the sup-norm in $\mathbb{R}^d$.

Interesting examples arise when one uses in (5.9) the functionals

$$\psi_1 = Lip(f)Lip(g)(|I| \wedge |J|) \quad \text{or} \quad \psi_2 = |I|Lip(f)\|g\|_\infty + |J|Lip(g)\|f\|_\infty.$$

Following [82] we will simply write (BL, θ) instead of (BL, ψ_1, θ) when no confusion can arise. Though this definition was given for a real-valued random field, it can be extended to more general systems. The most important among them are vector-valued random fields. The main task for extension here is to define test functions on a set of random vectors. The next generalization of association is proposed by Burton, Dabrowski and Dehling.

Definition 5.13. ([85]) A family of random vectors $\{X_t, t \in T\}$ with values in $\mathbb{R}^s$, $s \in \mathbb{N}$, is called *weakly associated* (or *negatively associated*), if for any disjoint finite sets $I, J \subset T$ and any functions $f \in M(s|I|)$, $g \in M(s|J|)$ one has

$$cov(f(X_I), g(X_J)) \geq 0 \quad (\text{resp. } cov(f(X_I), g(X_J)) \leq 0).$$

Definition 5.14. A vector-valued random field $X = \{X_j, j \in \mathbb{Z}^d\}$ with values in $\mathbb{R}^s$, $s \in \mathbb{N}$, is called (BL, ψ, θ)-*dependent* if there exists a nonincreasing sequence $\theta = (\theta_r)_{r \in \mathbb{N}}$, $\theta_r \to 0$ as $r \to \infty$, such that for all finite disjoint sets $I, J \subset \mathbb{Z}^d$ with $dist(I, J) = r \in \mathbb{N}$, and any functions $f \in BL(s|I|)$, $g \in BL(s|J|)$ inequality (5.9) holds where $\psi : \mathbb{N}^2 \times BL^2 \to \mathbb{R}_+$ is some specified functional.

See [68, 79, 85] for discussion and results for random vectors possessing dependence properties of the types introduced in Definitions 5.13 and 5.14.

Remark 5.15. One can view the Definitions above as an attempt to unify the ideas of using mixing coefficients and employing **PA** or **NA**. Other dependence coefficients were introduced in $[120, 121, 123, 124]$. For integrable random variables X on a probability space $(\Omega, \mathcal{F}, \mathsf{P})$ and σ-algebras $\mathcal{A} \subset \mathcal{F}$, set

$$\tau(X, \mathcal{A}) := \mathsf{E} \sup_{\substack{f \in BL(\mathbb{R}) \\ Lip(f) \leq 1}} |\mathsf{E}(f(X)|\mathcal{A}) - \mathsf{E}f(X)|.$$

It is proved in $[123]$ that this definition is correct, i.e. the expectation is applied to a measurable function. Dedecker and Prieur say that a random sequence $X = (X_n)_{n \in \mathbb{N}}$ is τ-*dependent* if

$$\tau_n := \sup_{k \in \mathbb{Z}} \tau(X_k, \mathcal{F}_{\leq k-n}) \to 0, \quad n \to \infty,$$

where, as above, $\mathcal{F}_{\leq m} = \sigma\{X_j, j \leq m\}$, $m \in \mathbb{Z}$. This notion of dependence admits a coupling result analogous to those known for mixing setup (see, e.g., $[138]$; some coupling techniques applications are provided also in Appendix A.6). It is, in a sense, more general than strong mixing; note that the random sequence appearing in Example 5.11 is τ-dependent with fast decreasing (τ_n), see $[123]$ for more complicated examples, e.g. based on autoregression processes and Markov kernels theory. As far as we know, this notion was not adapted to studying the multiindexed systems.

Now we provide three examples illustrating Definition 5.12. Before them, we provide a version of Theorem 1.8 for (BL, θ)-dependence.

Lemma 5.16. *(a) If* $\mathsf{E}X_j^2 < \infty$ *for each* $j \in \mathbb{Z}^d$, *then the Lipschitz functions* f *and* g *in Definition 5.12 with* $\psi = \psi_1$ *need not be bounded.*

(b) It suffices to prove (BL, θ)-*dependence assuming that* f *and* g *have bounded derivatives of all orders.*

(c) If $X^k = \{X_j^k, j \in \mathbb{Z}^d\}$ *is a sequence of* (BL, θ^k)-*dependent random fields such that finite-dimensional distributions of* X^k *weakly converge to the corresponding ones of* X *as* $k \to \infty$, *then the field* $X = \{X_j, j \in \mathbb{Z}^d\}$ *is* (BL, θ)-*dependent with* $\theta_r = \liminf_{k \to \infty} \theta_r^k$, $r \in \mathbb{N}$.

Proof. (a) and (b) are established by dominated convergence, (c) follows from the definition of convergence in law. $\square$

4°. Quasi-association of Gaussian systems.

Theorem 5.17. *($[139, 374]$) (a) A* **PA** *or* **NA** *random field* $X = \{X_j, j \in \mathbb{Z}^d\}$ *with values in* $\mathbb{R}^m$ $(m > 1)$ *is* (BL, θ)-*dependent whenever* $\mathsf{E}\|X_j\|^2 < \infty$ *for any* $j \in \mathbb{Z}^d$ *and the coefficients*

$$u_r = \sup_{k \in \mathbb{Z}^d} \sum_{\substack{j \in \mathbb{Z}^d \\ |j-k| \geq r}} \sum_{v,l=1}^m |cov(X_{kv}, X_{jl})| \qquad (5.10)$$

exist and tend to zero as $r \to \infty$. *In this case* $\theta_r = u_r$, $r \in \mathbb{Z}_+$.

(b) A Gaussian random field $X = \{X_j, j \in \mathbb{Z}^d\}$ *with values in* $\mathbb{R}^m$ $(m \geq 1)$ *is* (BL, θ)-*dependent, with* $\theta_r = u_r$, $r \in \mathbb{N}$, *whenever the quantities* $(u_r)_{r \in \mathbb{N}}$ *in* (5.10) *exist and vanish at infinity.*

Proof. (a) trivially follows from Theorem 5.3.

(b) Let $I, J \subset \mathbb{Z}^d$ be finite disjoint sets with $dist(I, J) = r \in \mathbb{N}$. We want to show that
$$cov|(f(X_I), g(X_J))| \leq (|I| \wedge |J|)Lip(f)Lip(g)u_r$$
for bounded Lipschitz functions $f : \mathbb{R}^{m|I|} \to \mathbb{R}$, $g : \mathbb{R}^{m|J|} \to \mathbb{R}$. Theorem 2.1 yields that one can consider only $m > 1$. By Lemma 5.16 we are able to take f and g which have continuous partial derivatives. Enumerate the components of $m(|I| + |J|)$-dimensional Gaussian random vector $\mathbf{X} = (X_I, X_J)$ in a row as follows
$$\mathbf{X} = (X_{i_1,1}, \ldots, X_{i_1,m}, \ldots, X_{i_r,1}, \ldots, X_{i_r,m}, X_{j_1,1}, \ldots, X_{j_1,m}, \ldots, X_{j_r,1}, \ldots, X_{j_r,m})$$
where $I = \{i_1, \ldots, i_r\}$ and $J = \{j_1, \ldots, j_m\}$. We may assume that $\mathbf{X} \sim N(0, \Sigma)$ with $\det \Sigma > 0$ (if $\det \Sigma = 0$, take vectors $\mathbf{X}^k \sim N(0, \Sigma + k^{-1}I)$ and then apply Lemma 5.16). Let $Z \sim N(0, \Sigma)$ be a Gaussian random vector independent of $\mathbf{X}$. Set $Y(\lambda) = \lambda\mathbf{X} + (1 - \lambda^2)^{1/2}Z$, $\lambda \in [0, 1]$, and consider $F(\lambda) = \mathsf{E}f(X)g(Y(\lambda))$, $\lambda \in [0, 1]$. This function is continuous and satisfies, by Lemma 2.3, the relation
$$F'(\lambda) = \int_{\mathbb{R}^q} \phi(x) \left(\sum_{k,t \in I \cup J} \sum_{v,l=1}^m cov(X_{kv}, X_{tl}) \frac{\partial f(x)}{\partial x_{kv}} \frac{\partial g(\lambda, x)}{\partial x_{tl}} \right) dx, \quad \lambda \in (0, 1).$$
Here ϕ and $g(\lambda, x)$ are the same as in the proof of Theorem 2.1, $q = m(|I| + |J|)$. Since f depends only on the first $m|I|$ coordinates and g depends only on the last $m|J|$ ones, the first double sum is taken in fact over $k \in I$ and $t \in J$. Also, for $t \in J$ and $l = 1, \ldots, m$, one has
$$\frac{\partial g(\lambda, x)}{\partial x_{tl}} = \frac{\partial}{\partial x_{tl}} \int_{\mathbb{R}^q} g(\lambda x - y)\phi_\lambda(y)dy \leq \left| \frac{\partial g(\lambda x - y)}{\partial x_{tl}} \right| \leq \lambda Lip(g),$$
since $\int_{\mathbb{R}^q} \phi_\lambda(y)dy = 1$. Consequently,
$$|F(1) - F(0)| \leq \sup_{\lambda \in (0,1)} |F'(\lambda)| \leq Lip(f)Lip(g) \sum_{k \in I, t \in J} \sum_{v,l=1}^m |cov(X_{kv}, X_{tl})|. \quad \Box$$

5°. Autoregression field. Let V be some finite subset of $\mathbb{Z}^d$ not containing 0, and $F : \mathbb{R}^{|V|} \to \mathbb{R}$ be some measurable function. We call $X = \{X_j, j \in \mathbb{Z}^d\}$ an *autoregression field* if
$$X_j = F(X_{j-v}, v \in V) + \varepsilon_j \tag{5.11}$$
where $\{\varepsilon_j, j \in \mathbb{Z}^d\}$ are some i.i.d. random variables with $\mathsf{E}\varepsilon_0 = 0$, $\mathsf{E}\varepsilon_0^2 = 1$. The classical *autoregression process* defined by the equation
$$X_n = \sum_{v=1}^k a_v X_{n-v} + \varepsilon_n, \quad n \in \mathbb{Z},$$
obviously satisfies the definition above with linear function F and $V = \{1, \ldots, k\}$.

Theorem 5.18. *Suppose that F is Lipschitz and $L := \sum_{i=1}^{|V|} Lip_i F < 1$. Then equation (5.11) has a strictly stationary solution which is (BL, θ)-dependent with*
$$\theta_r \leq L^{-1}(1 - L)^{-2}3^d \left(r^d L^{r/q} + \sum_{k=r}^\infty k^{d-1}L^{k/q} \right), \quad r \in \mathbb{N},$$
where $q = diam(V) = \max\{|i - j| : i, j \in V\}$.

Proof. Throughout the proof we will write $\rho(\cdot,\cdot)$ instead of $dist(\cdot,\cdot)$ in order to simplify the exponents. Define the strictly stationary random fields $X^{(k)} = \{X_i^{(k)}, i \in \mathbb{Z}^d\}, k \in \mathbb{Z}_+$, recursively. Set $X_i^{(0)} = \varepsilon_i$ and

$$X_i^{(k)} = F\left(X_{i-v}^{(k-1)}, v \in V\right) + \varepsilon_i, \quad k \in \mathbb{N}.$$

Then $X_i^{(k)}$ converges in quadratic mean to X_i as $k \to \infty$, for any $i \in \mathbb{Z}^d$. Indeed, for a fixed $i \in \mathbb{Z}^d$, using the association of the family $\{\varepsilon_{i-v}, v \in V\}$ one has

$$\begin{aligned}
\mathsf{E}(X_i^{(1)} - X_i^{(0)})^2 &= \mathsf{E}\left(F(0) + |F(\varepsilon_{i-v}, v \in V) - F(0)|\right)^2 \\
&\leq 2(F(0)^2 + |V|Lip^2(F)),
\end{aligned}$$

since $\sum_{v,w \in V} cov(\varepsilon_{i-v}, \varepsilon_{i-w}) = |V|$ (here $F(0)$ is evaluated at $0 \in \mathbb{R}^{|V|}$). For $k \in \mathbb{N}$

$$\begin{aligned}
\mathsf{E}(X_i^{(k+1)} - X_i^{(k)})^2 &\leq \mathsf{E}\left(\sum_{v \in V} Lip_v F |X_{i-v}^{(k)} - X_{i-v}^{(k-1)}|\right)^2 \\
&= \sum_{v,w \in V} Lip_v F Lip_w F \left(\mathsf{E}|X_{i-v}^{(k)} - X_{i-v}^{(k-1)}|^2 \mathsf{E}|X_{i-w}^{(k)} - X_{i-w}^{(k-1)}|^2\right)^{1/2} \\
&\leq L^2 \max_{v \in V} \mathsf{E}\left(X_{i-v}^{(k)} - X_{i-v}^{(k-1)}\right)^2.
\end{aligned}$$

Therefore,

$$\mathsf{E}(X_i^{(k+1)} - X_i^{(k)})^2 \leq 2L^{2k}(F(0)^2 + |V|Lip^2(F)), \quad k \in \mathbb{N}.$$

Thus $(X_i^{(k)})_{k \in \mathbb{N}}$ is a Cauchy sequence in $L^2(\Omega, \mathcal{F}, \mathsf{P})$. Hence the limit random field X is strictly stationary and $\mathsf{E}X_0^2 < \infty$. Moreover this field is a solution of (5.11) (as seen by passing to a subsequence converging a.s.). By Lemma 5.16, to prove the (BL, θ)-dependence it suffices to prove the same for any of the fields $X^{(k)}$. To accomplish this, for arbitrary set $U \subset \mathbb{Z}^d$ and $l \in \mathbb{Z}_+$, introduce $U^{(l)} = \{j \in \mathbb{Z}^d : \rho(j, U) \leq l\}$.

Lemma 5.19. *Let $I \subset \mathbb{Z}^d$ be a finite set. Then, for each $k \in \mathbb{Z}_+$, the random vector $(X_i^{(k)}, i \in I)$ admits a representation $(X_i^{(k)}, i \in I) = \phi_k(\varepsilon_j, j \in I^{(qk)})$ where $\phi : \mathbb{R}^{|I^{(qk)}|} \to \mathbb{R}^{|I|}$ is a Lipschitz map with $Lip_t \phi \leq (1 - L)^{-1} L^{\rho(t,I)/q}, \ t \in \mathbb{Z}^d$.*

Proof. We proceed by induction on k. For $k = 0$ the assertion is obvious as $X_i^{(0)} = \varepsilon_i, i \in \mathbb{Z}^d$. Suppose that it has been proved for the field $X^{(k)}$ and all finite subsets I of $\mathbb{Z}^d$. Fix a finite I. Clearly ϕ may only depend on values at points which are not farther from I than qk. To estimate the Lipschitz constant fix $t \in I^{(qk)}$. Let all the ε_j with $j \neq t$ be fixed and let $Y^{(k+1)}$ be the field obtained from $X^{(k+1)}$ by changing $\varepsilon_t(\omega)$ to some $\tau_t(\omega), \omega \in \Omega$. Then, by the definition of $X^{(k+1)}$,

$$\sum_{i \in I}\left|X_i^{(k+1)} - Y_i^{(k+1)}\right| \leq \mathbb{I}\{t \in I\}|\varepsilon_t - \tau_t| + \sum_{i \in I}\sum_{v \in V} Lip_v F \left|X_{i-v}^{(k)} - Y_{i-v}^{(k)}\right|$$

$$\leq \mathbb{I}\{t \in I\}|\varepsilon_t - \tau_t| + \sum_{v \in V} Lip_v F \sum_{i \in I}|X_{i-v}^{(k)} - Y_{i-v}^{(k)}|.$$

By induction hypothesis applied to the set $\{i - v, i \in I\}$ this upper bound does not exceed

$$|\varepsilon_t - \tau_t|\left(\mathbb{I}\{t \in I\} + (1 - L)^{-1}\sum_{v \in V} Lip_v F \exp\{(\rho(I, v + t)/q)\log L\}\right)$$

$$\leq |\varepsilon_t - \tau_t|\left(\mathbb{I}\{t \in I\} + L(1 - L)^{-1}\exp\{(\rho(I^{(q)}, t)/q)\log L\}\right).$$

If $t \in I$ then $1 + L(1 - L)^{-1} = (1 - L)^{-1}$ is the desired Lipschitz constant; if $t \notin I$, then, using the inequality $q + \rho(I^{(q)}, t) \geq \rho(I, t)$, we obtain the constant

$$L(1 - L)^{-1}L^{\rho(I^{(q)}, t)/q} \leq (1 - L)^{-1}L^{\rho(I, t)/q},$$

as stated. $\square$

Continuing the proof of Theorem, fix $k \in \mathbb{N}$ and take arbitrary finite sets $I, J \subset \mathbb{Z}^d$ with $|I| \leq |J|$ and $\rho(I, J) = r \in \mathbb{N}$. Consider any functions $f \in BL(|I|)$ and $g \in BL(|J|)$. The random variables $\{\varepsilon_t, t \in \mathbb{Z}^d\}$ are independent, hence associated. Therefore we see, on account of Theorem 5.3 and Lemma 5.19, that

$$|cov(f(X_I^{(k)}), g(X_J^{(k)}))| \leq (1 - L)^{-2}Lip(f)Lip(g)\sum_{x \in \mathbb{Z}^d} L^{(\rho(x, I) + \rho(x, J))/q}$$

$$\leq (1 - L)^{-2}Lip(f)Lip(g)|I|\max_{y \in I}\sum_{x \in \mathbb{Z}^d} L^{(\rho(x, y) + \rho(x, J))/q}.$$

It remains to prove that $(1 - L)^{-2}\sum_{x \in \mathbb{Z}^d} L^{(\rho(x, y) + \rho(x, J))/q} \leq \theta_r$ for $y \in I$. One has

$$\sum_{x : \rho(x, y) \geq r} L^{\rho(x, y)/q + \rho(x, J)/q} \leq \sum_{x \in \mathbb{Z}^d, \, |x| \geq r} L^{|x|/q} \leq 3^d\sum_{k=r}^{\infty} k^{d-1}L^{k/q}.$$

Note that $\rho(I, J) \leq \rho(x, y) + \rho(x, J)$. Thus

$$\sum_{x : \rho(x, y) < r} L^{(\rho(x, y) + \rho(x, J))/q} \leq L^{r/q}|\{x : \rho(x, y) < r\}| \leq 3^d r^d L^{r/q}. \tag{5.12}$$

From this estimate the Theorem follows. $\square$

6°. Interacting particle systems. The following example arises in the theory of interacting particle systems. For a detailed discussion of the subject one may consult the book by Liggett [265]. We only recall the basic facts which we will need. Let $\mathcal{S} = \{0, 1\}^{\mathbb{Z}^d}$ be the space of zero-one configurations on the lattice $\mathbb{Z}^d$ endowed with metric $\rho(\cdot, \cdot)$ making it a compact (e.g., $\rho(x, y) = \sum_{i \in \mathbb{Z}^d} 2^{-|i|}|x_i - y_i|$). For a function $f \in C(\mathcal{S}, \mathbb{R})$ set

$$\Delta_f(x) = \sup\{|f(\xi) - f(\eta)| : \xi, \eta \in \mathcal{S}, \xi(y) = \eta(y) \text{ if } y \neq x\}, \quad x \in \mathbb{Z}^d.$$

Consider a function $c : \mathbb{Z}^d \times \mathcal{S} \to \mathbb{R}_+$. One can construct a Markov process $\{X_s, s \geq 0\}$ defined on a probability space $(\Omega, \mathcal{F}, \mathsf{P})$ and taking values in $\mathcal{S}$ so that

$$\begin{aligned}
\mathsf{P}(X_s(x) \neq X_0(x)|X_0 = \eta) &= c(x, \eta)s + o(s), \quad s \to 0+, \\
\mathsf{P}(X_s(x) \neq X_0(x), X_s(y) \neq X_0(y)) &= o(s), \quad s \to 0+,
\end{aligned} \tag{5.13}$$

for all $x, y \in \mathbb{Z}^d$, $x \neq y$, and all $\eta \in \mathcal{S}$ (see [265]). That is, if the process has come to a configuration η, then it changes at point x with *intensity* $c(x, \eta)$, while the probability to change at two points almost simultaneously is negligibly small. Set

$$\gamma(x, z) = \mathbb{I}_{\{x \neq z\}} \sup_{\eta:\eta_1(y)=\eta_2(y),\, y\neq z} |c(x, \eta_1) - c(x, \eta_2)|, \quad x, z \in \mathbb{Z}^d,$$

$$M = \sup_{z\in\mathbb{Z}^d} \sum_{x\in\mathbb{Z}^d} \gamma(x, z), \quad M_1 = \sup_{x\in\mathbb{Z}^d} \sum_{z\in\mathbb{Z}^d} \gamma(x, z),$$

$$\varepsilon = \inf_{z\in\mathbb{Z}^d} \inf_{\substack{\eta_1(y)=\eta_2(y),\, y\neq z \\ \eta_1(z)\neq\eta_2(z)}} (c(x, \eta_1) + c(x, \eta_2)).$$

Next introduce a linear operator Γ (in general it may be unbounded) on

$$l_1(\mathbb{Z}^d) = \{a : \mathbb{Z}^d \to \mathbb{R}, \ \sum_{x\in\mathbb{Z}^d} |a(x)| < \infty\}$$

with infinite matrix $\gamma(x, z)$, i.e.

$$\Gamma\beta(z) = \sum_x \gamma(x, z)\beta(x), \quad \beta \in l_1(\mathbb{Z}^d).$$

We suppose that

$$D_0 := \sup_{x,\eta} c(x, \eta) < \infty \text{ and } M_1 < \infty.$$

These conditions ensure the existence of a Markov process with transition intensities $c(x, \eta)$ for every initial distribution on $\mathcal{S}$, see Theorem 3.9 in [265, Ch. 1]. The corresponding semigroup (see Appendix A.2) of Markov operators $(P_s)_{s\geq 0}$ is continuous (in the strong sense) and has generator G, defined on the linear subspace $D(\mathcal{S})$ of continuous functions $f : \mathcal{S} \to \mathbb{R}$ such that $\sum_{x\in\mathbb{Z}^d} \Delta_f(x) < \infty$, and acting as follows

$$Gf(\eta) = \sum_{x\in\mathbb{Z}^d} c(x, \eta)(f(\eta^x) - f(\eta)). \tag{5.14}$$

Here η^x stands for a configuration which is different from η at point x and coincides with η at other points. Note that if $M < \infty$, then Γ is a bounded linear operator defined on the (incomplete) space $(l_1(\mathbb{Z}^d), \|\cdot\|_\infty)$ having norm M. Here $\|z\|_\infty = \sup_{k\in\mathbb{Z}^d} |z_k|$, $z \in l_1(\mathbb{Z}^d)$. Besides, if *"the dynamics is translation-invariant"* (i.e. $c(x, \xi) = c(x + a, \zeta)$ for any $a \in \mathbb{Z}^d$, ζ being the configuration such that $\zeta(y) = \xi(y - a)$), then one has $\gamma(x, u) = \gamma(0, u - x)$ for all $x, u \in \mathbb{Z}^d$, so $M_1 = M$. The existence and uniqueness properties of our Markov process are gathered in the next statement.

Lemma 5.20. ([265, Ch. 1, Theorems 3.9 and 4.1]). *Suppose that $D_0 < \infty$ and $M_1 < \varepsilon < \infty$. Then for any $x \in \mathbb{Z}^d$, $f \in D(\mathcal{S})$ and $s \geq 0$ one has*

$$\Delta_{P_s f}(x) \leq e^{-\varepsilon s}(\exp(s\Gamma)\Delta_f)(x). \tag{5.15}$$

The corresponding Markov process $X = \{X_s, s \geq 0\}$ is ergodic, that is there exists an invariant distribution μ on $\mathcal{S}$ such that for any $f \in C(\mathcal{S})$ and $\eta \in \mathcal{S}$

$$P_s f(\eta) \to \int_S f(\eta)\mu(d\eta), \quad s \to \infty. \tag{5.16}$$

There are many examples of random fields meeting the conditions of Lemma 5.20. Moreover, it frequently happens that the intensities of change at point x do not depend on the value of configuration η at points which are far from x. This means that there is $q \in \mathbb{N}$ such that for any $x \in \mathbb{Z}^d$ and any pair of configurations η, ζ with $\eta(y) = \zeta(y)$ for all $y \in \mathbb{Z}^d$, $dist(x, y) \leq q$, one has $c(x, \eta) = c(x, \zeta)$. Clearly, if this holds, then the infinite matrix Γ is "$(2q+1)$-diagonal", i.e.

$$\gamma(x, z) = 0 \text{ for all } x, z \in \mathbb{Z}^d, \ dist(x, z) > q. \tag{5.17}$$

For example, this property is valid for Ising models corresponding to Gibbs fields with finite range potential, see, e.g., [265, Ch. 4].

We will assume a more general property than (5.17). Namely, for some $\lambda, c_0 > 0$

$$\gamma(x, u) \leq c_0 \exp\{-\lambda|x - u|\} \text{ for all } x, u \in \mathbb{Z}^d. \tag{5.18}$$

Throughout the proof of next theorem we will denote by $a_0, b_0, C_1, C_2, \ldots$ some positive constants which depend only on $d, D_0, M, M_1, \varepsilon, \lambda$, and c_0. We write also $\operatorname{Log} x = \log(x \vee e)$, $x > 0$. Now we are ready to state

Theorem 5.21. ([377]) *Suppose that $D_0 < \infty$ and $M, M_1 < \varepsilon < \infty$. If (5.18) holds, then the random field $\{\xi_t, t \in \mathbb{Z}^d\}$ having invariant distribution μ introduced above (see (5.16)) is (BL, θ)-dependent with*

$$\theta_r = C_1 \sum_{k=r}^{\infty} k^d \exp\left\{-C_2 \frac{k}{\operatorname{Log} k}\right\} \leq a_0 \exp\{-b_0 r\}, \quad r \in \mathbb{N}. \tag{5.19}$$

Proof resembles the ideas of Theorems 2.20 and 2.27. Let $\{X_t, t \geq 0\}$ be a Markov process which appears in (5.13) and Lemma 5.20, with initial distribution μ_0 concentrated at some configuration η_0. Take any finite disjoint sets $I, J \subset \mathbb{Z}^d$ such that $dist(I, J) = r$, $|I| \leq |J|$, and functions $f : \mathbb{R}^{|I|} \to \mathbb{R}$, $g : \mathbb{R}^{|J|} \to \mathbb{R}$. Since the values of f and g outside $\{0, 1\}^{|I|}$ (respectively $\{0, 1\}^{|J|}$) do not influence all the quantities under consideration, one can consider f and g as elements of the space $D(\mathcal{S})$ with $\|\Delta_f\|_\infty \leq Lip(f)$, $\|\Delta_g\|_\infty \leq Lip(g)$. By (5.16) we have

$$cov(f(X_I), g(X_J)) = \lim_{t \to \infty} (P_t(fg)(\eta_0) - P_t f(\eta_0) P_t g(\eta_0)).$$

Our goal is to estimate the expression under the sign of limit. To this end, fix $t \geq 0$ and set

$$h(s) = P_{t-s}((P_s f)(P_s g))(\eta_0), \quad 0 \leq s \leq t.$$

By Theorem A.4 the function $h(s)$ is continuous on $[0, t]$ and is differentiable in s when $0 < s < t$, and by (5.14)

$$-h'(s) = P_{t-s}G((P_s f)(P_s g))(\eta_0) - P_{t-s}((GP_s f)(P_s g))(\eta_0) - P_{t-s}((P_s f)(GP_s g))(\eta_0)$$

$$= P_{t-s}(G(P_s f, P_s g))(\eta_0). \tag{5.20}$$

Here

$$G(\phi, \psi)(\eta) = \sum_{x \in \mathbb{Z}^d} c(x, \eta)(\phi(\eta^x) - \phi(\eta))(\psi(\eta^x) - \psi(\eta))$$

for $\phi, \psi \in D(\mathcal{S})$. Note that if $f \in D(\mathcal{S})$ then $P_s f \in D(\mathcal{S})$ and $D(\mathcal{S})$ is closed under multiplication of functions, so (5.20) is correct. To estimate the right-hand side of (5.20) we need a bound for $\Delta_{P_s f}(x)$ and $\Delta_{P_s g}(x)$ at any point $x \in \mathbb{Z}^d$. Such estimate, similar to that of Lemma 5.19, will be given in the series of lemmas (we shall use the notation introduced above in this subsection).

Lemma 5.22. *Let $A : L \to L$ be a bounded linear operator on a normed linear space L and $\|A\| \leq K$, $K > 0$ (the operator norm is induced by the norm in L). Then, for any $m \in \mathbb{N}$, the following estimate holds*

$$\left\| \exp(A) - \sum_{k=0}^{m-1} \frac{A^k}{k!} \right\| \leq \frac{e^K K^m}{m!}.$$

Proof. This estimate is well-known and follows from application of the Taylor formula, see, e.g., [151, Ch. 1]. $\Box$

Lemma 5.23. *Suppose that $\lambda > 0$ and $x, y \in \mathbb{Z}^d$, $|x - y| = r \in \mathbb{N}$. Then, for any $m \in \mathbb{N}$, one has*

$$\sum_{j_1, \ldots, j_m \in \mathbb{Z}^d} \exp \left\{ -\lambda \sum_{k=0}^{m} |j_k - j_{k+1}| \right\} \leq \left((m+1)R(\lambda)^m + (2r)^{dm} \right) e^{-\lambda r} \qquad (5.21)$$

where $j_0 = x$, $j_{m+1} = y$ and $R(\lambda) = \sum_{v \in \mathbb{Z}^d} \exp\{-\lambda|v|\}$.

Proof. Let $x = j_0$, $y = j_{m+1}$. Introduce the sets

$$\mathbf{R} = \{(j_1, \ldots, j_m) : |j_i - j_{i+1}| < r, \ i = 0, \ldots, m\}, \ R_0 = \{(j_1, \ldots, j_m) : |j_0 - j_1| \geq r\},$$

$$R_1 = \{(j_1, \ldots, j_m) : |j_1 - j_2| \geq r\}, \ldots, R_m = \{(j_1, \ldots, j_m) : |j_m - j_{m+1}| \geq r\}.$$

Obviously $\mathbb{Z}^{dm} = \mathbf{R} \cup R_0 \cup \ldots \cup R_m$ and $r = |j_0 - j_{m+1}| \leq \sum_{k=0}^{m} |j_k - j_{k+1}|$. Then the sum in (5.21) taken over summands belonging to $\mathbf{R}$ contains no more than $(2r)^{dm}$ summands, each of them being no greater than $e^{-\lambda r}$. Observe that the sum in (5.21) taken over $(j_1, \ldots, j_m) \in R_0$ is bounded by

$$e^{-\lambda r} \sum_{j_m \in \mathbb{Z}^d} \exp\{-\lambda|j_m - j_{m+1}|\} \cdots \sum_{j_1 \in \mathbb{Z}^d} \exp\{-\lambda|j_1 - j_2|\} \leq e^{-\lambda r} R(\lambda)^m.$$

The remaining sums over $R_1, \ldots, R_m$ are estimated analogously to the sum over R_0. Summarizing, we come to (5.21). $\Box$

Lemma 5.24. *For any $q \in \mathbb{N}$ and $x \in \mathbb{Z}^d$ such that $\mathrm{dist}(x, I) = q$, one has*

$$\Delta_{P_s f}(x) \leq e^{(M-\varepsilon)s} \left(1 \wedge C_3 a(s) q^2 \exp \left\{ -\alpha q \left(1 - \frac{\log \log q}{\log q} \right) \right\} \right) Lip(f) \qquad (5.22)$$

for all $s \geq 0$ where $a(s) = s \vee s^{\alpha q / \log q + 1}$ $\alpha = \lambda / (d+1)$.

Proof. Fix $s \geq 0$. Taking the norm $\|\cdot\|_\infty$ in (5.15), we see that $\Delta_{P_s f}(x) \leq e^{(M-\varepsilon)s} Lip(f)$. Further, for $m \in \mathbb{N}$ and $x \in \mathbb{Z}^d$, by Lemma 5.22

$$e^{-\varepsilon s} \exp(s\Gamma)\Delta_f(x) \leq e^{-\varepsilon s} \left\| \exp(s\Gamma) - \sum_{j=0}^{m} \frac{s^j \Gamma^j}{j!} \right\|_\infty Lip(f) + \sum_{j=0}^{m} \frac{s^j \Gamma^j}{j!}\Delta_f(x)$$

$$\leq e^{(M-\varepsilon)s} \frac{(sM)^{m+1}}{(m+1)!} + \sum_{j=0}^{m} \frac{s^j}{j!}\Gamma^j \Delta_f(x).$$

Note that since Γ can be considered as infinite matrix and f does not depend on points outside I, for any $k \in \mathbb{N}$ such that $q \geq dk/\lambda$ one has

$$\Gamma^k \Delta_f(x) = \sum_{j_1,\ldots,j_{k-1} \in \mathbb{Z}^d,\, j_k \in I} \gamma(x,j_1)\gamma(j_1,j_2)\ldots\gamma(j_{k-1},j_k)\Delta_f(j_k)$$

$$\leq Lip(f)\, c_0^k \max_{y \in I} \sum_{j_1,\ldots,j_k \in \mathbb{Z}^d} \exp\{-\lambda(|x-j_1| + \ldots + |j_{k-1} - y|)\}$$

$$\leq Lip(f)\, c_0^k (kR(\lambda)^{k-1} + (2q)^{dk})e^{-\lambda q},$$

in view of Lemma 5.23. We used here that the sequence $(n^{dk}e^{-\lambda n})_{n \geq N_0}$ decreases when $N_0 \geq dk/\lambda$.

Now we have the estimate

$$\Delta_{P_s f}(x) \leq e^{(M-\varepsilon)s} \min_{m \leq q\lambda/d} \left(\frac{(sM)^{m+1}}{(m+1)!} + \sum_{k=0}^{m} \frac{c_0^k s^k}{k!}(kR(\lambda)^{k-1} + (2q)^{dk})e^{-\lambda q} \right) Lip(f).$$

After setting here $m = [\lambda q/(d+1)\log q]$ and invoking the Stirling formula one comes to the desired conclusion. $\square$

Let us proceed with the proof of Theorem 5.21. Denote the expression in large brackets in (5.22) by $\psi(s,q)$. Set $d_1 = d_1(x) = \rho(x,I)$, $d_2 = d_2(x) = \rho(x,J)$ for $x \in \mathbb{Z}^d$ (we write ρ instead of *dist*). We have

$$|P_t(fg)(\eta_0) - P_t f(\eta_0)P_t g(\eta_0)| = |h(t) - h(0)| = \left| \int_0^t h'(s)ds \right|$$

$$\leq D_0 \sum_{x \in \mathbb{Z}^d} \int_0^t e^{2(M-\varepsilon)s} Lip(f)Lip(g)\,(1 \wedge \psi(s,d_1))\,(1 \wedge \psi(s,d_2))\,ds$$

$$\leq D_0 Lip(f)Lip(g)|I| \max_{y \in I} \sum_{x \in \mathbb{Z}^d} \int_0^\infty e^{2(M-\varepsilon)s}\,(1 \wedge \psi(s,m))\,(1 \wedge \psi(s,d_2))\,ds \quad (5.23)$$

where $m = \rho(x,y)$ and the last inequality is due to the fact that $\rho(x,I) = \min_{i \in I} \rho(x,i)$, $x \in \mathbb{Z}^d$, $I \subset \mathbb{Z}^d$. Let us denote the integral over $\mathbb{R}_+$ in (5.23) by $I(y,x)$ and consider two cases separately.

Case 1: $\rho(x, y) > r$. Set $b_m(\alpha) = C_3 m^2 \exp\{-\alpha z_m(\operatorname{Log} m - \operatorname{Log}\operatorname{Log} m)\}$ where $z_m = m/\operatorname{Log} m$. Then

$$I(y, x) \le \int_0^\infty e^{2(M-\varepsilon)s}\left(1 \wedge (s \vee s^{\alpha z_m + 1})\right) b_m(\alpha)ds$$

$$\le b_m(\alpha) \int_1^{z_m/e} s^{\alpha z_m + 1} ds + C_3 e^{-C_4 m} \int_0^1 s\,ds + \int_{z_m/e}^\infty e^{2(M-\varepsilon)s} ds$$

$$\le C_3 e^{-C_4 m} + m^2 b_m(\alpha)\exp\{\alpha z_m \operatorname{Log}(z_m/e)\} + \frac{e^{2(M-\varepsilon)z_m/e}}{2(\varepsilon - M)} \le C_5 e^{-C_6 z_m}.$$

Case 2: $\rho(x, y) \le r$. We repeat the argument for Case 1 independently twice: the first time it is analogous to Case 1, whereas the next time one uses the second minimum in (5.23) instead of the first one, and the estimate $\rho(x, J) \ge r - \rho(x, y)$ which was employed in (5.12). Then we have

$$I(y, x) \le C_5 \exp\left\{-C_6\left(\frac{\rho(x, y)}{\operatorname{Log}\rho(x, y)} \vee \frac{d_2}{\operatorname{Log} d_2}\right)\right\}$$

$$\le C_5 \exp\left\{-C_6\left(\frac{\rho(x, y)}{\operatorname{Log}\rho(x, y)} \vee \frac{r - \rho(x, y)}{\operatorname{Log}(r - \rho(x, y))}\right)\right\} \le C_5 \exp\left\{-C_6\left(\frac{r}{2\operatorname{Log}(r/2)}\right)\right\}.$$

Summation (in (5.23)) over $x \in \mathbb{Z}^d$ leads to (5.19). Theorem 5.21 is proved. $\square$

Remark 5.25. The conditions ensuring that the field $\xi = \{\xi_j, j \in \mathbb{Z}^d\}$ is associated are also simple, they are provided in the following theorem by Harris (see, e.g., [265, Ch. IV]). The spin system is called *attractive* if, for any $x \in \mathbb{Z}^d$, the function $c(x, \eta)$ is increasing in η when $\eta_x = 0$, and decreasing in η otherwise. Indeed, one has

Theorem 5.26. *Let the spin system be attractive, and let its initial distribution be associated. Then its distribution at time t is associated, for any $t > 0$.*

7°. Further examples. For (BL, ψ_2, θ)-dependence, we give two examples due to Doukhan and Lang. For each $s \in \mathbb{N}$ let $\{a^{(s)}_{j_1,\ldots,j_s}, s \in \mathbb{N}, j_1, \ldots, j_s \in \mathbb{Z}^d\}$ be an array of real numbers such that $\sum_{s=1}^\infty \sum_{j_1,\ldots,j_s \in \mathbb{Z}^d} |a^{(s)}_{j_1,\ldots,j_s}| < \infty$. Let $\{\varepsilon_j, j \in \mathbb{Z}^d\}$ be a random field consisting of i.i.d. random variables with $\mathsf{E}\varepsilon_j = 0$, $\mathsf{E}\varepsilon_j^2 = 1$.

Definition 5.27. A *Chaotic Volterra field* $\{X_t, t \in \mathbb{Z}^d\}$ is defined by equations

$$X_t = \sum_{s=1}^\infty X_t^{(s)} \quad\text{where}\quad X_t^{(s)} = \sum_{j_1,\ldots,j_s \in \mathbb{Z}^d} a^{(s)}_{j_1,\ldots,j_s} \varepsilon_{t-j_1} \cdots \varepsilon_{t-j_s}, \quad t \in \mathbb{Z}^d,$$

assuming that the sum converges, e.g., in quadratic mean.

Theorem 5.28. ([139]) *A chaotic Volterra field is (BL, θ, ψ_2)-dependent with*

$$\theta_{2r} = 2\sum_{s=1}^\infty \sum_{v=1}^s \sum_{j_1,\ldots,j_s \in \mathbb{Z}^d, |j_v| \ge r} |a^{(s)}_{j_1,\ldots,j_s}|\,|\mathsf{E}|\varepsilon_0|^s, \quad r \ge 0,$$

provided that $\theta_0 < \infty$ and $\theta_r \to 0$, $r \to \infty$. Here it is assumed that $0 \cdot \infty = 0$ and $\theta_{2r+1} = \theta_{2r}$, $r \ge 0$. If the field is causal, that is, the indices j_i are taken from $\mathbb{N}^d$ instead of $\mathbb{Z}^d$, then the same equality holds for θ_r instead of θ_{2r}, $r \in \mathbb{N}$.

Proof. For $m \in \mathbb{N}$, let

$$Y_t^{(m)} = \sum_{s=1}^{\infty} \sum_{-m \le |j_1| \le m} \cdots \sum_{-m \le |j_s| \le m} a_{j_1,\ldots,j_s}^{(s)} \varepsilon_{t-j_1} \cdots \varepsilon_{t-j_s}, \quad t \in \mathbb{Z}^d.$$

Take disjoint finite sets $I, J \subset \mathbb{Z}^d$ with $dist(I, J) \ge 2r$. Let $f \in BL(|I|)$ and $g \in BL(|J|)$. One has

$$|cov(f(X_I), g(X_J))| \le |cov(f(X_I) - f(Y_I^{(r-1)}), g(X_J))|$$

$$+ |cov(f(Y_I^{(r-1)}), g(X_J) - g(Y_J^{(r-1)}))| + |cov(f(Y_I^{(r-1)}), g(Y_J^{(r-1)}))| \quad (5.24)$$

where $Y_I^{(m)} = (Y_i^{(m)}, i \in I)$ for $I \subset \mathbb{Z}^d$ and $m \in \mathbb{N}$. The last covariance in the right-hand side of (5.24) is zero, since the field $\{Y_t^{(m)}, t \in \mathbb{Z}^d\}$ is m-dependent. To estimate the first one, write

$$|cov(f(X_I) - f(Y_I^{(r-1)}), g(X_J))| \le 2\|g\|_{\infty} Lip(f)|I| \max_{i \in I} \mathsf{E}|X_i - Y_i^{(r-1)}|$$

$$\le 2\|g\|_{\infty} Lip(f)|I| \sum_{s=1}^{\infty} \sum_{v=1}^{s} \sum_{j_1,\ldots,j_s \in \mathbb{Z}^d, |j_v| \ge r} |a_{j_1,\ldots,j_s}^{(s)}| \mathsf{E}|\varepsilon_{t-j_1} \cdots \varepsilon_{t-j_s}|.$$

The second covariance is estimated analogously.

To prove the assertion related to causal fields, note that in that case one can repeat the argument above with the condition $dist(I, J) \ge 2r$ replaced by $dist(I, J) \ge r$. $\square$

Let $a > 0$, $\{b_j, j \in T\}$ be an array of real numbers, here $T = \mathbb{Z}_+^d \setminus \{0\}$, and let $\{\varepsilon_j, j \in \mathbb{Z}^d\}$ be a random field consisting of i.i.d. nonnegative random variables with $\mathsf{E}\varepsilon_j^2 = 1$. Consider a recursively defined random field

$$X_t = \left(a + \sum_{j \in T} b_j X_{t-j}\right) \varepsilon_t, \quad t \in T. \quad (5.25)$$

Suppose that $c := \mathsf{E}\varepsilon_0 \sum_{j \in T} b_j < 1$. Then there exists a stationary solution of (5.25) given by representation

$$X_t = a\varepsilon_t + a \sum_{s=1}^{\infty} \sum_{j_1,\ldots,j_s \in T} b_{j_1} \ldots b_{j_s} \varepsilon_{t-j_1} \cdots \varepsilon_{t-j_1-\ldots-j_s},$$

understood in mean-square convergence sense. Note that if all $b_j, j \in T$, are nonnegative, then $X = \{X_t, t \in T\}$ is (by Theorem 1.8, (c) and (d)) an associated random field, since all ε_t are independent and the product of nonnegative random variables is a nondecreasing function in them.

Now introduce

$$\rho(m) := \sum_{j \notin [0,m]^d} |b_j|, \quad m \in \mathbb{Z}_+.$$

Theorem 5.29. ([139]) *If $\rho(m) = 0$ for $m > p$, $p \in \mathbb{N}$, then the random field X defined by (5.25) is (BL, ψ_2, θ)-dependent with $\theta_r = c^{r/p}/(1 - c)$. If $\rho(m) = O(e^{-\lambda m})$ for some $\lambda > 0$, then X is (BL, ψ_2, θ)-dependent with $\theta_r = O(\exp(-\sqrt{-r\lambda \log c}))$, $r \to \infty$. If $\rho(m) = O(m^{-\lambda})$ for some $\lambda > 0$, then X is (BL, ψ_2, θ)-dependent with $\theta_r = O((\log r/r)^{\lambda})$, $r \to \infty$.*

Proof. Analogously to the proof of Theorem 5.28, for $L, m \in \mathbb{Z}_+$, consider the fields

$$X_t^{L,m} = a\varepsilon_t + a\sum_{s=1}^{L} \sum_{|j_1|\leq m,\ldots,|j_s|\leq m} b_{j_1}\ldots b_{j_s}\varepsilon_{t-j_1}\ldots\varepsilon_{t-j_1-\ldots-j_s}.$$

We have

$$|cov(f(X_I), g(X_J))| \leq |cov(f(X_I) - f(X_I^{L,m}), g(X_J))|$$

$$+|cov(f(X_I^{L,m}), g(X_J) - g(X_J^{L,m}))| + |cov(f(X_I^{L,m}), g(X_J^{L,m}))|. \qquad (5.26)$$

If $Lm < r$, then the last term in (5.26) is zero. For the first covariance we have

$$|cov(f(X_I) - f(X_I^{L,m}), g(X_J))| \leq 2\|g\|_\infty Lip(f)|I| \max_{i\in I} \mathsf{E}|X_i - X_i^{L,m}|$$

$$\leq 2\|g\|_\infty Lip(f)|I| \left(c^L(1-c)^{-1} + \rho(m)\right).$$

Similar estimate holds for the second covariance. Thus, one can take

$$\theta_r = \min_{L,m:Lm<r} \left(c^L(1-c)^{-1} + \rho(m)\right).$$

The optimization in L and m yields the result. $\square$

8°. Final remarks. We have made an overview of main types of covariance inequalities. It could be seen that there are many ways of describing systems dependent in such sense and different ideas are applied to different layouts. We have omitted a lot of notions which are adapted to some particular situations. For instance, in statistics and reliability one uses frequently the notions of *positively/negatively orthant dependent* (POD/NOD) random vectors.

Definition 5.30. A random vector $X = (X_1, \ldots, X_n)$ is POD if (1.10) holds, X is NOD if instead of (1.10) the inequality with opposite sign is satisfied.

This concept (and similar ones) is essentially due to Lehmann [256] for $n = 2$ and to papers [2, 9, 153] in general case. Note that POD is weaker than positive association (when $n > 2$) as is shown by

Example 5.31. ([160]) Let $X = (X_1, X_2, X_3)$ be a random vector such that

$$\mathsf{P}(X_1 = i, X_2 = j, X_3 = k) = p_{ijk} \quad \text{for} \quad i, j, k \in \{0, 1\}$$

and $p_{000} = 2/5$, $p_{001} = p_{010} = p_{100} = 0$, $p_{011} = p_{101} = p_{110} = 1/6$, $p_{111} = 1/10$. Then X is POD, but not **PA**, since $cov(X_1^+ X_2^+, X_3) < 0$.

It is necessary to mention also the notion of *linearly positively quadrant dependent* (LPQD) random variables ([308]). A family of random variables $\{X_t, t \in T\}$ is called LPQD if, for any finite disjoint sets $I, J \subset T$ and all nonnegative numbers $r_i, i \in I \cup J$, the random variables $\sum_{i\in I} r_i X_i$ and $\sum_{j\in J} r_j X_j$ are POD.

Return now to the percolation model (see Section 3). If A and B are increasing events then $\mathbb{P}_p(A \cap B) \geq \mathbb{P}_p(A)\mathbb{P}_p(B)$. In many problems it would be desirable to have an upper bound for $\mathbb{P}_p(A \cap B)$. Obviously it is impossible, in general, to reverse the sign in the above inequality. It turns out that one can specify the events to obtain the upper bound for their disjoint occurrence.

An event A is said to *occur* on the set S in the configuration ω if A occurs exercising only bonds in S and is independent of the values of the bonds in S^c. One denotes the collection of all such ω by $A|_S = \{\omega \in A : \widetilde{\omega}|_S = \omega|_S \Rightarrow \widetilde{\omega} \in A\}$, here, as usual, $\omega|_S$ is a restriction of a function ω to S. The events A_1, A_2 are said to *occur disjointly* (write $A_1 \circ A_2$) if there are two disjoint sets on which they occur:

$$A_1 \circ A_2 = \{\omega \in \Omega : \exists S_1, S_2 \subset \mathbb{B}_d, \, S_1 \cap S_2 = \varnothing, \omega \in A_1|_{S_1} \cap A_2|_{S_2}\}.$$

The following inequality is due to van den Berg, Kesten and Reimer.

Theorem 5.32. *For all* $A, B \in \mathcal{F}$ *depending on a finite number of bonds*

$$\mathbb{P}_p(A \circ B) \leq \mathbb{P}_p(A)\mathbb{P}_p(B). \tag{5.27}$$

Note that if A is increasing and B is decreasing, then $A \circ B = A \cap B$. Therefore in this case BKR-inequality (5.27) reduces to the FKG-inequality. The dependence of A and B on finite number of bonds can be relaxed in many examples of interest (see, e.g., [184]).

This inequality (5.27) was proved in 1985 by van den Berg and Kesten [33] for the case in which both A and B are increasing (or both decreasing) events. Then there were attempts to expand the validity of (5.27) to a larger class of events [31]. A general proof by Reimer [97] in 1995 confirmed the belief that inequality (5.27) holds in general situation. One should also refer to the important contributions by van den Berg [30], Talagrand [397] and Fishburn–Shepp [169].

Note that the FKG inequalities imply not only that the random variables X_i are associated, but also that their conditional distribution, given some part of them fixed, is also associated. In general the converse does not hold: an associated system can lose this property after conditioning over one element [170]. In this regard Liggett [266] introduces the idea of conditional association (see also [32]) and so-called downward FKG property (that is conditioning over sites where $X_i = 0$ is allowed), which is strictly between regular FKG and association.

In statistical physics one also invokes the *Griffiths inequalities* (see, e.g., [180, 183, 250]). In simple Ising situation the Griffiths inequalities

$$\mathsf{E}\sigma_V\sigma_W \geq \mathsf{E}\sigma_V\mathsf{E}\sigma_W \quad \text{for finite } V, W \subset \mathbb{Z}^d$$

are to hold whenever $J_{ij} \geq 0$ and $\mu_i \geq 0$ (and not when signs of different σ_i are different, even in the independent case). These inequalities can be also extended onto correlations (*Ursell functions*) of order higher than 2 (see, e.g., [359, §4.4]).

One can say that notions of standard, positive and negative association are general enough to admit a wide class of various random systems obeying them, but still strong enough to provide a large amount of limit theorems under plausible assumptions.

Chapter 2

Moment and Maximal Inequalities

In Chapter 2 we provide various inequalities for moments and maxima of partial sums (including multiindexed summands) for associated random systems and their modifications. Such estimates, known as moment and maximal inequalities, attract much attention in Probability Theory even in the case of independent random variables. They build important tools in studying the asymptotic behavior of random fields under consideration. For instance, we apply them in Chapter 5 to analysis of approximation of random fields by multiparameter Wiener process (both in distribution and almost surely) and in Chapter 8 to study fluctuations of the transformed solution of the multidimensional Burgers equation with random data. Recall that already for sequences of independent (or dependent in a sense) random variables there are deep results concerning the maximal inequalities for partial sums, such as the Kolmogorov, Doob, Burkholder – Davis – Gundy theorems etc. In this regard one can refer, e.g., to the books by Chow and Teicher ([101]), Petrov ([326]), Shiryaev ([383]), Khoshnevisan ([231]). However, for random fields the additional difficulties appear concerning the spatial configuration of index set of summands. Here the Móricz theorem is useful enabling us to prove the moment inequalities for partial sums instead of maximal ones. For associated random fields and partial sums we establish the sharp bound for their absolute moments of order $s > 2$ employing an analog of the classical Cox–Grimmett coefficient. We provide also recent results by the authors, for more general classes of weakly (so called (BL, θ)-)dependent random fields, obtained with the help of different approaches (the bisection method and the minimal graph techniques). A compact but important part of the discourse is based on supermodular functions, studied recently by Christofides and Vaggelatou in connection with positively and negatively associated random fields. In particular, it permits to deduce the Bernstein-type inequality for negatively associated random variables. Then we give a collection of Rosenthal-type inequalities established by Vronski, Bakhtin and Shashkin. The fruitful randomization idea introduced by Zhang and Wen (in negative association setup) is also employed, allowing to omit the assumption of existence of extra moments, which is usually difficult to avoid even in associated case. The last Section is devoted to inequalities involving not moments but distribution tails of (properly normalized) partial sums (and their maxima). Here we present a group of results by Newman, Wright, Bulinski and Keane. We also place there the inequality by Bakhtin for continuous-time random fields which is applied in the last Chapter to random PDE solutions.

1 Bounds for partial sums in the L^p space

1°. The Móricz Theorem. We assume that X is associated or satisfies other dependence conditions discussed in Chapter 1. Clearly it is also natural to impose some regularity conditions on moments of summands.

Let $\mathcal{U}$ be the class of blocks in $\mathbb{Z}^d$, that is, of the sets

$$U = (a, b] \cap \mathbb{Z}^d = ((a_1, b_1] \times \ldots \times (a_d, b_d]) \cap \mathbb{Z}^d, \ a_i < b_i, \ i = 1, \ldots, d.$$

We will say that a block W *belongs standardly* to a block U and denote this by $W \lhd U$ whenever $W \subset U$ and the minimal vertices of W and U (in the sense of the lexicographic order "$\prec$") coincide. Recall that $x = (x_1, \ldots, x_d) \prec y = (y_1, \ldots, y_d)$ in $\mathbb{R}^d$, $d > 1$, if either $x_1 < y_1$ or there exists $i \in \{1, \ldots, d-1\}$ such that $x_j = y_j$ for $j \leq i$ and $x_{i+1} < y_{i+1}$. For a random field $X = \{X_j, j \in \mathbb{Z}^d\}$ and $U \in \mathcal{U}$ set

$$S(U) = \sum_{j \in U} X_j, \quad M(U) = \sup_{W \lhd U} |S(W)|. \tag{1.1}$$

In the first Section we consider the problem of estimating the expectation $\mathsf{E}|S(U)|^p$ for $p > 2$ in terms of the cardinality $|U|$. The importance of such inequalities will be demonstrated in further Chapters. We start with Móricz version of the Erdös–Stechkin inequality for partial maxima. This is in fact an analytic result, which does not involve explicitly any dependence properties.

Definition 1.1. A function $\varphi : \mathcal{U} \to \mathbb{R}_+$ is called *superadditive*, if, for any blocks $U, U_1, U_2 \in \mathcal{U}$ such that $U = U_1 \cup U_2$ and $U_1 \cap U_2 = \varnothing$, one has

$$\varphi(U) \geq \varphi(U_1) + \varphi(U_2).$$

For example, a function $\varphi(U) = |U|^\tau$ is superadditive if $\tau \geq 1$.

Theorem 1.2. ([304]) *Let $\{X_j, j \in \mathbb{Z}^d\}$ be a random field such that for some $\gamma \geq 1, \alpha > 1$ and some superadditive function $\varphi : \mathcal{U} \to \mathbb{R}_+$ the estimate*

$$\mathsf{E}|S(U)|^\gamma \leq \varphi(U)^\alpha \tag{1.2}$$

holds for all $U \in \mathcal{U}$. Then, for any $U \in \mathcal{U}$,

$$\mathsf{E}M(U)^\gamma \leq C_0(d, \alpha, \gamma)\varphi(U)^\alpha, \quad C_0(d, \alpha, \gamma) = (5/2)^d(1 - 2^{(1-\alpha)/\gamma})^{-d\gamma}.$$

Proof is given in Appendix A.5.

Remark 1.3. Clearly, if $W, U \in \mathcal{U}$ and $W \subset U$, one can construct W by unions and differences of at most 2^d blocks standardly belonging to U. So, if the conditions of Theorem 1.2 are met, one has, for all $U \in \mathcal{U}$,

$$\mathsf{E} \sup_{W \subset U, \, W \in \mathcal{U}} |S(U)|^\gamma \leq C_1(d, \alpha, \gamma)\varphi(U)^\alpha$$

where $C_1(d, \alpha, \gamma) = 2^{d\gamma}C_0(d, \alpha, \gamma)$.

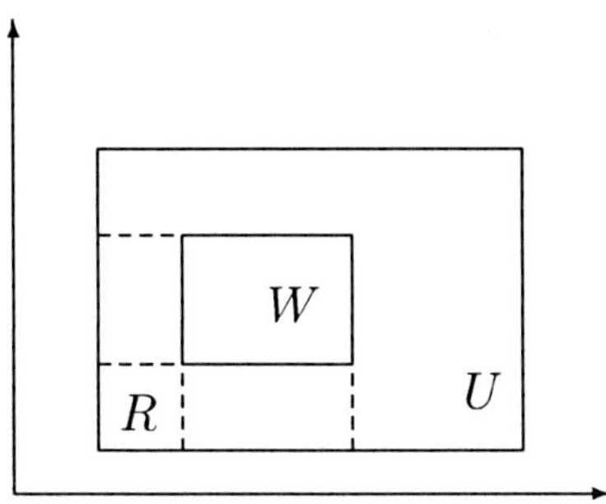

Fig. 2.1 $(d = 2,\ R \lhd U)$

Note that the inequality $\mathsf{E}|S(U)|^a \leq C|U|^{a/2}$, with $a > 2$ and C independent of U, ensures uniform boundedness of $\mathsf{E}M(U)^a/|U|^{a/2}, U \in \mathcal{U}$, since $\varphi(U) = |U|$ is a superadditive function. It is well-known (see, e.g., [326, Ch. II, §5]) that if $\{X_j, j \in \mathbb{Z}^d\}$ is a collection of independent centered random variables such that

$$\sup_{j \in \mathbb{Z}^d} \mathsf{E}|X_j|^r < \infty \quad \text{for some } r > 1, \tag{1.3}$$

then, for any finite set $U \subset \mathbb{Z}^d$,

$$\mathsf{E}|S(U)|^r \leq C|U|^{r/2} \tag{1.4}$$

where $S(U) = \sum_{j \in U} X_j$ and $C > 0$ does not depend on U. Moreover, the estimate (1.4) is sharp, that is, in general one can not take the exponent less than $r/2$ in the right-hand side of (1.4). Thus, in the case of independent summands satisfying (1.3) inequality (1.2) holds with $\gamma = r$ and $\alpha = r/2$.

For dependent random fields the exponent $r/2$ in the right-hand side of (1.4) need not be attainable. Besides, to have a "good" estimate, when $\mathsf{E}|S(U)|^r$ behaves as in the independent case, it is usually impossible to keep the same exponent r in the estimate (1.4) as the moment assumption (1.3) provides, but one has to decrease it slightly. Further we will explain this phenomenon.

$2°$. **Moment bounds for associated random fields.** As was mentioned in the previous Chapter, in describing the properties of an associated (**PA, NA**) field a prominent role is played by the Cox–Grimmett coefficient. It will appear in almost all limit theorems in this book, but now it will be convenient to introduce a function of more general character, still responsible for dependence. For $\tau > 0$, $n \in \mathbb{N}$ set

$$\Lambda_\tau(n) = \sup\left\{ \sum_{i \in U} \left(\sum_{j \notin U} |cov(X_i, X_j)| \right)^\tau : U \in \mathcal{U},\ |U| = n \right\}.$$

Given a block $U \in \mathcal{U}$, $r > 1$ and $n \in \mathbb{N}$, let

$$a_r(U) = \mathsf{E}\left| \sum_{j \in U} X_j \right|^r, \quad A_r(n) = \sup\left\{ a_r(U) : U \in \mathcal{U},\ |U| = n \right\}.$$

We will also often use the truncation functions

$$H_M(t) = (|t| \wedge M)sgn(t),\quad t \in \mathbb{R},\quad M > 0. \tag{1.5}$$

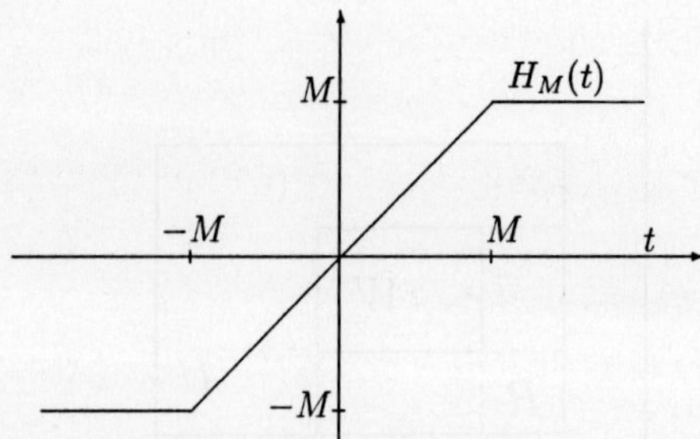

Now we are able to formulate the statement on the precise bound for absolute moment of partial sums of an associated (or more general **PA**) field.

Theorem 1.4. ([64]) *Let* $X = \{X_j, j \in \mathbb{N}^d\}$ *be a centered* **PA** *random field such that, for some* $r > 2, \delta > 0$ *and* $\mu \geq 0$, *one has*

$$A_{r+\delta}(1) = \sup_{j \in \mathbb{Z}^d} \mathsf{E}|X_j|^{r+\delta} < \infty,$$

$$\Lambda_{\delta/\varkappa}(n) = O(n^\mu) \quad \text{as} \quad n \to \infty, \tag{1.6}$$

here $\varkappa = \delta + (r + \delta)(r - 2)$. *Then* $A_r(n) = O(n^\tau)$ *where one can take*

$$\tau(r, \delta, \mu) = \begin{cases} r/2, & 0 \leq \mu < (1 + \delta/\varkappa)/2, \\ \varkappa(\mu \wedge 1)/(r + \delta - 2), & \text{otherwise.} \end{cases}$$

Proof. Let $U = V \cap \mathbb{Z}^d$ where $V = (a, b] \subset \mathbb{R}^d$ and $a, b \in \mathbb{Z}^d, a \leq b$. As everywhere, we denote the lengths of edges of a parallelepiped V by $l_1(V), \ldots, l_d(V)$. We write also $l_i(U)$ instead of $l_i(V)$, $i = 1, \ldots, d$. Set

$$\mathcal{U}' = \{U \in \mathcal{U} : l_k(U) = 2^{q_k}, \; q_k \in \mathbb{Z}_+, \; k = 1, \ldots, d\}, \quad l_0(U) = \max_k l_k(U).$$

For a block $U = V \cap \mathbb{Z}^d \in \mathcal{U}'$, $|U| > 1$, we introduce congruent blocks v, v', obtained from V by drawing a hyperplane orthogonal to some edge having length $l_0(U)$. Write

$$W_1 = \sum_{j \in v} X_j, \quad W_2 = \sum_{j \in v'} X_j. \tag{1.7}$$

By $b_r(v)$ we denote the supremum of $a_r(v')$ over all $v' \in \mathcal{U}'$ which are congruent to v. Passing to a series of auxiliary results, set as usual $f^+ = f\mathbb{I}\{f \geq 0\}$ and $f^- = -f\mathbb{I}\{f < 0\}$ for real-valued function f. So $f = f^+ - f^-$.

Lemma 1.5. *Let* (X, Y) *be a* **PA** *random vector such that* $cov(X, Y)$ *exists. Then*

$$0 \leq cov(X^+, Y) \leq cov(X, Y), \tag{1.8}$$

$$-cov(X, Y) \leq cov(X^-, Y) \leq 0. \tag{1.9}$$

If, moreover, $\mathsf{E}|X|^{r+\delta} < \infty$, $Y \geq 0$ and $\mathsf{E}Y^r < \infty$ where r, δ and $\varkappa$ are the same as in Theorem 1.4, then

$$0 \leq cov(X, Y^{r-1}) \leq (r+1)(\mathsf{E}|X|^{r+\delta})^{(r-2)/\varkappa}(\mathsf{E}Y^r)^{(r-2)(r+\delta-1)/\varkappa}(cov(X,Y))^{\delta/\varkappa}.$$

(1.10)

If, in addition to all that, $|X| \leq R < \infty$, then

$$cov(X, Y^{r-1}) \leq (r-1+2R)(\mathsf{E}Y^r)^{(r-2)/(r-1)}(cov(X,Y))^{1/(r-1)}.$$

(1.11)

Proof. The claim (1.8) is due to the fact that for $f(x) = x$ the functions x^+ and $x - x^+$ are nondecreasing; (1.9) follows analogously. As to (1.10), one has $cov(X, Y^{r-1}) \geq 0$ by association. To prove the second inequality in (1.10) write $Y^{r-1} = Z + T$, where $Z = H_M(Y^{r-1})$ for some $M > 0$ to be specified later. Note that for a function f, defined by $f(x) = H_M(x^{r-1})$ if $x \geq 0$ and $f(x) = 0$ if $x < 0$, one has $Lip(f) \leq (r-1)M^{r-2}$. Obviously $|T| \leq Y^{r-1}\mathbb{I}\{Y > M\}$. Next we observe that

$$cov(X, Y^{r-1}) = cov(X, Z) + cov(X, T) \leq (r-1)M^{r-2}cov(X,Y)$$

$$+ 2(\mathsf{E}|X|^{r+\delta})^{1/(r+\delta)}(\mathsf{E}Y^{(r+\delta)/(r+\delta-1)}\mathbb{I}\{Y > M\})^{(r+\delta-1)/(r+\delta)}$$

(1.12)

by Theorem 1.5.3, applied to the term $cov(X, Z)$, the Hölder inequality and the trivial estimate

$$\mathsf{E}Y^\gamma\mathbb{I}\{Y > M\} \leq M^{\gamma-r}\mathsf{E}Y^r \quad \text{for } 0 < \gamma < r, \ M > 0.$$

If $cov(X, Y) = 0$ then X and Y are independent. Hence X and Y^{r-1} are also independent and $cov(X, Y^{r-1}) = 0$. If $cov(X, Y) \neq 0$, take $M = (cov(X,Y))^\alpha(\mathsf{E}Y^r)^\beta(\mathsf{E}|X|^{r+\delta})^\nu$, then optimization of parameters α, β and ν yields (1.10). We again use (1.12) to prove (1.11). Namely,

$$cov(X, Y^{r-1}) \leq (r-1)M^{r-2}cov(X,Y) + 2RM^{-1}\mathsf{E}Y^r.$$

If $cov(X, Y) = 0$ then (1.11) holds. Otherwise take $M = (cov(X,Y))^\alpha(\mathsf{E}Y^r)^\beta$, then the appropriate choice of α and β implies (1.11). $\square$

Lemma 1.6. *Suppose that $X = \{X_j, j \in \mathbb{Z}^d\}$ is a **PA** random field with $A_r(1) < \infty$ and $r = l + \rho$ where $0 \leq \rho < 1$ and $l \in \mathbb{N}$, $l > 1$. Then*

$$\mathsf{E}|W_1 + W_2|^r \leq 2b_r(v) + 2^l b_r(v)^{1-\rho}\left((\mathsf{E}|W_1||W_2|^{r-1})^\rho + (\mathsf{E}|W_2||W_1|^{r-1})^\rho\right),$$

with W_1 and W_2 defined in (1.7).

Proof. We have

$$\mathsf{E}|W_1 + W_2|^r = \mathsf{E}|W_1 + W_2|^l|W_1 + W_2|^\rho \leq \mathsf{E}|W_1 + W_2|^l(|W_1|^\rho + |W_2|^\rho)$$

as $(x + y)^\rho \leq x^\rho + y^\rho$ for any $x, y \geq 0$ and $0 \leq \rho < 1$. Thus

$$\mathsf{E}|W_1 + W_2|^r \leq \sum_{m=0}^{l} \binom{l}{m}\left(\mathsf{E}|W_1|^{l-m+\rho}|W_2|^m + \mathsf{E}|W_1|^m|W_2|^{l-m+\rho}\right)$$

$$\leq 2b_r(v) + 2^l(\mathsf{E}|W_1|^\rho|W_2|^l + \mathsf{E}|W_1|^l|W_2|^\rho),$$

since $x^{l-m+\rho}y^m + x^m y^{l-m+\rho} \leq x^\rho y^l + x^l y^\rho$ for any $x, y \geq 0$, ρ, l under consideration and $m = 1, \ldots, l$ (clearly $x^\rho y^m(x^{l-m} - y^{l-m}) \leq y^\rho x^m(x^{l-m} - y^{l-m})$ in both cases $x \leq y$ and $x > y$). If $\rho = 0$, the Lemma is proved. Otherwise, applying the Hölder inequality

$$\mathsf{E}|W_1|^\rho|W_2|^l = \mathsf{E}|W_1|^\rho|W_2|^{\rho(r-1)}|W_2|^{l-\rho(r-1)}$$

$$\leq (\mathsf{E}|W_1||W_2|^{r-1})^\rho(\mathsf{E}|W_2|^{(l-\rho(r-1))/(1-\rho)})^{1-\rho} \leq b_r(v)^{1-\rho}(\mathsf{E}|W_1||W_2|^{r-1})^\rho$$

and estimating similarly another summand, one establishes the Lemma. $\square$

Lemma 1.7. *Suppose that* $A_r(1) < \infty$, $r > 2$, *and* $\Lambda_1(1) < \infty$. *Then*

$$\mathsf{E}|W_1||W_2|^{r-1} \leq 2^{r-2}\left(cov(W_1^+, (W_2^+)^{r-1}) + cov(W_1^-, (W_2^-)^{r-1})\right)$$

$$+ 2^r(\Lambda_1(1) + A_r(1)^{2/r})^{1/2}|v|^{1/2}b_r(v)^{(r-1)/r}.$$

Proof. Clearly $|W_i| = W_i^+ + W_i^-$, and

$$|W_i|^{r-1} = (W_i^+ + W_i^-)^{r-1} \leq 2^{r-2}\left((W_i^+)^{r-1} + (W_i^-)^{r-1}\right)$$

for $i = 1, 2$. Therefore,

$$\mathsf{E}|W_1||W_2|^{r-1} \leq 2^{r-2}\Big(cov(W_1^+, (W_2^+)^{r-1}) + cov(W_1^-, (W_2^-)^{r-1})$$

$$+ cov(W_1^+, (W_2^-)^{r-1}) + cov(W_1^-, (W_2^+)^{r-1}) + 4\mathsf{E}|W_1|\mathsf{E}|W_2|^{r-1}\Big). \tag{1.13}$$

Due to positive association $cov(W_1^+, (W_2^-)^{r-1}) \leq 0$ and $cov(W_1^-, (W_2^+)^{r-1}) \leq 0$. Furthermore, by the Lyapunov inequality, $\mathsf{E}X_j^2 \leq (\mathsf{E}|X_j|^r)^{2/r}$, $j \in \mathbb{Z}^d$, and one has

$$\mathsf{E}W_1^2 \leq (\Lambda_1(1) + A_r(1)^{2/r})|v|.$$

Thus, in view of the Hölder inequality, the last summand in (1.13) can be estimated as follows:

$$\mathsf{E}|W_1|\mathsf{E}|W_2|^{r-1} \leq (\Lambda_1(1) + A_r(1)^{2/r})^{1/2}|v|^{1/2}b_r(v)^{(r-1)/r}.$$

The Lemma is proved. $\square$

Lemma 1.8. *Suppose that* $A_{r+\delta}(1) < \infty$ *for some* $r > 2, \delta > 0$ *and let* $\Lambda_1(1) < \infty$. *Then*

$$cov(W_1^+, (W_2^+)^{r-1}) \leq (r+1)A_{r+\delta}(1)^{(r-1)/\varkappa}b_r(v)^{(r-2)(r+\delta-1)/\varkappa}\Lambda_{\delta/\varkappa}(v, v')$$

where $\Lambda_\gamma(U_1, U_2) := \sum_{i \in U_1}\left(\sum_{j \in U_2} cov(X_i, X_j)\right)^\gamma$, $U_1, U_2 \subset \mathbb{Z}^d$, $\gamma > 0$. *The same estimate holds for* $cov(W_1^-, (W_2^-)^{r-1})$.

Proof. By (1.8) we have

$$cov(W_1^+, (W_2^+)^{r-1}) \le cov(W_1, (W_2^+)^{r-1}) = \sum_{i \in v} cov\left(X_i, (W_2^+)^{r-1}\right).$$

Inequality (1.10) yields for $cov(X_i, (W_2^+)^{r-1})$ the following upper bound

$$(r+1)(\mathsf{E}|X_i|^{r+\delta})^{(r-2)/\varkappa}(\mathsf{E}|W_2^+|^r)^{(r-2)(r+\delta-1)/\varkappa}(cov(X_i, W_2^+))^{\delta/\varkappa}. \tag{1.14}$$

Note that $\mathsf{E}|W_2^+|^r \le b_r(v)$. Again by Lemma 1.5,

$$cov(X_i, W_2^+) \le cov(X_i, W_2) = \sum_{j \in v'} cov(X_i, X_j).$$

Combining these estimates leads to the desired bound for $cov(W_1^+, (W_2^+)^{r-1})$. Note that W_1 and $-(W_2^-)^{r-1}$ are **PA**. Therefore (1.9) implies

$$0 \ge cov(W_1^-, -(W_2^-)^{r-1}) = -cov(-W_1^-, -(W_2^-)^{r-1})$$

$$\ge -cov(W_1, -(W_2^-)^{r-1}) = cov(W_1, (W_2^-)^{r-1}).$$

Thus,

$$cov(W_1^-, (W_2^-)^{r-1}) \le \sum_{i \in v} cov(-X_i, (W_2^-)^{r-1}).$$

From (1.10) we infer an analogue of (1.14), and to complete the proof take into account that $cov(-X_i, W_2^-) \le cov(X_i, W_2)$ according to (1.9). $\square$

Lemma 1.9. *Suppose that* $\Lambda_1(1) < \infty$. *Then* $\Lambda_\tau(1) = \Lambda_1(1)^\tau$. *If* (1.6) *holds, then*

$$\Lambda_{\delta/\varkappa}(n) = O(n^{\mu \wedge 1}), \quad n \to \infty.$$

Proof. The equality for $\Lambda_\tau(1)$ is obvious. To check the asymptotic estimate, observe that

$$\sup_{U \in \mathcal{U}, |U|=n} \Lambda_{\delta/\varkappa}(U, \mathbb{Z}^d \setminus U) \le n \sup_{i \in \mathbb{Z}^d} \left(\sum_{j \ne i} cov(X_i, X_j)\right)^{\delta/\varkappa} = O(n). \quad \square$$

Now we return to the Theorem. Due to Lemmas 1.6—1.9, for any block $U \in \mathcal{U}'$ with $l_0(U) \ge 2$, we have

$$b_r(U) \le 2b_r(v) + L\left(|v|^{\gamma_1} b_r(v)^{1-\rho_1} + |v|^{\gamma_2} b_r(v)^{1-\rho_2}\right) \tag{1.15}$$

where v is obtained by partition of U, $L > 0$ is some positive constant and

$$\gamma_1 = \rho/2, \quad \rho_1 = \rho/r, \quad \gamma_2 = \rho(\mu \wedge 1), \quad \rho_2 = \rho(r+\delta-2)/(r-2).$$

At first we consider the case when $|U| = 2^q, q \in \mathbb{Z}_+$. Then the estimate

$$A_r(2^q) \le C \cdot 2^{\tau q} \tag{1.16}$$

follows from (1.15) by induction on q, if one takes $\tau > 1, \tau \ge \max_i(\gamma_i/\rho_i)$ and

$$C = 1 \vee A_r(1) \vee \left(\frac{L}{2^{\tau-1} - 1}\right)^{1/\rho_0}$$

with $\rho_0 = \rho_1 \wedge \rho_2$. Namely, if $q = 0$, then $A_r(1) \le C$ obviously. To perform the step of induction, assume, for $q \in \mathbb{N}$, that (1.16) holds for all blocks $v \in \mathcal{U}'$ with $|v| = 2^{q-1}$. Pick $U \in \mathcal{U}'$ such that $|U| = 2^q$. Then (1.15) and induction hypothesis imply that

$$b_r(U) \le 2^{1+\tau(q-1)}C + L\left(2^{\gamma_1(q-1)}C^{(1-\rho_1)}2^{(1-\rho_1)(q-1)} + 2^{\gamma_2(q-1)}C^{(1-\rho_2)}2^{(1-\rho_2)(q-1)}\right)$$
$$\le 2^{1+\tau(q-1)}C + LC^{1-\rho_0}\left(2^{(\gamma_1+1-\rho_1)(q-1)} + 2^{(\gamma_2+1-\rho_2)(q-1)}\right) \le 2^{\tau q},$$

due to the choice of C.

To finish the proof it remains to consider the case of an arbitrary block U. Let $l_1(U), \ldots, l_d(U)$ be, as above, lengths of edges of U (not necessarily ordered) and $q_1, \ldots, q_d$ be such nonnegative integers that

$$2^{q_k} \le l_k(U) < 2^{q_k+1}, \quad k = 1, \ldots, d.$$

Then $l_k(U) = \sum_{j=0}^{q_k} \mu_{kj} 2^{q_k-j}$ where $\mu_{kj} \in \{0,1\}$, $j = 0, \ldots, q_k$, $k = 1, \ldots, d$. Now define the set

$$J = J(U) = \{j = (j_1, \ldots, j_d) \in \mathbb{Z}^d : 0 \le j_k \le q_k, \ k = 1, \ldots, d\}.$$

It follows immediately that there exist pairwise disjoint blocks $U_j, j \in J$, such that

$$U = \bigcup_{j \in J} U_j, \text{ and } l_k(U_j) = \mu_{k,j_k} 2^{q_j - j_k}, \quad j \in J, \ k = 1, \ldots, d$$

(we stipulate that a block with an edge of zero length is an empty set). These blocks U_j can be constructed by drawing hyperplanes orthogonal to edges of U (i.e. of $V = (a, b] \subset \mathbb{R}^d$ where $U = V \cap \mathbb{Z}^d$). The cardinality of each $U_j \cap \mathbb{Z}^d$ is either 0 or a nonnegative power of 2. Therefore, by (1.16) and Minkowski's inequality we have

$$\mathsf{E}|S(U)|^r = \mathsf{E}\left|\sum_{j \in J} S(U_j)\right|^r \le \left(\sum_{j \in J}\left(\mathsf{E}|S(U_j)|^r\right)^{1/r}\right)^r \le \left(C^{1/r}\sum_{j \in J}|U_j|^{\tau/r}\right)^r.$$

Thus Theorem will be proved if we show that

$$\left(\sum_{j \in J}|U_j|^{\tau/r}\right)^r \le 2^{d\tau}(2^{\tau/r} - 1)^{-dr}|U|^\tau. \tag{1.17}$$

To this end we note that $2^{j_1+\ldots+j_d}|U_j| \le |U|$ for any $j \in J$, or, equivalently, $|U_j|^{\tau/r} \le 2^{-(j_1+\ldots+j_d)\tau/r}|U|^{\tau/r}$. Therefore,

$$\sum_{j \in J}|U_j|^{\tau/r} \le |U|^{\tau/r}\sum_{j \in J}2^{-(j_1+\ldots+j_d)\tau/r} \le |U|^{\tau/r}\sum_{j_1,\ldots,j_d=0}^{\infty}2^{-(j_1+\ldots+j_d)\tau/r}$$

$$= |U|^{\tau/r}\left(\sum_{m=0}^{\infty}2^{-m\tau/r}\right)^d = |U|^{\tau/r}\left(1 - 2^{-\tau/r}\right)^{-d}.$$

Taking the r-th power of both sides we see that the last estimate implies (1.17), hence the Theorem. $\square$

Remark 1.10. Analyzing the proof of Theorem 1.4 one sees that the condition $\Lambda_{\delta/\varkappa}(n) = O(n^\mu)$ can be replaced by a weaker assumption that $\Lambda_1(1) < \infty$ and

$$\sup\left\{\sum_{i \in v}\left(\sum_{j \in v'} cov(X_i, X_j)\right)^{\delta/\varkappa} : U \in \mathcal{U}', |v| = n\right\} = O(n^\mu), \tag{1.18}$$

here the blocks v and v' are obtained from U as in the proof of Theorem 1.4.

3°. Bakhtin's lemma. If the random field X in Theorem 1.4 is associated, then the final step of the proof (after (1.16) has been established) can be simplified by invoking the following lemma which is of independent interest. We remind that a function $f : \mathbb{R} \to \mathbb{R}$ is called *convex* if for any $x, y \in \mathbb{R}$ and $\alpha \in [0,1]$

$$f(\alpha x + (1 - \alpha)y) \leq \alpha f(x) + (1 - \alpha)f(y).$$

Lemma 1.11. ([15]) *Let (X, Y) be an associated random vector with values in $\mathbb{R}^2$ such that $\mathsf{E}X = 0$. Suppose that $f : \mathbb{R} \to \mathbb{R}$ is a convex function. Then*

$$\mathsf{E}f(Y) \leq \mathsf{E}f(X + Y) \tag{1.19}$$

if both expectations exist.

First of all we consider the end of the proof of Theorem 1.4. Let $U \in \mathcal{U}$ be a block with edges $l_1(U), \ldots, l_d(U)$. There are $q_i \in \mathbb{Z}_+$ such that $2^{q_i} \leq l_i(U) < 2^{q_i+1}$, $i = 1, \ldots, d$. Take a block $\Pi \in \mathcal{U}'$ with edges 2^{q_i+1}, $i = 1, \ldots, d$, such that $U \subset \Pi$. Then $S(\Pi) = S(U) + S(\Pi \setminus U)$ where $X = S(U)$ and $Y = S(\Pi \setminus U)$ are associated. Since $f(x) = |x|^r$ is a convex function, by Lemma 1.11 and (1.16)

$$\mathsf{E}\left|\sum_{j \in U} X_j\right|^r \leq A_r(|\Pi|) \leq A_r(2^{q_1 + \cdots + q_d + d}) \leq C \cdot 2^{r(q_1 + \cdots + q_d + d)} \leq C \cdot 2^{rd}|U|^r$$

where C is the same as in (1.16).

Proof of Lemma 1.11. We start with the case when f is bounded from below, nondecreasing and has a continuous derivative. It can be achieved (by adding a constant) that $f \geq 0$. Assume also at first that Y is a bounded random variable. Define a function $g : \mathbb{R}^2 \to \mathbb{R}$ by relation

$$g(x, y) = \begin{cases} \frac{f(x+y) - f(y)}{x}, & x \neq 0, \\ f'(y), & x = 0. \end{cases} \tag{1.20}$$

Then g is continuous and coordinate-wise nondecreasing. Indeed, for any $x \neq 0$, it has partial derivatives

$$\frac{\partial g(x, y)}{\partial y} = \frac{f'(x + y) - f'(y)}{x} \geq 0,$$

$$\frac{\partial g(x, y)}{\partial x} = \frac{f(x + y) - f(y) - xf'(y)}{x^2} \geq 0,$$

f' being a nondecreasing function (because of convexity of f). Note that $\mathsf{E}|g(X, Y)| < \infty$. Really,

$$\mathsf{E}|g(X, Y)| = \mathsf{E}|g(X, Y)|\mathbb{I}\{|X| \leq 1\} + \mathsf{E}|g(X, Y)|\mathbb{I}\{|X| > 1\}$$

$$\leq \mathsf{E}\sup_{x \in [-1,1]} f'(Y + x) + \mathsf{E}|f(X + Y)| + \mathsf{E}|f(Y)| < \infty,$$

as Y is bounded, and thus $\sup_{x \in [-1,1]} f'(Y + x)$ is also bounded. By association and since $\mathsf{E}X = 0$,

$$cov(g(X, Y), X) = \mathsf{E}g(X, Y)X = \mathsf{E}(f(X + Y) - f(Y)) \geq 0.$$

Now we treat the case when there are the same restrictions on f (it is assumed again that f is nonnegative) but Y is not bounded. For $n \in \mathbb{N}$, let $Y_n = H_n(Y)$ where H_n are the functions defined in (1.5) with $M = n$. Then for any n one has $f(Y_n) \leq f(Y)$ and $f(X + Y_n) \leq f(X + Y)$. Thus, by the dominated convergence theorem

$$\mathsf{E}f(X + Y) = \lim_{n \to \infty} \mathsf{E}f(X + Y_n) \geq \lim_{n \to \infty} \mathsf{E}f(Y_n) = \mathsf{E}f(Y).$$

The next step of the proof is to dispense with the restriction that f is differentiable. So let f be a continuous, nondecreasing, convex function, bounded from below. Having taken a sequence $(\chi_n)_{n \in \mathbb{N}}$ of infinitely differentiable nonnegative functions such that $\chi_n(x) = 0$ for $x \notin [0, 1/n]$ and $\int_{\mathbb{R}} \chi_n(t)dt = 1$, we define the functions

$$f_n(x) = (f * \chi_n)(x) = \int_{\mathbb{R}} f(t)\chi_n(x - t)dt = \int_{\mathbb{R}} f(x - t)\chi_n(t)dt, \quad x \in \mathbb{R}, \quad n \in \mathbb{N}.$$

Then f_n are bounded from below, nondecreasing and continuously differentiable. They are also convex because, for any $x, y \in \mathbb{R}$ and $\alpha \in [0, 1]$,

$$f_n(\alpha x + (1 - \alpha)y) = \int_{\mathbb{R}} f(\alpha(x - t) + (1 - \alpha)(y - t))\chi_n(t)dt$$

$$\leq \int_{\mathbb{R}} (\alpha f(x - t) + (1 - \alpha)f(y - t))\chi_n(t)dt = \alpha f_n(x) + (1 - \alpha)f_n(y).$$

Thus $\mathsf{E}f_n(X + Y) \geq \mathsf{E}f_n(Y)$, $n \in \mathbb{N}$. Since $f_n(x) \searrow f(x)$ as $n \to \infty$ for any $x \in \mathbb{R}$, the inequality $\mathsf{E}f(X + Y) \geq \mathsf{E}f(Y)$ follows by monotone convergence theorem.

Now, if f is a continuous, convex, nondecreasing function, we can take the approximating functions $(f_n)_{n \in \mathbb{N}}$, say $f_n(x) = f(x) \vee (-n)$, which are convex and bounded from below. Thus, again by monotone convergence, the Lemma is also true for such f.

Finally, if f is a continuous convex function which is not nondecreasing, it can be written as $f = f^{(+)} + f^{(-)}$, where $f^{(+)}$ is a continuous convex nondecreasing function and $f^{(-)}$ is continuous, convex and nonincreasing. By the previous argument we have $\mathsf{E}f^{(+)}(X + Y) \geq \mathsf{E}f^{(+)}(Y)$. It remains to define $h : \mathbb{R} \to \mathbb{R}$ by the relation $h(t) = f^{(-)}(-t)$, $t \in \mathbb{R}$, then $\mathsf{E}h(-Y) \leq \mathsf{E}h(-X - Y)$, since $(-X, -Y)$ is an associated vector. Therefore, $\mathsf{E}f(X + Y) \geq \mathsf{E}f(Y)$. $\square$

4°. Employing the Cox–Grimmett coefficient. The most useful applications of Theorem 1.4 occur when expectations of partial sums grow as in independent case, that is, there exists $r > 2$ for which $A_r(n) = O(n^{r/2})$. As was mentioned above, working with associated random fields it is convenient to express dependence in terms of the Cox–Grimmett coefficient, since its evaluation only requires that the covariance function is known. Here and subsequently we will denote this coefficient by $u(n)$ or u_n, $n \in \mathbb{N}$.

Corollary 1.12. *Let $X = \{X_j, j \in \mathbb{N}^d\}$ be an associated or* **PA** *random field such that $A_p(1) < \infty$ for some $p \in (2, 3]$ and $u(n) = O(n^{-\nu})$ for some $\nu > 0, n \to \infty$. Then there exists $r \in (2, p]$ such that $A_r(n) = O(n^{r/2}), n \to \infty$.*

Proof. If $u(n) = O(n^{-\nu})$ and $\nu < \varkappa/\delta$, then the left-hand side of (1.18) allows the following estimate:

$$\Lambda_{\delta/\varkappa}(v, v') \leq \frac{|U|}{l_0(U)} \sum_{k=1}^{l_0(U)/2} u(k)^{\delta/\varkappa} = O\Big(|U|l_0(U)^{-\nu\delta/\varkappa}\Big) = O(|U|^{1-\nu\delta/d\varkappa})$$

as $|U|$ grows to infinity. We used here that obviously $l_0(U) \geq |U|^{1/d}$. According to Theorem 1.4, we will have the desired result if it is possible to choose r and δ in such a way that, for $\nu < \varkappa/\delta$, i.e. $\nu\delta < \delta + (r + \delta)(r - 2)$, one has

$$1 - \frac{\nu\delta}{d(\delta + (r + \delta)(r - 2))} < \frac{1}{2}\left(1 + \frac{\delta}{\delta + (r + \delta)(r - 2)}\right).$$

These inequalities are equivalent to

$$2\nu d^{-1}\delta > (r - 2)(r + \delta) \quad \text{and} \quad (\nu - 1)\delta < (r - 2)(r + \delta).$$

If $\nu \leq 1$, one can easily satisfy them taking r sufficiently close to 2 and $\delta = p - r$; otherwise, letting also $\delta = p - r$, one can take $r \in ((\nu + 1)p(p + \nu - 1)^{-1}, p)$. $\square$

5°. Optimality of the moment bounds obtained. The moment bound provided by Theorem 1.4 is optimal in view of the following two theorems.

Theorem 1.13. ([64]) *For any $r > 2, \delta > 0, \mu \geq 0$ and $d \in \mathbb{N}$ there exists a random field $X = \{X_j, j \in \mathbb{Z}^d\}$ which satisfies all the conditions of Theorem 1.4 and such that*

$$a_r(U_n) \geq c|U_n|^\tau$$

where $U_n := (0, n]^d$, $\tau = \tau(r, \delta, \mu)$ was defined in Theorem 1.4 and $c > 0$ does not depend on n.

Proof. If $\tau(r, \delta, \mu) = r/2$ then the assertion is easy (e.g., one can take a field X consisting of independent $N(0, 1)$ random variables). Thus, in what follows, we prove the existence of X such that

$$a_r(U_n) \geq c|U_n|^{\varkappa(\mu\wedge 1)/(r+\delta-2)}. \tag{1.21}$$

Let $\rho := (\nu + d)(r + \delta - 2)^{-1}$ where $\nu \geq 0$ is a number to be specified later. For $\mu > 1$ the assertion coincides with the same for $\mu = 1$. If $\mu < (1 + \delta/\varkappa)/2$, then the exponent in (1.21) is less than $r/2$, and the statement is obvious (again one can take the field of i.i.d. $N(0, 1)$ random variables). So, one may assume without loss of generality that $(1 + \delta/\varkappa)/2 \leq \mu \leq 1$.

First we construct the probability space $(\Omega, \mathcal{F}, \mathsf{P})$ on which the desired random field X will be defined. Let $\Omega = \mathbb{Z}$, $\mathcal{F}$ be the σ-algebra of all subsets of Ω and probability measure P be such that $p_m = \mathsf{P}(\{m\}) = \mathsf{P}(\{-m\}) := zm^{-1-\rho(r+\delta)}$ for any $m \in \mathbb{N}$, $z > 0$ being the normalizing constant. Define the random variables

$$\xi_k(\omega) = \alpha_k(\mathbb{I}\{\omega \geq k\} - \mathbb{I}\{\omega \leq -k\}) \quad \text{where } \alpha_k = k^\rho, \ k \in \mathbb{N}, \ \omega \in \mathbb{Z}.$$

For $j \in \mathbb{N}^d$, set $X_j = \xi_{|j|}$ where $|j| = \max_{i=1,\ldots,d} |j_i|$. If $j \in \mathbb{Z}^d \setminus \mathbb{N}^d$, set $X_j := 0$.

Lemma 1.14. *The random field $X = \{X_j, j \in \mathbb{Z}^d\}$ is associated.*

Proof. Consider the random variable $\chi(\omega) = \omega$, $\omega \in \mathbb{Z}$. Then $\{\xi_k, k \in \mathbb{N}\} \in \mathbf{A}$ as all ξ_k are nondecreasing functions in χ, hence associated by Theorem 1.1.8, (d). The family $\{X_j, j \in \mathbb{Z}^d\}$ is the same as $\{\xi_k, k \in \mathbb{Z}_+\}$, if the almost surely equal random variables are identified. Thus, X is also associated. $\square$

Clearly the field X is centered. Writing down the exact expression for $\mathsf{E}|\xi_k|^{2+\delta}$, $k \in \mathbb{N}$, one sees that

$$A_{r+\delta}(1) = 2z \sup_{k\in\mathbb{N}} \alpha_k^{r+\delta} \sum_{m\geq k} p_m = 2z \sup_{k\in\mathbb{N}} k^{\rho(r+\delta)} \sum_{m\geq k} m^{-1-\rho(r+\delta)} < \infty. \qquad (1.22)$$

To achieve the requirement on the covariance function, we need one more auxiliary result.

Lemma 1.15. *For each $d \in \mathbb{N}$, any block $U \in \mathcal{U}$, every $\nu > 0$ and arbitrary $i \in U$ one has*

$$\sum_{j\in U, j\neq i} |i - j|^{-\nu} \leq c(d,\nu) f(|U|, d, \nu) \qquad (1.23)$$

where $c(d,\nu) > 0$ and

$$f(|U|, d, t) = \begin{cases} |U|^{1-t/d}, & 0 < t < d,\ t \notin \mathbb{N}, \\ (1 + \log|U|)|U|^{1-t/d}, & 0 < t \leq d,\ t \in \mathbb{N}, \\ 1, & t > d. \end{cases}$$

Proof. The case $d = 1$ is trivial. For $d \geq 2$, consider $U = (a, b] \cap \mathbb{Z}^d \in \mathcal{U}$. Without loss of generality we can assume that $l_1 \leq \ldots \leq l_d$ where $l_s = b_s - a_s$ and $s = 1, \ldots, d$. Set $h_d(s, k) = (2k+1)^{d-s} - (2k-1)^{d-s}$, $k \in \mathbb{N}$, $s = 0, \ldots, d-1$. It is easily seen that

$$\sum_{\substack{j\in U,\\ j\neq i}} |i - j|^{-\nu} = \sum_{k=1}^{\infty} k^{-\nu} \sum_{\substack{j\in\mathbb{Z}^d,\\ |j-i|=k}} \mathbb{I}\{j \in U\} = \sum_{s=0}^{d-1} \sum_{k=l_s+1}^{l_{s+1}} \prod_{1\leq m\leq s} l_m h_d(s, k) k^{-\nu}$$

$$\leq 3^d d \sum_{s=0}^{d-1} \prod_{1\leq m\leq s} l_m \sum_{k=l_s+1}^{l_{s+1}} k^{d-s-1-\nu},$$

where $l_0 = 0$ and a product over an empty set is equal to 1. Having used the well-known estimates for sums $\sum_{k=r_1}^{r_2} k^\gamma$ by means of corresponding integrals and the estimates

$$\Big(\prod_{1\leq m\leq s} l_m\Big) l^{d-s-\nu} \leq \Big(\prod_{1\leq m\leq d} l_m\Big)^{(1-\nu/d)\vee 0}$$

for $l = l_s$ and $l = l_{s+1}$, we come to (1.23). The Lemma is established. $\square$

It is seen at once that, for any positive integers i and j with $i \leq j$, one has

$$cov(\xi_i, \xi_j) = \mathsf{E}\xi_i\xi_j = 2(ij)^\rho \sum_{m\geq j} p_m \leq c_1 i^\rho j^{-\rho(r+\delta-1)}. \qquad (1.24)$$

Here and throughout the rest of the proof $c_1, c_2, \ldots$ are positive constants. From (1.24) we have that $\Lambda_{\delta/\varkappa}(n)$ admits an upper bound

$$c_2 \sup_{\substack{U \in \mathcal{U} \\ |U|=n}} \left\{ \sum_{i \in U} |i|^{\rho\delta/\varkappa} \left(\sum_{j:|j| \geq |i|} |j|^{-\rho(r+\delta-1)} \right)^{\delta/\varkappa} + \sum_{i \in U} |i|^{-\rho\delta(r+\delta-1)/\varkappa} \left(\sum_{j:|j|<|i|} |j|^{\rho} \right)^{\delta/\varkappa} \right\}$$

$$\leq c_3 \sup_{U \in \mathcal{U}, |U|=n} \sum_{i \in U} |i|^{\delta\nu/\varkappa} \leq c_4 n^{\mu} \tag{1.25}$$

where ν is chosen in such a way that $\mu = 1 - \delta\nu/(d\varkappa)$. The last estimate in (1.25) is obvious when $\mu = 1$; otherwise it follows from Lemma 1.15.

By Lemma 1.14, (1.22) and (1.25) the random field X satisfies the conditions of Theorem 1.4. At the same time, for any $n \in \mathbb{N}$, with $U_n = (0, n]^d$, we have

$$\mathsf{E} \left| \sum_{j \in U_n} X_j \right|^r \geq \sum_{m=n}^{\infty} \left(\sum_{k=1}^{n} k^{d-1+\rho} \right)^r p_m \geq c_5 n^{dr-\rho\delta}$$

$$= c_5 |U_n|^{r-\delta(\nu+d)/(r+\delta-2)d} = c_5 |U_n|^{\mu\varkappa/(r+\delta-2)}. \quad \square$$

Note that Birkel ([43]) was the first who obtained the upper bound of $\mathsf{E} |\sum_{j=1}^{n} X_j|^{2+\delta}$ for a sequence of associated random variables $(X_j)_{j \in \mathbb{N}} (r > 2, \delta > 0)$.

The second example which is due to Birkel shows that a slight loss in the power of moment (r instead of $r + \delta$) is unavoidable. It is proved that the existence of $A_r(1)$ doest not imply a nontrivial bound for $\mathsf{E}|S(U)|^r$, even if the Cox–Grimmett coefficient u_n decreases as fast as one wants (except when it becomes zero for large n; in that case the field is m-dependent for some $m \in \mathbb{N}$ and the moment bound is of the same type as in the case of independent summands). Here we give

Theorem 1.16. ([43]) *Let $(\gamma_n)_{n \in \mathbb{N}}$ be a monotone sequence of positive numbers such that $\gamma_n \to 0$ as $n \to \infty$. Then, for any $r > 2$, there exists a centered associated random sequence $X = \{X_j, j \in \mathbb{N}\}$ such that*
1) $A_r(1) < \infty$,
2) $u(n) = O(\gamma_n)$ *as $n \to \infty$,*
3) $a_r((0,n]) \geq cn^r$ *where $c > 0$ does not depend on n.*

Proof. We modify the probability space $(\Omega, \mathcal{F}, \mathsf{P})$ which was used in the proof of Theorem 1.13 letting now

$$\mathsf{P}(\{m\}) = \mathsf{P}(\{-m\}) = p_m = zm^{-1-r} \left(\frac{\gamma_m}{m} \right)^{r/(r-2)}, \quad m \in \mathbb{N},$$

$z > 0$ being the normalizing factor. Define the random variables

$$X_k(\omega) = kb_{|k|}(\mathbb{I}\{\omega \geq k\} - \mathbb{I}\{\omega \leq -k\})$$

where $k \in \mathbb{N}$, $\omega \in \mathbb{Z}$ and $b_j = (j/\gamma_j)^{1/(r-2)}$ for $j \in \mathbb{N}$.

Then the sequence X is centered and associated, which can be proved analogously to Lemma 1.14. Furthermore,

$$A_r(1) = \sup_{k \in \mathbb{N}} 2k^r \sum_{j=k}^{\infty} p_j b_j^r = \sup_{k \in \mathbb{N}} 2zk^r \sum_{j=k}^{\infty} j^{-1-r} < \infty,$$

$$cov(X_k, X_{k+v}) = 2k(k+v) \sum_{j=k+v}^{\infty} b_j^2 p_j = 2zk(k+v) \sum_{j=k+v}^{\infty} j^{-1-r}\frac{\gamma_j}{j} \leq c_1 \gamma_v v^{1-r},$$

here and in what follows $c_1, c_2, \ldots$ do not depend on n. Therefore,

$$u(n) = \sup_{k \in \mathbb{N}} \sum_{j:|j-k| \geq n} cov(X_k, X_j) \leq 2c_2 \sum_{j=n}^{\infty} \gamma_j j^{1-r} = O(\gamma_n)$$

because of monotonicity of $(\gamma_n)_{n \in \mathbb{N}}$. Finally, for $\omega \geq n \in \mathbb{N}$, we have

$$\sum_{j=1}^{n} X_j(\omega) = b_\omega \sum_{j=1}^{n} j \geq c_3 b_\omega n^2$$

and consequently

$$\mathsf{E}\left|\sum_{j=1}^{n} X_j\right|^r \geq c_4 n^{2r} \sum_{j=n}^{\infty} j^{-1-r} \geq c_5 n^r,$$

which is the last desired property. $\square$

6°. The study of (BL, θ)-dependent random fields. The next theorem explores the idea of bisection as Theorem 1.4 does. The result obtained is more involved, as association gives a possibility of more subtle estimating of the covariances of separate elements of the field. Here we will only pay attention to the conditions ensuring "independent"-type behavior for partial sums' moments.

Introduce a function

$$\psi(x) = \begin{cases} (x-1)(x-2)^{-1}, & 2 < x \leq 4, \\ (3 - \sqrt{x})(\sqrt{x}+1)/2, & 4 < x \leq t_0^2, \quad (1.26) \\ ((x-1)\sqrt{(x-2)^2 - 3} - x^2 + 6x - 11)(3x - 12)^{-1}, & x > t_0^2, \end{cases}$$

where $t_0 \approx 2.1413$ is the maximal root of the equation $t^3 + 2t^2 - 7t - 4 = 0$. Note that $\psi(x) \to 1$ as $x \to \infty$.

We use the collection $\mathcal{U}$ of blocks introduced above, and also the definitions of $S(U)$ and $M(U)$ given in the beginning of this Chapter.

Theorem 1.17. ([80]) *Let* $X = \{X_j, j \in \mathbb{Z}^d\}$ *be a centered,* (BL, θ)*-dependent random field, such that there are* $p > 2$ *and* $c_0 > 1$ *ensuring that* $D_p := A_p(1) < \infty$ *and*

$$\theta_r \leq c_0 r^{-\lambda}, \quad r \in \mathbb{N}, \tag{1.27}$$

for $\lambda > d\psi(p)$, *with* ψ *defined in* (1.26). *Then there exist* $\delta > 0$ *and* $C > 1$, *depending only on* d, p, D_p, c_0 *and* λ, *such that for any block* $U \in \mathcal{U}$ *one has*

$$\mathsf{E}|S(U)|^{2+\delta} \leq C|U|^{1+\delta/2}, \quad \mathsf{E}M(U)^{2+\delta} \leq AC|U|^{1+\delta/2}$$

where $A = 5^d(1 - 2^{\delta/(4+2\delta)})^{-d(2+\delta)}$.

Proof. We fix some $\delta \in (0,1]$, $\delta < p-2$. The exact value of δ will be specified later. Choose $A_\delta > 0$ (e.g., $A_\delta = 5$) to guarantee that

$$(x+y)^2(1+x+y)^\delta \leq x^{2+\delta} + y^{2+\delta} + A_\delta((1+x)^\delta y^2 + x^2(1+y)^\delta) \tag{1.28}$$

for any $x, y \geq 0$.

Let $h(n) = \min\{k \in \mathbb{Z}_+ : 2^k \geq n\}$, $n \in \mathbb{N}$. For any block $U \in \mathcal{U}$ having edges with lengths $l_1, \ldots, l_d$, we set $h(U) = h(l_1) + \ldots + h(l_d)$.

We will show that, for some $C > 2(D_p \vee 1)$ and all blocks $U \in \mathcal{U}$,

$$\mathsf{E} S^2(U)(1 + |S(U)|)^\delta \leq C|U|^{1+\delta/2}. \tag{1.29}$$

This is proved by induction on $h(U)$. For $h(U) = 0$ (i.e. when $|U| = 1$) inequality (1.29) is obviously true. Suppose now that (1.29) is verified for all U such that $h(U) \leq h_0$. Consider a block U having $h(U) = h_0 + 1$.

Let L be any of the longest edges of U. Denote its length by $l_0(U)$. Draw a hyperplane orthogonal to L dividing it into two intervals of length $[l_0(U)/2]$ and $l_0(U) - [l_0(U)/2]$, here $[\cdot]$ stands for integer part of a number. This hyperplane divides U into two blocks U_1 and U_2 with $h(U_1), h(U_2) \leq h_0$.

Lemma 1.18. *There exists a value $\tau_0 = \tau_0(\delta) < 1$ such that, for any block $U \subset \mathbb{Z}^d$ with $|U| > 1$, one has*

$$|U_1|^{1+\delta/2} + |U_2|^{1+\delta/2} \leq \tau_0 |U|^{1+\delta/2}. \tag{1.30}$$

Proof. We intend to show that for any $n \in \mathbb{N}$, $n > 1$,

$$[n/2]^{1+\delta/2} + (n - [n/2])^{1+\delta/2} \leq \tau_0 n^{1+\delta/2}.$$

If $n = 2k$ for $k \in \mathbb{N}$, then the desired bound obviously holds with $\tau_0 = 2^{-\delta/2}$. If $n = 2k + 1$, $k \in \mathbb{N}$, then one needs to check that

$$k^{1+\delta/2} + (k+1)^{1+\delta/2} \leq \tau_0(2k+1)^{1+\delta/2}, \quad k \in \mathbb{N}.$$

Thus it is sufficient to prove that $1 + (1+x)^{1+\delta/2} \leq \tau_0(2+x)^{1+\delta/2}$ if $0 < x \leq 1$. The last inequality is equivalent to the following one:

$$f_1(t) := 1 + t^{1+\delta/2} \leq \tau_0(1+t)^{1+\delta/2} =: f_2(t) \quad \text{for} \ \ t \in (1,2].$$

Note that $f_1(1) \leq f_2(1)$ if $\tau_0 \geq 2^{-\delta/2}$. Moreover,

$$f_1'(t) = (1+\delta/2)t^{\delta/2} \leq f_2'(t) = \tau_0(1+\delta/2)(1+t)^{\delta/2}$$

for $t \in [1,2]$ whenever $\tau_0 \geq (3/2)^{-\delta/2}$. Therefore in all cases (1.30) holds with $\tau_0 = (3/2)^{-\delta/2}$. $\square$

A very simple but important result is the following

Lemma 1.19. *Let $X = \{X_j, j \in \mathbb{Z}^d\}$ be a square-integrable, (BL, θ)-dependent random field. Then a bound*

$$\mathsf{Var} S(U) \leq \theta_1 |U| + \sum_{j \in U} \mathsf{Var} X_j$$

is true for any finite set $U \subset \mathbb{Z}^d$.

Proof. One has

$$\mathsf{Var}S(U) = \sum_{j\in U}\left(\mathsf{Var}X_j + cov\left(X_j, \sum_{i\in U,\,i\neq j}X_i\right)\right).$$

The last covariance is bounded by θ_1. $\square$

In our case Lemma 1.19 implies that

$$\mathsf{E}S^2(U) \leq (D_2 + c_0)|U| \tag{1.31}$$

for any block $U \subset \mathbb{Z}^d$. Here and in what follows $D_q = \sup_{j\in\mathbb{Z}^d}\mathsf{E}|X_j|^q$, $q \in (0, p]$.
Set $Q_k = S(U_k)$, $k = 1, 2$. By (1.28), induction hypothesis and Lemma 1.18,

$$\mathsf{E}S^2(U)(1 + |S(U)|)^\delta = \mathsf{E}(Q_1 + Q_2)^2(1 + |Q_1 + Q_2|)^\delta \leq C(|U_1|^{1+\delta/2} + |U_2|^{1+\delta/2})$$

$$+ A_\delta\mathsf{E}((1 + |Q_1|)^\delta Q_2^2 + (1 + |Q_2|)^\delta Q_1^2)$$

$$\leq C\tau_0|U|^{1+\delta/2} + A_\delta\mathsf{E}\left((1 + |Q_1|)^\delta Q_2^2 + (1 + |Q_2|)^\delta Q_1^2\right). \tag{1.32}$$

Our goal is to obtain upper bounds for $\mathsf{E}(1 + |Q_1|)^\delta Q_2^2$ and $\mathsf{E}(1 + |Q_2|)^\delta Q_1^2$. We proceed with the first estimate only, the second one being similar. To this end, let us take positive $\zeta < (1 - \tau_0)/(4A_\delta)$ and introduce a block

$$V = \{j \in U_2 : dist(\{j\}, U_1) \leq \zeta|U|^{1/d}\},$$

as usual, *dist* corresponds to the sup-norm. Note that the induction hypothesis applies to V because $V \subset U_2$. Using the Hölder inequality and (1.31) gives

$$\mathsf{E}(1 + |Q_1|)^\delta Q_2^2 \leq 2\mathsf{E}(1 + |Q_1|^\delta)S^2(V) + 2\mathsf{E}(1 + |Q_1|)^\delta S^2(U_2 \setminus V)$$

$$\leq 2(D_2 + c_0)|V| + 2(\mathsf{E}|Q_1|^{2+\delta})^{\delta/(2+\delta)}(\mathsf{E}|S(V)|^{2+\delta})^{2/(2+\delta)} + 2\mathsf{E}(1 + |Q_1|)^\delta S^2(U_2 \setminus V)$$

$$\leq 2(D_2 + c_0)|U| + 2C\zeta|U|^{1+\delta/2} + 2\mathsf{E}(1 + |Q_1|)^\delta S^2(U_2 \setminus V). \tag{1.33}$$

To obtain the last estimate we also used the inequality

$$|V| \leq \zeta|U|^{1/d}\frac{|U|}{l_0(U)} \leq \zeta|U|^{1/d}|U|^{(d-1)/d}$$

following from the choice of $l_0(U)$ and V. Fix any indices $i, j \in U_2 \setminus V$ and assume at first that $i \neq j$. Then $dist(\{j\}, \{i\}\cup U_1) = m > 0$. For any $y > 0$, let H_y be the function defined in (1.5), and for some $y, z \geq 1$ introduce the random variables

$$Q_1^I = H_y(Q_1), \quad Q_1^{II} = \left((1 + |Q_1|)^\delta - (1 + |Q_1^I|)^\delta\right)^{1/\delta},$$

$$X_i^I = H_z(X_i), \quad X_i^{II} = X_i - X_i^I.$$

To simplify the notation we do not write $Q_{1,y}^I, Q_{1,y}^{II}, X_{i,z}^I$ and $X_{i,z}^{II}$. Obviously,

$$|\mathsf{E}(1 + |Q_1|)^\delta X_i X_j|$$

$$\leq |\mathsf{E}(1 + |Q_1^I|)^\delta X_i^I X_j| + \mathsf{E}(1 + |Q_1|)^\delta|X_i^{II}X_j| + \mathsf{E}|Q_1^{II}|^\delta|X_i^I X_j|. \tag{1.34}$$

Note that $\Phi(v,w) = (1 + |H_y(v)|)^\delta H_z(w)$ is a bounded Lipschitz function with $Lip(\Phi) \le 2y^\delta + z$. Since X is a (BL,θ)-dependent centered field, we can write

$$|\mathsf{E}(1 + |Q_1^I|)^\delta X_i^I X_j| = |cov((1 + |Q_1^I|)^\delta X_i^I, X_j)| \le (2y^\delta + z)\theta_m. \tag{1.35}$$

Let q be a positive number such that $1/q + \delta/(2 + \delta) + 1/p = 1$, that is

$$q = p(2 + \delta)/(2p - 2 - \delta) < p.$$

By the Hölder and Lyapunov inequalities,

$$\mathsf{E}\left(1 + |Q_1|\right)^\delta |X_i^{II} X_j| \le (\mathsf{E}|X_i^{II}|^q)^{1/q} D_p^{1/p}\left(1 + (\mathsf{E}|Q_1|^{2+\delta})^{\delta/(2+\delta)}\right)$$

$$\le 2C^{\delta/(2+\delta)}|U|^{\delta/2}\left(\frac{D_p}{z^{p-q}}\right)^{1/q} D_p^{1/p},$$

the last estimate being due to induction hypothesis.

For $r \in (\delta, 2 + \delta)$ to be specified later,

$$\mathsf{E}|Q_1^{II}|^\delta |X_i^I X_j| \le z\mathsf{E}|Q_1^{II}|^\delta |X_j| \le z\left(|cov(|Q_1^{II}|^\delta, |X_j|)| + \mathsf{E}|X_j|\mathsf{E}|Q_1^{II}|^\delta\right)$$

$$\le z\delta y^{\delta-1}\theta_m + 2zy^{\delta-r}D_1 C^{r/(2+\delta)}|U|^{r/2}. \tag{1.36}$$

The last inequality follows by the induction hypothesis and the fact that the function $v \mapsto ((1 + |v|)^\delta - (1 + y)^\delta)\mathbb{I}\{|v| \ge y\}$ is Lipschitz.

Now from (1.27) and (1.34)—(1.36), with the notation $T = 2c_0(1 \vee D_p)$, we conclude that

$$|\mathsf{E}(1+|Q_1|)^\delta X_i X_j| \le TC^{r/(2+\delta)}\left((y^\delta+z)m^{-\lambda}+|U|^{\delta/2}z^{1-p/q}+zy^{\delta-r}|U|^{r/2}\right). \tag{1.37}$$

Let β, γ be positive parameters. Introduce $y = |U|^{1/2}m^{\beta\lambda}$, $z = m^{\gamma\lambda}$. Then in view of (1.37) we obtain

$$|\mathsf{E}(1 + |Q_1|)^\delta X_i X_j| \le TC^{r/(2+\delta)}\left(|U|^{\delta/2}\sum_{k=1}^{3}m^{-\lambda\nu_k} + m^{-\lambda\nu_4}\right)$$

where

$$\nu_1 = 1 - \delta\beta, \quad \nu_2 = \gamma\left(\frac{p}{q} - 1\right), \quad \nu_3 = (r - \delta)\beta - \gamma, \quad \nu_4 = 1 - \gamma.$$

Now let us pick r close enough to $2 + \delta$, and β, γ in such a way that $\lambda\nu_k > d$ for $k = 1, 2, 3$. To achieve this we need to satisfy restrictions

$$\frac{1}{2}(\gamma + \lambda^{-1}d) < \beta < \delta^{-1}(1 - d\lambda^{-1}) \quad \text{and} \quad 2\gamma\frac{p - 2 - \delta}{2 + \delta} > \frac{d}{\lambda}. \tag{1.38}$$

The solvability of this system is equivalent to the inequality

$$\frac{1}{2}(\gamma + \lambda^{-1}d) < \delta^{-1}(1 - d\lambda^{-1}).$$

One can verify that this is possible if

$$\lambda > \lambda_1(\delta, d) = d\,\frac{(2+\delta)(2p-4-\delta)}{4(p-2-\delta)}. \tag{1.39}$$

Moreover, to have simultaneously $\lambda\nu_4 > (1-\delta/2)d$, it suffices to require that

$$\gamma < 1 - (1-\delta/2)d/\lambda,$$

or (in view of (1.38)) that

$$\lambda > \lambda_2(\delta, d) = d\left(\frac{2+\delta}{2(p-2-\delta)} + 1 - \frac{\delta}{2}\right). \tag{1.40}$$

To understand whether inequalities (1.39) and (1.40) can be satisfied introduce (for d fixed) the continuous functions $f(\delta) = \lambda_2(\delta, d)/d$ and $g(\delta) = \lambda_1(\delta, d)/d$ defined for $\delta \in (0, p-2)$. A standard calculation shows that the minimum of f is attained at the point $\delta^* := p - \sqrt{p} - 2$, and that g is increasing if $\delta \in [0, 1]$, $\delta < p-2$. Let us consider three cases.

Case 1: $p \le 4$. In this case both functions increase when $\delta \in [0, (p-2) \wedge 1)$. Thus $\min(f(\delta) \vee g(\delta))$ is attained at $\delta = 0$ and equals $(p-1)/(p-2)$. Due to the condition of Theorem 1.17 we have $\lambda > d(p-1)/(p-2)$. Therefore, taking δ small enough guarantees that (1.39) and (1.40) hold.

Case 2: $4 < p \le t_0$. If $p > 4$ then it is easy to calculate that $f(1) < g(1)$. Since also $f(0) < g(0)$, there exists a root of equation $f(\delta) = g(\delta)$ in $(0, 1)$. This root is

$$\delta_* = \frac{2}{3}\left(p - 2 - \sqrt{(p-2)^2 - 3}\right).$$

If also $p \le t_0$, then $\delta^* \le \delta_* \le 1$. Therefore, the minimal value of f is attained to the left from δ_*.

Hence the function $f(\delta) \vee g(\delta)$ attains its minimum at the point $\arg\min_\delta(f(\delta) \vee g(\delta)) = \delta^*$, and (1.39) and (1.40) will be satisfied if $\delta = p - \sqrt{p} - 2$ and $\lambda/d > f(p - \sqrt{p} - 2) = (3 - \sqrt{p})(1 + \sqrt{p})/2$.

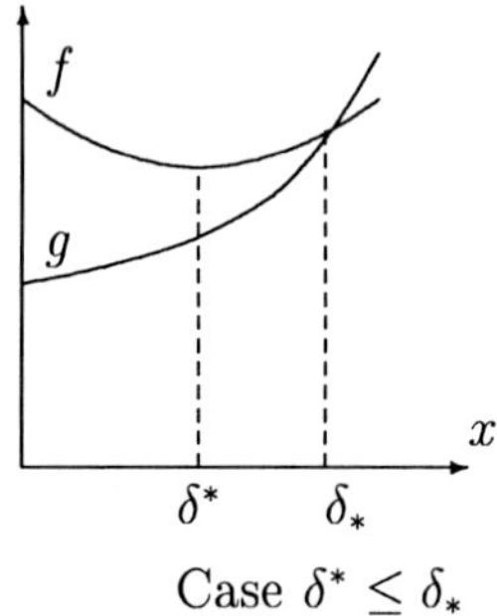

Case $\delta^* \le \delta_*$

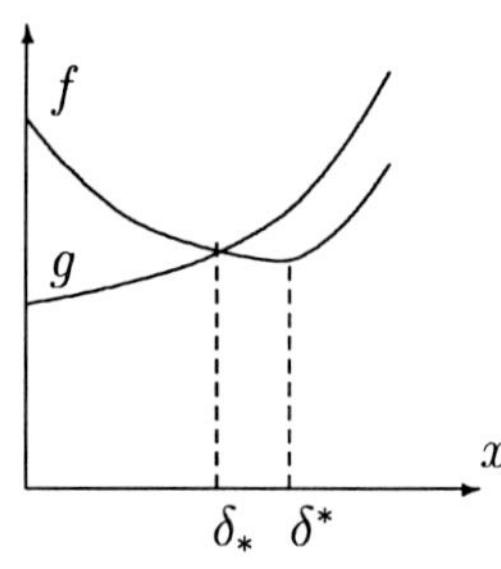

Case $\delta^* > \delta_*$

Fig. 2.2

Case 3: $p > t_0$. Now the root δ_* lies in the interval where f decreases, consequently $\arg\min_\delta(f(\delta) \vee g(\delta)) = \delta_*$. Thus the sufficient condition reads

$$\lambda/d > f(\delta_*) = \frac{(p-1)\sqrt{(p-2)^2 - 3} - p^2 + 6p - 11}{3p - 12}.$$

For arbitrary $i \in U_2 \setminus V$, set

$$\overline{U_2} = \{j \in U_2 \setminus V : |j - i| \geq \zeta |U|^{1/d}\}, \ \widehat{U_2} = \{j \in U_2 \setminus V : |j - i| < \zeta |U|^{1/d}\}.$$

By Lemma 1.15, for any $i \in U_2 \setminus V$, we have

$$\Big| \sum_{j \neq i, \, j \in U_2 \setminus V} \mathsf{E}(1 + |Q_1|)^\delta X_i X_j \Big| \leq \Big| \sum_{j \in \overline{U_2}} \mathsf{E}(1 + |Q_1|)^\delta X_i X_j \Big|$$

$$+ \Big| \sum_{j \in \widehat{U_2}, \, j \neq i} \mathsf{E}(1 + |Q_1|)^\delta X_i X_j \Big| \leq TC^{r/(2+\delta)} |U|^{\delta/2} \Big(4\zeta^{-\lambda\nu_0} + \sum_{k=1}^{4} c(d, \lambda\nu_k) \Big), \quad (1.41)$$

here $\nu_0 = \min_{k=1,\ldots,4} \nu_k$.

Now we treat the case of $i = j \in U_2 \setminus V$. Obviously one has $\delta p/(p - 2) < 2 + \delta$. Therefore, by Hölder's inequality and induction hypothesis we infer that

$$\mathsf{E}(1 + |Q_1|)^\delta X_i^2 \leq (\mathsf{E}|X_i|^p)^{2/p} \left(1 + \left(\mathsf{E}|Q_1|^{\delta p/(p-2)} \right)^{(p-2)/p} \right) \leq TC^{\delta/(2+\delta)} |U|^{\delta/2}.$$
$$(1.42)$$

From (1.41) and (1.42) one deduces that

$$\mathsf{E}(1 + |Q_1|)^\delta S^2(U_2 \setminus V) \leq \Big| \sum_{\substack{i,j \in U_2 \setminus V \\ i \neq j}} \mathsf{E}(1 + |Q_1|)^\delta X_i X_j \Big| + \sum_{i \in U_2 \setminus V} \mathsf{E}(1 + |Q_1|)^\delta X_i^2$$

$$\leq |U| \max_{i \in U_2 \setminus V} \Big| \sum_{j \neq i, j \in U_2 \setminus V} \mathsf{E}(1+|Q_1|)^\delta X_i X_j \Big| + TC^{\delta/(2+\delta)} |U|^{1+\delta/2} \leq MC^{r/(2+\delta)} |U|^{1+\delta/2}$$

where $M = T(1 + 4\zeta^{-\lambda\nu_0} + \sum_{k=1}^{4} c(d, \lambda\nu_k))$. Employing (1.32), (1.33) and the last inequality leads to

$$\mathsf{E}S^2(U)(1 + |S(U)|)^\delta \leq \Big(C\tau_0 + 4A_\delta(D_2 + c_0) + 4CA_\delta\zeta + 4C^{r/(2+\delta)} A_\delta M \Big) |U|^{1+\delta/2}.$$

The first assertion of the Theorem is now easily verified on account of (1.29). Namely, one has

$$\mathsf{E}S^2(U)(1 + |S(U)|)^\delta \leq C|U|^{1+\delta/2}$$

if C is so large that

$$(1 - \tau_0 - 4A_\delta\zeta)C > 4A_\delta MC^{r/(2+\delta)} + 4A_\delta(D_2 + c_0).$$

The second assertion follows from the first one and Theorem 1.2. $\square$

7°. The graph techniques application. The bisection method developed in the last two theorems allows to estimate the moment of any order greater than two. The situation becomes simpler if one is interested in the moment of order $2r$, when $r > 1$ is a positive integer, as the direct calculation for power of the sum is now possible. Following Bakhtin and Bulinski we show that it is useful to consider the estimates of the $cov(F(X_I), G(X_J))$ for power-type "test functions" F and G.

In the following we write $X(I)^n$ to denote the product $\prod_{i\in I} X_i^{n_i}$, for a finite $I \subset \mathbb{Z}^d$ and $n = (n_1,\ldots,n_{|I|}) \in \mathbb{N}^{|I|}$. Also $\|n\|_1 = \sum_{i\in I} |n_i|$.

Theorem 1.20. ([16]) *Let $X = \{X_j, j \in \mathbb{Z}^d\}$ be a centered random field such that*

$$D_{2r+\delta} := \sup_{j\in\mathbb{Z}^d} \mathsf{E}|X_j|^{2r+\delta} < \infty$$

for some $r \in \mathbb{N}$ and $\delta > 0$. Assume that, for any pair of finite disjoint sets $I, J \subset \mathbb{Z}^d$ and any $n \in \mathbb{N}^{|I|}$, $m \in \mathbb{N}^{|J|}$ with $\|n\|_1 + \|m\|_1 \leq 2r$, one has

$$|cov(X(I)^n, X(J)^m)| \leq D_{2r+\delta}^{(\|n\|_1+\|m\|_1)/(2r+\delta)} f\left(dist(I,J)\right) g(|I|,|J|) \qquad (1.43)$$

where $f : \mathbb{N} \to \mathbb{R}_+$ is nonincreasing and $g : \mathbb{N}^2 \to \mathbb{R}_+$ is nondecreasing. Then, for any block $U \subset \mathcal{U}$,

$$\mathsf{E}S(U)^{2r} \leq G(d,r)|U|^r\left(D_{2r} + D_{2r+\delta}^{2r/(2r+\delta)} g_0(2r) \sum_{k=1}^{diam(U)} k^{dr-1} f(k)\right),$$

here $g_0(2r) := \max_{k+l=2r} g(k,l)$ and $G(d,r) := r^2 2^{2r-1}(2r)^{2r} 3^{2dr} 2^{dr-1}(2r)!$

Proof. We have

$$\mathsf{E}S(U)^{2r} \leq (2r)! \sum_{q=1}^{2r} \sum{}' \sum{}'' |\mathsf{E}X(J)^n| \qquad (1.44)$$

where $\sum'$ is the sum over subsets $J \subset U$ with $|J| = q$, and $\sum''$ is over the multiindices $n \in \mathbb{N}^J$ with $\|n\|_1 = 2r$. By the Hölder inequality, $|\mathsf{E}X(J)^n| \leq D_{2r}$ if $\|n\|_1 = 2r$. The set $J \subset U$ with $|J| = q$ can be chosen in $\binom{|U|}{q}$ ways. The number of summands in $\sum''$ equals the number of possible arrangements of $2r$ balls in q boxes so that no box is empty, i.e., $\binom{2r-1}{q-1}$. Hence,

$$\sum_{q=1}^{r} \sum{}' \sum{}'' |\mathsf{E}X(J)^n| \leq D_{2r} \sum_{q=1}^{r} \binom{2r-1}{q-1} |U|^q \leq 2^{2r-1} D_{2r}|U|^r. \qquad (1.45)$$

Now suppose that $q \geq r+1$. Fix a set $J = \{j_1,\ldots,j_q\} \subset U$ and a multiindex $n \in \mathbb{N}^J$ describing one summand in the right-hand side of (1.44). Recall that *full graph* on a finite vertices set V is a graph (V, E) such that to every non-ordered pair of vertices $\{v_1, v_2\} \subset V$ one and only one edge $e = e(v_1, v_2)$ is incident. To each edge $j_s j_t$ of the full graph on vertices $\{j_1,\ldots,j_q\}$ we assign a weight $\mathcal{W}(j_s j_t) = dist(j_s, j_t)$. To each spanning tree[1] $T = T(J)$ of that graph we may thus assign a weight $\mathcal{W}(T)$ equal to the sum of weights of all the edges which are in T. Let T_0 be a minimal spanning tree (hence $\mathcal{W}(T_0) \leq \mathcal{W}(T)$ for any other spanning tree T). One may think that T_0 is uniquely determined by J. Namely, one may enumerate all edges $e_1, e_2, \ldots, e_{q(q-1)/2}$ of the full graph on vertices $\{j_1,\ldots,j_q\}$. Consider each tree as a set of edges ordered by that enumeration. Then, if T_1 and T_2 are different spanning trees, write $T_1 \prec T_2$ if $T_1 = \{e_{j_1},\ldots,e_{j_{q-1}}\}$ and $T_2 = \{e_{k_1},\ldots,e_{k_{q-1}}\}$, and either

---

[1] See Section 1.4, subsection 5.

$j_1 < k_1$ or there exists $l \in \{2, \ldots, q-1\}$ such that $j_1 = k_1, \ldots, j_{l-1} = k_{l-1}, j_l < k_l$. After ordering in this way all spanning trees we can take the least one among all candidates (all trees with minimal weight).

Let $m_1, \ldots, m_{q-1}$ be all the edges of T_0 ordered by increasing of weight. We need the following property.

Lemma 1.21. *There exists a vertex h of the tree T_0 such that $n(h) = 1$ and all edges incident to h belong to the set $\mathcal{R} = \{m_{q-r}, \ldots, m_{q-1}\}$.*

Proof. Let $z = \sharp\{j \in J : n(j) = 1\}$. Then $z + 2(q-z) \leq 2r$, i.e. $z \geq 2(q-r)$. Suppose that any such vertex is incident to an edge which does not belong to $\mathcal{R}$. Then the total number of such edges is not less than $z/2 \geq q - r$ (since one edge is incident to two vertices). However, in $\{m_1, \ldots, m_{q-r-1}\}$ there are only $q - r - 1$ elements. $\square$

Let $h(T_0)$ be the vertex given by Lemma 1.21 (if there are several ones, take, e.g., the least in the lexicographic order). Denote by $e_1, \ldots, e_p$ the edges incident to $h(T_0)$ (here, clearly, $p \leq |\mathcal{R}| = r$), ordered by increasing of weight. If one deletes any of the edges e_t ($t = 1, \ldots, p$) from T_0, the distance between two connected components of the remaining graph will be $\mathcal{W}(e_t)$, since T_0 was chosen to be a minimal tree. Let I_1 be the component of $T_0 \setminus e_1$, not containing h, and J_1 be the complementary one. Similarly we define the trees $I_2, J_2, \ldots, I_p, J_p = \{h\}$.

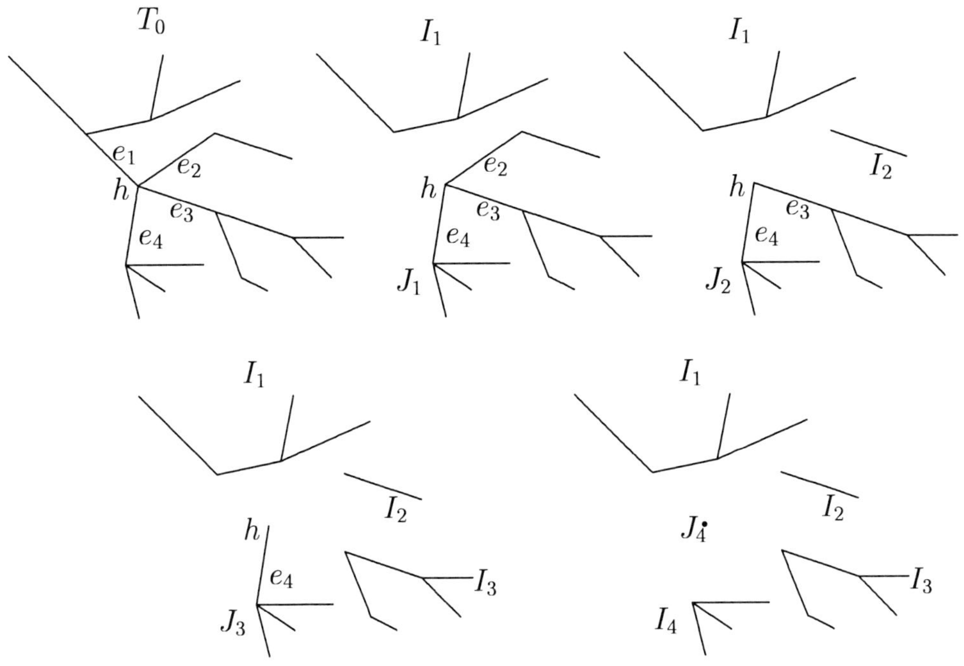

Fig. 2.3

Now it is convenient to adopt further notation. For $n = (n_{i_1}, \ldots, n_{i_q}) \in \mathbb{N}^q$, $\|n\|_1 = 2r$ and $I \subset J$, let $n(I) = (n_i, i \in I)$.

On account of the Hölder and Lyapunov inequalities, one can write

$$|\mathsf{E}X(I)^{n(I)}| \leq D_{2r+\delta}^{\|n(I)\|_1/(2r+\delta)} =: D_{r,\delta}(n(I)). \tag{1.46}$$

Having introduced

$$r_l := |cov(X(I_l)^{n(I_l)}, X(J_l)^{n(J_l)})|, \quad R_l := |\mathsf{E}X(I_l)^{n(I_l)}||\mathsf{E}X(J_l)^{n(J_l)}|$$

and using (1.46), one obtains the estimate, for $n \in \mathbb{N}^J$,

$$|\mathsf{E}X(J)^n| \leq r_1 + R_1 \leq r_1 + D_{r,\delta}(n(I_1))(r_2 + R_2) \leq \ldots$$

$$\leq r_1 + D_{r,\delta}(n(I_1))(r_2 + \ldots + D_{r,\delta}(n(I_2))(r_p + R_p)\ldots). \tag{1.47}$$

For any q such that $r + 1 \leq q \leq 2r$, from (1.47) it follows that

$$|\mathsf{E}X(J)^n| \leq g_0(q)D_{2r+\delta}^{2r/(2r+\delta)}(f(\mathcal{W}(e_1)) + \ldots + f(\mathcal{W}(e_p))).$$

Consequently,

$$\sum_{q=r+1}^{2r} \sum{}' \sum{}'' |\mathsf{E}X(J)^n| \leq g_0(2r)D_{2r+\delta}^{2r/(2r+\delta)} r \sum_{s=1}^{diam(U)} \sum_{q=r+1}^{2r} \sum{}''' \binom{2r-1}{j-1} f(s) \tag{1.48}$$

where $\sum{}'''$ is the sum over subsets $J \subset U$ having $|J| = q$ and such that the corresponding edge e_1 has length s. The vertex h can occupy no more than $|U|$ positions in U. For the second vertex h' of T_0 such that $dist(h, h') = s$, there are at most $2d(2s)^{d-1}$ possible places, and for other vertices in J there are totally $|U|^{d-2}$ possible configurations. We see that for a fixed positive integer $s \leq diam(U)$ and $q \in [r + 1, 2r]$ the sum $\sum{}'''$ contains no more than $2d(2s)^{d-1}|U|^{q-1}$ summands. If s is such that $(2s)^d \geq |U|$ then

$$2d(2s)^{d-1}|U|^{q-1} \leq d2^{dr}s^{dr-1}|U|^r.$$

Now consider the case when $(2s)^d < |U|$. Let $e_1(h(T_0(J))) = m_l$. Lemma 1.21 yields that $q - r \leq l \leq q - 1$.

The enumeration of the edges $m_1, \ldots, m_{q-1}$ can be done in no more than $(q-1)!$ ways. The edges $m_1, \ldots, m_l$ form a subgraph in $T_0(J)$ with connected components $T_1, \ldots, T_k$ where T_j has l_j edges, $j = 1, \ldots, k$.

Let us enumerate the edges $e_1^j, \ldots, e_{l_j}^j$ of the tree T_j in such a way that T_j can be drawn stepwise, by successive addition of the edge e_t^j, $1 \leq t \leq l_j$, making at every step the already drawn part of the tree connected. The (chosen uniquely[2]) vertex v_j, which is the initial for the drawing, could occupy at most $|U|$ positions. Every edge T_j different from m_l can be joined in at most $(2s + 1)^d$ ways, the edge m_l could be drawn maximum at $2d(2s)^{d-1}$ positions. The trees $T_1, \ldots, T_k$ have totally

---

[2] Also using lexicographic order.

l edges and k initial vertices $v_1, \ldots, v_k$. Thus the number of variants to place them in U is bounded by

$$|U|^k (2s+1)^{d(l-1)} 2d(2s)^{d-1} \leq 3^{2dr} |U|^k (2s)^{dl-1}.$$

There remain $q - \sum_{j=1}^{k}(l_j + 1) = q - l - k$ vertices of the tree $T_0(J)$ that are not in $T_1 \cup \ldots \cup T_k$. There are no more than $|U|^{q-l-k}$ positions for them. Thus the number of summands in $\sum'''$, corresponding to given s and q, does not exceed

$$(2r)^{2r} 3^{2dr} (2s)^{dl-1} |U|^{q-l} \leq (2r)^{2r} 3^{2dr} (2s)^{dr-1} |U|^r. \tag{1.49}$$

Combining (1.44), (1.45), (1.48) and (1.49) establishes the claim. $\square$

8°. Auxiliary results. The conditions of Theorem 1.20 hold for associated and for (BL, θ)-dependent random fields under appropriate moment restrictions and requirements on the dependence coefficient, as the following three lemmas show. As always, for a random variable X we write $\|X\|_p = (\mathsf{E}|X|^p)^{1/p}$, $p \geq 1$.

Lemma 1.22. ([405]) *Let* $Y_1, \ldots, Y_m$ *($m \in \mathbb{N}$, $m \geq 2$) be* **PA** *random variables having finite absolute moment of order $m + \delta$ for some $\delta > 0$. Set $Z_k = Y_1 \ldots Y_k$ and $W_k = Y_{k+1} \ldots Y_m$ where $k \in \{1, \ldots, m-1\}$. Then*

$$|cov(Z_k, W_k)| \leq 3 \left(\sum_{i=1}^{k} \sum_{j=k+1}^{m} cov(Y_i, Y_j) \right)^{\mu} \left(\prod_{i=1}^{m} \|Y_i\|_{m+\delta} \left(\sum_{j=1}^{m} \|Y_j\|_{m+\delta}^{\delta} \right) \right)^{1-\mu} \tag{1.50}$$

with $\mu = \delta/(m+\delta-2)$.

Proof. If $m = 2$, the assertion is obvious, thus in the following $m > 2$. For $M > 0$ to be specified later and the function $H_M(\cdot)$ defined in (1.5), write $h_M(t) = t - H_M(t)$, $t \in \mathbb{R}$. Then $Y_i = H_M(Y_i) + h_M(Y_i)$, $i \in \mathbb{N}$, and

$$cov(Z_k, W_k) = cov(H_M(Y_1) \ldots H_M(Y_k), H_M(Y_{k+1}) \ldots H_M(Y_m))$$

$$+ cov(h_M(Y_1) H_M(Y_2) \ldots H_M(Y_k), H_M(Y_{k+1}) \ldots H_M(Y_m)) + \ldots$$

$$+ cov(Y_1 \ldots Y_k, Y_{k+1} \ldots h_M(Y_{m-1}) H_M(Y_m))$$

$$+ cov(Y_1 \ldots Y_k, Y_{k+1} \ldots Y_{m-1} h_M(Y_m)) = J_0 + J_1 + \ldots + J_m. \tag{1.51}$$

Actually relation (1.51) is established by moving "from the right to the left". By Theorem 1.5.3 we have

$$|J_0| \leq M^{m-2} \sum_{i=1}^{k} \sum_{j=k+1}^{m} cov(Y_i, Y_j).$$

An obvious inequality $|h_M(t)| \leq M^{-\delta}|t|^{1+\delta}$ (for $t \in \mathbb{R}$, $\delta > 0$) and the Hölder inequality imply that for $1 \leq r \leq m$

$$|J_r| \leq 2M^{-\delta} \left(\mathsf{E}|Y_r|^{(1+\delta)\frac{m+\delta}{1+\delta}} \right)^{(1+\delta)/(m+\delta)} \leq 2M^{-\delta} \|Y_r\|_{m+\delta}^{\delta} \prod_{i=1}^{m} \|Y_i\|_{m+\delta}.$$

Thus the left-hand side of (1.51) has upper bound

$$M^{m-2} \sum_{i=1}^{k} \sum_{j=k+1}^{m} cov(Y_i, Y_j) + 2M^{-\delta} \prod_{i=1}^{m} \|Y_i\|_{m+\delta} \sum_{j=1}^{m} \|Y_j\|_{m+\delta}^{\delta}.$$

If $\Lambda := \sum_{i=1}^{k} \sum_{j=k+1}^{m} cov(Y_i, Y_j) = 0$, then $(Y_1, \ldots, Y_k)$ and $(Y_{k+1}, \ldots, Y_m)$ are independent and the assertion is true. If $\Lambda \neq 0$, taking $M = \Lambda^\alpha R^\beta$ with $R = \prod_{i=1}^{m} \|Y_i\|_{m+\delta} \sum_{j=1}^{m} \|Y_j\|_{m+\delta}^{\delta}$ we come to (1.50) (after appropriate choice of α and β, namely $\beta = -\alpha = 1/(m+\delta-2)$). $\square$

We can rephrase Lemma 1.22 in the following manner.

Lemma 1.23. *Let* $X = \{X_j, j \in \mathbb{Z}^d\} \in \mathbf{PA}$ *and* $\mathsf{E}|X_j|^{m+\delta} < \infty$ *for some integer* $m \geq 2$, *some* $\delta > 0$ *and each* $j \in \mathbb{Z}^d$. *For any finite disjoint nonempty sets* $I, J \subset \mathbb{Z}^d$ *with* $|I| + |J| = m$ *put* $X(I) = \prod_{i \in I} X_i$ *and* $X(J) = \prod_{j \in J} X_j$, *then*

$$|cov(X(I), X(J))| \leq 3\left(\sum_{i \in I} \sum_{j \in J} cov(X_i, X_j)\right)^{\mu} \left(\prod_{i \in I \cup J} \|X_i\|_{m+\delta} \left(\sum_{j \in I \cup J} \|X_j\|_{m+\delta}^{\delta}\right)\right)^{1-\mu}.$$

Lemma 1.24. *Assume that in the previous Lemma the condition* $X \in \mathbf{PA}$ *is replaced by the hypothesis* X *is* (BL, θ)-*dependent. In this case*

$$|cov(X(I), X(J))| \leq 3\left((|I| \wedge |J|)\theta_r\right)^{\mu} \left(\prod_{i \in I \cup J} \|X_i\|_{m+\delta} \left(\sum_{j \in I \cup J} \|X_j\|_{m+\delta}^{\delta}\right)\right)^{1-\mu}.$$

Proof is the same as for Lemmas 1.22 and 1.23, with the amendments that $\sum_{i \in I} \sum_{j \in J} cov(X_i, X_j)$ is replaced with $(|I| \wedge |J|)\theta_r$ and instead of Theorem 1.5.3 one uses the definition of (BL, θ)-dependent random field. $\square$

$9°$. **The fourth moment of partial sums.** Theorem 1.20 is still far from optimality. The application of covariance inequality can be done repeatedly, which helps to invoke moments of smaller order and to decrease the exponent of $|U|$. Due to complexity of this method we only give a result for $r = 2$ (i.e. $p = 4$).

For $n \in \mathbb{N}$ and $\tau \in \mathbb{R}$ set

$$B(n, \tau) = \begin{cases} n^\tau, & \tau > 0, \\ \log(n \vee e), & \tau = 0, \\ 1, & \tau < 0. \end{cases}$$

Theorem 1.25. ([375]) *Let* $X = \{X_j, \ j \in \mathbb{Z}^d\}$ *be a centered* (BL, θ)-*dependent random field satisfying* (1.27). *Assume that* $D_p < \infty$ *for some* $p > 4$. *Then for any* $U \in \mathcal{U}$ *one has the estimate*

$$\mathsf{E}S(U)^4 \leq \sum_{j \in U} \mathsf{E}X_j^4 + 12|U|^2 (D_2 + \theta_1)^2 + |U|C_1(d, \lambda, p)c_0^v D_p^{2/(p-2)} B(|U|, \tau) \quad (1.52)$$

where

$$v = \frac{p-4}{p-2}, \quad \tau = 3 - \frac{\lambda v}{d}, \quad C_1(d, \lambda, p) = 192d^2 3^{3d}(p-4)^{2/(p-2)} h(\tau)v^{-1}, \quad (1.53)$$

$$h(\tau) = \begin{cases} \tau^{-1} \vee 1, & \tau > 0, \\ 2, & \tau = 0, \\ 2(|\tau|^{-1} \vee 1), & \tau < 0. \end{cases}$$

If X is quasi-associated and

$$|cov(X_i, X_j)| \leq c_1 |i - j|^{-\kappa} \tag{1.54}$$

for any $i \neq j$, with some c_1, $\kappa > 0$, the same estimate (1.52) holds upon replacement of c_0 with c_1 and λ with κ.

Proof. If $c_0 = 0$ in (1.27) then the field X consists of independent random variables and

$$\mathsf{E}S(U)^4 = \sum_{j \in U} \mathsf{E}X_j^4 + 3 \sum_{i,j \in U, i \neq j} \mathsf{E}X_i^2 \mathsf{E}X_j^2 \leq \sum_{j \in U} \mathsf{E}X_j^4 + 3\left(\sum_{j \in U} \mathsf{E}X_j^2\right)^2.$$

Thus we will further assume that $c_0 > 0$. We have

$$\mathsf{E}S(U)^4 = \sum_{j \in U} \mathsf{E}X_j^4 + \sum_{J_4 \in \Lambda} \mathsf{E}X_{j_1} X_{j_2} X_{j_3} X_{j_4} = S^{(1)} + S^{(2)},$$

here the second sum is taken over the set Λ of ordered 4-tuples of indices $J_4 = (j_1, j_2, j_3, j_4) \in U^4$ having at least one pair of distinct indices.

To estimate $S^{(2)}$ we represent the set Λ as a union[3] of sets Λ_r, $r \in \mathbb{N}$, in the following way: a 4-tuple $J_4 = (t_1, t_2, t_3, t_4) \in \Lambda_r$, if $\max_{Q \subset J_4} dist(Q, J_4 \setminus Q) = r$.

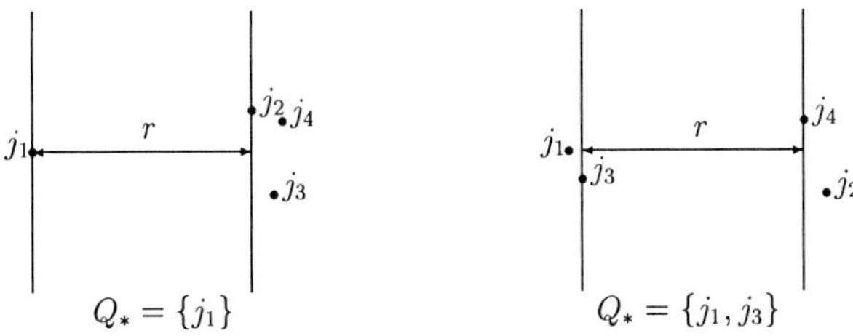

Fig. 2.4

We may assume (employing the lexicographic order) that there is a uniquely determined set $Q_* = Q_*(J_4)$ such that

$$\max_{Q \subset J_4} dist(Q, J_4 \setminus Q) = dist(Q_*, J_4 \setminus Q_*). \tag{1.55}$$

Without loss of generality one may think that lengths of the edges of U obey the relations $l_1 \leq \ldots \leq l_d$. Set $l_0 = 0$. Then to each $r = 1, \ldots, l_d$ we can assign a unique number $k = k(r)$, $0 \leq k \leq d - 1$, in the following way. Let $k(r) = 0$ if $l_1 = \ldots = l_d$ and otherwise $k(r) := \max\{k : l_k < r \leq l_{k+1}\}$. Let us show that

$$|\Lambda_r| \leq 144d \left(\prod_{i=1}^{k(r)} l_i\right)^3 (3r)^{3(d-k)-1} |U|, \tag{1.56}$$

---

[3] Note that $\Lambda_r = \varnothing$ if $r > diam(U)$.

here and below the product over an empty set is 1. For $J_4 = (t_1, t_2, t_3, t_4) \in \Lambda_r$, there is $Q_*(J_4)$ such that the distance in (1.55) is exactly r. The 4-tuple obtained from J_4 by a permutation also belongs to Λ_4; for a non-ordered set $\{t_1, t_2, t_3, t_4\}$ there are at most 24 such 4-tuples. Suppose that $\{t_1, t_2, t_3, t_4\}$ is a non-ordered set. Divide all such sets into d subclasses $\widetilde{\Lambda}_{r,l}$, $l = 1, ..., d$, where l is the minimal coordinate number $i \in \{1, \ldots, d\}$, where the distance between subsets of the 4-tuple is attained (that is, l-th axis is parallel to the arrows on Fig. 1). Then $|\Lambda_r| \leq 24d \max_l |\widetilde{\Lambda}_{r,l}|$. One can certainly assume that $dist(t_1, t_2) = r$ and $|(t_2)_l - (t_1)_l| = r$, so that t_2 belongs to the intersection of U with a $(d-1)$-dimensional ball, of radius r, which has center point $t_1 + (0, ..., 0, r, 0, ..., 0)$. Thus we have $|U|$ positions for t_1, whereas for t_2 there are

$$l_1 ... l_k (2r + 1)^{d-k-1} \leq l_1 ... l_k (3r)^{d-k-1}$$

positions. The point t_3 belongs to U intersected with one of the balls, with the same radius r, centered at t_1 and t_2, hence it has at most $2l_1 ... l_k (2r + 1)^{d-k}$ possible positions; the point t_4 is inside one of balls with radii r centered at t_1, t_2, t_3, thus for t_4 the number of positions does not exceed $3l_1 ... l_k (2r + 1)^{d-k}$. In total $|\Lambda_r| \leq 24d \cdot 6(l_1 ... l_k)^3 (3r)^{3(d-k)-1} |U| = 144d(l_1 ... l_k)^3 (3r)^{3(d-k)-1} |U|$. Thus (1.56) is proved.

Fix $J_4 = (t_1, t_2, t_3, t_4) \in \Lambda_r$, $r = 1, ..., l_d$. At first consider the case when $|Q_*(J_4)| = 1$. There is no loss of generality in assuming $dist(\{t_1\}, T_4 \setminus \{t_1\}) = r$. To simplify the notation we will further write X_q instead of X_{t_q}, $q = 1, 2, 3, 4$.

For $M > 0$ consider the function $H_M(\cdot)$ defined in (1.5). Then $|H_M(t)| \leq M$, $t \in \mathbb{R}$, and $Lip(H_M) = 1$. For $u = 2, 3, 4$ one has $X_u = X'_u + X''_u$ where $X'_u = H_M(X_u)$, $X''_u = X_u - X'_u$. It is clear that

$$|EX_1 X_2 X_3 X_4| \leq |cov(X_1, X'_2 X'_3 X'_4)| + |EX_1 X''_2 X'_3 X'_4|$$

$$+ |EX_1 X_2 X''_3 X'_4| + |EX_1 X_2 X_3 X''_4|.$$

Condition (1.27) yields

$$|cov(X_1, X'_2 X'_3 X'_4)| \leq 3c_0 M^2 r^{-\lambda}, \tag{1.57}$$

and by Chebyshev's and Hölder's inequalities

$$|EX_1 X''_2 X'_3 X'_4| \leq M^2 E|X_1 X''_2| \leq M^2 (E|X_1|^p)^{1/p} \left(E|X''_2|^{p/(p-1)} \right)^{(p-1)/p}$$

$$\leq M^2 D_p^{1/p} \left(\frac{D_p}{M^{p-p/(p-1)}} \right)^{1-1/p} = D_p M^{4-p}.$$

In the same way we estimate $|EX_1 X_2 X''_3 X'_4|$ and $|EX_1 X_2 X_3 X''_4|$.

Summing up the bounds obtained gives

$$|cov(X_1, X_2 X_3 X_4)| \leq 3c_0 M^2 r^{-\lambda} + 3D_p M^{4-p}. \tag{1.58}$$

Now let $Q_* = \{t_1, t_2\}$. Again with $X_q = X_{t_q}$, we have, for $M > 0$,

$$\mathsf{E}X_1 X_2 X_3 X_4 = \mathsf{E}X_1 X_2 \mathsf{E}X_3 X_4 + cov(X_1' X_2', X_3' X_4') + cov(X_1'' X_2', X_3' X_4')$$

$$+ cov(X_1 X_2'', X_3' X_4') + cov(X_1 X_2, X_3'' X_4') + cov(X_1 X_2, X_3 X_4''). \tag{1.59}$$

So, by definition of (BL, θ)-dependence, $|cov(X_1' X_2', X_3' X_4')| \le 4c_0 M^2 r^{-\lambda}$. After estimating all the summands in (1.59), except the first and the second one, analogously to (1.57), (1.58) we conclude that

$$|\mathsf{E}X_1 X_2 X_3 X_4| \le |\mathsf{E}X_1' X_2' \mathsf{E}X_3' X_4'| + 4c_0 M^2 r^{-\lambda} + 8D_p M^{4-p}.$$

Set $F(M) := 4c_0 M^2 r^{-\lambda} + 8D_p M^{4-p}$. Optimization in M shows that the minimal value of F is

$$F(M) = 4c_0^v v^{-1} \big((p-4) D_p \big)^{2/(p-2)} r^{-\lambda v}$$

where v was introduced in (1.53).

Now it remains to estimate the quadruple sum $S^{(4)} := \Big| \sum_{J_4} \mathsf{E}X_{j_1} X_{j_2} \mathsf{E}X_{j_3} X_{j_4} \Big|$, taken over all sets J_4 with $Q_*(J_4) = \{j_1, j_2\}$. Obviously

$$S^{(4)} \le \sum_{j_1, j_2, j_3, j_4 \in U} |\mathsf{E}X_{j_1} X_{j_2}||\mathsf{E}X_{j_3} X_{j_4}| \le 3 \Big(\sum_{j_1, j_2 \in U} |\mathsf{E}X_{j_1} X_{j_2}| \Big)^2. \tag{1.60}$$

We fix index $j_1 \in U$ and estimate the corresponding sum over the free index j_2. Introduce the following notation:

$$J(j_1)^+ = \{j_2 \in U : \mathsf{E}X_{j_1} X_{j_2} \ge 0\}, \quad J(j_1)^- = \{j_2 \in U : \mathsf{E}X_{j_1} X_{j_2} < 0\}. \tag{1.61}$$

We consider only the sum $\sum^{(+)}$ over $j_2 \in J(j_1)^+$, other sum being estimated similarly.

$$\sum{}^{(+)} |\mathsf{E}X_{j_1} X_{j_2}| = \sum{}^{(+)} \mathsf{E}X_{j_1} X_{j_2} = \mathsf{E}X_1^2 + cov\Big(X_{j_1}, \sum_{q \in J(j_1)^+, q \ne j_1} X_q \Big) \le D_2 + \theta_1$$

by (BL, θ)-dependence.

Now (1.60) and (1.61) imply that $S^{(4)} \le 12(D_2 + \theta_1)^2 |U|$. Thus due to (1.56) one has

$$|S^{(2)}| \le R(U) + 4a \sum_{r=1}^{l_d} |\Lambda_r| r^{-\lambda v}$$

$$\le R(U) + 4 \cdot 144 d |U| a \sum_{k=0}^{d-1} \Big(\prod_{i=1}^{k} l_i \Big)^3 \sum_{r:k(r)=k} 3^{3d-1} r^{3(d-k)-1-\lambda v}$$

$$\le R(U) + 192 d |U| 3^{3d} a \sum_{k=0}^{d-1} \Big(\prod_{i=1}^{k} l_i \Big)^\tau \sum_{r=l_k+1}^{l_{k+1}} r^{(d-k)\tau - 1}$$

where $a = c_0^v v^{-1} \big((p-4)D_p\big)^{2/(p-2)}$ and τ is the same as in (1.53), whereas $R(U) = 12(D_2 + \theta_1)^2|U|^2$. From this one easily deduces that

$$|S^{(2)}| \leq 12(D_2 + \theta_1)^2|U|^2 + 192d^2|U|3^{3d}v^{-1}c_0^v((p-4)D_p)^{2/(p-2)}h(\tau)|U|B(|U|,\tau).$$

To prove the second assertion it suffices to notice that in quasi-association case all the estimates above are true if λ is replaced with κ. $\square$

Remark 1.26. The analysis of the proof shows that in fact θ_1 at the right-hand side of (1.52) can be replaced with the value

$$\vartheta_1 = \vartheta_1(X) := \sup_{j \in \mathbb{Z}^d} \sup_{V \not\ni j} \sup_\gamma \left| \sum_{k \in V} \gamma_k \mathsf{E} X_j X_k \right|$$

(the second supremum is over all finite subsets of $\mathbb{Z}^d$ not containing j, and the third one is over all functions $\gamma : \mathbb{Z}^d \to \{0,1\}$).

We also note that the proof of Theorem 1.25 demonstrates application of the dependence condition (1.5.9) when $|I| + |J| \leq 4$, for the subsets I and J of $\mathbb{Z}^d$ appearing in Definition 1.5.12.

Corollary 1.27. *Suppose that the conditions of Theorem 1.25 are satisfied with* $\lambda \geq 2d(p-2)/(p-4)$. *Then*

$$\mathsf{E} M(U)^4 \leq \gamma^d \left(D_4 + 3(D_2 + \theta_1)^2 + C_1(d, \lambda, p)c_0^{\frac{p-4}{p-2}} D_p^{\frac{2}{p-2}} \right) |U|^2 \qquad (1.62)$$

where $M(U)$ was introduced in (1.1) and $\gamma = (5/2)(1 - 2^{-1/4})^{-4}$. In quasi-association case the same claim is true with κ instead of λ.

Proof. Note that $\tau = 3 - \lambda v/d$ in the formula (1.52) and $v = (p-4)/(p-2)$. Therefore $\tau \leq 1 \Leftrightarrow \lambda \geq 2d(p-4)/(p-2)$. Thus the estimate (1.62) follows from (1.52) and Theorem 1.2. $\square$

2   Results based on supermodular order

In this Section we consider an approach to moment and maximal inequalities based on the theory of stochastic order. The key idea is to reduce the problems for initial random variables to the corresponding ones for independent variables.

1°. Supermodular functions. We begin with an important

Definition 2.1. A function $f : \mathbb{R}^n \to \mathbb{R}$ is called *supermodular* if it is bounded on any bounded set in $\mathbb{R}^n$ and $\exp\{f\}$ is MTP_2 (see Section 1.4, **2°**).

In other words,

$$f(x \vee y) + f(x \wedge y) \geq f(x) + f(y), \quad x, y \in \mathbb{R}^n.$$

A function f is called *submodular* if $(-f)$ is supermodular.

Lemma 2.2. *The following functions* $f : \mathbb{R}^n \to \mathbb{R}$ *are supermodular:*

(a) $\sum_{i=1}^{n} x_i$;

(b) $\max_{k=1,\dots,n} \sum_{i=1}^{k} x_i$;

(c) $-\max_{k=1,\dots,n} x_k$;

(d) $-\sum_{k=1}^{n}(x_k - \bar{x})^2$, *where* $\bar{x} = n^{-1}\sum_{i=1}^{n} x_i$;

(e) *the composition* $f(x) = g(h_1(x),\dots,h_n(x))$ *where* $g : \mathbb{R}^n \to \mathbb{R}$ *is supermodular and* $h_i : \mathbb{R} \to \mathbb{R}$ *are nondecreasing functions for* $i = 1,\dots,n$;

(f) *the composition* $f(x) = g(h(x))$, *where* $h : \mathbb{R}^n \to \mathbb{R}$ *is supermodular coordinatewise nondecreasing function and* $g : \mathbb{R} \to \mathbb{R}$ *is a nondecreasing convex function.*

Proof. (a) is obviously true. If $x, y \in \mathbb{R}^n$, then, for any $l, m \in \{1,\dots,n\}$,

$$\max_{k=1,\dots,n} \sum_{i=1}^{k}(x_i \vee y_i) + \max_{k=1,\dots,n} \sum_{i=1}^{k}(x_i \wedge y_i)$$

$$\geq \sum_{i=1}^{l\wedge m}(x_i \wedge y_i) + \sum_{i=1}^{l\vee m}(x_i \vee y_i) = \sum_{i=1}^{l\wedge m}(x_i \vee y_i + x_i \wedge y_i) + \sum_{i=l\wedge m+1}^{l\vee m}(x_i \vee y_i)$$

$$= \sum_{i=1}^{l\wedge m}(x_i + y_i) + \sum_{i=l\wedge m+1}^{l\vee m}(x_i \vee y_i) \geq \sum_{i=1}^{l} x_i + \sum_{i=1}^{m} y_i, \tag{2.1}$$

$$\max_{k=1,\dots,n} x_i + \max_{k=1,\dots,n} y_i \geq x_l \vee y_m + x_l \wedge y_m. \tag{2.2}$$

Taking the maxima in l, m on the right-hand side in (2.1) and (2.2) proves the claim for (b) and (c). For (d) we observe that $\partial^2 f/(\partial x_i \partial x_j) \leq 0$ whenever $i \neq j$ and apply Lemma 1.4.6. The case (e) is verified directly. For (f) take $x, y \in \mathbb{R}^n$ and note that $h(x), h(y) \in [h(x \wedge y), h(x \vee y)]$, as h is coordinatewise nondecreasing (clearly, one can only consider the case when $h(x) \leq h(y)$). If $a \leq b \leq c \leq d$ and $a + d \geq b + c$, then

$$g(a) + g(d) \geq g(b) + g(c). \tag{2.3}$$

To see this let Y be a random variable equal to b or c with probabilities $1/2$, and suppose that $X = a - b$ if $Y = b$ and $X = b - a$ if $Y = c$. Then the vector (X, Y) is associated (as X is a nondecreasing function in Y) and $\mathsf{E}X = 0$. Thus Lemma 1.20 yields $\mathsf{E}g(X+Y) \geq \mathsf{E}g(Y)$, which implies (2.3) as g is nondecreasing and $d \geq b+c-a$. Application of (2.3) with $a = h(x \wedge y)$, $b = h(x)$, $c = h(y)$, $d = h(x \vee y)$ implies the result. $\square$

Definition 2.3. Given random vectors $X = (X_1,\dots,X_n)$ and $Y = (Y_1,\dots,Y_n)$, we say that X is less than Y in the *supermodular order*, and write it as $X \leq_{\mathrm{sm}} Y$, if

$$\mathsf{E}\varphi(X) \leq \mathsf{E}\varphi(Y) \tag{2.4}$$

for any measurable bounded supermodular function $\varphi : \mathbb{R}^n \to \mathbb{R}$.

We establish an analogue of Theorem 1.1.5 showing that in (2.4) one can use a subclass of supermodular functions. Namely, we employ the following result by Müller and Scarsini.

Lemma 2.4. ([305]) *Let W and U be random vectors in $\mathbb{R}^n$ having the same marginal distributions. Suppose that $\mathsf{E}\varphi(W) \leq \mathsf{E}\varphi(U)$ for any supermodular function $\varphi : \mathbb{R}^n \to \mathbb{R}$ which is bounded, coordinatewise nondecreasing and twice differentiable. Then $W \leq_{\mathrm{sm}} U$.*

Proof passes through a series of steps. We prove in succession that the conditions of Lemma imply the inequality $\mathsf{E}f(W) \leq \mathsf{E}f(U)$ for supermodular f from the following classes of functions:

    1) bounded continuous nondecreasing,
    2) bounded right-continuous[4] nondecreasing,
    3) bounded nondecreasing,
    4) bounded.

Now we shall perform this plan.

1) Let $f : \mathbb{R}^n \to \mathbb{R}$ be a continuous bounded (i.e. $\sup_{x\in\mathbb{R}^n} |f(x)| < \infty$) nondecreasing supermodular function. Let

$$\chi_k(x) = (k/2\pi)^{n/2} \exp(-\|x\|^2 k/2), \quad k \in \mathbb{N},$$

be the functions introduced in the proof of Theorem 1.1.5, (c). Then the convolutions

$$f_k(x) := (f * \chi_k)(x) = \int_{\mathbb{R}^n} f(x - y)\chi_k(y)dy, \quad x \in \mathbb{R}^n,$$

are nondecreasing twice differentiable functions, bounded uniformly in $k \in \mathbb{N}$ and converging pointwise to f. Besides that, each f_k is supermodular, this follows directly from linearity of the integral. Hence $\mathsf{E}f_k(W) \leq \mathsf{E}f_k(U)$ for any $k \in \mathbb{N}$, and by the dominated convergence theorem $\mathsf{E}f(W) \leq \mathsf{E}f(U)$.

2) Suppose that $f : \mathbb{R}^n \to \mathbb{R}$ is bounded and right-continuous. Let ϱ_k, $k \in \mathbb{N}$, be the function equal to k^n on the cube $[(-k^{-1}, \ldots, -k^{-1}), 0] \subset \mathbb{R}^n$ and vanishing off this cube. Then $g_k := f * \varrho_k$ are bounded, continuous and pointwisely converge from above to f as $k \to \infty$ (to see that these functions are continuous in $x \in \mathbb{R}$, take a sequence $(x_m)_{m\in\mathbb{N}} \subset \mathbb{R}^n$ such that $x_m \to x$ as $m \to \infty$ and apply the dominated convergence theorem). Thus by the monotone convergence theorem $\mathsf{E}g_k(U) - \mathsf{E}g_k(W) \to \mathsf{E}f(U) - \mathsf{E}f(W)$ as $k \to \infty$. If f is supermodular, then f_g are also supermodular and $\mathsf{E}f(U) - \mathsf{E}f(W) \geq 0$.

3) Let $f : \mathbb{R}^n \to \mathbb{R}$ be a bounded nondecreasing supermodular function. For $i = 1, \ldots, n$ consider the functions

$$f^{(i)}(t) := \lim_{x_j \to +\infty, \, j\neq i} f(x_1, \ldots, x_{i-1}, t, x_{i+1}, \ldots, x_n), \quad t \in \mathbb{R}.$$

---

[4] A function $f : \mathbb{R}^n \to \mathbb{R}$ is right-continuous at $x = (x_1, \ldots, x_n) \in \mathbb{R}^n$ if $f(x_1 + h_1, \ldots, x_n + h_n) \to f(x)$ as $h_i \to 0+$, $i = 1, \ldots, n$.

These functions are obviously well defined and belong to $\mathcal{M}(1)$. For any $i = 1, \ldots, n$ and each $k \in \mathbb{N}$, introduce a set $S_k^{(i)}$ as follows. Put $s_{m,k}^{(i)} = \inf\{t : f^{(i)}(t) \geq 2^{-k}m\}$ for $m \in \mathbb{Z}$ and define

$$S_k^{(i)} = \left\{ s_{m,k}^{(i)}, \ m \in \mathbb{Z} \right\} \bigcup \left\{ m2^{-k}, \ m \in \mathbb{Z} \right\}.$$

Further take $S_k = S_k^{(1)} \times \ldots \times S_k^{(n)}$, $k \in \mathbb{N}$, and

$$f_k(x) := \sup\{f(z) : z \in S_k, \ z \leq x\}, \ x \in \mathbb{R}^n.$$

Then $f_k(x) = f(g^{(1)}(x_1), \ldots, g^{(n)}(x_n))$, for $x = (x_1, \ldots, x_n) \in \mathbb{R}^n$ where

$$g^{(i)}(y) = \sup\{t \in S_k^{(i)} : t \leq y\}, \ y \in \mathbb{R}, \quad i = 1, \ldots, n.$$

These functions $g^{(i)}$ are nondecreasing, hence by Lemma 2.2, e), f_k is a nondecreasing supermodular function. It is also right-continuous, which is easily verified. Moreover, for any $x \in \mathbb{R}^n$ the sequence $(f_k(x))_{k \in \mathbb{N}}$ is nondecreasing (because $S_k \subset S_{k+1}$, $k \in \mathbb{N}$) and $f_k(x) \leq f(x)$. Therefore, there exists a monotone pointwise limit $f_\infty(x) = \lim_{k \to \infty} f_k(x)$. Since f is supermodular, by definition of $f^{(i)}$ one has

$$f(x_1, \ldots, x_{i-1}, x_i + y, x_{i+1}, \ldots, x_n) - f(x) \leq f^{(i)}(x_i + y) - f^{(i)}(x_i) \qquad (2.5)$$

for any $i = 1, \ldots, n$ and all $y > 0$.

Take arbitrary $\varepsilon > 0$ and $x \in \mathbb{R}$. Let $k \in \mathbb{N}$ be large enough so that $2^{-k}n < \varepsilon$, and take $t = (t_1, \ldots, t_n)$ where $t_i = \sup\{t : t \in S_k, \ t \leq x\}$. Then by (2.5)

$$f(x) - f(t) = f(x_1, \ldots, x_n) - f(t_1, x_2, \ldots, x_n) + \ldots + f(t_1, \ldots, t_{n-1}, x_n) - f(t_1, \ldots, t_n)$$

$$\leq \sum_{i=1}^{n} (f^{(i)}(x_i) - f^{(i)}(t_i)) \leq \varepsilon,$$

as $f^{(i)}(x_i) - f^{(i)}(t_i) \leq 2^{-k}$, $i = 1, \ldots, n$. Therefore, we have $f_k(x) = f(t) \geq f(x) - \varepsilon$, which implies that $f_\infty(x) \geq f(x) - \varepsilon$. Since ε was arbitrary, we infer that $\lim_{k \to \infty} f_k(x) = f(x)$. The inequality $\mathsf{E}f(W) \leq \mathsf{E}f(U)$ follows now by the monotone convergence theorem.

4) Finally we dispense with the (coordinatewise) monotonicity of f. Suppose that $f : \mathbb{R}^n \to \mathbb{R}$ is a bounded supermodular function and $\|f\|_\infty = \sup_{x \in \mathbb{R}^n} |f(x)| < \infty$. For arbitrary $a \in \mathbb{R}$, introduce

$$f_a(x) = f(x) - \sum_{i=1}^{n} f(x_i e_i + a\mathbf{1}) + (n-1)f(a, \ldots, a)$$

where $e_i \in \mathbb{R}^n$ is the i-th coordinate vector and $\mathbf{1} = (1, \ldots, 1) \in \mathbb{R}^n$. Then f_a is supermodular, bounded by $2n\|f\|_\infty$, nonnegative and coordinatewise nondecreasing on the set $[a, +\infty)^n$. Indeed, for all $y > 0, x_1 \geq a, \ldots, x_n \geq a$ and any $i = 1, \ldots, n$, setting $\vec{a}_i = (a, \ldots, a) \in \mathbb{R}^i$ one has

$$f_a(x_1, \ldots, x_i + y, \ldots, x_n) - f_a(x_1, \ldots, x_i, \ldots, x_n) \geq f_a(a, x_2, \ldots, x_i + y, \ldots, x_n)$$

$$- f_a(a, x_2, \ldots, x_i, \ldots, x_n) \geq \ldots \geq f_a(\vec{a}_{i-1}, x_i + y, \vec{a}_{n-i}) - f_a(\vec{a}_{i-1}, x_i, \vec{a}_{n-i}) = 0.$$

For any $i = 1, \ldots, n$, from the equality $Law(W_i) = Law(U_i)$ it follows that

$$\Delta := \mathsf{E}f(U) - \mathsf{E}f(W) = \mathsf{E}f_a(U) - \mathsf{E}f_a(W). \tag{2.6}$$

Let $\varepsilon > 0$ be arbitrary and let $a = a(\varepsilon) \in \mathbb{R}$ be such that

$$\mathsf{P}(U \ngeq \mathbf{a}) + \mathsf{P}(W \ngeq \mathbf{a}) \leq \frac{\varepsilon}{2n\|f\|_\infty} \tag{2.7}$$

where $\mathbf{a} = (a, \ldots, a) \in \mathbb{R}^n$ and $\{x \ngeq \mathbf{a}\} := \mathbb{R}^n \setminus \{x \geq \mathbf{a}\}$. For each $\varepsilon > 0$ the function

$$g_\varepsilon(x_1, \ldots, x_n) := \mathbb{I}\{x \geq \mathbf{a}\} f_{a(\varepsilon)}(x_1, \ldots, x_n)$$

is nondecreasing and supermodular. This fact does not follow directly from Lemma 2.2 but is anyway easily seen. Indeed, if $x, y \in \mathbb{R}^n$, then either $(x \wedge y) \geq \mathbf{a}$ or $(x \wedge y) \ngeq \mathbf{a}$. In the first case $x, y, x \vee y \geq \mathbf{a}$, so

$$g_\varepsilon(x \vee y) + g_\varepsilon(x \wedge y) - g_\varepsilon(x) - g_\varepsilon(y) = f_a(x \vee y) + f_a(x \wedge y) - f_a(x) - f_a(y) \geq 0$$

since f is supermodular. In the second case one can find $i \in \{1, \ldots, n\}$ such that $(x_i \wedge y_i) < a$. Without loss of generality we may think that $i = 1$ and $x_1 < a$. Then

$$g_\varepsilon(x \vee y) + g_\varepsilon(x \wedge y) - g_\varepsilon(x) - g_\varepsilon(y) = g_\varepsilon(x \vee y) - g_\varepsilon(y).$$

If $y \geq \mathbf{a}$, then the right-hand side is equal to $f_a(x \vee y) - f_a(y)$ and is nonnegative by monotonicity of f_a; otherwise $g_\varepsilon(y) = 0$ and it remains to recall that the functions f_a are also nonnegative.

Set $A = \{W \geq \mathbf{a}, U \geq \mathbf{a}\}$, $B = \{W \ngeq \mathbf{a}, U \geq \mathbf{a}\}$ and $C = \{W \ngeq \mathbf{a}, U \ngeq \mathbf{a}\}$. Consequently, by (2.7)

$$\Delta = \mathsf{E}(f_a(U) - f_a(W))\mathbb{I}\{A\} + \mathsf{E}f_a(U)\mathbb{I}\{B\} - \mathsf{E}f_a(W)\mathbb{I}\{C\}$$

$$\geq \mathsf{E}(g_\varepsilon(U) - g_\varepsilon(W)) - 2n\|f\|_\infty(\mathsf{P}(U \ngeq \mathbf{a}) + \mathsf{P}(W \ngeq \mathbf{a})) \geq -\varepsilon.$$

Due to (2.6) we now have $\mathsf{E}f(U) \geq \mathsf{E}f(W)$ as required, since the choice of $\varepsilon > 0$ was arbitrary. $\square$

2°. Decoupled version of a random vector.

Definition 2.5. For a random vector $X = (X_1, \ldots, X_n)$, its *decoupled version* is a vector $Y = (Y_1, \ldots, Y_n)$ such that $Law(X_i) = Law(Y_i)$, $i = 1, \ldots, n$, and the components of Y are mutually independent.

Now we give a proof of a powerful result on positive and negative association by Christofides and Vaggelatou [103]. It shows that a (weakly) associated random vector is in some intuitively clear sense "greater" than its decoupled version, while for a negatively associated random vector the situation is converse.

Theorem 2.6. ([103]) *Let $Y = (Y_1, \ldots, Y_n)$ be the decoupled version of a random vector $X = (X_1, \ldots, X_n)$. If $X \in \mathbf{PA}$ then $X \geq_{sm} Y$, if $X \in \mathbf{NA}$ then $X \leq_{sm} Y$.*

Proof. The proofs in both cases are similar, thus we only deal with the **PA** case. We may assume from the beginning that X is independent of Y. Take a bounded, increasing and twice differentiable supermodular function $h : \mathbb{R}^n \to \mathbb{R}$. At first we will prove that

$$\mathsf{E}h(Y_1, \vec{X}) \leq \mathsf{E}h(X_1, \vec{X}) \tag{2.8}$$

where $\vec{X} = (X_2, \ldots, X_n)$. Set $g(x_1, \vec{x}) = \partial h / \partial x_1(x_1, \vec{x})$, $x = (x_2, \ldots, x_n) \in \mathbb{R}^{n-1}$. Then, for any $x, y \in \mathbb{R}$,

$$h(x, \vec{X}) = h(y, \vec{X}) + \int_y^x g(t, \vec{X})dt = h(y, \vec{X}) + \int_{\mathbb{R}} (\mathbb{I}\{x > t\} - \mathbb{I}\{y > t\})g(t, \vec{X})dt,$$

and consequently

$$h(X_1, \vec{X}) - h(Y_1, \vec{X}) = \int_{\mathbb{R}} (\mathbb{I}\{X_1 > t\} - \mathbb{I}\{Y_1 > t\})g(t, \vec{X})dt.$$

Invoking the Fubini theorem and independence of Y_1 and $\vec{X}$, we can conclude that

$$\mathsf{E}h(X) - \mathsf{E}h(Y) = \int_{\mathbb{R}} \mathsf{E}(\mathbb{I}\{X_1 > t\} - \mathbb{I}\{Y_1 > t\})g(t, \vec{X})dt$$

$$= \int_{\mathbb{R}} \left(\mathsf{E}\mathbb{I}\{X_1 > t\}g(t, \vec{X}) - \mathsf{E}\mathbb{I}\{Y_1 > t\}\mathsf{E}g(t, \vec{X}) \right) dt = \int_{\mathbb{R}} cov(\mathbb{I}\{X_1 > t\}, g(t, \vec{X})dt.$$

The last expression is nonnegative by **PA**, which yields (2.8).

Now the Theorem is proved by induction on n. For $n = 1$ the assertion is trivially true (with equality in (2.4)). Suppose that it holds for random vectors in $\mathbb{R}^{n-1}$. Note that the function

$$h_1(\vec{x}) := \mathsf{E}h(Y_1, \vec{x})$$

on $\mathbb{R}^{n-1}$ is supermodular. It is also bounded, twice differentiable and nondecreasing provided that h has all these properties. By the Fubini theorem and induction hypothesis

$$\mathsf{E}h(Y_1, \ldots, Y_n) = \mathsf{E}h_1(Y_2, \ldots, Y_n) \leq \mathsf{E}h_1(\vec{X}) = \mathsf{E}h(Y_1, \vec{X}).$$

From this inequality and (2.8) the assertion follows. $\square$

The statement converse to Theorem 2.6 is not true, as was shown by Hu, Müller and Scarsini who proposed the following

Example 2.7. ([203]) Let $X = (X_1, X_2, X_3, X_4)$ be a binary random vector in $\mathbb{R}^4$ such that $\mathsf{P}(X = 0) = 1/6$, $\mathsf{P}(A_{1,2}) = 5/12$, $\mathsf{P}(A_{3,4}) = 5/12$ where

$$A_{1,2} = \{0 < X_1 + X_2 + X_3 + X_4 \leq 2\}, \quad A_{3,4} = \{X_1 + X_2 + X_3 + X_4 > 2\}$$

and all 10 possible outcomes constituting events $A_{1,2}$ have equal probability, as well as 5 events constituting $A_{3,4}$. Then $X \geq_{sm} Y$, where Y is the decoupled version of X, but $X \notin$ **PA**.

Indeed, let $f : \mathbb{R}^4 \to \mathbb{R}$ be a supermodular function. Since the distribution of X is symmetric, proving the "$\leq_{sm}$" property we may and will confine ourselves to the case when f is symmetric (see the argument establishing Theorem 1.1.26). We may also assume that $f(0,0,0,0) = 0$. Note that Y_i ($i = 1,2,3,4$) are i.i.d. random variables taking values 0 and 1 with equal probability. Thus

$$\mathsf{E}f(X) - \mathsf{E}f(Y) = \frac{1}{6}f(1,0,0,0) + \frac{1}{4}f(1,1,0,0) + \frac{1}{3}f(1,1,1,0) + \frac{1}{12}f(1,1,1,1)$$

$$-\frac{1}{4}f(1,0,0,0) - \frac{3}{8}f(1,1,0,0) - \frac{1}{4}f(1,1,1,0) - \frac{1}{16}f(1,1,1,1)$$

$$= \frac{1}{12}\left(f(1,1,1,0) - f(1,1,0,0) - f(1,0,0,0) + f(0,0,0,0)\right)$$

$$+\frac{1}{48}\left(f(1,1,1,1) - f(1,1,0,0) - f(0,0,1,1) + f(0,0,0,0)\right) \geq 0$$

by symmetry and supermodularity of f. To see that $X \notin \mathbf{PA}$ write

$$\mathsf{P}(X_i = 1, i = 1,2,3,4) = \frac{1}{12} < \left(\frac{1}{4} + \frac{1}{24}\right)^2 = \mathsf{P}(X_1 = X_2 = 1)\mathsf{P}(X_3 = X_4 = 1).$$

Corollary 2.8. *Let $X = \{X_t, t \in T\}$ be a countable nonnegatively correlated Gaussian random system[5] and $Y = \{Y_t, t \in T\}$ be independent random variables such that $Law(Y_t) = Law(X_t)$, $t \in T$. Then, for any nondecreasing convex function $f : \mathbb{R} \to \mathbb{R}$,*

$$\mathsf{E}f\left(\sup_{t \in T} X_t\right) \leq \mathsf{E}f\left(\sup_{t \in T} Y_t\right)$$

if both expectations exist. If the system is nonpositively correlated, the inequality above is valid with reversed sign.

Proof. At first assume that $|T| < \infty$. For any $k \in \mathbb{N}$ let H_k be the function defined in (1.5). The function $(x_t, t \in T) \mapsto f(\max_{t \in T} H_k(x_t))$ is bounded (as a convex function defined on $\mathbb{R}$ is bounded on any compact) and submodular by Lemma 2.2, (c), (e) and (f). Hence by Theorems 1.2.1 and 2.6 one has

$$\mathsf{E}f\left(\sup_{t \in T} H_k(X_t)\right) \leq \mathsf{E}f\left(\sup_{t \in T} H_k(Y_t)\right).$$

Letting $k \to \infty$ and using the dominated convergence theorem yields the result. If T is infinite, take a sequence of finite sets $(T_n)_{n \in \mathbb{N}}$ such that $T_n \subset T_{n+1}$, $n \in \mathbb{N}$, and $\cup_{n \in \mathbb{N}} T_n = T$. For any $n \in \mathbb{N}$ one has, by the already proved statement,

$$\mathsf{E}f\left(\sup_{t \in T_n} X_t\right) \leq \mathsf{E}f\left(\sup_{t \in T_n} Y_t\right)$$

and, since f is nondecreasing, the result follows by the monotone convergence theorem. $\square$

---

[5] In view of Theorem 1.2.1 one can say that $X \in \mathbf{A}$.

3°. Applications to sums and maxima of partial sums. Now we can obtain the result by Shao.

Corollary 2.9. ([371]) *Let* $X_1, X_2, \ldots$ *be a sequence of* **PA** *random variables,* $(Y_n)_{n\in\mathbb{N}}$ *be its decoupled version* [6] *and* $T_n = \sum_{j=1}^{n} Y_j$. *Then, for any* $n \in \mathbb{N}$ *and any convex function* $f : \mathbb{R} \to \mathbb{R}$,

$$\mathsf{E}f(S_n) \geq \mathsf{E}f(T_n), \tag{2.9}$$

whenever the expectations exist. If, moreover, f *is nondecreasing, then*

$$\mathsf{E}f(M_n) \geq \mathsf{E}f(M_n^*) \tag{2.10}$$

if both expectations exist, where $M_n = \max_{j=1,\ldots,n} S_j$ *and* $M_n^* = \max_{j=1,\ldots,n} T_j$. *If the sequence is* **NA**, *both assertions are valid with reversed signs.*

Proof. We proceed with **PA** case, the **NA** case being similar. At first we require that $|X_j| \leq k$ a.s. for some fixed $k \in \mathbb{N}$ and any $j \in \mathbb{N}$. If f is nondecreasing, both inequalities follow from Theorem 2.6 and Lemma 2.2, (b), (d) and (f) (note that all functions under expectations are bounded as we consider fixed $k \in \mathbb{N}$). Otherwise one can write $f(x) = f^{(+)}(x) + f^{(-)}(x)$, where $f^{(+)}$ is nondecreasing convex and $f^{(-)}$ is nonincreasing convex function. Namely, put

$$f^{(+)}(x) = f(c) + (f(x) - f(c))\mathbb{I}\{x \geq c\}$$

where $c \in \mathbb{R}$ is such that $f(c) = \inf_{x\in\mathbb{R}} f(x)$. Then by (2.9) we have

$$\mathsf{E}f_+(S_n) \geq \mathsf{E}f_+(T_n). \tag{2.11}$$

Moreover, the sequence $(-X_n)_{n\in\mathbb{N}}$ is **PA** and its decoupled version is $(-Y_n)_{n\in\mathbb{N}}$. Therefore, taking $g(x) = f_-(-x)$, by (2.9) one obtains the relation

$$\mathsf{E}f_-(S_n) = \mathsf{E}g(-S_n) \geq \mathsf{E}g(-T_n) = \mathsf{E}f_-(T_n). \tag{2.12}$$

It remains to add (2.11) and (2.12).

To avoid the requirement that random variables X_j are uniformly bounded, consider the function H_k ($k \in \mathbb{N}$) defined in (1.5). Due to Theorem 1.1.8, (d), the sequence $(H_k(X_n))_{n\in\mathbb{N}}$ is **PA**. For its decoupled version $(H_k(Y_n))_{n\in\mathbb{N}}$, by the already proved part of assertion, we have

$$\mathsf{E}f\left(\sum_{j=1}^{n} H_k(X_j)\right) \geq \mathsf{E}f\left(\sum_{j=1}^{n} H_k(Y_j)\right)$$

instead of (2.9) and similar inequality instead of (2.10). It remains to use the dominated convergence theorem, which applies as f is convex. $\square$

Specializing f we provide an analogue of the Bernstein inequality. Note that the general way of proving inequalities for random vectors with the help of their decoupled versions is described in the book by de la Peña and Giné [327]. The results

---

[6] That is $Y_1, Y_2, \ldots$ are independent random variables and $Law(Y_n) = Law(X_n)$ for any $n \in \mathbb{N}$.

constructing decoupled versions, also known as reconstructions, help to approximate a given random vector (having dependent components) with one consisting of independent random variables. That is, these two vectors are built on the same probability space, possibly different from the initial one, and they are close in some sense (say, in probability or in mean). Important contributions in this domain were made by Strassen [391], Dobrushin [137], Berkes and Philipp [37], Bryc [56], Berbee [29], Peligrad [319].

Corollary 2.10. *Let* $(X_1, \ldots, X_n)$ *be negatively associated centered random variables such that* $B_n := \sum_{j=1}^{n} \operatorname{Var} X_j < \infty$. *Suppose that there exists some* $R > 0$ *such that* $|X_k| \leq R$ *a.s,* $k = 1, \ldots, n$. *Then, for any* $x \geq 0$ *and all* $n \in \mathbb{N}$,

$$
\mathsf{P}(S_n \geq x) \vee \mathsf{P}(S_n \leq -x) \leq \begin{cases} \exp\{-\frac{x^2}{2B_n}\}, & x \in [0, B_n/R], \\ \exp\{-\frac{x}{2R}\}, & x \geq B_n/R. \end{cases} \tag{2.13}
$$

Proof. We only prove the estimate for $\mathsf{P}(S_n \geq x)$, since the other one is similar (as $(-X_j)_{j=1,\ldots,n}$ is the random vector satisfying the conditions of the Corollary). For any $t \in [0, 1/R]$ and $k = 1, \ldots, n$, due to the Fubini theorem one has

$$
\mathsf{E} e^{tX_k} = \mathsf{E} \sum_{m=0}^{\infty} \frac{t^m X_k^m}{m!} = 1 + \sum_{m=2}^{\infty} \frac{t^m \mathsf{E} X_k^m}{m!} \leq 1 + t^2 \operatorname{Var} X_k \sum_{m=2}^{\infty} \frac{(tR)^{m-2}}{(m-2)!m}
$$

$$
\leq 1 + \frac{t^2}{2} \operatorname{Var} X_k \sum_{m=2}^{\infty} \frac{1}{(m-2)!} = 1 + \frac{t^2}{2} \operatorname{Var} X_k \leq \exp\{(t^2/2) \operatorname{Var} X_k\} \tag{2.14}
$$

because $1 + y \leq e^y$, $y \in \mathbb{R}$. Let $(Y_1, \ldots, Y_n)$ be the decoupled version of $(X_1, \ldots, X_n)$ and $T_n = Y_1 + \ldots + Y_n$. Then, by Corollary 2.9 (or by Corollary 1.1.10), for any $t > 0$ and $x \geq 0$ we have

$$
\mathsf{P}(S_n \geq x) \leq e^{-tx} \mathsf{E} e^{tS_n} \leq e^{-tx} \mathsf{E} e^{tT_n} = e^{-tx} \prod_{k=1}^{n} \mathsf{E} e^{tX_k} \leq \exp\{-tx + B_n t^2/2\}.
$$

$$
\tag{2.15}
$$

Now, if $x \leq B_n/R$, take $t = x/B_n \leq 1/R$; if $x > B_n/R$, then take $t = R^{-1}$. Substitution of such t in (2.15) yields (2.13). $\square$

Remark 2.11. Clearly, the inequality (2.9) is applicable to random fields as well, since it does not involve the partial order of the parameter set. However, for the maxima of the partial sums such generalization is not possible, since the function

$$
f(x_1, \ldots, x_n) = \sup_{I} \sum_{i \in I} x_i,
$$

for a given system of subsets $I \subset \{1, \ldots, n\}$, in general, is not supermodular. Namely, we have a counterexample to Corollary 2.9 (**NA** case) in the multidimensional situation.

As usual, for a field $X = \{X_j, j \in \mathbb{N}^d\}$ let $Y = \{Y_j, j \in \mathbb{Z}^d\}$ be its decoupled version and M_n (resp. M_n^*) denote the maxima of the partial sums of X (resp. Y) over blocks standardly belonging to $(0, n]$, $n \in \mathbb{N}^d$.

Theorem 2.12. ([82]) *Let $f : \mathbb{R} \to \mathbb{R}$ be a function such that $f(1) > f(0)$ (in particular, strictly increasing). Then, for any $d > 1$, there exists a **NA** random field $X = \{X_j, j \in \mathbb{Z}^d\}$ and a multiindex $n \in \mathbb{N}^d$ such that*

$$\mathsf{E}f(M_n) > \mathsf{E}f(M_n^*).$$

Proof. It suffices to consider $d = 2$ and $n = (2,2)$ (we can take other random variables to be a.s. zero). Let η be a random variable such that $\mathsf{P}(\eta = -1) = \mathsf{P}(\eta = 1) = 1/2$, and define

$$X_{1,1} = 0, \quad X_{1,2} = \eta, \quad X_{2,1} = -\eta, \quad X_{2,2} = -3.$$

Then $(X_{1,1}, X_{1,2}, X_{2,1}, X_{2,2}) \in \mathbf{NA}$ by Theorem 1.1.8 (d). For $n = (2,2)$ one has $M_n = 1$ a.s., since always either $S_{2,1} = 1$ or $S_{1,2} = 1$, other partial sums being at the same time nonpositive. Therefore $\mathsf{E}f(M_n) = f(1)$.

The random variables $Y_{1,2}$ and $Y_{2,1}$ are independent and distributed as η. Thus, since $Y_{1,1} = 0$ and $Y_{1,2} + Y_{2,1} + Y_{2,2} < 0$, we see that $M_{(2,2)}^*$ equals 0 if and only if both $Y_{1,2}$ and $Y_{2,1}$ take the value -1. Otherwise $M_{(2,2)}^*$ equals 1. Hence

$$\mathsf{E}f(M_n^*) = \frac{1}{4}f(0) + \frac{3}{4}f(1) < f(1). \qquad \square$$

3 Rosenthal-type inequalities

For $p \geq 1$ and a real-valued random variable X we write $\|X\|_p = (\mathsf{E}|X|^p)^{1/p}$. Let $\{X_t, t \in T\}$ be a family of (real-valued) random variables. For a finite set $U \subset T$ and real numbers $a, b \geq 1$ define the functions

$$S_U = S(U) = \sum_{t \in U} X_t, \quad Q(U, a, b) = \sum_{t \in U} \|X_t\|_a^b. \tag{3.1}$$

The classical Rosenthal inequality (see, e.g., [326, Ch. III, §5,6]) states that, if X_t, $t \in T$, are independent centered random variables such that $\mathsf{E}|X_t|^p < \infty$ for some $p > 1$, then for any finite $U \subset T$ one has

$$\mathsf{E}|S(U)|^p \leq 2^{p^2} \left(Q(U, p, p) \vee (Q(U, 1, 1))^p \right), \tag{3.2}$$

$$\mathsf{E}|S(U)|^p \geq 2^{-p}(Q(U, p, p) \vee Q(U, 2, 2)^{p/2}). \tag{3.3}$$

In this Section we give several theorems extending the above inequalities to weakly dependent random systems. The truncation of the random variables and employment of randomization technique are the main tools here. Moreover, the weighted sums, shot-noise and cluster random fields are studied as well. The former will be used in Chapter 8.

1°. The moment bounds for PA and NA random fields. The next result is due to Vronski.

Theorem 3.1. ([405]) *Let a centered random field $X = \{X_j, j \in \mathbb{Z}^d\} \in \mathbf{PA}^7$ and*
1) $a < \mathsf{E}X_j^2 < \infty$ *for some $a > 0$ and all $j \in \mathbb{Z}^d$;*
2) *the finite susceptibity condition (see (1.5.4)) holds;*
3) *there exist an even integer $k \geq 2$ and some $\delta > 0$ such that*

$$\mathsf{E}|X_j|^{k+\delta} < \infty \text{ for any } j \in \mathbb{Z}^d,$$

$$\sum_{n=1}^{\infty} u_n^{\delta/(k+\delta-2)} n^{d(k-1)-1} < \infty \tag{3.4}$$

where u_n is the Cox–Grimmett coefficient (1.5.4).
Then there exists $C = C(k)$ determined by d, k, δ and the sequence $(a^{-2}u_n)_{n\in\mathbb{N}}$ such that

$$\mathsf{E}|S(U)|^k \leq C(k)\left(Q(U, k+\delta, k) \vee Q(U, 2+\delta, 2)^{k/2}\right)$$

for any finite $U \subset \mathbb{Z}^d$.

Proof. The proof is based on the expansion of $\mathsf{E}S_U^k$ into the sum of expectations over the k-tuples $(j_1, \ldots, j_k)$ where $j_i \in U$, $i = 1, \ldots, k$. We rely on Lemmas 1.22 and 1.23, devoted to estimation of the covariance between products of X_j's, and need one more auxiliary result.

Lemma 3.2. ([138]) *Let $X_t, t \in U$, be a set of random variables and U be a finite subset of $\mathbb{Z}^d$. Assume that, for some $\delta > 0$, and some integers $m > 1$ and $n > 1$, one has $\mathsf{E}|X_t|^{m+n+\delta} < \infty$, $t \in U$. Then*

$$\left(Q(U, m+\delta, m) \vee Q(U, 2+\delta, 2)^{m/2}\right)\left(Q(U, n+\delta, n) \vee Q(U, 2+\delta, 2)^{n/2}\right)$$

$$\leq \left(Q(U, m+n+\delta, m+n) \vee Q(U, 2+\delta, 2)^{(m+n)/2}\right).$$

Proof. If all $X_t, t \in U$, are almost surely zero, the assertion is obviously true. Otherwise, we can normalize them in such a way that

$$\sum_{t\in U} \|X_t\|_{2+\delta}^2 = Q(U, 2+\delta, 2) = 1. \tag{3.5}$$

Thus we have to verify the inequality

$$A(U, m+\delta, m)A(U, n+\delta, n) \leq A(U, m+n+\delta, m+n)$$

where $A(U, n+\delta, v) := Q(U, n+\delta, v) \vee 1$. We write

$$c = m+n, \; u = \frac{(c+\delta)(m-2)}{c-2}, \; v = \frac{(2+\delta)n}{c-2}, \; r = \frac{mu}{c(m+\delta)}, \; s = \frac{mn(2+\delta)}{2(c-2)(m+\delta)}.$$

---

[7]The proof given in [405] for associated fields remains valid in **PA** case.

Then, by the Hölder inequality, for any random variable Z,

$$\mathsf{E}|Z|^{m+\delta} \leq (\mathsf{E}|Z|^{c+\delta})^{u/(c+\delta)}(\mathsf{E}|Z|^{2+\delta})^{v/(2+\delta)} = \|Z\|_{c+\delta}^{u}\|Z\|_{2+\delta}^{v}.$$

Therefore, again by Hölder's inequality,

$$Q(U, m+\delta, m) = \sum_{t\in T}\|X_t\|_{m+\delta}^{m} \leq \sum_{t\in T}\|X_t\|_{c+\delta}^{mu/(m+\delta)}\|X_t\|_{2+\delta}^{mv/(m+\delta)}$$

$$\leq Q(U, m+n+\delta, m+n)^{r} Q(U, 2+\delta, 2s/(1-r))^{1-r}.$$

In view of (3.5) and inequality $s \geq 1-r$ we have $Q(U, 2+\delta, 2s/(1-r)) \leq 1$. Hence,

$$Q(U, m+\delta, m) \leq Q(U, m+n+\delta, m+n)^{r}$$

$$\leq Q(U, m+n+\delta, m+n)^{m/c} \vee 1 = A(U, m+n+\delta, m+n)^{\frac{m}{c}},$$

here we have used the fact that $r \leq m/c$. Changing the roles of m and n in the argument we see that an analogous relation holds for $Q(U, n+\delta, n)$. Finally we get

$$A(U, m+\delta, m)A(U, n+\delta, n)$$

$$\leq A(U, m+n+\delta, m+n)^{m/c} A(U, m+n+\delta, m+n)^{n/c} = A(U, m+n+\delta, m+n),$$

as required. $\square$

Now we are able to start proving Theorem 3.1. By $[U]^p$, $p \in \mathbb{N}$, we will denote the collection of non-ordered p-tuples consisting of points of U. For a set $S \in [U]^p$, we introduce a "product" variable $X(S) = X_{s_1}\ldots X_{s_p}$ and $\Pi(S) = \prod_{s\in R(S)} n(s)!$, here $R(S)$ is the set of different components of S and $n(s)$ is the multiplicity of s in S. For a finite set $V \subset \mathbb{Z}^d$, define a function $c(V)$ (the size of a *"maximal hole"* in V) as follows. If $|V| = 1$, then $c(V) := 0$. Otherwise, $c(V) := r$, where $r \in \mathbb{N}$ is the maximal number chosen so that it is possible to divide V into nonempty disjoint subsets V_1 and V_2 with $dist(V_1, V_2) = r$. One can view V_1 and V_2 as set-valued functions of V. For example, one can select V_1 satisfying the requirement that the union of all balls in $\mathbb{R}^d$ of radius r centered at points of V_1 forms a connected set. To make the choice of V_1 unique we define $\min(W)$ for $W \subset \mathbb{Z}^d$ as a point which is minimal in the lexicographic sense. Among all possible choices of $V_1 = V_1(V)$ we fix that one for which $\min(V_1)$ is minimal (also in the lexicographic sense) for these choices. If there are V_1' and V_1'' with $s = \min(V_1') = \min(V_1'')$, we consider in V_1' and V_1'' the points t' and t'' respectively which are the closest to s. If $t' < t''$, set $V_1 = V_1'$. If $t' = t''$ continue in the same way. In case when all the points in V_1' were compared with their counterparts in V_1'' in this manner and coincided, take $V_1 = V_1'$ if $|V_1'| < |V_1''|$. We have then

$$\mathsf{E}S(U)^k \leq A_k = \sum_{S\in[U]^k} \frac{k!}{\Pi(S)}|\mathsf{E}X(S)|.$$

For any positive integer $m \leq k$

$$A_m \leq \sum_{r=1}^{diam(U)} \sum_{q=1}^{m-1} {\sum_{S_1,S_2}}' \frac{m!}{\Pi(S)}|\mathsf{E}X(S_1)X(S_2)| + \sum_{j\in U}|\mathsf{E}X_j^m| \tag{3.6}$$

where $\sum'$ is the sum taken over $S_1 \in [U]^q$, $S_2 \in [U]^{m-q}$ such that

$$dist(R(S_1), R(S_2)) = c(R(S_1) \cup R(S_2)) = r. \tag{3.7}$$

The first sum on the right-hand side of (3.6) is not greater than

$$\sum {}'' \frac{m!}{\Pi(S)} |EX(S_1)||EX(S_2)| + \sum {}'' \frac{m!}{\Pi(S)} |cov(X(S_1), X(S_2))| =: I_1 + I_2$$

where $\sum''$ means that summation is carried over the same r, q, S_1 and S_2 that were used in (3.6). If S_1 and S_2 are such sets that $R(S_1) \cap R(S_2) = \varnothing$, then, clearly, $\Pi(S_1 S_2) = \Pi(S_1)\Pi(S_2)$. Therefore,

$$I_1 \leq \sum_{q=1}^{m-1} \binom{m}{q} \sum_{S_1 \in [U]^q} |EX(S_1)| \sum_{S_2 \in [U]^{m-q}} |EX(S_2)| = \sum_{q=1}^{m-1} \binom{m}{q} A_q A_{m-q}.$$

Set $\mu = \delta/(m+\delta-2)$. Note that under condition (3.7) for $j \in R(S_1)$ and $s \in R(S_2)$ one has $cov(X_j, X_s) \leq u_r$. Thus by Lemma 1.22

$$I_2 \leq m! \sum {}'' 3m^{2\mu} u_r^{\mu} F(S_1 \cup S_2) = 3m!m^{2\mu} \sum_{r=1}^{diam(U)} u_r^{\mu} \sum_{S \in [U]^m} {}''' F(S).$$

Here

$$F(S) = \left(\prod_{i=1}^{m} \|X_{s_i}\|_{m+\delta} \sum_{j=1}^{m} \|X_{s_j}\|_{m+\delta}^{\delta} \right)^{1-\mu}$$

and $\sum'''$ is the sum over $S = \{s_1, \ldots, s_m\} \in [U]^m$ such that: a) $c(R(S)) = r$, b) the union $S_1 \cup S_2$ belongs to $[U]^m$. Furthermore,

$$F(S) \leq m^{1-\mu} \max_{s \in R(S)} \|X_s\|_{m+\delta}^{(m+\delta)(1-\mu)}.$$

Hence

$$I_2 \leq 3m!m^{1+\mu} \sum_{r=1}^{diam(U)} u_r^{\mu} \sum_{s \in U} \sum_{V \subset U} {}^{(s)} \|X_s\|_{m+\delta}^{(m+\delta)(1-\mu)}$$

where the sum $\sum^{(s)}$ is taken over subsets $V \subset U$ such that $s \in V$, $c(V) = r$ and $2 \leq |V| \leq m$. An equivalent relation is

$$I_2 \leq 3m!m^{1+\mu} \sum_{r=1}^{diam(U)} u_r^{\mu} \sum_{s \in U} \|X_s\|_{m+\delta}^{(m+\delta)(1-\mu)} \sum_{V \subset U} {}^{(s)} 1.$$

To estimate the last sum, let $V = V_1 \cup V_2$ be a partition of V into disjoint sets with $dist(V_1, V_2) = r$, $|V_1| = t$, $|V_2| = p$. If $t > 1$ then the ball $B_r(s)$ of radius r with center s contains a point $s^{(1)}$ from U_1, $s^{(1)} \neq s$. If $t > 2$ then the set $B_r(s) \cup B_r(s^{(1)})$ contains a point $s^{(2)} \in V_1 \setminus \{s, s^{(1)}\}$. Proceeding with this process we can exhaust V_1. Then within the union of spheres of radii r with centers at all points of V_1 there is a point $s^{(t+1)}$ from V_2. If $p > 1$ then the union of that spheres and the ball with center at $s^{(t+1)}$ contains a point from $V_2 \setminus \{s^{(t+1)}\}$. So we come to the following estimate for $\sum_{V \subset \mathbb{Z}^d}^{(s)} 1$:

$$\sum_{i=1}^{m} \sum_{q=1}^{i-1} b_r 2b_r \ldots (q-1)b_r q s_r (q s_r + b_r) \ldots (q s_r + (i-q-1)b_r) \leq m! m^2 s_r b_r^{m-2}$$

where $b_r = (2r+1)^d$ and $s_r = (2r+1)^d - (2r-1)^d$ are the respective cardinalities of the ball and the sphere of radius r in $\mathbb{Z}^d$ (instead of the condition $V \subset U$ we used the assumption $V \subset \mathbb{Z}^d$). Thus, with the notation $G_{m,\delta} := 3(m!)^2 m^2 m^{1+\mu}$, we can write

$$I_2 \leq G_{m,\delta} \sum_{r=1}^{diam(U)} u_r^{\mu} s_r b_r^{m-2} \sum_{s \in U} \|X_s\|_{m+\delta}^{(m+\delta)(1-\mu)}$$

$$\leq G_{m,\delta} \sum_{r=1}^{diam(U)} (a^{-2} u_r)^{\mu} s_r b_r^{m-2} \sum_{s \in U} \|X_s\|_{m+\delta}^{(m+\delta)(1-\mu)} \|X_s\|_2^{2\mu}$$

$$\leq G_{m,\delta} \sum_{r=1}^{diam(U)} (a^{-2} u_r)^{\mu} s_r b_r^{m-2} \sum_{s \in U} \|X_s\|_{m+\delta}^{m}$$

in view of the bound $\mathsf{E} X_j^2 \geq a^2$, $j \in \mathbb{Z}^d$, and the Lyapunov inequality. So

$$A_m \leq \sum_{q=1}^{m-1} \binom{m}{q} A_q A_{m-q} + H_{m,\delta} \sum_{s \in U} \|X_s\|_{m+\delta}^{m} \tag{3.8}$$

where

$$H_{m,\delta} = 1 + G_{m,\delta} \sum_{r=1}^{\infty} (a^{-2} u_r)^{\mu} s_r b_r^{m-2}, \tag{3.9}$$

and the series in (3.9) converges in view of (3.4). Notice that $A_1 = \sum_{j \in U} |\mathsf{E} X_j| = 0$ and $A_2 \leq H_{2,\delta} Q(U, 2+\delta, 2)$ by (3.8). Now set

$$L(U,n) = Q(U, n+\delta, n) \vee Q(U, 2+\delta, 2)^{n/2}, \quad n \in \mathbb{N}.$$

For $k > 2$, we apply (3.8) with $m = 3, \ldots, k$ to get, by virtue of Lemma 3.2,

$$A_k \leq \sum_{q=1}^{k-1} \binom{k}{q} C(q) C(k-q) L(U,q) L(U, k-q) + H_{k,\delta} L(U,k)$$

$$\leq L(U,k) \Big(\sum_{q=1}^{k-1} \binom{k}{q} C(q) C(k-q) + H_{k,\delta} \Big). \tag{3.10}$$

The proof is complete. $\square$

The study of negatively associated random fields is easier due to the supermodular functions approach by Christofides and Vaggelatou.

Theorem 3.3. ([103]) *Let* $X = \{X_j, j \in \mathbb{Z}^d\}$ *be a* **NA** *random field such that* $\mathsf{E} X_j = 0$ *and* $\mathsf{E}|X_j|^p < \infty$ *for some* $p > 2$. *Then, for any finite* $U \subset \mathbb{Z}^d$,

$$\mathsf{E}|S_U|^p \leq 2^{p^2} \Big(Q(U, p, p) \vee Q(U, 2, 2)^{p/2} \Big)$$

where the function Q *is introduced in* (3.1).

Proof. Let $\{Y_j, j \in U\}$ be a decoupled version of $\{X_j, j \in U\}$, and let $T_U = \sum_{j \in U} Y_j$. The function $x \mapsto |x|^p$ is convex, therefore by Corollary 2.9 we have $\mathsf{E}|S_U|^p \leq \mathsf{E}|T_U|^p$. The classical Rosenthal inequality (3.2) implies that

$$\mathsf{E}|T_U|^p \leq 2^{p^2}\left(\sum_{j \in U} \mathsf{E}|Y_j|^p \vee \left(\sum_{j \in U} \mathsf{E}Y_j^2\right)^{p/2}\right) = 2^{p^2}\left(Q(U, p, p) \vee Q(U, 2, 2)^{p/2}\right). \quad \square$$

For **PA** random variables, the same approach yields the lower bound.

Theorem 3.4. ([103]) *Let* $X = \{X_j, j \in \mathbb{Z}^d\}$ *be a* **PA** *random field such that* $\mathsf{E}X_j = 0$ *and* $\mathsf{E}|X_j|^p < \infty$ *for some* $p > 2$. *Then, for any finite* $U \subset \mathbb{Z}^d$,

$$\mathsf{E}|S_U|^p \geq 2^{-p}\left(Q(U, p, p) \vee Q(U, 2, 2)^{p/2}\right).$$

2°. Weighted sums. An important restriction imposed in Theorem 3.1 is that the variances must be uniformly separated from zero. A modification of that result, devoid of such restriction, permits to estimate the weighted sums of associated random variables, which will help to establish a moment inequality for integrals over associated measures in Ch. 8. Recall that a polynomial function $P : \mathbb{R}^n \to \mathbb{R}$ is called a *homogeneous polynomial of degree s* (here $s \in \mathbb{Z}_+$) if

$$P(x_1, \ldots, x_n) = \sum c_{j_1, \ldots, j_n} x_1^{j_1} \ldots x_n^{j_n},$$

where the sum is carried over all n-tuples of nonnegative integers $(j_1, \ldots, j_n)$ with $j_1 + \ldots + j_n = s$, and $c_{j_1, \ldots, j_n} \in \mathbb{R}$.

The following result was established by Bakhtin.

Theorem 3.5. ([15]) *Let* $X = \{X_j, j \in \mathbb{Z}^d\}$ *be a centered associated random field. Suppose that there exists an even positive integer* k *such that* $\|X_j\|_{k+\delta} \leq D$ *for some* $\delta > 0$, $D \geq 1$ *and any* $j \in \mathbb{Z}^d$. *Assume also that* (3.4) *holds. Then there exists a homogeneous polynomial* $P = P_{k,\delta,u,D}$ *of degree* $k/2$ *such that*

$$\mathsf{E}\left(\sum_{j \in \mathbb{Z}^d} a_j X_j\right)^k \leq \|a\|_\infty^{k/2} P(\|a\|_1, \|a\|_\infty) \tag{3.11}$$

for any function $a : \mathbb{Z}^d \to \mathbb{R}$ *having* $\|a\|_1 = \sum_{j \in \mathbb{Z}^d} |a_j| < \infty$, *here* $\|a\|_\infty = \sup_{j \in \mathbb{Z}^d} |a_j|$. *In particular, the sum on the left-hand side is a.s. finite if* $\|a\|_\infty < \infty$.

Proof. For $m \in \mathbb{N}$, denote the set $\{1, \ldots, m\}$ by V_m and introduce

$$A_m = \left|\sum_{f:V_m \to \mathbb{Z}^d} \mathsf{E}\prod_{l=1}^m a_{f(l)} X_{f(l)}\right|, \quad m \leq k$$

(the sum is over all functions on V_m with values in $\mathbb{Z}^d$; a priori it might be infinite). Next, to any map $f : V_m \to \mathbb{Z}^d$ one can naturally attribute

$$c(f) = \max\{r : \text{ there exists } U \subset V_m, \,, 0 < |U| < m, \, dist(f(U), f(V_m \setminus U)) = r\}$$

where, as usual, *dist* is the distance due to the norm $|\cdot|$ ($|z| = \max_l |z_l|$ for $z = (z_1, \ldots, z_d)$) and the maximum over an empty set is zero. It is convenient to think that the partition of V_m realizing this maximum is uniquely determined by f. For example, one may use the lexicographic ordering as in the proof of Theorem 3.1. In this case we write $V_m = U_1(f) \cup U_2(f)$. Set

$$Y_1(f) = \prod_{l \in U_1(f)} a_{f(l)} X_{f(l)}, \quad Y_2(f) = \prod_{l \in U_2(f)} a_{f(l)} X_{f(l)}$$

and let $\sum'$ stands for summation over $r \in \mathbb{N}$, $q \in \{1, \ldots, m-1\}$ and $f \in H$ where $H = \{f : V_m \to \mathbb{Z}^d \text{ having } c(f) = r \text{ and } |U_1(f)| = q\}$. Then

$$A_m \leq \left| \sum{}' \mathsf{E} Y_1(f) Y_2(f) \right| \leq \sum{}' |\mathsf{E} Y_1(f)||\mathsf{E} Y_2(f)| + \sum{}' |cov(Y_1(f), Y_2(f))|$$

$$+ \sum_{j \in \mathbb{Z}^d} |\mathsf{E} a_j^m X_j^m| =: I_1 + I_2 + I_3.$$

An argument similar to that used to prove Theorem 3.1 (after formulas (3.6) and (3.7)) shows that

$$I_1 \leq m! \sum_{q=1}^{m-1} A_q A_{m-q}. \tag{3.12}$$

Clearly,

$$I_3 \leq D^m \sum_{j \in \mathbb{Z}^d} |a_j|^m \leq D^m \|a\|_1 \|a\|_\infty^{m-1}. \tag{3.13}$$

To estimate I_2 we apply Lemma 1.22. Set $\mu = \delta/(m + \delta - 2)$. By that Lemma and by linearity of covariances in each argument, we have

$$|cov(Y_1(f), Y_2(f))| = \prod_{l=1}^{m} |a_{f(l)}| \left| cov\left(\prod_{l \in U_1(f)} X_{f(l)}, \prod_{l \in U_2(f)} X_{f(l)} \right) \right|$$

$$\leq 3 \prod_{l=1}^{m} |a_{f(l)}| \left(\sum_{l_1 \in U_1(f)} \sum_{l_2 \in U_2(f)} cov(X_{f(l_1)}, X_{f(l_2)}) \right)^\mu$$

$$\times \left(\prod_{l=1}^{m} \|X_{f(l)}\|_{m+\delta} \sum_{n=1}^{m} \|X_{f(n)}\|_{m+\delta}^\delta \right)^{1-\mu} \leq 3 m u_{c(f)}^\mu \prod_{l=1}^{m} |a_{f(l)}| D^{(m+\delta)(1-\mu)}.$$

Therefore,

$$I_2 \leq 3 m D^{(m+\delta)(1-\mu)} \sum_{r=1}^{\infty} u_r^\mu \sum_{s \in \mathbb{Z}^d} \sum_{q=1}^{m-1} \sum{}'' |a_s| \|a\|_\infty^{m-1} \tag{3.14}$$

where $\sum''$ denotes summation performed over all maps $f : V_m \to \mathbb{Z}^d$ having $c(f) = r$, $|U_1(f)| = q$ and over $s \in f(U_1(f))$. We are left with the task of estimating the total number of maps having the mentioned properties. Using the lexicographic order, to each map f we can assign a permutation G of V_m such that the following is true:

1. $f(G(1)) = s$;
2. $\{G(1),\ldots,G(q)\} = U_1(f)$;
3. $\{G(q+1),\ldots,G(m)\} = U_2(f)$;
4. For $i \notin \{1, q+1\}$ one has $dist(f(G(i)), f(\{G(1),\ldots,G(i-1)\})) \le r$;
5. $dist(f(G(q+1)), f(U_1(f))) = r$.

The total number of possible permutations is $m!$ The choice of $f(G(1)) = s$ is determined uniquely. For $f(G(2))$ there are b_r possibilities, for $f(G(3))$ one has $2b_r$ choices, $\ldots$, for $f(G(q))$ they amount to $(q-1)b_r$, for $f(G(q+1))$ there are qs_r, for $f(G(q+2))$ there are $(q+1)b_r$, $\ldots$, and for $f(G(m))$ there are $(m-1)b_r$ possibilities. Here, as in the proof of Theorem 3.1, b_r and s_r denote the number of points in a ball and a sphere of radius r respectively. The overall number of maps f is at most $m!b_r \cdot 2b_r \ldots (q-1)b_q qs_r(q+1)b_r \ldots (m-1)b_r$. Therefore, we write $Q = Q(m, D, \delta) = 3m^2(m!)^2 D^{(m+\delta)(1-\mu)}$ and obtain from (3.14) the inequality

$$I_2 \le Q \sum_{r=1}^{\infty} u_r^\mu b_r^{m-2} s_r \sum_{s\in\mathbb{Z}^d} |a_s| \cdot \|a\|_\infty^{m-1} \le K(m, \delta, u, D)\|a\|_1\|a\|_\infty^{m-1} \qquad (3.15)$$

with some $K(m, \delta, u, D) > 0$ (here we used (3.4)). The bounds (3.12), (3.13) and (3.15) together imply that, for any $m \ge 2$, there exists $C(m, \delta, u, D) > 0$ such that

$$A_m \le C(m, \delta, u, D)\Big(\sum_{q=1}^{m-1} A_q A_{m-q} + \|a\|_1\|a\|_\infty^{m-1}\Big). \qquad (3.16)$$

Furthermore, on account of (3.4) and the moment restriction $\|X_j\|_{k+\delta} \le D$,

$$A_1 = 0 \text{ and } A_2 \le \sum_{j,v\in\mathbb{Z}^d} |a_j a_v| cov(X_j, X_v)$$

$$\le \sum_{j\in\mathbb{Z}^d} |a_j| \sum_{v\in\mathbb{Z}^d} \|a\|_\infty cov(X_j, X_v) \le u_0\|a\|_1\|a\|_\infty \qquad (3.17)$$

where $u_0 = \sup_{j\in\mathbb{Z}^d} \sum_{v\in\mathbb{Z}^d} cov(X_j, X_v)$. From (3.16) and (3.17) one deduces by induction on m that, for any $m = 1,\ldots,k$, the number A_m does not exceed the value of some homogeneous polynomial $Q_{m,\delta,u,D}(\|a\|_1, \|a\|_\infty)$ of degree m. Moreover, for any monomial constructing this polynomial, the degree of $\|a\|_\infty$ is no less than that of $\|a\|_1$. Therefore, we can write

$$Q_{k,\delta,u,D}(\|a\|_1, \|a\|_\infty) = \|a\|_\infty^{k/2} P_{k,\delta,u,D}(\|a\|_1, \|a\|_\infty).$$

Finally the Theorem follows from the obvious fact that the left-hand side of (3.11) is not greater than A_k. $\square$

3°. Results for shot-noise and cluster fields. Now we return to examples of point and cluster random fields studied in Section 1.3. We will give sufficient conditions for the existence of moments of order higher than two and estimate the rate of decrease of covariance function. These conditions will ensure the applicability of main limit theorems proved in Chapters 3—8. In other words, the next two theorems are counterparts of Corollary 1.2.15.

As in Section 1.3, $U = \{x_i\}$ is the Poisson spatial process in $\mathbb{R}^n$ with σ-finite intensity measure Λ. Consider a shot-noise field $Y = \{Y(u), u \in \mathbb{R}^n\}$ defined in (1.3.10) with nonnegative ξ_i and ψ. Set

$$I_r(x, \psi) := \int_{\mathbb{R}^n} \psi(x - t)^r \Lambda(dt), \quad x \in \mathbb{R}^n, \quad r \geq 1.$$

Theorem 3.6. *Let Y be a shot-noise field introduced above and $\mathsf{E}\xi_1^2 < \infty$, then*

$$cov(Y(x), Y(y)) = (\mathsf{Var}\xi_1) \int_{\mathbb{R}^n} \psi(x - t)\psi(y - t)\Lambda(dt), \quad x, y \in \mathbb{R}^n, \tag{3.18}$$

provided that the integral converges. If, moreover, $\mathsf{E}\xi_1^p < \infty$ for some $p > 2$, then

$$\mathsf{E}|X(x)|^p \leq C(p)\mathsf{E}|\xi_1|^p \left(I_1(x, \psi)^p + I_2(x, \psi)^{p/2} + I_p(x, \psi) \right) \tag{3.19}$$

for any $x \in \mathbb{R}^n$, with $C(p)$ depending only on p.

Proof. At first we consider the case $\Lambda(\mathbb{R}^n) < \infty$. If $\Lambda(\mathbb{R}^n) = 0$ then there is nothing to prove, so we omit this case. Let $\tau \sim Pois(\Lambda(\mathbb{R}^n))$ and $X_1, X_2, \ldots$ be i.i.d. random vectors with distribution $\Lambda(\cdot)/\Lambda(\mathbb{R}^n)$, independent from τ. By Theorem A.5, the random finite collection of points $\{X_1, X_2, \ldots, X_\tau\}$ constitutes a Poisson random field with intensity measure Λ. In what follows we assume that this representation takes place.

By the Fubini theorem

$$\mathsf{E}X(x) = \mathsf{E}\sum_i \xi_i \psi(x - x_i) = \sum_{N=1}^{\infty} \mathsf{P}(\tau = N) \sum_{i=1}^{N} \mathsf{E}\xi_i \psi(x - X_i)$$

$$= e^{-\Lambda(\mathbb{R}^n)} \sum_{N=1}^{\infty} \frac{\Lambda(\mathbb{R}^n)^N}{N!\Lambda(\mathbb{R}^n)} N(\mathsf{E}\xi) \int_{\mathbb{R}^n} \psi(x - t)\Lambda(dt) = (\mathsf{E}\xi)I_1(x, \psi). \tag{3.20}$$

In the same way

$$\mathsf{E}X(x)X(y) = \mathsf{E}\sum_{N=1}^{\infty} \mathbb{I}\{\tau = N\} \sum_{i,j} \xi_i \psi(x - x_i)\xi_j \psi(y - x_j)$$

$$= (\mathsf{E}\xi_1^2) \int_{\mathbb{R}^n} \psi(x - t)\psi(y - t)\Lambda(dt) + (\mathsf{E}\xi_1)^2 I_1(x, \psi)I_1(y, \psi) \tag{3.21}$$

since

$$\sum_{N=1}^{\infty} e^{-a}a^N N(N - 1)/N! = a^2 \quad \text{for} \quad a > 0.$$

On subtracting (3.21) from (3.20) one comes to (3.18).

To establish (3.19) we set $b = \mathsf{E}\xi_1$ and write

$$\mathsf{E}\left(\sum_i \xi_i \psi(x - x_i)\right)^p = \sum_{N=1}^{\infty} e^{-\Lambda(\mathbb{R}^n)} \frac{\Lambda(\mathbb{R}^n)^N}{N!} \mathsf{E}\left(\sum_{i=1}^{N} \xi_i \psi(x - X_i)\right)^p$$

$$\leq 2^p \sum_{N=1}^{\infty} e^{-\Lambda(\mathbb{R}^n)} \frac{\Lambda(\mathbb{R}^n)^N}{N!} \mathsf{E}\Big(\sum_{i=1}^{N} (\xi_i \psi(x - X_i) - b\mathsf{E}\psi(x - X_i))\Big)^p$$

$$+2^p \sum_{N=1}^{\infty} e^{-\Lambda(\mathbb{R}^n)} \frac{\Lambda(\mathbb{R}^n)^N}{N!} \Big(Nb\mathsf{E}\psi(x - X_i)\Big)^p =: \sum_{N=1}^{\infty} e^{-\Lambda(\mathbb{R}^n)} \frac{\Lambda(\mathbb{R}^n)^N}{N!} (E_{1,N} + E_{2,N}).$$

We apply the Rosenthal inequality (see the beginning of Section 2.3) and the Cauchy–Bunyakowski–Schwarz inequality to obtain

$$\mathsf{E}\Big(\sum_{i=1}^{N} \eta_i \psi(x - X_i)\Big)^p \leq 2^{p^2} \Big(N\mathsf{E}|\eta_1|^p \psi^p(x - X_1) + N^{p/2}\Big(\mathsf{Var}(\xi_1 \psi(x - X_1))\Big)^{p/2}\Big)$$

$$\leq 2^{p^2} \Big(2^{p+1} N\mathsf{E}\xi_1^p \Lambda(\mathbb{R}^n)^{-1} I_p(x, \psi) + N^{p/2}\mathsf{E}\xi_1^p \Big(\Lambda(\mathbb{R}^n)^{-1} I_2(x, \psi)\Big)^{p/2}\Big).$$

Thus $E_{1,N} + E_{2,N}$ does not exceed

$$2^{p^2+1}\mathsf{E}\xi_1^p \Big(\frac{N}{\Lambda(\mathbb{R}^n)} I_p(x, \psi) + \frac{N^{p/2}}{\Lambda(\mathbb{R}^n)^{p/2}} \Big(I_2(x, \psi)\Big)^{p/2} + \frac{N^p}{\Lambda(\mathbb{R}^n)^p} \Big(I_1(x, \psi)\Big)^p\Big), \quad N \in \mathbb{N}.$$

It remains to recall that if $\tau_a \sim Pois(a)$, then $\sup_{a>0} \mathsf{E}\tau^q/a^q < \infty$ for any $q > 1$.

If $\Lambda(\mathbb{R}^n) = +\infty$, consider a sequence of compacts $(K_n)_{n\in\mathbb{N}}$ such that $\cup_n K_n = \mathbb{R}^n$ and $\Lambda_n(\mathbb{R}^n) < \infty$ where $\Lambda_n(B) := \Lambda(B \cap K_n)$, $n \in \mathbb{N}$, $B \in \mathcal{B}(\mathbb{R}^n)$. Then the desired statement follows by the monotone convergence theorem. $\square$

Corollary 3.7. *Suppose that a shot-noise field Y meets conditions of the previous Theorem and, moreover, the function ψ is bounded, integrable and satisfies the condition $\psi(t) \leq \varphi(|t|)$ for some function $\varphi : \mathbb{R}_+ \to \mathbb{R}_+$ which vanishes at infinity. Then*

$$cov(Y(x), Y(y)) \leq 2(\mathsf{Var}\xi_1)\varphi\left(\frac{\|x - y\|}{2}\right) \sup_{z\in\mathbb{R}} \int_{z\in\mathbb{R}^n} \psi(z - t)\Lambda(dt), \quad x, y \in \mathbb{R}^n.$$

Proof. Decompose the integral over $\mathbb{R}^n$ in (3.18) into the sum of two ones, taken over the set $\{t : \|t - x\| \geq \|t - y\|\}$ and its complement. Now we only have to note that if, for instance, $\|t - x\| \geq \|t - y\|$ then $\|t - x\| \geq \|x - y\|/2$. $\square$

The next result for a cluster field $X = \{X(B), B \in \mathcal{B}_0(\mathbb{R}^n)\}$ is analogous to Theorem 3.6, so the proof is omitted.

Theorem 3.8. *Let X be a cluster field defined in (1.3.13). Then*

$$cov(X(B), X(C)) = \int_{\mathbb{R}^n} \mathsf{E}M(B + t)M(C + t)\Lambda(dt), \quad B, C \in \mathcal{B}_0(\mathbb{R}^n),$$

if this integral is well-defined. Moreover, for each $p > 2$ and any $B \in \mathcal{B}_0(\mathbb{R}^n)$,

$$\mathsf{E}|X(B)|^p \leq C_*(p) \Big(J_1(B, M)^p + J_2(B, M)^{p/2} + J_p(B, M)\Big),$$

with $C_*(p)$ depending only on p and $J_r(B, \psi) = \int_{\mathbb{R}^n} \mathsf{E}M^r(B + t)\Lambda(dt), r \geq 1$.

4°. Randomization techniques. For random fields with more general dependence conditions than association we have a theorem which can be considered as a generalization of Rosenthal's inequalities. It is especially helpful in situations when the random variables of the field under consideration have no moments higher than the second order. The idea of additional randomization ascends to papers by Peligrad [317], Zhang and Wen [431]. We shall use the following result by Shashkin.

Theorem 3.9. ([375]) *Let X be a centered random field such that $D_2 = A_2(1) < \infty$.*

(a) *Assume that X is (BL,θ)-dependent and (1.27) holds. Then, for any $m \in \mathbb{N}$ and any block $U \in \mathcal{U}$, one has*

$$\mathsf{E}M_U^2 \le 3m^d \left(\left(\sum_{j \in U} \mathsf{E}|X_j| \right)^2 + 18 \sum_{j \in U} \mathsf{E}X_j^2 + 16c_0|U|m^{-\lambda} \right). \tag{3.22}$$

(b) *If the field is quasi-associated and (1.54) holds, then (3.22) is true with λ and c_0 replaced by $\varkappa$ and $c_1 C(\varkappa)$ respectively where*

$$C(\varkappa) = \sum_{k \in \mathbb{Z}^d, k \ne 0} |k|^{-\varkappa}. \tag{3.23}$$

(c) *If X is* **NA**, *then, for any block $U \in \mathcal{U}$,*

$$\mathsf{E}M_U^2 \le 2 \left(\sum_{j \in U} \mathsf{E}|X_j| \right)^2 + 100 \sum_{j \in U} \mathsf{E}X_j^2.$$

Remark 3.10. In general this Theorem provides only trivial estimate of moments of partial maxima, i.e. $\mathsf{E}M_U^2 = O(|U|^2)$. However it will be applied to "tails" of random variables arising after appropriate truncation (for specified arrays). Thus combined with Theorem 1.25 it will provide an "independent-type" bound, see Theorem 3.11 below.

Proof. (a) We may assume that $(\Omega, \mathcal{F}, \mathsf{P}) = (\Omega_1, \mathcal{F}_1, \mathsf{P}_1) \otimes (\Omega_2, \mathcal{F}_2, \mathsf{P}_2)$ and the random variables $\{X_j(\omega_1), j \in U, \omega_1 \in \Omega_1\}$ are defined on $(\Omega_1, \mathcal{F}_1, \mathsf{P}_1)$, whereas $(\Omega_2, \mathcal{F}_2, \mathsf{P}_2)$ supports a random field $\{\varepsilon_j(\omega_2), j \in \mathbb{Z}^d, \omega_2 \in \Omega_2\}$ such that ε_i are i.i.d. taking values ± 1 with equal probability $1/2$. For random variables

$$Y(\omega_1, \omega_2) = F(X_j(\omega_1), \varepsilon_j(\omega_2), j \in U),$$

$$Y'(\omega_1, \omega_2) = G(X_j(\omega_1), \varepsilon_j(\omega_2), j \in U)$$

where $F : \mathbb{R}^{2|U|} \to \mathbb{R}$ and $G : \mathbb{R}^{2|U|} \to \mathbb{R}$ are Borel functions, we write

$$\mathsf{E}_1 Y = \int_{\Omega_1} Y \, d\mathsf{P}_1, \quad \mathsf{E}_2 Y = \int_{\Omega_2} Y \, d\mathsf{P}_2,$$

$$cov_1(Y, Y') = \mathsf{E}_1 YY' - \mathsf{E}_1 Y \mathsf{E}_1 Y', \quad \|Y\| = \left(\mathsf{E}_1 Y^2 \right)^{1/2}.$$

Clearly, if $\mathsf{E}Y$ exists then $\mathsf{E}_1 \mathsf{E}_2 Y = \mathsf{E}_2 \mathsf{E}_1 Y = \mathsf{E}Y$ by the Fubini theorem. Moreover, if, for any $\omega_2 \in \Omega_2$, F and G are Lipschitz functions, one can use Theorem 1.5.3 to estimate the covariance $cov_1(Y, Y')$.

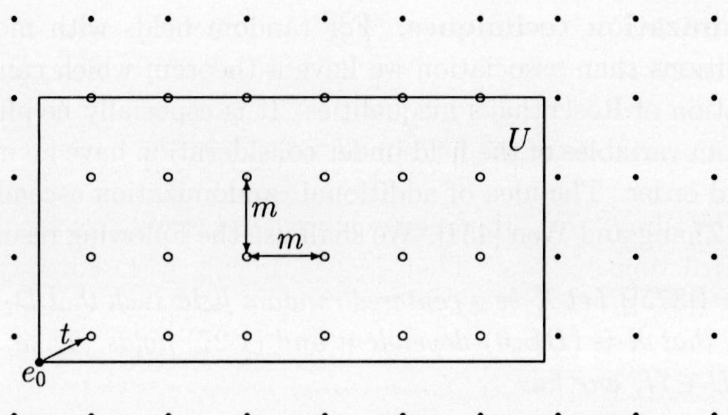

Fig. 2.5 $(d = 2$, white circles correspond to $U(t))$

Fix $m \in \mathbb{N}$. Let $e_0 \in U$ be the minimal element of U in the sense of lexicographic order. Consider the set $\Gamma = \{0, ..., m - 1\}^d$. For any $t \in \Gamma$ we introduce an integer lattice

$$\mathbb{Z}^d(t) = \mathbb{Z}^d(t, m) = \{j \in \mathbb{Z}^d : j = e_0 + t + (mz_1, ..., mz_d),\ z_i \in \mathbb{Z},\ i = 1, \ldots, d\}$$

and let $U(t) = U \cap \mathbb{Z}^d(t)$, see 3.

Then obviously

$$\sum_{t \in \Gamma} |U(t)| = |U| \quad \text{and} \quad \min_{\substack{i, j \in U(t) \\ i \neq j}} dist(i, j) = m \ \text{ if } \ U(t) \neq \varnothing. \tag{3.24}$$

Set $M_U(t) = \max_{V \lhd U} |S_{V \cap U(t)}|$, using the symbol $\lhd$ introduced before (1.1). Having taken into account that, for any $a_i \in \mathbb{R}$, $i = 1, \ldots, N$,

$$(a_1 + \ldots + a_N)^2 \leq N(a_1^2 + \ldots + a_N^2),$$

one observes that

$$\mathsf{E} M_U^2 \leq m^d \mathsf{E} \sum_{t \in \Gamma} \max_{V \lhd U} \left(\sum_{j \in V \cap U(t)} X_j \right)^2 \leq m^d \sum_{t \in \Gamma} \mathsf{E} M_U^2(t).$$

For fixed $t \in \Gamma$ such that $U(t) \neq \varnothing$ we will estimate $\mathsf{E} M_U^2(t)$. Following Zhang and Wen [431], for $V \lhd U$ write

$$S_{1,V} = \sum_{j \in V \cap U(t):\varepsilon_j = 1} X_j, \quad S_{-1,V} = \sum_{j \in V \cap U(t):\varepsilon_j = -1} X_j, \quad \widetilde{S}_V = S_{1,V} - S_{-1,V},$$

$$M = M_U(t), \quad M_1 = \max_{V \lhd U} |S_{1,V}|, \quad M_{-1} = \max_{V \lhd U} |S_{-1,V}|, \quad \widetilde{M} = \max_{V \lhd U} |\widetilde{S}_V|.$$

Then we have

$$2M_{\pm 1} = \max_{V \lhd U} |S_{\pm 1,V} + S_{\pm 1,V}| \leq \max_{V \lhd U} |S_{1,V} + S_{-1,V}| + \max_{V \lhd U} |S_{1,V} - S_{-1,V}| \leq M + \widetilde{M},$$

therefore $M_1 + M_{-1} \leq M + \widetilde{M}$. Notice also that

$$M \leq M_1 + M_{-1}, \quad M_1 \leq M_{-1} + \widetilde{M}, \quad M_{-1} \leq M_1 + \widetilde{M}, \quad |M_1 - M_{-1}| \leq \widetilde{M}$$

(the last inequality follows from the second and third ones). Since M is independent of $\{\varepsilon_j, j \in U\}$, one has $(\mathsf{E}M^2)^{1/2} = \|M\| \le \|M_1\| + \|M_{-1}\|$. Let

$$\xi = \frac{M_1}{\|M_1\| \vee \beta}, \quad \eta = \frac{M_{-1}}{\|M_{-1}\| \vee \beta}$$

where $\beta = \left(c_0 |U(t)| m^{-\lambda}\right)^{1/2}$. If some denominator here equals zero, we regard the corresponding fraction as zero. Then $\|\xi\| \le 1$, $\|\eta\| \le 1$. For any $\omega_2 \in \Omega_2$ the sets $\{j \in U(t) : \varepsilon_j = 1\}$ and $\{j \in U(t) : \varepsilon_j = -1\}$ are disjoint, the distance between them is m. The random variables M_1 and M_{-1} are Lipschitz functions of disjoint (nonoverlapping) sets of random variables X_j, $j \in U(t)$. The corresponding Lipschitz constants equal 1. Hence

$$\mathsf{E}_1 \xi M_{-1} = cov_1(\xi, M_{-1}) + \mathsf{E}_1 \xi \mathsf{E}_1 M_{-1}$$

$$\le |cov_1(\xi, M_{-1})| + \|\xi\| \mathsf{E}_1 M_{-1} \le \beta + \mathsf{E}_1 M_{-1},$$

by (BL, θ)-dependence, Lyapunov inequality and (3.24). Similarly we conclude that $\mathsf{E}_1 \eta M_1 \le \beta + \mathsf{E}_1 M_1$. We also notice that if $\|M_1\| \ge \beta$ then $\mathsf{E}_1 \xi M_1 = \|M_1\|$, and if $\|M_{-1}\| \ge \beta$ then $\mathsf{E}_1 \eta M_{-1} = \|M_{-1}\|$. Really, when $\|M_1\| \ge \beta$ one has $\xi = M_1 / \|M_1\|$, and consequently $\mathsf{E}\xi M_1 \|M_1\| = \|M_1\|^2$.

We consider four possible cases separately.

Case 1: $\|M_1\| \ge \beta$, $\|M_{-1}\| \ge \beta$. By the already proved bounds

$$(\mathsf{E}M^2)^{1/2} = \|M\| \le \|M_1\| + \|M_{-1}\| = \mathsf{E}_1 \xi M_1 + \mathsf{E}_1 \eta M_{-1} = \mathsf{E}_1 (\xi - \eta)(M_1 - M_{-1})$$

$$+ \mathsf{E}_1 \xi M_{-1} + \mathsf{E}_1 \eta M_1 \le \|\xi - \eta\| \|M_1 - M_{-1}\| + 2\beta + \mathsf{E}_1(M_1 + M_{-1}) \qquad (3.25)$$

$$\le 2\|\widetilde{M}\| + 2\beta + \mathsf{E}_1(M + \widetilde{M}) \le 3\|\widetilde{M}\| + 2\beta + \mathsf{E}_1 M.$$

Case 2: $\|M_1\| \ge \beta$, $\|M_{-1}\| < \beta$. Then $|\mathsf{E}_1 \eta M_{-1}| \le \beta$, therefore

$$(\mathsf{E}M^2)^{1/2} \le \|M_1\| + \|M_{-1}\| \le \mathsf{E}_1 \xi M_1 + \beta + \mathsf{E}_1 \eta M_{-1} - \mathsf{E}_1 \eta M_{-1}$$

$$\le |\beta - \mathsf{E}_1 \eta M_{-1}| + \mathsf{E}_1 \xi M_1 + \mathsf{E}_1 \eta M_{-1} \le 2\beta + \mathsf{E}_1 \xi M_1 + \mathsf{E}_1 \eta M_{-1} \le 3\|\widetilde{M}\| + 4\beta + \mathsf{E}_1 M$$

analogously to the first case.

Case 3: $\|M_1\| < \beta$, $\|M_{-1}\| \ge \beta$ is analogous to case 2.

Case 4: $\|M_1\| < \beta$, $\|M_{-1}\| < \beta$. Then

$$(\mathsf{E}M^2)^{1/2} \le \|M_1\| + \|M_{-1}\| < 2\beta.$$

In any case we have $(\mathsf{E}M^2)^{1/2} \le 3\|\widetilde{M}\| + 4\beta + \mathsf{E}M$. Thus

$$\mathsf{E}M^2 \le 3 \left(9\mathsf{E}\widetilde{M}^2 + (\mathsf{E}M)^2 + 16c_0 |U(t)| m^{-\lambda}\right). \qquad (3.26)$$

Now we estimate $\mathsf{E}\widetilde{M}^2$. To this end we make use of the structure of the field $\{\varepsilon_j, j \in \mathbb{Z}^d\}$. For any fixed $\omega_1 \in \Omega_1$,

$$\widetilde{M} = \max_{V \triangleleft U} \left| \sum_{j \in V \cap U(t)} \zeta_j \right|$$

where $\zeta_j = \varepsilon_j X_j$ are independent symmetric random variables on $(\Omega_2, \mathcal{F}_2, \mathsf{P}_2)$ with $\mathsf{E}_2 \zeta_j = 0$ and $\mathsf{E}_2 \zeta_j^2 = X_j^2$, $j \in U$. We claim that

$$\mathsf{E}_2 \widetilde{M}^2 \leq 2\mathsf{E}_2 \Big(\sum_{j \in U(t)} \zeta_j \Big)^2 \leq 2 \sum_{j \in U(t)} X_j^2. \tag{3.27}$$

This is seen by a tedious application of the reflection principle for symmetric independent random variables. Namely, since

$$\mathsf{P}(\varepsilon_j = 1) = \mathsf{P}(\varepsilon_j = -1) = 1/2$$

for each j, the probability space Ω_2 is a union of $2^{|U(\gamma)|}$ events (configurations) $A_j = \{\varepsilon(j) = \delta_j,\ j \in U\}$, where $\delta_j \in \{\pm 1\}$ and all events A_j, $j \in U$, have equal probability. Take arbitrary $a > 0$ and a block $V \lhd U$. We call a configuration with $\widetilde{M} \geq a$ a V-configuration, if V is the minimal block standardly belonging to U and such that

$$|\widetilde{S}_{V \cap U(\gamma)}| \geq a, \tag{3.28}$$

i.e. its maximal vertex has the least lexicographic order among all blocks $V \lhd U$ satisfying the property (3.28). For any fixed $V \lhd U$ all the V-configurations can be split into pairs obtained by the change of sign of those ε_j that have $j \notin V$. Then in each pair either $|\widetilde{S}_{U(\gamma)}| \geq a$ for both configurations or one configuration has $|\widetilde{S}_{U(\gamma)}| < a$, but then another one has $|\widetilde{S}_{U(\gamma)}| \geq a$. So the number of configurations with $\widetilde{M} \geq a$ is no greater than twice the number of configurations for which $|\widetilde{S}_{U(\gamma)}| \geq a$. Consequently,

$$\mathsf{P}_2(\widetilde{M} \geq a) \leq 2\mathsf{P}_2(\widetilde{S} \geq a),$$

hence

$$\mathsf{E}_2 \widetilde{M}^2 = \int_0^\infty \mathsf{P}_2(\widetilde{M}^2 \geq x)dx = \int_0^\infty \mathsf{P}_2\left(\widetilde{M} \geq \sqrt{x} \right) dx$$

$$\leq 2\int_0^\infty \mathsf{P}_2\left(|\widetilde{S}| \geq \sqrt{x} \right) dx = 2\int_0^\infty \mathsf{P}_2(\widetilde{S}^2 \geq x)dx = 2\mathsf{E}_2\widetilde{S}^2 = 2 \sum_{j \in U(\gamma)} X_j^2. \tag{3.29}$$

Thus $\mathsf{E}\widetilde{M}^2 = \mathsf{E}_1\mathsf{E}_2\widetilde{M}^2 \leq 2\sum_{j \in U(\gamma)} \mathsf{E}X_j^2$. Due to (3.24)—(3.27) we conclude that

$$\mathsf{E}M_U^2 \leq 3m^d \sum_{t \in \Gamma} \Big(18 \sum_{j \in U(t)} \mathsf{E}X_j^2 + \Big(\sum_{j \in U(t)} \mathsf{E}|X_j| \Big)^2 + 16c_0|U(t)|m^{-\lambda} \Big)$$

$$\leq 3m^d \Big(18 \sum_{j \in U} \mathsf{E}X_j^2 + \Big(\sum_{j \in U} \mathsf{E}|X_j| \Big)^2 + 16c_0|U|m^{-\lambda} \Big).$$

The first assertion of Theorem is proved.

For proving the second assertion we repeat the previous argument with the following amendment. Take $\beta = (c_1 C(\varkappa)|U(t)|m^{-\varkappa})^{1/2}$. Then, by definition of quasi-association,

$$|cov(\xi, M_{-1})| \leq \frac{1}{\|M_1\| \vee \beta} \sum_{\substack{j \in U(t) \\ \varepsilon_j = 1}} \sum_{\substack{v \in U(t) \\ \varepsilon_v = -1}} c_1 |j - v|^{-\varkappa}$$

$$\leq \frac{1}{\|M_1\| \vee \beta} |U(t)| c_1 \sum_{k \in \mathbb{Z}^d,\, k \neq 0} |mk|^{-\varkappa} = \frac{1}{\|M_1\| \vee \beta} c_1 |U(t)| m^{-\varkappa} \sum_{k \in \mathbb{Z}^d,\, k \neq 0} |k|^{-\varkappa},$$

since the distances between nonequal elements of $U(t)$ are divisible by m; here, as usual, $mk = (mk_1, \ldots, mk_d) \in \mathbb{Z}^d$ for $m \in \mathbb{N}$ and $k \in \mathbb{Z}^d$.

It remains to examine the case when the field X is **NA**, which may be handled in the same way as the first two cases. The only distinction is that now we invoke the covariances of monotone functions of the random variables under consideration, and these functions need not be Lipschitz. Let us take $m = 1$, introduce the field $\{\varepsilon_j, j \in \mathbb{Z}^d\}$ and construct $M, M_1, M_{-1}, \widetilde{M}$ as in the proof of (a). In addition, we define the following auxiliary random variables:

$$M_{\pm 1}^{(+)} := \max_{V \lhd U} S_{1,V}^+, \quad M_{\pm 1}^{(-)} := \max_{V \lhd U} (-S_{1,V})^+.$$

Then

$$\|M_1\| \leq \|M_1^{(+)}\| + \|M_1^{(-)}\|, \quad \|M_{-1}\| \leq \|M_{-1}^{(+)}\| + \|M_{-1}^{(-)}\|. \tag{3.30}$$

Now let

$$\xi = M_1^{(+)} / \|M_1^{(+)}\|, \quad \eta = M_{-1}^{(+)} / \|M_{-1}^{(+)}\|.$$

As above, the fraction $0/0$ is treated as zero. Thus, for a fixed $\omega_2 \in \Omega_2$, the vectors $(\xi, M_{-1}^{(+)})$ and $(\eta, M_1^{(+)})$ are 2-dimensional **NA** random vectors. Therefore

$$\mathsf{E}_1 \xi M_{-1}^{(+)} \leq \mathsf{E}_1 \xi \mathsf{E}_1 M_{-1}^{(+)} \leq \mathsf{E} M_{-1}^{(+)}, \quad \mathsf{E}_1 \eta M_1^{(+)} \leq \mathsf{E}_1 \eta \mathsf{E}_1 M_{11}^{(+)} \leq \mathsf{E}_1 M_1^{(+)}. \tag{3.31}$$

Consequently, rewriting the estimate (3.25) (with $M_1^{(+)}$ instead of M_1 and $M_{-1}^{(+)}$ instead of M_{-1}), we have

$$\|M_1^{(+)}\| + \|M_{-1}^{(+)}\| = \mathsf{E}_1 \xi M_1^{(+)} + \mathsf{E}_1 \eta M_{-1}^{(+)}$$

$$= \mathsf{E}_1 (\xi - \eta)(M_1^{(+)} - M_{-1}^{(+)}) + \mathsf{E}_1 \xi M_{-1}^{(+)} + \mathsf{E}_1 \eta M_1^{(+)}$$

$$\leq 2\|M_1^{(+)} - M_{-1}^{(+)}\| + \mathsf{E}_1 \left(M_{-1}^{(+)} + M_1^{(+)} \right) \leq 2\|\widetilde{M}\| + \mathsf{E}_1 \left(M_{-1}^{(+)} + M_1^{(+)} \right) \tag{3.32}$$

because $|M_1^{(+)} - M_{-1}^{(+)}| \leq |M_1 - M_{-1}|$. Similarly,

$$\|M_1^{(-)}\| + \|M_{-1}^{(-)}\| \leq 2\|\widetilde{M}\| + \mathsf{E}_1 \left(M_{-1}^{(-)} + M_1^{(-)} \right). \tag{3.33}$$

From (3.32) and (3.33), using (3.30), we may infer that

$$(\mathsf{E} M^2)^{1/2} \leq \|M_1\| + \|M_{-1}\| \leq \|M_1^{(+)}\| + \|M_1^{(-)}\| + \|M_{-1}^{(+)}\| + \|M_{-1}^{(-)}\|$$

$$\leq 4\|\widetilde{M}\| + \mathsf{E}_1\left(M_{-1}^{(-)} + M_1^{(-)} + M_{-1}^{(+)} + M_1^{(+)}\right)$$

$$\leq 4\|\widetilde{M}\| + \mathsf{E}_1\left(M_{-1} + M_1\right) \leq 4\|\widetilde{M}\| + \mathsf{E}_1\left(M + \widetilde{M}\right) \leq 5\|\widetilde{M}\| + \mathsf{E}_1 M.$$

Thus, the analogue of (3.26) is

$$\mathsf{E}M^2 \leq 2\left(25\mathsf{E}\widetilde{M}^2 + (\mathsf{E}M)^2\right).$$

The proof can be finished as in the case (a) by taking into account (3.29). $\square$

The following corollary of the previous theorem was established by Zhang and Wen for **NA** and by Shashkin for (BL,θ)-dependent random fields.

Theorem 3.11. ([375, 431]) *Let X be a (BL,θ)-dependent centered random field such that, for some $q \geq 2$, $D_q < \infty$.*

(a) Suppose that (1.27) holds with some $\lambda > 3d$. Then, for any $U \in \mathcal{U}$, one has

$$|U|^{-1}\mathsf{E}M_U^2 \leq \Delta = \left(96c_0 + 24D_2^2 + 432D_2 + 2\gamma^{d/2}(4D_q^{1/2} + D_2 + c_0 + C_2(d,\lambda))\right)$$

where γ was introduced in (1.62) and

$$C_2(d,\lambda) = 16\sqrt{6}d \cdot 3^{3d/2}\left((\lambda d^{-1} - 3)^{-1/2} \vee 1\right).$$

If, moreover, the random field $\{X_j^2, \ j \in \mathbb{Z}^d\}$ is uniformly integrable (in particular, if X is strictly stationary and the second moment exists), then the random family $\{|U|^{-1}M_U^2, \ U \in \mathcal{U}\}$ is also uniformly integrable.

(b) If the field X is quasi-associated then assertion of (a) holds upon replacement of λ with $\varkappa$ and c_0 with $c_1 C(\varkappa)$, where $\varkappa, c_1$ were introduced in (1.54) and $C(\varkappa)$ in (3.23).

*(c) Let the field X be **NA**. Then*

$$|U|^{-1}\mathsf{E}M_U^2 \leq \left(800D_2 + 16D_2 + 32\gamma^{d/2}(4D_2 + D_2^2)^{1/2}\right).$$

If X_j^2, $j \in \mathbb{Z}^d$, are uniformly integrable random variables, then $\{|U|^{-1}M_U^2, \ U \in \mathcal{U}\}$ is also uniformly integrable.

Proof. (a) Consider transformed variables

$$X_j' = H_{\sqrt{|U|}}(X_j) - \mathsf{E}H_{\sqrt{|U|}}(X_j), \ X_j'' = X_j - X_j', \ j \in U,$$

where the function H_M was defined in (1.5). Take

$$S_U' = \sum_{j \in V} X_j', \ S_U'' = \sum_{j \in V} X_j'', \ M_U' = \max_{V \vartriangleleft U}|S_V'|, \ M_U'' = \max_{V \vartriangleleft U}|S_V''|.$$

To estimate moments of X_j' and X_j'' we use a simple inequality

$$\mathsf{E}|Z - \mathsf{E}Z|^p \leq 2^p\mathsf{E}|Z|^p, \tag{3.34}$$

valid for any $p \geq 1$ and any random variable Z with $\mathsf{E}|Z|^p < \infty$. Note also that, for any $U \in \mathcal{U}$,

$$\mathsf{E}\left|X_j - H_{\sqrt{|U|}}(X_j)\right| \leq \mathsf{E}|X_j|\mathbb{I}\{|X_j| \geq \sqrt{|U|}\} \leq \frac{1}{\sqrt{|U|}}\mathsf{E}X_j^2\mathbb{I}\{|X_j| \geq \sqrt{|U|}\}. \tag{3.35}$$

We estimate $\mathsf{E}(M_U'')^2$ with the help of (3.34), (3.35) and the first assertion of Theorem 3.9, applied for $m = 1$, as follows:

$$\mathsf{E}(M_U'')^2 \leq 12\left(\left(|U|\max_{j \in U}\mathsf{E}|X_j|\mathbb{I}\{|X_j| \geq \sqrt{|U|}\}\right)^2\right.$$

$$\left. + 18|U|\max_{j \in U}\mathsf{E}X_j^2\mathbb{I}\{|X_j| \geq \sqrt{|U|}\} + 4c_0|U|\right) \leq 12|U|(D_2^2 + 18D_2 + 4c_0). \tag{3.36}$$

Now estimate $\mathsf{E}M_U'^4$ by Theorem 1.25. The random variables X_j' are bounded. For any $p \geq q$, (3.34) implies that

$$D_p' = \sup_{j \in \mathbb{Z}^d} \mathsf{E}|X_j'|^p \leq 2^p|U|^{(p-q)/2}D_q. \tag{3.37}$$

Let $p > 4$, $p \geq q$. By Theorem 1.25 and (3.37)

$$\mathsf{E}S_U'^4 \leq |U|\left((D_2 + \theta_1)^2|U| + D_4' + C_1c_0^v D_p'^{2/(p-2)}B(|U|,\tau)\right)$$

$$\leq |U|\left((D_2 + \theta_1)^2|U| + 16D_q|U|^{(2-q/2)\vee 0} + C_1c_0^v(2^p|U|^{(p-q)/2}D_q)^{2/(p-2)}B(|U|,\tau)\right)$$

where v, τ and $C_1 = C_1(d,\lambda,p)$ are defined in (1.53).

For all p large enough we have $\tau < 0$ and, consequently, $B(|U|,\tau) = 1$. Letting p tend to infinity we come to an estimate

$$\mathsf{E}S_U'^4 \leq \left((D_2 + \theta_1)^2 + 16D_q + 1536d^2 \cdot 3^{3d}\left((\lambda d^{-1} - 3)^{-1} \vee 1\right)c_0\right)|U|^2.$$

Again by Theorem 1.2, for $\lambda > 3d$ one has the inequality

$$\mathsf{E}(M_U')^4 \leq \gamma^d\left((D_2 + \theta_1)^2 + 16D_q + 1536d^2 \cdot 3^{3d}\left((\lambda d^{-1} - 3)^{-1} \vee 1\right)c_0\right)|U|^2. \tag{3.38}$$

Due to (3.36), (3.38) and the obvious fact that $M_U \leq M_U' + M_U''$, we have

$$\mathsf{E}|U|^{-1}M_U^2 \leq 2|U|^{-1}(\mathsf{E}M_U'^2 + \mathsf{E}M_U''^2) \leq 24(D_2^2 + 18D_2 + 4c_0)$$

$$+ \gamma^{d/2}\left(2\sqrt{6}(D_2 + \theta_1) + 8D_q^{1/2} + C_2(d,\lambda)c_0^{1/2}\right) \leq \Delta.$$

Let now the family $\{X_j^2, \ j \in \mathbb{Z}^d\}$ be uniformly integrable. By Theorem 3.9 and (3.34), for any $m \in \mathbb{N}$

$$\mathsf{E}M_U''^2 \leq 3m^d|U|\left(|U|\max_{j \in U}(\mathsf{E}|X_j''|)^2 + 18\max_{j \in U}\mathsf{E}X_j''^2 + 16c_0|U|m^{-\lambda}\right)$$

$$\leq 12m^d|U|\left(\max_{j \in U}\left(\mathsf{E}X_j^2\mathbb{I}\{|X_j| > \sqrt{|U|}\}\right)^2 + 18\max_{j \in U}\mathsf{E}X_j^2\mathbb{I}\{|X_j| > \sqrt{|U|}\} + 4c_0|U|m^{-\lambda}\right).$$

By uniform integrability the last formula yields

$$\limsup_{|U|\to\infty} |U|^{-1} \mathsf{E} M_U''^2 \le 4c_0 m^{-\lambda}.$$

Since m can be chosen arbitrarily large, this amounts to the property

$$|U|^{-1}\mathsf{E} M_U''^2 \to 0 \text{ as } |U| \to \infty.$$

For an arbitrary $\mu > 0$ one has

$$|U|^{-1}\mathsf{E} M_U^2 \mathbb{I}\{M_U^2 \ge \mu|U|\} \le 6|U|^{-1}\mathsf{E}(M_U'')^2$$

$$+2|U|^{-1}\mathsf{E}(M_U')^2 \mathbb{I}\left\{(M_U')^2 \ge \frac{\mu|U|}{4}\right\} \le 6|U|^{-1}\mathsf{E}(M_U'')^2 + 8\mu^{-1}|U|^{-2}\mathsf{E}(M_U')^4 \to 0$$

as $|U| \to \infty$ and $\mu \to \infty$. Thus $\{|U|^{-1}M_U^2, \ U \in \mathcal{U}\}$ is uniformly integrable.

(b) The proof in case of quasi-association is completely similar.

(c) Let $\{X_j', j \in \mathbb{Z}^d\}$ and $\{X_j'', j \in \mathbb{Z}^d\}$ be the random fields introduced as in the proof of (a). By Theorem 3.3 and (3.37), for any block $U \in \mathcal{U}$ one has

$$\mathsf{E} S_U'^{\,4} \le 2^{16}\left(\sum_{j\in U}\mathsf{E} X_j'^{\,4} + \left(\sum_{j\in U}\mathsf{E} X_j^2\right)^2\right)$$

$$\le 2^{16}\left(\sum_{j\in U} 4D_2|U| + \left(D_2|U|\right)^2\right) \le 2^{16}(4D_2 + D_2^2)|U|^2.$$

Theorem 1.2 ensures that instead of (3.38) we have now the bound

$$\mathsf{E}(M_U')^4 \le 2^{16}(4D_2 + D_2^2)\gamma^d|U|^2 \tag{3.39}$$

where γ is defined in Theorem 1.2. The rest of the proof goes as for (a). The claim concerning the uniform integrability also follows as in (a), with the help of (3.39). The Theorem is proved. $\square$

4   Estimates for the distribution functions of partial maxima

In this Section we collect results providing upper bounds for the probability $\mathsf{P}(M(U) \ge x)$ when $|U|$ and x are large enough. Of course, any bound for $\mathsf{E} M(U)^p$ established in previous sections yields an estimate for $\mathsf{P}(M(U) > x)$ via Markov inequality. However sometimes that would not suffice. For example, to prove a law of the iterated logarithm an adequate tool would be an exponential bound for $\mathsf{P}(M(U) > x)$. A group of theorems below, such as Newman–Wright and Bulinski–Keane inequalities, allow to replace, inside the probability, $M(U)$ with $S(U)$ and then use the results on normal approximation.

1°. The study of demimartingales. For a demisubmartingale[8] $S_1, \ldots, S_n$ (in particular, for partial sums of **PA** random variables), set $L_k = \max_{j=1,\ldots,k} S_j$ and $T_k = L_k \vee 0$. As before, $M_k = \max_{j=1,\ldots,k} |S_j|$.

---

[8]See Definition 1.1.17.

Theorem 4.1. ([310]) *Let $(S_1, \ldots, S_n)$ be a demisubmartingale and $m : \mathbb{R} \to \mathbb{R}$ be a nondecreasing right-continuous function such that $m(0) = 0$. Then*

$$\mathsf{E} \int_0^{T_n} u\, dm(u) \leq \mathsf{E} S_n m(T_n).$$

In particular, for any $x > 0$, one has

$$x\, \mathsf{P}(T_n \geq x) \leq \mathsf{E} S_n \mathbb{I}\{T_n \geq x\}.$$

Proof. We have

$$S_n m(T_n) = \sum_{k=0}^{n-1} S_{k+1}(m(T_{k+1}) - m(T_k)) + \sum_{k=0}^{n-1} m(T_k)(S_{k+1} - S_k)$$

where, as usual, $T_0 := 0$. The k-th summand in the first sum is not less than $\int_{T_k}^{T_{k+1}} u\, dm(u)$. The second sum is nonnegative as $m(x) \geq 0$ for $x \geq 0$ and due to the definition of demisubmartingale (when one takes $f \equiv 1$) we have $\mathsf{E}(S_{k+1} - S_k) \geq 0$ for all $k = 1, \ldots, n - 1$. The first relation is verified, which entails the second one by substituting $m(u) = \mathbb{I}\{u \geq x\}$. $\square$

Corollary 4.2. ([310]) *Let $S_1, \ldots, S_n$ be a square-integrable demisubmartingale. If $\sigma_n = (\mathsf{E} S_n^2)^{1/2} > 0$, then, for $0 < x_1 < x_2$,*

$$\mathsf{P}(L_n \geq x_2 \sigma_n) \leq \frac{1}{x_2 - x_1} \sqrt{\mathsf{P}(S_n \geq x_1 \sigma_n)}.$$

If, in addition, $S_1, \ldots, S_n$ form a demimartingale then

$$\mathsf{P}(M_n \geq x_2 \sigma_n) \leq \frac{1}{x_2 - x_1} \sqrt{\mathsf{P}(|S_n| \geq x_1 \sigma_n)}.$$

Proof. By the second assertion of Theorem 4.1 with $x = x_2 \sigma_n$, we have

$$x_2 \sigma_n \mathsf{P}(L_n \geq x_2 \sigma_n) = x_2 \sigma_n \mathsf{P}(T_n \geq x_2 \sigma_n) \leq \mathsf{E} S_n \mathbb{I}\{T_n \geq x_2 \sigma_n\}$$

$$= \mathsf{E} S_n \mathbb{I}\{L_n \geq x_2 \sigma_n\} \leq \mathsf{E} S_n (\mathbb{I}\{S_n > x_1 \sigma_n\} + \mathbb{I}\{S_n \leq x_1 \sigma_n, L_n \geq x_2 \sigma_n\})$$

$$\leq \mathsf{E} S_n \mathbb{I}\{S_n > x_1 \sigma_n\} + x_1 \sigma_n \mathsf{P}(L_n \geq x_2 \sigma_n).$$

Therefore,

$$\mathsf{P}(L_n \geq x_2 \sigma_n) \leq \frac{\mathsf{E} S_n \mathbb{I}\{S_n > x_1 \sigma_n\}}{(x_2 - x_1)\sigma_n}. \tag{4.1}$$

The first desired estimate follows from (4.1) by the Cauchy-Bunyakovski-Schwarz inequality. For the second one note that, applying (4.1) to the random sequence $(-S_1, \ldots, -S_n)$ which is also a demimartingale, we can write

$$\mathsf{P}\left(\max_{j=1,\ldots,n}(-S_j) \geq x_2 \sigma_n\right) \leq \frac{\mathsf{E}|S_n| \mathbb{I}\{S_n < -x_1 \sigma_n\}}{(x_2 - x_1)\sigma_n}. \tag{4.2}$$

From (4.1) and (4.2) one easily infers that

$$\mathsf{P}(M_n \geq x_2 \sigma_n) \leq \frac{\mathsf{E}|S_n| \mathbb{I}\{|S_n| > x_1 \sigma_n\}}{(x_2 - x_1)\sigma_n}$$

and it remains to use the Cauchy–Bunyakowski–Schwarz inequality again. $\square$

The partial sums of associated (or even **PA**) integrable random variables form a demimartingale, hence possess the property given in the above Corollary.

2°. Extensions for two-dimensional array. An important application of the demimartingale notion is to establish a maximal inequality for sums of two-parameter associated array, though no natural complete ordering is possible there. For a random field $X = \{X_j, j \in \mathbb{N}^2\}$, put

$$S_{m,n} = \sum_{1 \leq j \leq m, 1 \leq k \leq n} X_{jk}, \quad S_j^{(m)} = \max_{v=1,\ldots,m} S_{(v,j)}, \quad m, n, j \in \mathbb{N}.$$

Lemma 4.3. *Let $X = \{X_{j,k}, j, k \in \mathbb{N}\}$ be a **PA** centered random field defined on $\mathbb{N}^2$. Then, for any $m \in \mathbb{N}$, the sequence $(S_j^{(m)})_{j \in \mathbb{N}}$ is a demisubmartingale.*

Proof. Let $m, j \in \mathbb{N}$ and $f \in \mathcal{M}(j)$ be a nonnegative function (the class $\mathcal{M}(j)$ was introduced in Section 1.1). Set $Y = f(S_1^{(m)}, \ldots, S_j^{(m)})$. We only need to show that

$$\mathsf{E}(S_{j+1}^{(m)} - S_j^{(m)})Y \geq 0.$$

Let us introduce a random variable

$$K_j = \min\{k \in \mathbb{N} : S_{k,j} = \max\{S_{1,j}, \ldots, S_{m,j}\}\} = \min\{k \in \mathbb{N} : S_{k,j} = S_j^{(m)}\}.$$

Then, with this notation,

$$S_{j+1}^{(m)} - S_j^{(m)} = S_{j+1}^{(m)} - S_{K_j,j} \geq S_{K_j,j+1} - S_{K_j,j} = \sum_{k=1}^{K_j} X_{k,j+1} = \sum_{k=1}^{m} X_{k,j+1}\mathbb{I}\{K_j \geq k\}.$$

Since Y was taken nonnegative, we have

$$\mathsf{E}(S_{j+1}^{(m)} - S_j^{(m)})Y \geq \sum_{k=1}^{m} \mathsf{E}X_{k,j+1}\mathbb{I}\{K_j \geq k\}Y = \sum_{k=1}^{m} cov(X_{k,j+1}, \mathbb{I}\{K_j \geq k\}Y),$$

so the proof will be completed as soon as we establish that $\mathbb{I}\{K_j \geq k\}$ (and then $\mathbb{I}\{K_j \geq k\}Y$) is a nondecreasing function in $X_{tv}, t \leq m, v \leq j$. But this fact is easy, since if one adds some positive quantities to X_{tv}, with $t \leq m, v \leq j$, then S_{lj} increases no more than S_{rj} does, for $r > l$. $\square$

Using the same notation as in subsection 1° set

$$L_{m,n} = \max_{k \leq m, j \leq n} S_{j,k} \quad \text{and} \quad M_{m,n} = \max_{k \leq m, j \leq n} |S_{j,k}|.$$

Lemma 4.3 entails

Corollary 4.4. ([310]) *Let $X = \{X_{j,k}, j, k \in \mathbb{N}\}$ be a **PA** centered random field, and let $m, n \in \mathbb{N}$ be such that $\mathsf{E}X_{j,k}^2 > 0$ for at least one pair of indices $(j, k) \leq (m, n)$. Then, for any $x_1, x_2 > 0$ such that $x_2 > x_1$, one has*

$$\mathsf{P}(M_{m,n} \geq x_2\sigma_n) \leq 4\frac{3^{3/2}}{(x_2 - x_1)^{3/2}}\mathsf{P}(|S_{m,n}| \geq x_1\sigma_n)^{1/4}.$$

Proof. It is clear that

$$L_{m,n} = \max_{k \leq m, j \leq n} S_{j,k} = \max_{k \leq m} \max_{j \leq n} S_{j,k} = \max_{k \leq m} S_k^{(n)}.$$

Thus, by Corollary 4.2 and Lemma 4.3

$$P(L_{m,n} \geq x_2 \sigma_n) \leq \frac{1}{x_2 - x} P(S_m^{(n)} \geq x \sigma_n)^{1/2}$$

for any $x \in (x_1, x_2)$. Since $S_m^{(n)} = \max_{j \leq n} S_{j,m}$ is the maximal value of partial sums of positively associated random variables, the last probability can be in turn estimated by Corollary 4.2, namely

$$P(|S_m^{(n)}| \geq x \sigma_n) \leq \frac{1}{x - x_1} P(|S_{m,n}| \geq x_1 \sigma_n)^{1/2},$$

and consequently

$$P(L_{m,n} \geq x_2 \sigma_n) \leq \frac{1}{(x_2 - x)(x - x_1)^{1/2}} P(S_{m,n} \geq x_1 \sigma_n)^{1/4}. \tag{4.3}$$

The maximum of the function $x \mapsto (x_2 - x)(x - x_1)^{1/2}$ is attained when $x = (2x_1 + x_2)/3$; on inserting this value into (4.3) and repeating the same argument for the random field $\{-X_{j,k}, j, k \in \mathbb{N}\}$ we come to the desired conclusion. $\square$

3°. **Further generalizations.** In the case when $d = 1$ and the random variables are associated, the inequality in Corollary 4.2 can be modified.

Theorem 4.5. ([309]) *Suppose that* $X_1, \ldots, X_n$ *are* **PA**, *square-integrable, centered random variables. Then* $\mathsf{E}L_n^2 \leq \mathsf{E}S_n^2$. *If* $(X_1, \ldots, X_n) \in \mathbf{A}$, *then, for any* $\lambda > 0$,

$$P(M_n \geq \lambda(\mathrm{Var}S_n)^{1/2}) \leq 2P\left(|S_n| \geq (\lambda - \sqrt{2})(\mathrm{Var}S_n)^{1/2}\right).$$

Proof. The first assertion is proved by induction on n. For $n = 1$ it is obviously true. For general n write

$$K_n = \min\{S_n - S_m : m = 1, \ldots, n\}, \ R_n = \max\{S_m - X_1 : m = 1, \ldots, n\}.$$

Then $K_n = X_2 + \ldots + X_n - R_n$ is a nondecreasing function in $X_2, \ldots, X_n$, and $L_n = X_1 + R_n$. Consequently,

$$\begin{aligned}
\mathsf{E}(L_n)^2 &= \mathsf{E}(X_1 + R_n)^2 = \mathrm{Var}X_1 + 2cov(X_1, R_n) + \mathsf{E}R_n^2 \\
&= \mathrm{Var}X_1 + 2cov(X_1, X_2 + \ldots + X_n) - 2cov(X_1, K_n) + \mathsf{E}J_n^2 \\
&\leq \mathrm{Var}X_1 + 2cov(X_1, X_2 + \ldots + X_n) + \mathsf{E}R_n^2 \\
&\leq \mathrm{Var}X_1 + 2cov(X_1, X_2 + \ldots + X_n) + \mathsf{E}(X_2 + \ldots + X_n)^2 = \mathsf{E}S_n^2
\end{aligned}$$

where the last inequality is due to induction hypothesis.

To prove the second assertion observe that we can write, for positive λ_1, λ_2 such that $\lambda_1 < \lambda_2$,

$$P(L_n \geq \lambda_2) \leq P(S_n \geq \lambda_1) + P(L_{n-1} \geq \lambda_2, L_{n-1} - S_n > \lambda_2 - \lambda_1) \leq P(S_n \geq \lambda_1)$$

$$+ P(L_{n-1} \geq \lambda_2)P(L_{n-1} - S_n > \lambda_2 - \lambda_1) \leq P(S_n \geq \lambda_1) + P(L_n \geq \lambda_2)\frac{\mathsf{E}(L_{n-1} - S_n)^2}{(\lambda_2 - \lambda_1)^2},$$

using the same idea as above, i.e. observing that L_{n-1} and $S_n - L_{n-1}$ are nondecreasing functions in $X_1, \ldots, X_n$. The first assertion, applied to random variables $(-X_1, \ldots, -X_n)$, also implies that

$$E\left(\max\{S_1, \ldots, S_{n-1}\} - S_n\right)^2 = E\left(\max\{S_1 - S_n, \ldots, S_{n-1} - S_n\}\right)^2$$
$$= E\max\{-X_n, -X_n - X_{n-1}, \ldots, -X_n - \cdots - X_1\}^2 \leq ES_n^2.$$

Thus

$$P(L_n \geq \lambda_2) \leq \left(1 - \frac{\mathrm{Var}\, S_n}{(\lambda_2 - \lambda_1)^2}\right)^{-1} P(S_n \geq \lambda_1),$$

provided that $(\lambda_2 - \lambda_1)^2 > \mathrm{Var}\, S_n$. Therefore, taking $\lambda_1 = (\lambda - \sqrt{2})(\mathrm{Var}\, S_n)^{1/2}$, $\lambda_2 = \lambda(\mathrm{Var}\, S_n)^{1/2}$ and considering in a similar way $(-X_1, \ldots, -X_n)$ we come to the desired relation. $\square$

To prove the maximal inequality for random fields we need notation analogous to that given above for sequences. Namely, for a random field $X = \{X_j, j \in \mathbb{Z}^d\}$ and an integer block U we set

$$L(U) = \max\{S(V) : V \in \mathcal{U}, V \subset U\}, \quad M(U) = \max\{|S(V)| : V \in \mathcal{U}, V \subset U\}.$$

Theorem 4.6. ([72]) *Let X be a centered, associated random field such that $D_p := \sup_{j \in \mathbb{Z}^d} E|X_j|^p < \infty$ and $u(m) = O(m^{-\lambda})$ as $m \to \infty$, for some $\lambda > 0$. Then, for any $\tau \in (0,1)$, there exists $x_0 > 0$ such that for all $U \in \mathcal{U}$ and $x \geq x_0$*

$$P(M(U) \geq x|U|^{1/2}) \leq 2P(|S(U)| \geq \tau x|U|^{1/2}).$$

Here x_0 does not depend on U.

Proof. For $y, z > 0$ we have

$$P(L(U) \geq y) \leq P(S(U) \geq z) + P(L(U) \geq y, L(U) - S(U) \geq y - z)$$

$$\leq P(S(U) \geq z) + P(L(U) \geq y)P(L(U) - S(U) \geq y - z), \tag{4.4}$$

since $L(U)$ and $S(U) - L(U)$ are nondecreasing in $X_j, j \in U$. For $y > z$ and $r < p$, by the Markov inequality

$$P(L(U) - S(U) \geq y - z) \leq \frac{2^r E|L(U)|^r}{(y - z)^r}. \tag{4.5}$$

In view of Corollary 1.12, one can choose $r > 2$ in such a way that $E|L(U)|^r \leq C|U|^{r/2}$, where $C > 0$ does not depend on U. From (4.4) and (4.5) we obtain that

$$(1 - 2^r C(y - z)^{-r}|U|^{r/2})P(L(U) \geq y) \leq P(S(U) \geq z).$$

Taking $y = x|U|^{1/2}$ and $z = y/2$ one has

$$(1 - 2^r(1 - \tau)^{-r}x^{-r}C)P(L(U) \geq x|U|^{1/2}) \leq P(S(U) \geq \tau x|U|^{1/2}).$$

The same inequality holds for the random field $\{-X_j, j \in \mathbb{N}^d\}$. Thus, for all x such that $2^{r+1}C < (x - x\tau)^r$, the claimed estimate is true. $\square$

Now we provide the Vronski inequality.

Theorem 4.7. ([405]) *Suppose that* $X = \{X_j, j \in \mathbb{Z}^d\}$ *is a strictly stationary centered associated random field. Assume that*
1) $\mathsf{E}\exp\{h|X_0|^\beta\} < \infty$ *for some* $h, \beta > 0$;
2) *for some* $\lambda, \gamma > 0$, *the Cox–Grimmett coefficient* $u_r = O(e^{-\lambda r^\gamma})$, $r \to \infty$.
Then there exist $A > 0$ *and* $B > 0$ *determined by* $d, \mathsf{E}X_0^2, h, \beta, \mathsf{E}\exp\{h|X_0|^\beta\}$ *and* $(u_r)_{r\in\mathbb{N}}$ *such that, for any* $\alpha > 1/2, \varepsilon \in (0,1)$ *and any finite set* $U \subset \mathbb{Z}^d$, *one has*

$$\mathsf{P}(|S(U)| \geq \varepsilon|U|^\alpha) \leq A|U|^{2\alpha-1} \exp\left\{-B\varepsilon^\mu |U|^{(\alpha-1/2)\mu}\right\}$$

where $\mu := (2 + \beta^{-1} + \gamma^{-1})^{-1}$.

Proof. Throughout the proof we denote positive factors not depending on U by $c_1, c_2, \ldots$ Assume that $a^2 = \mathsf{E}X_0^2 > 0$, otherwise the assertion is trivial.

Benefitting from the proof of Theorem 3.1 (formula (3.10)), we take $\delta = k - 2$ for any $k > 2$ ($k \in \mathbb{N}$) and obtain the estimate

$$A_k \leq C(k)\left(Q(U, k+\delta, k) \vee Q(U, 2+\delta, 2)^{k/2}\right)$$

where $C(1) = 0$ and

$$C(k) = \sum_{p=2}^{k-2} \frac{k!}{p!(k-p)!} C(p)C(k-p) + H(k), \quad k > 1, \tag{4.6}$$

with

$$H(k) = 1 + 3(k!)^2 k^4 \sum_{r=1}^\infty (a^{-2}u_r)^{\delta/(k+\delta-2)} r^{k-2} = 1 + 3(k!)^2 k^4 \sum_{r=1}^\infty (a^{-2}u_r)^{1/2} r^{k-2}. \tag{4.7}$$

It is easily seen that $\sum_{r=1}^\infty \exp\{-\lambda r^\gamma/2\}r^{k-2} \leq c_1^k k^{k/\gamma}$, $k \in \mathbb{N}$. From this, using in (4.7) the Stirling formula, we infer that $H(k) \leq c_2^k k^{\nu k}$ where $\nu = 2+1/\gamma$. Therefore, (4.6) and induction on k prove that

$$C(k) \leq c_3^k k^{\nu k}, \quad k \in \mathbb{N}. \tag{4.8}$$

By Chebyshev's inequality

$$\mathsf{P}(|S(U)| \geq \varepsilon|U|^\alpha) \leq \varepsilon^{-k}|U|^{-k\alpha}\mathsf{E}|S_U|^k.$$

If k is an even number, then, by Theorem 3.1 and strict stationarity,

$$\mathsf{E}S(U)^k \leq C(k)\left((|U|\||X_0\|_{2k}^k) \vee (|U|^{k/2}\|X_0\|_k^k)\right).$$

Denote by $F(x)$ the distribution function of the random variable $|X_0|$. For any positive integer k, the existence of exponential moment (condition 1)) implies that

$$\mathsf{E}|X_0|^k = \int_0^\infty x^k dF(x) = \int_0^\infty x^k \exp\{-hx^\beta\}\exp\{hx^\beta\}dF(x)$$

$$\leq \sup_{x>0} x^k \exp\{hx^{-\beta}\}\mathsf{E}\exp\{h|X_0|^\beta\} \leq c_4^k k^{k/\beta}. \tag{4.9}$$

Relations (4.8), (4.9) show that, for any finite $U \in \mathcal{U}$ and positive even integer k,

$$P(|S(U)| \geq \varepsilon |U|^\alpha) \leq c_5^k k^{k(\nu + \beta^{-1})} |U|^{k(1/2 - \alpha)}.$$

It remains to optimize the last expression in k. Namely, take k to be the smallest positive even number exceeding the quantity

$$\exp \left\{ -1 + \frac{1}{\nu + \beta^{-1}} \log \left(c_5 \varepsilon^{-1} |U|^{1/2 - \alpha} \right) \right\}. \quad \square$$

Now we present an inequality by Bakhtin for continuous-parameter random fields. Note that this inequality does not require any dependence assumptions.

Theorem 4.8. ([15]) *Let $X = \{X_t, t \in \mathbb{R}_+^d\}$ be a random field with (almost surely) continuous trajectories. Suppose that $T \in \mathbb{R}_+^d$ is a positive vector and there are some $K, \gamma > 0$, $\alpha > d$ such that, for any $z > 0$ and any pair $x, y \in \mathbb{R}_+^d$, $0 \leq x \leq y \leq T$,*

$$P(X_y - X_x > z) \leq K z^{-\gamma} |y - x|^\alpha.$$

Then there exist $\tau > 1$ and $C > 0$, depending on K, α, γ, d, such that for any $\varepsilon > 0$

$$P(L_T > \tau \varepsilon) \leq P(X_0 > \varepsilon) + P(X_T > \varepsilon) + C \varepsilon^{-\gamma} |T|^\alpha$$

where $L_T = \sup_{t \in [0,T]} X_t$.

Proof. Let $\mathcal{P}_{-1} := \{0, T\} \subset \mathbb{R}_+^d$ and

$$\mathcal{P}_k := \{(2^{-k} T_1 l_1, \ldots, 2^{-k} T_d l_d), \ 0 \leq l_i \leq 2^k \text{ for } i = 1, \ldots, d\}, \ k \in \mathbb{Z}_+.$$

For any k set $L_T^{(k)} = \max_{t \in \mathcal{P}_k} X_t$. Note that, for any $\rho \in (0, 1)$,

$$P\left(L_T > \frac{\varepsilon}{1 - \rho}\right) \leq P(X_0 \vee X_T > \varepsilon) + \sum_{k=0}^{\infty} P(L_T^{(k)} - L_T^{(k-1)} > \varepsilon \rho^{k+1}).$$

For any $x \in \mathcal{D}_k := \mathcal{P}_k \setminus \mathcal{P}_{k-1}$, define $p(x) \in \mathcal{P}_{k-1}$ to be the closest to x point such that $p(x) \leq x$. On the event $\{L_T^{(k)} - L_T^{(k-1)} > \varepsilon \rho^{k+1}\}$, for the (random) point $x \in \mathcal{D}_k$ where $X_x = L_T^{(k)}$, one has

$$P\left(L_T > \frac{\varepsilon}{1 - \rho}\right) \leq P(X_0 \vee X_T > \varepsilon) + \sum_{k=0}^{\infty} \sum_{x \in \mathcal{D}_k} P(X_x - X_{p(x)} > \varepsilon \rho^{k+1})$$

$$\leq P(X_0 \vee X_T > \varepsilon) + \sum_{k=0}^{\infty} \sum_{x \in \mathcal{D}_k} K \varepsilon^{-\gamma} \rho^{-(k+1)\gamma} |x - p(x)|^\alpha. \tag{4.10}$$

Since for any $x \in \mathcal{D}_k$ the estimate $|x - p(x)| \leq 2^{-k} |T|$ holds and $|\mathcal{D}_k| \leq 2^{(k+1)d}$, inequality (4.10) can be continued as

$$P\left(L_T > \frac{\varepsilon}{1 - \rho}\right) \leq P(X_0 \vee X_T > \varepsilon) + 2^d K \varepsilon^{-\gamma} d^{\alpha/2} T^\alpha \sum_{k=0}^{\infty} 2^{k(d-\alpha)} \rho^{-\gamma(k+1)}.$$

To prove the Theorem, pick ρ such that $2^{d-\alpha} \rho^{-\gamma} < 1$ and $\tau = 1/(1 - \rho)$. $\square$

Theorem 4.9. ([15]) *Let $X = \{X_t, t \in \mathbb{R}_+^d\}$ be a random field with (almost surely) continuous trajectories. Suppose that $T \in \mathbb{R}_+^d$ and there are some $K, \gamma > 0$, $\alpha > d/2$ such that, for any $z > 0$ and any three points $x, t, y \in \mathbb{R}_+^d$, $0 \leq x \leq t \leq y \leq T$,*

$$\mathsf{P}(X_t - X_x > z, X_t - X_y > z) \leq K z^{-\gamma} (|y - t||t - x|)^\alpha.$$

Then there exist $\tau > 1$ and $C > 0$, depending on K, α, γ, d, such that for any $\varepsilon > 0$

$$\mathsf{P}(L_T > \tau\varepsilon) \leq \mathsf{P}(X_0 > \varepsilon) + \mathsf{P}(X_T > \varepsilon) + C\varepsilon^{-\gamma}|T|^{2\alpha}.$$

Proof is analogous to that of Theorem 4.8. The only significant difference is that instead of a point $p(x)$ one takes two points $p(x)$ and $n(x)$ where $n(x)$ is the closest to x point in $\mathcal{P}_{k-1}$ for which $n(x) \geq x$. $\square$

Remark 4.10. The conditions of Theorem 4.8 can be verified with the help of Markov inequality. If the increments of a random field X over disjoint rectangles are associated, then conditions of Theorem 4.9 are checked analogously. Indeed, for $x \leq t \leq y \leq T$, by association,

$$\mathsf{P}(X_t - X_x > z, X_t - X_y > z) = \mathsf{P}(X_t - X_x > z, X_y - X_t < -z)$$

$$\leq \mathsf{P}(X_t - X_x > z)\mathsf{P}(X_y - X_t < -z).$$

4°. Final remarks. The useful inequalities (in particular, of exponential type) are established in [213, 249, 273, 274, 277, 289, 301, 312, 333, 395, 408]. The proofs of these results involve various combinations of such tools used above as, e.g., the truncations of random variables, inequalities similar to (1.5.1) and division of the index set into appropriate blocks. For example, we formulate the following result by Oliveira.

Theorem 4.11. [312] *Suppose that $X = (X_n)_{n \in \mathbb{Z}}$ is a centered, strictly stationary, associated random sequence such that $\sup_{t:|t| \leq \delta} \mathsf{E} e^{tX_0} \leq M_\delta$ for some $\delta, M_\delta > 0$. Assume that the Cox–Grimmett coefficient of X admits the estimate $u_r \leq C_0 e^{-\lambda r^\mu}$ as $r \to \infty$, for some $C_0, \lambda, \mu > 0$. Then, for any $\varepsilon > 0$ and all $n \in \mathbb{N}$ large enough,*

$$\mathsf{P}(|S_n| > n\varepsilon) \leq C_1 (1 + \varepsilon^{-2} M_\delta n^3) \exp\left\{-C_2 \varepsilon^2 n^{\mu\alpha}\right\}$$

where $S_n = \sum_{i=1}^n X_i$, the value $\alpha \in (0, 1)$ can be chosen arbitrarily close to 1, and positive C_1 and C_2 are determined by C_0, μ, λ and α.

The consideration of associated random systems without the second order moment leads to appearance of a more complicated function at the right-hand side of the estimate. By a theorem by Louhichi [275] it follows that if $X_1, \ldots, X_n$ are identically distributed associated random variables, then for any $M > 0$ and $x > 0$

$$\mathsf{P}(L_n > x) \leq 4nx^{-2}\mathsf{E}X_1^2 \mathbb{I}\{|X_1| < M\} + 4nx^{-1}\mathsf{E}X_1 \mathbb{I}\{|X_1| \geq M\}$$

$$+ 4nM^2 x^{-2} \mathsf{P}(|X_1| \geq M) + 8x^{-2} \sum_{1 \leq i < j \leq n} cov\left(H_M(X_i), H_M(X_j)\right)$$

where $H_M(\cdot)$ is the same as in (1.5) and L_n is defined before Theorem 4.1.

The last inequality can serve (see [275]) to establish the rate of convergence in the law of large numbers (LLN); in particular, it yields the strong LLN for linear processes which is stronger than that proved via mixing.

There are also inequalities of the Rosenthal-type for the sequences obtained by taking smooth functions of associated random variables. These are important for the study of empirical processes [273, 373]. An account on maximal inequalities for **NA** random systems is given in the article by Zhang [430].

The approach we presented in Section 2 is a sub-stream of rapidly developing theory of stochastic ordering and its connections with probability metrics, one can refer, e.g., to books [127],[434], [338]. These two instruments provide complementary tools to study the behavior of a random vector by approximating it with a vector modified to be simpler (e.g. the decoupled version). The most useful stochastic order to work with positive and negative dependence is the convex one since the supermodularity of a function is a weakened convexity assumption. The applications of stochastic orders to positively dependent (not necessarily associated, but also POD and LPQD) random systems was studied (at first for integer-valued random variables) by Boutsikas and Koutras [51], Boutsikas and Vaggelatou [52] (with application to normal and Poisson approximation), Denuit, Lefèvre and Utev [129]. The general results connecting positive orders with Zolotarev's *ideal metrics* help to prove the following theorem.

Theorem 4.12. ([52]) *Let the assumptions of Corollary 2.9 be true, and all the random variables X_n be square-integrable. Then, for any $n \in \mathbb{N}$, one has*

$$\sup_{x \in \mathbb{R}} D(x) \le |R_n|^{1/2}, \quad \int_{\mathbb{R}} D(x)dx = |R_n|,$$

where $D(x) = |\mathsf{E}((S_n - x)^+) - \mathsf{E}((T_n - x)^+)|$ and $R_n = \sum_{1 \le i < j \le n} cov(X_i, X_j)$.

The function $\sup_{x \in \mathbb{R}} D(x)$ is the *stop-loss metric* used in actuarial mathematics. Interesting generalizations appear when one considers stochastic orders described by s-convex functions (which are defined by differences of orders higher than 2) [52, 128]. They are also in close relation with TP systems to which we paid some attention in previous Chapter.

It would be desirable to build analogs of Theorem 2.6 describing the proximity of random vectors to their decoupled versions in terms of some probability metrics. Useful approaches here are proposed by Louhichi [278], Li and Shao [262].

Note in passing that we do not tackle a problem of finding the optimal constants in the right-hand sides of the inequalities. In this regard we refer, e.g., to the paper by Hitczenko [198] where the optimal constant in the Burkholder–Davis–Gundy inequalityfor martingales is obtained.

To finish the Section we stress again that the study of moment and maximal inequalities for partial sums of (especially dependent and multiindexed) random variables is very fruitful. Various applications will be provided in the next Chapters.

Chapter 3

Central Limit Theorem

Chapter 3 is devoted to CLT for random fields. This theorem belongs to the principal ones of the Probability Theory and has important statistical applications which we consider further in Chapter 7. The present Chapter is aimed at examining normal approximation for partial sums taken over the regularly growing (in a specified sense) finite subsets of the lattice $\mathbb{Z}^d$. First of all, we discuss the concept of sequence of sets increasing in $\mathbb{R}^d$ in the Van Hove sense and the discrete analog of this notion for subsets of $\mathbb{Z}^d$. After that the asymptotic behavior of the variances of sums over such sets is investigated. We study thoroughly the remarkable Newman CLT, for associated random fields and their extensions. The characteristic functions technique and the Stein method are employed to estimate the rate of convergence to the Gaussian law under the Lindeberg condition. In particular we give a simple direct proof of CLT for the power functions of Ornstein–Uhlenbeck random field. Theorems by Lewis and Vronski, concerning the asymptotic normality under dependence assumptions weaker than standard ones, are proved as well. Here one assumes that the covariance function does not satisfy the finite susceptibility condition by Newman (i.e. is not summable), whereas the partial sums increase as a slowly varying function in the Karamata sense. In 1980 Newman set up the hypothesis that the latter property and the existence of second moment imply the CLT for strictly stationary associated random fields. In 1984 Herrndorf constructed a counterexample in the particular case and in 2005 Shashkin gave a complete negative answer to the hypothesis, which is treated in Section 2 by means of a new integral representation for slowly varying functions. Finally, Section 3 is devoted to the asymptotically sharp analog of the Berry–Esseen estimate for associated random fields proved by Bulinski via the Stein–Tikhomirov techniques.

In this Chapter we study the (normalized) partial sums generated by associated and similar random fields. More exactly, we consider the problem of approximation of their laws by the Gaussian laws. Further generalizations concerning the weak and strong invariance principles will be investigated in Chapter 5.

1   Sufficient conditions for normal approximation

As usual, we study a centered square integrable random process or field $X = \{X_j, j \in \mathbb{Z}^d\}, d \geq 1$, and, for a sequence of finite sets $U_n \subset \mathbb{Z}^d$, consider the normalized sums $S(U_n)/\sqrt{\mathrm{Var}S(U_n)}$ (or $S(U_n)/\sqrt{|U_n|}$), $n \in \mathbb{N}$. If their distributions converge weakly to the normal one when $U_n \to \infty$ (in a certain sense), then we say that a sequence $(S(U_n)/\sqrt{\mathrm{Var}S(U_n)})_{n \in \mathbb{N}}$ satisfies the central limit theorem (CLT). We also say that X satisfies CLT if the above mentioned property holds for any sequence of finite sets U_n growing to infinity in a fixed manner.

The first result of such type for associated fields appeared in 1980. It was the beautiful CLT by Newman ([307]) for strictly stationary random fields where blocks $U_n \in \mathcal{U}$ were used[1]. A sequence of blocks is growing to infinity if all the edges' lengths tend to infinity.

There are several ways for possible generalizations of that limit theorem. One can consider more general sets of summation, non-stationary processes or fields, random fields having various dependence structure (not only weak dependence but also long-range dependence, see, e.g., [143]), vector-valued random fields. It is also interesting to estimate the convergence rates in the CLT. We are going to pay attention to all these generalizations.

1°. Regularly growing sets. First of all we recall some concepts of "regular growth" for a family of subsets in $\mathbb{R}^d$ and $\mathbb{Z}^d$. Let $a = (a_1, \ldots, a_d) \in \mathbb{R}^d$ be a vector with positive components (one writes $a > 0$). Introduce

$$\Pi_0(a) = \{x \in \mathbb{R}^d : 0 < x_i \leq a_i, \, i = 1, \ldots, d\}$$

and for $j \in \mathbb{Z}^d$ define the shifted blocks

$$\Pi_j(a) = \Pi_0(a) + ja = \{x \in \mathbb{R}^d : j_i a_i < x_i \leq (j_i + 1)a_i, \, i = 1, \ldots, d\}.$$

Thus $\{\Pi_j(a), j \in \mathbb{Z}^d\}$ form a partition of $\mathbb{R}^d$. For a set $V \subset \mathbb{R}^d$, put

$$J_-(V, a) = \{j : \Pi_j(a) \subset V\}, \quad J_+(V, a) = \{j : \Pi_j(a) \cap V \neq \varnothing\},$$

$$V^-(a) = \bigcup_{j \in J_-(V,a)} \Pi_j(a), \quad V^+(a) = \bigcup_{j \in J_+(V,a)} \Pi_j(a).$$

From now on, in this Chapter, for a measurable set $B \subset \mathbb{R}^d$ its Lebesgue measure[2] is denoted by $|B|$.

Definition 1.1. (see, e.g., [359]) A sequence of sets $V_n \subset \mathbb{R}^d$ ($n \in \mathbb{N}$) tends to infinity in the *Van Hove sense* (or is VH-*growing*) if, for any $a \in \mathbb{R}^d$ ($a > 0$), one has

$$|V_n^-(a)| \to \infty, \quad |V_n^-(a)|/|V_n^+(a)| \to 1 \text{ as } n \to \infty. \tag{1.1}$$

---

[1] See the beginning of Section 2.1 for the definition of $\mathcal{U}$.

[2] For a finite set J, $|J|$ also often stands for its cardinality, occasionally written as $\sharp(J)$ to avoid misunderstanding.

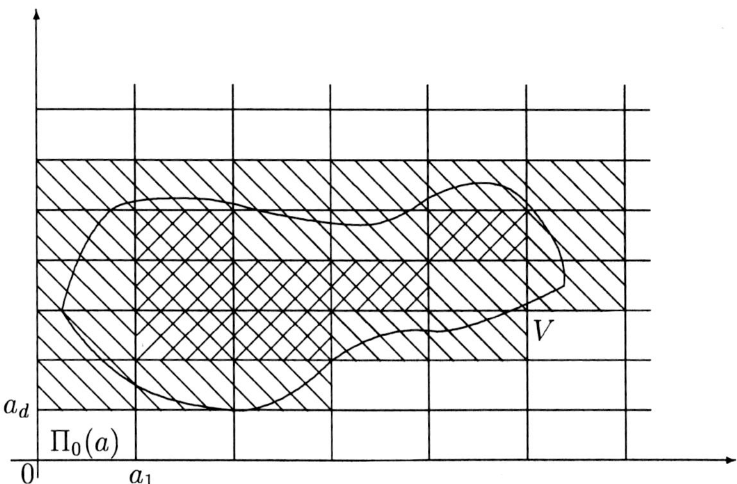

Fig. 3.1   ($d = 2$, crossed hachures mark $V^-(a)$ and single $V^+(a) \setminus V^-(a)$)

It is easy to show (see also (1.2) below) that a family of blocks

$$V_n = (a^{(n)}, b^{(n)}] = \{x \in \mathbb{R}^d : a_i^{(n)} < x_i \leq b_i^{(n)}, i = 1, \ldots, d\} \to \infty$$

in the Van Hove sense if and only if $\min_{1 \leq i \leq d}(b_i^{(n)} - a_i^{(n)}) \to \infty$ as $n \to \infty$.

For $\varepsilon > 0$ and $V \subset \mathbb{R}^d$, we define (using the Euclidean distance ρ) the neighborhood of V as previously:

$$V^\varepsilon = \{x \in \mathbb{R}^d : \rho(x, V) := \inf\{\rho(x, y) : y \in V\} < \varepsilon\}.$$

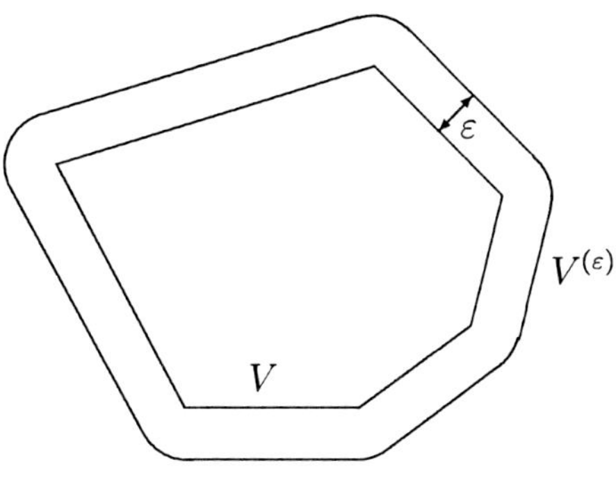

Fig. 3.2

Recall that the boundary of a set $V \subset \mathbb{R}^d$ is a set ∂V consisting of such points $z \in \mathbb{R}^d$ that in every neighbourhood of z there exist a point $x \in V$ and a point $y \notin V$. In the book by Ruelle [359] one can find (without proof) the following statement.

Lemma 1.2. *Let $(V_n)_{n \in \mathbb{N}}$ be a sequence of bounded measurable[3] sets in $\mathbb{R}^d$. Then $V_n \to \infty$ in the Van Hove sense if and only if, for any $\varepsilon > 0$,*

$$|(\partial V_n)^\varepsilon|/|V_n| \to 0 \quad as \ n \to \infty. \tag{1.2}$$

Proof. Necessity. Note that $(\partial V_n)^\varepsilon$ is an open (as union of open balls) bounded set for any $n \in \mathbb{N}$. Thus $(\partial V_n)^\varepsilon$ has finite Lebesgue measure for each $\varepsilon > 0$ and all $n \in \mathbb{N}$. Let $V_n \to \infty$ in the Van Hove sense. Take $a = (t, \ldots, t) \in \mathbb{R}^d$, $t > 2\varepsilon$, and introduce

$$\Pi_0(a)_\varepsilon = \{x \in \mathbb{R}^d : \varepsilon < x_i \le a_i - \varepsilon, i = 1, \ldots, d\}.$$

Then $|\Pi_0(a)^\varepsilon| = (t + 2\varepsilon)^d$ and $|\Pi_0(a)^\varepsilon \setminus \Pi_0(a)_\varepsilon| = (t + 2\varepsilon)^d - (t - 2\varepsilon)^d \le 2^{d+1} d\varepsilon t^{d-1}$. Clearly,

$$|(\partial V_n)^\varepsilon| \le \sharp(J_-(V_n, a)) |\Pi_0(a)^\varepsilon \setminus \Pi_0(a)_\varepsilon| + \sharp(J_+(V_n, a) \setminus J_-(V_n, a))|\Pi_0(a)^\varepsilon|.$$

Equally obviously, $\sharp(J_-(V_n, a)) < \infty$ and $\sharp(J_+(V_n, a) < \infty$, as $diam(V_n) < \infty$. Thus $|V_n^-(a)| > 0$ for all n large enough and

$$\frac{|(\partial V_n)^\varepsilon|}{|V_n|} \le \frac{|(\partial V_n)^\varepsilon|}{|V_n^-(a)|} \le \frac{|\Pi_0(a)^\varepsilon \setminus \Pi_0(a)_\varepsilon|}{|\Pi_0(a)|} + \frac{|\Pi_0(a)^\varepsilon|}{|\Pi_0(a)|} \frac{(|V_n^+(a)| - |V_n^-(a)|)}{|V_n^-(a)|} \le$$

$$\le \frac{d \, 2^{d+1} \varepsilon}{t} + \left(1 + \frac{2\varepsilon}{t}\right)^d \left(\frac{|V_n^+(a)|}{|V_n^-(a)|} - 1\right).$$

Due to (1.1), one has

$$\limsup_{n \to \infty} \frac{|(\partial V_n)^\varepsilon|}{|V_n|} \le \frac{d \, 2^{d+1} \varepsilon}{t}.$$

Since t can be taken arbitrary large, we come to (1.2).

Sufficiency. Suppose that (1.2) holds. Let $B_r(x) = \{y \in \mathbb{R}^d : dist(x, y) < r\}$. Evidently, for any $\varepsilon > 0$ and $n \in \mathbb{N}$, $|(\partial V_n)^\varepsilon| \ge |B_\varepsilon(0)|$. Therefore $|V_n| \to \infty$ as $n \to \infty$. For any fixed $a \in \mathbb{R}^d$ $(a > 0)$ consider a partition of $\mathbb{R}^d$ by a family $\{\Pi_j(a), j \in \mathbb{Z}^d\}$. If $J_-(V_n, a) = J_+(V_n, a)$, then, clearly, $|V_n^-(a)| = |V_n|$ and (1.1) is valid. Now let $J_+(V_n, a) \setminus J_-(V_n, a) \ne \varnothing$. For each $j \in J_+(V_n, a) \setminus J_-(V_n, a)$ there exists $x_{n_j} \in \Pi_j(a)$ such that $x_{n_j} \in \partial V_n$.

Then, for any $\varepsilon > 0$, one has

$$\left|(\partial V_n)^\varepsilon\right| \le \sum_{j \in J_n(a)} \left|\Pi_j(a) \cap B_\varepsilon(x_{n_j})\right| \ge C(\varepsilon, a)\sharp(J_n(a)) \tag{1.3}$$

where $J_n(a) = J_+(V_n, a) \setminus J_-(V_n, a)$ and

$$C(\varepsilon, a) = \inf\{|\Pi_0(a) \cap B_\varepsilon(x)| : x \in \Pi_0(a)\} > 0.$$

For instance, if $\varepsilon < \min_{1 \le i \le d} a_i$ then $C(\varepsilon, a) = 2^{-d}|B_\varepsilon(0)|$. Note that $J_-(V_n, a) \ne \varnothing$ for all n large enough, as otherwise (1.3) would imply that

$$|\partial V_n)^\varepsilon|/|V_n| \ge C(\varepsilon, a)\sharp J_+(V_n, a)/(|\Pi_0(a)|\sharp J_+(V_n, a)) = C(\varepsilon, a)/|\Pi_0(a)| > 0,$$

---

[3]That is V_n is Lebesgue measurable and $diam(V_n) := \sup\{\rho(x, y) : x, y \in V_n\} < \infty$, $n \in \mathbb{N}$.

which contradicts (1.3). Using a trivial estimate $|V_n| \leq \sharp(J_+(V_n, a))|\Pi_0(a)|$, we see that

$$\frac{|(\partial V_n)^\varepsilon|}{|V_n|} \geq \frac{C(\varepsilon, a)}{|\Pi_0(a)|}\left(1 - \frac{|V_n^-(a)|}{|V_n^+(a)|}\right). \tag{1.4}$$

So the second relation in (1.1) follows from (1.2).

Since $|V_n^-(a)| \leq |V_n| \leq |V_n^+(a)|$ and $|V_n| \to \infty$, we conclude on account of (1.2) and (1.4) that $|V_n^-(a)| \to \infty$, $n \to \infty$. Therefore (1.1) is established. $\square$

Remark 1.3. For a sequence $(V_n)_{n \in \mathbb{N}}$ of bounded measurable subsets of $\mathbb{R}^d$, in view of Lemma 1.2 the concept of growing in the Van Hove sense does not depend on the choice of coordinate system in $\mathbb{R}^d$.

For finite sets $U_n \subset \mathbb{Z}^d, n \in \mathbb{N}$, one can use an analog of condition (1.2). Given $U \subset \mathbb{Z}^d$ with $|U| < \infty$, write

$$U^p = \{j \in \mathbb{Z}^d : dist(j, U) := \inf\{dist(i, j) : i \in U\} \leq p\}$$

where $p \in \mathbb{N}$ and $dist$ is the metric corresponding to the sup-norm in $\mathbb{R}^d$. Put $\delta U = U^1 \setminus U$. That is

$$\delta U = \{j \in \mathbb{Z}^d \setminus U : dist(j, U) = 1\}.$$

Thus for a finite set $U \subset \mathbb{Z}^d$ we use the symbol δU as an analog of ∂V for $V \subset \mathbb{R}^d$.

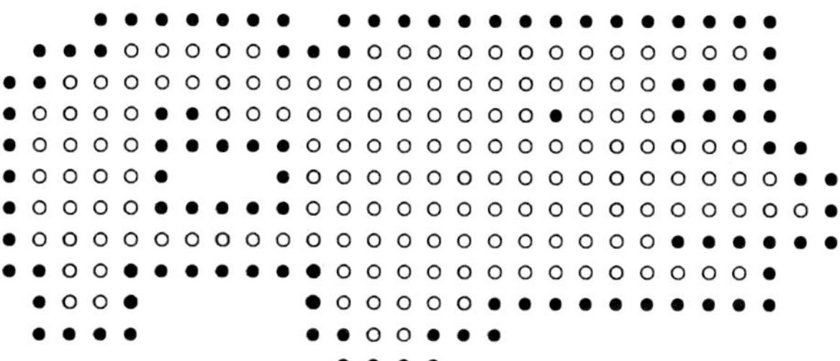

Fig. 3.3   (black circles correspond to points of δU and white ones to U)

Definition 1.4. A sequence of finite sets $U_n \subset \mathbb{Z}^d$ ($n \in \mathbb{N}$) is called *regularly growing* (to infinity) if

$$|U_n| \to \infty \quad \text{and} \quad |\delta U_n|/|U_n| \to 0 \text{ as } n \to \infty. \tag{1.5}$$

One also says that $U_n \to \infty$ in a *regular manner* as $n \to \infty$.

Lemma 1.5. *Let $(V_n)_{n \in \mathbb{N}}$ be a sequence of bounded sets in $\mathbb{R}^d$ such that $V_n \to \infty$ in the Van Hove sense as $n \to \infty$. Then $U_n := V_n \cap \mathbb{Z}^d$ ($n \in \mathbb{N}$) form a regularly growing sequence.*

Proof. Take $a = (m, \ldots, m) \in \mathbb{N}^d$ such that $m > 2$. Obviously $V_n^-(a) \cap \mathbb{Z}^d \subset V_n \cap \mathbb{Z}^d = U_n$ and $\sharp(V_n^-(a) \cap \mathbb{Z}^d) = |V_n^-(a)|$. Thus $|U_n| \to \infty$ as $n \to \infty$. Introduce

$$\Gamma_j(a) := \{x \in \mathbb{R}^d : mj_i - 1 < x_i \le m(j_i + 1) + 1, i = 1, \ldots, d\}.$$

Then by definition

$$\delta U_n \subset \left(\bigcup_{j \in J_-(V_n, a)} (\Gamma_j(a) \setminus \Pi_j(a)) \right) \bigcup \left(\bigcup_{j \in J_+(V_n, a) \setminus J_-(V_n, a)} \Gamma_j(a) \right).$$

Note that $|\Gamma_j(a)| = (m+2)^d$ and $|\Gamma_j(a) \setminus \Pi_j(a)| = (m+2)^d - m^d \le 2d(m+2)^{d-1} \le d2^d m^{d-1}$ because $m > 2$. So we come to the inequality

$$d\, 2^d\, m^{d-1} |J_-(V_n, a)| + |\delta U_n| \le (m+2)^d (|J_+(V_n, a)| - |J_-(V_n, a)|)$$

$$= d\, 2^d |V_n^-(a)|/m + (m+2)^d (|V_n^+(a)| - |V_n^-(a)|)/|\Pi_0(a)|.$$

Taking into account that $\sharp(U_n) \ge |V_n^-(a)|$, we have

$$\frac{|\delta U_n|}{|U_n|} \le \left(1 + \frac{2}{m}\right)^d \left(\frac{|V_n^+(a)|}{|V_n^-(a)|} - 1\right) + \frac{d\, 2^d}{m}.$$

Due to (1.1),

$$\limsup_{n \to \infty} \frac{|\delta U_n|}{|U_n|} \le \frac{d\, 2^d}{m}.$$

As m can be taken arbitrary large, we come to (1.5). $\square$

Let $Q_j := \{x \in \mathbb{R}^d : j_i < x_i \le j_i + 1, i = 1, \ldots, d\}$, $j = (j_1, \ldots, j_d) \in \mathbb{Z}^d$, be a unit cube with a "lower vertex" j.

Lemma 1.6. *If $(U_n)_{n \in \mathbb{N}}$ is a sequence of finite subsets of $\mathbb{Z}^d$, regularly growing to infinity, then $V_n := \cup_{j \in U_n} Q_j \to \infty$ in the Van Hove sense as $n \to \infty$.*

Proof. Take arbitrary $m \in \mathbb{N}$ and consider the partition of $\mathbb{R}^d$ formed by a family of blocks $\Pi_j(a)$, $j \in \mathbb{Z}^d$, where $a = (m, \ldots, m) \in \mathbb{N}^d$. Using the same proof as for Lemma 1.2 we obtain that, for every $\varepsilon > 0$,

$$\frac{|(\partial V_n)^\varepsilon|}{|V_n|} \le \frac{2^{d+1} d\varepsilon}{m} + \left(1 + \frac{2\varepsilon}{m}\right)^d \left(\frac{|J_+(V_n, a)|}{|J_-(V_n, a)|} - 1\right). \tag{1.6}$$

Observe that

$$|\Pi_0(a)||J_-(V_n, a)| \ge |U_n| - |\delta U_n|(2m + 1)^d,$$

$$|\Pi_0(a)|(|J_+(V_n, a)| - |J_-(V_n, a)|) \le |\delta U_n|(2m + 1)^d.$$

Evidently, $|J_-(V_n, a)| > 0$ for all n large enough. Indeed, if $J_-(V_n, a) = \varnothing$ then

$$\frac{|\delta U_n|}{|U_n|} \ge \frac{|J_+(V_n, a)|}{|\Pi_0(a)||J_+(V_n, a)|} = \frac{1}{|\Pi_0(a)|},$$

which contradicts (1.5). Therefore (1.5) shows that, for all n large enough (such that the denominator in the formula below is positive),

$$0 \le \frac{|J_+(V_n, a)| - |J_-(V_n, a)|}{|J_-(V_n, a)|} \le \frac{|\delta U_n|}{|U_n|(2m+1)^{-d} - |\delta U_n|} \to 0 \text{ as } n \to \infty.$$

Consequently, in view of (1.6) we come to (1.2). $\square$

Remark 1.7. Lemmas 1.5 and 1.6 are important because they show that the study of additive random functions, defined on a system of bounded measurable subsets of $\mathbb{R}^d$ growing to infinity in the Van Hove sense, and investigation of sums of multiindexed random variables $X_j, j \in \mathbb{Z}^d$, taken over regularly growing finite sets $U_n \subset \mathbb{Z}^d$ ($n \in \mathbb{N}$), are in a certain sense equivalent.

2°. Variances of partial sums. Before we start studying limit theorems it will be useful to consider asymptotical behavior of variances for partial sums $S(U_n)$ when $U_n \to \infty$ in a regular manner as $n \to \infty$. If the field X is (wide-sense) stationary and the Cox–Grimmett coefficient is finite, then this behavior is simply determined by the covariance function. Recall that a field $X = \{X_j, j \in \mathbb{Z}^d\}$, consisting of square integrable random variables, is called a *wide sense stationary* if, for all $i, j \in \mathbb{Z}^d$, one has

$$\mathsf{E} X_j = c \, (= \text{const}) \quad \text{and} \quad cov(X_i, X_j) =: \mathsf{R}(\mathsf{i} - \mathsf{j}).$$

The following result by Bolthausen concerns arbitrary (wide-sense) stationary fields.

Theorem 1.8. ([49]) *Suppose that X is a wide-sense stationary random field such that the series (1.5.3) is absolutely convergent. Then*

$$\frac{\mathsf{Var} S(U_n)}{|U_n|} \to \sigma^2 \text{ as } n \to \infty \tag{1.7}$$

for any sequence $U_n \to \infty$ in a regular manner.

Proof. Take arbitrary $p \in \mathbb{N}$ and set $G_n = U_n \cap (\delta U_n)^p$, $W_n = U_n \setminus G_n$. Then

$$\sigma^2 |U_n| - \mathsf{Var} S(U_n) = \sum_{j \in U_n} \sum_{k \notin U_n} cov(X_j, X_k)$$

$$= \sum_{j \in G_n} \sum_{k \notin U_n} cov(X_j, X_k) + \sum_{j \in W_n} \sum_{k \notin U_n} cov(X_j, X_k) =: R_{1,n} + R_{2,n}.$$

Obviously $|G_n| \le |(\delta U_n)^p| \le (2p+1)^d |\delta U_n|$, and by (1.5)

$$\frac{R_{1,n}}{|U_n|} \le \frac{|G_n|}{|U_n|} \sum_{j \in \mathbb{Z}^d} |cov(X_0, X_j)| \le c_0 (2p+1)^d (|\delta U_n|/|U_n|) \to 0, \, n \to \infty,$$

where $c_0 = \sum_{j \in \mathbb{Z}^d} |cov(X_0, X_j)|$. Since $dist(W_n, \mathbb{Z}^d \setminus U_n) \ge p$ and $|W_n| \le |U_n|$, we have

$$\limsup_{n \to \infty} \frac{R_{2,n}}{|U_n|} \le \sum_{j \in \mathbb{Z}^d : |j| \ge p} |cov(X_0, X_j)|.$$

On account of the condition $\sum_j |cov(X_0, X_j)| < \infty$ the right-hand side of the last inequality can be made arbitrarily small when p is large enough. Thus (1.7) is established. $\square$

Sometimes it is helpful to notice that, for non-positively correlated random variables, the conditions of Theorem 1.8 can be simplified as the following result by Newman shows.

Lemma 1.9. ([308]) *Let* $X = \{X_j, j \in \mathbb{Z}^d\}$ *be a square-integrable, wide-sense stationary random field such that* $cov(X_i, X_j) \leq 0$ *for any* $i, j \in \mathbb{Z}^d$, $i \neq j$. *Then the series (1.5.3) is absolutely convergent and its sum is nonnegative.*

Proof. Let $U_n = [-n, n]^d \cap \mathbb{Z}^d$, $n \in \mathbb{N}$. Assume that Lemma's first assertion is false. Then there exist some $n_0 \in \mathbb{N}$ and $b < 0$ such that

$$\sum_{j \in \mathbb{Z}^d : |j| \leq n} cov(X_0, X_j) \leq b \tag{1.8}$$

for all $n \geq n_0$ (as all the summands except for one with $j = 0$ are nonpositive).

We apply a device which is similar to that used in the proof of Theorem 1.8. Fix some number $\alpha \in (0, 1/(d+1))$ and let $p_n = [n^\alpha]$, $n \in \mathbb{N}$. For any $n \in \mathbb{N}$ such that $p_n > n_0$, define the sets $H_n = U_n \cap (\delta U_n)^{p_n}$, $M_n = U_n \setminus H_n$. Then we have

$$0 \leq \mathsf{Var} S(U_n) = \sum_{j \in H_n} \sum_{k \in U_n} cov(X_j, X_k) + \sum_{j \in M_n} \sum_{k \in U_n} cov(X_j, X_k). \tag{1.9}$$

As $cov(X_j, X_k) \leq 0$ whenever $j \neq k$, the right-hand side of (1.9) will not decrease if in both double sums we take the sum over k belonging not to U_n but to a subset of U_n containing j. Thus $\mathsf{Var} S(U_n)$ admits the following upper bound

$$\sum_{j \in H_n} \sum_{\substack{k \in U_n: \\ |k-j| \leq p_n}} cov(X_j, X_k) + \sum_{j \in M_n} \sum_{\substack{k \in \mathbb{Z}^d: \\ |k-j| \leq p_n}} cov(X_j, X_k) =: T_{1,n} + T_{2,n}. \tag{1.10}$$

By wide-sense stationarity and (1.8) we have

$$T_{2,n} = |M_n| \sum_{j : |j| \leq p_n} cov(X_0, X_j) \leq b|M_n|.$$

Because of trivial estimate $|cov(X_j, X_k)| \leq \mathsf{Var} X_0$, for $j, k \in \mathbb{N}$, we have

$$|T_{1,n}| \leq |H_n|(2p_n + 1)^d \mathsf{Var} X_0.$$

Hence from (1.10) we infer that

$$0 \leq \mathsf{Var} S(U_n) \leq b|M_n| + |H_n|(2p_n + 1)^d \mathsf{Var} X_0$$

$$= b(2n + 1 - 2p_n)^d + \left((2n + 1)^d - (2n + 1 - 2p_n)^d\right)(2p_n + 1)^d \, \mathsf{Var} X_0,$$

but the choice of p_n with $\alpha \in (0, 1/(d+1))$ implies that the right-hand side is negative for all n large enough, which is a contradiction. $\square$

Remark 1.10. If we assume that the field under consideration is (BL, θ)-dependent, then the finite susceptibility condition is certainly satisfied. Really, introduce the auxiliary random field $Y = \{Y_j, j \in \mathbb{Z}^d\}$ where $Y_j = \gamma_j X_j$ with $\gamma_j = sgn(cov(X_0, X_j))$, $j \in \mathbb{Z}^d$. Clearly, $cov(Y_0, Y_j) = |cov(X_0, X_j)|$ for all $j \in \mathbb{Z}^d$. In view of Lemma 1.5.2 the field Y is (BL, θ)-dependent with the same sequence $(\theta_r)_{r \in \mathbb{N}}$ as for the field X. For any finite set $U \subset \mathbb{Z}^d$ with $dist(U, \{0\}) \geq r$ one has

$$\sum_{j \in U} |cov(X_0, X_j)| = \sum_{j \in U} cov(Y_0, Y_j) = cov\left(Y_0, \sum_{j \in U} Y_j\right) \leq \theta_r. \qquad (1.11)$$

Hence, by the Cauchy criterion the series $\sum_{j \in \mathbb{Z}^d} |cov(X_0, X_j)|$ converges.

We will also make use of

Lemma 1.11. *Suppose that $X = \{X_j, j \in \mathbb{Z}^d\}$ is a wide-sense stationary random field such that $\sum_{j \in \mathbb{Z}^d} |cov(X_0, X_j)| < \infty$. If*

$$\sigma^2 := \sum_{j \in \mathbb{Z}^d} cov(X_0, X_j) > 0,$$

then $\rho_0 := \inf_{U \in \mathcal{U}} \operatorname{Var}S(U)/|U| > 0$ where $S(U) = \sum_{j \in U} X_j$. The value of ρ_0 is determined by the covariance function of X.

Proof. By Theorem 1.8 there exists some $m \in \mathbb{N}$ such that $\operatorname{Var}S(U)/|U| > \sigma^2/2$ whenever U is a block having all edges longer than m. Suppose that Lemma's assertion does not hold for the field X, and let $\varepsilon \in (0, \sigma^2 m^{-d}/2)$. Then there exists a block $B_{(0)} = (a^0, b^0] \in \mathcal{U}$ such that $\operatorname{Var}S(B_{(0)}) < \varepsilon|B_{(0)}|$. Because of wide-sense stationarity we can think that $a^0 = 0$. Moreover, for any multiindex $j \in \mathbb{Z}^d$ the block $B_{(j)} := (j_1 b_1^0; (j_1 + 1)b_1^0] \times \ldots \times (j_d b_d^0; (j_d + 1)b_d^0]$ is obtained from $B_{(0)}$ by a translation. Thus, again by stationarity, one has $\operatorname{Var}S(B_{(j)}) < \varepsilon|B_{(0)}|$ and $cov(S(B_{(j)}), S(B_{(k)})) < \varepsilon|B_{(0)}|$, $j \neq k$. Now the block

$$\widehat{U} := \bigcup_{0 < j_v \leq m,\, v=1,\ldots,d} U_{(j)}$$

has all its edges longer than m, but at the same time

$$\operatorname{Var}S\left(\widehat{U}\right) = \sum_{1 \leq j,k \leq m\mathbf{1}} cov(S(B_{(j)}), S(B_{(k)})) < \varepsilon m^{2d}|B_{(0)}| = \varepsilon m^d|\widehat{U}| < \frac{\sigma^2}{2}|\widehat{U}|,$$

contrary to the choice of m. $\square$

3°. CLT for random fields by means of characteristic functions. Now we will formulate the CLT. For strictly stationary random field, due to Newman [307], one may only require the existence of second moment and the finite susceptibility condition. The CLT for associated random fields when summation is carried over VH-growing sets was proved by Bulinski and Vronski [83]. Here we impose the condition of (BL, θ)-dependence instead of association, since the method of proof applies in that generality.

Theorem 1.12. *Let $X = \{X_j, j \in \mathbb{Z}^d\}$ be a (BL, θ)-dependent strictly stationary centered square-integrable random field. Then, for any sequence of regularly growing sets $U_n \subset \mathbb{Z}^d$, one has*

$$S(U_n)/\sqrt{|U_n|} \to N(0, \sigma^2) \text{ in law as } n \to \infty, \tag{1.12}$$

here σ^2 is that defined in (1.5.3).

    Proof. First of all we show that, in order to establish (1.12), instead of $S(U_n)/\sqrt{|U_n|}$ one can operate with normalized sums taken over unions of fixed blocks which form a partition of $\mathbb{R}^d$. For any bounded $V \subset \mathbb{R}^d$ let

$$S(V) = \sum_{j \in V \cap \mathbb{Z}^d} X_j,$$

then obviously $S(V) = S(U)$ where $U = V \cap \mathbb{Z}^d$.

    If $(U_n)_{n \in \mathbb{N}}$ is a sequence of regularly growing sets in $\mathbb{Z}^d$ (see Definition 1.4) then, by virtue of Lemma 1.6, $V_n := \cup_{j \in U_n} Q_j \to \infty$ in the Van Hove sense as $n \to \infty$, where $Q_j = (j, j+1] \subset \mathbb{R}^d$, $j \in \mathbb{Z}^d$. Thus we have to prove that, for each $t \in \mathbb{R}$,

$$\mathsf{E} \exp\{it S(V_n)/\sqrt{|V_n|}\} \to \exp\{-\sigma^2 t^2/2\} \quad \text{as } n \to \infty, \tag{1.13}$$

here $i^2 = -1$, $|V_n|$ is the Lebesgue measure of V_n. Take $a = (a_1, \ldots, a_d) \in \mathbb{N}^d$ with $a_k = mr$, $k = 1, \ldots, d$, where $m, r \in \mathbb{N}$ will be specified later. Consider $V_n^-(a)$ introduced before the Definition 1.1. Clearly, for any $x, y \in \mathbb{R}$ one has $|e^{ix} - e^{iy}| \le |x - y|$. Therefore

$$\left| \mathsf{E} \exp\{it |V_n|^{-1/2} S(V_n)\} - \mathsf{E} \exp\{it |V_n^-(a)|^{-1/2} S(V_n^-(a))\} \right|$$

$$\le |t| |V_n|^{-1/2} \mathsf{E}|S(V_n) - S(V_n^-(a))| + |t| \left| |V_n|^{-1/2} - |V_n^-(a)|^{-1/2} \right| \mathsf{E}|S(V_n^-(a))|. \tag{1.14}$$

The estimate (1.11) yields the bound

$$\mathsf{E}|S(V_n) - S(V_n^-(a))| \le (\mathsf{E}(S(V_n \setminus V_n^-(a)))^2)^{1/2} \le ((\mathsf{E}X_0^2 + \theta_1)|V_n \setminus V_n^-(a)|)^{1/2}$$

and, since V_n are VH-growing,

$$|V_n \setminus V_n^-(a)|/|V_n| \to 0, \quad n \to \infty.$$

Note also that

$$\left| |V_n|^{-1/2} - |V_n^-(a)|^{-1/2} \right| = \frac{|V_n| - |V_n^-(a)|}{(|V_n||V_n^-(a)|)^{1/2}(|V_n|^{1/2} + |V_n^-(a)|^{1/2})} \le \frac{|V_n| - |V_n^-(a)|}{|V_n^-(a)|^{3/2}}.$$

Furthermore,

$$\mathsf{E}|S(V_n^-(a))| \le (\mathsf{E}(S(V_n^-(a)))^2)^{1/2} \le ((\mathsf{E}X_0^2 + \theta_1)|V_n^-(a)|)^{1/2}. \tag{1.15}$$

Relations (1.14), (1.15) show that to prove (1.13) it suffices to verify that, for $t \in \mathbb{R}$,

$$\mathsf{E} \exp\{it |V_n^-(a)|^{-1/2} S(V_n^-(a))\} \to \exp\{-\sigma^2 t^2/2\} \quad \text{as } n \to \infty.$$

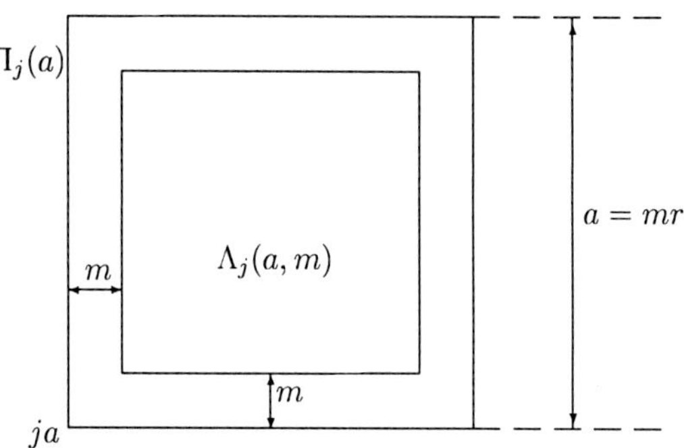

Fig. 3.4

Now we make the next step. For each $j \in \mathbb{Z}^d$ introduce a block $\Lambda_j(a, m) \subset \Pi_j(a)$ in such a way that, for all $q, j \in \mathbb{Z}^d$, $q \ne j$,

$$dist(\Lambda_j(a, m), \Lambda_q(a, m)) \ge m \tag{1.16}$$

where $dist$ is, as usual, the distance corresponding to sup-norm in $\mathbb{R}^d$. Namely, let $r > 2$ and

$$\Lambda_j(a, m) = \{x \in \mathbb{R}^d : mrj_k + m < x_k \le mr(j_k + 1) - m, \ k = 1, \ldots, d\}.$$

Set $T_n = |\widetilde{V_n}|^{-1/2} S(\widetilde{V_n})$ where

$$\widetilde{V_n} = \widetilde{V_n}(a, m) := \bigcup_{j \in J^-(V_n, a)} \Lambda_j(a, m).$$

It means that instead of $V_n^-(a)$ we will consider a set $\widetilde{V_n}(a, m) \subset V_n^-(a)$ consisting of separated blocks $\Gamma_j(a, m)$.

So, for each fixed $t \in \mathbb{R}$ and $m \in \mathbb{N}$, we show that $\mathsf{E} \exp\{itT_n\}$, for all $n \in \mathbb{N}$, is close to

$$\mathsf{E} \exp\{it|V_n^-(a)|^{-1/2} S(V_n^-(a))\}$$

if r is taken large enough. Obviously,

$$\left| V_n^-(a) \setminus \widetilde{V_n} \right| = |J^-(V_n, a)| \left(|\Pi_0(a)| - |\Lambda_0(a, m)| \right) = |V_n^-(a)| \frac{|\Pi_0(a)| - |\Lambda_0(a, m)|}{|\Pi_0(a)|}$$

$$= |V_n^-(a)| \frac{(rm)^d - ((r - 2)m)^d}{(rm)^d} \le |V_n^-(a)| \frac{2d}{r}.$$

The same reasoning as that used to prove (1.14) and (1.15) leads to the following estimate. For every fixed $t \in \mathbb{R}$, any $\varepsilon > 0$ and all $m, n \in \mathbb{N}$,

$$\left| \mathsf{E} \exp\left\{it|V_n^-(a)|^{-1/2} S(V_n^-(a))\right\} - \mathsf{E} \exp\left\{itT_n\right\} \right| \le 2(2d(\mathsf{E}X_0^2 + \theta_1)/r)^{1/2} < \varepsilon$$

if r is large enough.

Write $N_n = |J^-(V_n, a)|$ and enumerate a family of random variables

$$\{|\Lambda_j(a, m)|^{-1/2} S(\Lambda_j(a, m)),\ j \in J^-(V_n, a)\}$$

to obtain a collection $\xi_{n,1}, \ldots, \xi_{n,N_n}$ (clearly, $N_n = N_n(a)$ and $\xi_{n,k} = \xi_{n,k}(a, m)$). Then $|\widetilde{V_n}| = N_n |\Lambda_0(a, m)|$ and we see that

$$\mathsf{E} \exp\{itT_n\} = \mathsf{E} \exp\left\{itN_n^{-1/2} \sum_{k=1}^{N_n} \xi_{n,k}\right\}.$$

By Corollary 1.5.5, in view of (1.16) one has, for any $\varepsilon > 0$, $t \in \mathbb{R}$, $n \in \mathbb{N}$ and $r > 2$,

$$\left| \mathsf{E} \exp\left\{it \sum_{k=1}^{N_n} \xi_{n,k} N_n^{-1/2}\right\} - \prod_{i=1}^{N_n} \mathsf{E} \exp\left\{itN_n^{-1/2}\xi_{n,k}\right\} \right|$$

$$\leq \sum_{q=1}^{N_n-1} \left| cov\left(\exp\left\{itN_n^{-1/2}\xi_{n,q}\right\}, \exp\left\{-itN_n^{-1/2} \sum_{k=q+1}^{N_n} \xi_{n,k}\right\}\right) \right|$$

$$\leq 4t^2 N_n \frac{\theta_m}{N_n |\Lambda_0(a, m)|} = \frac{4t^2 \theta_m}{|\Lambda_0(a, m)|} \leq 4t^2 \theta_m < \varepsilon$$

under appropriate choice of m. Taking independent random variables $\eta_{n,1}, \ldots, \eta_{n,N_n}$ with $Law(\xi_{n,k}) = Law(\eta_{n,k})$, $k = 1, \ldots, N_n$ we see that, for $t \in \mathbb{R}$, $n \in \mathbb{N}$, one has

$$\prod_{i=1}^{N_n} \mathsf{E} \exp\left\{itN_n^{-1/2}\xi_{n,k}\right\} = \mathsf{E} \exp\left\{itN_n^{-1/2} \sum_{k=1}^{N_n} \eta_{n,k}\right\}.$$

For an array of square-integrable i.i.d. random variables $\{\eta_{n,k}, n \in \mathbb{N}, 1 \leq k \leq N_n\}$ we can apply the CLT (see, e.g., [326, Ch. IV]). Therefore, in the sense of convergence in law,

$$N_n^{-1/2} \sum_{k=1}^{N_n} \eta_{n,k} \to Z \sim N(0, \Sigma^2(a, m)) \quad \text{as } n \to \infty.$$

Here

$$\Sigma^2(a, m) = \Sigma^2(rm, m) = \mathsf{Var}(\eta_{1,1}) = \mathsf{Var} S(\Lambda_0(a, m))/|\Lambda_0(a, m)|.$$

Theorem 1.8 yields that, for every $m \in \mathbb{N}$, $\Sigma^2(a, m) \to \sigma^2$ as $r \to \infty$. Using the trivial fact that, for every $t \in \mathbb{R}$,

$$\exp\{-\Sigma^2(a, m)t^2/2\} \to \exp\{-\sigma^2 t^2/2\}, \quad m \to \infty,$$

we complete the proof. $\square$

Corollary 1.13. *Let $X = \{X_j, j \in \mathbb{Z}^d\}$ be a (BL, θ)-dependent, strictly stationary, centered, square-integrable vector-valued random field taking values in $\mathbb{R}^k$. Then, for any sequence of regularly growing sets $U_n \subset \mathbb{Z}^d$, one has*

$$|U_n|^{-1/2} \sum_{j \in U_n} X_j \to N(0, C) \quad \text{in law} \quad \text{as } n \to \infty.$$

Here C is the $k \times k$ matrix having the elements

$$C_{lm} = \sum_{j \in \mathbb{Z}^d} cov(X_{0,l}, X_{j,m}), \quad l, m = 1, \ldots, k. \tag{1.17}$$

Proof. The convergence of all k^2 series in (1.17) is proved in the same way as the convergence of the series defining σ^2 in one-dimensional case. The convergence in law is proved via the Cramér–Wald device. More precisely, it suffices to show that, for any fixed $v \in \mathbb{R}^k$, the sum of inner products $|U_n|^{-1/2} \sum_{j \in U_n} (X_j, v)$ converges in law to a Gaussian random variable with mean zero and variance (Cv, v). The field $\{(X_j, v), j \in \mathbb{Z}^d\}$ meets the conditions of Theorem 1.12, since taking the inner product with v is a Lipschitz function in $\mathbb{R}^k$. Therefore such field satisfies the CLT with asymptotic variance

$$\sigma_v^2 = \sum_{j \in \mathbb{Z}^d} cov((X_0, v), (X_j, v)),$$

but the last expression is the same as (Cv, v). $\square$

Remark 1.14. Assume that conditions of Theorem 1.12 are fulfilled and $\sigma^2 = \sum_{j \in \mathbb{Z}^d} cov(X_0, X_j) > 0$. Then $\mathsf{Var}S(U) > 0$ for any block $U \in \mathcal{U}$ by Lemma 1.11. In this case it is possible to reformulate Theorem 1.12 for normalization by $\sqrt{\mathsf{Var}S(U)}$. Namely, one has

Corollary 1.15. *Let* $X = \{X_j, j \in \mathbb{Z}^d\}$ *be a* (BL, θ)*-dependent strictly stationary centered random field such that* $\sigma > 0$ *where* σ^2 *is defined in* (1.5.3). *Then, for any sequence of regularly growing sets* $U_n \subset \mathbb{Z}^d$, *one has*

$$S(U_n)/\sqrt{\mathsf{Var}S(U_n)} \to N(0, 1)$$

in law as $n \to \infty$.

Proof is immediate in view of Theorem 1.12 and Theorem 1.8. $\square$

The classical Newman's CLT is a consequence of the result above. Because of its significance we state it separately. For $k = (k_1, \ldots, k_d) \in \mathbb{Z}^d$ and $n \in \mathbb{N}$, let

$$B_k^n = \{j : nk_l < j_l \le n(k_l + 1)\}, \quad Y_{k,n} = n^{-d/2}S(B_k^n).$$

Corollary 1.16. ([307]) *Let* $X = \{X_j, j \in \mathbb{Z}^d\}$ *be a centered, strictly stationary, associated random field such that* $\mathsf{E}X_0^2 < \infty$ *and* $\sigma^2 = \sum_{j \in \mathbb{Z}^d} cov(X_0, X_j) < \infty$. *Then the finite-dimensional distributions of the field* $\{Y_{k,n}, k \in \mathbb{Z}^d\}$ *converge, as* $n \to \infty$, *to the corresponding ones of the field* $Z = \{Z_k, k \in \mathbb{Z}^d\}$ *consisting of independent* $N(0, \sigma^2)$ *random variables.*

Proof. Let $m \in \mathbb{N}$ and $k^1, \ldots, k^m \in \mathbb{Z}^d$. For $t = (t_1, \ldots, t_m) \in \mathbb{R}^m$ with $\|t\|^2 = t_1^2 + \ldots + t_m^2$, writing down the characteristic function of the vector $(Y_{k^1,n}, \ldots, Y_{k^m,n})$ gives

$$\left| \mathsf{E} \exp\left\{ i \sum_{l=1}^m t_l Y_{k^l,n} \right\} - \exp\left\{ -\frac{\sigma^2 \|t\|^2}{2} \right\} \right|$$

$$\le \left| \mathsf{E} \exp\left\{ i \sum_{l=1}^m t_l Y_{k^l,n} \right\} - \prod_{l=1}^m \mathsf{E} \exp\left\{ it_l Y_{k^l,n} \right\} \right| + \left| \prod_{l=1}^m \mathsf{E} \exp\left\{ it_l Y_{k^l,n} \right\} - \exp\left\{ -\frac{\sigma^2 \|t\|^2}{2} \right\} \right|.$$

The second term on the right-hand side of the last inequality converges to zero, since $Y_{k,n} \to Z_k$ in law for any $k \in \mathbb{Z}^d$ ($n \to \infty$) due to Theorem 1.12. For the first term by Theorem 1.5.3 we have

$$\left| \mathsf{E} \exp\left\{ i \sum_{l=1}^{m} t_l Y_{k^l,n} \right\} - \prod_{l=1}^{m} \mathsf{E} \exp\left\{ i t_l Y_{k^l,n} \right\} \right| \le 4\|t\|_1^2 \sum_{1 \le v,l \le n, v \ne l} cov(Y_{k^v,n}, Y_{k^l,n})$$

$$= 4\|t\|_1^2 n^{-d} \left(\mathsf{Var} \sum_{l=1}^{m} S(B_{k^l}^n) - \sum_{l=1}^{m} \mathsf{Var} S(B_{k^l}^n) \right).$$

By virtue of Theorem 1.8 the right-hand side tends to zero as $n \to \infty$. $\square$

Remark 1.17. Corollary 1.16 holds also for (BL, θ)-dependent random fields. In that case it does not suffice to apply directly Theorem 1.5.3, but a device analogous to that used in the proof of Theorem 1.8 helps to avoid this difficulty. See Lemma 5.1.8.

4°. The Stein techniques application. The proofs considered were based on the characteristic functions theory. Another powerful method, well-adapted to dependent random systems, is the celebrated Stein method ([388, 389]). We will briefly review this matter and note that at a small additional cost it enables us to provide an estimate of the convergence rate in the CLT for random fields under consideration.

Let $X = \{X_j, j \in \mathbb{Z}^d\}$ be such random field that $\mathsf{E}X_j^2 < \infty$, $j \in \mathbb{Z}^d$. For a finite $U \subset \mathbb{Z}^d$ with $B := \sum_{j \in U} \mathsf{E}X_j^2 > 0$, define

$$W = B^{-1} \sum_{j \in U} X_j, \quad R = B^{-2}|U|\theta_1,$$

where θ_1 is the weak dependence coefficient (see Definition 1.5.12). To simplify the notation we do not write $W = W(U)$, $B = B(U)$, etc. Introduce also, for $\varepsilon > 0$, the *Lindeberg function*

$$\mathcal{L}_\varepsilon = B^{-2} \sum_{j \in U} \mathsf{E}X_j^2 \mathbb{I}\{|X_j| > \varepsilon B\}.$$

If, moreover, $\mathsf{E}|X_j|^s < \infty$ for $s > 2$, define the *Lyapunov fraction*

$$L_s = B^{-s} \sum_{j \in U} \mathsf{E}|X_j|^s.$$

The idea of the Stein method is the following. Suppose that $Z \sim N(0,1)$. If $g : \mathbb{R} \to \mathbb{R}$ is a bounded continuous function, then the function

$$f(w) = e^{w^2/2} \int_{-\infty}^{w} (g(t) - \mathsf{E}g(Z)) e^{-t^2/2} dt \qquad (1.18)$$

is the unique bounded solution of the differential equation

$$f'(w) - w f(w) = g(w) - \mathsf{E}g(Z). \qquad (1.19)$$

On substituting the random variable W instead of w into (1.19) and taking the expectation, one obtains at the right-hand side the accuracy of normal approximation for $\mathsf{E}g(W)$. If g is smooth enough, $\mathsf{E}f'(W) - \mathsf{E}Wf(W)$ can be examined to find that accuracy.

Lemma 1.18. ([21, 159, 389]) *Let* $g : \mathbb{R} \to \mathbb{R}$ *be a bounded function having bounded continuous derivative. Then for f defined by (1.18) the following relations hold*

$$\|f\|_\infty \leq \sqrt{\pi/2}\|K\|_\infty, \quad Lip(f) \leq 2\|K\|_\infty, \quad Lip(f') \leq \sqrt{2\pi}\|K\|_\infty + 2Lip(K) \quad (1.20)$$

with $K(t) = g(t) - \mathsf{E}g(Z)$. *Obviously,* $\|K\|_\infty \leq 2\|g\|_\infty$ *and* $Lip(K) = Lip(g)$.

Proof. If $w \leq 0$, then

$$|f(w)| \leq \|K\|_\infty \int_{-\infty}^{w} e^{w^2/2 - t^2/2} dt = \|K\|_\infty \int_{-\infty}^{0} e^{w^2/2 - (u+w)^2/2} du$$

$$= \|K\|_\infty \int_{-\infty}^{0} e^{-u^2/2 - uw} du \leq \|K\|_\infty \int_{-\infty}^{0} e^{-u^2/2} du = \|K\|_\infty \sqrt{\pi/2},$$

$$|f'(w)| = \left| w \int_{-\infty}^{w} e^{w^2/2 - t^2/2} K(t) dt + g(w) - \mathsf{E}g(Z) \right|.$$

Evidently, $|f'(0)| \leq \|K\|_\infty$. For $w < 0$ the function $|f'(w)|$ admits the upper bound

$$-\|K\|_\infty e^{w^2/2} \int_{-\infty}^{w} t e^{-t^2/2} dt + \|K\|_\infty = 2\|K\|_\infty.$$

If $w > 0$, note that on account of (1.18)

$$f(w) = e^{w^2/2} \int_{w}^{+\infty} (\mathsf{E}g(Z) - g(t)) e^{-t^2/2} dt,$$

hence analogous estimates are valid for $w > 0$. Thus first two relations in (1.20) are verified. The third estimate is slightly harder. Again we only consider the case $w < 0$, since the case $w > 0$ is similar and the case $w = 0$ is treated by continuity. In view of (1.18),

$$f''(w) = (K(w) + wf(w))' = K'(w) + wK(w) + (1 + w^2)f(w).$$

Therefore, by the already proved part of the Lemma,

$$|f''(w)| \leq Lip(K) + \sqrt{\pi/2}\|K\|_\infty + \sup_{w<0} |wK(w) + w^2 f(w)|. \qquad (1.21)$$

Further, after integrating by parts we have

$$wK(w) + w^2 f(w) = wK(w) + w^2 e^{w^2/2} \int_{-\infty}^{w} e^{-t^2/2} K(t) dt$$

$$= wK(w) + w^2 e^{w^2/2} \int_{-\infty}^{w} t e^{-t^2/2} \frac{K(t)}{t} dt$$

$$= wK(w) + w^2 e^{w^2/2} \left(e^{-t^2/2} \frac{K(t)}{t} \Big|_w^{-\infty} + \int_{-\infty}^w e^{-t^2/2} \frac{K'(t)t - K(t)}{t^2} dt \right)$$

$$= w^2 e^{w^2/2} \int_{-\infty}^w e^{-t^2/2} \frac{K'(t)t - K(t)}{t^2} dt.$$

Hence,

$$|wK(w) + w^2 f(w)| \leq |w| e^{w^2/2} \int_{-\infty}^w e^{-t^2/2} |K'(t)| dt$$

$$+ e^{w^2/2} \int_{-\infty}^w e^{-t^2/2} |K(t)| dt \leq Lip(K) + \sqrt{\pi/2} \|K\|_\infty.$$

This relation and (1.21) yield the desired bound. $\square$

Theorem 1.19. ([82]) *Let* $X = \{X_j, j \in \mathbb{Z}^d\}$ *be a centered,* (BL, θ)-*dependent random field such that* $\mathsf{E}X_j^2 < \infty$ *for all* $j \in \mathbb{Z}^d$. *Then, for any finite subset* U *of* $\mathbb{Z}^d$ *such that* $B(U) > 0$ *and arbitrary positive* ε, γ, *one has*

$$\sup_{x \in \mathbb{R}^n} |\mathsf{P}(W \geq x) - \mathsf{P}(Z \geq x)| \leq \gamma + (\sqrt{2\pi} + 4\gamma^{-1}) A(\varepsilon, R, \mathcal{L}_\varepsilon) + B(\varepsilon, R, \mathcal{L}_\varepsilon) \quad (1.22)$$

with $A(\varepsilon, R, \mathcal{L}_\varepsilon) = \frac{3}{2}\varepsilon + 2\varepsilon R + \varepsilon \mathcal{L}_\varepsilon$, $B(\varepsilon, R, \mathcal{L}_\varepsilon) = 2R + 8\mathcal{L}_\varepsilon + 4\varepsilon^{-1}\mathcal{L}_\varepsilon$, $Z \sim N(0,1)$. *If, in addition,* $\mathsf{E}|X_j|^{2+\delta} < \infty$ *for some* $\delta \in (0,1]$, $j \in \mathbb{Z}^d$, *then for any* $\gamma > 0$

$$\Delta(U) \leq \gamma + 2(\sqrt{2\pi} + 4\gamma^{-1})(R + L_{2+\delta}) + 2R + 12L_{2+\delta} \quad (1.23)$$

where $\Delta(U)$ *stands for the left-hand side of* (1.22).

Remark 1.20. Using optimization in γ we find instead of (1.22) that for any $\varepsilon > 0$

$$\Delta(U) \leq \sqrt{2\pi} A(\varepsilon, R, \mathcal{L}_\varepsilon) + B(\varepsilon, R, \mathcal{L}_\varepsilon) + 5A(\varepsilon, R, \mathcal{L}_\varepsilon)^{1/2} \quad (1.24)$$

and instead of (1.23) we derive:

$$\Delta(U) \leq 2(\sqrt{2\pi} + 1)R + 2(\sqrt{2\pi} + 6)L_{2+\delta} + 5\sqrt{2}(R + L_{2+\delta})^{1/2}. \quad (1.25)$$

Thus, if $\theta_1 = 0$ then $R = 0$, and the estimates (1.33) and (1.25), for $\varepsilon \in (0,1]$, $\mathcal{L}_\varepsilon \leq 1$ and $L_{2+\delta} \leq 1$, take respectively the forms $c_1(\varepsilon^{1/2} + 3\varepsilon^{-1}\mathcal{L}_\varepsilon)$ and $c_2 L_{2+\delta}^{1/2}$ where c_1 and c_2 are positive constants. If $\theta_1 \neq 0$ then, for a sequence of finite sets $U_n \subset \mathbb{Z}^d$ ($n \in \mathbb{N}$), $R(U_n) \to 0$ whenever the growth of $B^2(U_n)$ (as $n \to \infty$) is more rapid than that generic for stationary random fields (see Theorem 1.8).

Proof. Fix $x \in \mathbb{R}$ and $\gamma > 0$. Define a three times differentiable nondecreasing function $g(w) = g_{x,\gamma}(w)$ in such a way that $g(w) = 0$ if $w < x$, $g(w) = 1$ if $w > x + \gamma$ and $g'(w) \leq 2\gamma^{-1}$, for all $w \in \mathbb{R}$. Obviously

$$\mathbb{I}\{w \geq x + \gamma\} \leq g(w) \leq \mathbb{I}\{w \geq x\}, \quad w \in \mathbb{R}. \quad (1.26)$$

Then, by Stein's method,

$$\mathsf{E}g(W) - \mathsf{E}g(Z) = \mathsf{E}f'(W) - \mathsf{E}Wf(W)$$

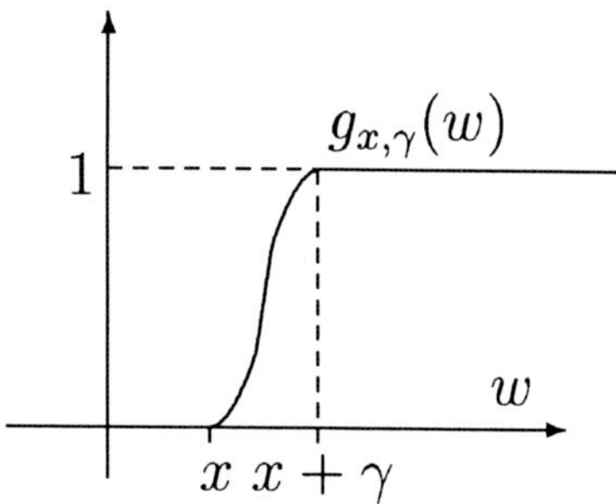

Fig. 3.5

where f is a function given in (1.18). Note that $|g(w) - \mathsf{E}g(Z)| \le 1$, $w \in \mathbb{R}$. Therefore, Lemma 1.18 shows that

$$\|f\|_\infty \le \sqrt{\pi/2}, \quad Lip(f) \le 2, \quad Lip(f') \le \sqrt{2\pi} + 4\gamma^{-1} =: c_\gamma. \tag{1.27}$$

Let $H_\varepsilon(t)$ be the truncation function defined in (2.1.5). For any $j \in U$, set

$$Y_j = X_j/B, \quad T_j = H_\varepsilon(Y_j), \quad V_j = Y_j - T_j, \quad W^{(j)} = W - Y_j.$$

Then

$$\mathsf{E}Wf(W) = \sum_{j \in U} \mathsf{E}Y_j f(W) = \sum_{l=1}^{4} Q_l$$

where

$$Q_1 = \sum_{j \in U} \mathsf{E}Y_j f(W^{(j)}), \quad Q_2 = \sum_{j \in U} \mathsf{E}V_j (f(W) - f(W^{(j)})),$$

$$Q_3 = \sum_{j \in U} \mathsf{E}T_j (f(W) - f(W^{(j)} + T_j)), \quad Q_4 = \sum_{j \in U} \mathsf{E}T_j (f(W^{(j)} + T_j) - f(W^{(j)})).$$

Note that

$$|T_j| \le \varepsilon \wedge |Y_j|, \quad |V_j| \le |Y_j| \mathbb{I}\{|Y_j| > \varepsilon\} \le \varepsilon^{-1} Y_j^2 \mathbb{I}\{|Y_j| > \varepsilon\}. \tag{1.28}$$

By (BL, θ)-dependence, definition of R, (1.27) and (1.28) we have

$$|Q_1| \le \sum_{j \in U} |cov(T_j, f(W^{(j)}))| + 2\|f\|_\infty \sum_{j \in U} \mathsf{E}|V_j| \le 2(R + 2\varepsilon^{-1}\mathcal{L}_\varepsilon).$$

$$|Q_2| \le \sum_{j \in U} |\mathsf{E}V_j (f(W^{(j)} + T_j + V_j) - f(W^{(j)} + T_j))| + \sum_{j \in U} |\mathsf{E}V_j (f(W^{(j)} + T_j) - f(W^{(j)}))|$$

$$\le 2\sum_{j \in U} (\mathsf{E}V_j^2 + \mathsf{E}|T_j V_j|) \le 2\left(\sum_{j \in U} \mathsf{E}Y_j^2 \mathbb{I}\{|Y_j| > \varepsilon\} + \varepsilon \sum_{j \in U} \mathsf{E}|Y_j| \mathbb{I}\{|Y_j| > \varepsilon\}\right) \le 4\mathcal{L}_\varepsilon.$$

In the same way

$$|Q_3| \le 2\sum_{j \in U} \mathsf{E}|T_j V_j| \le 2\mathcal{L}_\varepsilon.$$

Invoking the Taylor formula yields

$$f(W^{(j)} + T_j) - f(W^{(j)}) = f'(W^{(j)})T_j + \frac{1}{2}f''(\eta_j)T_j^2$$

where $\eta_j = \eta_j(\omega)$ is a point (depending on the elementary outcome ω) between $W^{(j)}(\omega)$ and $W^{(j)}(\omega) + T_j(\omega)$. Thus, taking into account (1.28) and the relation $\sum_{j\in U}\mathsf{E}T_j^2 \leq \sum_{j\in U}\mathsf{E}Y_j^2 = 1$, we see that

$$Q_4 = \sum_{j\in U}\mathsf{E}T_j^2 f'(W^{(j)}) + \Delta_1$$

where

$$|\Delta_1| \leq \frac{1}{2}Lip(f')\sum_{j\in U}\mathsf{E}|T_j|^3 \leq \frac{1}{2}c_\gamma\varepsilon. \tag{1.29}$$

Furthermore,

$$\sum_{j\in U}\mathsf{E}T_j^2 f'(W^{(j)}) = \sum_{j\in U}cov(T_j^2, f'(W^{(j)})) + \sum_{j\in U}\mathsf{E}T_j^2\mathsf{E}f'(W^{(j)}). \tag{1.30}$$

Clearly, $Lip(H_\varepsilon^2(\cdot)) = 2\varepsilon$. By (BL,θ)-dependence and (1.27),

$$\left|\sum_{j\in U}cov(T_j^2, f'(W^{(j)}))\right| \leq 2c_\gamma\varepsilon R. \tag{1.31}$$

For the second summand in (1.30) we can write

$$\sum_{j\in U}\mathsf{E}T_j^2\mathsf{E}f'(W^{(j)}) = \sum_{j\in U}\mathsf{E}T_j^2\mathsf{E}f'(W) + \sum_{j\in U}\mathsf{E}T_j^2(\mathsf{E}f'(W^{(j)}) - \mathsf{E}f'(W)) \tag{1.32}$$

and

$$\left|\sum_{j\in U}\mathsf{E}T_j^2(\mathsf{E}f'(W^{(j)}) - \mathsf{E}f'(W))\right| \leq \Delta_2 + \Delta_3,$$

where

$$\Delta_2 = \sum_{j\in U}\mathsf{E}T_j^2|\mathsf{E}(f'(W^{(j)} + T_j + V_j) - f'(W^{(j)} + T_j))|,$$

$$\Delta_3 = \sum_{j\in U}\mathsf{E}T_j^2|\mathsf{E}(f'(W^{(j)} + T_j) - f'(W^{(j)}))|.$$

Due to (1.28) we have $|\Delta_2| \leq c_\gamma$ and

$$\sum_{j\in U}\mathsf{E}T_j^2\mathsf{E}|V_j| \leq c_\gamma\varepsilon^2\sum_{j\in U}\mathsf{E}|V_j| \leq c_\gamma\varepsilon\mathcal{L}_\varepsilon, \tag{1.33}$$

$$|\Delta_3| \leq c_\gamma\sum_{j\in U}\mathsf{E}T_j^2\mathsf{E}|T_j| \leq c_\gamma\varepsilon. \tag{1.34}$$

Using again that $\sum_{j\in U}\mathsf{E}Y_j^2 = 1$, we get

$$\mathsf{E}f'(W)\sum_{j\in U}\mathsf{E}T_j^2 = \mathsf{E}f'(W) + \mathsf{E}f'(W)\sum_{j\in U}(\mathsf{E}T_j^2 - \mathsf{E}Y_j^2),$$

but

$$\left| \mathsf{E} f'(W) \sum_{j \in U} (\mathsf{E} T_j^2 - \mathsf{E} Y_j^2) \right| \le Lip(f) \sum_{j \in U} \mathsf{E} |T_j^2 - Y_j^2| \le 2 \sum_{j \in U} \mathsf{E} Y_j^2 \mathbb{I}\{|Y_j| > \varepsilon\} = 2\mathcal{L}_\varepsilon.$$

$$(1.35)$$

Hence (1.29)—(1.35) imply that

$$Q_4 = \mathsf{E} f'(W) + \Delta_4, \quad |\Delta_4| \le c_\gamma \left(\frac{\varepsilon}{2} + 2\varepsilon R + \varepsilon \mathcal{L}_\varepsilon + \varepsilon \right) + 2\mathcal{L}_\varepsilon.$$

Finally, gathering all estimates, we obtain, in view of (1.26), the bound

$$\mathsf{P}(W \ge x) - \mathsf{P}(Z \ge x + \gamma) \ge \mathsf{E} g(W) - \mathsf{E} g(Z) = \mathsf{E} f'(W) - \mathsf{E} W f(W)$$

$$\ge -(Q_1 + Q_2 + Q_3 + \Delta_4) \ge -c_\gamma \left(\frac{3}{2}\varepsilon + 2\varepsilon R + \varepsilon \mathcal{L}_\varepsilon \right) - (2R + 8\mathcal{L}_\varepsilon + 4\varepsilon^{-1}\mathcal{L}_\varepsilon).$$

Taking instead of g the function $G(t) = g_{x-\gamma,\gamma}(t)$, $t \in \mathbb{R}$, leads to

$$\mathsf{P}(W \ge x) - \mathsf{P}(Z \ge x - \gamma) \le \mathsf{E} G(W) - \mathsf{E} G(Z).$$

As $\|G\|_\infty = \|g\|_\infty$ and $Lip(G) = Lip(g)$, the estimates for $\mathsf{E} G(W) - \mathsf{E} G(Z)$ and for $\mathsf{E} g(W) - \mathsf{E} g(Z)$ coincide. The first assertion of Theorem follows by observation that $\mathsf{P}(|Z - x| \le \gamma) \le \gamma$.

For the second assertion, use the proof of the first one with $\varepsilon = 1$ and notice that $\mathcal{L}_1 \le L_{2+\delta}$. The only amendment needed is the estimation of Δ_1 and Δ_3. These two procedures are similar; for example, the last one gives

$$|\Delta_3| \le Lip(f') \sum_{j \in U} \mathsf{E} T_j^2 \mathsf{E} |T_j| \le Lip(f') \sum_{j \in U} \mathsf{E} |T_j|^{2+\delta} \le Lip(f') L_{2+\delta}.$$

by the Lyapunov inequality. $\square$

If $X = \{X_j, j \in \mathbb{Z}^d\}$ is a centered strictly stationary (BL, θ)-dependent random field with values in $\mathbb{R}^k$ and one sets

$$\Sigma_{ij} = \sum_{t \in \mathbb{Z}^d} cov(X_{0i}, X_{tj}),$$

then Corollary 1.13 is true and the random vectors $S(U_n)/\sqrt{|U_n|}$ converge in law to Gaussian vector with mean zero and covariance matrix Σ. The non-stationary case can be also treated, though in a more tedious way. To find some rate of convergence is, nevertheless, considerably more difficult, since the accuracy of approximation should be controlled on a rather wide class of sets (see, e.g. [79]).

The Stein method provides such an opportunity. For a random vector $W = (W_1, \ldots, W_n)$ we study the accuracy of approximation of $\mathsf{E} h(W)$ by $\mathsf{E} h(Z)$ where h is a smooth enough function and $Z \sim N(0, I_n)$. For two random elements ξ and η taking values in a space S more general than $\mathbb{R}$, the idea of Stein's method is as follows (see, e.g., [8]). Let P and Q be the distributions of ξ and η respectively. Suppose that $\mathbf{X} = \{\mathbf{X}_s, s \ge 0\}$ is a Markov process with infinitesimal operator $\mathcal{A}$ and stationary distribution P. Then $\mathsf{E}(\mathcal{A}h)(\xi) = 0$ for a relatively wide class of functionals h defined on the domain of $\mathcal{A}$. Thus, the proximity of P and Q can be estimated through estimation of $\mathsf{E}(\mathcal{A}h)(\eta)$. For the space $S = \mathbb{R}^k$ and $\xi \sim N(0, I_n)$, the corresponding generator is that of the Ornstein–Uhlenbeck process, i.e.

$$\mathcal{A}(f)(x) = \sum_{i=1}^{n} \frac{\partial^2 f(x)}{\partial x_i^2} - \sum_{i=1}^{n} x_i \frac{\partial f(x)}{\partial x_i}. \tag{1.36}$$

Thus, taking sufficiently regular function $h : \mathbb{R}^k \to \mathbb{R}$, one may solve the equation $\mathcal{A}f(x) = h(x) - \mathsf{E}h(Z)$ and then examine the expectation of the right-hand side of (1.36) where x is replaced with W.

5°. The Bernstein method. We have mentioned in the previous subsection that if the variances of partial sums behave in a regular way (i.e. $\mathsf{Var}S(U) \sim \sigma^2|U|$ as $U \to \infty$) then the value $R = R(U)$ does not approach zero as the index set grows and the estimate might fail to provide a normal approximation. Thus it is reasonable to apply methods of sectioning, proposed initially by Bernstein for stochastic sequences, in order to extract $\theta_{r(U)}$, $r(U) \to \infty$, instead of θ_1. Moreover, this approach gives the rate of convergence in the CLT.

Set

$$\Delta_U = \sup_{x \in \mathbb{R}} \left| \mathsf{P}\left(\frac{S(U)}{\sqrt{\mathsf{Var}S(U)}} \leq x \right) - \mathsf{P}(Z \leq x) \right|, \quad U \in \mathcal{U}.$$

Theorem 1.21. *Let* $X = \{X_j, j \in \mathbb{Z}^d\}$ *be a wide-sense stationary, centered,* (BL, θ)*-dependent random field such that*

1) $D_{2+\delta} := \sup_{j \in \mathbb{Z}^d} \mathsf{E}|X_j|^{2+\delta} < \infty$ *for some* $\delta \in (0, 1]$;

2) $\sum_{j \in \mathbb{Z}^d} cov(X_0, X_j) = \sigma^2 > 0$.

Assume, moreover, that, for some positive c_0, λ *and all* $r \in \mathbb{N}$, *one of the following conditions holds:*

(a) $\theta_r \leq c_0(\mathrm{Log}\, r)^{-\lambda}$;

(b) $\theta_r \leq c_0 r^{-\lambda}$;

(c) $\theta_r \leq c_0 e^{-r\lambda}$.

Then in these cases, for any $U \in \mathcal{U}$, *one has the following respective estimates:*

$$\begin{aligned}
&\text{(a)} \quad \Delta_U \leq C(\mathrm{Log}\,|U|)^{-\lambda/2}, \\
&\text{(b)} \quad \Delta_U \leq C'|U|^{-\mu}, \quad \mu = \delta/4(1 + d(1 + \delta)(\lambda^{-1} + 3/2)), \\
&\text{(c)} \quad \Delta_U \leq C''|U|^{-\kappa}(\mathrm{Log}\,|U|)^{d(1+\delta)}, \quad \kappa = \delta/2(2 + 3d(1 + \delta)),
\end{aligned} \tag{1.37}$$

where C, C', C'' *depend only on* $d, \lambda, c_0, D_{2+\delta}$ *and the covariance function of* X.

Proof. First of all, note that by Lemma 1.11 one has $\mathsf{Var}S(U) > \rho_0|U|$ for some $\rho_0 > 0$ and any block $U \subset \mathbb{Z}^d$.

The proof employs the Bernstein method (for details of its application see, e.g., [210], [84]). We proceed with the case (b) as most used in the sequel and only indicate which amendment works in other cases. Let α, β, ν be positive numbers such that $\alpha < \beta < 1$. Their exact values will be chosen later. Take functions $p(U) = [|U|^\beta]$, $q(U) = [|U|^\alpha]$. If U is large enough divide every edge which is not shorter than $p(U)$ into segments of lengths $p(U), q(U), \ldots, p(U), q(U), \widehat{p(U)}$, where $p(U) \leq \widehat{p(U)} \leq 3p(U)$ (i.e., if the last segment is of length not equal to $p(U)$, we unite the two or three last segments to make the resulting last segment not shorter than $p(U)$, and the previous one be of length $q(U)$). If an edge has length $u \geq p(U)$ then the total number of long (longer than $q(U)$) segments in it equals

$$\left[\frac{u - p(U)}{p(U) + q(U)}\right] + 1 \leq \frac{u}{p(U)}.$$

This division induces the division of U into blocks of (at most) 2^d types. Call a block "large" if any of its edges either is not less than $p(U)$ or has the same length as the parallel edge of the whole block U. As in the proof of Theorem 1.12, we denote "large" blocks by $U^{(1)}, \ldots, U^{(m)}$, with $m = m(U)$. As above, $U^{(0)} = U \setminus (\cup_{j=1}^{m} U^{(j)})$. To simplify the writings we sometimes write q, p, m instead of $q(U), p(U), m(U)$.

It is easy to verify that

$$\frac{|U^{(0)}|}{|U|} \leq \frac{dq}{p}, \quad \frac{|U^{(0)}|}{|U|} \leq \frac{1}{2}. \tag{1.38}$$

We use the following elementary

Lemma 1.22. *Let ξ, η, Z be random variables such that $\mathsf{E}\eta^2 < \infty$ and $Z \sim N(0,1)$. Then, for any number $\tau > 0$,*

$$\sup_{x \in \mathbb{R}} |\mathsf{P}(\xi + \eta \leq x) - \mathsf{P}(Z \leq x)| \leq \sup_{x \in \mathbb{R}} |\mathsf{P}(\xi \leq x) - \mathsf{P}(Z \leq x)| + \tau + \frac{\mathsf{E}\eta^2}{\tau^2}.$$

Proof. For any $x \in \mathbb{R}$ one can write

$$\mathsf{P}(\xi + \eta \leq x) - \mathsf{P}(Z \leq x) \leq \mathsf{P}(\xi \leq x + \tau) - \mathsf{P}(Z \leq x + \tau) + \mathsf{P}(x < Z \leq x + \tau) + \mathsf{P}(|\eta| > \tau),$$

$$\mathsf{P}(Z \leq x) - \mathsf{P}(\xi + \eta \leq x) \leq \mathsf{P}(Z \leq x - \tau) - \mathsf{P}(\xi \leq x - \tau) + \mathsf{P}(x - \tau < Z \leq x) + \mathsf{P}(|\eta| > \tau).$$

Employing the estimate $\mathsf{P}(a < Z < b) < b - a$, $a < b$, and the Chebyshev inequality, we come to the desired conclusion. $\square$

Now let

$$\zeta_j = S(U^{(j)}), \quad j = 1, \ldots, m, \quad B^2 = \sum_{j=1}^{m} \mathsf{Var}\zeta_j, \quad \xi = \sum_{j=1}^{m} \frac{\zeta_j}{B}, \quad \eta = \frac{S(U)}{\sqrt{\mathsf{Var}S(U)}} - \xi.$$

The system $\{\zeta_j\}$ is (BL, θ)-dependent and satisfies the hypotheses of Theorem 1.19. Note that

$$|\mathsf{Var}S(U) - B^2| = \left|\mathsf{Var}S(U^{(0)}) + \sum_{\substack{1 \leq j,k \leq m \\ j \neq k}} cov(\zeta_j, \zeta_k) + \sum_{j=1}^{m} cov(S(U^{(0)}), \zeta_j)\right|$$

$$\leq \theta_q \sum_{j=1}^{m} |U^{(j)}| + (\mathsf{E}X_0^2 + c_0)|U^{(0)}| \leq \theta_q|U| + (\mathsf{E}X_0^2 + c_0)|U|\frac{dq}{p} \tag{1.39}$$

by (BL, θ)-dependence and (1.38). Furthermore, by the second assertion of the mentioned Theorem 1.19 and by Lemma 1.22, applied with $\gamma = |U|^{-\nu}$ and $\tau = |U|^{-\nu}$ respectively,

$$\sup_{x \in \mathbb{R}} \left| P\left(\frac{S(U)}{\sqrt{\mathrm{Var}S(U)}} \leq x\right) - P(Z \leq x) \right| \leq \sup_{x \in \mathbb{R}} |P(\xi \leq x) - P(Z \leq x)| + |U|^{-\nu} + |U|^{2\nu} \mathsf{E}\eta^2$$

$$\leq 2|U|^{-\nu} + |U|^{2\nu} \mathsf{E}\eta^2 + C_1 |U|^\nu (R(U) + L_{2+\delta}(U)) \tag{1.40}$$

where $R(U)$ and $L_{2+\delta}(U)$ are constructed for the family $\{\zeta_j\}$. That is,

$$R(U) \leq m(U) B^{-2} \theta_q \max_{j=1,\ldots,m} |U^{(j)}| \leq 3^d \rho_0^{-1} \theta_q, \tag{1.41}$$

since $\mathrm{Var}\zeta_j \geq \rho_0 |U_n^{(j)}|$ by Lemma 1.11. Let us denote lengths of the edges of U by $u_1, u_2, \ldots, u_d$; rearranging the coordinates we may assume that there exists $k = k(U) \in \{1, \ldots, d\}$ such that $u_j \geq p(U)$ for $j \leq k$ and $u_j < p(U)$ whenever $j > k$ (possibly $k = d$). Recall that $B^2 \geq \rho_0 |U \setminus U^{(0)}| \geq \rho_0 |U|/2$ by Lemma 1.11 and (1.38). Thus

$$L_{2+\delta}(U) \leq B^{-2-\delta} \sum_{j=1}^m \mathsf{E}|\zeta_j|^{2+\delta} \leq 2^{9/2} \rho_0^{-1-\delta/2} (u_1 \ldots u_d)^{-1-\delta/2} D_{2+\delta} \sum_{j=1}^m |U^{(j)}|^{2+\delta}$$

$$\leq 2^{9/2} 3^{d(2+\delta)} \rho_0^{-1-\delta/2} D_{2+\delta} (u_1 \ldots u_k)^{-1-\delta/2} m(U) p(U)^{k(2+\delta)} (u_{k+1} \ldots u_d)^{1+\delta/2}$$

$$\leq 2^{9/2} 3^{3d} \rho_0^{-1-\delta/2} D_{2+\delta} (u_1 \ldots u_k)^{-\delta/2} p(U)^{k(1+\delta)} (u_{k+1} \ldots u_d)^{1+\delta/2}$$

$$= 2^{9/2} 3^{3d} \rho_0^{-1-\delta/2} D_{2+\delta} |U|^{-\delta/2} p(U)^{d(1+\delta)} \tag{1.42}$$

(as everywhere, the product over an empty index set equals 1). Here we used the trivial estimate $\mathsf{E}|S(U^{(j)})|^{2+\delta} \leq 2^{2+\delta} |U^{(j)}|^{2+\delta} D_{2+\delta}$, the definition of index $k(U)$, and the fact that

$$m(U) p(U)^k u_{k+1} \ldots u_d \leq u_1 \ldots u_k u_{k+1} \ldots u_d = |U|$$

ensuing from the bound $|U^{(j)}| \geq p(U)^k u_{k+1} \ldots u_d$.

To estimate $\mathsf{E}\eta^2$, observe that

$$\eta = \frac{S(U)}{\sqrt{\mathrm{Var}S(U)}} - \sum_{j=1}^m \frac{\zeta_j}{B} = \frac{S(U)}{\sqrt{\mathrm{Var}S(U)}} \left(1 - \sqrt{\frac{\mathrm{Var}S_U}{B^2}}\right) + \frac{S(U_n^{(0)})}{B} =: \eta_1 + \eta_2.$$

Then one has

$$\mathsf{E}\eta_1^2 = \left(1 - \sqrt{\frac{\mathrm{Var}S_U}{B^2}}\right)^2 \leq \frac{(\mathrm{Var}S(U) - B^2)^2}{B^4} \leq 4\rho_0^{-2}\left(\theta_q + (\mathsf{E}X_0^2 + c_0)\frac{dq}{p}\right)^2 \tag{1.43}$$

in view of (1.38), (1.39), Lemma 1.11 and since $|\sqrt{t} - 1| \leq |t - 1|$, $t \geq 0$. Further, due to the same facts,

$$\mathsf{E}\eta_2^2 \leq 2d\rho_0^{-1}(\mathsf{E}X_0^2 + c_0)\frac{q}{p}. \tag{1.44}$$

Finally, if the condition (b) holds, from (1.40)—(1.44) one deduces that (1.37) holds with

$$\mu = \max_{\alpha,\beta,\nu} \min \left\{ \nu, -\nu + \lambda\alpha, -\nu + \frac{\delta}{2} - d\beta(1+\delta), -2\nu + \beta - \alpha \right\}.$$

The maximum is attained when all four expressions inside the minimum sign are equal. Solving the linear system one obtains the announced value of μ.

If condition (a) holds instead of (b), take $q(U) = [|U|^\alpha]$, $p(U) = [|U|^\beta]$ for some $0 < \alpha < \beta < 1$, and $\gamma = \gamma(U) = \tau = \tau(U) = (\text{Log}\,|U|)^{-\lambda/2}$.

If condition (c) is given, then take

$$q(U) = 1 + [\alpha\text{Log}\,|U|], \quad p(U) = 1 + [q(U)e^{(3/2)\lambda q(U)}] \quad \text{and} \quad \gamma = \tau = [e^{-\lambda q(U)/2}],$$

with $\alpha = \delta/(2\lambda + 3\lambda d(1 + \delta))$. $\square$

Remark 1.23. The analysis of the proof above shows that in case of condition (b) the factor C' at the right-hand side of (1.37) can be written in the form

$$\left(1 + 2^\delta \rho_0^{-1} D_{2+\delta} + (\mathsf{E}X_0^2 + c_0)^2 \rho_0^{-2}\right) A_0$$

where $A_0 > 0$ depends on d only. Similar assertions are true for other rates of convergence of $(\theta_r)_{r\in\mathbb{N}}$ studied above.

Remark 1.24. The assumption of wide-sense stationarity is actually unessential and can be replaced, e.g., by the following condition: there exist positive A_1 and A_2 such that

$$A_1|U| \le \mathsf{Var}S(U) \le A_2|U|,$$

this also implies condition 2) of Theorem 1.21. The proof is exactly the same.

$6°$. **The study of multiparameter Ornstein–Uhlenbeck process.** As was proved in Chapter 1, the dependence between Lipschitz functions in Gaussian random variables can be determined in terms of (BL, θ)-dependence. We are not able to prove this for non-Lipschitz functions of Gaussian random variables; but they may be well approximated by Lipschitz ones (e.g. using appropriate truncation) and thus the theory of weak dependence still applies. In particular, one can give an easy and direct proof of CLT which is usually proved with the help of mixing theory (the Kolmogorov–Rozanov theorem on the mixing conditions for Gaussian processes, see [210, Ch. 18]). In the next theorem we carry out this idea for Wiener process with functions $f(x) = |x|^s$, $s > 0$. By *multiparameter Ornstein–Uhlenbeck process* we mean a random field $U = \{U_t, t \in \mathbb{R}_+^d\}$ where

$$U_t = \exp\{-(t_1 + \ldots + t_d)/2\}W(e^t), \quad t = (t_1, \ldots, t_d) \in \mathbb{R}_+^d$$

and W is a d-parameter Wiener process, i.e. a centered Gaussian random field with a.s. continuous trajectories, defined on $\mathbb{R}_+^d$ and having the covariance function

$$cov(W(z), W(t)) = \prod_{i=1}^{d} \min\{z_i, t_i\} \quad \text{for} \quad z, t \in \mathbb{R}_+^d.$$

Here and in what follows the algebraic operations over multiindices are defined coordinatewise, in particular, $e^t = (e^{t_1}, \ldots, e^{t_d})$. Clearly, U_t is a d-parameter stochastic process with continuous trajectories. As above, for positive real T we write $(0, T\mathbf{1}]$ to denote the cube with "lower" and "upper" points 0 and $T\mathbf{1} = (T, \ldots, T)$ respectively. The proof of the following result seems to be new one.

Theorem 1.25. *For any $s > 0$ one has*

$$\frac{\int_{(0,T\mathbf{1}]} |U_t|^s dt - T^d \mathsf{E}|Z|^s}{T^{d/2}} \to N(0, A_{d,s}^2) \quad \text{in law} \tag{1.45}$$

as $T \to \infty$. Here $Z \sim N(0,1)$ and $A_{d,s} > 0$.

    Proof. Define the functions

$$f_s(x) = |x|^s sgn(x), \quad x \in \mathbb{R}.$$

At first we dwell on the following fact. As usual, $|t - z| = \max_{i=1,\ldots,d} |t_i - z_i|$.

Lemma 1.26. *The field $X = \{X_t, t \in \mathbb{R}_+^d\} := \{f_s(U_t), t \in \mathbb{R}_+^d\}$ is strictly stationary and associated. Its covariance function decreases exponentially, i.e. $cov(X_t, X_z) \leq C e^{-|t-z|\lambda}$ for all $t, z \in \mathbb{R}_+^d$ and some positive C and λ depending only on s.*

    Proof. Since the field U is Gaussian, the strict stationarity for it is the same as wide-sense stationarity. Clearly $\mathsf{E}U_t = 0$, $t \in \mathbb{R}_+^d$. Moreover, for any $t, z \in \mathbb{R}_+^d$,

$$cov(U_t, U_z) = \mathsf{E}U_t U_z = \exp\left\{-\frac{1}{2}\sum_{i=1}^d (t_i + z_i)\right\} \mathsf{E}W(e^t)W(e^z)$$

$$= \exp\left\{-\frac{1}{2}\sum_{i=1}^d (t_i + z_i) + \sum_{i=1}^d (t_i \wedge z_i)\right\} = \exp\left\{-\frac{1}{2}\sum_{i=1}^d |t_i - z_i|\right\} > 0. \tag{1.46}$$

Thus the field U is strictly stationary and associated by Theorem 1.2.1. The field X is obtained from U by the formula $X_t = f_s(U_t), t \in \mathbb{R}_+^d$. Thus X is strictly stationary and since f_s is a nondecreasing function, it is associated as well.

    It remains to estimate the covariances. Let $M > 0$ be arbitrary and $H_M(\cdot)$ be the truncation function defined in (2.1.5). Set

$$V_t = f_s(H_M(U_t)), \quad Y_t = X_t - V_t, \quad t \in \mathbb{R}_+^d.$$

For any $t, z \in \mathbb{R}_+^d$,

$$0 \leq cov(X_t, X_z) = \mathsf{E}X_t X_z \leq cov(V_t, V_z) + cov(Y_t, Y_z) + \mathsf{E}|Y_t||V_z| + \mathsf{E}|V_t||Y_z|, \tag{1.47}$$

all the random variables introduced above being symmetric.

    If $s \geq 1$, then by Theorem 1.5.3 one has $cov(V_t, V_z) \leq s^2 M^{2(s-1)} cov(U_t, U_z)$. Consequently, (1.46) and (1.47) give for $cov(X_t, X_z)$ the following upper bound

$$s^2 M^{2(s-1)} \exp\left\{-\frac{1}{2}\|t - z\|_1\right\} + 2\mathsf{E}|X_t||X_z|\mathbb{I}\{|U_t| \geq M\} + \mathsf{E}|X_t||X_z|\mathbb{I}\{|U_z| \geq M\},$$

since $|Y_t| \vee |V_t| \leq |X_t|$ and $Y_t = 0$ if $|U_t| \leq M$. By the Hölder inequality

$$cov(X_t, X_z) \leq s^2 M^{2(s-1)} e^{-|t-z|/2} + 3(\mathsf{E}|Z|^{2s})^{1/2} \left(\mathsf{E}|Z|^{2s}\mathbb{I}\{|Z| > M\}\right)^{1/2}$$

$$\leq s^2 M^{2(s-1)} e^{-|t-z|/2} + 3(\mathsf{E}|Z|^{2s}\mathsf{E}|Z|^{6s-4})^{1/2} M^{2(1-s)}.$$

Here $Z \sim N(0,1)$. Let M be such that $M^{2(s-1)} = \sqrt{3}(\mathsf{E}|Z|^{2s}\mathsf{E}|Z|^{6s-4})^{1/4} s^{-1} e^{|t-z|/4}$, then we obtain

$$cov(X_t, X_z) \leq 2\sqrt{3}(\mathsf{E}|Z|^{2s}\mathsf{E}|Z|^{6s-4})^{1/4} s e^{-|t-z|/4}. \tag{1.48}$$

If $0 < s < 1$, then $|V_t| \leq M^s$ and Y_t is a Lipschitz function of U_t with Lipschitz constant sM^{s-1}, $t \in \mathbb{R}_+^d$. Therefore

$$cov(X_t, X_z) \leq cov(Y_t, Y_z) + M^s \mathsf{E}|Y_t| + M^s \mathsf{E}|Y_z| + M^{2s}$$

$$\leq s^2 M^{2(s-1)} e^{-|t-z|/2} + 2M^s \mathsf{E}|Z|^s + M^{2s}.$$

Taking

$$M = \left(\frac{s^2 e^{-|t-z|/2}}{2\mathsf{E}|Z|^s}\right)^{1/(2-s)}$$

yields

$$cov(X_t, X_z) \leq C(s) \exp\left\{-\frac{s}{2(2-s)}|t-z|\right\}. \tag{1.49}$$

From (1.48) and (1.49) we infer that the covariances of X decrease exponentially as the distance between points grows. It remains to take $\lambda = 1/4$ if $s \geq 1$ and $\lambda = s/(4-2s)$ if $s < 1$. $\square$

We did not present any limit results for integrals over associated or related to them random fields with continuous parameter, though such extension of the scope can be readily achieved in view of Remark 1.7. Instead we will prove the theorem approximating the integrals with sums.

Recall the useful Slutsky lemma, which can be found in every elementary course on probability and statistics.

Lemma 1.27. *Let ξ, ξ_n, η_n $(n \in \mathbb{N})$ be random variables defined on a probability space $(\Omega, \mathcal{F}, \mathsf{P})$ such that $\xi_n \to \xi$ in law and $\eta_n \to a$ in probability as $n \to \infty$, for some $a \in \mathbb{R}$. Then*

$$\xi_n + \eta_n \to \xi + a \quad \text{and} \quad \xi_n \eta_n \to a\xi \quad \text{in law}, \quad n \to \infty.$$

Lemma 1.28. *The convergence in law in (1.45) follows from the relation*

$$\frac{\int_{(0,[T]\mathbf{1}]} |U_t|^s dt - [T]^d \mathsf{E}|Z|^s}{[T]^{d/2}} \to N(0, A_{d,s}^2), \quad T \to \infty;$$

$[\cdot]$ *stands for an integer part of a number.*

Proof. Note that any power of the random variable $\sup_{t\in B}|W(t)|$ is integrable for any rectangle $B \subset \mathbb{R}^d_+$ and a d-parameter Brownian motion $\{W(t), t \in \mathbb{R}^d_+\}$. This follows from Theorem 2.4.8; namely, to find that $\mathsf{E}\sup_{t\in[0,K]}|W(t)|^p < \infty$ $(K \in \mathbb{R}^d_+, p > 2d)$ apply that Theorem with arbitrary $z = \varepsilon > 0$, $\gamma > p$ and $\alpha = \gamma/2$. The condition of Theorem 2.4.8 is true since

$$\mathsf{P}(W(y) - W(x) > z) \le z^{-\gamma}\mathsf{E}|W(y) - W(x)|^\gamma$$

$$= z^{-\gamma}\mathsf{E}|Z|^\gamma(\langle y\rangle - \langle x\rangle)^{\gamma/2} \le d|T|^{d-1}|y - x|^{\gamma/2}z^{-\gamma}\mathsf{E}|Z|^\gamma$$

if $0 \le x \le y \le T$ and $z > 0$, here $\langle z\rangle := z_1 \ldots z_d$, $z \in \mathbb{R}^d_+$. Consequently, for some $\tau > 1$ and $C > 0$ independent of $z > 0$ one has

$$\mathsf{P}(L_T > \tau z) \le \mathsf{P}(W(T) > z) + C|T|^\alpha z^{-\gamma}$$

where L_T was defined in Theorem 2.4.8. One can also apply the maximal inequality of Wichura ([414]).

Thus the random variable $\int_{(0,[T]\mathbf{1}]}|U_t|^s dt$ is integrable. The assertion of Lemma is now easily checked with the help of Lemma 1.27, since $[T]/T \to 1$ as $T \to \infty$. Moreover, by the Fubini theorem

$$[T]^{-d/2}\mathsf{E}\int_{([T]\mathbf{1},T\mathbf{1}]}|U_t|^s dt = [T]^{-d/2}\mathrm{mes}(([T]\mathbf{1},T\mathbf{1}])\mathsf{E}|Z|^s$$

$$= [T]^{-d/2}(T - [T])^d\mathsf{E}|Z|^s \to 0, \quad T \to \infty. \qquad \square$$

Lemma 1.29. *The random field $\xi = \{\xi_j, j \in \mathbb{N}^d\}$, defined by way of*

$$\xi_j = \int_{(j-\mathbf{1},j]}|X_t|dt$$

(where X_t was introduced in Lemma 1.26) is strictly stationary, square-integrable and (BL, θ)-dependent with $\theta_r = O(e^{-\varkappa r})$, $r \in \mathbb{N}$, where $\varkappa \in (0, \lambda)$ and λ is the same as in Lemma 1.26.

Proof. The fact that $\mathsf{E}\xi_j^2 < \infty$, $j \in \mathbb{Z}^d$, was actually established in the proof of previous Lemma. Furthermore, the field $\{|X_t|, t \in \mathbb{R}^d_+\}$ is continuous in $L^2(\Omega, \mathcal{F}, \mathsf{P})$. This follows (cf. [412, §1.2, Pr.9]) from the fact that the function $(t, u) \mapsto \mathsf{E}X_t X_u$ is continuous on $(\mathbb{R}^d_+)^2$. Indeed, standard calculation of expectations yields

$$\mathsf{E}X_t X_u = \iint_{\mathbb{R}^2} e^{-(a^2+b^2)/2}f_s\left(a^2\exp\left\{\frac{1}{2}\sum_{i=1}^d(2t_i \wedge u_i - t_i - u_i)\right\}\right.$$

$$\left. +ab\left(1 - \exp\left\{\sum_{i=1}^d(2t_i \wedge u_i - t_i - u_i)\right\}\right)^{1/2}\right)dadb,$$

but the last integral is continuous in variables (t, u) by the dominated convergence theorem.

For $k \in \mathbb{N}$ let

$$\xi_j^{(k)} = k^{-d} \sum_{0 \leq v_1,\ldots,v_d < k} \left| X_{j_1-1+v_1/k,\ldots,j_d-1+v_d/k} \right|, \quad j \in \mathbb{N}^d,$$

which is the Riemann sum approximating the integral ξ_j. Since X is a field which is continuous in quadratic mean, one has $\xi_j^{(k)} \to \xi_j$ in $L^2(\Omega, \mathcal{F}, \mathsf{P})$ as $k \to \infty$ (see [412, §1.2, Pr.16]). Obviously $\{\xi_j^{(k)}, j \in \mathbb{N}^d\}$ is a strictly stationary field, since X is strictly stationary. Thus, $\{\xi_j, j \in \mathbb{N}^d\}$ is also strictly stationary. If we prove that $\{\xi_j^{(k)}, j \in \mathbb{N}^d\}$ is (BL, θ)-dependent for any $k \in \mathbb{N}$ with sequence $(\theta_r)_{r \in \mathbb{N}}$ independent of k, then $\{\xi_j, j \in \mathbb{N}^d\}$ will be (BL, θ)-dependent with the same sequence $(\theta_r)_{r \in \mathbb{N}}$ by Lemma 1.5.2.

Let $I, J \subset \mathbb{N}^d$ be finite disjoint sets with $dist(I, J) \geq r \in \mathbb{N}$. Take any bounded Lipschitz functions $F : \mathbb{R}^{|I|} \to \mathbb{R}$ and $G : \mathbb{R}^{|J|} \to \mathbb{R}$. Then, by Theorem 1.5.3 and Lemma 1.26

$$|cov(F(\xi_I^{(k)}), G(\xi_J^{(k)}))| \leq Lip(F)Lip(G)k^{-2d}$$

$$\times \sum_{\substack{i \in I, \\ j \in J}} \sum_{\substack{0 \leq v_1,\ldots,v_d < k, \\ 0 \leq m_1,\ldots,m_d < k}} cov(X_{i_1-1+v_1/k,\ldots,i_d-1+v_d/k}, X_{j_1-1+m_1/k,\ldots,j_d-1+m_d/k})$$

$$\leq Lip(F)Lip(G) \sum_{i \in I, j \in J} e^{-\lambda|i-j|+\lambda} \leq Lip(F)Lip(G)(|I| \wedge |J|) \sum_{j \in \mathbb{Z}^d, |j| \geq r} e^{\lambda-\lambda|j|}.$$

We have used that $|t - z| \geq |i - j| - 1$ for any two points $i, j \in \mathbb{N}^d$ and arbitrary $t \in (i-1, i], z \in (j-1, j]$. Thus, $\{\xi_j^{(k)}, j \in \mathbb{N}^d\}$ is a weakly dependent random field with $\theta_r = \sum_{j \in \mathbb{Z}^d, |j| \geq r} e^{\lambda-\lambda|j|}$. Consequently the limiting field $\{\xi_j, j \in \mathbb{N}^d\}$ inherits this property. $\square$

Now we can prove the Theorem. Lemma 1.29 allows to apply Theorem 1.12 to the field $\{\xi_j, j \in \mathbb{N}^d\}$, therefore

$$n^{-d/2} \sum_{j \in \mathbb{N}^d, j \leq n\mathbf{1}} (\xi_j - \mathsf{E}\xi_j) = n^{-d/2} \left(\int_{(0,n\mathbf{1}]} |X_t| dt - n^d \mathsf{E}|Z|^s \right) \to N(0, A_{d,s}^2)$$

where

$$A_{d,s}^2 = \lim_{n \to \infty} n^{-d} \mathsf{Var} \sum_{j \in \mathbb{N}^d, j \leq n\mathbf{1}} \xi_j = \lim_{n \to \infty} n^{-d} \sum_{j,m \in \mathbb{N}^d, j,m \leq n\mathbf{1}} cov(\xi_j, \xi_m). \quad (1.50)$$

It remains to prove that $A_{d,s}^2 > 0$. We use a lemma which is of independent interest, as we obtain an example of associated random vector and non-monotone transformation of its components such that the new vector is associated as well.

Lemma 1.30. *Let (X, Y) be a Gaussian random vector with mean zero and $cov(X, Y) > 0$. Then $(|X|, |Y|)$ is an associated random vector.*

Proof. Multiplying X and Y by appropriate positive constants (which cannot change neither the conditions of the Lemma nor the assertion) we can assume that the covariance matrix of (X, Y) is

$$\begin{pmatrix} 1 & \rho \\ \rho & 1 \end{pmatrix},$$

with some $\rho > 0$. If $\rho = 1$, then $X = Y$ a.s. and the assertion of the Lemma is obvious due to Theorem 1.1.8, (a). Otherwise let $p(x, y)$ be the density of $(|X|, |Y|)$. If $x \wedge y \le 0$, then $p(x, y) = 0$. For positive x and y we have

$$p(x, y) = \frac{\partial^2}{\partial x \partial y} \mathsf{P}(|X| \le x, |Y| \le y) = \frac{\partial^2}{\partial x \partial y} \int_{-x}^{x} \int_{-y}^{y} \phi(u, v) du dv$$

$$= \phi(x, y) - \phi(-x, y) - \phi(x, -y) + \phi(-x, -y) = 2(\phi(x, y) - \phi(-x, y))$$

$$= \frac{2}{\pi \sqrt{1 - \rho^2}} \exp\{-(x^2 + y^2)/2\} \sinh(axy)$$

where $\phi(x, y) = (2\pi\sqrt{1 - \rho^2})^{-1} \exp(-(x^2 - \rho xy + y^2)/(2 - 2\rho^2))$ is the density of (X, Y) and $a = \rho/(2 - 2\rho^2)$. The function $p(x, y)$ is MTP_2 (see Section 1.4). To check this we will show that for any $x, y, z, w \in \mathbb{R}$ such that $x \ge z$ and $y \le w$

$$p(x, y)p(z, w) \le p(x, w)p(y, z). \tag{1.51}$$

If $z \le 0$ or $y \le 0$, then (1.51) is obvious. If $x, y, z, w > 0$, then (1.51) is equivalent to the inequality

$$\sinh(axy) \sinh(azw) \le \sinh(axw) \sinh(ayz).$$

This is in turn equivalent (see Lemma 1.4.6) to the relation

$$\frac{\partial^2}{\partial x \partial y} \sinh(axy) = a \cosh(axy) + a^2 xy \sinh(axy) \ge 0 \quad \text{for} \quad x, y > 0,$$

which is obviously true. Whence (1.51) holds, and so by Theorem 1.4.13 (or Theorem 1.4.7 and the Remark 1.4.8) the random vector $(|X|, |Y|)$ is associated. $\square$

Lemma 1.30 ensures that $cov(|X_t|, |X_z|) \ge 0$ for any $t, z \in \mathbb{R}_+^d$ because $|U_t|$ and $|U_z|$ are associated whereas $X_t = f_s(U_t)$ and $f_s(\cdot)$ is a nondecreasing function. Therefore, for any $k \in \mathbb{N}$ and all $j, l \in \mathbb{Z}_+^d$ one has $cov(\xi_j^{(k)}, \xi_l^{(k)}) \ge 0$. Since $\xi_j^{(k)} \to \xi_l$ as $k \to \infty$ in quadratic mean, we have

$$cov(\xi_j^{(k)}, \xi_l^{(k)}) \to cov(\xi_j, \xi_l), \quad j, l \in \mathbb{Z}_+^d.$$

Thus, the value of expression inside the limit sign in (1.50) is no less than

$$\mathsf{Var}\xi_0 = \mathsf{Var} \int_{(0,1]} |X_t| dt = \int_{(0,1] \times (0,1]} cov(|X_u|, |X_v|) du dv > 0$$

as continuous function $F(u, v) := cov(|X_u|, |X_v|) \ge 0$ for all $u, v \in [0, 1]$ and $F(v, v) > 0$ for any $v \in [0, 1]$ ($X_v = f_s(U_v)$, thus $|X_v| \ne const$ a.s.). Consequently, $A_{d,s}^2 > 0$. $\square$

2   The Newman conjecture

We turn to more careful investigation of the necessary and sufficient conditions for validity of the CLT in the association case. Here we will confine ourselves to one-dimensional parameter set, that is we will only examine $X = \{X_j, j \in T\}$ with $T = \mathbb{N}$ or $T = \mathbb{Z}$. We also assume that all the random sequences will be strictly stationary.

Recall that Newman's CLT considered in the previous Section comprises the elementary CLT for i.i.d. random variables having finite second moment. A natural question in this respect is whether the key hypothesis of finite susceptibility can be weakened. In this Section we deal with the *Newman conjecture* ([307]).

1°. Slowly varying functions and variances of partial sums. We start with auxiliary results involving the following classical notion.

Definition 2.1. (a) A function $L : \mathbb{R}_+ \to \mathbb{R} \setminus \{0\}$ is called *slowly varying* (at infinity) in the Karamata sense if, for any $a > 0$,

$$\frac{L(ax)}{L(x)} \to 1 \text{ as } x \to \infty. \tag{2.1}$$

In such case we write $L \in \mathcal{L}$.

(b) A function $L : \mathbb{N} \to \mathbb{R} \setminus \{0\}$ is *slowly varying* (at infinity) if (2.1) holds for any $a \in \mathbb{N}$ as $x \to \infty$, $x \in \mathbb{N}$. We write $L \in \mathcal{L}(\mathbb{N})$.

Obviously one can only consider $a \in (0, 1)$ in (2.1). If $G : \mathbb{R}_+ \to \mathbb{R} \setminus \{0\}$ and $G \sim L$ where $L \in \mathcal{L}$, then it is clear that $G \in \mathcal{L}$, the same is true for $G : \mathbb{N} \to \mathbb{R} \setminus \{0\}$ and $L \in \mathcal{L}(\mathbb{N})$.

We refer to [210, App. 1] and [366] for basic information on slowly varying functions. Note in passing that if $L \in \mathcal{L}$ then the restriction of L onto $\mathbb{N}$ is slowly varying as well. In general a slowly varying function $L : \mathbb{N} \to \mathbb{R} \setminus \{0\}$ does not admit an extension to $\mathbb{R}_+$ which satisfies (2.1). However, if L is nondecreasing, such extension is possible, e.g., one can set $L(t) = L([t])$ where $[\cdot]$ is the integer part of a number and take $L(0) = L(1)$. Thus we have to show that, for any $a > 1$,

$$\frac{L([ax])}{L([x])} \to 1 \text{ as } x \to \infty. \tag{2.2}$$

Note that, for all $x > x_0(a)$, one has $[a][x] \le [ax] \le ([a]+2)[x]$. Taking into account that L is nondecreasing on $\mathbb{N}$ and, for any $m \in \mathbb{N}$, $L(nm)/L(n) \to 1$ as $n \to \infty$, we come to (2.2). Then we will also write $L \in \mathcal{L}$ (as for the extension of L to $\mathbb{R}_+$).

Let $X = (X_j)_{j \in \mathbb{Z}}$ be a sequence of square-integrable random variables. Set

$$\mathsf{K}_X(n) := \sum_{|j| \le n} cov(X_0, X_j), \quad n \in \mathbb{N},$$

and fix a slowly varying nondecreasing function $L \in \mathcal{L}(\mathbb{N})$.

Lemma 2.2. *Let $X = (X_j)_{j \in \mathbb{Z}}$ be a wide-sense stationary random sequence with nonnegative covariance function. If $\mathsf{K}_X(\cdot) \in \mathcal{L}(\mathbb{N})$, then $\mathsf{Var} S_n \sim n\mathsf{K}_X(n)$, $n \to \infty$. If $\mathsf{Var} S_n \sim nL(n)$ as $n \to \infty$, with $L \in \mathcal{L}(\mathbb{N})$, then $L(n) \sim \mathsf{K}_X(n)$, $n \to \infty$.*

Proof. Suppose that $K_X(\cdot) \in \mathcal{L}(\mathbb{N})$. Due to stationarity $cov(X_i, X_j) =: R(i-j)$, so for $S_n = \sum_{j=1}^{n} X_j$ we get an estimate

$$\mathrm{Var} S_n = \sum_{i,j=1}^{n} cov(X_i, X_j) = \sum_{i,j=1}^{n} R(i-j) = \sum_{m=-(n-1)}^{n-1} (n - |m|)R(m)$$

$$\leq n \sum_{m=-(n-1)}^{n-1} R(m) \leq n K_X(n), \qquad (2.3)$$

since R is a nonnegative function. For any $a \in (0,1)$ and $n \geq 1/(1-a)$ (i.e. $na \leq n-1$) one has

$$\mathrm{Var} S_n = \sum_{|m| \leq n-1} (n - |m|)R(m) \geq \sum_{|m| \leq na} (n - |m|)R(m)$$

$$\geq n(1-a) \sum_{|m| \leq na} R(m) = n(1-a)K_X([na]). \qquad (2.4)$$

Thus, in view of (2.2)

$$(1-a)nK_X([na]) \sim (1-a)nK_X(n), \quad n \to \infty. \qquad (2.5)$$

Relations (2.3)—(2.5) yield that $\mathrm{Var} S_n \sim nK_X(n), n \to \infty$ (as a could be taken arbitrarily close to 0).

Now let $\mathrm{Var} S_n \sim nL(n), n \to \infty$, where $L \in \mathcal{L}(\mathbb{N})$. Then, for any $\varepsilon \in (0,1)$ and all n large enough, (2.3) implies that

$$K_X(n) \geq \frac{\mathrm{Var} S_n}{n} \geq (1-\varepsilon)L(n). \qquad (2.6)$$

Now, given $q \in \mathbb{N}, q > 1, n \in \mathbb{N}$ and $m \in \mathbb{Z}$ such that $|m| \leq n$, we see that

$$\frac{q}{q-1}\left(1 - \frac{|m|}{nq}\right) \geq \frac{q}{q-1}\left(1 - \frac{n}{nq}\right) = 1.$$

Therefore

$$K_X(n) \leq \frac{q}{q-1} \sum_{|m| \leq n} \frac{(nq - |m|)}{nq} R(m) \leq \frac{q}{(q-1)nq} \sum_{|m| \leq nq} (nq - |m|)R(m)$$

$$= \frac{q}{q-1} \frac{\mathrm{Var} S_{nq}}{nq} \sim \frac{q}{q-1} L(nq), \quad n \to \infty. \qquad (2.7)$$

As q can be taken arbitrarily large and $L \in \mathcal{L}(\mathbb{N})$, combining (2.6) and (2.7) completes the proof. $\square$

Remark 2.3. For a wide-sense stationary random sequence $X = (X_j)_{j \in \mathbb{N}}$ with nonnegative covariance function one can introduce the sequence

$$\widetilde{K}_X(n) := \mathrm{Var} X_1 + 2 \sum_{1 < j \leq n} cov(X_1, X_j), \quad n \in \mathbb{N},$$

and observe that an analogue of Lemma 2.2 holds in this case.

Note that ([210, Ch. 18]), for uniformly strongly mixing sequences, the behavior of variances of partial sums as $nL(n)$, with $L \in \mathcal{L}$, is necessary and sufficient for CLT.

Following [197], for wide-sense stationary sequences we will consider the spectral density defined on the segment $[-1/2, 1/2]$. More precisely, a wide-sense stationary sequence $(X_n)_{n \in \mathbb{N}}$ has spectral density $f(t)$, $t \in [-1/2, 1/2]$, if, for any $n \in \mathbb{Z}_+$,

$$cov(X_0, X_n) = \int_{-1/2}^{1/2} \exp\{2\pi i n t\}\, f(t)\, dt.$$

Lemma 2.4. *Suppose that a centered wide-sense stationary random sequence* $X = (X_n)_{n \in \mathbb{N}}$ *has an even spectral density* $f(t)$, $t \in [-1/2, 1/2]$, *such that* f *is bounded outside any neighborhood of zero and* $f(1/t) \sim L(t)$ *as* $t \to +\infty$ *where* $L(t)$ *is a nondecreasing slowly varying function defined on* $\mathbb{R}_+$. *Then*

$$\mathsf{K}_X(n) \sim L(n) \text{ as } n \to \infty.$$

Proof. Applying the formula for the variance of a partial sum of a stationary process (see [210, Theorem 18.2.1]), and making the change of variable $s = nt$ one has

$$\frac{\mathrm{Var} S_n}{nL(n)} = \int_{-1/2}^{1/2} \frac{\sin^2 n\pi t}{nL(n)\sin^2 \pi t} f(t)\, dt = 2 \int_0^{n/2} \frac{\sin^2 \pi s}{n^2 \sin^2(\pi s/n)} \frac{f(s/n)}{L(n)}\, ds. \quad (2.8)$$

If $1 \le s \le n/2$, then $f(s/n) \le C_1 L(n/s) \le C_1 L(n)$ where

$$C_1 := \sup_{t \ge 2} \frac{f(1/t)}{L(t)} < \infty$$

by Lemma's assumptions. If $0 < s < 1$, then, taking into account the integral representation (2.26) for the function $L(t)$, we conclude that

$$\frac{f(s/n)}{L(n)} \le C_1 \frac{L(n/s)}{L(n)} = C_1 \exp\left\{ \int_n^{n/s} \frac{\varepsilon(u)}{u}\, du \right\}$$

where $\varepsilon(u) \to 0$ as $u \to \infty$. Let u_0 be such a number that $\varepsilon(u) < 1/2$ if $u \ge u_0$. Then, for $n > u_0$ and $0 < s < 1$,

$$\frac{f(s/n)}{L(n)} \le C_1 s^{-1/2}.$$

Moreover, an elementary estimate shows that $n^2 \sin^2(\pi s/n) \ge 4s^2$ for $0 < s \le n/2$ and all $n \in \mathbb{N}$. Thus, for $n > u_0$ the integrand on the right-hand side of (2.8) can be majorized by the integrable function $C_1 \max\{s^{-1/2}, 1\} s^{-2} \sin^2 \pi s$. Therefore, one may pass to a limit as $n \to \infty$ under the sign of integral on the right-hand side of (2.8):

$$\lim_{n \to \infty} \frac{\mathrm{Var} S_n}{nL(n)} = 2 \int_0^{\infty} \lim_{n \to \infty} \frac{f(s/n)\sin^2 \pi s}{n^2 L(n)\sin^2(\pi s/n)}\, ds = 2 \int_0^{\infty} \frac{\sin^2 \pi s}{\pi^2 s^2}\, ds = 1.$$

From this fact the assertion of the Lemma follows by Lemma 2.2. $\square$

2°. CLT without the finite susceptibility condition. Now we will give two groups of sufficient conditions for CLT applicable in the absence of finite susceptibility.

Recall that a sequence of (real-valued) random variables $(\xi_n)_{n\in\mathbb{N}}$ is called *uniformly integrable* if

$$\lim_{c\to+\infty}\sup_{n\in\mathbb{N}}\mathsf{E}|\xi_n|\mathbb{I}\{|\xi_n|\ge c\}=0.$$

Clearly if $(\xi_n)_{n\in\mathbb{N}}$ is uniformly integrable then $\sup_n \mathsf{E}|\xi_n|<\infty$ and, for any sequence $(c_n)_{n\in\mathbb{N}}$ of real numbers such that $c_n\to\infty$ as $n\to\infty$, one has

$$\mathsf{E}|\xi_n|\mathbb{I}\{|\xi_n|\ge c_n\}\to 0,\quad n\to\infty. \tag{2.9}$$

The following result by De la Valle Poussin provides a criterion of uniform integrability, see, e.g., [383, Ch. II, §6.4].

Theorem 2.5. *A sequence $(\xi_n)_{n\in\mathbb{N}}$ is uniformly integrable if and only if there exists a nondecreasing function $H:\mathbb{R}_+\to\mathbb{R}_+$ such that*

$$\lim_{x\to\infty}\frac{H(x)}{x}=\infty\quad\text{and}\quad\sup_{n\in\mathbb{N}}\mathsf{E}H(|\xi_n|)<\infty.$$

The next theorem is due to Lewis.

Theorem 2.6. ([259]) *Let $X=(X_j)_{j\in\mathbb{Z}}$ be a strictly stationary **PA** random sequence such that $0<\mathsf{E}X_0^2<\infty$ and $\mathsf{K}_X(\cdot)\in\mathcal{L}(\mathbb{N})$. Then X satisfies the CLT, that is*

$$\frac{S_n-\mathsf{E}S_n}{\sqrt{\mathsf{Var}S_n}}\to Z\sim N(0,1)\quad\text{in law as}\quad n\to\infty, \tag{2.10}$$

if and only if the random sequence $((S_n-\mathsf{E}S_n)^2/(n\mathsf{K}_X(n)))_{n\in\mathbb{N}}$ is uniformly integrable.

Proof. Necessity. Assume that (2.10) holds. Without loss of generality we stipulate that $\mathsf{E}X_0=0$. Then $S_n^2/\mathsf{Var}S_n\to Z^2$ in law as $n\to\infty$ (if $\xi_n\to\xi$ in law and h is a continuous function then $h(\xi_n)\to h(\xi)$ in law). We quote the following well-known result without proof.

Lemma 2.7. (see, e.g., [39, §5, Th. 5.4]) *Let $(\xi_n)_{n\in\mathbb{N}}$ be a sequence of uniformly integrable random variables. Suppose that $\xi_n\to\xi$ in law as $n\to\infty$. Then ξ is integrable and*

$$\mathsf{E}\xi_n\to\mathsf{E}\xi,\quad n\to\infty. \tag{2.11}$$

Moreover, if $(\xi_n)_{n\in\mathbb{N}}$ is a sequence of nonnegative random variables such that $\xi_n\to\xi$ in law as $n\to\infty$, ξ_n $(n\in\mathbb{N})$ and ξ are integrable and (2.11) holds then the sequence $(\xi_n)_{n\in\mathbb{N}}$ is uniformly integrable.

Obviously, $S_n^2/\mathrm{Var}S_n \geq 0$, $\mathsf{E}(S_n^2/\mathrm{Var}S_n) = \mathsf{E}Z^2 = 1$. Therefore a sequence $(S_n^2/\mathrm{Var}S_n)_{n\in\mathbb{N}}$ is uniformly integrable. In view of Lemma 2.2 the sequence $(S_n^2/n\mathsf{K}_X(n))_{n\in\mathbb{N}}$ is uniformly integrable as well. The necessity part of Theorem is established.

Sufficiency. We can assume that K_X is unbounded, as otherwise Theorem follows from Theorem 1.12. First of all we show that there exists a sequence $(q_n)_{n\in\mathbb{N}}$ of positive integers such that

$$q_n \to \infty, \; q_n/n \to 0 \;\; \text{and} \;\; \mathsf{K}_X(q_n)/\mathsf{K}_X(n) \to 1 \;\; \text{as} \;\; n \to \infty. \tag{2.12}$$

Set $\mathsf{K}_X(t) := \mathsf{K}_X([t])$ for $t \in \mathbb{R}_+$. Then, in view of (2.2), for any $\varepsilon > 0$ one has $\mathsf{K}_X(\varepsilon n)/\mathsf{K}_X(n) \to 1$, $n \to \infty$. So it is possible to construct a sequence $\varepsilon_n \to 0+$ such that $\mathsf{K}_X(\varepsilon_n n)/\mathsf{K}_X(n) \to 1$, $n \to \infty$, and to take $q_n = [\varepsilon_n n] + 1$ ($n \in \mathbb{N}$). Note that then $q_n \to \infty$ when $n \to \infty$. Indeed, if not, one could extract a subsequence along which $K_X(\varepsilon_n n)$ tends to some finite limit; but this is impossible as K_X is unbounded. We can also construct a sequence $(p_n)_{n\in\mathbb{N}}$ of positive integers such that

$$q_n \leq p_n \leq n, \;\; q_n/p_n \to 0 \;\; \text{and} \;\; p_n/n \to 0, \;\; n \to \infty. \tag{2.13}$$

Now we use the well-known sectioning device introduced by Bernstein. Let us take a partition of $\mathbb{R}_+$ with points $0, p_n, p_n+q_n, 2p_n+q_n, 2p_n+2q_n, \ldots$. For n large enough let $U_1^{(1)}, \ldots, U_n^{(m_n)}$ be the blocks (intervals) of length p_n, where $m_n = [n/(p_n+q_n)]$, and $U_n^{(0)} := (0, n] \setminus (\cup_{j=1}^{m_n} U_n^{(j)})$. More precisely, we consider

$$U_n^{(j)} = ((j-1)(p_n + q_n); ((j-1)q_n + jp_n)] \cap \mathbb{Z},$$

$$V_n^{(j)} = ((j-1)q_n + jp_n; j(p_n + q_n)] \cap \mathbb{Z}, \;\; j = 1, \ldots, m_n,$$

and let $U_n^{(0)} = \left(\cup_{j=1}^{m_n} V_n^{(j)}\right) \cup (m_n(p_n + q_n) \wedge n, n]$.

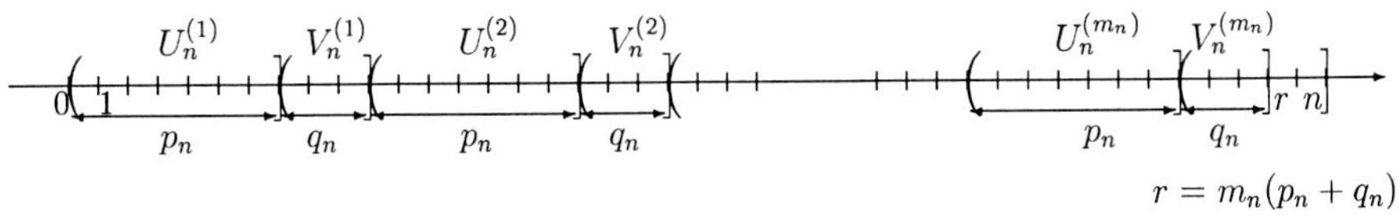

Fig. 3.6

It is a basic fact (which we used in Section 3.1) that for random variables the convergence in law is equivalent (see, e.g., [383, Ch. III, §3.2]) to pointwise convergence of their characteristic functions. Note that, for each $t \in \mathbb{R}$, with the notation

$$S(U) := \sum_{j\in U} X_j, \;\; U \subset \mathbb{Z}, \; |U| < \infty, \; v_n = \sqrt{n\mathsf{K}_X(n)}, \; n \in \mathbb{N},$$

we have

$$\left| \mathsf{E}\exp\{itS_n/v_n\} - e^{-t^2/2} \right| \leq \left| \mathsf{E}\exp\{itS_n/v_n\} - \mathsf{E}\exp\{it \sum_{j=1}^{m_n} S(U_n^{(j)})/v_n\} \right|$$

$$+\left|\mathsf{E}\exp\left\{it\sum_{j=1}^{m_n}S(U_n^{(j)})/v_n\right\}-\prod_{j=1}^{m_n}\mathsf{E}\exp\left\{itS(U_n^{(j)})/v_n\right\}\right|$$

$$+\left|\prod_{j=1}^{m_n}\mathsf{E}\,exp\left\{itS(U_n^{(j)})/v_n\right\}-e^{-t^2/2}\right|=:Q_1+Q_2+Q_3. \qquad (2.14)$$

Here $i^2=-1$, $Q_k=Q_k(n,t)$, $k=1,2,3$, $n\in\mathbb{N}$, $t\in\mathbb{R}$. As $|e^{ix}-e^{iy}|\le|x-y|$ for $x,y\in\mathbb{R}$, one may write, using the Lyapunov inequality,

$$Q_1\le|t|\mathsf{E}|S(U_n^{(0)})|/v_n\le|t|\left(\mathsf{E}S(U_n^{(0)})^2\right)^{1/2}/v_n.$$

By **PA**, the covariance function is nonnegative. Thus, in view of (wide-sense) stationarity, we have

$$\mathsf{E}S(U_n^{(0)})^2\le\sum_{j\in U_n^{(0)}}\sum_{1\le k\le n}cov(X_j,X_k)\le|U_n^{(0)}|\mathsf{K}_X(n)\le(m_nq_n+p_n+q_n)\mathsf{K}_X(n).$$

Due to (2.12) and (2.13),

$$\frac{\mathsf{E}S(U_n^{(0)})^2}{n\mathsf{K}_X(n)}\le\frac{[n/(p_n+q_n)]q_n+p_n+q_n}{n}\to0,\ \ n\to\infty.$$

Therefore, for each $t\in\mathbb{R}$, one has $Q_1(n,t)\to0,\ \ n\to\infty$.

To estimate Q_2 note that random variables $S(U_n^{(j)}),j=1,\dots,m_n$, are **PA** by Theorem 1.1.8, (d). Theorem 1.5.3 implies that

$$Q_2\le\sum_{k=1}^{m_n-1}\left|cov\left(\exp\left\{itS(U_n^{(k)})/v_n\right\},\exp\left\{-it\sum_{j=k+1}^{m_n}S(U_n^{(j)})/v_n\right\}\right)\right|$$

$$\le\frac{4t^2}{v_n^2}\sum_{\substack{1\le k,r\le m_n\\k\ne r}}cov\left(S(U_n^{(k)}),S(U_n^{(r)})\right)\le\frac{4t^2}{n\mathsf{K}_X(n)}\sum_{j=1}^{n}\sum_{\substack{1\le k\le n,\\|k-j|>q_n}}cov(X_j,X_k).$$

Obviously, for any $j\in\{1,\dots,n\}$,

$$\{k:1\le k\le n,|k-j|>q_n\}\subset\{j-n\le k\le j+n\}\setminus\{k:|k-j|\le q_n\}.$$

Consequently,

$$\sum_{j=1}^{n}\sum_{\substack{1\le k\le n,\\|k-j|>q_n}}cov(X_j,X_k)\le n(\mathsf{K}_X(n)-\mathsf{K}_X(q_n)),$$

and we conclude in view of (2.12) that $Q_2(n,t)\to0$ as $n\to\infty$ (for every $t\in\mathbb{R}$).

To handle the term Q_3 consider independent random variables $Z_n^{(1)},\dots,Z_n^{(m_n)}$ such that each $Z_n^{(j)}$ is distributed as $S(U_n^{(j)})/\sqrt{n\mathsf{K}_X(n)}$. By Lemma 2.2, for each j we have $\mathsf{Var}Z_n^{(j)}=\mathsf{Var}Z_n^{(1)}\sim p_n\mathsf{K}_X(p_n)/n\mathsf{K}_X(n),n\to\infty$. Therefore,

$$\sum_{j=1}^{m_n}\mathsf{Var}Z_n^{(j)}=m_n\mathsf{Var}Z_n^{(1)}\to1,\ \ n\to\infty,$$

because (2.12) and (2.13) yield that $m_n p_n = [n/(p_n + q_n)]p_n \sim n$ and $K_X(p_n)/K_X(n) \to 1$ as $n \to \infty$. Furthermore, for any $\varepsilon > 0$, using the strict stationarity of X, one has

$$\sum_{j=1}^{m_n} \mathsf{E}(Z_n^{(j)})^2 \mathbb{I}\left\{|Z_n^{(j)}| > \varepsilon\right\} = \frac{m_n}{n K_X(n)} \mathsf{E} S(U_n^{(1)})^2 \mathbb{I}\{S(U_n^{(1)})^2 > \varepsilon^2 n K_X(n)\}$$

$$= \frac{m_n p_n K_X(p_n)}{n K_X(n)} \mathsf{E} \frac{S(U_n^{(1)})^2}{p_n K_X(p_n)} \mathbb{I}\left\{\frac{S(U_n^{(1)})^2}{p_n K_X(p_n)} > \varepsilon^2 \frac{n K_X(n)}{p_n K_X(p_n)}\right\} \to 0 \quad \text{as} \quad n \to \infty$$

in view of (2.9) since $n K_X(p_n)/p_n K_X(p_n) \to \infty$ and the family $\{S(U_n^{(1)})^2/(p_n K_X(p_n))\}$ is uniformly integrable. Indeed, $(S_{p_n}^2/p_n K_X(p_n))_{n \in \mathbb{N}}$ is a subfamily of $(S_n^2/(n K_X(n)))_{n \in \mathbb{N}}$. Thus, by the Lindeberg theorem (see, e.g., [39, §7, Th. 7.2]), the following convergence in law holds

$$\sum_{j=1}^{m_n} Z_n^{(j)} \to Z \sim N(0, 1), \ n \to \infty.$$

Consequently, for each $t \in \mathbb{R}$ one has $Q_3(n, t) \to 0$ as $n \to \infty$. $\square$

Following Vronski [405] we introduce

Definition 2.8. Call a function $L : \mathbb{R}_+ \to \mathbb{R} \setminus \{0\}$ *very slowly varying* if for any $a > 0$ it satisfies the condition

$$\frac{L(x^{1+a})}{L(x)} \to 1 \text{ as } x \to \infty \tag{2.15}$$

(e.g., $L(x) = \log\log x$, $x > e$). Denote by $\mathcal{L}_s$ the class of nondecreasing very slowly varying functions.

Theorem 2.9. ([405]) *Suppose that* $X = \{X_j, j \in \mathbb{N}\}$ *is a wide-sense stationary centered associated* (**PA**) *sequence such that* $0 < \mathsf{E}|X_1|^{2+\delta} < \infty$ *for some* $\delta \in (0, 1]$ *and* $K_X(n) \sim L(n)$ *where* $L \in \mathcal{L}_s$. *Then* X *satisfies the CLT.*

Proof. Take numbers α, β such that $0 < \alpha < \beta < \delta/(2 + 2\delta)$. Let $p_n = [n^\beta]$, $q_n = [n^\alpha]$ $(n \in \mathbb{N})$ and construct the sets $U_n^{(0)}, U_n^{(1)}, \ldots, U_n^{(m_n)}$ in the same way as in the proof of Theorem 2.6. Note that in that case $m_n \sim n^{1-\beta}$. For any $t \in \mathbb{R}$, introduce the terms $Q_i = Q_i(n, t)$, $i = 1, 2, 3$ in the same way as in (2.14). Again one can show that $Q_1 + Q_2 + Q_3 \to 0$, $n \to \infty$. We omit proving that $Q_1 + Q_2 \to 0$ as $n \to \infty$, it follows likewise in Theorem 2.6 (now for estimating Q_2 apply (2.15) to infer that $L(q_n)/L(n) \to 1$). It remains to prove that

$$\sum_{j=1}^{m_n} Z_n^{(j)} \to Z \sim N(0, 1) \tag{2.16}$$

in law as $n \to \infty$, where $Z_n^{(j)}$ are i.i.d. (for each $n \in \mathbb{N}$) random variables distributed as $S(U_n^{(j)})/\sqrt{nL(n)}$, $j = 1, \ldots, m_n$.

Having noticed that Minkowski's inequality implies

$$(\mathsf{E}|\xi_1 + \ldots + \xi_k|^{2+\delta})^{1/(2+\delta)} \leq \sum_{j=1}^{k}(\mathsf{E}|\xi_j|^{2+\delta})^{1/(2+\delta)},$$

we obtain an estimate

$$\mathsf{E}|Z_n^{(j)}|^{2+\delta} = (nL(n))^{-1-\delta/2}\mathsf{E}|S(U_n^{(j)})|^{2+\delta}$$

$$\leq (nL(n))^{-1-\delta/2}\mathsf{E}|X_1|^{2+\delta}|U_n^{(j)}|^{2+\delta} \leq \frac{n^{\beta(2+\delta)}}{(nL(n))^{1+\delta/2}}\mathsf{E}|X_1|^{2+\delta}. \tag{2.17}$$

Instead of applying the Lindeberg theorem, we use the Berry–Esseen result. Namely, let $\xi_1, \ldots, \xi_n$ be independent centered random variables such that $\mathsf{E}|\xi_j|^{2+\delta} < \infty$ for some $\delta \in (0, 1]$ and all $j = 1, \ldots, n$. Let $F_n(x)$ be the distribution function of $\sigma^{-1}\sum_{j=1}^{n}\xi_j$ where $\sigma_n^2 := \sum_{j=1}^{n}\mathsf{Var}\xi_j$. Then if $\sum_{j=1}^{n}\mathsf{Var}\xi_j \neq 0$ (see, e.g., [326, Ch. V, §3]) then one has

$$\sup_{x \in \mathbb{R}}\left|F_n(x) - \frac{1}{\sqrt{2\pi}}\int_{-\infty}^{x}e^{-u^2/2}du\right| \leq c_0\sigma_n^{-2-\delta}\sum_{j=1}^{n}\mathsf{E}|\xi_j|^{2+\delta} \tag{2.18}$$

where c_0 is an absolute positive constant.

Now consider i.i.d. random variables $Z_n^{(1)}, \ldots, Z_n^{(m_n)}$ introduced in the proof of Theorem 2.6. Then, if $G_n(x)$ is the distribution function of $\sum_{j=1}^{m_n}Z_n^{(j)}$, in view of (2.17), (2.18) and Lemma 2.2 we have

$$\sup_{x \in \mathbb{R}}\left|G_n(x) - \frac{1}{\sqrt{2\pi}}\int_{-\infty}^{x}e^{-u^2/2}du\right| \leq 2c_0m_n\frac{n^{\beta(2+\delta)}\mathsf{E}|X_1|^{2+\delta}}{(nL(n))^{1+\delta/2}}\left(\frac{nL(n)}{m_np_nL(p_n)}\right)^{1+\delta/2},$$

for all n large enough, because

$$\mathsf{Var}Z_n^{(j)} = \mathsf{Var}S(U_n^{(1)})/(nL(n)) \sim p_nL(p_n)/(nL(n)), \quad j = 1, \ldots, m_n \quad (n \to \infty).$$

Obviously,

$$\frac{nL(n)}{m_np_nL(p_n)} \to 1, \quad \frac{m_nn^{\beta(2+\delta)}}{(nL(n))^{1+\delta/2}} \to 0, \quad n \to \infty,$$

since $m_n \sim n^{1-\beta}$ and $\beta < \delta/(2 + 2\delta)$. Thus (2.16) holds. The proof is complete. $\square$

In Theorems 2.6 and 2.9 the CLT was established without the assumption of finite susceptibility (1.5.3) used in the classical Newman's Theorem 1.12. However, besides strict (or wide) stationarity, **PA** and existence of second moments for summands, some additional conditions were imposed concerning the uniform integrability of $((S_n - \mathsf{E}S_n)^2/\mathsf{Var}S_n)$ in Theorem 2.6, or existence of moments of order $2 + \delta$ for random variables under consideration and very slow behavior of $K_X(\cdot)$ in Theorem 2.9. We mention also that in paper by Louhichi and Soulier ([279]) a CLT for associated sequences (possibly without assumption (1.5.3)) was proved for the non-classical normalization. That is, the normalizing factor different from $(nK_X(n))^{-1/2}$ (in particular, $(\mathsf{E}|S_n|)^{-1}$ used by Philipp et al. [126] for mixing random variables) was considered there.

3°. The generalization of the Herrndorf example. In 1980 Newman ([307]) conjectured that for strictly stationary associated random field $X = \{X_j, j \in \mathbb{Z}^d\}$ instead of the condition of finite susceptibility it is sufficient for CLT (when partial sums are taken over growing sets) to use the hypothesis that

$$\mathsf{K}_X(n) = \sum_{|j| \leq n} cov(X_0, X_j), \ n \in \mathbb{N},$$

is slowly varying function. Unfortunately, this elegant hypothesis is not true. We give below two results providing counterexamples even in the case $d = 1$.

Theorem 2.10. ([197]) *There exists a strictly stationary, centered, associated random sequence $X = (X_j)_{j \in \mathbb{N}}$ with $\mathsf{K}_X(n) \sim \log n$, as $n \to \infty$, such that normalized partial sums $S_n / \sqrt{n \mathsf{K}_X(n)}$ do not have any nondegenerate limit distribution as $n \to \infty$.*

This Herrndorf's theorem can be obtained as a corollary from the following result by Shashkin.

Theorem 2.11. ([378]) *Let L be a nondecreasing slowly varying function such that $L(n) \to \infty$ as $n \to \infty$. Then, for any positive unbounded sequence $(b(n))_{n \in \mathbb{N}}$, there exists a strictly stationary associated random sequence $X = (X_j)_{j \in \mathbb{N}}$ such that*
   a) $\mathsf{E}X_1 = 0$ *and* $\mathsf{E}X_1^2 = 1$;
   b) $\mathsf{K}_X(n) \sim L(n)$ *as* $n \to \infty$;
   c) *the normalized partial sums $(S_n / \sqrt{n b(n)})_{n \in \mathbb{N}}$ do not have any non-degenerate limit in law as $n \to \infty$.*

Remark 2.12. If the function L is not equivalent to some nondecreasing function, then, clearly, associated sequence $X = (X_j)_{j \in \mathbb{N}}$ such that $\mathsf{K}_X(n) \sim L(n)$ does not exist. Thus, the assumption that L is nondecreasing is natural.

Proof. To start with, recall the following

Lemma 2.13. ([197]) *If $h : [0, 1/2] \to \mathbb{R}_+$ is a bounded nonincreasing convex function, then*

$$\int_0^{1/2} h(t) \cos(2\pi nt)dt \geq 0 \quad \text{for all } n \in \mathbb{N}.$$

Proof. Let $k = [n/2]$ and $C_n(t) := \cos(2\pi nt)$ where $n \in \mathbb{N}$, $t \in \mathbb{R}$. Write the integral as

$$\int_0^{1/2} h(t)C_n(t)dt = \sum_{m=0}^{k-1} \int_{m/n}^{(m+1)/n} h(t)C_n(t)dt + \int_{k/n}^{1/2} h(t)C_n(t)dt. \tag{2.19}$$

For any $m = 0, \ldots, k - 1$, divide the interval $(m/n, (m + 1)/n]$ into four equal intervals. By the change of variable, using the periodicity of cosine, one obtains

$$\int_{m/n}^{(m+1)/n} h(t)C_n(t)dt = \int_0^{1/(4n)} h_{m,n}(t)C_n(t)dt \tag{2.20}$$

where for $t \in [0, 1/4n]$

$$h_{m,n}(t) = h\left(\frac{m}{n} + t\right) - h\left(\frac{2m+1}{2n} - t\right) - h\left(\frac{2m+1}{2n} + t\right) + h\left(\frac{m+1}{n} - t\right).$$

The convexity of h in view of relation (2.2.3) implies that the right-hand side of (2.20) is nonnegative.

For even n the last summand in the right-hand side of (2.19) is equal to 0. Thus the Lemma is proved for even n.

For the odd n, it remains to consider the last integral in (2.19).

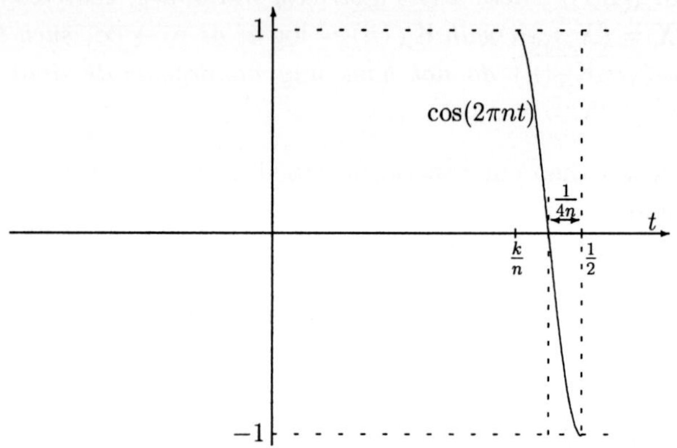

Fig. 3.7 $(n = 2r + 1,\ k/n = r/(2r+1))$

Again by the change of variable and monotonicity of h, taking into account that $C_n(t) \geq 0$ for $t \in \left[\frac{k}{n}, \frac{k}{n} + \frac{1}{4n}\right]$ and $C_n(t) \leq 0$ for $t \in \left[\frac{k}{n} + \frac{1}{4n}, \frac{1}{2}\right]$, one has

$$\int_{k/n}^{1/2} h(t)C_n(t)dt = \int_0^{1/(4n)} h\left(\frac{k}{n} + t\right)C_n(t)dt + \int_{1/(4n)}^{1/(2n)} h\left(\frac{k}{n} + t\right)C_n(t)dt$$

$$\geq h\left(\frac{k}{n} + \frac{1}{4n}\right)\left(\int_0^{1/(4n)} C_n(t)dt + \int_{1/(4n)}^{1/(2n)} C_n(t)dt\right) = 0. \quad \square$$

Without loss of generality we may assume that $L(n) > 0$ for any $n \in \mathbb{N}$ (otherwise we would consider the function $L(n) + c$ instead of $L(n)$ where c is a positive constant). Since L is nondecreasing, it can be extended (see (2.2)) to a nondecreasing slowly varying function $L(x)$ defined on $\mathbb{R}_+$ and such that $L(x) \to \infty$, $x \to \infty$.

The construction of example uses ideas of [197]. It will be shown that

$$L\left(\frac{1}{t}\right) \sim f(t) = \sum_{k=1}^{\infty}(P_k(t))^2 \tag{2.21}$$

as $t \searrow 0$, where $P_k(t)$ are some even trigonometric polynomials with nonnegative coefficients. Then the desired sequence X will be constructed as a sum of a series of independent moving average processes, each of them having spectral density $(P_k(t))^2$. The main difficulty arising in the generalization of [197] is to prove the possibility of representation (2.21) for an arbitrary nondecreasing slowly varying function L (since in [197] this part of proof was based on the Taylor expansion of the function $L(1/t) = -\log t$). The facts needed to solve this task are given below in Lemmas 2.14—2.18. For a process X we employ also the connection between the behavior of $K_X(n)$ as $n \to \infty$ and of its spectral density $f_X(t)$ as $t \to 0+$, see Lemma 2.4. Finally, in the last part of proof we show that some subsequence of normalized sums which were introduced in assertion c) of Theorem 2.11 converge in probability to zero.

We need some notation. Set

$$r(m) = \min\{k \in \mathbb{N} \colon 2^k > 5^m\}, \quad m \in \mathbb{Z}_+. \tag{2.22}$$

The integer and fractional part of a number x will be denoted, as usual, by $[x]$ and $\{x\}$ respectively. Define for $m \in \mathbb{Z}_+$ the sets

$$\Delta_m = \left[\{\log_5 2^{r(m)}\}, \{\log_5 2^{r(m)+1}\}\right], \quad T_m = [0,1] \setminus \Delta_m. \tag{2.23}$$

Note that, for any m, $\Delta_m \subset (0,1)$. Let $(\phi_m(s))_{m \in \mathbb{Z}_+}$ be a sequence of continuous functions on $[0,1]$ such that:

(a) $\phi_m(s) \geq 0$ for $s \in [0,1]$, $\phi_m(0) = \phi_m(1) = 0$;

(b) the function ϕ_m restricted to the segment Δ_m is constant and equal to $\max_{[0,1]} \phi_m(s)$;

(c) $\int_0^1 \phi_m(s)\, ds = 1$;

(d) if $m > 0$, then $\int_{T_m} \phi_m(s)\, ds < m^{-2} \min\{1, (L(5^{m+2}))^{-1}\}$.

Show that the length of any of the segments Δ_m is equal to $\log_5 2$. Let for some $k = k(m) \in \mathbb{N}$, $k < \log_5 2^{r(m)} \leq k+1$ or equivalently $5^k < 2^{r(m)} \leq 5^{k+1}$. From (2.22) one has $2^{r(m)-1} \leq 5^m$. Thus $2^{r(m)+1} = 2^{r(m)-1} \cdot 4 \leq 4 \cdot 5^{k+1}$ and $k < \log_5 2^{r(m)+1} < k+1$. If $x, x + \Delta \in [k, k+1)$ then $\{x + \Delta\} - \{x\} = \Delta$. Consequently $\{\log_5 2^{r(m)} + \log_5 2\} - \{\log_5 2^r(m)\} = \log_5 2$. Therefore, under the requirements introduced, the estimate $\max_{s \in [0,1]} \phi_m(s) \leq \log_2 5$, $m \in \mathbb{Z}_+$, holds.

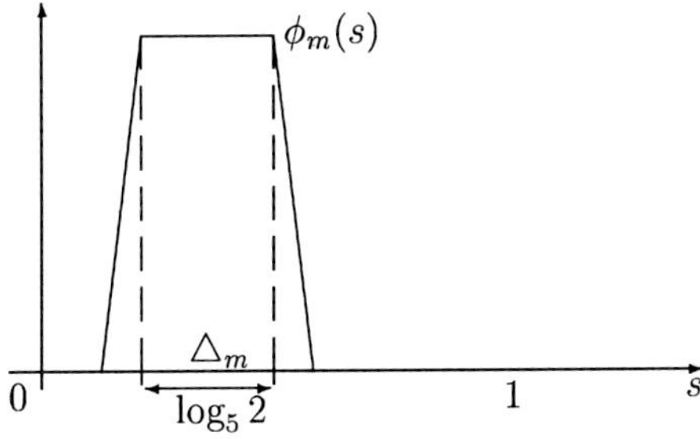

Fig. 3.8

Set $\nu_m = (\log_5 e)\log(L(5^{m+1})/L(5^m))$, $m \in \mathbb{Z}_+$. Note that $\nu_m \to 0$ as $m \to \infty$. Consider the functions

$$L_1(x) = \exp\Big\{\log L(5^m) + \log\Big(L(5^{m+1})/L(5^m)\Big)\int_0^{\{\log_5 x\}} \phi_m(s)\,ds\Big\}, \qquad (2.24)$$

$$\varepsilon(x) = \nu_m \phi_m(\{\log_5 x\}) \qquad (2.25)$$

where $x \geq 1$, $m = m(x) = [\log_5 x]$. For $x \in [0,1)$ set $L_1(x) = L_1(1)$. Note that $L_1(5^m) = L(5^m)$ for all $m \in \mathbb{Z}_+$. One can also easily see that $\varepsilon(x) \geq 0$ is a continuous function on $\mathbb{R}_+$ and $\varepsilon(x) \to 0$ as $x \to \infty$.

We will say that a positive integer k belongs to a set K, if there exists $m \in \mathbb{Z}_+$ such that $k = r(m)$ and $L(5^{m+1}) > L(5^m)$. The notation $\lim_K$ means that the limit is taken over $k \in K$, $k \to \infty$.

The next Lemma summarizes the useful properties of the function $L_1(x)$.

Lemma 2.14. *The following claims are valid:*

1) for $\varepsilon(x)$ defined in (2.25),

$$L_1(x) = L(1)\exp\Big\{\int_1^x \frac{\varepsilon(u)}{u}\,du\Big\}, \quad x \geq 1, \qquad (2.26)$$

2) for all $k \in K$ large enough, the function $L_1(x)$ on the segment $[2^k, 2^{k+1}]$ has positive derivative and is concave;

3) $\lim_K L_1'(2^k)/L_1'(2^{k+1}) = 2$;

4) $L_1(2^{k+1}) - L_1(2^k) \leq d_k$ for any $k \in \mathbb{N}\setminus K$, here $d_k \geq 0$ and $\sum_{k \in \mathbb{N}\setminus K} d_k < \infty$.

Remark 2.15. The property 1) means that (2.24) is a modification of the classical Karamata integral representation for slowly varying functions. Thus, formula (2.26) shows that the function L_1 is nondecreasing and belongs to $\mathcal{L}$. For any $x \geq 1$ one can find $n(x) \in \mathbb{Z}_+$ such that $5^{n(x)} \leq x < 5^{n(x)+1}$. Therefore, for all $x \geq 1$,

$$\frac{L(5^{n(x)})}{L(5^{n(x)+1})} = \frac{L_1(5^{n(x)})}{L_1(5^{n(x)+1})} \leq \frac{L_1(x)}{L(x)} \leq \frac{L_1(5^{n(x)+1})}{L(5^{n(x)})} = \frac{L(5^{n(x)+1})}{L(5^{n(x)})}$$

as $L_1(5^m) = L(5^m)$, $m \in \mathbb{Z}_+$. Consequently, $L_1(x) \sim L(x)$ as $x \to \infty$. Representations similar to (2.24)—(2.26) were considered, e.g., in [421, §1.4].

Proof. If $x \geq 1$, let $m = m(x) = [\log_5 x]$ and $A_k = \log(L(5^{k+1})/L(5^k))$, $k \in \mathbb{N}$. Then one can write, using the property (c) and making the substitution $u = 5^s$,

$$L_1(x) = \exp\Big\{\log L(1) + \sum_{k=0}^{m-1} A_k + A_m \int_0^{\{\log_5 x\}} \phi_m(s)\,ds\Big\}$$

$$= L(1)\exp\Big\{\sum_{k=0}^{m-1} A_k \int_k^{k+1} \phi_k(s-k)\,ds + A_m \int_m^{m+\{\log_5 x\}} \phi_m(s-m)\,ds\Big\}$$

$$= L(1)\exp\Big\{\sum_{k=0}^{m-1} \nu_k \int_{5^k}^{5^{k+1}} \phi_k(\log_5 u - k)u^{-1}du + \nu_m \int_{5^m}^x \phi_m(\log_5 u - m)u^{-1}du\Big\}$$

$$= L(1)\exp\left\{\sum_{k=0}^{m-1}\int_{5^k}^{5^{k+1}}\frac{\varepsilon(u)}{u}\,du + \int_{5^m}^{x}\frac{\varepsilon(u)}{u}\,du\right\} = L(1)\exp\left\{\int_{1}^{x}\varepsilon(u)u^{-1}\,du\right\},$$

as $\varepsilon(x) = \nu_m\phi_m(\log_5 x - m)$ when $m = [\log_5 x]$, with ν_m defined before (2.24). Thus 1) is verified. Now (2.26) implies that

$$L_1'(x) = L_1(x)\frac{\varepsilon(x)}{x}, \quad x \geq 1. \tag{2.27}$$

So, if $k \in K$ (i.e. $k = r(m)$ for some $m \in \mathbb{Z}_+$), then for any $x \in [2^k, 2^{k+1}]$ one has

$$L_1'(x) = L_1(x)\frac{\nu_m}{x} > 0 \tag{2.28}$$

as $\phi_m(\{\log_5 x\}) = 1$ for such x. From (2.27) one also sees that $L_1'(x)$ is a continuous function differentiable off a countable set of points (namely, off points with $\{\log_5 x\} \in Q_m$, where $Q_m \subset [0,1]$ is the set of all points where ϕ_m does not have a derivative, $m = [\log_5 x]$). Hence, to check that L_1 is concave on $[2^k, 2^{k+1}]$, it suffices to prove that its second derivative is nonpositive when it is well-defined. Evidently, for $k \in K$, $k = r(m)$, and $x \in (2^k, 2^{k+1})$, by (2.28),

$$L_1''(x) = \nu_m\frac{L_1'(x)x - L_1(x)}{x^2} = \nu_m L_1(x)\frac{L_1(x)\nu_m - L_1(x)}{x^2} \geq 0,$$

if $k \in K$ is large enough, since $\nu_m \to 0$ as $m \to \infty$. Moreover, for thus chosen k, by construction, $\varepsilon(2^k) = \varepsilon(2^{k+1})$, hence from (2.28) we deduce that

$$\lim_{K}\frac{L_1'(2^k)}{L_1'(2^{k+1})} = \lim_{K}\frac{L_1(2^k)\,\varepsilon(2^k)\,2^{k+1}}{L_1(2^{k+1})\,\varepsilon(2^{k+1})\,2^k} = 2.$$

This establishes 3).

Now let us check the last property. If $k = r(m)$ for some $m \in \mathbb{Z}_+$, but $k \notin K$, then $L_1(5^m) = L_1(5^{m+1})$. Since $[2^k, 2^{k+1}] \subset [5^m, 5^{m+1}]$, it follows that $L_1(2^k) = L_1(2^{k+1})$ and we may take $d_k = 0$. If, on the contrary, $k \neq r(m)$ for all m, then, letting $m = [\log_5 2^k]$, one has

$$5^m < 2^{r(m)} < 2^k < 5^{m+1}.$$

Further,

$$L_1(2^{k+1}) - L_1(2^k) = L_1(2^k)\left(\exp\left\{\int_{2^k}^{2^{k+1}}\frac{\varepsilon(u)}{u}\,du\right\} - 1\right). \tag{2.29}$$

Observe that $x \in (2^k, 2^{k+1})$ entails $\{\log_5 x\} \in T_m \cup T_{m+1}$ where $m = m(x) = [\log_5 x]$ and the sets T_m were introduced in (2.23). Write the integral in (2.29) as the sum of integrals over the sets $(2^k, \min\{5^{m+1}, 2^{k+1}\})$ and $(\min\{5^{m+1}, 2^{k+1}\}, 2^{k+1})$ (the latter set might be empty). In the first integral make a change of variable $s = \log_5 u - m$, and in the second one $s = \log_5 u - m - 1$. Then, as $u = 5^{s+m}$, we have

$$\int_{2^k}^{2^{k+1}\wedge 5^{m+1}}\frac{\varepsilon(u)}{u}\,du = \int_{2^k}^{2^{k+1}\wedge 5^{m+1}}\frac{\nu_m\varepsilon(\{\log_5 u\})}{u}\,du \leq \int_{T_m}(\log 5)\nu_m\phi_m(s)\,ds$$

and analogously

$$\int_{2^{k+1}\wedge 5^{m+1}}^{2^{k+1}} \frac{\varepsilon(u)}{u}\,du \le \int_{T_{m+1}} (\log 5)\nu_{m+1}\phi_{m+1}(s)ds.$$

By the property (d) of functions $\phi_m(s)$, from (2.29) for $k > 2$ we obtain

$$L_1(2^{k+1}) - L_1(2^k)$$

$$\le L(5^{m+2})\left(\exp\left\{\int_{T_m}(\log 5)\,\nu_m\phi_m(s)\,ds + \int_{T_{m+1}}(\log 5)\,\nu_{m+1}\phi_{m+1}(s)\,ds\right\} - 1\right)$$

$$< em^{-2} = e[k\log_5 2]^{-2},$$

if $\max\{\nu_m,\nu_{m+1}\} < \frac{1}{2}\log_5 e$. Since $\nu_m \to 0$ as $m \to \infty$, the assertion 4) is valid. $\square$

So, proving the Theorem we will assume that the function $L(x) = L_1(x)$, i.e. the representation (2.24) holds, implying all the assertions of Lemma 2.14. In what follows we also employ an elementary Lemma, its proof being obvious.

Lemma 2.16. *Suppose that $(\alpha_n)_{n\in\mathbb{N}}, (\beta_n)_{n\in\mathbb{N}}$ are two sequences of real numbers such that $\alpha_n \to 0$ and $\beta_n \nearrow \infty$ as $n \to \infty$. Then*

$$\sum_{k=1}^{n-1} \alpha_k(\beta_{k+1} - \beta_k) = o(\beta_n) \qquad as \quad n \to \infty.$$

Continuing the proof of Theorem 2.11, set $h(t) = L(1/t)$, $t \in (0,1/2]$. For any $k \in \mathbb{N}$ and $t \in [0,\ 1/2]$ define a function by

$$H_k(t) = \frac{1}{2}L'(2^k)\left(2^{3k}t^2 - 2^{2k}(2+\sqrt{2})\,t + 2^k(1+\sqrt{2})\right) + L(2^k). \qquad (2.30)$$

It is easy to check that $H_k(2^{-k}) = L(2^k)$. For $k \in \mathbb{N}$ set

$$g_k(t) = \begin{cases} H_{k+1}(t) - H_k(t), & t \in (0, 2^{-k-1}], \\ h(t) - H_k(t), & t \in (2^{-k-1}, 2^{-k}], \\ 0, & t \in (2^{-k}, 1/2]. \end{cases} \qquad (2.31)$$

By construction, we have $h(t) = H_1(t) + \sum_{k=1}^{\infty} g_k(t)$.

Lemma 2.17. *The following relation holds:*

$$G(t, K) := \sum_{k\notin K} g_k(t) = o(h(t)) \quad as \ \ t \to 0 + .$$

Proof. For any $t \in (0, 1/2]$ and for $q \in \mathbb{N}$ such that $t \in (2^{-q-1}, 2^{-q}]$ one has

$$G(t, K) = (h(t) - H_q(t))I_{\{q\notin K\}} + \sum_{1\le k<q,\,k\notin K}(H_{k+1}(t) - H_k(t)). \qquad (2.32)$$

The functions $h(t)$ and $H_q(t)$ increase when $t \in (2^{-q-1}, 2^{-q}]$ (for the second function this follows from the easily verifiable inequality $H_q'(2^{-q}) \leq 0$). Hence, for any $q \in \mathbb{N}$,

$$h(t) - H_q(t) \leq h(2^{-q-1}) - H_l(2^{-q}) = L(2^{q+1}) - L(2^q) = o(L(2^q)),$$

$$H_q(t) - h(t) \leq H_q(2^{-q-1}) - h(2^{-q}) = (2^{-3} + 2^{-3/2}) 2^q L'(2^q) \qquad (2.33)$$

$$= (2^{-3} + 2^{-3/2}) \varepsilon(2^q) L(2^q) = o(L(2^q))$$

as $q \to \infty$. Now we denote by $\sum$ summation over $k \notin K$, $1 \leq k \leq q$, and decompose the last sum in (2.32) into four sums Q_i by way of

$$\sum (H_{k+1}(t) - H_k(t)) = \sum (L(2^{k+1}) - L(2^k))$$

$$+ \left(\frac{1}{2} + \frac{1}{\sqrt{2}} \right) \sum \left(\varepsilon(2^{k+1}) L(2^{k+1}) - \varepsilon(2^k) L(2^k) \right)$$

$$- \left(1 + \frac{1}{\sqrt{2}} \right) \sum \left(2^{k+1} \varepsilon(2^{k+1}) L(2^{k+1}) - 2^k \varepsilon(2^k) L(2^k) \right) t$$

$$+ \frac{1}{2} \sum \left(2^{2k+2} \varepsilon(2^{k+1}) L(2^{k+1}) - 2^{2k} \varepsilon(2^k) L(2^k) \right) t^2 =: \sum_{i=1}^{4} Q_i.$$

Due to assertion 4) of Lemma 2.14 we have $0 \leq Q_1 \leq \sum_{k=1}^{\infty} d_k < \infty$. To estimate Q_2, write

$$\left(\frac{1}{2} + \frac{1}{\sqrt{2}} \right)^{-1} Q_2 = \left(\sum_{1 \leq k < q} - \sum_{1 \leq k < q,\, k \in K} \right) \left(\varepsilon(2^{k+1}) L(2^{k+1}) - \varepsilon(2^k) L(2^k) \right)$$

$$= \varepsilon(2^q) L(2^q) - \varepsilon(2) L(2) - \sum_{1 \leq k < q,\, k \in K} \left(\varepsilon(2^{k+1}) L(2^{k+1}) - \varepsilon(2^k) L(2^k) \right)$$

$$= o(L(2^q)) - \sum_{1 \leq k < q,\, k \in K} \varepsilon(2^k)(L(2^{k+1}) - L(2^k)),$$

since by construction $\varepsilon(2^k) = \varepsilon(2^{k+1})$ for $k \in K$. As L is nondecreasing, the sum subtracted in the last relation is nonnegative; on the other hand the same sum is no greater than

$$\sum_{1 \leq k < q} \varepsilon(2^k)(L(2^{k+1}) - L(2^k)) = o(L(2^q)),$$

by Lemma 2.16 applied to $\alpha_q = \varepsilon(2^q)$ and $\beta_q = L(2^q)$. The sums Q_3 and Q_4 are estimated analogously (in that cases one should apply Lemma 2.16 taking $\beta_q = 2^q L(2^q)$ and $\beta_q = 2^{2q} L(2^q)$ respectively). So, $G(t, K) = o(h(2^{-q}))$, $t \searrow 0$, where $q = [-\log_2 t]$. Now the assertion of Lemma 2.17 is obvious. $\square$

Lemma 2.18. *There exists a family $\{h_k(t),\ t \in (0, 1/2],\ k \in \mathbb{N}\}$ of nonnegative bounded functions such that*

1) one has $h(t) \sim \sum_{k=1}^{\infty} h_k(t)$ as $t \to 0+$;

2) for any $k \in \mathbb{N}$, the function $h_k(t)$ does not increase and is differentiable for all $t \in (0, 1/2]$, except, possibly, for one point;

3) for any $k \in \mathbb{N}$, the function $\sqrt{h_k(t)}$ is convex on the set $(0, 1/2]$.

Proof. From Lemma 2.17 it follows that

$$h(t) \sim \sum_{k \in K} g_k(t), \quad t \to 0+,$$

where g_k are defined in (2.31). For $k \in K$, $k \geq k_0$ ($k_0 \in \mathbb{N}$), the following inequalities hold: $H_{k+1}(t) \geq H_k(t)$, $t \in \mathbb{R}$, and $H'_{k+1}(2^{-k-1}) - H'_k(2^{-k-1}) < 0$. One can verify them applying the assertions 2) and 3) of Lemma 2.14. For these k we define nonnegative functions $v_k(t)$ in the following way. If $0 < t \leq 2^{-k-1}$, let $v_k(t) = g_k(t)$. If $t > 2^{-k-1}$, we require that

$$\sqrt{v_k(t)} = \max\{y'_k(2^{-k-1})\,t + y_k(2^{-k-1}) - 2^{-k-1}y'_k(2^{-k-1}), 0\}$$

where $y_k(t) = \sqrt{g_k(t)}$, and the derivative at the point $t = 2^{-k-1}$ is understood as the left derivative. Then the functions $v_k(t)$ for $k \in K$, $k \geq k_0$, have desired properties 2) and 3) of $h_k(t)$. Again applying Lemma 2.14, claims 2) and 3), and evaluating $y_k(2^{-k-1})$, $y'_k(2^{-k-1})$, one can show that there exists $p \in \mathbb{N}$ (depending on h but not on k) such that the inequality $v_k(t) \neq g_k(t)$, $k \in K$, is possible only when $t \in (2^{-k-1}, 2^{-k+p}]$. Besides that, due to (2.33) one has

$$|v_k(t) - g_k(t)| \leq |g_k(t)| + v_k(2^{-k-1}) = |g_k(t)| + g_k(2^{-k-1})$$

$$\leq 2\max\big(L(2^{k+1}) - L(2^k),\ \varepsilon(2^k)\,L(2^k)\big) =: \tau_k = o(L(2^k)), \quad \text{as } k \to \infty. \tag{2.34}$$

Thus, for any point $t \in (2^{-q-1}, 2^{-q}]$, $q \in \mathbb{N}$, there are at most $p + 1$ functions $v_k(t) - g_k(t)$, $k \in K$, whose values at t are non-zero. Namely, it can happen when $2^{k+1} > 1/t \geq 2^{p-k}$, i.e. for $k \in (-\log_2 t - 1, -\log_2 t + p] =: I(t)$. We write

$$Y(t) = \sum_{k \in I(t) \cap K} (v_k(t) - g_k(t))$$

and take $q = q(t) = [-\log_2 t]$ for $t \in (0, 1/2]$. Then, because of (2.34), $Y(t)$ satisfies the inequality $|Y(t)| \leq \sum_{j=q}^{q+p+1} \tau_j$. Therefore, for any small enough $t > 0$ we have

$$\frac{|Y(t)|}{h(t)} \leq \frac{\tau_q + \ldots + \tau_{q+p+1}}{L(2^q)} \to 0, \quad t \to 0+. \tag{2.35}$$

Finally,

$$h(t) \sim \sum_{k < k_0, k \in K} g_k(t) + Y(t) + \sum_{k \geq k_0, k \in K} v_k(t), \quad t \to 0+,$$

and from (2.35) and the fact that the functions $g_k(t)$ are bounded we infer that

$$\sum_{1 \leq k < k_0, k \in K} g_k(t) + Y(t) = o(h(t)), \quad t \to 0+.$$

Now it only remains to re-enumerate the functions $\{v_k(t), k \in K, k \geq k_0\}$. Lemma is proved. $\square$

Now we turn to the construction of the desired $(X_n)_{n \in \mathbb{Z}}$. Let $\{h_k(t), k \geq 1\}$ be the functions which appeared in the formulation of Lemma 2.18. We use the same symbols h_k for their continuous even extensions to the segment $[-1/2, 1/2]$. Then, for any $k \in \mathbb{N}$, the function $\sqrt{h_k(t)}$ is continuous and even on $[-1/2, 1/2]$. This segment can be divided into four ones $[x_{j-1}, x_j]$ $(j = 1, 2, 3, 4,\ x_0 = -1/2,\ x_4 = 1/2)$ such that the derivative of $\sqrt{h_k(t)}$ is continuous on (x_{j-1}, x_j) and has finite one-side limits at the ends of this interval. Therefore, by the well-known result (see, e.g., [108, Ch. IX, §5, 1,3]) it can be expanded into a uniformly convergent Fourier series

$$\sqrt{h_k(t)} = a_{0,k} + 2 \sum_{n=1}^{\infty} a_{n,k} \cos(2\pi nt), \qquad |t| \leq 1/2.$$

Due to Lemma 2.13, monotonicity and convexity of $\sqrt{h_k(t)}$ when $t > 0$ imply that $a_{n,k} \geq 0$ for any $n \in \mathbb{Z}_+,\ k \in \mathbb{N}$. For $M \in \mathbb{N},\ k \in \mathbb{N}$, we set

$$\varphi_{M,k}(t) = a_{0,k} + 2 \sum_{n=1}^{M} a_{n,k} \cos(2\pi nt), \qquad f_{M,k}(t) = \varphi_{M,k}^2(t).$$

For any $k \in \mathbb{N}$, let $N(k)$ be a positive integer such that

$$\sup_{|t| \leq 1/2} |\sqrt{h_k(t)} - \varphi_{N(k),k}(t)| \leq 2^{-k} \min\{1, \lambda_k^{-1/2}\}$$

where $\lambda_k = \sup_{|t| \leq 1/2} |h_k(t)|$. Then we see that

$$\sup_{|t| \leq 1/2} |h_k(t) - (\varphi_{N(k),k}(t))^2|$$

$$\leq 2^{-k} \min\{1, \lambda_k^{-1/2}\}(2\lambda_k^{1/2} + 2^{-k} \min\{1, \lambda_k^{-1/2}\}) \leq 3 \cdot 2^{-k}, \qquad k \in \mathbb{N}.$$

Therefore, the series

$$f(t) = \sum_{k=1}^{\infty} f_{N(k),k}(t)$$

converges and $|f(t) - \tilde{h}(t)| \leq 3$ for any $t \in (0, 1/2]$, here $\tilde{h}(t) = \sum_{k=1}^{\infty} h_k(t)$. Thus, $f(t) \sim \tilde{h}(t) \sim L(1/t)$ as $t \to 0 +$.

Let $(n_k)_{k \in \mathbb{N}}$ be an increasing sequence of positive integers such that $k/b(n_k) \to 0$ and $R(2^{k+1})/b(n_k) \to 0$ as $k \to \infty$, where $b(n)$ appeared in the formulation of Theorem 2.11 and

$$R(x) = L(x) + \frac{x}{2}(1 + \sqrt{2})\,L'(x), \qquad x \geq 1. \tag{2.36}$$

Take $\alpha_k = 2^k n_k(1 + 2N(k))$. Consider a family $\{\xi_{n,k},\ n \in \mathbb{Z},\ k \in \mathbb{N}\}$ of independent random variables such that every $\xi_{n,k}$ takes the values $\pm\sqrt{\alpha_k}$, each with probability $1/(2\alpha_k)$, and the value 0 with probability $1 - 1/\alpha_k$. Then $\mathsf{E}\xi_{n,k} = 0$, $\mathsf{Var}\xi_{n,k} = 1$ for all k, n. It turns convenient to introduce, for any $k, n \in \mathbb{N}$, random variables

$$X_{n,k} = \sum_{j=-N(k)}^{N(k)} a_{|j|,k}\xi_{j+n,k}.$$

For each $k \in \mathbb{N}$ the sequence $(X_{n,k})_{n \in \mathbb{N}}$ is strictly stationary and centered, and its spectral density equals $f_k(t)$, $t \in [-1/2, 1/2]$. For any $n \in \mathbb{N}$, the random variables $X_{n,k}$, $k \in \mathbb{N}$, are independent, and

$$\sum_{k=1}^{\infty} \mathrm{Var}\, X_{n,k} = \sum_{k=1}^{\infty} \int_{-1/2}^{1/2} f_k(t)\, dt = \int_{-1/2}^{1/2} f(t)\, dt < \infty.$$

The second equality here is due to the fact that $f_k(t) \geq 0$ for all $k \in \mathbb{N}$ and $t \in [-1/2, 1/2]$, and also since $f(t) \sim L(1/t)$ when $t \to 0+$. The last relation guarantees that there exists such $C_2 > 0$ that $f(t) \leq C_2 t^{-1/2}$ for all $t \in (0, 1/2]$, so f is integrable with respect to the Lebesgue measure on $[-1/2, 1/2]$. Hence, every series $\sum_{k=1}^{\infty} X_{n,k}$ converges in quadratic mean and almost surely to a random variable X_n, and the arising sequence $X = (X_n)_{n \in \mathbb{N}}$ is centered, strictly stationary, and has spectral density $f(t)$.

As usual, set $S_n = \sum_{j=1}^{n} X_j$. By Lemma 2.4 we have $\mathsf{K}_X(n) \sim L(n)$ as $n \to \infty$. Furthermore, as $a_{n,k} \geq 0$ for all n and k, the random sequence $(\sum_{j=1}^{k} X_{n,j})_{n \in \mathbb{N}}$ is associated for any k, therefore, the same is true for X. We will show that $S_{n_k} / \sqrt{n_k b(n_k)} \to 0$ in probability as $k \to \infty$. For $n, r \in \mathbb{N}$ write $\sigma_{n,r}^2 = \mathsf{Var} \sum_{j=1}^{n} X_{j,r}$.

Lemma 2.19. *For any $n, r \in \mathbb{N}$ one has $\sigma_{n,r}^2 \leq n(3 + R(2^{r+1}) - R(2^r))$ where the function $R(x)$ was introduced in (2.36).*

Proof. We have

$$\sigma_{n,r}^2 = \int_{-1/2}^{1/2} \frac{\sin^2 n\pi t}{\sin^2 \pi t} f_r(t)\, dt \leq n \sup_t f_r(t) \leq n\left(3 + \sup_t h_r(t)\right).$$

From this, taking into account that $h_r(t)$ is an even function, nonincreasing when $t \in [0, 1/2]$, and $h_r(0) = H_{r+1}(0) - H_r(0)$ (see (2.30) for H_r) we obtain the desired conclusion. $\square$

Further on, let $S_{n_k} = V_k + W_k$ where

$$V_k = \sum_{r=1}^{k} \sum_{j=1}^{n_k} X_{j,r}, \quad W_k = \sum_{r=k+1}^{\infty} \sum_{j=1}^{n_k} X_{j,r}, \quad k \in \mathbb{N}.$$

We see that

$$\mathsf{E} V_k = 0, \quad \frac{\mathsf{Var}\, V_k}{n_k b(n_k)} = \frac{\sum_{r=1}^{k} \sigma_{n_k,r}^2}{n_k b(n_k)} \leq \frac{3k + R(2^{k+1})}{b(n_k)} \to 0$$

as $k \to \infty$, due to Lemma 2.19 and the choice of sequence $(n_k)_{k \in \mathbb{N}}$. To estimate the second summand W_k, write:

$$\mathsf{P}(W_k \neq 0) \leq \sum_{r=k+1}^{\infty} \mathsf{P}\left(\sum_{j=1}^{n_k} X_{j,r} \neq 0\right) \leq \sum_{r=k+1}^{\infty} n_k \mathsf{P}(X_{1,r} \neq 0)$$

$$\leq \sum_{r=k+1}^{\infty} n_r (1 + 2N(r)) \alpha_r^{-1} \leq 2^{-k} \to 0, \quad k \to \infty.$$

Thus, $(V_k + W_k)/\sqrt{n_k b(n_k)} \to 0$ in probability, $k \to \infty$. The Theorem is proved. $\square$

3   Sharp rates of normal approximation

In this Section we study, in general non-stationary, associated random field and give a Berry–Esseen type estimate which is asymptotically sharp. In Section 1 we concentrated on the estimates of the convergence rates to the normal law for normalized partial sums (in terms of the Lindeberg function) under lower moment assumptions for summands. As a by-product the Lyapunov-type estimates were obtained. These estimates were not optimal for i.i.d. random variables with $\mathsf{E}|X_0|^3 < \infty$. Namely, in that case the error bound had the form $O(|U|^{-1/4})$ where $U \subset \mathbb{Z}^d$ was a finite subset of $\mathbb{Z}^d$ over which the summation was carried. However, in the mentioned case of i.i.d. random variables it is well-known that the optimal estimate has the form $O(|U|^{-1/2})$, see, e.g., [326, Ch. V, § 2].

One feasible way to achieve the better estimate is to adapt Stein's equation to non-continuous function g (for such approach to i.i.d. random variables see, e.g., [22, 99]). An alternative way is a modification of the Stein techniques due to Tikhomirov [401] who proposed (for mixing random sequences) to employ the analogue of the basic differential equation to characteristic functions, see also [28, 61, 182, 188, 396].

1°. The upper bound for rate in the CLT. For a centered random field $X = \{X_j, j \in \mathbb{Z}^d\}$ and finite set $U \subset \mathbb{Z}^d$ we write, as usual, $S(U) = \sum_{j \in U} X_j$. Assume that $\sigma(U) := \sqrt{\mathsf{Var}S(U)} > 0$ and introduce

$$\Delta(U) = \sup_{x \in \mathbb{R}} |\mathsf{P}(S(U) \le x\sigma(U)) - \Phi(x)|$$

where Φ is the distribution function of a standard normal random variable.

Theorem 3.1. *([67]) Let $X = \{X_j, j \in \mathbb{Z}^d\}$ be a **PA** or **NA** centered random field such that, for some $s \in (2, 3]$ and $\lambda > 0$,*

$$D_s = \sup_j \mathsf{E}|X_j|^s < \infty,$$

$$u_k = O(k^{-\lambda}), \ k \to \infty,$$

where the Cox–Grimmett coefficient u_k was introduced in (1.5.4). Then, for every finite $U \subset \mathbb{Z}^d$ such that $\sigma(U) > 0$, one has

$$\Delta(U) \le C|U|\sigma(U)^{-s+\gamma}$$

with any $\gamma > \gamma^* = (1/(s-1) + (\lambda+d)/(ds(s-2)))^{-1}$ and $C > 0$ independent of U.

Proof. We begin with a short notation list. Given a finite set $U \subset \mathbb{Z}^d$ having $\sigma = \sigma(U) > 0$, for $j \in U$, $m, l, r \in \mathbb{N}$ and $t \in \mathbb{R}$ write

$$f(t) = \mathsf{E}\exp\{it\sigma^{-1}S(U)\},$$

$$U_j^{(0)} = U, \ U_j^{(l)} = \{q \in U : |q - j| > lm\}, \ W_j^{(l)} = U_j^{(l-1)} \setminus U_j^{(l)},$$

$$S_j^{(0)} = \sigma^{-1}S(U), \quad S_j^{(l)} = \sigma^{-1}S(U_j^{(l)}), \quad Z_j^{(l)} = S(W_j^{(l)}),$$

$$\xi_j^{(l)} = \exp\{it\sigma^{-1}Z_j^{(l)}\} - 1, \quad \zeta_j^{(r)} = X_j \prod_{l=1}^{r} \xi_j^{(l)}.$$

These expressions depend also on t, m, U, which will be frequently omitted as arguments to simplify the formulas. As usual the sum (resp. product) over empty set is equal to zero (resp. one).

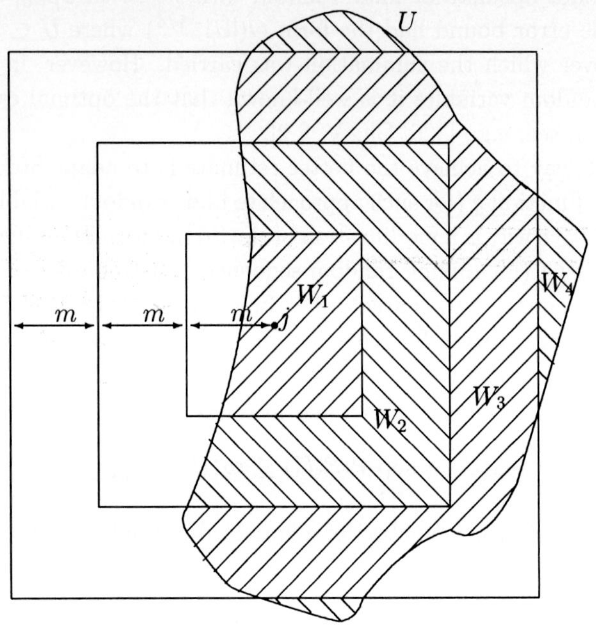

Fig. 3.9

Note that, for any $w \in \mathbb{C}$, $k \in \mathbb{N}$ and $j \in \mathbb{Z}^d$,

$$e^{wS(U)} = e^{wS_j^{(1)}} + e^{wS_j^{(1)}}(e^{wZ_j^{(1)}} - 1) = e^{wS_j^{(1)}} + e^{wS_j^{(1)}}\xi_j^{(1)}$$

$$= \ldots = e^{wS_j^{(1)}} + \sum_{r=1}^{k} e^{wS_j^{(r+1)}} \prod_{l=1}^{r} \xi_j^{(l)} + e^{wS_j^{(k+1)}} \prod_{l=1}^{k+1} \xi_j^{(l)}.$$

Therefore, for any $k \in \mathbb{N}$ and all $t \in \mathbb{R}$ one has

$$f'(t) = i\sigma^{-1} \sum_{j \in U} \mathsf{E}X_j \exp\{it\sigma^{-1}S(U)\} = i\sigma^{-1} \sum_{j \in U} \mathsf{E}X_j \exp\{itS_j^{(1)}\}$$

$$+ i\sigma^{-1} \sum_{r=1}^{k} \sum_{j \in U} \mathsf{E}\zeta_j^{(r)}(\exp\{itS_j^{(r+1)}\} - f(t)) + i\sigma^{-1}f(t) \sum_{r=1}^{k} \sum_{j \in U} \mathsf{E}\zeta_j^{(r)}$$

$$+ i\sigma^{-1} \sum_{j \in U} \mathsf{E}\zeta_j^{(k+1)} \exp\{itS_j^{(k+1)}\}. \tag{3.1}$$

Set

$$b_s(m,r) = \max_{j\in U}\max_{1\le l\le r}\|Z_j^{(l)}\|_s, \quad s\ge 1, \quad m,r\in\mathbb{N}.$$

Let $\theta_l(t)$ $(t\in\mathbb{R}, l\in\mathbb{Z}_+)$ be the complex-valued functions with $|\theta_l(t)|\le 1$. They may depend on other dropped parameters.

Lemma 3.2. *For any $t\in\mathbb{R}$ and $m\in\mathbb{N}$, we have*

$$i\sigma^{-1}\sum_{j\in U}\mathsf{E}\zeta_j^{(1)} = -t+\theta_1(t)tu_m|U|\sigma^{-2}+2\theta_2(t)D_s^{1/s}(|t|b_s(m,1))^{s-1}|U|\sigma^{-s}, \quad (3.2)$$

$$i\sigma^{-1}\sum_{j\in U}\mathsf{E}X_j\exp\{itS_j^{(1)}\} = 2\theta_3(t)|t|u_m|U|\sigma^{-2}. \tag{3.3}$$

Proof. To verify (3.2), apply the relation $e^{ix}-1 = ix+2\theta_0(x)|x|^{s-1}$, $x\in\mathbb{R}$, the Hölder inequality for the term $\mathsf{E}|X_j||Z_j^{(1)}|^{s-1}$ and recall that $\sum_{j\in U}\mathsf{E}X_jS_j^{(0)} = \sigma^2$. The equality (3.3) follows by Theorem 1.5.3. $\square$

For $m\in\mathbb{N}$ put

$$Q(m,1) = Q(m,2) = 0,$$

$$Q(m,r) = d\cdot(3m)^d r^{d-1}u_m, \quad r\in\mathbb{N}, \quad r>2.$$

Let $\nu\in(0,1/2)$, $a = a(d)>0$, $m,k\in\mathbb{N}$ be some fixed numbers to be specified later. A standing assumption on $t\in\mathbb{R}$ in the proof of Theorem is

$$a|t|D_s^{1/s}m^d(k+1)^{d-1}\le\sigma\nu. \tag{3.4}$$

Lemma 3.3. *For any t satisfying (3.4), $j\in U$ and $r = 1,\ldots,k+1$ one has*

$$\|\zeta_j^{(1)}\|_1\le 3^{d/2}D_s^{1/s}(D_s^{2/s}+u_1)^{1/2}\tau m^{d/2} := E_{m,1}, \tag{3.5}$$

$$\|\zeta_j^{(r)}\|_1\le c\left\{\prod_{l=1}^r\|\xi^{(l)}\|_s + 2^{r/2}\tau^{2/s}Q^{1/s}\left(\prod_{l=1}^r{}'\|\xi_j^{(l)}\|_s + \prod_{l=1}^r{}''\|\xi_j^{(l)}\|_s\right)\right.$$

$$\left.+2^r\tau^{4/s}Q^{2/s}+(\tau u_m)^{(s-2)/(s-1)}\left(2^r\tau^{2/s}Q^{1/s}+2^{r/2}\prod_{l=1}^r{}'\|\xi_j^{(l)}\|_s\right)\right\} =: E_{m,r} \tag{3.6}$$

where $Q = Q(m,r)$, $r = 2,\ldots,k+1$, $\prod_{l=1}^r{}'$ and $\prod_{l=1}^r{}''$ mean respectively products over odd and even $l\in\{1,\ldots,r\}$, c depends only on d and $\tau = |t|/\sigma$.

Proof. For $\mathsf{E}|\zeta_j^{(1)}|$ we have the following upper bound

$$\tau\mathsf{E}|X_jZ_j^{(1)}|\le\tau D_s^{1/s}(|W_j^{(1)}|(\sup_{j\in\mathbb{Z}^d}\mathsf{E}X_j^2+u_1))^{1/2}\le D_s^{1/s}(D_s^{2/s}+u_1)^{1/2}\tau(2m+1)^{d/2}.$$

This gives (3.5). For $r\ge 2$ let us consider

$$A'(j,r) := \prod_{l=1}^r{}'\xi_j^{(l)}, \quad A''(j,r) := \prod_{l=1}^r{}''\xi_j^{(l)}.$$

Set $\alpha = s/(s-1)$. Further, using the Hölder inequality we find that $\|\zeta_j^{(r)}\|_1$ admits an upper bound

$$\|A'(j,r)\|_s \left\{ \left|cov(|X_j|^\alpha, A''(j,r)^\alpha)\right|^{1/\alpha} + \|X_j\|_\alpha \|A''(j,r)\|_\alpha \right\}. \tag{3.7}$$

Observe that

$$|\xi_j^{(l)}|^s = 2^{s/2}\left(1 - \cos(\sigma^{-1}tZ_j^{(l)})\right)^{s/2}.$$

Denote by $Odd(p,r)$ the cardinality of odd numbers subset in $\{p,\ldots,r\}$. Then by virtue of Theorem 1.5.3 we obtain

$$\mathsf{E}\left|A'(j,r)\right|^s = cov\left(\left|\xi_j^{(1)}\right|^s, \left|\prod_{l=3}^r{}'\xi_j^{(l)}\right|^s\right) + \mathsf{E}\left|\xi_j^{(1)}\right|^s \mathsf{E}\left|\prod_{l=3}^r{}'\xi_j^{(l)}\right|^s$$

$$\leq (\tau s 2^{s-2})^2 Q(m,r)\left\{ 2^{s(Odd(1,r)-2)} + \mathsf{E}\left|\xi_j^{(1)}\right|^s 2^{s(Odd(3,r)-2)} + \ldots \right.$$

$$\left. + \mathsf{E}\left|\xi_j^{(1)}\right|^s \ldots \mathsf{E}\left|\xi_j^{(p(r)-2)}\right|^s 2^{s(Odd(p(r)-2,p(r))-2)} \right\} + \prod_{l=1}^r{}' \mathsf{E}\left|\xi_j^{(l)}\right|^s \tag{3.8}$$

where $p(r)$ is the greatest odd number not exceeding r. It is easy to see that

$$\mathsf{E}\left|\xi_j^{(l)}\right|^s \leq \tau^s \mathsf{E}|Z_j^{(l)}|^s \leq (a\tau D_s^{1/s} m^d l^{d-1})^s, \quad a = d \cdot 3^d. \tag{3.9}$$

From (3.4), (3.8) and (3.9) it follows that

$$\mathsf{E}\left|A'(j,r)\right|^s \leq (\tau s 2^{s-2})^2 2^{s(Odd(1,r)-2)} Q(m,r) \sum_{n=0}^{Odd(1,r)} \left(\frac{\nu}{2}\right)^{ns} + \prod_{l=1}^r{}' \mathsf{E}\left|\xi_j^{(l)}\right|^s$$

$$\leq \left(1 - \frac{\nu^s}{2^s}\right)^{-1}\left(\frac{\tau s}{4}\right)^2 2^{s\,Odd(1,r)} Q + \prod_{l=1}^r{}' \mathsf{E}\left|\xi_j^{(l)}\right|^s.$$

Likewise we consider the product over even numbers and, letting $Even(1,r) = r - Odd(r)$ be the cardinality of even numbers subset in $\{1,\ldots,r\}$. Finally, for all $r = 1,\ldots,k+1$ we have

$$\left\|A'(j,r)\right\|_s \leq \tau^{2/s} 2^{Odd(1,r)} Q^{1/s} + \prod_{l=1}^r{}' \left\|\xi_j^{(l)}\right\|_s, \tag{3.10}$$

$$\left\|A''(j,r)\right\|_s \leq \tau^{2/s} 2^{Even(1,r)} Q^{1/s} + \prod_{l=1}^r{}'' \left\|\xi_j^{(l)}\right\|_s. \tag{3.11}$$

To estimate the covariance in (3.7) take $N > 0$ and let $H_N : \mathbb{R} \to \mathbb{R}$ be the function defined in (2.1.5). We assume that $u_m > 0$, since otherwise that covariance would be zero. Setting $\gamma = s(s-2)/(s-1)$, we obtain for any $x \in \mathbb{R}$

$$|x|^\alpha - |H_N(x)|^\alpha \leq 2^\gamma N^{-\gamma}|x|^s.$$

Therefore, taking into account that $|\xi_j^{(l)}| \leq 2$, one may write

$$\left| cov\left(|X_j|^\alpha - |H_N(X_j)|^\alpha, |A'(j,r)|^\alpha \right) \right|^{1/\alpha} \leq 2^{s-1} N^{2-s} 2^{[1,r]} D_s^{1/\alpha}. \qquad (3.12)$$

Theorem 1.5.3 provides us with the estimate

$$\left| cov\left(|H_N(X_j)|^\alpha, |A'(j,r)|^\alpha \right) \right|^{1/\alpha} \leq 2N^{1/s} 2^{r/2} (\tau u_m)^{1/\alpha}.$$

Due to this estimate and (3.12), the optimization in N leads to the bound

$$\left| cov\left(|X_j|^\alpha, |A'(j,r)|^\alpha \right) \right|^{1/\alpha} \leq c_1 2^{r/2} 2^{r/2} (\tau u_m)^{(s-2)/(s-1)} \qquad (3.13)$$

here c_1 depends only on D_s. From (3.7), (3.10), (3.11) and (3.13) we infer (3.6). $\square$

Lemma 3.4. *Under condition* (3.4), *for* $j \in U$, $r = 1, \ldots, k+1$ *the following estimates are valid:*

$$\left| cov\left(\zeta_j^{(r+1)}, \exp\{-itS_j^{(r+1)}\} \right) \right|$$

$$\leq c_2 4^r \left(\tau u_m + (3^d r^d m^d \tau^2 u_m)^{1/\alpha} \right) =: H_{m,r}, \quad c_2 = c_2(D_s), \qquad (3.14)$$

$$\left| \sum_{j \in U} \mathsf{E}\zeta_j^{(r)} \left(\exp\{itS_j^{(r+1)}\} - f(t) \right) \right|$$

$$\leq |U| H_{m,r} + 5^{d/2} |U| (D_s^{2/s} + u_1)^{1/2} (rm)^{d/2} E_{m,r} \tau |f(t)|$$

$$+ 3^{2d} |U|^{1/2} (D_s^{2/s} + u_1)^{1/2} (rm)^d E_{m,r} \tau. \qquad (3.15)$$

Proof. As in the proof of Lemma 3.3, we consider a number $N > 0$ to be specified later and the truncation function H_N. Then

$$\left| cov\left(\zeta_j^{(r+1)}, \exp\{-itS_j^{(r+1)}\} \right) \right| \leq \left| cov\left(H_N(X_j) \prod_{l=1}^{r+1} \xi_j^{(l)}, \exp\{-itS_j^{(r+1)}\} \right) \right|$$

$$+ \left| cov\left((X_j - H_N(X_j)) \prod_{l=1}^{r+1} \xi_j^{(l)}, \exp\{-itS_j^{(r+1)}\} \right) \right| =: I_1 + I_2.$$

Set

$$\delta(m,r) = \max_{j \in U} \sum_{l=1}^{r} \sum_{n=r+2}^{\infty} cov(Z_j^{(l)}, Z_j^{(n)}).$$

An obvious calculation shows that

$$\delta(m,r) \leq (3rm)^d u_m. \qquad (3.16)$$

To estimate I_1 we write all complex exponents appearing at both terms of covariance as a sum of real and imaginary parts, thus obtaining 2^{r+1} summands. Then apply to each summand Theorem 1.5.3. As all the random variables $S_j^{(l)}, Z_j^{(l)}$ are **PA** or **NA**, the mentioned Theorem implies that I_1 admits the upper bound

$$4^{r+1}t\,cov(X_j, S_j^{(r+1)}) + 4^{r+1}\sum_{l=1}^{r+1} N\sigma^{-1}t^2 cov(Z_j^{(l)}, S_j^{(r+1)}) = 4^{r+1}\left(\tau u_m + N\tau^2\delta(m,r)\right).$$

The inequalities $|\xi^{(l)}| \le 2$ and $|X_j - H_N(X_j)| \le |X_j|\mathbb{I}\{|X_j| \ge N\} \le N^{1-s}D_s$ allow us to estimate I_2 :

$$I_2 \le 2^{r+2}N^{1-s}D_s.$$

Thus taking into account (3.16) one has

$$I_1 + I_2 \le 4^{r+1}\left(\tau u_m + N\tau^2(3rm)^d u_m\right) + 2^{r+2}N^{1-s}D_s. \tag{3.17}$$

If $u_m = 0$ then $\delta(m,r) = 0$ due to (3.16). Thus in this case letting $N \to \infty$ in (3.17) one arrives at (3.14). If $u_m > 0$, then (3.14) is obtained by minimization of the right-hand side of (3.17) in N.

The proof of (3.15) is more difficult. An elementary estimate yields

$$\left|\sum_{j\in U}\mathsf{E}\zeta_j^{(r)}\left(\exp\{itS_j^{(r+1)}\} - f(t)\right)\right| \le \sum_{j\in U}\left|cov\left(\zeta_j^{(r)}, \exp\{itS_j^{(r+1)}\}\right)\right|$$

$$+ \sum_{j\in U}\left|\mathsf{E}\zeta_j^{(r)}\left(\mathsf{E}\exp\{itS_j^{(0)}\} - \exp\{itS_j^{(r+1)}\}\right)\right|. \tag{3.18}$$

By (3.14) we have

$$\sum_{j\in U}\left|cov\left(\zeta_j^{(r)}, \exp\{itS_j^{(r+1)}\}\right)\right| \le |U|H_{m,r}. \tag{3.19}$$

Furthermore,

$$\sum_{j\in U}\left|\mathsf{E}\zeta_j^{(r)}\left(\mathsf{E}\exp\{itS_j^{(0)}\} - \exp\{itS_j^{(r+1)}\}\right)\right| \le \left|f(t)\sum_{j\in U}h_j^{(r)}\mathsf{E}\eta_j^{(r)}\right|$$

$$+ \left|\mathsf{E}\left(\exp\{it\sigma^{-1}S(U)\}T_r(U)\right)\right| \tag{3.20}$$

where

$$T_r(U) = \sum_{j\in U}h_j^{(r)}\left(\eta_j^{(r)} - \mathsf{E}\eta_j^{(r)}\right), \quad h_j^{(r)} = \mathsf{E}\zeta_j^{(r)} \quad \text{and} \quad \eta_j^{(r)} = 1 - \exp\left\{-it(S_j^{(0)} - S_j^{(r)})\right\}.$$

The inequality $|e^{ix} - 1| \le |x| \ (x \in \mathbb{R})$ permits to observe that for $j \in U$

$$\mathsf{E}|\eta_j^{(r)}| \le |t|\mathsf{E}|S_j^{(0)} - S_j^{(r)}| \le \tau(\mathsf{E}S^2(B_j))^{1/2} \le \tau\left(L(5rm)^d\right)^{1/2},$$

here $B_j = \{q \in U : |q - j| \le rm\}$, $L = D_s^{2/s} + u_1$ and Lemma 2.1.20 together with Lyapunov inequality was used to bound the expectation. Consequently, from Lemma 3.3 one deduces the estimate

$$\left|f(t)\sum_{j\in U}h_j^{(r)}\mathsf{E}\eta_j^{(r)}\right| \le 5^{d/2}|U|L^{1/2}(rm)^{d/2}E_{m,r}\tau|f(t)|. \tag{3.21}$$

For the second summand in (3.20) we write, first of all, that

$$\left|\left(\mathsf{E}\exp\{it\sigma^{-1}S(U)\}T_r(U)\right)\right| \le (\mathsf{E}|T_r(U)|^2)^{1/2} \tag{3.22}$$

since $|\exp\{it\sigma^{-1}S(U)\}| = 1$. Recall that $|z|^2 = (\operatorname{Re} z)^2 + (\operatorname{Im} z)^2 = z\bar{z}$, where $z \in \mathbb{C}$ and $\bar{z}$ denotes the complex conjugate to z. Thus, by Lemma 3.3,

$$\mathsf{E}|T_r(U)|^2 \le \sum_{j,q\in U} E^2_{m,r}\left|cov\left(\eta_j^{(r)} - \mathsf{E}\eta_j^{(r)}, \eta_q^{(r)} - \mathsf{E}\eta_q^{(r)}\right)\right| =: \sum_{j,q\in U} E^2_{m,r}\Delta_{j,q}^{(r)}. \tag{3.23}$$

For $j \in U$ introduce the "close to j" set $V(j,r,m) := \{v \in U : |j - v| \le 3rm\}$. Note that for any $j \in U$ and $q \in V(j,r,m)$, applying the Cauchy-Bunyakovski-Schwarz inequality and the bound $|e^{ix} - 1| \le |x|$ $(x \in \mathbb{R})$ we have

$$\Delta_{j,q}^{(r)} \le \left|cov\left(\exp\{i\tau S(U \setminus U_j^{(r)})\} - 1, \exp\{i\tau S(U \setminus U_q^{(r)})\} - 1\right)\right|$$

$$\le \tau^2 \max_{v\in U} \mathsf{E}S(U \setminus U_v^{(r)})^2 \le \tau^2 (D_s^{2/s} + u_1) \max_{v\in U} |U \setminus U_v^{(r)}| \le \tau^2 (3rm)^d L. \tag{3.24}$$

If $j \in U$ and $q \notin V(j,r,m)$, then $(U \setminus U_j^{(r)}) \cap (U \setminus U_q^{(r)}) = \varnothing$. Thus Theorem 1.5.3 yields for such j a bound

$$\sum_{q\notin V(j,r,m)} \Delta_{j,q}^{(r)} \le 4\tau^2 \sum_{\substack{v\in U: \\ |v-j|\le rm}} \sum_{w\in U} \sharp\{q \in U : |q - w| \le rm\}|cov(X_v, X_w)|$$

$$\le 4\tau^2 (3rm)^{2d} \max_{v\in U} \sum_{w\in U} |cov(X_v, X_w)| \le 4\tau^2 (3rm)^{2d} L. \tag{3.25}$$

Of course the absolute value signs at the covariances are essential if the field is **NA**. Clearly, $|V(j,r,m)| \le (7rm)^d$, $j \in U$. Consequently, the relations (3.24) and (3.25) show that

$$\sum_{j,q\in U} \left|cov\left(\eta_j^{(r)} - \mathsf{E}\eta_j^{(r)}, \eta_q^{(r)} - \mathsf{E}\eta_q^{(r)}\right)\right| \le 3^{4d}|U|\tau^2(rm)^{2d} L. \tag{3.26}$$

Now (3.15) follows from (3.18)—(3.25).

Lemmas 3.2—3.4 together with (3.1) imply that for t satisfying (3.4)

$$f'(t) = -tf(t) + \theta_4(t)A(t)f(t) + \theta_5(t)B(t),$$

where

$$A(t) = 5^{d/2}|U|\sigma^{-2}|t|m^{d/2}L^{1/2}\Sigma_1 + |U|\sigma^{-1}\Sigma_2 + 2D_s^{1/s}|U|\sigma^{-s}(|t|b_s(m,1))^{s-1},$$

$$B(t) = 3|U|\sigma^{-2}|t|u_m + |U|\sigma^{-1}\Sigma_3 + 2^{5d}|t||U|^{1/2}\sigma^{-2}m^d L^{1/2}\Sigma_4 + |U|\sigma^{-1}E_{m,k+1}$$

where

$$\Sigma_1 = \sum_{r=1}^{k} r^{d/2}E_{m,r}, \quad \Sigma_2 = \sum_{r=2}^{k} E_{m,r}, \quad \Sigma_3 = \sum_{r=1}^{k} H_{m,r}, \quad \Sigma_4 = \sum_{r=1}^{k} r^d E_{m,r}.$$

Note also that

$$b_s(m,r) \leq 3^d d D_s^{1/s} r^{d-1} m^d.$$

Thus, applying the condition on decrease of $(u_r)_{r\in\mathbb{N}}$ and restriction (3.4) on t, we establish, for all k and m such that $k \leq (m \wedge |U|^{1/2})M$, M being some constant, that

$$f'(t) = -tf(t) + c\theta_6(t)f(t)|U|\sum_{l=1}^{4} a_l|t|^{\beta_l} + C\theta_7(t)|U|\sum_{n=1}^{11} b_n|t|^{\gamma_n}. \qquad (3.27)$$

Here c and C do not depend on t and U, with the notation $y = (s-2)/(s-1)$ we have $\beta_1 = s-1$, $\beta_2 = 2$, $\beta_3 = 1+2/s$, $\beta_4 = 1+y$, $a_1 = \sigma^{-s}m^{d(s-1)}$, $a_2 = \sigma^{-3}m^{2d}$, $a_3 = \sigma^{-2-2/s}m^{d+(d-\lambda)/s}$, $a_4 = \sigma^{-2-y}m^{d-\lambda y}$, $\gamma_1 = 2$, $\gamma_2 = 2 + 2/s$, $\gamma_3 = 3$, $\gamma_4 = 1$, $\gamma_5 = 2(s-1)/s$, $\gamma_6 = 4/s$, $\gamma_7 = 1 + 2/s$, $\gamma_8 = 2$, $\gamma_9 = (s^2 - 2)/(s^2 - s)$, $\gamma_{10} = 1 + y$, $\gamma_{11} = 2 + y$, $b_1 = |U|^{-1/2}\sigma^{-3}m^{3d/2}$, $b_2 = |U|^{-1/2}\sigma^{-3-2/s}m^{2d+(d-\lambda)/s}$, $b_3 = |U|^{-1/2}\sigma^{-4}m^{3d}$, $b_4 = 4^k\sigma^{-2}m^{-\lambda}$, $b_5 = 4^k\sigma^{-1-2(s-1)/s}m^{(d-\lambda)(s-1)/s}k^{d(s-1)/s}$, $b_6 = 2^k\sigma^{-1-4/s}m^{2(d-\lambda)/s}k^{2(d-1)/s}$, $b_7 = 2^{k/2}\sigma^{-2-2/s}m^{d+(d-\lambda)/s}k^{(d-1)/s}\nu^{(k-3)/2}$, $b_8 = \sigma^{-3}m^{2d}\nu^{k-2}$, $b_9 = 2^k\sigma^{-1-(s^2-2)(s^2-s)}m^{-\lambda(s-2)/(s-1)+(d-\lambda)/s}k^{(d-1)/s}$, $b_{10} = 2^{k/2}\sigma^{-2-y}m^{d-\lambda y}\nu^{(k-2)/2}$, $b_{11} = |U|^{-1/2}\sigma^{-3-y}m^{2d-\lambda y}$.

Writing down an explicit formula for the solution of (3.27), we arrive at the bound for $|f(t) - \exp\{-t^2/2\}|$ where t meets (3.4). In addition choose $k = [\varepsilon \log m] + 1$ where $\varepsilon > 0$ and $m = [\sigma^\mu] + 1$ with any $\mu > (d + (\lambda + d)(s-1)/(s(s-2)))^{-1}$.

The classical Esseen inequality (see, e.g., [326]) provides for any distribution function F, every $T > 0$ and each $b > (2\pi)^{-1}$ the following estimate

$$\sup_{x\in\mathbb{R}} |F(x) - \Phi(x)| \leq b \int_{-T}^{T} \left| \frac{g(t) - \exp\{-t^2/2\}}{t} \right| dt + (2\pi)^{-1/2} r(b) T^{-1}$$

where g is the characteristic function corresponding to F and $r(b) > 0$ depends on b only.

The application of this inequality to estimate the difference between the distribution function of the normalized sum and one of a standard normal law leads to the desired statement. $\square$

2°. The lower bound for rate in the CLT. The rate given by Theorem 3.1 is almost optimal in a sense, as the following example shows.

Theorem 3.5. ([66]) *There exists a random field satisfying all the conditions of Theorem 3.1 and a family of sets U_n, $|U_n| = n$, $n \in \mathbb{N}$, such that*

$$\Delta(U_n) \geq c|U_n|\sigma(U_n)^{-s+\gamma} \qquad (3.28)$$

for any $\gamma < \gamma_* = (s-1)(s-2)/(\lambda + s - 1)$ with $c > 0$ independent of $(U_n)_{n\in\mathbb{N}}$.

Proof. Let ξ_k, $k \in \mathbb{N}$, be i.i.d. random variables such that for some positive α specified later

$$P(\xi_1 = A_k) = P(\xi_1 = -A_k) = ck^{-3-2\alpha-\lambda}, \quad A_k = \sum_{j=1}^{k} j^\alpha. \qquad (3.29)$$

As usual, $c, C, C_0, C_1, \ldots$ denote positive nonrandom quantities depending only on α, λ and sometime on s as well. Let us introduce

$$X_j = \sum_{i=1}^{j} i^\alpha \left(\mathbb{I}\left\{ \xi_{j-i+1} > A_{i-1} \right\} - \mathbb{I}\left\{ \xi_{j-i+1} < -A_{i-1} \right\} \right), \quad j \in \mathbb{N}, \tag{3.30}$$

and set $X_j = 0$ for $j \in \mathbb{Z}^d \setminus \mathbb{N}$, where we identify the sets $\mathbb{N}$ and $\mathbb{N} \times \mathbb{Z}^{d-1}$ (a sum over an empty set is equal to zero as usual). Then $\{X_j, j \in \mathbb{Z}^d\}$ is a centered associated random field by Theorem 1.1.8, (d).

For $s \in (2, 3]$, any $j \in \mathbb{N}$ and all $a_1, \ldots, a_j \in \mathbb{R}$ a standard calculation yields

$$|a_1 + \ldots + a_j|^s \le (a_1 + \ldots + a_j)^2 |a_1 + \ldots + a_j|^{s-2} \le \sum_{i,r=1}^{j} a_i a_r \sum_{m=1}^{j} |a_m|^{s-2}.$$

Consequently,

$$\mathsf{E}|X_j|^s \le \sum_{i,r,m=1}^{j} (irm^{s-2})^\alpha P_{i,r,m} \tag{3.31}$$

where $P_{i,r,m} = \mathsf{P}\left(|\xi_{j-i+1}| > A_{i-1}, |\xi_{j-r+1}| > A_{r-1}, |\xi_{j-m+1}| > A_{m-1} \right)$.

Let $i = r = m$, then for all $j \in \mathbb{N}$

$$\sum_{i=1}^{j} i^{\alpha s} P_{i,i,i} = \sum_{i=1}^{j} i^{\alpha s} \mathsf{P}(|\xi_1| > A_{i-1}) = 2c \sum_{i=1}^{j} i^{\alpha s} \sum_{k=i}^{\infty} k^{-3-2\alpha-\lambda}$$

$$\le C_0 \sum_{i=1}^{\infty} i^{\alpha s - 2 - 2\alpha - \lambda} < \infty$$

if

$$\alpha < (1 + \lambda)/(s - 2). \tag{3.32}$$

Here we used, for $\gamma > 1$, the simple relation

$$\sum_{k=r}^{\infty} k^{-\gamma} \sim \int_{r}^{\infty} x^{-\gamma} dx = \frac{r^{1-\gamma}}{\gamma - 1}, \quad r \to \infty.$$

Introduce a set J consisting of $i, r, m \in \{1, \ldots, j\}$ such that $i \ne r$, $i \ne m$ and $r \ne m$. Using the independence of a sequence $(\xi_k)_{k \in \mathbb{N}}$ we have

$$\sum_{i,r,m \in J} (irm^{s-2})^\alpha P_{i,r,m} \le (2c)^3 \left(\sum_{i=1}^{\infty} i^\alpha \sum_{q=i}^{\infty} q^{-3-2\alpha-\lambda} \right)^2 \sum_{m=1}^{\infty} m^{\alpha(s-2)} \sum_{q=m}^{\infty} q^{-3-2\alpha-\lambda}.$$

The last expression is finite under condition (3.32) as $\alpha < \alpha s$ and $\alpha(s - 2) < \alpha s$.

In a similar way we consider other summands in (3.31) to conclude that $D_s = \sup_j \mathsf{E}|X_j|^s < \infty$ whenever (3.32) holds.

Furthermore,

$$\mathsf{E}X_j^2 = 2c \sum_{r=1}^{j} r^{2\alpha} \sum_{k=r}^{\infty} k^{-3-2\alpha-\lambda}, \quad j \in \mathbb{N}.$$

Therefore, $\sigma_1^2 := \inf_{j\in\mathbb{N}} \mathsf{E}X_j^2 > 0$. For any random variable ξ and for all $0 \le a \le b$,

$$(\mathbb{I}\{\xi > a\} - \mathbb{I}\{\xi < -a\})(\mathbb{I}\{\xi > b\} - \mathbb{I}\{\xi < -b\}) = \mathbb{I}\{|\xi| > b\}.$$

Thus, for $k, j \in \mathbb{N}$,

$$cov(X_j, X_{j+k}) = \sum_{i=1}^{j} i^\alpha (i + k)^\alpha \mathsf{P}\left(|\xi_{j-i+1}| > A_{i+k-1}\right)$$

$$= 2c \sum_{i=1}^{j} i^\alpha (i + k)^\alpha \sum_{v=i+k}^{\infty} v^{-3-2\alpha-\lambda}$$

$$\le C_1 \sum_{i=1}^{j} i^\alpha (i + k)^{-2-\alpha-\lambda} \le C_1 \sum_{r=k}^{\infty} r^{-2-\lambda} \le C_2 k^{-1-\lambda}.$$

This estimate implies that $u_n = O(n^{-\lambda})$, $n \to \infty$. Now we can write, for all $n \in \mathbb{N}$,

$$n\sigma_1^2 \le \sigma_n^2 \le u(0)n \tag{3.33}$$

where $\sigma_n^2 := \mathsf{E}S_n^2$. Note that, for $r \in \mathbb{N}$,

$$\xi_r = \sum_{i=1}^{\infty} A_i \left(\mathbb{I}\{\xi_r = A_i\} - \mathbb{I}\{\xi_r = -A_i\}\right) = \sum_{i=1}^{\infty} i^\alpha \left(\mathbb{I}\{\xi_r > A_{i-1}\} - \mathbb{I}\{\xi_r < -A_{i-1}\}\right).$$

Using the change of summation order we have

$$S_n = \sum_{j=1}^{n} X_j = \sum_{j=1}^{n}\sum_{i=1}^{j} i^\alpha \left(\mathbb{I}\{\xi_{j-i+1} > A_{i-1}\} - \mathbb{I}\{\xi_{j-i+1} < -A_{i-1}\}\right)$$

$$= \sum_{i=1}^{n}\sum_{j=i}^{n} i^\alpha \left(\mathbb{I}\{\xi_{j-i+1} > A_{i-1}\} - \mathbb{I}\{\xi_{j-i+1} < -A_{i-1}\}\right)$$

$$= \sum_{i=1}^{n}\sum_{r=1}^{n-i+1} i^\alpha \left(\mathbb{I}\{\xi_r > A_{i-1}\} - \mathbb{I}\{\xi_r < -A_{i-1}\}\right)$$

$$= \sum_{r=1}^{n}\sum_{i=1}^{n-r+1} i^\alpha \left(\mathbb{I}\{\xi_r > A_{i-1}\} - \mathbb{I}\{\xi_r < -A_{i-1}\}\right).$$

Thus $S_n = \sum_{r=1}^{n} \xi_r - T_n$ where

$$T_n = \sum_{r=1}^{n} \sum_{i=n-r+2}^{\infty} i^\alpha \left(\mathbb{I}\{\xi_r > A_{i-1}\} - \mathbb{I}\{\xi_r < -A_{i-1}\}\right), \quad n \in \mathbb{N}.$$

For $t > 0$ we have an obvious inequality

$$\mathsf{P}(S_n > t) \ge \mathsf{P}\left(\sum_{i=1}^{n} \xi_i \ge 2t\right) - \mathsf{P}\left(T_n \ge t\right). \tag{3.34}$$

If $\xi_1, \xi_2, \ldots$ are symmetric i.i.d. random variables then, for any $x > 0$ and all $n \in \mathbb{N}$, the following well-known inequality holds (see, e.g., [167, V. 2, Ch. 5, §5]):

$$\mathsf{P}(|\xi_1 + \ldots + \xi_n| > x) \ge \frac{1}{2}\mathsf{P}\left(\max_{1\le j\le n} |\xi_j| > x\right).$$

Note that

$$P\left(\max_{1\le j\le n}|\xi_j| > x\right) = 1 - P\left(\bigcap_{j=1}^{n}\{|\xi_j| \le x\}\right) = 1 - (1 - P(|\xi_1| > x))^n.$$

For any $p \in [0,1]$ and all $n \in \mathbb{N}$,

$$(1-p)^n \le 1 - np + \frac{n(n-1)}{2}p^2.$$

Therefore,

$$P(|\xi_1 + \ldots + \xi_n| > x) \ge \frac{n}{2}P(|\xi_1| > x)\left(1 - \frac{n-1}{2}P(|\xi_1| > x)\right).$$

Using the symmetry of distribution of the random variable ξ_1 one has

$$P(|\xi_1| > x) = 2P(\xi_1 > x), \quad x > 0.$$

Consequently, for any $x > 0$,

$$P\left(\sum_{i=1}^{n}\xi_i \ge x\right) \ge \frac{n}{2}P(\xi_1 \ge x)(1 - (n-1)P(\xi_1 \ge x)). \tag{3.35}$$

From (3.29), taking into account (3.33), we have

$$C_3\psi(n) \le P(\xi_1 \ge 2\sigma_n \log n) \le C_4\psi(n) \tag{3.36}$$

for all n large enough, where $\psi(n) = n^{-1-\lambda/(2+2\alpha)}(\log n)^{-2-\lambda/(1+\alpha)}$. For all $n \in \mathbb{N}$

$$E|T_n| \le \sum_{r=1}^{n}\sum_{i=n-r+2}^{\infty} 2ci^\alpha \sum_{k=i+1}^{\infty} k^{-3-2\alpha-\lambda} \le C_5 \sum_{r=1}^{n}\sum_{i=n-r+2}^{\infty} i^{-2-\alpha-\lambda}$$

$$\le C_6 \sum_{r=1}^{n}(n-r+1)^{-1-\alpha-\lambda} \le C_7 \sum_{i=1}^{\infty} i^{-1-\alpha-\lambda} < \infty$$

and the Chebyshev inequality guarantees the estimate

$$P\left(T_n \ge \sigma_n \log n\right) \le \frac{C_8}{\sigma_n \log n}, \quad n > 1. \tag{3.37}$$

Pick now α in such a way that (3.32) holds and

$$\lambda(1+\alpha)^{-1} < 1. \tag{3.38}$$

From (3.34)—(3.38) we have

$$P(S_n \ge \sigma_n \log n) \ge C_9 n^{-\lambda/(2+2\alpha)}(\log n)^{-2-\lambda/(1+\alpha)}. \tag{3.39}$$

The well-known relation

$$P(Z \ge x) \sim (x\sqrt{2\pi})^{-1}\exp\{-x^2/2\} \text{ as } x \to \infty,$$

where $Z \sim N(0,1)$, yields

$$P(Z \ge \log n) = o(n^{-\log n}), \quad n \to \infty. \tag{3.40}$$

From (3.39) and (3.40) we get for all n large enough and the sets $U_n = \{1, \ldots, n\}$

$$\Delta(U_n) \geq C_{10} n^{-\lambda/(2+2\alpha)} (\log n)^{-2-\lambda/(1+\alpha)}.$$

Thus the proof will be done if we select α under (3.32) and (3.38) so that

$$\frac{\lambda}{2(1+\alpha)} < -1 + \frac{s-\gamma}{2}.$$

Consequently, $\gamma < s - 2 - \lambda/(\lambda+1)$. Taking α close to $(\lambda+1)/(s-2)$ we conclude that (3.28) is satisfied and such choice leads to the inequality

$$\gamma < (s-1)(s-2)/(\lambda+s-1) = \gamma_*.$$

The proof is complete. $\square$

3°. Final remarks. There is a vast literature devoted to CLT and its applications for various classes of random systems (including martingales, Markov and mixing processes). Besides the papers discussed above in this Chapter it is worth to provide the following references. The τ-dependent random sequences were studied by Dedecker and Prieur [124]. The triangular array of (BL, θ)-dependent and similar random variables were investigated by Coulon-Prieur and Doukhan [106]. Asymptotic normality of nonstationary **NA** random fields was studied by Roussas [352] and later, under minimal moment assumptions, by Hu, Su and Yuan [205]. The first paper concerning the Berry-Esseen type estimates for random fields seems to be that of Shao [370]. The paper by Louhichi [276] contains the Berry-Esseen type estimate for a random sequence of quasi-associated random variables (in the sense noted in our Section 1.5). The convergence rate of order $O(n^{-1/2})$ to the standard Gaussian law of normalized partial sums S_n $(n \to \infty)$ for a sequence of bounded associated random variables is established by Matula [287], and an "almost" sharp bound for **NA** random variables is due to Pan [314]. Matula and Rychlik [295] treated also the analogs of CLT for products of associated random variables. Also merit mentioning the papers by Zhang, giving further generalizations of CLT for **NA** random fields [409, 427, 429], and a result on large deviations by Tang [398] for **NA** random variables.

For associated sequences Dabrowski and Jakubowski [118] considered the convergence of normalized partial sums to stable distributions with parameter $\alpha \in (0, 2)$. The asymptotical normality of different statistics was studied, e.g., by Dewan and Rao [133] and Huang and Zhang [209] (the rank statistics), see also Section 7.2 and the corresponding references. A local limit theorem is treated by Wood [419], and the Poisson approximation to associated random sequences proved via the Stein–Chen method ([98]) was the object of the paper by Ferreira [168]. Results for linearly transformed associated and **NA** random variables are proved by Peligrad and Utev [324] and by and Yun-xia and Zhang [263].

The Newman CLT was applied to the proofs of CLTs for density of infinite clusters and for the random paintings of cluster appearing in percolation model, see Garet [174] and [175] respectively. The relations between the validity of CLT and the choice of the summation regions in $\mathbb{Z}^d$ were studied by Perera [323]. In [70] for random field $X = \{X_t, t \in \mathbb{R}^d\}$ the analog of (BL, θ)-dependence condition was proposed and the CLT was established when the observations belong to VH-growing subsets of $\mathbb{R}^d$ and at the same time form more and more dense grid, i.e. two scaling procedures was studied.

The development of the Stein, Stein-Chen and Tikhomirov techniques and their modifications is the object of intensive studies. In this regard we refer, e.g., to [19, 20, 23, 181, 182, 345]. The Stein technique (more exactly, the system of partial differential equations) was used in [79] to estimate the convergence rate (e.g., over the collection of convex sets) to a normal law for normalized sums of (BL, θ)-dependent random vectors . In [340] the analogous problem was studied using the Laplace operator. The asymptotic expansions of the Edgeworth type for distributions of normalized partial sums were studied by Rinott and Rotar [345] when the summands poses some kind of local dependence. As far as we know there are no similar expansions for (BL, θ)-dependent random fields.

Note that Levèvre and Utev [255] studied exact norm of the Stein operator with applications to the CLT (cf. Lemma 1.18). Namely, let ϕ be a density of a random variable $Z \sim N(0, 1)$ and

$$Sh(x) := \frac{1}{\phi(x)} \int_{-\infty}^{x} (h(u) - \mathsf{E}h(Z))\phi(u)du$$

be the Stein operator providing the solution to the equation $(S)'(x) - xSh(x) = h(x) - \mathsf{E}h(Z)$. Introduce the usual differential operator $Df := f'$ and let $\|D^k S\|_{[k]} := \sup_{h:\|h\|_{[k]}=1} \|D^k Sh\|_\infty$ where $\|h\|_{[k]} := \frac{1}{k!}\|D^k h\|_{TV}$ ($\|\cdot\|_{TV}$ is the total variation norm and $\|\cdot\|_\infty$ is the sup-norm). Then $\|D^k S\|_{[k]} = (\mathsf{E}|Z|^k)^2 \sqrt{\pi/8}$, $k \in \mathbb{Z}_+$.

An interesting approach to CLT is based on the information-theoretic arguments due to Linnik [270] and developed further by other researchers, see, e.g. Johnson and Suhov [222]. Recall that if a random variable Y with finite variance σ^2 has a differentiable positive density f, then its *Fisher information* $J(Y) := \mathsf{E}\rho^2(Y)$ where $\rho(u) = f'(u)/f(u)$, whereas the *standardized Fisher information*

$$J_{\mathrm{st}}(Y) := \sigma^2 J(Y) - 1 = \mathsf{E}(\sigma\rho(Y) + Y/\sigma)^2.$$

Observe also that if $Y \sim N(0, \sigma^2)$ then $J_{\mathrm{st}}(Y) = 0$. Shimizu [382] proved that if Y has variance σ^2, density f and distribution function F then

$$\sup_{x \in \mathbb{R}} |F(x) - \Phi_{0,\sigma^2}(x)| \leq 4\sqrt{3J_{\mathrm{st}}}$$

where Φ_{0,σ^2} is a distribution function of $N(0, \sigma^2)$.

Recently Johnson [220] adapted the arguments concerning information-theoretic convergence in CLT from the independent case to associated random fields. The difficulties arose here because in the independent case the Fisher information is a sub-additive quantity on convolutions whereas in the dependent one it is "almost sub-additive" and one has to control the error term. Consider a stationary associated random field $X = \{X_j, j \in \mathbb{Z}^d\}$ with mean zero satisfying the finite susceptibility condition. Let $Z^{(\tau)} = \{Z_j^{(\tau)}, j \in \mathbb{Z}^d\}$ be a family consisting of independent $N(0, \tau)$ random variables such that X and $Z^{(\tau)}$ are independent. Introduce a random field $Y^{(\tau)} = \{Y_j^{(\tau)} = X_j + Z_j^{(\tau)}, j \in \mathbb{Z}^d\}$ perturbed by normal random variables. In [220] the counterparts of Theorem 2.6 are established concerning the uniform integrability of normalized (auxiliary) partial sums and asymptotical behavior of $J_{\mathrm{st}}(V_i^{(\tau)})$ (as $|i| \to \infty$, $i \in \mathbb{Z}^d$) where $V_i^{(\tau)} = \sum_{j \in B(i)} Y_j^{(\tau)} / |B(i)|$ and $B(i) = [0, i]$ is a block belonging to $\mathcal{U}$. The relative entropy distance between density functions was used their also.

Let $(Z_1, \ldots, Z_n)$ be a sample from a random process $(Z_i)_{i \in \mathbb{N}}$ where $Z_1, Z_2, \ldots$ are i.i.d. random variables with values in a finite alphabet $\mathcal{A}$. An interesting problem having applications in number theory and cryptography is to estimate the entropy H of the process. In [221] the following approach is used. Consider a partition of (Z_i) into blocks of size l, that is $X_i = Z_{(i-1)l+1}^{il}$, $i \in \mathbb{N}$, where $Z_a^b := (Z_a, \ldots, Z_b)$. Given the finite blocks $X_1, \ldots, X_k$, we can count how long it takes for these blocks to reappear. Introduce the "return time for the j-th block"

$$S_j := \min\{t \geq 1 : X_{j+t} = X_j\} \quad j = 1, \ldots, k.$$

Note that there are related definitions, e.g., the Grassberg prefix. It turns possible (under certain conditions) to approximate the S_i by means of the auxiliary almost independent random variables R_i and on this way to construct an asymptotically normal and consistent estimate for H. In a special case $(R_i)_{i \in \mathbb{N}} \in \mathbf{NA}$, which simplifies the asymptotic analysis.

To conclude this Section we remark that it would be desirable to continue the study of random fields without finite susceptibility condition, see the book [143] where the problems of long-range dependence are considered. An interesting research domain is the summation of random number of random variables (see [179] and references therein). This theory was built mainly for summation of sequences of independent random variables with random index independent of the examined sequence. Thus the investigation of various functionals constructed for dependent summands belonging to random sets would be a natural extension.

Chapter 4

Almost Sure Convergence

In Chapter 4 we are concerned with the law of large numbers (LLN) which is the cornerstone result ascending to the origin of Probability Theory. There is a long way from the classical Bernoulli LLN to modern results describing almost sure behavior of normalized sums of random variables. It is worth mentioning the powerful tools which were created to study the fluctuations of partial sums of independent or dependent (in a sense) random summands. Namely, the moment and maximal inequalities play an important role here as well as the study of subsequences. We prove an analog of Etemadi's theorem for associated and similar random sequences. The main attention is paid here to the extensions, due to Vronski and Mikusheva, of the famous Baum–Katz theorem, concerning the convergence rate in the LLN, to positively and negatively associated fields (in the last case the necessary and sufficient conditions are provided to have a given rate of convergence to the mean). We also tackle the problem of establishing the almost sure CLT, originating from the papers by Brosamler, Schatte and recently developed by Ibragimov and Lifshits. It is shown that the theorems proved have a nice application to the limit properties of order statistics.

1 Strong law of large numbers

LLN established by Bernoulli in 1713 was the origin of the modern Probability Theory based now on the solid axiomatic foundation proposed by Kolmogorov in 1933. There are a number of brilliant results concerning the law of large numbers in weak and strong forms (usual abbreviations are LLN and SLLN). Among them one can mention, e.g., the Kolmogorov theorem for i.i.d. random variables, the results by Khintchine, Feller, Birkhoff, Prohorov, Petrov, Martikainen, Gut and many other researchers. The general trend is to extend the classical results by analysis of dependent summands, invoking random fields and vector-valued random variables. In this regard we refer to papers by Etemadi [162, 163], Gaposhkin [173], Klesov and Fazekas [165, 236], Volodin [403]. In Section 1 we consider a sequence of random variables $X = (X_n)_{n \in \mathbb{N}}$ and examine the sums $S_n = X_1 + \ldots + X_n$, $n \in \mathbb{N}$.

1°. The application of the Etemadi method.

Theorem 1.1. *Let $X = (X_n)_{n \in \mathbb{N}}$ be a sequence of square-integrable random variables such that*

$$\sup_{n \in \mathbb{N}} \mathsf{E}|X_n| < \infty. \tag{1.1}$$

(a) *If the sequence is* **PA**, *assume in addition that*

$$\sum_{n=1}^{\infty} n^{-2} cov(X_n, S_n) < \infty. \tag{1.2}$$

(b) *If the sequence is either* **NA** *or* (BL, θ)-*dependent, assume that*

$$\sum_{n \in \mathbb{N}} n^{-2} \mathsf{Var} X_n < \infty. \tag{1.3}$$

Then

$$\frac{S_n - \mathsf{E}S_n}{n} \to 0 \quad \text{a.s., } n \to \infty. \tag{1.4}$$

Proof. We treat the mentioned three types of dependence on equal foot by adapting the idea from Etemadi's proof of the SLLN for pairwise independent identically distributed random variables, see [162, 163]. Provisionally we require that $X_j \geq 0$, $j \in \mathbb{N}$. Having fixed some $\alpha > 1$, at first we will prove that

$$\frac{S_{[\alpha^k]} - \mathsf{E}S_{[\alpha^k]}}{[\alpha^k]} \to 0 \quad \text{a.s., } k \to \infty. \tag{1.5}$$

For any $\varepsilon > 0$ and all $k \in \mathbb{N}$, by the Chebyshev inequality

$$\mathsf{P}\left(\frac{S_{[\alpha^k]} - \mathsf{E}S_{[\alpha^k]}}{[\alpha^k]} > \varepsilon\right) \leq \frac{\mathsf{Var}S_{[\alpha^k]}}{\varepsilon^2 [\alpha^k]^2}.$$

Therefore,

$$\sum_{k=1}^{\infty} \mathsf{P}\left(\frac{S_{[\alpha^k]} - \mathsf{E}S_{[\alpha^k]}}{[\alpha^k]} > \varepsilon\right) \leq \varepsilon^{-2} \sum_{k=1}^{\infty} [\alpha^k]^{-2} \mathsf{Var}S_{[\alpha^k]}. \tag{1.6}$$

Set $L(j) = \inf\{k : [\alpha^k] \geq j\}$. In the case (a), due to the positive association,

$$\mathsf{Var}S_n = \sum_{j=1}^{n} \mathsf{Var}X_j + 2\sum_{j=1}^{n} cov(X_j, S_{j-1}) \leq 2\sum_{j=1}^{n} cov(X_j, S_j), \ n \in \mathbb{N}.$$

Using (1.2), (1.6) and changing the order of summation leads to the observation

$$\sum_{k=1}^{\infty} [\alpha^k]^{-2} \sum_{j=1}^{[\alpha^k]} cov(X_j, S_j) = \sum_{j=1}^{\infty} cov(X_j, S_j) \sum_{k=L(j)}^{\infty} [\alpha^k]^{-2}$$

$$\leq 4\sum_{j=1}^{\infty} cov(X_j, S_j) \sum_{k=L(j)}^{\infty} \alpha^{-2k} = 4(1 - \alpha^{-2})^{-1} \sum_{j=1}^{\infty} cov(X_j, S_j) \alpha^{-2L(j)} < \infty,$$

because $\alpha^k \leq 2[\alpha^k]$ and $\alpha^{L(j)} \geq j$, for all $k, j \in \mathbb{N}$. Consequently, in view of (1.2) the series in the right-hand side of (1.6) converges. Hence by the Borel–Cantelli lemma relation (1.5) follows, as $\varepsilon > 0$ was arbitrary.

In the case (b), if the sequence is **NA**, just notice that $\mathrm{Var} S_n \leq \sum_{j=1}^{n} \mathrm{Var} X_j$. If the sequence is (BL, θ)-dependent, apply Lemma 2.1.20 yielding the bound

$$\mathrm{Var} S_n \leq \sum_{j=1}^{n} \mathrm{Var} X_j + n\theta_1.$$

Then the further argument leading to (1.5) is a similar change of summation order.

Now, for any $n \in \mathbb{N}$ large enough, we have, as all random variables X_j are nonnegative,

$$\frac{S_{[\alpha^{L(n)-1}]} - \mathsf{E}S_{[\alpha^{L(n)-1}]}}{[\alpha^{L(n)}]} - \frac{\mathsf{E}S_{[\alpha^{L(n)}]} - \mathsf{E}S_{[\alpha^{L(n)-1}]}}{[\alpha^{L(n)}]}$$

$$\leq \frac{S_n - \mathsf{E}S_n}{n} \leq \frac{S_{[\alpha^{L(n)}]} - \mathsf{E}S_{[\alpha^{L(n)}]}}{[\alpha^{L(n)-1}]} + \frac{\mathsf{E}S_{[\alpha^{L(n)}]} - \mathsf{E}S_{[\alpha^{L(n)-1}]}}{[\alpha^{L(n)-1}]}.$$

Moreover, for any $n \in \mathbb{N}$ we have

$$\mathsf{E}S_{[\alpha^{L(n)}]} - \mathsf{E}S_{[\alpha^{L(n)-1}]} \leq \left([\alpha^{L(n)}] - [\alpha^{L(n)-1}]\right) \sup_{n \in \mathbb{N}} \mathsf{E}X_n.$$

Therefore, according to (1.5) and since $[\alpha^{L(n)}]/[\alpha^{L(n)-1}] \to 1$ as $n \to \infty$, we see that almost surely

$$-(\alpha-1)\sup_{n \in \mathbb{N}} \mathsf{E}X_n \leq \liminf_{n \to \infty} \frac{S_n - \mathsf{E}S_n}{n} \leq \limsup_{n \to \infty} \frac{S_n - \mathsf{E}S_n}{n} \leq (\alpha-1)\sup_{n \in \mathbb{N}} \mathsf{E}X_n. \quad (1.7)$$

As $\alpha > 1$ could be chosen arbitrarily close to 1, relation (1.7) entails (1.4). So the Theorem is proved under additional assumption of nonnegativity of $X_j, j \in \mathbb{N}$.

For general $(X_n)_{n \in \mathbb{N}}$ let $S_n^{(+)} = \sum_{j=1}^{n} X_j^+$ and $S_n^{(-)} = \sum_{j=1}^{n} X_j^-$, $n \in \mathbb{N}$. Both sequences $(X_n^+)_{n \in \mathbb{N}}$ and $(X_n^-)_{n \in \mathbb{N}}$ have the same type of dependence as the sequence $(X_n)_{n \in \mathbb{N}}$ itself (by Theorem 1.1.8, (d) and Lemma 1.5.16). If $(X_n)_{n \in \mathbb{N}}$ is **PA** and (1.2) is true then (1.2) holds also for $(X_n^+)_{n \in \mathbb{N}}$ and $(X_n^-)_{n \in \mathbb{N}}$ (by the first assertion of Lemma 2.1.5). Obviously, $0 \leq X_n^+ \leq |X_n|$ and $0 \leq X_n^- \leq |X_n|$, $n \in \mathbb{N}$. Therefore, $\sup_{n \in \mathbb{N}} \mathsf{E}X_n^+ < \infty$ and $\sup_n \mathsf{E}X_n^- < \infty$ if (1.1) holds. Moreover, X_n^+ and X_n^- are square-integrable, as $X_n \in L^2(\Omega, \mathcal{F}, \mathsf{P})$ and

$$\mathrm{Var} X_n = \mathrm{Var} X_n^+ + \mathrm{Var} X_n^- - 2cov(X_n^+, X_n^-), \quad n \in \mathbb{N}.$$

If $(X_n)_{n \in \mathbb{N}} \in \mathbf{NA}$, then $cov(X_n^+, X_n^-) \geq 0$ and

$$\sum_{n \in \mathbb{N}} n^{-2} \mathrm{Var} X_n^+ < \infty, \quad \sum_{n \in \mathbb{N}} n^{-2} \mathrm{Var} X_n^- < \infty. \quad (1.8)$$

If $(X_n)_{n \in \mathbb{N}}$ is (BL, θ)-dependent, then

$$\mathrm{Var} X_n^+ + \mathrm{Var} X_n^- \leq \mathrm{Var} X_n + 2\theta_1, \quad n \in \mathbb{N}.$$

Consequently, in view of (1.3), one has (1.8). Thus, by the already proved part of the Theorem

$$\frac{S_n - \mathsf{E}S_n}{n} = \frac{S_n^{(+)} - \mathsf{E}S_n^{(+)}}{n} - \frac{S_n^{(-)} - \mathsf{E}S_n^{(-)}}{n} \to 0 \text{ a.s., } n \to \infty. \ \square$$

Note that (1.3) implies the classical Kolmogorov SLLN for independent summands with finite second moments under extra condition (1.1). For positively associated sequences without that property we can prove an analogous theorem established by Birkel using the Newman–Wright maximal inequality (Theorem 2.4.5).

Theorem 1.2. ([42]) *Let* $X = (X_n)_{n \in \mathbb{N}}$ *be a square-integrable* **PA** *sequence of random variables such that (1.2) holds. Then (1.4) is valid.*

Proof. Clearly it suffices to consider the case when $\mathsf{E}X_n = 0$, $n \in \mathbb{N}$. For arbitrary $\varepsilon > 0$ and all $k \in \mathbb{N}$, combining Theorem 2.4.5 and the Chebyshev inequality gives

$$\mathsf{P}\left(\max_{j:2^k < j \le 2^{k+1}} \frac{|S_j|}{j} > \varepsilon\right) \le \varepsilon^{-2} 2^{-2k} \mathsf{E}\left(\max_{j:j \le 2^{k+1}} |S_j|\right)^2 \le 2\frac{\mathsf{E}S_{2^{k+1}}^2}{\varepsilon^2 2^{2k}}.$$

Therefore,

$$\sum_{k=1}^{\infty} \mathsf{P}\left(\max_{j:2^k < j \le 2^{k+1}} \frac{|S_j|}{j} > \varepsilon\right) \le 4\varepsilon^{-2} \sum_{k=1}^{\infty} 2^{-2k} \sum_{j=1}^{2^{k+1}} cov(X_j, S_j)$$

since

$$\mathsf{E}S_n^2 = \sum_{j=1}^{n} \mathsf{E}X_j^2 + 2\sum_{j=1}^{n} cov(X_j, S_{j-1}) \le 2\sum_{j=1}^{n} cov(X_j, S_j),$$

due to positive association. The proof is finished by changing the order of summation as in the proof of Theorem 1.1. $\square$

2°. The study of stationary sequences. In the strictly stationary case, the conditions of Theorems 1.1 and 1.2 can be relaxed as was shown by Newman for **PA** or **NA** systems.

Theorem 1.3. ([308]) *Let* $X = (X_n)_{n \in \mathbb{N}}$ *be a square-integrable, strictly stationary sequence of random variables which is either* **PA**, **NA**, *or* (BL, θ)*-dependent. In the* **PA** *case suppose also that* $n^{-1}cov(X_n, S_n) \to 0$ *as* $n \to \infty$. *Then (1.4) is valid.*

Proof. We may assume without loss of generality that $\mathsf{E}X_j = 0$, $j \in \mathbb{N}$. By the Birkhoff–Khintchine ergodic theorem for stationary sequences with finite expectation (see, e.g., [383, Ch. V, §3.3]) there exists a random variable ξ such that $S_n/n \to \xi$ a.s., $n \to \infty$.

If the sequence is **NA** or (BL, θ)-dependent, then $\mathsf{E}S_n^2 \le n(\mathsf{E}X_1^2 + 2\theta_1)$ in view of Lemma 2.1.20. Thus, in these cases $S_n/n \to 0$ in $L^2(\Omega, \mathcal{F}, \mathsf{P})$, hence in probability. As the limit in probability is unique we conclude that $\xi = 0$ a.s.

To make the same conclusion in **PA** case note that for any $n \in \mathbb{N}$, due to stationarity and positive association,

$$\frac{\mathsf{E}S_n^2}{n^2} = \frac{1}{n^2} \sum_{j=1}^{n} cov(X_j, S_n) = \frac{1}{n^2} \sum_{j=1}^{n} \left(cov(X_j, S_j) + cov(X_j, S_n - S_j) \right)$$

$$\leq \frac{1}{n^2} \sum_{j=1}^{n} \left(cov(X_j, S_j) + cov(X_{n-j+1}, S_{n-j+1}) \right) \leq \frac{2}{n} \sum_{j=1}^{n} \frac{cov(X_j, S_j)}{j}.$$

Evidently, if $(a_n)_{n \in \mathbb{N}}$ is a sequence of numbers having limit zero, the Cesaro averages $n^{-1} \sum_{j=1}^{n} a_j$ converge to zero as $n \to \infty$. Consequently $S_n/n \to 0$ in probability as $n \to \infty$ which means that $\xi = 0$ a.s. $\square$

2 Rate of convergence in the LLN

In this section some analogs of the Baum–Katz theorem are studied. Let $X_1, X_2, \ldots$ be i.i.d. random variables such that $\mathsf{E}X_1 = 0$, and suppose that $p \geq 1$. Then the famous Baum–Katz theorem (see, e.g., [326, Ch. VI]) states that the following three conditions are mutually equivalent:

1) $\mathsf{E}|X_1|^p < \infty$,

2) for any $\varepsilon > 0$ and $\alpha > 1/2$ such that $\alpha p \geq 1$,

$$\sum_{n=1}^{\infty} n^{\alpha p - 2} \mathsf{P}(|S_n| \geq \varepsilon n^{\alpha}) < \infty, \tag{2.1}$$

3) for any $\varepsilon > 0$ and $\alpha > 1/2$ such that $\alpha p \geq 1$,

$$\sum_{n=1}^{\infty} n^{\alpha p - 2} \mathsf{P}\left(\sup_{j \geq n} \frac{|S_j|}{j^{\alpha}} \geq \varepsilon \right) < \infty.$$

We mention in passing that for independent multiindexed summands further generalizations of this theorem were established recently by Dilman [135].

1°. The results for associated families. As in Chapter 2, we set $M_n = \max\{|S_1|, \ldots, |S_n|\}$, $n \in \mathbb{N}$. The following result is due to Vronski.

Theorem 2.1. ([404]) *Let a centered strictly stationary sequence of random variables $X = (X_n)_{n \in \mathbb{Z}} \in \mathbf{A}$. Assume that its Cox–Grimmett coefficients $(u_r)_{r \in \mathbb{N}}$ are finite and $u_r \to 0$ as $r \to \infty$. Suppose that for some $p \geq 2, \delta > 0, \alpha \in (1/2, 1]$ and even $k > (p\alpha - 1)/(\alpha - 1/2)$ one has*

$$\mathsf{E}|X_0|^{p+\delta} < \infty, \tag{2.2}$$

$$\sum_{r=1}^{\infty} u_r^{\delta/(k-2+\delta)} r^{k-2} < \infty. \tag{2.3}$$

Then, for any $\varepsilon > 0$,

$$\sum_{n=1}^{\infty} n^{\alpha p - 2} \mathsf{P}(M_n \geq \varepsilon n^{\alpha}) < \infty.$$

Remark 2.2. Thus for associated random variables, assuming (2.2) instead of condition 1) above and specifying the decrease rate of u_r by means of (2.3), one can obtain (for certain α) a counterpart of relation (2.1) for maxima of absolute values of partial sums.

Proof. If $\mathsf{E}X_0^2 = 0$ then $M_n = 0$ a.s. for any $n \in \mathbb{N}$, and the assertion is trivially true. Thus in what follows we suppose that $\mathsf{E}X_0^2 > 0$. For a positive a and $j \in \mathbb{N}$ we will write $X_j(a) = H_a(X_j) - \mathsf{E}H_a(X_j)$ where H_a are the truncation functions defined in (2.1.5). We also denote by $S_n(a)$ and $M_n(a)$ the sums and maxima of their absolute values respectively. Clearly

$$M_n \leq \max_{j=1,\dots,n} \left| S_j - \sum_{q=1}^{j} \mathsf{E}H_a(X_q) \right| + \max_{j=1,\dots,n} \left| \sum_{q=1}^{j} \mathsf{E}H_a(X_q) \right|. \tag{2.4}$$

Since X_q's are centered, we have

$$|\mathsf{E}H_a(X_q)| = |\mathsf{E}(X_q - H_a(X_q))| \leq a^{-1}\mathsf{E}X_q^2, \quad q = 1,\dots,n.$$

Thus the second summand in (2.4) is no greater than $na^{-1}\mathsf{E}X_1^2$. On the event

$$A = A_{a,n} := \{\omega : X_q = H_a(X_q), q = 1,\dots,n\}$$

one has

$$S_j - \sum_{q=1}^{j} \mathsf{E}H_a(X_q) = S_j(a),$$

and the first summand in the right-hand side of (2.4) coincides with $M_n(a)$. Further, by (2.4),

$$\mathsf{P}(M_n \geq 2\varepsilon a) \leq \mathsf{P}\left(A \cap \{M_n(a) + na^{-1}\mathsf{E}X_1^2 \geq 2\varepsilon a\}\right) + \mathsf{P}(A^c).$$

Note that $\mathsf{P}(A^c) \leq n\mathsf{P}(|X_1| \geq a)$. Consequently, for all n large enough

$$\mathsf{P}(M_n \geq 2\varepsilon a) \leq \mathsf{P}(M_n(a) \geq \varepsilon a) + n\mathsf{P}(|X_1| \geq a). \tag{2.5}$$

Applying Theorem 2.4.6 we have, for any $k \in \mathbb{N}$,

$$\mathsf{P}(M_n(a) \geq \varepsilon a) \leq 2\mathsf{P}(|S_n(a)| \geq a\varepsilon/2) \leq \frac{2^{k+1}}{(\varepsilon a)^k}\|S_n(a)\|_k^k.$$

On the other hand, $\mathsf{E}(X_1(a))^2/\mathsf{E}X_1^2 > 1/2$ for all a large enough. So, in view of Theorem 2.3.1 one has

$$\|S_n(a)\|_k^k \leq C_1 \max\left\{ n\|X_1(a)\|_{k+\delta}^k, n^{k/2}\|X_1(a)\|_{2+\delta}^k \right\}. \tag{2.6}$$

Here $C_1 = 2C$ and $C = C(k, \delta, ((\mathsf{E}X_1^2)^{-1}u_r)_{r\in\mathbb{N}})$ appeared in the formulation of Theorem 2.3.1. We took into account that for large enough a, by Theorem 1.5.3,

$$\frac{u_a(r)}{\mathsf{E}X_1(a)^2} \leq 2\frac{u_r}{\mathsf{E}X_1^2}$$

where $u_a(r) = 2\sum_{j>r} cov(X_0(a), X_j(a))$. Take $a = n^\alpha$, $\alpha > 0$, and estimate the quantity $\sum_{n=1}^{\infty} n^{\alpha p - 2 - \alpha k}\|S_n(n^\alpha)\|_k^k$, with k satisfying the condition in the formulation of Theorem. According to (2.6) this sum does not exceed

$$C_1 \sum_{n=1}^{\infty} n^{\alpha p - 2 - \alpha k} \left(n\|X_1(n^\alpha)\|_{k+\delta}^k + n^{k/2}\|X_1(n^\alpha)\|_{2+\delta}^k \right) =: I_1 + I_2,$$

the separation into I_1 and I_2 being due to summands in brackets.

Note that $|X_1(n^\alpha)| \leq |X_1| + |\mathsf{E}H_a(X_1)| \leq |X_1| + 1$ for all n large enough as $\mathsf{E}X_1 = 0$. Thus, for any $b \in [1, p + \delta]$, one has $\|X_1(n^\alpha)\|_b \leq \|X_1\|_b + 1$. Therefore,

$$I_2 = O(\sum_{n=1}^{\infty} n^{\alpha p - 2 - \alpha k + k/2}) < \infty$$

due to the choice of k.

In estimating I_1 it is convenient to consider two cases.

Case 1: $\mathsf{E}|X_1|^{k+\delta} < \infty$. Note that $(p\alpha - 1)/(\alpha - 1/2) \geq p$ as $p \geq 2$. Hence $k > p$, therefore $\alpha p - 2 - \alpha k + 1 < -1$ and $I_1 = O\left(\sum_{n=1}^{\infty} n^{\alpha p - 2 - \alpha k + 1}\right) < \infty$.

Case 2: $\mathsf{E}|X_1|^{k+\delta} = \infty$. Then $\|X_1(a)\|_{k+\delta}^k \geq 1$ for all a large enough (i.e. for all n large enough as $a = n^\alpha$, $\alpha > 0$). Indeed, if there is a sequence $(a_m)_{m\in\mathbb{N}}$ such that $\|X_1(a_m)\|_{k+\delta}^k < 1$, $m \in \mathbb{N}$, and $a_m \to \infty$ as $m \to \infty$, then the Fatou lemma implies that $\|X_1\|_{k+\delta}^k \leq 1$.

Obviously $\|X_1(a)\|_{k+\delta}^k \leq \|X_1(a)\|_{k+\delta}^{k+\delta}$. Set

$$p_j := \mathsf{P}(j - 1 \leq |X_1| < j), \quad j \in \mathbb{N},$$

then

$$I_1 = O\left(\sum_{n=1}^{\infty} n^{\alpha p - 1 - \alpha k}\left(\sum_{j=1}^{[n^\alpha]+1} j^{k+\delta} p_j + n^{\alpha(k+\delta)}\sum_{j>n^\alpha} p_j\right)\right)$$

$$= O\left(\sum_{j=1}^{\infty} j^{k+\delta} p_j \sum_{n \geq (j-1)^{1/\alpha}} n^{\alpha p - 1 - \alpha k} + \sum_{j=1}^{\infty} p_j \sum_{n \leq j^{1/\alpha}} n^{\alpha(p+\delta) - 1}\right) = O(\sum_{j=1}^{\infty} p_j j^{p+\delta}) < \infty$$

due to (2.2). In view of (2.5) it remains to show that

$$S := \sum_{n=1}^{\infty} n^{p\alpha - 1}\mathsf{P}(|X_1| \geq n^\alpha) < \infty.$$

It is clear that

$$S \leq \sum_{n=1}^{\infty} n^{p\alpha - 1}\sum_{j \geq n^\alpha} p_j \leq \sum_{j=1}^{\infty} p_j \sum_{n \leq j^{1/\alpha}} n^{p\alpha - 1} = O(\sum_{j=1}^{\infty} j^p p_j) < \infty$$

since (2.2) holds. The proof is complete. $\square$

To formulate an analog of the above result for random fields let us recall that, for a block U in $\mathbb{Z}^d$, the maximum $M(U)$ is defined in (2.1.1).

Theorem 2.3. ([405]) *Let $X = \{X_j, j \in \mathbb{Z}^d\}$ be a field of associated identically distributed random variables such that $\mathsf{E}X_0 = 0$ and $\mathsf{E}X_0^2 < \infty$. Assume that its Cox–Grimmett coefficients $(u_r)_{r \in \mathbb{N}}$ are finite and $u_r \to 0, r \to \infty$. Suppose that there exist $p \geq 2, \delta > 0, \alpha \in (1/2, 1]$ and even $k > (p\alpha - 1)/(\alpha - 1/2)$ such that $\mathsf{E}|X_0|^{p+\delta} < \infty$ and*

$$\sum_{r=1}^{\infty} u_r^{\delta/(k-2+\delta)} r^{d(k-1)-1} < \infty.$$

Then, for any $\varepsilon > 0$ and any sequence of blocks U_n in $\mathbb{Z}^d$ such that $U_n \to \infty$ and $|U_n| \nearrow \infty$ as $n \to \infty$, one has

$$\sum_{n=1}^{\infty} |U_n|^{\alpha p - 2} \mathsf{P}(M(U_n) \geq \varepsilon|U_n|^{\alpha}) < \infty.$$

Proof is very close to that of Theorem 2.1. Namely, for $n \in \mathbb{N}$ consider the random field $\{X_j(v_n), j \in \mathbb{Z}^d\}$, where $v_n = |U_n|$, and define the corresponding sums and maxima $S(U_n, v_n)$ and $M(U_n, v_n)$ in the same way as before. Then, for any $\varepsilon > 0$ and all n large enough,

$$\mathsf{P}(M(U_n) \geq 2\varepsilon v_n^{\alpha}) \leq \mathsf{P}(M(U_n, v_n) \geq \varepsilon v_n^{\alpha}) + v_n \mathsf{P}(|X_0| \geq v_n^{\alpha}).$$

Theorems 2.4.6 and 2.3.1 imply that

$$\mathsf{P}(M(U_n, v_n) \geq \varepsilon v_n^{\alpha}) \geq 2\mathsf{P}(|S(U_n, v_n)| \geq \varepsilon v_n^{\alpha}/2) \leq 2^{k+1} \frac{\|S(U_n, v_n)\|_k^k}{\varepsilon^k v_n^{k\alpha}}$$

$$\leq 2^{k+1} \frac{2C}{\varepsilon^k v_n^{k\alpha}} \max\left\{ v_n \|X_0(v_n)\|_{k+\delta}^k, v_n^{k/2} \|X_0(v_n)\|_{2+\delta}^k \right\}$$

if n is so large that $\mathsf{E}X_0(v_n)^2 / \mathsf{E}X_0^2 \geq 1/2$.

To complete the proof we consider two cases, as in the proof of Theorem 2.1. $\square$

2°. The estimates involving cardinalities of auxiliary sets. Introduce

$$R_m = \#\{n = (n_1, \ldots, n_d) \in \mathbb{N}^d : \langle n \rangle = n_1 \ldots n_d \leq m\}, \quad m \in \mathbb{N}.$$

Lemma 2.4. *R_m is estimated by way of*

$$A_1 m \mathrm{Log}^{d-1} m \leq R_m \leq A_2 m \mathrm{Log}^{d-1} m, \quad m \in \mathbb{N},$$

for some positive A_1, A_2 depending only on d.

Proof. Consider the sets

$$K_t(d) = \{x \in \mathbb{R}^d : x_1 \geq 1, \ldots, x_d \geq 1, \ x_1 x_2 \ldots x_d \leq t\}, \quad d \in \mathbb{N}; \ t > 1.$$

Then the respective Lebesgue measures are

$$\mathrm{mes}(K_t(1)) = t, \tag{2.7}$$

and for $d > 1$, by the Fubini theorem,

$$\mathrm{mes}(K_t(d)) = \int_{K_t(d)} 1 dx_1 \ldots dx_d = \int_1^t \mathrm{mes}(K_{t/x_1}(d-1)) dx_1. \tag{2.8}$$

Now we claim that

$$\operatorname{mes}(K_t(d)) = \frac{1}{(d-1)!} t \log^{d-1} t, \quad d \in \mathbb{N}, \quad t > 1. \tag{2.9}$$

This is verified by induction on d. Indeed, for $d = 1$ we return to (2.7), which is obvious. If $d > 1$ and (2.9) is true for the dimension $d - 1$, then, by (2.8) and the induction hypothesis,

$$\operatorname{mes}(K_t(d)) = \int_1^t \operatorname{mes}(K_{t/x}(d-1))dx = \frac{1}{(d-2)!} \int_1^t \frac{t}{x} \log^{d-2}\left(\frac{t}{x}\right) dx$$

$$= -\frac{t}{(d-2)!} \int_1^t \log^{d-2}\left(\frac{t}{x}\right) d\left(\log \frac{t}{x}\right) = \frac{t}{(d-1)!} \log^{d-1} t,$$

as required.

Now, for fixed $d \in \mathbb{N}$ and $m \in \mathbb{N}$, let L_m be the union of cubes of the kind $\{(j, j+1]; j \in K_m(d)\}$. Then we have

$$R_m = \operatorname{mes}(L_m) \geq \operatorname{mes}(K_m(d)). \tag{2.10}$$

On the other hand, if $j \in \mathbb{N}^d$ then

$$\langle j + 1 \rangle = (j_1 + 1) \dots (j_d + 1) \leq 2^d j_1 \dots j_d = 2^d \langle j \rangle,$$

and therefore, for $j \in \mathbb{N}^d \cap K_m(d)$,

$$R_m = \operatorname{mes}(L_m) \leq K_{2^d m}(d) = \frac{2^d m \log^{d-1}(2^d m)}{(d-1)!} \sim \frac{2^d}{(d-1)!} m \log^{d-1} m \tag{2.11}$$

as $m \to \infty$. The Lemma follows from (2.10) and (2.11). $\square$

Set $d_1 = 1$ and $d_m = R_m - R_{m-1}$ for $m > 1, m \in \mathbb{N}$, i.e. d_m is the number of points $n \in \mathbb{N}^d$ with $\langle n \rangle = m$.

Lemma 2.5. *For each $d \in \mathbb{N}$ and any $\nu \in \mathbb{R}$ there exist some $B_i = B_i(d, \nu) > 0$, $i = 1, 2$, such that for all $j \in \mathbb{N}$ the following claims hold:*

1) *if* $\nu > -1$ *then* $B_1 j^{\nu+1} \operatorname{Log}^{d-1} j \leq \sum_{k=1}^{j} k^\nu d_k \leq B_2 j^{\nu+1} \operatorname{Log}^{d-1} j;$ (2.12)

2) *if* $\nu = -1$ *then* $B_1 \operatorname{Log}^d j \leq \sum_{k=1}^{j} k^\nu d_k \leq B_2 \operatorname{Log}^d j;$ (2.13)

3) *if* $\nu < -1$ *then* $B_1 \leq \sum_{k=1}^{j} k^\nu d_k \leq B_2.$ (2.14)

Thus $\sum_{k=1}^{j} k^\nu d_k < \infty$ if and only if $\nu < -1$.

Proof. 1) Let $\nu > -1$. Consider separately the cases $\nu \geq 0$ and $-1 < \nu < 0$.

Case 1: $\nu \geq 0$. The function k^ν, $k \in \mathbb{N}$, is nondecreasing and by virtue of Lemma 2.4 we have

$$\sum_{k=1}^{j} k^\nu d_k \leq j^\nu \sum_{j=1}^{k} d_k = j^\nu R_j \leq A_2 j^{\nu+1} \mathrm{Log}^{d-1} j,$$

which is the upper bound in (2.12).

To establish the lower bound let us choose $c \in (0,1)$ and take $q_j = [cj]$, $j \in \mathbb{N}$. Then, for all j large enough, $q_j < j$ and we have

$$\sum_{k=1}^{j} k^\nu d_k \geq \sum_{k=q_j}^{j} k^\nu d_k \geq q_j^\nu \sum_{k=q_j}^{j} d_k = q_j^\nu (R_j - R_{q_j}). \tag{2.15}$$

One can pick c in such a way that $b := A_1 - cA_2 > 0$ (A_1 and A_2 are the same as in Lemma 2.4). Then

$$R_j - R_{q_j} \geq A_1 j \mathrm{Log}^{d-1} j - A_2 q_j \mathrm{Log}^{d-1} j \geq b j \mathrm{Log}^{d-1} j. \tag{2.16}$$

Using (2.15), (2.16) and the relation $[cj] \sim cj$ as $j \to \infty$ we come to the lower bound in (2.12).

Case 2: $-1 < \nu < 0$. Now the function k^ν, $k \in \mathbb{N}$, is decreasing. Set $R_0 = 0$. For $j \in \mathbb{N}$, the Abel formula (i.e. the discrete analog of integration by parts) yields

$$\sum_{k=1}^{j} k^\nu d_k = \sum_{k=1}^{j} k^\nu (R_k - R_{k-1}) = \sum_{k=1}^{j} R_k (k^\nu - (k+1)^\nu) + R_j (j+1)^\nu. \tag{2.17}$$

Due to Lemma 2.5 one has

$$R_j (j+1)^\nu \leq A_2 j^{\nu+1} \mathrm{Log}^{d-1} j. \tag{2.18}$$

Note that $k^\nu - (k+1)^\nu \leq -\nu k^{\nu-1} \leq k^{\nu-1}$, $k \in \mathbb{N}$. Thus,

$$\sum_{k=1}^{j} R_k (k^\nu - (k+1)^\nu) \leq \sum_{k=1}^{j} R_k k^{\nu-1} \leq A_2 \sum_{k=1}^{j} k^\nu \mathrm{Log}^{d-1} k$$

$$\leq A_2 \mathrm{Log}^{d-1} j \sum_{k=1}^{j} k^\nu \leq A_2 \mathrm{Log}^{d-1} j \left(1 + \int_1^j x^\nu \, dx \right) \leq \frac{A_2}{\nu+1} j^{\nu+1} \mathrm{Log}^{d-1} j. \tag{2.19}$$

Formulas (2.17)—(2.19) lead to the upper bound in (2.12).

For any $c \in (0,1)$ and $q_j = [cj]$, in view of (2.17), one has for all j large enough

$$\sum_{k=1}^{j} k^\nu d_k \geq \sum_{k=1}^{j} R_k (k^\nu - (k+1)^\nu)$$

$$\geq \min_{q_j \leq k \leq j} R_k \sum_{k=q_j}^{j} (k^\nu - (k+1)^\nu) \geq \frac{cA_1}{2} j \mathrm{Log}^{d-1} j (q_j^\nu - j^\nu).$$

Using appropriate constant c we come to the lower bound in (2.12) in a similar way as in Case 1.

2) Let $\nu = -1$. Then applying (2.17), (2.18) and the equality

$$\int_1^j x^{-1} dx = \log j, \quad j \in \mathbb{N},$$

in (2.19) we obtain the upper bound in (2.13). In view of Lemma 2.4 and due to (2.17) one has

$$\sum_{k=1}^j k^{-1} d_k \geq \sum_{k=1}^j R_k(k^{-1} - (k+1)^{-1}) = \sum_{k=1}^j \frac{R_k}{k(k+1)} \geq \frac{A_1}{2} \sum_{k=1}^j \frac{\mathrm{Log}^{d-1} k}{k}$$

$$\geq \frac{A_1}{2}\left(1 + \int_1^j x^{-1} \log^{d-1} x\, dx\right) = \frac{A_1}{2}\left(1 + \frac{\mathrm{Log}^d j}{d}\right).$$

Thus the lower bound in (2.13) is established.

3) Let $\nu < -1$. Then (2.17) and Lemma 2.4 yield

$$0 < \sum_{k=1}^j k^\nu d_k \leq A_2 \sum_{k=1}^j k^\nu \mathrm{Log}^{d-1} k + A_2 j^{\nu+1} \mathrm{Log}^{d-1} j.$$

It remains to observe that, for $\nu < -1$,

$$j^{\nu+1} \mathrm{Log}^{d-1} j \to 0, \; j \to \infty \text{ and } \sum_{k=1}^\infty k^\nu \mathrm{Log}^{d-1} j < \infty. \quad \square$$

3°. The results for NA systems. For negatively associated random fields one may attempt to obtain conditions necessary and sufficient for the desired rate of convergence. Moreover, it appears that the decrease rate of covariance function does not play a role here. In other words, this is an illustration of the idea that **NA** random systems behave almost as independent ones.

For $k \in \mathbb{N}^d$ we set

$$S_k = \sum_{j \in \mathbb{N}^d : j \leq k} X_j \text{ and } M_k = \max_{j \in \mathbb{N}^d : j \leq k} |S_j|. \tag{2.20}$$

All the sums in multiindices are taken only over $j \in \mathbb{N}^d$, unless otherwise stated. Throughout this Section, we write $\langle n \rangle = n_1 \dots n_d$ for $n \in \mathbb{N}^d$. As usual, set $\mathrm{Log}\,(x) = \log(x \vee e)$, $x \in \mathbb{R}$.

Theorem 2.6. ([299]) *Let $\{X_n, n \in \mathbb{Z}^d\}$ be a negatively associated random field consisting of identically distributed random variables. Then, for any $p \in \mathbb{N}$ and $\alpha \in (1/2, 1]$ such that $p\alpha > 1$, the following three conditions are equivalent:*

$$\mathsf{E}|X_0|^p (\mathrm{Log}\,|X_0|)^{d-1} < \infty \text{ and } \mathsf{E}X_0 = 0, \tag{2.21}$$

$$\sum_{n \in \mathbb{N}^d} \langle n \rangle^{p\alpha-2} \mathsf{P}(M_n \geq \varepsilon \langle n \rangle^\alpha) < \infty \text{ for any } \varepsilon > 0, \tag{2.22}$$

$$\sum_{l=1}^\infty l^{\alpha p - 2} \mathsf{P}\left(\sup_{k \in \mathbb{N}^d : \langle k \rangle \geq l} \frac{|S_k|}{\langle k \rangle^\alpha} \geq \varepsilon\right) < \infty \text{ for any } \varepsilon > 0. \tag{2.23}$$

Proof. Before proving the Theorem we dwell on a series of lemmas.

Lemma 2.7. *Let ξ be a random variable and $s > 0$ be some number. Then*

$$\sum_{n \in \mathbb{N}^d} \langle n \rangle^{s-1} \mathsf{P}(|\xi| \geq \langle n \rangle) < \infty$$

if and only if $\mathsf{E}|\xi|^s (\mathrm{Log}\,|\xi|)^{d-1} < \infty.$

Proof. Note that

$$\sum_{n \in \mathbb{N}^d} \langle n \rangle^{s-1} \mathsf{P}(|\xi| \geq \langle n \rangle) = \sum_{k=1}^{\infty} \sum_{n \in \mathbb{N}^d,\,\langle n \rangle = k} k^{s-1} \mathsf{P}(|\xi| \geq k)$$

$$= \sum_{k=1}^{\infty} d_k k^{s-1} \mathsf{P}(|\xi| \geq k) = \sum_{k=1}^{\infty} d_k k^{s-1} \sum_{j=k}^{\infty} \mathsf{P}(j \leq |\xi| < j+1)$$

$$= \sum_{j=1}^{\infty} \mathsf{P}(j \leq |\xi| < j+1) \sum_{k=1}^{j} d_k k^{s-1}. \tag{2.24}$$

The finiteness of $\mathsf{E}|\xi|^s (\mathrm{Log}\,|\xi|)^{d-1}$ is equivalent to convergence of the series

$$\sum_{j=1}^{\infty} \mathsf{P}(j \leq |\xi| < \jmath + 1) j^s \mathrm{Log}^{\,d-1} j.$$

Therefore, in view of (2.24) and by Lemma 2.5 with $\nu = s - 1 > -1$ we come to the statement of the Lemma. $\square$

Lemma 2.8. *Let $\{X_j, j \in \mathbb{Z}^d\}$ be a random field of negatively associated, identically distributed, centered random variables. Suppose that $\mathsf{E}|X_0|^\gamma < \infty$ for some $\gamma > 2$. Then*

$$\mathsf{E}M_n^\gamma \leq \left(a_\gamma \langle n \rangle \mathsf{E}|X_0|^\gamma + b_\gamma (\mathsf{E}X_0^2)^{\gamma/2} \langle n \rangle^{\gamma/2} \right) ([\mathrm{Log}\,n_1] \ldots [\mathrm{Log}\,n_d])^\gamma, \quad n \in \mathbb{N}^d,$$

where a_γ, b_γ depend only on γ and d.

Proof. On account of the Rosenthal inequality (Theorem 2.3.3) we have

$$\mathsf{E}|S_n|^\gamma \leq 2^{\gamma^2} (\langle n \rangle \mathsf{E}|X_0|^\gamma + (\mathsf{E}X_0^2)^{\gamma/2} \langle n \rangle^{\gamma/2}).$$

This estimate implies the Lemma according to Corollary A.14, see Appendix A.5.
$\square$

Further in this Section we tacitly consider only $n \in \mathbb{N}^d$ without further notice.

Lemma 2.9. *Suppose that X is a random field satisfying all the conditions of Theorem 2.6. Then, for any $\varepsilon > 0$ and all α, p such that $p \in \mathbb{N}$, $\alpha \in (1/2, 1]$ and $p\alpha > 1$, one has the bound*

$$\sum_{r=1}^{\infty} r^{\alpha p - 2} P_{\alpha, \varepsilon}(r) \leq \sum_{r=1}^{\infty} \frac{2^{(r+1)(\alpha p - 1)}}{\alpha p - 1} \sum_{n:\langle n \rangle = 2^{r+d}} \mathsf{P}\left(\sup_{k:k \leq n} |S_k| \geq \langle n \rangle^\alpha \varepsilon \right)$$

where $P_{\alpha, \varepsilon}(r) = \mathsf{P}\left(\sup_{n:\langle n \rangle \geq r} \frac{|S_n|}{\langle n \rangle^\alpha} \geq 2^{\alpha(1+d)} \varepsilon \right).$

Proof. It is easily seen that

$$\sum_{r=1}^{\infty} r^{\alpha p-2} P_{\alpha,\varepsilon}(r) \leq \sum_{m=0}^{\infty} \sum_{r=2^m}^{2^{m+1}-1} r^{\alpha p-2} P_{\alpha,\varepsilon}(r) \leq \sum_{m=0}^{\infty} \frac{2^{(m+1)(\alpha p-1)}}{\alpha p - 1} P_{\alpha,\varepsilon}(2^m)$$

$$\leq \sum_{m=0}^{\infty} \frac{2^{(m+1)(\alpha p-1)}}{\alpha p - 1} \sum_{r=m+1}^{\infty} \mathsf{P}\left(\max_{n:2^{r-1}\leq\langle n\rangle<2^r} \frac{|S_n|}{\langle n\rangle^\alpha} \geq 2^{\alpha(1+d)}\varepsilon\right),$$

and the probability in the last expression is no greater than

$$\sum_{n:\langle n\rangle=2^{r+d}} \mathsf{P}\left(\sup_{k\leq n} |S_k| \geq \langle n\rangle^\alpha \varepsilon\right).$$

It remains to change the order of summation to obtain the Lemma. $\square$

To prove the Theorem we show that $(2.21) \Rightarrow (2.22) \Rightarrow (2.23) \Rightarrow (2.21)$. Assume that (2.21) holds. Writing $X_j = X_j^+ - X_j^-$ for any $j \in \mathbb{Z}_+^d$, we see that to prove (2.22) it will be enough to consider the case when X_j are bounded from below (and centered). Take some

$$q \in \left(\frac{\alpha p + 1}{2p}, \alpha\right) \tag{2.25}$$

and let $X_j(n^q) = X_j \wedge \langle n\rangle^q$ for $j \in \mathbb{Z}_+^d$. Define $S(n^q)_k$ and $M(n^q)_k$, $k \in \mathbb{Z}_+^d$, as in the proof of previous theorem. Let

$$A_j = \{\omega : X_j > \langle n\rangle^q \text{ and } X_t \leq \langle n\rangle^q \text{ for all } t \in \mathbb{N}^d \setminus \{q\}, t \leq n\}, \quad j \leq n.$$

Then the events $\{A_j, j \leq n\}$ are pairwise disjoint and, as $M_n \leq X_j + M(n^q)_n$ on the event A_j,

$$A_j \cap \{M_n \geq \varepsilon\langle n\rangle^\alpha\} \subset \{X_j > \varepsilon\langle n\rangle^\alpha/2\} \cup \{M(n^q)_n \geq \varepsilon\langle n\rangle^\alpha/2\}.$$

Therefore

$$\mathsf{P}(M_n \geq \varepsilon\langle n\rangle^\alpha) \leq \mathsf{P}\left(\max_{k\leq n}|S(n^q)_k| \geq \varepsilon\langle n\rangle^\alpha/2\right) + \mathsf{P}\left(X_j > \varepsilon\langle n\rangle^\alpha/2 \text{ for some } j \leq n\right)$$

$$+ \mathsf{P}\left(X_j > \langle n\rangle^q, X_t > \langle n\rangle^q \text{ for some } j \neq t \text{ with } j, t \leq n\right)$$

$$\leq \mathsf{P}\left(\max_{k\leq n}|S_k(n^q)| \geq \frac{\varepsilon\langle n\rangle^\alpha}{2}\right) + \langle n\rangle \mathsf{P}\left(|X_0| > \frac{\varepsilon\langle n\rangle^\alpha}{2}\right) + \langle n\rangle^2 \mathsf{P}\left(|X_0| > \varepsilon\langle n\rangle^q\right)^2 \tag{2.26}$$

where the last inequality is due to **NA** property. Clearly, using the fact that

$$\mathsf{E}X_0 = 0 = \mathsf{E}X_0\mathbb{I}\{|X_0| > \langle n\rangle^q\} + \mathsf{E}X_0\mathbb{I}\{|X_0| \leq \langle n\rangle^q\},$$

we have

$$\langle n\rangle^{1-\alpha}|\mathsf{E}X_0(n^q)| \leq \langle n\rangle^{1-\alpha+q}\mathsf{P}(|X_0| > \langle n\rangle^q) + \langle n\rangle^{1-\alpha}|\mathsf{E}X_0\mathbb{I}\{X_0 > \langle n\rangle^q\}|$$

$$\leq \langle n\rangle^{1-\alpha+q-pq}\mathsf{E}|X_0|^p + \langle n\rangle^{1-\alpha+q(1-p)}\mathsf{E}\left|X_0^p\mathbb{I}\{X_0 > \langle n\rangle^q\}\right| \to 0$$

as $\langle n \rangle \to \infty$. In other words, $\mathsf{E}S_k(n^q) = o(\langle n \rangle^\alpha)$, $\langle n \rangle \to \infty$. Therefore, for any $\gamma > 2$ and all n, having $\langle n \rangle$ large enough,

$$\mathsf{P}\left(\max_{k \leq n} |S_k(n^q)| \geq \varepsilon \langle n \rangle^\alpha / 2\right) \leq \mathsf{P}\left(\max_{k \leq n} |(S_k(n^q) - \mathsf{E}S_k(n^q))| \geq \varepsilon \langle n \rangle^\alpha / 4\right)$$

$$\leq C\varepsilon^{-\gamma} \langle n \rangle^{-\gamma\alpha} \left(\mathsf{E}|X_0(n^q)|^\gamma \langle n \rangle + \langle n \rangle^{\gamma/2} (\mathsf{E}X_0(n^q)^2)^{\gamma/2}\right) \prod_{i=1}^{d} (\mathrm{Log}\, n_i)^\gamma \qquad (2.27)$$

by Lemma 2.8, here $C = C_\gamma = 2^\gamma (a_\gamma \vee b_\gamma)$. The value $\gamma > p$ will be picked later.

Note that condition (2.21) entails

$$\mathsf{E}|X_0(n^q)|^\gamma \leq \langle n \rangle^{q(\gamma - p)} \mathsf{E}|X_0|^p \qquad (2.28)$$

for $\gamma > p$ and all n large enough (more precisely, for n such that $X_0 \geq -\langle n \rangle^q$ with probability one). This fact and (2.26)—(2.27) imply that, for any $\varepsilon > 0$,

$$\sum_n \langle n \rangle^{\alpha p - 2} \mathsf{P}(M_n \geq \varepsilon \langle n \rangle^\alpha) \leq \sum_n \langle n \rangle^{\alpha p - 1} \mathsf{P}(|X_0| \geq \varepsilon \langle n \rangle^\alpha / 2)$$

$$+ \sum_n \langle n \rangle^{\alpha p} \mathsf{P}(|X_0| > \langle n \rangle^q)^2 + C_1 \varepsilon^{-\gamma} \sum_n \langle n \rangle^{\alpha p - 1 - \alpha\gamma + q(\gamma - p)} \left(\prod_{i=1}^{d} \mathrm{Log}\, n_i\right)^\gamma$$

$$+ C_1 \sum_n \langle n \rangle^{\alpha p - 2 - \gamma\alpha + \gamma/2} (\mathsf{E}X_0(n^q)^2)^{\gamma/2} \left(\prod_{i=1}^{d} \mathrm{Log}\, n_i\right)^\gamma. \qquad (2.29)$$

Here C_1 depends on γ and q.

The convergence of the first series in (2.29) is equivalent to (2.21). Namely,

$$\sum_n \langle n \rangle^{\alpha p - 1} \mathsf{P}\left(|X_0| \geq \varepsilon \langle n \rangle^\alpha / 2\right) = \sum_{m=1}^{\infty} m^{\alpha p - 1} d_m \mathsf{P}\left((2|X_0|\varepsilon^{-1})^{1/\alpha} \geq m\right)$$

$$= \sum_{m=1}^{\infty} m^{\alpha p - 1} d_m \sum_{j=m}^{\infty} \mathsf{P}\left(j \leq (2|X_0|\varepsilon^{-1})^{1/\alpha} < j + 1\right)$$

$$= \sum_{j=1}^{\infty} \mathsf{P}\left(j \leq (2|X_0|\varepsilon^{-1})^{1/\alpha} < j + 1\right) \sum_{m=1}^{j} m^{\alpha p - 1} d_m.$$

In view of (2.12) the convergence of the last series is equivalent to the relation

$$\sum_{j=1}^{\infty} \mathsf{P}\left(j \leq (2|X_0|\varepsilon^{-1})^{1/\alpha} < j + 1\right) j^{\alpha p} \log_+^{d-1} j < \infty.$$

Clearly this series converges if and only if $\mathsf{E}|X_0|^p \log_+^{d-1} |X_0| < \infty$, which amounts to (2.21).

Since by (2.21) one has $\mathsf{P}(X_0 > n^q) \leq \mathsf{E}|X_0|^p n^{-qp}$, the sum of the second series in (2.29) does not exceed $\mathsf{E}|X_0|^p \sum_n \langle n \rangle^{\alpha p - 2pq} < \infty$ by our choice of q. The third and fourth series are convergent for large enough γ, provided that $\mathsf{E}X_0^2 < \infty$, i.e. if $p \geq 2$. If that is not the case, apply (2.28) with 2 instead of γ to establish that $\mathsf{E}X_0(n^q)^2 \leq \langle n \rangle^{q(2-p)}$ if $\langle n \rangle$ is large enough. Therefore, the exponent for $\langle n \rangle$ in any element of the fourth series in (2.29) is no greater than $\gamma(1/2 - \alpha + q(2-p)/2) + \alpha p$. But the choice (2.25) implies that the factor at γ in the last expression is negative. Indeed, we have

$$q(2-p) < \alpha(2-p) \leq 2\alpha - 1$$

as $p < 2$. Consequently, taking γ large enough allows us to verify the convergence of all series in (2.29). Thus one obtains (2.22).

Assume now that (2.22) holds. We will prove (2.23). Lemma 2.9 ensures that instead we may prove that

$$\sum_{r=1}^{\infty} 2^{(r+1)(\alpha p - 1)} \sum_{n:\langle n \rangle = 2^{r+d}} \mathsf{P}\left(M_n \geq \langle n \rangle^{\alpha} \varepsilon\right) < \infty \quad \text{for any } \varepsilon > 0. \tag{2.30}$$

To this end define the function $h : \mathbb{N}^d \to \mathbb{Z}^d$ as follows. For $n = (n_1, \ldots, n_d) \in \mathbb{N}^d$ we set

$$h(n) = (2^{j_1}, \ldots, 2^{j_d}) \quad \text{with } j_i = [\log_2 n_i], \quad i = 1, \ldots, d.$$

Then $\langle h(n) \rangle \leq \langle n \rangle \leq 2^d \langle h(n) \rangle$ for any $n \in \mathbb{N}^d$. This double inequality and (2.22) imply that

$$\sum_{n \in \mathbb{N}^d} \langle h(n) \rangle^{\alpha p - 2} \mathsf{P}\left(\frac{M_{h(n)}}{\langle h(n) \rangle^{\alpha}} \geq \varepsilon\right) < \infty \quad \text{for any } \varepsilon > 0. \tag{2.31}$$

Denote by $\mathbb{N}^d(2)$ the set of elements of $\mathbb{N}^d$ whose components are nonnegative powers of 2. For any $k \in \mathbb{N}^d(2)$ the total number of points $n \in \mathbb{N}^d$ such that $k = h(n)$ is $\langle k \rangle$. Therefore, by (2.31),

$$\sum_{k \in \mathbb{N}^d(2)} \langle k \rangle^{\alpha p - 1} \mathsf{P}\left(\frac{M_k}{\langle k \rangle^{\alpha}} \geq \varepsilon\right) < \infty,$$

but the last series can be rewritten as

$$\sum_{k \in \mathbb{N}^d(2)} \langle k \rangle^{\alpha p - 1} \mathsf{P}\left(\frac{M_k}{\langle k \rangle^{\alpha}} \geq \varepsilon\right) = \sum_{m=0}^{\infty} (2^m)^{(\alpha p - 1)} \sum_{k \in \mathbb{N}^d(2):\langle k \rangle = 2^m} \mathsf{P}\left(\frac{M_k}{\langle k \rangle^{\alpha}} \geq \varepsilon\right)$$

$$\geq \sum_{m=0}^{\infty} (2^{m-d+1})^{(\alpha p - 1)} \sum_{k \in \mathbb{N}^d(2):\langle k \rangle = 2^{m+d}} \mathsf{P}\left(\frac{M_k}{\langle k \rangle^{\alpha}} \geq \varepsilon\right),$$

which clearly entails (2.30).

The only implication left is that (2.21) follows from (2.23). So, let us assume that (2.23) is valid. Then, obviously,

$$\sum_{n \in \mathbb{N}^d} \langle n \rangle^{\alpha p - 2} \mathsf{P}\left(\sup_{k \in \mathbb{N}^d : k \geq n} \frac{|S_k|}{\langle k \rangle^{\alpha}} \geq \varepsilon\right) < \infty \quad \text{for any } \varepsilon > 0. \tag{2.32}$$

    Limit Theorems for Associated Random Fields and Related Systems

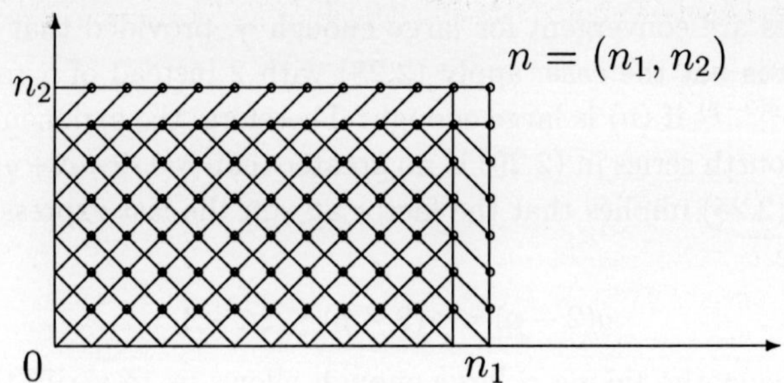

Fig. 4.1 $(d = 2;\ X_n = S_n - S_{n_1, n_2 - 1} - S_{n_1 - 1, n_2} + S_{n_1 - 1, n_2 - 1})$

Denote by e_i the multiindex $(0, \ldots, 0, 1, 0, \ldots, 0)$ where 1 stands at the i-th position, $i = 1, \ldots, d$. Then, for $n \in \mathbb{N}^d$, one has

$$X_n = \sum_{I \subset \{1, \ldots, d\}} (-1)^{|I|} S\left(\left(0, n - \sum_{r \in I} e_r\right] \right)$$

(the multiindices are added and subtracted as vectors in $\mathbb{R}^d$).

Consequently, for any $\varepsilon > 0$ and $m \in \mathbb{N}$ large enough,

$$\mathsf{P}\left(\sup_{k: \langle k \rangle \geq m} \frac{|X_k|}{\langle k \rangle^\alpha} \geq 2^d \varepsilon \right) \leq \mathsf{P}\left(\sup_{k: \langle k \rangle \geq m} \frac{|S_k|}{\langle k \rangle^\alpha} \geq \varepsilon \right).$$

Therefore, by (2.32),

$$\sum_n \langle n \rangle^{\alpha p - 2} \mathsf{P}\left(\sup_{k \geq n} \frac{|X_k|}{\langle k \rangle^\alpha} \geq \varepsilon \right) < \infty \quad \text{for any } \varepsilon > 0. \tag{2.33}$$

Let $\{Z_j, j \in \mathbb{Z}^d\}$ be a random field of independent identically distributed random variables distributed as X_0. For any finite set $U \subset \mathbb{Z}^d$ and any $x \geq 0$, on account of Corollary 1.1.10 we have

$$\mathsf{P}\left(\max_{k \in U} \frac{|X_k|}{\langle k \rangle^\alpha} \geq x \right) = 1 - \mathsf{P}\left(\bigcap_{k \in U} \left\{ \frac{|X_k|}{\langle k \rangle^\alpha} < x \right\} \right)$$

$$\geq 1 - \prod_{k \in U} \mathsf{P}\left(\frac{|X_k|}{\langle k \rangle^\alpha} < x \right) = \mathsf{P}\left(\max_{k \in U} \frac{|Z_k|}{\langle k \rangle^\alpha} \geq x \right). \tag{2.34}$$

Thus, by (2.33) and (2.34),

$$\sum_n \langle n \rangle^{\alpha p - 2} \mathsf{P}\left(\sup_{k \geq n} \frac{|Z_k|}{\langle k \rangle^\alpha} \geq \varepsilon \right) < \infty \quad \text{for any } \varepsilon > 0.$$

Due to independence and stationarity,

$$\mathsf{P}\left(\sup_{k \geq n} \frac{|Z_k|}{\langle k \rangle^\alpha} \geq \varepsilon \right) \geq \mathsf{P}\left(\max_{n < k \leq 2n} |Z_k| \geq \varepsilon \langle n \rangle^\alpha \right) = \mathsf{P}\left(\max_{0 < k \leq n} |Z_k| \geq \varepsilon \langle n \rangle^\alpha \right),$$

here, as usual, $2n = (2n_1, \ldots, 2n_d)$, $n \in \mathbb{N}^d$. As all Z_k have the same distribution,

$$\infty > \sum_n \langle n \rangle^{\alpha p - 2} \mathsf{P}\left(\max_{0 < k \leq n} |Z_k| \geq \varepsilon \langle n \rangle^\alpha\right) = \sum_n \langle n \rangle^{\alpha p - 2}\left(1 - \prod_{0 < k \leq n} \mathsf{P}\left(|Z_k| < \varepsilon \langle n \rangle^\alpha\right)\right)$$

$$= \sum_n \langle n \rangle^{\alpha p - 2}\left(1 - \left(\mathsf{P}\left(|Z_0| < \varepsilon \langle n \rangle^\alpha\right)\right)^{\langle n \rangle}\right) \quad \text{for any } \varepsilon > 0. \tag{2.35}$$

Fix $\varepsilon > 0$. The last series in (2.35) can be rewritten as

$$\sum_{m=1}^{\infty} m^{\alpha p - 2} d_m \left(1 - (1 - \mathsf{P}(|Z_0| \geq \varepsilon m^\alpha))^m\right) < \infty. \tag{2.36}$$

Set $b_m = \mathsf{P}(|Z_0| \geq \varepsilon m^\alpha)$, $m \in \mathbb{N}$. Since $\log(1 - x) \leq -x$ for $x \in [0, 1)$ and $d_m \geq 1$ for $m \in \mathbb{N}$, it can be seen that (2.36) implies the inequality

$$\sum_{m=1}^{\infty} m^{\alpha p - 2} \left(1 - e^{-m b_m}\right) < \infty \tag{2.37}$$

(note that $b_m < 1$ for all m large enough). We claim that

$$m \mathsf{P}(|Z_0| \geq \varepsilon m^\alpha) \leq 1 \tag{2.38}$$

for all m large enough (i.e. $b_m \leq 1/m$ for all m large enough). Suppose on the contrary that this is not true. Then there exists a sequence $(m_r)_{r \in \mathbb{N}}$ of positive integers such that $m_r \to \infty$ as $r \to \infty$ and $m_r b_{m_r} > 1$, $r \in \mathbb{N}$.

For any $r \in \mathbb{N}$ and all $j \in \mathbb{N}$ with $m_r/2 < j \leq m_r$ one clearly has the inequality $j b_j \geq b_{m_r} m_r/2$, as $(b_m)_{m \in \mathbb{N}}$ is a nonincreasing sequence. Therefore, for all $r \in \mathbb{N}$,

$$\sum_{j=[m_r/2]+1}^{m_r} j^{\alpha p - 2} \left(1 - \exp\{-j b_j\}\right) \geq \sum_{j=[m_r/2]+1}^{m_r} j^{\alpha p - 2} \left(1 - \exp\{-m_r b_{m_r}/2\}\right)$$

$$\geq \sum_{j=[m_r/2]+1}^{m_r} j^{\alpha p - 2} \left(1 - e^{-1/2}\right).$$

The last expression tends to infinity as $r \to \infty$, since

$$\sum_{j=[m/2]+1}^{m} j^{\alpha p - 2} \sim \frac{m^{\alpha p - 1}}{\alpha p - 1}\left(1 - \frac{1}{2^{\alpha p - 1}}\right) \to \infty, \quad m \to \infty,$$

if $\alpha p > 1$. However by the Cauchy criterion this contradicts the convergence of the series (2.37). The obtained contradiction shows the validity of (2.38).

For any $m \in \mathbb{N}$ and $x \in [0, 1]$, we have the well-known inequality

$$1 - (1 - x)^m \geq mx - \frac{m^2 x^2}{2}$$

which is easily checked by calculating the derivatives of both parts and invoking the fact that $(1 - x)^m \geq 1 - xm$. Applying this inequality with $x = b_m$, we infer from (2.36) that

$$\sum_{m=1}^{\infty} m^{\alpha p - 1} d_m \mathsf{P}(|Z_0| \geq \varepsilon m^\alpha)\left(1 - \frac{m}{2}\mathsf{P}(|Z_0| \geq \varepsilon m^\alpha)\right) < \infty \tag{2.39}$$

(more precisely, all elements of the series are positive, except for a finite number, and the series converges). Due to (2.38), for all m large enough, the expression which is inside the big brackets in (2.39) is not less than $1/2$. Thus, from (2.38) and (2.39) we conclude that, for some $m_0 > 1$,

$$\sum_{m=m_0}^{\infty} m^{\alpha p-1} d_m \mathsf{P}\left(\varepsilon^{-1}|Z_0|^{1/\alpha} \geq m\right)$$

$$\leq 2 \sum_{m=m_0}^{\infty} m^{\alpha p-1} d_m \mathsf{P}\left(|Z_0| \geq \varepsilon m^{\alpha}\right)\left(1 - \frac{m}{2}\mathsf{P}\left(|Z_0| \geq \varepsilon m^{\alpha}\right)\right) < \infty.$$

Now, as $\varepsilon > 0$ was arbitrary, (2.12) implies that $\mathsf{E}|Z_0|^p \mathrm{Log}^{d-1}|Z_0| = \mathsf{E}|X_0|^p \mathrm{Log}^{d-1}|X_0| < \infty$.

To establish (2.21) it remains to prove that $\mathsf{E}X_0 = 0$. Since $p > 1$, the first moment exists. Suppose that $\mathsf{E}X_0 = a \neq 0$. Then, by assumption (2.23) and the already proved implication (2.21)$\Rightarrow$(2.23) applied to $\{X_n - a, n \in \mathbb{N}^d\}$, for any $\varepsilon > 0$ we obtain

$$\sum_{r=1}^{\infty} r^{\alpha p-2}\left(\mathsf{P}\left(\sup_{k:\langle k\rangle \geq r}\frac{|S_k|}{\langle k\rangle^{\alpha}} \geq \varepsilon/2\right) + \mathsf{P}\left(\sup_{k:\langle k\rangle \geq r}\frac{|S_k - \langle k\rangle a|}{\langle k\rangle^{\alpha}} \geq \varepsilon/2\right)\right) < \infty. \quad (2.40)$$

Take $\varepsilon < |a|$. As $\alpha \leq 1$, for any $k \in \mathbb{N}^d$ we have $\langle k\rangle^{1-\alpha}|a| > \varepsilon$. So if $|S_k| < \varepsilon\langle k\rangle^{\alpha}/2$, then

$$|S_k - a\langle k\rangle|/\langle k\rangle^{\alpha} \geq a\langle k\rangle^{1-\alpha} - |S_k|/\langle k\rangle^{\alpha} \geq \varepsilon - \varepsilon/2 = \varepsilon/2.$$

Thus the sum of two probabilities in the right-hand side of (2.40) is not less than one, for any $r \in \mathbb{N}$. Hence the series sum majorizes $\sum_{r=1}^{\infty} r^{\alpha p-2}$. But since $\alpha p > 2$, the last series diverges, which contradicts (2.40). The contradiction shows that $\mathsf{E}X_0 = 0$. The Theorem is proved. $\square$

3   Almost sure Gaussian approximation

This short Section (not divided into subsections) is devoted to the strong central limit theorem for associated sequences. The pioneering results in this field were established by Brosamler [58] and Schatle [364].

Theorem 3.1. ([321, 406]) *Let* $X = (X_j)_{j\in\mathbb{Z}}$ *be a strictly stationary, associated, centered, square-integrable random sequence and* $S_n = X_1 + \ldots + X_n$, $n \in \mathbb{N}$. *Suppose that* X *satisfies the finite susceptibility condition* (1.5.3). *Then, for any Lipschitz function* $f : \mathbb{R} \to \mathbb{R}$,

$$\frac{1}{\mathrm{Log}\, n}\sum_{j=1}^{n}\frac{1}{j}f\left(\frac{S_j}{\sqrt{j}}\right) \to \mathsf{E}f(Z) \quad a.s., \quad n \to \infty. \quad (3.1)$$

Here $Z \sim N(0, \sigma^2)$ *and* σ^2 *is given in* (1.5.3).

Proof. Set $b_n = \sum_{j=1}^n 1/j$, $n \in \mathbb{N}$, and consider

$$Z_n(f) = \frac{1}{b_n} \sum_{j=1}^n \frac{1}{j} f\left(\frac{S_j}{\sqrt{j}}\right). \tag{3.2}$$

Note that $b_n \sim \log n$ as $n \to \infty$. Thus it obviously suffices to prove that

$$Z_n(f) \to \mathsf{E}f(Z) \text{ a.s., } n \to \infty.$$

Define a nonnegative sequence $(d_n)_{n \in \mathbb{N}}$ as follows:

$$d_n^2 = \frac{1}{b_n^2} \sum_{i,j=1}^n (ij)^{-3/2} cov(S_i, S_j).$$

Then

$$d_n^2 \le \frac{2}{b_n^2} \sum_{1 \le i \le j \le n} (ij)^{-3/2} cov(S_i, S_j) \le \frac{2\sigma^2}{b_n^2} \sum_{i=1}^n i^{-1/2} \sum_{j=i}^n j^{-3/2} \le \frac{4\sigma^2}{b_n} \to 0, \quad n \to \infty.$$

Here we have used the relations

$$cov(S_i, S_j) = \sum_{q=1}^i cov(X_q, S_j) \le i\sigma^2$$

for all $1 \le i \le j$ and $\sum_{j=i}^\infty j^{-3/2} \le 2i^{-1/2}$ for $i \in \mathbb{N}$. Moreover, it is possible to extract such increasing sequence $(n_k)_{k \in \mathbb{N}}$, $n_k \to \infty$ as $k \to \infty$, that $\sum_{k=1}^\infty d_{n_k}^2 < \infty$, but at the same time

$$b_{n_k}/b_{n_{k+1}} \to 1 \text{ as } k \to \infty. \tag{3.3}$$

For example, one can take $n_k = [\exp(k^2)]$, $k \in \mathbb{N}$.

We claim that

$$Z_{n_k}(f) \to \mathsf{E}f(Z) \text{ a.s., } \quad k \to \infty. \tag{3.4}$$

Indeed, for an arbitrary $\varepsilon > 0$, by Theorem 1.5.3,

$$\sum_{k=1}^\infty \mathsf{P}\left(|Z_{n_k}(f) - \mathsf{E}Z_{n_k}(f)| > \varepsilon\right) \le \varepsilon^{-2} \sum_{k=1}^\infty \mathsf{Var}\, Z_{n_k}(f)$$

$$\le Lip^2(f)\varepsilon^{-2} \sum_{k=1}^\infty b_{n_k}^{-2} \sum_{i,j=1}^{n_k} (ij)^{-1} cov\left(\frac{S_i}{\sqrt{i}}, \frac{S_j}{\sqrt{j}}\right) = Lip^2(f)\varepsilon^{-2} \sum_{k=1}^\infty d_{n_k}^2 < \infty.$$

Thus, by the Borel–Cantelli lemma,

$$Z_{n_k}(f) - \mathsf{E}Z_{n_k}(f) \to 0 \text{ a.s., } k \to \infty. \tag{3.5}$$

Besides that, $S_n/\sqrt{n} \to Z$ in law (by Theorem 3.1.12) and the sequence $(f(S_n/\sqrt{n}))_{n \in \mathbb{N}}$ is uniformly integrable, since

$$|f(S_n/\sqrt{n})| \le |f(0)| + \frac{|S_n|}{\sqrt{n}} Lip(f)$$

and $\mathsf{E}(S_n^2/n) \le \sigma^2$. Consequently, $\mathsf{E}f(S_n/\sqrt{n}) \to \mathsf{E}f(Z)$ as $n \to \infty$.

Let $(c_j)_{j\in\mathbb{N}}$ be a sequence of positive numbers such that $r_n := \sum_{j=1}^{n} c_j \to \infty$ as $n \to \infty$. Then the Toeplitz lemma entails, for any sequence $(x_n)_{n\in\mathbb{N}}$ such that $x_n \to x$ as $n \to \infty$, the relation

$$\frac{1}{r_n} \sum_{j=1}^{n} c_j x_j \to x, \quad n \to \infty.$$

Consequently, $\mathsf{E}Z_n(f) \to \mathsf{E}f(Z)$ as $n \to \infty$. From this and (3.5) it follows that (3.4) holds. It remains to observe that

$$\sup_{n_k < n \le n_{k+1}} |Z_{n_k}(f) - Z_n(f)| \le \sup_{n_k < n \le n_{k+1}} \left| \frac{b_{n_k}}{b_n} Z_{n_k}(f) - Z_{n_k}(f) \right|$$

$$+ \sup_{n_k < n \le n_{k+1}} \left| \frac{b_{n_k}}{b_n} Z_{n_k}(f) - Z_n(f) \right|. \tag{3.6}$$

Indeed, we have

$$\sup_{n_k < n \le n_{k+1}} \left| \frac{b_{n_k}}{b_n} Z_{n_k}(f) - Z_{n_k}(f) \right| = \left(1 - \frac{b_{n_k}}{b_{n_{k+1}}} \right) |Z_{n_k}(f)| \to 0 \text{ a.s.}$$

as $k \to \infty$, since $Z_{n_k}(f)$ is, with probability 1, a bounded sequence (by (3.4)), and $(b_{n_k})_{k\in\mathbb{N}}$ satisfies (3.3). To estimate the second summand in (3.6) write that, for $n_k < n \le n_{k+1}$,

$$\left| \frac{b_{n_k}}{b_n} Z_{n_k}(f) - Z_n(f) \right| = b_n^{-1} \left| \sum_{j=n_k+1}^{n} \frac{1}{j} f\left(\frac{S_j}{\sqrt{j}} \right) \right| \le b_n^{-1} \sum_{j=n_k+1}^{n_{k+1}} \frac{1}{j} \left| f\left(\frac{S_j}{\sqrt{j}} \right) \right|$$

$$\le b_{n_k}^{-1} \left(b_{n_{k+1}} Z_{n_{k+1}}(|f|) - b_{n_k} Z_{n_k}(|f|) \right).$$

Note that $Z_{n_k}(|f|) \to \mathsf{E}|f(Z)|$ a.s. $(k \to \infty)$ by (3.4) because $|f|$ is a Lipschitz function. Therefore, in view of (3.3) the examined last term in (3.6) tends a.s. to zero as $k \to \infty$. Thus the left-hand side of (3.6) also tends to zero almost surely as $k \to \infty$. We conclude that (3.2) holds. The proof of Theorem is complete. $\square$

The analysis of the argument above shows that it is not necessary to consider only normal approximation for weighted sums. The proof remains valid for general weak convergence of associated random variables provided that all the steps of proof can be performed. Moreover, the stationarity assumption can be easily removed as well. In fact we have the following result. As before, $b_n = \sum_{j=1}^{n} 1/j$, $n \in \mathbb{N}$.

Theorem 3.2. *Let $(Y_n)_{n\in\mathbb{N}}$ be an associated random sequence such that*

1) $\mathsf{E}Y_j^2 < \infty$ *for any $j \in \mathbb{N}$,*

2) $d_n^2 := b_n^{-2} \sum_{i,j=1}^{n} (ij)^{-1} cov(Y_i, Y_j) \to 0$, $n \to \infty$, *and $\sum_{k=1}^{\infty} d_{n_k}^2 < \infty$ for some increasing sequence $(n_k)_{k\in\mathbb{N}}$ such that $b_{n_{k+1}}/b_{n_k} \to 1$, $k \to \infty$;*

3) $(Y_n)_{n\in\mathbb{N}}$ *is uniformly integrable and there exists a random variable Y such that $Y_n \to Y$ in law as $n \to \infty$.*

Then for any Lipschitz function $f : \mathbb{R} \to \mathbb{R}$

$$b_n^{-1} \sum_{j=1}^{n} j^{-1} f(Y_j) \to \mathsf{E}f(Y) \text{ a.s.}, \quad n \to \infty.$$

Proof is the same as that of Theorem 3.1. The property $\mathsf{E}|f(Y)| < \infty$ holds since $(f(Y_n))_{n \in \mathbb{N}}$ is a uniformly integrable random sequence. $\square$

Corollary 3.3. *Let* $U_1, U_2, \ldots$ *be i.i.d. random variables uniformly distributed on* $[0,1]$. *Set* $Y_k = k \min\{U_1, \ldots, U_k\}$, $k \in \mathbb{N}$. *Then*

$$\frac{1}{\operatorname{Log} n} \sum_{j=1}^{n} \frac{1}{j} f(Y_j) \to \int_0^\infty f(t) e^{-t} dt \ \text{a.s.}, \quad n \to \infty, \tag{3.7}$$

where $f : \mathbb{R} \to \mathbb{R}$ *is any Lipschitz function.*

Proof. The random variables $Y_1, Y_2, \ldots$ are associated by Theorem 1.1.8, (c) and (d). So we only need to check the conditions of Theorem 3.2 taking Y exponentially distributed with parameter 1. The condition 1) is trivial. As to 3), note that, for any $t > 0$ and $n > t$,

$$\mathsf{P}(Y_n > t) = \mathsf{P}\left(\min_{i=1,\ldots,n} U_i > \frac{t}{n}\right) = \mathsf{P}\left(U_1 > \frac{t}{n}, \ldots, U_n > \frac{t}{n}\right)$$

$$= \left(1 - \frac{t}{n}\right)^n \to e^{-t} = \mathsf{P}(Y > t), \quad n \to \infty.$$

Hence $Y_n \to Y$ in law as $n \to \infty$. The random variables Y_n are nonnegative and

$$\mathsf{E}Y_n = \int_0^\infty \mathsf{P}(Y_n > t)dt = \int_0^n \left(1 - \frac{t}{n}\right)^n dt = \frac{n}{n+1} \to 1 = \mathsf{E}Y \tag{3.8}$$

as $n \to \infty$. Thus, by Lemma 3.2.6 a sequence $(Y_n)_{n \in \mathbb{N}}$ is uniformly integrable.

To verify the condition 2) let us calculate the covariance $cov(Y_m, Y_n)$ where $1 \le m \le n$. Let $Z_k = Y_k/k$, $k \in \mathbb{N}$. Then by linearity of the covariance and the Hoeffding formula (see Appendix A.1)

$$\mathsf{E}Y_m Y_n - \mathsf{E}Y_m \mathsf{E}Y_n = mn\, cov(Z_m, Z_n)$$

$$= mn \iint_{\mathbb{R}^2} (\mathsf{P}(Z_m > t, Z_n > u) - \mathsf{P}(Z_m > t)\mathsf{P}(Z_> u))\, dt du$$

$$= mn \int_0^1 \int_0^1 \mathsf{P}(Z_m > t, Z_n > u)dt du - \frac{mn}{(m+1)(n+1)}$$

$$= mn \left(\iint_{\{t \ge u\} \cap Q} + \iint_{\{t < u\} \cap Q}\right) - \frac{mn}{(m+1)(n+1)},$$

since the expectations were evaluated in (3.8), here $Q = [0,1]^2$. In the domain $G := Q \cap \{t \ge u\}$ we have

$$\iint_G \mathsf{P}(Z_m > t, Z_n > u)dt du = \int_0^1 \int_0^t \mathsf{P}(X_1 \wedge \ldots \wedge X_m > t, X_{m+1} \wedge \ldots \wedge X_n > u)du dt$$

$$= \iint_G (1-t)^m (1-u)^{n-m} du dt = \frac{1}{n-m+1} \int_0^1 (1-t)^m (1 - (1-t)^{n-m+1})dt$$

$$= \frac{1}{(m+1)(n-m+1)} - \frac{1}{(n+2)(n-m+1)} = \frac{1}{(m+1)(n+2)}.$$

The second integral is

$$\iint_{\{t<u\}\cap Q} \mathsf{P}(Z_m > t, Z_n > u)\,dt\,du$$

$$= \iint_{\{t<u\}\cap Q} \mathsf{P}(X_1 > u, \dots, X_m > u, X_{m+1} > u, \dots, X_n > u)\,du\,dt$$

$$= \iint_{\{t<u\}\cap Q} (1-u)^n\,dt\,du = \int_0^1 u(1-u)^n\,du = \frac{1}{(n+1)(n+2)}.$$

Therefore

$$cov(Y_m, Y_n) = mn\left(\frac{1}{(m+1)(n+2)} + \frac{1}{(n+1)(n+2)} - \frac{1}{(n+1)(m+1)}\right)$$

$$= \frac{m^2 n}{(m+1)(n+1)(n+2)} \le \frac{m}{n}, \quad n \ge m.$$

From this estimate one infers that

$$d_n^2 := b_n^{-2} \sum_{i,j=1}^n (ij)^{-1} cov(Y_i, Y_j) \le 2b_n^{-2} \sum_{1 \le i \le j \le n} (ij)^{-1} cov(Y_i, Y_j)$$

$$\le 2b_n^{-2} \sum_{i=1}^n \sum_{j=i}^n j^{-2} \le 4b_n^{-2} \sum_{i=1}^n i^{-1} = 4b_n^{-1} = O((\mathrm{Log}\,n)^{-1}), \quad n \to \infty.$$

Thus, the condition 2) holds with $n_k = [\exp(k^2)]$. Clearly the integral in the right-hand side of (3.7) is equal to $\mathsf{E}f(Y)$. The proof of (3.7) is complete. $\square$

Final remarks. Note that a powerful approach to study the relation of the type (3.1) was proposed by Ibragimov and Lifshits ([211, 212]). That method is based on the technique of empirical characteristic functions. We mention also the general theory of almost sure limit theorems studied in [35] and [321]. Associated (and **NA**) random systems and limit theorems with probability one were considered in [288, 291]. The version of Theorem 3.1 for (BL, θ)-dependent processes was established in [149], and for PQD sequences in [294].

The convergence of sums of **NA** (or possessing similar types of negative dependence) random variables is the object of works [7, 40, 234, 238, 243, 286, 288, 297]. Demimartingales were studied in [102], τ-dependent processes in [121]. Non-linearly transformed associated random variables are considered in [293]. The Marcinkiewicz–Zygmund theorem for associated sequences is proved in [275].

The Baum-Katz theorem is in close connection with the so-called complete convergence of random sequences and with logarithmic laws for partial sums (see Chapter 6). In the independent case there is the Gut–Spataru theorem characterizing how often the partial sums exceed the threshold $\varepsilon\sqrt{n\log n}$. This result is extended in the context of random fields and **NA** sequences by Dil'man [135] and Zhao [432] respectively. The necessary and sufficient conditions for **NA** sequence to converge completely are due to Mikusheva [297].

Chapter 5

Invariance Principles

In Chapter 5 the functional limit theorems are considered. This branch of the modern Probability Theory has arisen due to the fundamental investigations by Kolmogorov, Erdös, Kac, Prohorov, Skorokhod, Borovkov, Strassen and other scientists. We are interested in the problems of approximation for partial sums processes, generated by associated random fields and their modifications, by means of multiparameter Wiener process (Brownian motion). Two Sections are devoted to approximation in law (weak invariance principle) and almost surely (strong invariance principle) respectively. We dwell on the versions of the functional CLT obtained by Newman, Wright, Bulinski, Keane, Zhang, Wen and Shashkin. Here we discuss various conditions imposed on the moments of summands, dependence structure of a field $\{X_j, j \in \mathbb{Z}^d\}$ and the dimension d. It is worth mentioning that we study approximation in law not only in the Skorokhod space $D([0,1]^d)$ but also in the space $L^2([0,1]^d)$, following the paper by Oliveira and Suquet. The strong invariance principle is more involved. For associated random processes it was proved by Yu in 1996. The corresponding result for associated random fields was established quite recently, in 2005, by Balan. The previous studies are comprised now by the new strong invariance principle for (BL, θ)- dependent random fields given in Section 2. Here we also rely on the results by Csörgő, Révész, Philipp, Strassen, Berkes, Morrow and on the results by the authors presented in Chapters 2 and 3.

1 Weak invariance principle

The study of asymptotic behavior for distributions of stochastic processes and random fields belongs to the mainstream and frontline of the Modern Probability. One can estimate the impact of contributions related to these problems considering the development of Probability Theory in the XXth century (recall, e.g., the construction of Brownian motion by Wiener, the famous Kolmogorov test in mathematical statistics, the formation of the theory of stochastic differential equations initiated by Itô etc.). However, one says that the origin of the theory of functional limit theorems was the paper by Erdös and Kac [158] where the authors examined the

251

asymptotic behavior of four functionals in partial sums processes (e.g., maximum of partial sums) and established the weak convergence for distributions of such functionals to the corresponding ones for a Wiener process (Brownian motion). The basis for the theory of weak convergence of probability measures was created by Kolmogorov, Alexandrov, Prohorov, Skorokhod, Borovkov, Strassen, Varadarajn. Further development was due to Zolotarev, Dudley, Talagrand and other scientists. One can refer, e.g., to the books by Billingsley [39], Kallenberg [224], Jacod and Shiryaev [216], Pollard [332].

$1°$. **Weak convergence and tightness of probability measures.** Recall the following

Definition 1.1. A sequence $(Q_n)_{n\in\mathbb{N}}$ of *probability measures* defined on the Borel σ-algebra $\mathcal{B}(S)$ of a metric space S *converges weakly* to a probability measure Q (defined on $\mathcal{B}(S)$) if

$$\int_S f(x)Q_n(dx) \to \int_S f(x)Q(dx)$$

for any bounded continuous function $f : S \to \mathbb{R}$.

Let $X_n\,(n \in \mathbb{N})$, X be random elements taking values in S. We say that X_n converge in distribution (or *in law*) to X if the sequence P_{X_n} weakly converges to P_X on $\mathcal{B}(S)$ as $n \to \infty$. If $X_n = \{X_n(t),\, t \in T\}$, $n \in \mathbb{N}$, and $X = \{X(t),\, t \in T\}$ are random functions defined on some set T, then typically one can regard them as random elements with values in a functional metric space (elementary outcomes ω are identified with trajectories). In the most part of applications one uses processes with continuous trajectories or with trajectories belonging to the Skorokhod space.

Definition 1.2. Given $d \in \mathbb{N}$, a *Skorokhod space* $D([0,1]^d)$ is a space of functions $f : [0,1]^d \to \mathbb{R}$ such that f is right-continuous (or "upper-continuous" for $d > 1$) at every point $x \in [0,1)^d$ and for any $x \in (0,1]^d$ there exists a left-hand limit (or "limit from below" for $d > 1$). That is, there exist

$$\lim_{x \searrow x_0} f(x) = f(x_0) \quad \text{for any} \ \ x_0 \in [0,1)^d,$$

$$\lim_{x \nearrow x_0} f(x) \quad \text{for each} \ \ x_0 \in (0,1]^d.$$

Here $x \searrow x_0$ (resp. $x \nearrow x_0$) means that $x_i \searrow x_{0,i}$ (resp. $x_i \nearrow x_{0,i}$), $i = 1,\ldots,d$, where $x = (x_1,\ldots,x_d)$ and $x_0 = (x_{0,1},\ldots,x_{0,d})$.

Quite similarly one can define the space $D([0,\infty)^d)$ (see, e.g., [302]). These spaces are Polish under appropriate choice of metric (see, e.g., [414]). Therefore due to the fundamental Prohorov Theorem (see, e.g., [39, §6, Th. 6.1,6.2]), the necessary and sufficient condition for the weak convergence of a sequence of processes $(W_n)_{n\in\mathbb{N}}$ where $W_n = \{W_n(t), t \in [0,1]^d\}$ is the tightness of the sequence of their distributions in the Skorokhod space $D([0,1]^d)$ and weak convergence of the finite-dimensional distributions. The *tightness* means that, for any $\varepsilon > 0$, it is possible to find a compact subset $K \subset D([0,1]^d)$ such that $\mathsf{P}(W_n \in K) \geq 1 - \varepsilon$ for any n.

For a random field $X = \{X_j, j \in \mathbb{Z}^d\}$ introduce the *partial sums process* (corresponding to X) by the following formula

$$W_n(t) = \frac{1}{\sqrt{\langle n \rangle}} \sum_{1 \leq j_k \leq n_k t_k, \, k=1,\ldots,d} X_j, \quad t \in [0,1]^d, \; n \in \mathbb{N}^d, \tag{1.1}$$

where $\langle n \rangle = n_1 \ldots n_d$ for $n = (n_1, \ldots, n_d) \in \mathbb{N}^d$. Let $\mathcal{U}$ be the same system of blocks as in Ch. 2 and $M(U)$ be defined by the relation (2.1.1), for $U \in \mathcal{U}$. We also write M_U instead of $M(U)$.

Note that the convergence of multiindexed family $\{W_n, n \in \mathbb{N}^d\}$ as $n \to \infty$ is understood in sequential sense, i.e. one assumes the convergence to the same limit for any sequence $\{W_{n_k}, k \in \mathbb{N}\}$ if $n_k \to \infty$ as $k \to \infty$. We also call the family $\{W_n, n \in \mathbb{N}^d\}$ *sequentially tight* if, for any sequence of points $(n_k)_{k \in \mathbb{N}}$ in $\mathbb{N}^d$ such that $n_k \to \infty$ as $k \to \infty$, the family $\{W_{n_k}, k \in \mathbb{N}\}$ is tight.

We remind an easy condition ensuring the sequential tightness in $D([0,1]^d)$ for the laws of processes defined in (1.1). Namely, one can check that, for any $\varepsilon > 0$,

$$\lim_{\delta \to 0+} \limsup_{n \to \infty} \mathsf{P}\left(\sup_{s,t \in [0,1]^d : |s-t| < \delta} |W_n(s) - W_n(t)| \geq \varepsilon \right) = 0 \tag{1.2}$$

where $n \in \mathbb{N}^d$ and $|\cdot|$ is the sup-norm in $\mathbb{R}^d$ (see [39, §15, Th. 15.5], [414]; though conditions given in [39] are applicable directly when $d = 1$, they extend to the multiparameter case without difficulties). We also mention that (1.2) and the convergence of finite-dimensional distributions of $(W_n)_{n \in \mathbb{N}^d}$ imply both the weak convergence of laws of W_n in the Skorokhod space endowed with uniform metric and the almost sure continuity of the limit process.

Theorem 1.3. *Let $X = \{X_j, j \in \mathbb{N}^d\}$ be a random field with corresponding partial sums process W_n, $n \in \mathbb{N}^d$. Assume that, for any sequence of blocks $U_{n_k} \in \mathcal{U}, k \in \mathbb{N}$, growing to infinity (as $k \to \infty$), the sequence $(M(U_{n_k})^2 / |U_{n_k}|)_{k \in \mathbb{N}}$ is uniformly integrable. Then the family $\{W_n, n \in \mathbb{N}^d\}$ is sequentially tight in the space $D([0,1]^d)$.*

Proof. We will check the condition (1.2). Take $\delta \in (0,1)$. The event inside the probability sign in (1.2) is contained in the union over $i = 1, \ldots, d$ of events

$$C_n(i,\delta) = \left\{ \sup_{t \in [0,1]^d : t_i \leq 1-\delta, \gamma \in (0,\delta)} |W_n(t_1, \ldots, t_i + \gamma, \ldots, t_d) - W_n(t)| \geq \varepsilon d^{-1} \right\}.$$

For a fixed $i \in \{1, \ldots, d\}$ introduce the sets

$$A_m(i,\delta) = (0,1] \times \ldots \times ((m-1)\delta, m\delta \wedge 1] \times \ldots \times (0,1],$$

so that the half-interval of length not exceeding δ stands here on the i-th place for $m = 1, \ldots, q = q(\delta) = [\delta^{-1}] + 1$, and next define in $\mathbb{R}^d_+$ the blocks

$$U_{m,n} = \{([n_1 t_1], \ldots, [n_d t_d]), \, t = (t_1, \ldots, t_d) \in A_m(i,\delta)\}, \quad m = 1, \ldots, q, \; n \in \mathbb{N}^d.$$

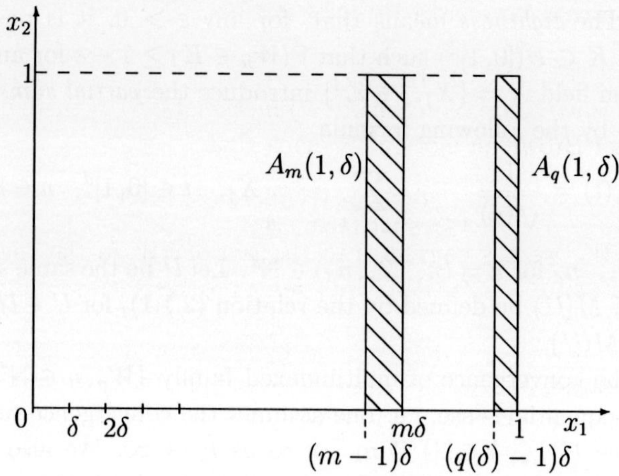

Fig. 5.1　$(d=2)$

For each $i \in \{1,\dots,d\}$, using for blocks the notation $\lhd$ introduced in Section 2.1, we have

$$P(C_n(i,\delta)) \leq \sum_{m=1}^{q} P\left(\sup_{\substack{t,s\in A_m(i,\delta) \\ t_r=s_r,\, r\neq i}} |W_n(t) - W_n(s)| \geq \frac{\varepsilon}{2d} \right)$$

$$\leq \sum_{m=1}^{q} P\left(\sup_{V \lhd U_{m,n}} \frac{|S_V|}{\sqrt{\langle n\rangle}} \geq \frac{\varepsilon}{4d} \right) \leq \sum_{m=1}^{q} P\left(\frac{M_{U_{m,n}}}{\sqrt{\langle n\rangle}} \geq \frac{\varepsilon}{4d} \right)$$

$$\leq \frac{16d^2}{\varepsilon^2} \sum_{m=1}^{q} E\left(\frac{M_{U_{m,n}}^2}{\langle n\rangle} \mathbb{I}\left\{ \frac{M_{U_{m,n}}}{\sqrt{\langle n\rangle}} \geq \frac{\varepsilon}{4d} \right\} \right) = \frac{16d^2}{\varepsilon^2} \sum_{m=1}^{q} \Delta_m.$$

To obtain the last relation we employed the Chebyshev inequality.

Clearly $|U_{m,n}|/\langle n\rangle \to \mathrm{mes}(A_m(i,\delta)) \leq \delta$ as $n \to \infty$. Consequently, for all n large enough, we can continue the estimation and obtain

$$\sum_{m=1}^{q} \Delta_m = \sum_{m=1}^{q} \frac{|U_{m,n}|}{\langle n\rangle} E\left(\frac{M_{U_{m,n}}^2}{|U_{m,n}|} \mathbb{I}\left\{ \frac{M_{U_{m,n}}}{\sqrt{|U_{m,n}|}} \sqrt{\frac{|U_{m,n}|}{\langle n\rangle}} \geq \frac{\varepsilon}{4d} \right\} \right)$$

$$\leq \sum_{m=1}^{q} 2\delta E\left(\frac{M_{U_{m,n}}^2}{|U_{m,n}|} \mathbb{I}\left\{ \frac{M_{U_{m,n}}}{\sqrt{|U_{m,n}|}} \geq \frac{\varepsilon}{4d\sqrt{2\delta}} \right\} \right)$$

$$\leq 2\delta(1+\delta^{-1}) \max_m E\left(\frac{M_{U_{m,n}}^2}{|U_{m,n}|} \mathbb{I}\left\{ \frac{M_{U_{m,n}}}{\sqrt{|U_{m,n}|}} \geq \frac{\varepsilon}{4d\sqrt{2\delta}} \right\} \right).$$

Obviously it is sufficient to consider $\varepsilon \in (0,1)$. Taking $\delta = \varepsilon^4$ we conclude, due to the uniform integrability of a family $\{M(U_{m,n})^2/|U_{m,n}|\}$, that (1.2) holds. $\square$

2°. The functional central limit theorem.

Definition 1.4. One says that a random field $X = \{X_j, j \in \mathbb{Z}^d\}$ satisfies the (classical) *weak invariance principle* or[1] a *functional CLT* if there exist $\sigma^2 \geq 0$ and a multiparameter Brownian motion[2] W such that

$$W_n \Rightarrow \sigma W, \quad n \to \infty,$$

i.e. the law of W_n weakly converges to the law of σW in the space $D([0,1]^d)$ as $n \to \infty$. Here W_n $(n \in \mathbb{N}^d)$ is the partial sums process corresponding to X.

Now we may apply this concept to dependent random systems under consideration. We will gather all the conditions sufficient for FCLT in one theorem, as the proofs will be similar.

Theorem 1.5. ([72, 310, 376, 431]) *Let* $X = \{X_j, j \in \mathbb{Z}^d\}$ *be a square-integrable, centered random field. Let any of the following groups of conditions be satisfied. In all conditions except for* (a) *there are no requirements on the lattice dimension d.*
   (a) $d \leq 2$, *a strictly stationary* $X \in$ **PA** *and satisfies condition* (1.5.3);
   (b) X *is wide-sense stationary and* **A**, *for some positive* δ *and* λ *one has*

$$\sup_{j \in \mathbb{Z}^d} \mathsf{E}|X_j|^{2+\delta} < \infty, \quad u_r = O(r^{-\lambda}) \text{ as } r \to \infty$$

where u_r $(r \in \mathbb{N})$ *is the Cox–Grimmett coefficient[3];*
   (c) X *is strictly stationary and* **NA***;*
   (d) X *is strictly stationary,* (BL, θ)*-dependent,* $\theta_r = O(r^{-\lambda})$ *for some* $\lambda > 3d$;
   (e) X *is wide-sense stationary,* (BL, θ)*-dependent and satisfies the conditions of either Theorem 2.1.18 or Corollary 2.1.28.*
   Then X *satisfies the FCLT.*

The assertion (a) is due to Newman and Wright [310], (b) is established by Bulinski and Keane, (c) is proved by Zhang and Wen, (d) is obtained by Shashkin, (e) belongs to Bulinski and Shashkin.

Proof. At first we prove the tightness of the family of random processes under consideration.

Lemma 1.6. *Each of the conditions* (b)—(e) *implies the condition of Theorem 1.3.*

Proof. Case (b). Corollary 2.1.13 yields that there is some $\eta > 0$ such that, for all $U \in \mathcal{U}$, $\mathsf{E}M(U)^{2+\eta} \leq C|U|^{1+\eta/2}$ where C does not depend on U This estimate implies uniform integrability since

$$\sup_{U \in \mathcal{U}} \mathsf{E}\frac{M(U)^2}{|U|}\mathbb{I}\left\{\frac{M(U)^2}{|U|} \geq x\right\} \leq \sup_{U \in \mathcal{U}} \frac{\mathsf{E}M(U)^{2+\eta}}{|U|^{1+\eta/2}x^{\eta/2}} \to 0, \quad x \to \infty.$$

---

[1] One writes FCLT.
[2] We consider the restriction of a Brownian motion defined on $\mathbb{R}^d_+$ to $[0,1]^d$.
[3] See (1.5.4).

Cases (c) and d). If X is strictly stationary and square-integrable, then obviously $\{X_j^2, j \in \mathbb{Z}^d\}$ is a uniformly integrable system. Thus the desired conclusion is the second assertion of Theorem 2.3.11, (a).

Case (e). This case is analogous to (b), because of validity of Theorem 2.1.18 or Corollary 2.1.28. $\square$

For the case (a) we have the following analogue of the previous Lemma:

Lemma 1.7. *Let $(n_k)_{k\in\mathbb{N}}$ be a sequence of points in $\mathbb{Z}^d$ ($d \le 2$) such that $n_k \to \infty$, $k \to \infty$. Then under condition (a) the sequence of random processes $W_{n_k} = \{W_{n_k}(t), t \in [0,1]^d\}$, $k \in \mathbb{N}$, is tight in $D([0,1]^d)$.*

Proof. If $\sigma^2 = \sum_{j\in\mathbb{Z}^d} cov(X_0, X_j) = 0$, then all processes W_n are identically zero and the Lemma is true. Hence we can assume that $\sigma^2 > 0$. In particular that means that $\mathsf{Var}S(U) > 0$ for any block $U \subset \mathbb{Z}^d$ (as all the covariances are nonnegative). Fix a sequence $(n_k)_{k\in\mathbb{N}} \subset \mathbb{Z}^d$ and a number $\varepsilon > 0$. Arguing as in the proof of Theorem 1.3 (and preserving the notation), we see that it suffices to prove

$$\lim_{\delta\to 0+}\ \limsup_{n\to\infty} \sum_{m=1}^{q} \mathsf{P}\left(\frac{M_{U_{m,n}}}{\sqrt{\langle n\rangle}} \ge \frac{\varepsilon}{4d}\right) = 0 \tag{1.3}$$

where $q = q(\delta) = [\delta^{-1}] + 1$. Note that here $n = n_k$, but for simplicity we omitted the index k. Set $a_{m,n} = (\langle n\rangle/\mathsf{Var}S(U_{m,n}))^{1/2}$. Thus, for any small enough $\delta > 0$ and all $n = n_k$,

$$\sum_{m=1}^{q} \mathsf{P}\left(\frac{M_{U_{m,n}}}{\sqrt{\langle n\rangle}} \ge \frac{\varepsilon}{4d}\right) = \sum_{m=1}^{q} \mathsf{P}\left(\frac{M_{U_{m,n}}}{\sqrt{\mathsf{Var}S(U_{m,n})}} \ge \frac{\varepsilon}{4d}\sqrt{\frac{\langle n\rangle}{\mathsf{Var}S(U_{m,n})}}\right)$$

$$\le b_\varepsilon \sum_{m=1}^{q} a_{m,n}^{-3/2}\mathsf{P}\left(\frac{S_{U_{m,n}}}{\sqrt{\mathsf{Var}S(U_{m,n})}} \ge \frac{\varepsilon}{8d}a_{m,n}\right)^{1/4} = b_\varepsilon T_q(m,n) \tag{1.4}$$

in view of Corollary 2.4.4, here $b_\varepsilon = 2(24d/\varepsilon)^{3/2}$. Theorem 3.1.8 implies that

$$\lim_{n\to\infty} \frac{\mathsf{Var}S(U_{m,n})}{\langle n\rangle} = \begin{cases} \delta, & m < q, \\ 1 - (q-1)\delta \le \delta, & m = q. \end{cases}$$

Thus, if we denote $Z \sim N(0,1)$, then due to the CLT for associated random fields (Theorem 3.1.12),

$$T_q(m,n) \le \sum_{m=1}^{q-1} \delta^{3/4}\mathsf{P}\left(Z \ge \frac{\varepsilon}{8d}\delta^{-1/2}\right)^{1/4} + \delta^{3/4}$$

$$\le \delta^{-1/4}\left(\frac{1}{\sqrt{2\pi}}\int_{\varepsilon\delta^{-1/2}/8d}^{+\infty} e^{-u^2/2}du\right)^{1/4} + 2\delta^{3/4} \to 0, \quad \delta \to 0+ .$$

This relation and (1.4) prove (1.3), hence the Lemma. $\square$

By Lemma 1.7 in case (a), by Lemma 1.6 and Theorem 1.3 in other cases, each sequence $W_{n_k}(\cdot)$, where $n_k \in \mathbb{N}^d$, $n_k \to \infty$, contains a subsequence which converges in law to some process $X(\cdot)$. Moreover, this process is a.s. continuous (see the remark after (1.2)). So we have to prove that it is a d-parameter Brownian motion multiplied by a factor σ, i.e. $X(t) = \sigma W(t)$, $t \in [0,1]^d$.

Let us denote by $X(B)$ the *difference of the process X over a rectangle B*. That is, for $B = (a,b] = (a_1,b_1] \times \ldots \times (a_d,b_d] \subset \mathbb{R}^d$ set

$$X(B) := \sum_{I \subset \{1,\ldots,d\}} (-1)^{|I|} X(t(I)), \tag{1.5}$$

here $t(I) = (t_1(I),\ldots,t_d(I))$ is such that $t_i(I) = a_i$ if $i \in I$ and $t_i(I) = b_i$ otherwise. Recall that "$\subset$" means "$\subseteq$".

We show that, for any $q \in \mathbb{N}$ and any rectangles $B_1,\ldots,B_q \subset [0,1]^d$ with $B_i \cap B_j = \varnothing$ whenever $i \neq j$, the random vector $(X(B_1),\ldots,X(B_q))$ has Gaussian distribution with mean zero and covariance matrix $\sigma^2 diag(\mathrm{mes}(B_1),\ldots,\mathrm{mes}(B_q))$.

For a rectangle $B = (a,b] \subset [0,1]^d$ and any $n \in \mathbb{N}^d$ define a block

$$U(n,B) = \{([n_1 t_1],\ldots,[n_d t_d]),\ t = (t_1,\ldots,t_d) \in B\} \subset \mathbb{Z}^d.$$

Lemma 1.8. *For any rectangle $B = (a,b] \subset [0,1]^d$,*

$$S(U(n,B))/\sqrt{\langle n \rangle} \to N(0,\sigma^2 \mathrm{mes}(B)) \ \text{ in law as } \ n \to \infty.$$

Proof. Since $U(n,B) \to \infty$ in a regular manner as $n \to \infty$, we have $S(U(n,B))/\sqrt{|U(n,B)|} \to N(0,\sigma^2)$ by Theorem 3.1.12. It remains to note that $|U(n,B)|/\langle n \rangle \to \mathrm{mes}(B)$, $n \to \infty$, in view of the definition of $U(n,B)$. $\square$

Lemma 1.9. *For any $q \in \mathbb{N}$ and any pairwise disjoint rectangles $B_1,\ldots,B_q \subset [0,1]^d$ the random variables $X(B_1),\ldots,X(B_q)$ are independent.*

Proof. We prove this by induction on q. Let $B_1,\ldots,B_q,B_{q+1} \subset [0,1]^d$ be arbitrary rectangles with $\mathrm{mes}(B_i \cap B_j) = 0$ whenever $i \neq j$, $i,j = 1,\ldots,q+1$.

Let $\varphi_{n,1}(z_1,\ldots,z_q)$, $\varphi(z)$, $\varphi_{n,q+1}(z_{q+1})$ be the characteristic functions of the random vectors $(W_n(B_1),\ldots,W_n(B_q))$, $(W_n(B_1),\ldots,W_n(B_q),W_n(B_{q+1}))$ and $W_n(B_{q+1})$ respectively, here $z \in \mathbb{R}^{q+1}$. Then by Lemma 1.8 if $q = 1$ and also by the induction hypothesis if $q > 1$ we have

$$\varphi_{q,1}(z_1,\ldots,z_q)\varphi_{n,q+1}(z_{q+1}) \to \exp\left\{-\frac{1}{2}\sum_{k=1}^{q+1}\sigma^2 \mathrm{mes}(B_k)z_k^2\right\}, \quad n \to \infty.$$

Thus it suffices to prove that, for any fixed $z \in \mathbb{R}^{q+1}$,

$$|\varphi(z) - \varphi_{n,1}(z_1,\ldots,z_q)\varphi_{n,q+1}(z_{q+1})|$$

$$= \left|cov\left(\exp\left\{i\sum_{k=1}^{q} z_k W_n(B_k)\right\}, \exp\left\{-iz_{q+1}W_n(B_{q+1})\right\}\right)\right| \to 0, \quad n \to \infty.$$

Select a function $r(n) : \mathbb{N}^d \to \mathbb{N}$ such that $r(n) \to \infty$ and $r(n)/\min\{n_1, \ldots, n_d\} \to 0$ as $n \to \infty$. Introduce the blocks

$$V(n, B_{q+1}) = \{j \in U(n, B_{q+1}) : dist(\{j\}, \mathbb{N}^d \setminus U(n, B_{q+1})) \geq r_n\}$$

and define random variables $Y_n = \langle n \rangle^{-1/2} S_{V(n,B_{q+1})}$. Then by Theorem 1.5.3 or by the definition of (BL, θ)-dependence we infer that

$$\left| cov\left(\exp\left\{ i \sum_{k=1}^{q} z_k W_n(B_k) \right\}, \exp\left\{ -iz_{q+1} Y_n \right\} \right) \right| \leq 4 \max_{k=1,\ldots,q+1} z_k^2 \theta_{r(n)} \to 0,$$

as $n \to \infty$ (for **PA** or **NA** cases θ_r is the Cox–Grimmett coefficient). Since, for any $x, y \in \mathbb{R}$, one has $|e^{ix} - e^{iy}| \leq |x - y|$, we see that for all n large enough

$$\left| cov\left(\exp\left\{ i \sum_{k=1}^{q} z_k W_n(B_k) \right\}, \exp\left\{ -iz_{q+1} W_n(B_{q+1}) \right\} - \exp\left\{ -iz_{q+1} Y_n \right\} \right) \right|$$

$$\leq 2\langle n \rangle^{-1/2} |z_{q+1}| |U(n, B_{q+1}) \setminus V(n, B_{q+1})| \to 0, \quad n \to \infty. \quad \square$$

From Lemma 1.9 it follows that the $(X(B_1), \ldots, X(B_q))$ is a Gaussian vector with independent components. Moreover, $X(B) \sim N(0, \sigma^2 \text{mes}(B))$ for any rectangle $B \subset [0, 1]^d$.

Note that $W_n(t) = W_n((0, t])$ for $t \in [0, 1]^d$. If $r \in \mathbb{N}$ and $t_1, \ldots, t_r \in [0, 1]^d$, one can write $W_n(t_1), \ldots, W_n(t_r)$ as sums of some random variables taken from the set $W_n(B_i)$, $i = 1, \ldots, L$, where the rectangles $B_i = (a_i, b_i]$ are pairwise disjoint and L does not depend on n. As weak convergence survives under continuous maps, we conclude that

$$(W_n(t_1), \ldots, W_n(t_r)) \to (W(t_1), \ldots, W(t_r)) \text{ in law as } n \to \infty.$$

Therefore $X = \sigma W = \{\sigma W_t, t \in [0, 1]^d\}$ is the d-parameter Wiener process on $[0, 1]^d$ multiplied by σ. $\square$

3°. FCLT in the space $\mathbf{L^2([0, 1]^d)}$. In this case the situation is simpler than in the previous one. Here we prove the extension of the result due to Oliveira and Suquet [313] (for associated random fields, in fact even for LPQD) obtained by Shashkin.

Theorem 1.10. ([379]) *Let $X = \{X_j, j \in \mathbb{Z}^d\}$ be a centered, strictly stationary, (BL, θ)-dependent random field with $\mathsf{E}X_0^2 < \infty$. Then $\{W_n(t), t \in [0, 1]^d, n \in \mathbb{N}^d\}$ converge weakly in the Hilbert space $L^2([0, 1]^d)$, as $n \to \infty$, to the random process σW where $W = \{W(t), t \in [0, 1]^d\}$ is the d-parameter Brownian motion and σ is given by (1.5.3).*

Proof. Again we begin with establishing the relative compactness. Recall the classical lemma from calculus known as Dini's theorem.

Lemma 1.11. *Let $n \in \mathbb{N}$ and $K \subset \mathbb{R}^n$ be a compact set. Suppose that $(f_k)_{k \in \mathbb{N}}$ is a sequence of continuous nonnegative functions on K such that $f_k(x) \searrow 0$, $k \to \infty$, for any $x \in K$. Then $f_k(x) \to 0$ uniformly in $x \in K$, as $k \to \infty$.*

Proof. Let $\varepsilon > 0$ be arbitrary and fixed, then, for any $x \in K$, there exists a number $k_0(\varepsilon, x) \in \mathbb{N}$ such that $f_{k_0(\varepsilon,x)} < \varepsilon/2$. As the functions f_k are continuous, to any $x \in K$ we may find an open set $B(\varepsilon, x)$ such that $x \in B(\varepsilon, x)$ and $f_{k_0(\varepsilon,x)}(y) < \varepsilon$ for any $y \in B(\varepsilon, x)$. Since K is compact, we can select a finite number of points $x_1, \ldots, x_N \in K$ so that $K \subset \cup_{j=1}^{N} B(\varepsilon, x_j)$, with some $N = N(\varepsilon) \in \mathbb{N}$. Take $k_0 = \max_{j=1,\ldots,N} k_0(\varepsilon, x_j)$. Then, for any $y \in K$ and all $k > k_0$, by monotonicity we have

$$0 \leq f_k(y) \leq f_{k_0}(y) \leq f_{k_0(\varepsilon, x_m)}(y) < \varepsilon$$

where $m = m(y) \in \{1, \ldots, N\}$ is such that $y \in B(\varepsilon, x_m)$. Thus f_k tend uniformly to zero as $k \to \infty$. $\square$

Lemma 1.12. *The family of distributions of $W_n = \{W_n(t), t \in [0,1]^d, n \in \mathbb{Z}^d\}$ is tight in $L^2([0,1]^d)$.*

Proof. Let $(e_i)_{i \in \mathbb{N}}$ be some orthonormal basis in $L^2([0,1]^d)$. Set

$$f_i(s) = \int_{s_1}^{1} \cdots \int_{s_d}^{1} e_i(t)dt = \int_{[0,1]^d} e_i(t)\mathbb{I}_{[s,1]}(t)dt, \quad s \in [0,1]^d.$$

Changing the summation and integration order we have, for any $i \in \mathbb{N}$,

$$\int_{[0,1]^d} W_n(t)e_i(t)dt = \frac{1}{\sqrt{\langle n \rangle}} \int_{[0,1]^d} \sum_{j_k \leq n_k t_k,\, k=1,\ldots,d} X_j e_i(t)dt$$

$$= \frac{1}{\sqrt{\langle n \rangle}} \int_{[0,1]^d} \sum_{j \leq n} \mathbb{I}\{j_k \leq n_k t_k,\, k = 1, \ldots, d\} X_j e_i(t)dt =$$

$$\frac{1}{\sqrt{\langle n \rangle}} \sum_{j \leq n} X_j \int_{[0,1]^d} \mathbb{I}\{j_k \leq n_k t_k,\, k = 1, \ldots, d\} e_i(t)dt = \frac{1}{\sqrt{\langle n \rangle}} \sum_{j \leq n} X_j f_i\left(\frac{j}{n}\right).$$

Here

$$\frac{j}{n} = \left(\frac{j_1}{n_1}, \ldots, \frac{j_d}{n_d}\right), \quad j, n \in \mathbb{N}^d.$$

Thus, for any $N \in \mathbb{N}$,

$$\sup_{n \in \mathbb{N}^d} \mathsf{E} \sum_{i=N}^{\infty} \left(\int_{[0,1]^d} W_n(t)e_i(t)dt \right)^2 = \sup_{n \in \mathbb{N}^d} \mathsf{E} \sum_{i=N}^{\infty} \frac{1}{\langle n \rangle} \left(\sum_{j \leq n} f_i\left(\frac{j}{n}\right) X_j \right)^2$$

$$= \sup_{n \in \mathbb{N}^d} \frac{1}{\langle n \rangle} \mathsf{E} \sum_{i=N}^{\infty} \sum_{j,k \leq n} f_i\left(\frac{j}{n}\right) f_i\left(\frac{k}{n}\right) X_j X_k = \sup_{n \in \mathbb{N}^d} \frac{1}{\langle n \rangle} \mathsf{E} \sum_{j,k \leq n} X_j X_k a_{jk},$$

with notation $a_{jk} := \sum_{i=N}^{\infty} f_i(j/n)f_i(k/n)$. All the functions f_i are continuous and $\sum_{i=1}^{\infty} f_i^2(x) = \|\mathbb{I}_{[x,1]}\|_{L^2}^2 = (1 - x_1) \ldots (1 - x_d)$ on account of the Parceval identity,

$\|\cdot\|_{L^2}$ being the norm in $L^2([0,1]^d)$. Therefore, by Lemma 1.11, $\sum_{i=N}^{\infty} f_i^2(x) \to 0$ uniformly in $x \in [0,1]^d$ as $N \to \infty$. In view of the (BL, θ)-dependence,

$$0 \leq \frac{1}{\langle n \rangle} \sum_{j,k \leq n} \mathsf{E} X_j X_k a_{jk} \leq C \sup_{j,k} |a_{jk}|$$

for some positive C independent of n and N. Thus,

$$\sup_{n \in \mathbb{N}^d} \mathsf{E} \sum_{i=N}^{\infty} (W_n, e_i)^2 \to 0, \quad N \to \infty. \tag{1.6}$$

We claim that (1.6) implies the desired statement. Indeed, (1.6) is the *Prohorov condition* ([334]) *of tightness in a separable Hilbert space* H for $(W_n)_{n \in \mathbb{N}^d}$. For the sake of completeness we reproduce the proof. Let $\varepsilon > 0$ be arbitrary. First of all observe that there exists $M > 0$ such that for any $n \in \mathbb{Z}^d$

$$\mathsf{P}(\|W_n\| > M) \leq M^{-2} \sup_{n \in \mathbb{N}^d} \mathsf{E}\|W_n\|^2 = M^{-2} \sup_{n \in \mathbb{N}^d} \sum_{i=N}^{\infty} (W_n, e_i)^2 < \varepsilon.$$

Using (1.6) we may choose a sequence of positive integers $(N_k)_{k \in \mathbb{N}}$ verifying

$$\sup_{n \in \mathbb{N}^d} \mathsf{E} \sum_{i=N_k}^{\infty} (W_n, e_i)^2 < \frac{\varepsilon}{2^{2k}}, \quad k \in \mathbb{N}. \tag{1.7}$$

By the Chebyshev inequality and (1.7),

$$\sup_{n \in \mathbb{N}^d} \mathsf{P}\Big(\Big\| \sum_{i=N_k}^{\infty} (W_n, e_i)e_i \Big\|_{L^2} > 2^{-k/2} \Big) \leq 2^k \mathsf{E}\Big(\sum_{i=N_k}^{\infty} (W_n, e_i)^2 \Big) < \frac{\varepsilon}{2^k}. \tag{1.8}$$

Here we also used that $\{e_i\}_{i \in \mathbb{N}}$ is an orthonormal system. Introduce in $L^2([0,1]^d)$ the set

$$K = \Big\{ f \in L^2([0,1]^d) : \|f\| \leq M \text{ and } \sum_{i=N_k}^{\infty} (f, e_i)^2 \leq 2^{-k} \text{ for any } k \in \mathbb{N} \Big\}.$$

Then K is a compact in $L^2([0,1]^d)$, as follows by demonstrating that K is closed and completely bounded[4]. For any $n \in \mathbb{N}^d$, from (1.8) we have the estimate

$$\mathsf{P}(W_n \notin K) \leq \mathsf{P}(\|W_n\| > M) + \sum_{k=1}^{\infty} \mathsf{P}\Big(\sum_{i=N_k}^{\infty} (W_n, e_i)^2 > 2^{-k} \Big)$$

$$\leq \varepsilon + \sum_{k=1}^{\infty} \mathsf{P}\Big(\Big\| \sum_{i=N_k}^{\infty} (W_n, e_i)e_i \Big\|_{L^2} > 2^{-k/2} \Big) \leq \varepsilon + \sum_{k=1}^{\infty} \frac{\varepsilon}{2^k} = 2\varepsilon,$$

which shows the desired tightness of $\{W_n, n \in \mathbb{N}^d\}$. The Lemma is proved. $\square$

According to [334] the weak convergence of Y_n to a random element Y in the Hilbert space H (as $n \to \infty$) is equivalent to the tightness of $\{Y_n\}$ and the relation

$$\mathsf{E} \exp(i(Y_n, g)) \to \mathsf{E} \exp(i(Y, g)) \quad \text{for any } g \in H. \tag{1.9}$$

---

[4]See [239, Ch. II, §7, 2]

Indeed, the *characteristic functional* $\mathsf{E}\exp\{i(Y,g)\}$ determines the distribution of a random element Y uniquely (since it determines the distributions of all projections of Y onto finite-dimensional subspaces of H). Therefore the existence of two different partial limit (as $n \to \infty$) distributions of the family $\{Y_n\}$ contradicts (1.9).

Lemma 1.13. *Suppose that the family $\{Y_n\}$ is tight. Then it suffices to check (1.9) for $g \in M$, where M is a dense subset in H.*

Proof. Suppose that (1.9) holds for any $g \in M$ but the convergence in law does not hold. Then there exist (by tightness) a random element L with distribution different from that of Y and a sequence of points $n_k \in \mathbb{N}^d$, such that $n_k \to \infty$ as $k \to \infty$ and $Y_{n_k} \to L$ in law, $k \to \infty$. If ξ is a random element in H, then the function $v \mapsto \mathsf{E}\exp\{i(\xi,v)\}$ is continuous[5] in H. Thus, since the characteristic functionals of L and Y coincide on M and are both continuous, they coincide on H. That means that the laws of L and Y are the same. $\square$

Take $M = C([0,1]^d)$. We are going to prove that whenever $g \in C[0,1]^d$, the convergence in law occurs,

$$S_n(g) = \int_{[0,1]^d} W_n(t)g(t)dt \to \sigma \int_{[0,1]^d} W(t)g(t)dt, \quad n \to \infty. \qquad (1.10)$$

The right-hand side has normal distribution with parameters $(0, \sigma^2 \int_{[0,1]^d} h^2(s)ds)$ where $h(s) = \int_s^1 g(u)du$. Indeed, the fact that the random variable at the right-hand side of (1.10) is Gaussian and centered is seen easily, at first for step functions h and then, by an appropriate approximation, for continuous ones. Further, the random field W has a.s. continuous trajectories, so (see, e.g., [412, §2.1, 4])

$$\mathsf{Var}\int_{[0,1]^d} W(t)g(t)dt = \iint_{s,u\in[0,1]^d} \prod_{i=1}^{d}(s_i \wedge u_i)g(s)g(u)dsdu$$

$$= \iiint_{t,s,u\in[0,1]^d} \mathbb{I}\{t \leq s \wedge u\}g(s)g(u)dsdudt$$

$$= \int_{t\in[0,1]^d} \iint_{s,u\in[0,1]^d} g(s)\mathbb{I}\{t \leq s\}\mathbb{I}\{t \leq u\}g(u)dsdudt = \int_{t\in[0,1]^d} h^2(t)dt.$$

The left-hand side is equal to

$$\frac{1}{\sqrt{\langle n \rangle}} \sum_{j \leq n} h\left(\frac{j}{|n|}\right) X_j$$

as was shown in the proof of Lemma 1.12. Note that since $g \in C[0,1]^d$, one has $Lip(h) < \infty$.

Let $p \in \mathbb{N}$. Divide the d-dimensional unit cube into p^d cubes $B_1, \dots, B_{p^d}$ by dividing each of its edges into p equal parts and drawing through the division

---

[5]This is proved exactly as the analogous statement for $H = \mathbb{R}^n$.

points hyperplanes orthogonal to the corresponding axis. Let $h_p : [0,1]^d \to \mathbb{R}$ be such function that it is constant inside any of the "small" cubes $B_1, \ldots, B_{p^d}$ and its value at the center of each cube equals the value of $h(t)$ at the same point. Since

$$|h_p(x)| \leq \sup_{t \in [0,1]^d} |h(t)| + Lip(h), \quad x \in [0,1]^d,$$

by the dominated convergence theorem we have

$$\int_{[0,1]^d} h_p^2(t)dt \to \int_{[0,1]^d} h^2(t)dt, \quad p \to \infty. \tag{1.11}$$

Applying Lemmas 1.6 and 1.9 to the family of cubes $B_1, \ldots, B_{p^d}$, we deduce, taking into account the properties of h_p, that

$$S_n(h_p) = \frac{1}{\sqrt{\langle n \rangle}} \sum_{k \leq n} h_p\left(\frac{k}{n}\right) X_k \to N\left(0, \sigma^2 \int_{[0,1]^d} h_p^2(t)dt\right), \quad n \to \infty.$$

Since $Lip(h) < \infty$, one has a bound $|h(x) - h_p(x)| \leq Lip(h)p^{-1}$, $x \in [0,1]^d$. Therefore, for any $p \in \mathbb{N}$ and all n large enough, we have

$$\mathsf{E}(S_n(h) - S_n(h_p))^2 = \frac{1}{\langle n \rangle} \sum_{j,k \leq n} \left(h\left(\frac{j}{n}\right) - h_p\left(\frac{j}{n}\right)\right)\left(h\left(\frac{k}{n}\right) - h_p\left(\frac{k}{n}\right)\right)\mathsf{E}X_j X_k$$

where for all j, k

$$\left|h\left(\frac{j}{n}\right) - h_p\left(\frac{j}{n}\right)\right|\left|h\left(\frac{k}{n}\right) - h_p\left(\frac{k}{n}\right)\right| \leq (Lip(h))^2 p^{-2}.$$

Consequently, by (BL, θ)-dependence (as in the last step of proof of Lemma 1.12)

$$\mathsf{E}(S_n(h) - S_n(h_p))^2 \leq \theta_1 p^{-2}(Lip(h))^2. \tag{1.12}$$

Take in $\mathbb{N}^d$ an arbitrary sequence $n_k \to \infty$. In order to check that $S_{n_k}(h) \to Z \sim N(0, \sigma^2 \int h^2(t)dt)$, as $k \to \infty$, we shall employ an elementary

Lemma 1.14. *Let for an array $\{\xi_{k,p}, k, p \in \mathbb{Z}_+\}$ of integrable random variables*
   1) $\xi_{k,p} \to \xi_{0,p}$ in law for any $p \in \mathbb{N}$ as $k \to \infty$,
   2) $\xi_{0,p} \to \xi_{0,0}$ in law as $p \to \infty$,
   3) $\mathsf{E}|\xi_{k,p} - \xi_{k,0}| \to 0$ uniformly in $k \in \mathbb{N}$ as $p \to \infty$.
   Then $\xi_{k,0} \to \xi_{0,0}$ in law as $k \to \infty$.

Proof. Let $\varphi_{k,p}(\cdot)$ be a characteristic function of a random variable $\xi_{k,p}$ with $k, p \in \mathbb{Z}_+$. For an arbitrary $\lambda \in \mathbb{R}$ we show that $\varphi_{k,0}(\lambda) \to \varphi_{0,0}(\lambda)$, $k \to \infty$. Since $|\mathsf{E}\exp(i\lambda\eta) - \mathsf{E}\exp(i\lambda\zeta)| \leq |\lambda|\mathsf{E}|\eta - \zeta|$ for integrable random variables η and ζ and all $\lambda \in \mathbb{R}$, we see that $\varphi_{k,p}(\lambda) \to \varphi_{k,0}(\lambda)$ uniformly in $k \in \mathbb{N}$, as $p \to \infty$. Take arbitrary $\varepsilon > 0$ and pick p so large that $|\varphi_{k,p}(\lambda) - \varphi_{k,0}(\lambda)| < \varepsilon$ for all $k \in \mathbb{Z}_+$. Then the upper limit (as $k \to \infty$) of the right-hand side of the inequality

$$|\varphi_{k,0}(\lambda) - \varphi_{0,0}(\lambda)| \leq |\varphi_{k,p}(\lambda) - \varphi_{k,0}(\lambda)| + |\varphi_{k,p}(\lambda) - \varphi_{0,p}(\lambda)| + |\varphi_{0,p}(\lambda) - \varphi_{0,0}(\lambda)|$$

does not exceed 2ε. Consequently, it is zero. $\square$

Now take $\xi_{k,p} = S_n(h_p)$, $\xi_{k,0} = S_n(h)$, here $n = n_k$, $k, p \in \mathbb{N}$ and

$$\xi_{0,p} \sim N(0, \sigma^2 \int h_p^2(t)dt), \quad \xi_{0,0} \sim N(0, \sigma^2 \int h^2(t)dt).$$

From (1.11) and (1.12) we infer that the conditions of Lemma 1.14 hold and therefore $S_n(h) \to Z$ as $n_k \to \infty$. Theorem is proved. $\square$

Remark 1.15. Along with the mentioned above, the weak functional Gaussian approximation was treated in the papers by Birkel [41], Burton and Kim [86] (where the random measures were also considered), Matula and Rychlik [296] (the non-stationary case). The weak approximation of associated and related random systems by Gaussian ones is still elaborated intensively, see Kim and Ko [232, 233], Doukhan and Lang [139]. We refer the reader to the works of Burton, Dabrowski and Dehling [85], Baek and Kim [10] concerning the generalizations for vector-valued random systems. The Barbour – Götze – Stein techniques to FCLT for associated random fields were employed by Bulinski and Shabanovich [78].

2 Strong invariance principle

In the previous Section we were concerned with the problem of weak convergence to the distribution of a Wiener process in the Skorokhod space for distributions of partial sums processes $\{W_n, n \in \mathbb{N}^d\}$ (as $n \to \infty$). Now we shall consider the problem of the almost sure approximation of the trajectories of W_n by those of limiting process. This research direction, called *strong invariance principle* (as approximation is almost sure), was founded in the classical papers by Skorokhod [387] and Strassen [390]. Note that a key idea here is to redefine initial stochastic processes on a new probability space, together with limiting processes, to have their trajectories close in a sense. At first important results were obtained for the partial sums processes generated by sequences of independent random variables, see the papers by Csörgő and Révész [113, 114], Kómlos, Major, Tusnady [240, 241], Borovkov [50], Sakhanenko [360] and other researchers. Various techniques were proposed (Skorokhod embedding, quantile transformation by Csörgő and Révész, the reconstruction method by Berkes and Philipp, the employment of entropy etc.). We mention in passing that further generalizations for vector-valued partial sum processes with independent increments are due to Zaitsev [425] and Sakhanenko [146]. Later one began the investigation of partial sum processes corresponding to the sequences of dependent random variables, see, e.g., works by Strassen [393], Philipp and Stout [328], Yu [424], Jacod and Shiryaev [216], Revus and Yor [343]. In 1981 Berkes and Morrow [36] studied the strong invariance principle for mixing random fields, whereas for associated random fields the deep result was established only recently by Balan [17]. Note that strong invariance principle enables us to obtain important corollaries. On these lines we will investigate the problems related to the law of the iterated logarithm.

1°. The main result. Suppose that $X = \{X_j, j \in \mathbb{Z}^d\}$ is a (BL, θ)-dependent random field and

$$D_p := \sup_{j \in \mathbb{Z}^d} \mathsf{E}|X_j|^p < \infty \text{ for some } p > 2, \tag{2.1}$$

$$\theta_r \leq c_0 e^{-\lambda r}, r \in \mathbb{N}, \text{ for some } c_0 > 1 \text{ and } \lambda > 0, \tag{2.2}$$

and

$$\sigma^2 := \sum_{j \in \mathbb{Z}^d} cov(X_0, X_j) \neq 0. \tag{2.3}$$

Set, as usual,

$$S(V) = \sum_{j \in V \cap \mathbb{Z}^d} X_j, \quad V \subset \mathbb{R}^d,$$

therefore, $S(V) = S(V \cap \mathbb{Z}^d)$. We write $S_N = S((0, N])$ for $N \in \mathbb{N}^d$ where

$$(0, N] = (0, N_1] \times \ldots \times (0, N_d], \quad N = (N_1, \ldots, N_d).$$

We use the collection $\mathcal{U}$ of blocks introduced in Section 2.1. Recall that (BL, θ)-dependence entails the convergence of the series in (2.3) for a field X having $\mathsf{E}X_j^2 < \infty$, $j \in \mathbb{Z}^d$ (see Remark 1.10). By Theorem 3.1.8,

$$\mathsf{Var}(S_N) \sim \sigma^2 \langle N \rangle, \text{ as } N \to \infty,$$

where $N \in \mathbb{N}^d$. If $d > 1$ then, following [36], for any $\tau > 0$ we introduce the set

$$G_\tau = \bigcap_{s=1}^{d} \left\{ j \in \mathbb{N}^d : j_s \geq \left(\prod_{s' \neq s} j_{s'} \right)^\tau \right\}. \tag{2.4}$$

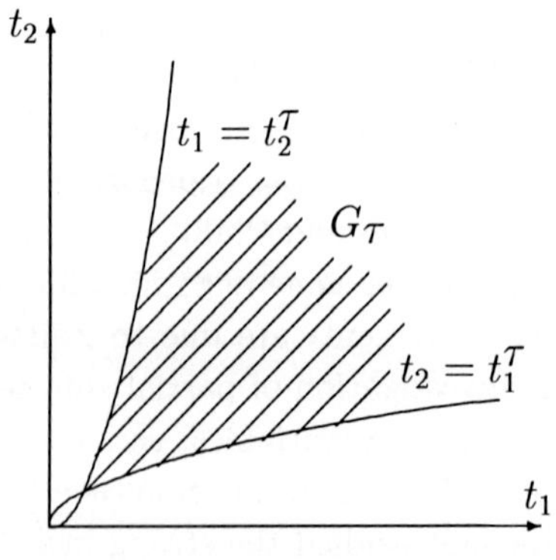

Fig. 5.2 $(d = 2)$

In the case $d = 1$ we set $G_\tau = \mathbb{N}$ for any $\tau > 0$. Note that $G_\tau = \varnothing$ for $\tau > 1/(d-1)$, $d > 1$.

Theorem 2.1. ([80]) *Suppose that X is a wide-sense stationary, (BL, θ)-dependent, centered random field satisfying (2.1), (2.2) and (2.3). Then, for any $\tau > 0$, one can redefine X, without changing its distribution, on a new probability space together with a d-parameter Wiener process $W = \{W_t, t \in [0, \infty)^d\}$, so that for some $\varepsilon > 0$ the following relation holds:*

$$S_N - \sigma W_N = O(\langle N \rangle^{1/2 - \varepsilon}) \text{ a.s.} \tag{2.5}$$

as $N \to \infty$, $N \in G_\tau$. The value ε in (2.5) depends on a field X. More precisely, ε is determined by τ, the covariance function of X and parameters d, p, D_p, c_0, λ.

Proof. The proof adapts the approach of [17] and [36]. However, as the random field under consideration possesses a dependence property more general than association, we have to involve other results on normal approximation for partial sums and use different moment and maximal inequalities. We also give simplified proofs of some steps employed in [17] for associated random fields.

$2°$. **Construction of auxiliary blocks.** Let $\alpha > \beta > 1$ be integers specified later. Introduce

$$n_0 = 0, \quad n_l := \sum_{i=1}^{l} (i^\alpha + i^\beta), \quad l \in \mathbb{N}.$$

For $k \in \mathbb{N}^d$, put $k - 1 = (k_1 - 1, \ldots, k_d - 1)$ and $\mathsf{N}_k = (n_{k_1}, \ldots, n_{k_d})$. Let

$$B_k = (\mathsf{N}_{k-1}, \mathsf{N}_k], \quad H_k = \prod_{s=1}^{d} (n_{k_s - 1}, n_{k_s - 1} + k_s^\alpha], \quad I_k = B_k \setminus H_k, \tag{2.6}$$

$$u_k = S(H_k), \quad \sigma_k^2 = \mathsf{Var}(u_k), \quad v_k = S(I_k), \quad \tau_k^2 = \mathsf{Var}(v_k).$$

Thus the sets I_k form the "corridors" between the "rooms" (i.e. blocks H_k, $k \in \mathbb{N}^d$).

We can redefine the random field $\{X_k, k \in \mathbb{N}^d\}$ on another probability space together with a random field $\{w_k, k \in \mathbb{N}^d\}$, consisting of independent random variables such that

$$w_k \sim N(0, \tau_k^2), \quad k \in \mathbb{N}^d$$

and the fields $\{u_k\}$ and $\{w_k\}$ are independent.

Consider now a new set G_ρ with $\rho = \tau/8$, L be the set of all indices i corresponding to the ("good") blocks $B_i \subset G_\rho$, and H be the union of all good blocks. For each $N = (N_1, \ldots, N_d) \in H$, let $N^{(1)}, \ldots, N^{(d)}$ be the points in $\mathbb{N}^d$ defined as follows:

$$N_{s'}^{(s)} := N_{s'}, \ s' \neq s \quad \text{and} \quad N_s^{(s)} := \min\left\{n_s : n \in H, \ n_{s'} = N_{s'}, s' \neq s\right\}.$$

We also consider the sets

$$R_k = (M_k, \mathsf{N}_k] \quad \text{where} \quad M_k = \left((\mathsf{N}_k^{(1)})_1, \ldots, (\mathsf{N}_k^{(d)})_d\right), \quad L_k = \{i : B_i \subset R_k\}.$$

So $R_k = (M_k, \mathsf{N}_k]$ is the largest block of the type $(a, \mathsf{N}_k]$ which is contained in H.

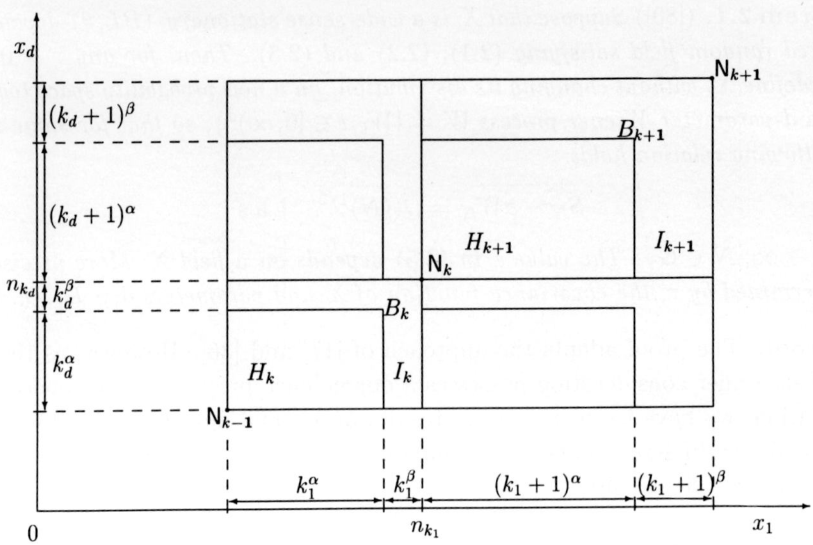

Fig. 5.3 $(d = 2)$

If $U = (a, b] \cap \mathbb{Z}_+^d$ is a block in $\mathbb{Z}_+^d$ for some $a, b \in \mathbb{Z}_+^d$, and $\overline{U} = (a, b] \subset \mathbb{R}_+^d$, then let $W(U) := W(\overline{U})$. Here $W(\overline{U})$ is defined as usual (i.e. the signed sum of $W(t_l)$ where t_l, $l = 1, \ldots, 2^d$, are the vertices of $\overline{U}$, see (1.5)).

We are going to estimate $|S_N - \sigma W_N|$ for $N \in G_\tau$ and $\mathsf{N}_k < N \le \mathsf{N}_{k+1}$, $k \in \mathbb{N}^d$. Note that

$$|S_N - \sigma W_N| \le |S((0, N]) - S((0, \mathsf{N}_k])| + |S((0, \mathsf{N}_k]) - S(R_k)|$$

$$+|S(R_k) - \sigma W(R_k)| + |\sigma W(R_k) - \sigma W((0, \mathsf{N}_k])| + |\sigma W((0, \mathsf{N}_k]) - \sigma W((0, N])|.$$

All the terms involving the Wiener process can be considered as sums of independent identically distributed normal random variables, thus, forming a weakly dependent field, and we will proceed only with the partial sums generated by X.

3°. Variances of partial sums. Further on, we will denote by C any positive factor which might depend only on $d, c_0, \lambda, p, D_p, \tau$ and the covariance function of the field X, unless otherwise stated. Occasionally, when C is a positive random variable, we write[6] $C(\omega)$. Now we pass to a number of lemmas.

By wide-sense stationarity and due to (2.3), we have $\sigma_k^2 > 0$ for any $k \in \mathbb{N}^d$ (see Lemma 3.1.11). Also there exists $k_0 \in \mathbb{N}$ such that $\tau_k^2 > 0$ for all $k \in \mathbb{N}^d$ with $\min_{s=1,\ldots,d} k_s \ge k_0$. This is an immediate consequence of the following

Lemma 2.2. *Let X be a random field such that $D_2 < \infty$ and (2.2) holds. Then, for any finite union U of pairwise disjoint blocks $U_q \in \mathcal{U}$, i.e. $U = \cup_{q=1}^{m(U)} U_q$ where $1 \le m(U) \le m_0$ for some $m_0 \in \mathbb{N}$, one has*

$$\sigma^2 - \frac{1}{|U|} \mathsf{Var} S(U) = O(l(U)^{-1/2}) \quad \text{as} \quad l(U) \to \infty, \tag{2.7}$$

---

[6] A point ω, in general, belongs to an extension of the initial probability space.

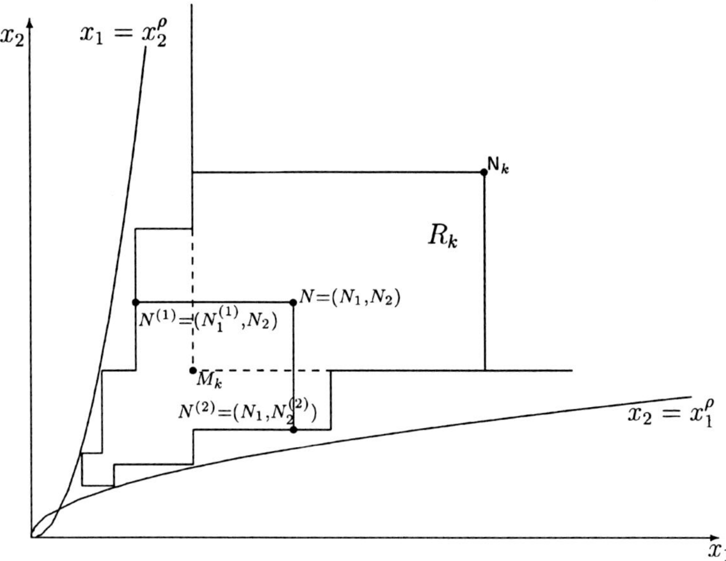

Fig. 5.4 $(d = 2)$

here $l(U)$ is the minimal edge of all blocks U_q. If instead of (2.2) one only has the bound $\theta_r = O(r^{-\lambda})$ as $r \to \infty$ (for some $\lambda > 0$), then (2.7) holds with right-hand side replaced with $O(l(U)^{-\lambda/(1+\lambda)})$.

Proof. Let $l_{q,i}$, $i = 1,\ldots,d$, be the list of edges' length for U_q. Set $l_{q,0} = \min_{1 \le i \le d} l_{q,i}$. Thus $l(U) = \min_{1 \le q \le m(U)} l_{q,0}$. We have

$$\left| \sigma^2 |U| - \mathsf{Var}S(U) \right| = \left| \sum_{j \in U, k \notin U} cov(X_j, X_k) \right| \le \sum_{q=1}^{m(U)} (|T_{q,1}| + |T_{q,2}|)$$

where $T_{q,1}$ is the sum of $cov(X_j, X_k)$ over those $k \notin U$ and $j \in U_q$ for which $dist(\{j\}, \mathbb{Z}^d \setminus U_q) \ge l_{q,0}^{1/2}$, $T_{q,2}$ denoting the analogous sum over $k \notin U$ and the rest of $j's$ in U_q $(q = 1, \ldots, m(U))$.

By (2.2) one has

$$|T_{q,1}| \le c_0 |U_q| \exp\{-\lambda l_{q,0}^{1/2}\} \le c_0 |U_q| \exp\{-\lambda l(U)^{-1/2}\}, \quad 1 \le q \le m(U).$$

Furthermore, let us show that

$$|T_{q,2}| \le 2dC |U_q| l(U)^{-1/2}, \quad 1 \le q \le m(U), \tag{2.8}$$

with $C = \sum_{k \in \mathbb{Z}^d} |cov(X_0, X_k)| < \infty$. For $d = 1$

$$l_q - (l_q - 2[l_q^{1/2}]) = 2[l_q^{1/2}] \le 2l_q^{1/2} = 2l_q/l_q^{1/2} = 2|U_q| l(U)^{-1/2}.$$

For $d > 1$ and $x_1, \ldots, x_d > 0$, $0 < x_0 \le \min_{1 \le i \le d} x_i$ note that

$$\prod_{i=1}^{d} x_i - \prod_{i=1}^{d} (x_i - x_0) = \prod_{i=1}^{d} x_i - (x_1 - x_0) \prod_{i=2}^{d} x_i + (x_1 - x_0) \prod_{i=2}^{d} x_i - \ldots$$

$$+ \left(\prod_{i=1}^{d-1} (x_i - x_0) \right) x_d - \prod_{i=1}^{d} (x_i - x_0) \le dx_0 \prod_{i=1}^{d} x_i \left(\min_{1 \le i \le d} x_i \right)^{-1}.$$

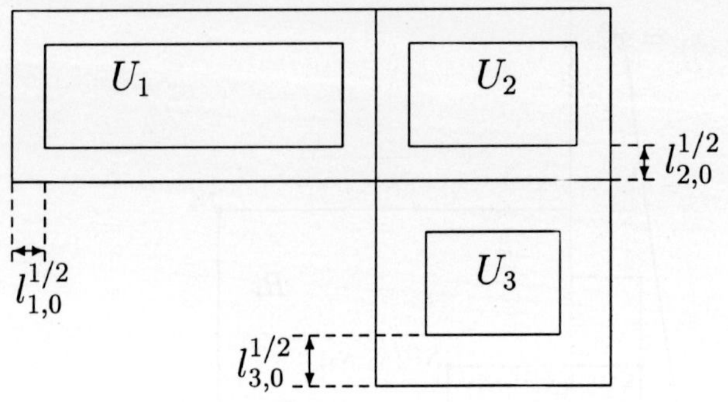

Fig. 5.5 $(d = 2,\ m(U_n) = 3)$

Thus

$$\left|\{j \in U_q : dist(\{j\}, \mathbb{Z}^d \setminus U_q) < [l_{q,0}^{1/2}]\}\right| \leq 2d[l_{q,0}^{1/2}]\frac{|U_q|}{l_{q,0}} \leq 2d|U_q|l_{q,0}^{-1/2}. \qquad (2.9)$$

Using (2.9) we come to (2.8).

For the second assertion, take some $\varkappa > 0$ to be specified later and repeat the argument, letting now $T_{q,1}$ stand for the sum over all $k \notin U$ and $j \in U_q$ such that $dist(\{j\}, \mathbb{Z}^d \setminus U_q) > l(U)^\varkappa$. Then $|T_{q,1}| \leq C|U|l(U)^{-\varkappa\lambda}$, $|T_{q,2}| \leq 2dC|U_q|l_{q,0}^{\varkappa-1}$. It remains to choose $\varkappa = (\lambda + 1)^{-1}$. $\square$

Remark 2.3. Let, for some $c_0, \lambda > 0$, the Cox–Grimmett coefficients admit the bound $u_r \leq c_0 r^{-\lambda}$, $r \in \mathbb{N}$. As seen from the proof, in association case one can write

$$\sigma^2 - |U|^{-1}\mathrm{Var}S(U) \leq m(U)(c_0 + 2d\sigma^2)l(U)^{-\lambda/(1+\lambda)}.$$

4°. Employment of the quantile transform by Csörgő–Révész. In what follows, we will consider only k "large enough", that is having $\min_s k_s \geq k_0$. For such k and $x \in \mathbb{R}$, let

$$\xi_k = \frac{u_k + w_k}{\sqrt{\sigma_k^2 + \tau_k^2}}, \quad F_k(x) = \mathsf{P}(\xi_k \leq x). \qquad (2.10)$$

Then ξ_k has a density $f_k(x)$, because of the used convolution with a Gaussian random variable w_k having density.

Analogously to [17] we introduce the random variables

$$\eta_k = \Phi^{-1}(F_k(\xi_k)), \quad e_k = \sqrt{\sigma_k^2 + \tau_k^2}(\xi_k - \eta_k), \quad k \in \mathbb{N}^d, \qquad (2.11)$$

where

$$\Phi(x) = (2\pi)^{-1/2}\int_{-\infty}^{x} e^{-t^2/2}dt, \quad x \in \mathbb{R}.$$

Obviously $F_k(\xi_k)$ has uniform distribution on $[0, 1]$. Therefore, $\eta_k \sim N(0, 1)$ for all k large enough.

Clearly,

$$S(R_k) = \sum_{t \in L_k} S(B_t) = \sum_{t \in L_k} (u_t + w_t) - \sum_{t \in L_k} w_t + \sum_{t \in L_k} v_t$$

$$= \sum_{t \in L_k} e_t + \sum_{t \in L_k} \eta_t \sqrt{\sigma_t^2 + \tau_t^2} - \sum_{t \in L_k} w_t + \sum_{t \in L_k} v_t$$

$$= \sum_{t \in L_k} e_t + \sum_{t \in L_k} \sqrt{|B_t|} \left(\sqrt{\frac{\sigma_t^2 + \tau_t^2}{|B_t|}} - \sigma \right) \eta_t$$

$$+ \sum_{t \in L_k} \sigma \sqrt{|B_t|} \eta_t - \sum_{t \in L_k} w_t + \sum_{t \in L_k} v_t. \tag{2.12}$$

We need also the following result on perturbation of a field X.

Lemma 2.4. *Suppose that $X = \{X_j, j \in \mathbb{Z}^d\}$ is a (BL, θ)-dependent random field and $Y = \{Y_j, j \in \mathbb{Z}^d\}$ is a field consisting of independent random variables and independent of X. Let $I, J \subset \mathbb{Z}^d$ be disjoint finite sets, $f : \mathbb{R}^{|I|} \to \mathbb{R}$ and $g : \mathbb{R}^{|J|} \to \mathbb{R}$ be bounded Lipschitz functions. Then[7]*

$$|cov(f(X_I + Y_I), g(X_J + Y_J))| \le Lip(f)Lip(g)(|I| \wedge |J|)\theta_r, \quad r = dist(I, J). \tag{2.13}$$

Proof. If the Lemma is true for the vector (Y_I, Y_J) having density, then it is true for a general field Y satisfying its conditions. Indeed, take a field $Z = \{Z_j, j \in \mathbb{Z}^d\}$ consisting of independent standard Gaussian variables, independent from X and Y. Define random fields $Y^{(n)} = \{Y_j + n^{-1}Z_j, j \in \mathbb{Z}^d\}$, $n \in \mathbb{N}$. Clearly $Y^{(n)}$ does not depend on X and the random vector $(Y_I^{(n)}, Y_J^{(n)})$ has a density. It remains to notice that $(Y_I^{(n)}, Y_J^{(n)})$ converge in distribution to (Y_I, Y_J) as $n \to \infty$.

Suppose that the random vector (Y_I, Y_J) has a density $q(t^1, t^2) = q_I(t^1)q_J(t^2)$, here $t^1 \in \mathbb{R}^{|I|}, t^2 \in \mathbb{R}^{|J|}$. Evidently (2.13) is true for a field Y consisting of some constants. Thus by independence hypothesis, the Fubini theorem and the definition of (BL, θ)-dependence we have

$$|cov(f(X_I + Y_I), g(X_J + Y_J))| = \left| \iint cov(f(X_I + y_I), g(X_J + y_J))q(y_I, y_J)dy_I dy_J \right|$$

$$\le Lip(f)Lip(g)(|I| \wedge |J|)\theta_r \iint q(y_I, y_J)dy_I dy_J = Lip(f)Lip(g)(|I| \wedge |J|)\theta_r$$

where the double integral is taken over $\mathbb{R}^{|I|} \times \mathbb{R}^{|J|}$. $\square$

Lemma 2.5. *If (2.1), (2.2) and (2.3) are satisfied, then, for all $k \in L$ large enough,*

$$\sup_{x \in \mathbb{R}} |F_k(x) - \Phi(x)| \le C\langle k \rangle^{-\alpha\mu}$$

where $\mu > 0$ does not depend on k (F_k was introduced in (2.10)).

---

[7]X_I, X_J, Y_I and Y_J are defined in accordance with Definition 1.1.1.

Proof. If X is a weakly dependent random field and we replace some of the variables X_j with zeros, then the random field $X' = \{X'_j, j \in \mathbb{N}^d\}$ obtained is again weakly dependent with the same sequence $\theta = (\theta_r)_{r \in \mathbb{N}}$ (e.g., by Lemma 1.5.16). Let $Y_j \sim N(0, \tau_k^2/|I_k|)$ for $j \in I_k$ and $Y_j = 0$ a.s. for other $j \in H_k$. Consider the random variables $T_j = X'_j + Y_j$, $j \in \mathbb{Z}^d$. Then

$$\frac{1}{|B_k|} \mathsf{Var}\left(\sum_{j \in B_k} T_j \right) = \frac{\sigma_k^2 + \tau_k^2}{|B_k|} \geq \frac{\sigma_k^2 + \tau_k^2}{2^d \langle k \rangle^\alpha} \tag{2.14}$$

for any B_k, $k \in \mathbb{N}^d$, as $k_i^\alpha + k_i^\beta \leq 2k_i^\alpha$, $i = 1, \ldots, d$. Now the desired result follows from Lemma 2.4 and Theorem 3.1.21, applied to the field $T = \{T_j, j \in \mathbb{Z}^d\}$. $\square$

Lemma 2.6. *Let $(a_n)_{n \in \mathbb{N}}$, $(b_n)_{n \in \mathbb{N}}$ be two sequences of points taken from the interval $(0, 1)$ so that $a_n \to 1$, $b_n \to 1$ and $(1 - a_n)/(1 - b_n) \to 1$ as $n \to \infty$. Then*

$$(\Phi^{-1}(a_k))^2 - (\Phi^{-1}(b_k))^2 \to 0, \quad n \to \infty.$$

Proof. Without loss of generality we can assume that $a_n \leq b_n$ for all n (letting $\tilde{a}_n = a_n \wedge b_n$ and $\tilde{b}_n = a_n \vee b_n$, $n \in \mathbb{N}$). Set $x_n = \Phi^{-1}(a_n)$, $y_n = \Phi^{-1}(b_n)$. Then $0 < x_n \leq y_n$ and $x_n \to \infty$, $y_n \to \infty$ as $a_n \to 1$, $b_n \to 1$ $(n \to \infty)$. One has

$$\frac{1 - a_n}{1 - b_n} = \frac{\int_{x_n}^{+\infty} e^{-x^2/2} dx}{\int_{y_n}^{+\infty} e^{-x^2/2} dx} \sim \frac{y_n e^{-x_n^2/2}}{x_n e^{-y_n^2/2}} = e^{(y_n^2 - x_n^2)/2} \frac{y_n}{x_n}, \quad n \to \infty.$$

We know that $(1 - a_n)/(1 - b_n) \to 1$ as $n \to \infty$. Therefore

$$e^{(y_n^2 - x_n^2)/2} \frac{y_n}{x_n} \to 1, \quad n \to \infty.$$

Both factors here are not less than 1. Consequently, $\exp\{y_n^2 - x_n^2\} \to 1$ as $n \to \infty$. Thus, $y_n^2 - x_n^2 \to 0$, which is our claim. $\square$

Lemma 2.7. *Under conditions* (2.1), (2.2) *and* (2.3), *for any $K \in (0, \sqrt{2\alpha\mu})$ and for all $k \in L$ large enough, one has*

$$|\Phi^{-1}(F_k(x)) - x| \leq C\langle k \rangle^{-\alpha\mu + K^2/2}$$

if $|x| \leq K\sqrt{\mathrm{Log}\,\langle k \rangle}$. Here C may also depend on K.

Proof. First of all, notice that since $K^2/2 < \alpha\mu$ and

$$1 - \Phi(K\sqrt{\log\langle k \rangle}) \sim (2\pi)^{-1/2} \frac{e^{-(K^2/2)\log\langle k \rangle}}{K\sqrt{\log\langle k \rangle}}, \quad \langle k \rangle \to \infty,$$

one has $\Phi(K\sqrt{\log\langle k \rangle}) + C\langle k \rangle^{-\alpha\mu} < 1$ for all $\langle k \rangle$ large enough, where C is the same as in Lemma 2.6. In the rest of the proof we consider only such k. For $u \in (0, 1)$, by a standard calculus formula,

$$\frac{d\Phi^{-1}(u)}{du} = \frac{1}{\Phi'(\Phi^{-1}(u))} = \sqrt{2\pi}\exp\left((\Phi^{-1}(u))^2/2\right). \tag{2.15}$$

Now the Lagrange formula shows that, for any $x \in \mathbb{R}$,

$$\Delta_k(x) := |\Phi^{-1}(F_k(x)) - x| \le |F_k(x) - \Phi(x)| \sup_{F_k(x) \wedge \Phi(x) \le u \le F_k(x) \vee \Phi(x)} \left| \frac{d\Phi^{-1}(u)}{du} \right|.$$

If $|x| \le K\sqrt{\log\langle k \rangle}$ then, on applying Lemma 2.5, one has

$$\Delta_k(x) \le C\langle k \rangle^{-\alpha\mu} \sup_{u \in M(k)} \exp\left\{ (\Phi^{-1}(u))^2/2 \right\},$$

as the interval $[F_k(x) \wedge \Phi(x), F_k(x) \vee \Phi(x)] \subset M(k) := \{u : |u| \le \Phi(K\sqrt{\log\langle k \rangle}) + C_1\langle k \rangle^{-\alpha\mu}\} \subset (-1, 1)$. Therefore

$$\sup_{|x| \le K\sqrt{\log\langle k \rangle}} \Delta_k(x) \le C\langle k \rangle^{-\alpha\mu}\sqrt{2\pi} \sup_{u \in M(k)} \exp\{(\Phi^{-1}(u))^2/2\}. \qquad (2.16)$$

Let $a_{\langle k \rangle} = \Phi(K\sqrt{\log\langle k \rangle})$, $b_{\langle k \rangle} = \Phi(K\sqrt{\log\langle k \rangle}) + C_1\langle k \rangle^{-\alpha\mu}$. Our goal is to estimate $\exp\left((\Phi^{-1}(b_{\langle k \rangle}))^2/2\right)$. We have $a_{\langle k \rangle} \to 1$, $b_{\langle k \rangle} \to 1$ and

$$(1 - a_{\langle k \rangle})/(1 - b_{\langle k \rangle}) \to 1$$

as $\langle k \rangle \to \infty$ (the last relation is true because $\langle k \rangle^{-\alpha\mu} = o(1 - \Phi(K\sqrt{\log\langle k \rangle}))$, $\langle k \rangle \to \infty$). On account of Lemma 2.6 we have

$$\exp\left((\Phi^{-1}(b_{\langle k \rangle}))^2/2\right) = \exp\left\{ (\Phi^{-1}(b_{\langle k \rangle}))^2/2 - (\Phi^{-1}(a_{\langle k \rangle}))^2/2 \right\} \exp\left((\Phi^{-1}(a_{\langle k \rangle}))^2/2\right)$$

$$\sim \exp\left\{ (K^2/2)\log\langle k \rangle \right\}.$$

The Lemma follows from this relation and (2.16). $\square$

Lemma 2.8. *If* (2.1), (2.2) *and* (2.3) *hold, then, for all* $k \in \mathbb{N}^d$ *large enough,*

$$\mathsf{E}e_k^2 \le C\langle k \rangle^{\alpha - \varepsilon_0}$$

with $\varepsilon_0 = 2\alpha\mu\delta/(4 + 3\delta)$ *where* $\delta > 0$ *is the same as in Theorem 2.1.18 and* μ *is the same as in Lemma 2.5.*

Proof. Since $\sigma_k^2 + \tau_k^2 \le 2^d(\mathsf{E}X_0^2 + c_0)\langle k \rangle^{\alpha}$, $k \in \mathbb{N}^d$, it suffices to show that $\mathsf{E}(\eta_k - \xi_k)^2 \le C\langle k \rangle^{-\varepsilon_0}$. To this end we take K from Lemma 2.7 and write

$$\mathsf{E}(\eta_k - \xi_k)^2 \mathbb{I}\{|\xi_k| \le K\sqrt{\log\langle k \rangle}\}$$

$$= \mathsf{E}\left(\Phi^{-1}(F_k(\xi_k)) - \xi_k\right)^2 \mathbb{I}\{|\xi_k| \le K\sqrt{\log\langle k \rangle}\} \le C\langle k \rangle^{-2\alpha\mu + K^2}.$$

Moreover, the Hölder and Minkowsky inequalities yield

$$\mathsf{E}(\eta_k - \xi_k)^2 \mathbb{I}\{|\xi_k| > K\sqrt{\log\langle k \rangle}\} \le (\mathsf{E}|\eta_k - \xi_k|^{2+\delta})^{2/(2+\delta)} (\mathsf{P}(|\xi_k| > K\sqrt{\log\langle k \rangle}))^{\delta/(2+\delta)}$$

$$\le 2((\mathsf{E}|\eta_k|^{2+\delta})^{2/(2+\delta)} + (\mathsf{E}|\xi_k|^{2+\delta})^{2/(2+\delta)})(\mathsf{P}(|\xi_k| > K\sqrt{\log\langle k \rangle}))^{\delta/(2+\delta)}.$$

Note that $\|\eta_k\|_{2+\delta}$ does not depend on k, as $\eta_k \sim N(0, 1)$. By Theorem 2.1.18 one can select such $\delta > 0$ that $\mathsf{E}|\xi_k|^{2+\delta} \le C$ for all $k \in \mathbb{N}^d$, with C independent of α

and β. Then for that $\delta > 0$ and any k, taking into account Lemmas 2.5 and 2.7, one has

$$\mathsf{E}(\eta_k - \xi_k)^2 = \mathsf{E}(\eta_k - \xi_k)^2 \mathbb{I}\{|\xi_k| \le K\sqrt{\log\langle k\rangle}\} + \mathsf{E}(\eta_k - \xi_k)^2 \mathbb{I}\{|\xi_k| > K\sqrt{\log\langle k\rangle}\}$$

$$\le C\langle k\rangle^{-2\alpha\mu + K^2} + C(\mathsf{P}(|\eta_k| > K\sqrt{\log\langle k\rangle}))^{\delta/(2+\delta)}$$

$$+ C\left|\mathsf{P}(|\xi_k| > K\sqrt{\log\langle k\rangle}) - \mathsf{P}(|\eta_k| > K\sqrt{\log\langle k\rangle})\right|^{\delta/(2+\delta)}$$

$$\le C(\langle k\rangle^{-2\alpha\mu + K^2} + \langle k\rangle^{-K^2\delta/2(2+\delta)} + \langle k\rangle^{-\alpha\mu\delta/(2+\delta)}).$$

To optimize the last expression in K we take $K^2 = 4\alpha\mu(2+\delta)/(4+3\delta) < 2\alpha\mu$. Thus one obtains $\varepsilon_0 = 2\alpha\mu\delta/(4+3\delta)$. $\square$

Lemma 2.9. *If* (2.1), (2.2), (2.3) *hold and* $\alpha - \beta \le \varepsilon_0/4$, *then, for all* $k \in \mathbb{N}^d$ *large enough,*

$$\sup_{x\in\mathbb{R}} |f_k(x) - f(x)| \le C\langle k\rangle^{2(\alpha-\beta)-\varepsilon_0/2}$$

where $f(x) = (2\pi)^{-1/2}e^{-x^2/2}$, f_k *appeared after* (2.11) *and* ε_0 *is the same as in Lemma 2.8.*

Proof. Let $\varphi_k(t) = \mathsf{E}\exp\{it\xi_k\}$, $\widehat{\varphi}_k(t) = \mathsf{E}\exp\{itu_k/\sqrt{\sigma_k^2 + \tau_k^2}\}$, $\varphi(t) = e^{-t^2/2}$ where $t \in \mathbb{R}$, $i^2 = -1$. Note that

$$\varphi_k(t) = \widehat{\varphi}_k(t)\mathsf{E}\exp\left\{itw_k/\sqrt{\sigma_k^2 + \tau_k^2}\right\} = \widehat{\varphi}_k(t)\exp\left\{-\tau_k^2 t^2/2(\sigma_k^2 + \tau_k^2)\right\}.$$

By Lemma 2.8, for any $t \in \mathbb{R}$,

$$|\varphi_k(t) - \varphi(t)| \le \mathsf{E}|\exp\{it\xi_k\} - \exp\{it\eta_k\}| \le |t|\mathsf{E}|\xi_k - \eta_k| \le |t|\frac{(\mathsf{E}e_k^2)^{1/2}}{\sigma_k^2 + \tau_k^2} \le C|t|\langle k\rangle^{-\varepsilon_0/2}.$$

Therefore for any $T > 0$ and any $x \in \mathbb{R}$, by the inversion formula for characteristic functions (see, e.g., [383, Ch. II, §12.5]) one has, with notation $\nu_k = \tau_k^2/(\sigma_k^2 + \tau_k^2)$,

$$|f_k(x) - f(x)| \le \frac{1}{2\pi}\int_{\mathbb{R}} |\varphi_k(t) - \varphi(t)|dt \le \frac{1}{2\pi}\int_{|t|\le T} |\varphi_k(t) - \varphi(t)|dt$$

$$+ \frac{1}{2\pi}\int_{|t|\ge T} \left(|\widehat{\varphi}_k(t)|e^{-t^2\nu_k/2} + \varphi(t)\right)dt \le CT^2\langle k\rangle^{-\varepsilon_0/2} + C\int_T^\infty e^{-t^2/2}dt$$

$$+ C\int_T^\infty e^{-\nu_k t^2/2}dt \le C\left(T^2\langle k\rangle^{-\varepsilon_0/2} + \frac{1}{T}e^{-T^2/2} + \frac{1}{T\sqrt{\nu_k}}\exp\left\{-\frac{T^2\sqrt{\nu_k}}{2}\right\}\right).$$

Since $\sigma_k^2 = O(\langle k\rangle^\alpha)$ and $\tau_k^2 \ge C\langle k\rangle^\beta$, the Lemma follows if we take $T = \langle k\rangle^{\alpha-\beta}$. $\square$

Remark 2.10. Clearly the condition $\alpha - \beta \le \varepsilon_0/4$ is equivalent to

$$(\alpha/\beta)(1 - \mu\delta/(8 + 6\delta)) \le 1. \tag{2.17}$$

To satisfy (2.17) one has to select α and β large enough so that the ratio α/β be close to 1. Note that in Lemmas 2.8, 2.9 and Lemma 2.11 below one can replace condition (2.2) on λ with that used in Theorem 2.1.18.

Lemma 2.11. *Suppose that* (2.1), (2.2), (2.3) *hold and* α *is so large that* $\varepsilon_0 > 2$. *Then there exists* $\varepsilon_1 > 0$ *such that*

$$\sum_{t \in L_k} |e_t| \leq C(\omega) \langle N_k \rangle^{1/2 - \varepsilon_1} \text{ a.s., } k \in \mathbb{N}^d.$$

Proof. For $q > 0$, invoking Lemma 2.8 we see that

$$\mathsf{P}(|e_t| \geq \langle t \rangle^q) \leq \langle t \rangle^{-2q} \mathsf{E} e_t^2 \leq C \langle t \rangle^{\alpha - \varepsilon_0 - 2q}.$$

If $2q > \alpha - \varepsilon_0 + 1$, then by the Borel–Cantelli lemma $|e_t| \leq C(\omega)\langle t \rangle^q$ a.s., therefore

$$\sum_{t \in L_k} |e_t| \leq \sum_{t \in L_k} C(\omega)\langle t \rangle^q \leq C(\omega)\langle k \rangle^{q+1} \leq C(\omega)\langle N_k \rangle^{(q+1)/(\alpha+1)},$$

as $\langle N_k \rangle \sim C\langle k \rangle^{\alpha+1}$. We have also used that $L_k \subset \{t : B_t \subset (0, N_k]\}$, hence $t \leq k$ for any $t \in L_k$. The Lemma will be established if we take q in such a way that $2q > \alpha - \varepsilon_0 + 1$ but $(q+1)/(\alpha+1) < 1/2$. When $\varepsilon_0 > 2$, this is possible. $\square$

5°. The bounds for auxiliary random variables.

Lemma 2.12. *Assume that* $D_2 < \infty$. *If* (2.2) *holds and* $\alpha - \beta > 6/\rho$ *then, for some* $\varepsilon_2 > 0$,

$$\sum_{t \in L_k} (|v_t| + |w_t|) \leq C(\omega)\langle N_k \rangle^{1/2 - \varepsilon_2} \text{ a.s., } k \in \mathbb{N}^d.$$

Proof. We study only $\sum_{t \in L_k} |v_t|$; the argument for $\sum_{t \in L_k} |v_t|$ is completely analogous. One has $I_t = \cup_{s=1}^d I_t(s)$ where $I_t(s) \in \mathcal{U}$ with $|I_t(s)| \leq C t_s^\beta \prod_{l \neq s} t_l^\alpha$. Consequently one may write

$$v_t = \sum_{s=1}^d v_t(s), \quad v_t(s) = \sum_{j \in I_t(s)} X_j.$$

Then, in view of the estimate $\mathsf{E}|v_t(s)|^2 \leq C|I_t(s)|$ and the Chebyshev inequality, we obtain

$$\mathsf{P}(|v_t(s)| \geq \langle t \rangle^{\alpha/2-1}) \leq C\langle t \rangle^{-\alpha+2}|I_t(s)| \leq C\langle t \rangle^{-\alpha+2} t_s^\beta \prod_{l \neq s} t_l^\alpha = C\langle t \rangle^2 t_s^{-\alpha+\beta}.$$

Since $t_s \geq C(\prod_{l \neq s} t_l)^\rho$, one has $t_s \geq C\langle t \rangle^{\rho/(1+\rho)}$ and $t_s \geq C\langle t \rangle^{\rho/2}$, because $\rho = \tau/8 < 1$. Thus

$$\mathsf{P}(|v_t(s)| \geq \langle t \rangle^{\alpha/2-1}) \leq C\langle t \rangle^{2 - \rho/2(\alpha - \beta)}.$$

So, the series $\sum_{t \in L_k} \mathsf{P}(|v_t(s)| \geq \langle t \rangle^{\alpha/2-1})$ converges if $\alpha - \beta > 6/\rho$. Hence by the Borel–Cantelli lemma, one has, with probability one,

$$\sum_{t \in L_k} v_t(s) \leq C(\omega) \sum_{t \in L_k} \langle t \rangle^{\alpha/2-1} \leq C(\omega)\langle k \rangle^{\alpha/2} \leq C(\omega)\langle N_k \rangle^{\alpha/(2\alpha+2)} = C(\omega)\langle N_k \rangle^{1/2 - \varepsilon_1}$$

as $\alpha/(2\alpha + 2) < 1/2$. $\square$

Lemma 2.13. *Suppose that* $D_2 < \infty$. *If, in addition,* (2.2), (2.3) *hold and* $\beta > 6/\rho$, *then, for some* $\varepsilon_3 > 0$,

$$\sum_{t \in L_k} \sqrt{|B_t|} \left(\sigma - \sqrt{\frac{\sigma_t^2 + \tau_t^2}{|B_t|}} \right) |\eta_t| \leq C(\omega)\langle N_k \rangle^{1/2 - \varepsilon_3} \text{ a.s., } k \in \mathbb{N}^d.$$

Proof. Let us apply Lemma 2.2 to to the union of blocks $\{H_t, I_t(1), \ldots, I_t(d)\}$ (see Lemma 2.12 for the notation). The minimum of the edges of these blocks has by definition the length $\min_{s=1,\ldots,d} t_s^\beta \geq C\langle t \rangle^{\beta\rho/2}$, see Lemma 2.12. Set $b_t = ((\sigma_t^2 + \tau_t^2)/|B_t|)^{1/2}$ By the estimate

$$|\sigma - b_t| \leq \sigma^{-1}\left|\sigma^2 - \frac{1}{|B_t|}\mathsf{Var}(u_t + w_t)\right| \leq C\langle t \rangle^{-\beta\rho/4}$$

ensuing from Lemma 2.2, we have

$$\mathsf{P}\left(\sqrt{|B_t|}\left(\sigma - b_t\right)|\eta_t| \geq \langle t \rangle^{\alpha/2-1}\right) \leq \langle t \rangle^{2-\alpha}|B_t|\left|\sigma - b_t\right|^2 \leq C\langle t \rangle^{2-\beta\rho/2}.$$

If $\beta > 6/\rho$, then $\beta\rho/2 > 3$, hence the Borel–Cantelli lemma yields

$$\sum_{t\in L_k} \sqrt{|B_t|}\left(\sigma - b_t\right)|\eta_t| \leq \sum_{t\leq k} C(\omega)\langle t \rangle^{\alpha/2-1} \leq C(\omega)\langle k \rangle^{\alpha/2} \leq C(\omega)\langle N_k \rangle^{\alpha/(2\alpha+2)} \quad \text{a.s.}$$

because $\langle N_k \rangle \sim C\langle k \rangle^{\alpha+1}$ as $\langle k \rangle \to \infty$. $\square$

In the following Lemma we consider only $d > 1$, since in the case $d = 1$ the enumeration of L by increase is trivial. We remind that $L = \{t \in \mathbb{N}^d : B_t \subset G_\rho\}$ where $\rho = \tau/8$, see (2.4).

Lemma 2.14. ([17]) *Suppose that $L \neq \varnothing$. Then there exists a bijection $\psi : \mathbb{N} \to L$ with the following properties:*

    (a) if $l < m$ then there exists $s^* = s^*(l,m) \in \{1,\ldots,d\}$ such that

$$\psi(l)_{s^*} \leq \psi(m)_{s^*}; \tag{2.18}$$

    (b) if $\gamma_0 > (1+1/\rho)(1-1/d)$, one can find $m_0 \in \mathbb{N}$ such that for all $m \geq m_0$

$$m \leq \langle \psi(m) \rangle^{\gamma_0}. \tag{2.19}$$

Proof. Choose $m \in \mathbb{N}$, $m > 1$, to ensure that $(m,\ldots,m) \in L$ and $k = (k_1,\ldots,k_{d-1},m) \in L$ satisfy the condition $k_s \geq m$, for any $s < d$. Whence all the vertices of B_k will be in G_ρ. In particular $n_m \geq n_{k_s}^\rho$, $s < d$. Since m is fixed, this can only happen for a finite number of the k_s. It follows that, for each $s = 1,\ldots,d-1$, there exists $k_s^*(m) \geq m$ such that $k_s \leq k_s^*(m)$, and one has $k_s^*(m) \leq Cm^{1/\rho}$ for all m large enough. Therefore, we have at most

$$k^*(m) := \prod_{s=1}^{d-1}(k_s^*(m) - m + 1)$$

points of the form $(k_1,\ldots,k_{d-1},m)$ in L with $k_s > m$, $s < d$.

By symmetry we may repeat this argument for each of d axes. Let

$$L_s(m) := \{k = (k_1,\ldots,k_{s-1},m,k_{s+1},\ldots,k_d), \ m \leq k_l \leq k_l^*(m), \ l < d\},$$

for every $s = 1,\ldots,d$, and

$$L(m) = \bigcup_{s=1}^{d} L_s(m).$$

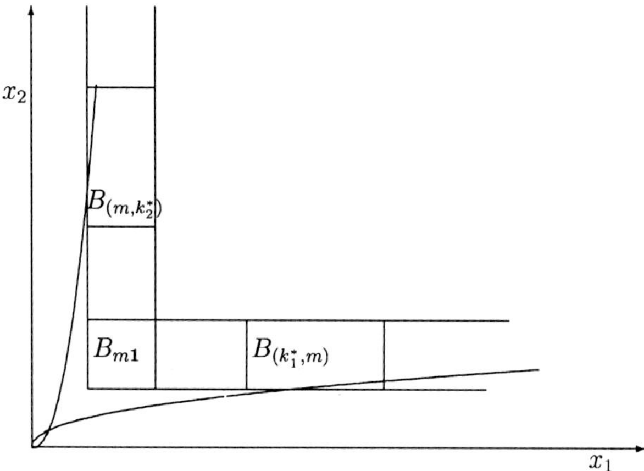

Fig. 5.6 $(d = 2, \, m\mathbf{1} = (m, m))$

Note that $|L(m)| \leq dk^*(m)$ and the inclusion $k \in L(m)$ implies $\langle k \rangle \geq m^d$. Clearly, $L \subset \cup_{m \in \mathbb{N}} L(m)$ (if $B_{m\mathbf{1}} \not\subset G_\rho$ then $L(m) = \varnothing$). Moreover, for each $z \in \mathbb{N}$

$$\sum_{l=1}^{m-1} |L(l)| < z \leq \sum_{l=1}^{m} |L(l)| \Rightarrow \psi(z) \in L(m). \tag{2.20}$$

Now (2.20) implies

$$z \leq d \sum_{l=2}^{m} \prod_{s=1}^{d-1} (k_s^*(l) - l + 1) \leq d \prod_{s=1}^{d-1} \sum_{l=2}^{m} (k_s^*(l) - l + 1) \leq d \prod_{s=1}^{d-1} \left(\sum_{l=2}^{m} k_s^*(l) \right)$$

$$\leq Cm^{(1+1/\rho)(d-1)} \leq Cm^{d\gamma_0} \leq C\langle \psi(z) \rangle^{\gamma_0}$$

for m large enough and arbitrary $\gamma_0 > (1 + 1/\rho)(1 - 1/d)$. Thus (2.19) follows. $\square$

Set $Y_m = \eta_{\psi(m)}$, $m \in \mathbb{N}$, using η_k determined in (2.12).

Lemma 2.15. *If* (2.1), (2.2), (2.3) *and* (2.17) *hold, then for every* $m \in \mathbb{N}, m > 1$, *and all* $t = (t_1, \ldots, t_m) \in \mathbb{R}^m$, *one has*

$$\varphi_m(t) := \left| \mathsf{E} \exp\left\{ i \sum_{l=1}^{m} t_l Y_l \right\} - \mathsf{E} \exp\left\{ i \sum_{l=1}^{m-1} t_l Y_l \right\} \mathsf{E} \exp\{ it_m Y_m \} \right| \leq Cm\Delta^\gamma \tag{2.21}$$

where $i^2 = -1$,

$$\Delta = \langle \psi(m) \rangle^\alpha \theta_r \|t\|^2, \quad r = dist\left(H_{\psi(m)}, \bigcup_{l=1}^{m-1} H_{\psi(l)} \right), \quad \gamma = \begin{cases} 1, & \Delta > 1, \\ 1/3, & \Delta \leq 1, \end{cases} \tag{2.22}$$

$\|t\|^2 = t_1^2 + \ldots + t_m^2$, *the blocks* H_k *being defined in* (2.6).

Proof. Take a number $M > 0$ to be specified later and let $H_M(t)$ be the function defined in (2.1.5). We set $Y_{j,M} = H_M(Y_j)$, $j \in \mathbb{N}$. Note that

$$|\exp\{itY_{j,M}\} - \exp\{itY_j\}| \le 2I\{|Y_j| > M\}, \quad t \in \mathbb{R}.$$

Therefore, for the left-hand side of (2.21) we have

$$\varphi_m(t) \le \left| cov\left(\exp\left\{ i \sum_{l=1}^{m-1} t_l Y_{l,M} \right\}, \exp\{-it_m Y_{m,M}\} \right) \right| + 4 \sum_{l=1}^{m} \mathsf{P}(|Y_l| > M). \quad (2.23)$$

Since $Y_l \sim N(0,1)$, $l \in \mathbb{N}$, by the Markov inequality one has

$$\mathsf{P}(Y_l > M) \le e^{-sM} \mathsf{E} e^{sY_l} = e^{s^2/2 - tM}, \quad s, M > 0.$$

On inserting $s = M$ one comes to the estimate $\mathsf{P}(|Y_l| > M) \le 2e^{-M^2/2}$, $M > 0$.

To estimate the first summand in the right-hand side of (2.23), we notice that, for any $k \in \mathbb{N}^d$, the random variable $\eta_{k,M}$ is a Lipschitz function of ξ_k. Indeed, $Lip(F_k) \le C$ by Lemma 2.9 and $\eta_{k,M} = H_M(\Phi^{-1}(F_k(\xi_k))) = g_M(F_k(\xi_k))$ where

$$g_M(x) = \left(|\Phi^{-1}(x)| \wedge M \right) sgn(\Phi^{-1}(x)), \quad x \in (0,1).$$

Formula (2.15) implies that $Lip(g_M) \le \sqrt{2\pi} e^{M^2/2}$. By Lemma 2.4, one can estimate the covariance in the same way as if the Gaussian variables w_k, $k \in \mathbb{N}^d$, were constants, and with the help of (2.14) we obtain

$$\left| cov\left(\exp\left\{ i \sum_{l=1}^{m-1} t_l Y_{l,M} \right\}, \exp\{-it_m Y_{m,M}\} \right) \right| \le Cm\theta_r e^{M^2} \langle \psi(m) \rangle^\alpha \|t\|^2, \quad (2.24)$$

here we also use that $t_m(\sum_{l=1}^{m-1} t_l) \le \sum_{l=1}^{m-1} (t_l^2 + t_m^2)/2 \le m \sum_{l=1}^{m} t_l^2$ and

$$Y_{m,M} = g_M\left(F_k\left(\frac{S(H_{\psi(m)}) + w_{\psi(m)}}{\sqrt{\sigma_{\psi(m)}^2 + \tau_{\psi(m)}^2}} \right) \right), \quad |H_{\psi(m)}| \le C \langle \psi(m) \rangle^\alpha.$$

Thus, from (2.23) and (2.24) we see that

$$\left| cov\left(\exp\left\{ i \sum_{l=1}^{m-1} t_l Y_l \right\}, \exp\{-it_m Y_m\} \right) \right| \le Cm\left(e^{-M^2/2} + e^{M^2} \langle \psi(m) \rangle^\alpha \theta_r \|t\|^2 \right).$$

The result follows now by optimization in $M > 0$. $\square$

6°. The employment of the Berkes–Philipp techniques.

Lemma 2.16. *Suppose that* (2.1), (2.2), (2.3) *and* (2.17) *hold and* $\beta > 2\gamma_0/\rho$ *where* γ_0 *appears in the formulation of Lemma 2.14. Then we can redefine the random field* X, *without changing its distribution, on a new probability space together with a d-parameter Wiener process* $W = \{W_t, t \in [0,\infty)^d\}$, *such that for some* $\varepsilon_4 > 0$

$$\sum_{t \in L_k} \sigma \sqrt{|B_t|} \left| \eta_t - \frac{W(B_t)}{\sqrt{|B_t|}} \right| \le C(\omega) \langle \mathsf{N}_k \rangle^{1/2 - \varepsilon_4} \quad \text{a.s.}$$

Proof. By the Berkes–Philipp strong approximation result (see Appendix A.6I, Theorem A.15) it is enough to establish the following fact. There exist numbers $\varkappa_m > 0$ and $z_m > 0$ ($m \in \mathbb{N}$, $m > 1$) such that

1) for any $m \in \mathbb{N}, m > 1$, and all $t = (t_1, \ldots, t_m) \in \mathbb{R}^m$ with $\|t\|^2 \leq z_m^2$ one has

$$\left| \mathsf{E} \exp\left\{ i \sum_{l=1}^{m} t_l Y_l \right\} - \mathsf{E} \exp\left\{ i \sum_{l=1}^{m-1} t_l Y_l \right\} \mathsf{E} \exp\{ i t_m Y_m \} \right| \leq \varkappa_m,$$

2) $m z_m^{-2} + m^3 z_m^{2m-1} \exp\{ -m^{-6} z_m^2 / 2 \} + \varkappa_m z_m^{2m} = O(m^{-2}), \quad m \to \infty.$

Assume at first that such $\varkappa_m$ and z_m have been found. Then, by the mentioned result, we can redefine the sequence $\{Y_m\}_{m \in \mathbb{N}}$ on a new probability space together with a sequence $Z = (Z_m)_{m \in \mathbb{N}}$ of independent standard Gaussian variables such that

$$|Y_m - Z_m| = O(m^{-2}) \quad \text{a.s.} \tag{2.25}$$

By extending the probability space one may assume that the field X and the sequences Y and Z are defined together and all their properties are preserved (due to Lemma A.18; see Theorems A.15 and A.22 in Appendix A.6 for a more careful argument).

One has, with probability 1,

$$\sum_{t \in L_k} \sigma \sqrt{|B_t|} \left| \eta_t - Z_{\psi^{-1}(t)} \right| \leq C \langle k \rangle^{\alpha/2} \sum_{t \in L_k} \left(\psi^{-1}(t) \right)^{-2} \leq C \langle k \rangle^{\alpha/2} \leq C \langle \mathsf{N}_k \rangle^{\alpha/2(\alpha+1)},$$

in view of (2.25) and as ψ is a bijection of $\mathbb{N}$ to L.

Therefore, by Theorem A.22 the field X and the sequence Z can be redefined together on a richer probability space supporting a d-parameter Wiener process $W = \{ W_t, t \in \mathbb{R}_+^d \}$ such that $Z_m = W(B_{\psi(m)}) / \sqrt{|B_{\psi(m)}|}$, $m \in \mathbb{N}$. Then

$$\sum_{t \in L_k} \sigma \sqrt{|B_t|} \left| \eta_t - \frac{W(B_t)}{\sqrt{|B_t|}} \right| \leq C \langle \mathsf{N}_k \rangle^{\alpha/2(\alpha+1)} \quad \text{a.s.}$$

Thus the Lemma will be proved if we succeed in finding the appropriate sequences (z_m) and $(\varkappa_m)$. We take $z_m = m^q, q > 8$. Then it suffices to prove that one can take $\varkappa_m = O(\exp\{-m^R\}), m \to \infty$, for some $R > 1$.

The distance r between H_m and any of the blocks $H_1, \ldots, H_{m-1}$ is, by construction, not less than $\min_{s=1,\ldots,d}(\psi(m)_s - 1)^\beta$. Since $\psi(m) \in L$, by Lemma 2.14 for $m > m_0$ we have

$$r \geq C \min_{s=1,\ldots,d} \psi(m)_s^\beta \geq C \langle \psi(m) \rangle^{\rho\beta/2} \geq C m^{\rho\beta/2\gamma_0} \tag{2.26}$$

(the second inequality is true, because if $r \in G_\rho$, then $\min_{s=1,\ldots,d} r_s \geq \langle r \rangle^{\rho/2}$; but for $t \in L$ we have $\mathsf{N}_{k+1} \in G_\rho$, and $(\mathsf{N}_k)_s \sim k_s^{\alpha+1} / (\alpha + 1)$).

From Lemma 2.14 one can also easily see that for all m large enough

$$\langle \psi(m) \rangle \leq C \min_{s=1,\ldots,d} \psi(m)_s^{2/\rho} \leq C m^{2/\rho}. \tag{2.27}$$

Obviously, $\Delta \leq Cm^{2q+2\alpha/\rho} \exp\{-C\lambda m^{\rho\beta/(2\gamma_0)}\} < 1$ for all m large enough. There-fore, for such m, on account of Lemma 2.15, (2.2), (2.26) and (2.27), we have

$$\left| \mathsf{E} \exp\left\{ i \sum_{l=1}^{m} t_l Y_l \right\} - \mathsf{E} \exp\left\{ i \sum_{l=1}^{m-1} t_l Y_l \right\} \mathsf{E} \exp\{it_m Y_m\} \right| \leq Cm\Delta_1^{1/3}$$

where Δ and γ are defined in (2.22). Thus one can take $R \in (1, \rho\beta/(2\gamma_0))$. The Lemma is proved. $\square$

7°. The maxima of partial sums and the proof completion. Induction on d yields

$$|S((0, \mathsf{N}_k] \setminus R_k)| \leq \sum_{s=1}^{d} 2^{d-s} \mathsf{M}_s(\mathsf{N}_k)$$

where $\mathsf{M}_s(N) = \max_{n \leq N^{(s)}} |S_n|$. Furthermore, with the notation

$$I_k^{(J)}(N) = \prod_{s \in J} (n_{k_s}, N_s] \times \prod_{s \notin J} (0, n_{k_s}]$$

where $\mathsf{N}_k < N \leq \mathsf{N}_{k+1}$, $k \in \mathbb{N}^d$ and J is a nonempty subset of $\{1, \ldots, d\}$, we have $(0, N] \setminus (0, \mathsf{N}_k] = \bigcup_J I_k^{(J)}(N)$. Therefore

$$\max_{\mathsf{N}_k < N \leq \mathsf{N}_{k+1}} |S_N - S((0, \mathsf{N}_k])| \leq \sum_J M_k^{(J)} \quad \text{where} \quad M_k^{(J)} = \sup |S(I_k^{(J)}(N))|,$$

the supremum being taken over all N such that $n_{k_s} < N_s \leq n_{k_s+1}$ for $s \in J$ and $N_s = n_{k_s}$ for $s \notin J$.

Lemma 2.17. *Suppose that conditions of Theorem 1.17 are satisfied. Then there exists $\delta > 0$ such that, for any $x > 0$ and any block $U \in \mathcal{U}$,*

$$\mathsf{P}(M(U) \geq x\sqrt{|U|}) \leq Cx^{-2-\delta}.$$

Proof follows from the second assertion of Theorem 2.1.18 and the Markov inequality. $\square$

Lemma 2.18. *Let conditions of Theorem 1.17 hold. If (2.3) is true then there exists $\gamma_1 > 0$ such that, for any block $U = (m, m + n]$ with $n \in G_\rho$ and $m \in \mathbb{Z}_+^d$,*

$$\mathsf{P}(M(U) \geq |U|^{1/2}(\log |U|)^{d+1}) \leq C|U|^{-\gamma_1}$$

where C does not depend on m and n.

Proof. The proof is the same as that for the second inequality of Lemma 7 in [36], and relies on Berry–Esseen type estimates. If suffices to consider $V = (0, n]$, $n \in \mathbb{N}^d$, $n \in G_\rho$, $\langle n \rangle \geq 4^d$. Let $N \in \mathbb{N}^d$ be such that $2^{N_s-1} \leq n_s \leq 2^{N_s}$, $s = 1, \ldots, d$. For points $\nu \in \mathbb{Z}_+^d$, $k \in \mathbb{N}^d$ consider events

$$A(\nu, k) = \left\{ \omega : |S((\nu 2^k, \nu 2^k + 2^k \mathbf{1}])| \geq \langle n \rangle^{1/2} \log\langle n \rangle \right\},$$

$$A = \bigcup_{k \leq N} \bigcup_{\nu < 2^{N-k}} A(\nu, k).$$

By Theorem 2.1.18

$$\mathsf{P}(A(\nu, k)) \leq C \langle 2^{N-k} \rangle^{-1-\delta/2} \tag{2.28}$$

(this probability does not exceed $\mathsf{P}(|S((\nu 2^k, \nu 2^k + 2^k])| \geq \langle 2^k \rangle^{1/2}(\langle 2^{N-1} \rangle / \langle 2^k \rangle)^{1/2}))$).
On applying the obvious modification of Lemma 2.5 (or Theorem 3.1.21) for all $k \leq N$ one has

$$\mathsf{P}(A(\nu, k)) \leq \left| \mathsf{P}\left(\frac{|S((\nu 2^k, \nu 2^k + 2^k])|}{\sqrt{\mathsf{Var} S((\nu 2^k, \nu 2^k + 2^k])}} \geq C \log\langle n \rangle \right) - \mathsf{P}(|Z| \geq C \log\langle n \rangle) \right|$$

$$+ \mathsf{P}(|Z| \geq C \log\langle n \rangle) \leq C \langle 2^k \rangle^{-\mu} + C e^{-C(\log\langle n \rangle)^2/2} \tag{2.29}$$

where $Z \sim N(0, 1)$. Set $J = J(l, N) = \{\nu, k : k \leq N, \sum_i k_i = l, \nu < 2^{N-k}\}$ and

$$p_l = \sum_{\nu, k \in J} \mathsf{P}(A(\nu, k)).$$

If $l \leq \|N\|_1$, then, due to (2.28),

$$p_l \leq C \sum_{\nu, k \in J} \langle 2^{N-k} \rangle^{-1-\delta/2} \leq C \sum_{\substack{k : k \leq N, \\ \sum_i k_i = l}} \langle 2^{N-k} \rangle^{-\delta/2} \leq C (2^{-l} \langle 2^N \rangle)^{-\delta/2} \|N\|_1^d \tag{2.30}$$

as there are at most $\|N\|_1^d$ d-tuples k with $\sum k_i = l \leq \|N\|_1^d$. Since $n \in G_\rho$, for large enough n ($\langle n \rangle \geq 4^d$) one has $N_i \geq \rho \|N\|_1/4$, $i = 1, \ldots, d$. Indeed, this is equivalent to the relation $2^{N_i} \geq \left(\prod_{j=1}^N 2^{N_j} \right)^{\rho/4}$.

Consequently, if $k \in \mathbb{N}^d$, $k \leq N$ and $k_i < N_i/2$ for some i, then

$$k_1 + \ldots + k_d < \|N\|_1 - N_i/2 \leq (1 - \rho/8)\|N\|_1.$$

Thus, for $l \in [(1 - \rho/8)\|N\|_1, \|N\|_1]$ the sum defining p_l contains only the summands with $k_i \geq N_i/2 \geq \rho \|N\|_1/8 \geq \rho \sum k_i/8$. In other words, $2^k \in G_{\rho/8}$. Thus by (2.29)

$$p_l \leq \|N\|_1^q 2^{(\|N\|_1 - l)} C (e^{-\mu_1 (\log\langle n \rangle)^2} + 2^{-\mu l}) \tag{2.31}$$

with some fixed $\mu_1 > 0$. Let $c \in (0, 1)$. For $l \leq c\|N\|_1$ we will use (2.30), for $c\|N\|_1 \leq l \leq \|N\|_1$ we use (2.31). Then

$$\mathsf{P}(A) \leq \sum_{l=1}^{\|N\|_1} p_l \leq C \left(\|N\|_1^d 2^{(c-1)N\delta/2} + \|N\|_1^d e^{-\mu_2 \|N\|_1^2} + \|N\|_1^d 2^{-(c+c\mu-1)\|N\|_1} \right)$$

$$\leq C e^{-\mu_3 \|N\|_1} \leq C \langle n \rangle^{-\mu_3}$$

for some $\mu_2, \mu_3 > 0$ independent of n. For any $m \in \mathbb{N}$, $m \leq n$, at most $N_1 \ldots N_d$ sums of the kind $S((\nu 2^k, (\nu + 1) 2^k])$ constitute the sum $S((0, m])$, hence, on the event A^c,

$$M((0, n]) \leq \langle n \rangle^{1/2} \log\langle n \rangle N_1 \ldots N_d \leq \langle n \rangle^{1/2} (\log\langle n \rangle)^{d+1}. \quad \square$$

Lemma 2.19. *Let conditions of Theorem 2.1.18 be satisfied and* $\alpha > 8/(3\tau) - 1$. *Then*

$$\max_{s=1,\ldots,d} M_s(N_k) \leq C(\omega)\langle N_k \rangle^{1/2 - \varepsilon_5} \quad \text{a.s.}$$

for some $\varepsilon_5 > 0$ *and every* $N_k \in G_\tau$.

Proof. We proceed with $s = 1$, other cases following by re-enumerating the coordinates. For $N \in \mathbb{N}^d$, $N \in G_\tau$, we have $N_1^{(1)} \leq C(N_2 \ldots N_d)^\rho$, because $\lim_{N \to \infty,\, N \in G_\tau} N_1^{(1)}/(N_2 \ldots N_d)^\rho = 1$. At the same time $n_1 \geq (n_2 \ldots n_d)^\tau$, so $n_1 \geq \langle n \rangle^{\tau/2}$. By Lemma 2.17, for any $\varepsilon_5 \in (0, 1)$,

$$P(M_1(N_k) > \langle N_k \rangle^{1/2 - \varepsilon_5}) = P\left(M_1(N_k) > \langle N_k^{(1)} \rangle^{1/2}\, \frac{\langle N_k \rangle^{1/2 - \varepsilon_5}}{\langle N_k^{(1)} \rangle^{1/2}} \right)$$

$$\leq C\left(\frac{\langle N_k^{(1)} \rangle}{\langle N_k \rangle^{1 - 2\varepsilon_5}} \right)^{1 + \delta/2}. \tag{2.32}$$

The last fraction in brackets is estimated as follows:

$$\frac{(N_k^{(1)})_1 (N_k)_2 \ldots (N_k)_d}{((N_k)_1 (N_k)_2 \ldots (N_k)_d)^{1 - 2\varepsilon_5}} = \frac{(N_k^{(1)})_1 ((N_k)_2 \ldots (N_k)_d)^{2\varepsilon_5}}{((N_k)_1)^{1 - 2\varepsilon_5}}$$

$$\leq C\frac{((N_k)_2 \ldots (N_k)_d)^{2\varepsilon_5 + \rho}}{((N_k)_1)^{1 - 2\varepsilon_5}} \leq \frac{((N_k)_2 \ldots (N_k)_d)^{2\varepsilon_5 + \rho}}{\langle N_k \rangle^{\tau/2(1 - 2\varepsilon_5)}}$$

$$= C((N_k)_1)^{-\tau/2(1 - 2\varepsilon_5)} ((N_k)_2 \ldots (N_k)_d)^{2\varepsilon_5 + \rho - \tau/2 + \tau\varepsilon_5}.$$

Since $(N_k)_s \sim k_s^{\alpha+1}/(\alpha + 1)$, $s = 1, \ldots, d$, and $\rho - \tau/2 = -3\tau/8$, we have

$$\frac{\langle N_k^{(1)} \rangle}{\langle N_k \rangle^{1 - 2\varepsilon_5}} \leq C k_1^{-\tau/2(\alpha+1)(1 - 2\varepsilon_5)} (k_2 \ldots k_d)^{-3\tau(\alpha+1)/8 + (\alpha+1)(2 + \tau)\varepsilon_5}. \tag{2.33}$$

One can take $\varepsilon_5 > 0$ in such a way that the exponents in (2.33) are less than -1. Thus the probabilities in (2.32) are majorized by members of a convergent series. By the Borel–Cantelli lemma the desired result holds. $\square$

Lemma 2.20. *Suppose that conditions of Theorem 2.1.18 and (2.3) hold true. Let* γ_1 *be the constant given by Lemma 2.18. If* $\alpha > 2/\gamma_1$ *then one has*

$$\max_J M_k^{(J)} \leq C(\omega)\langle N_k \rangle^{1/2 - \varepsilon_6} \quad \text{a.s.}$$

for some $\varepsilon_6 > 0$ *and every* $N_k \in G_\rho$.

Proof. We need to estimate the maximum of sums over the block U which has one edge (without loss of generality, the first one) of length k_1^α whereas all other edges lengths are no greater than $C k_s^{\alpha+1}$, $s = 2, \ldots, d$.

For any $\varepsilon_6 \in (0, 1/2)$, clearly,

$$P(M(U) \geq \langle N_k \rangle^{1/2 - \varepsilon_6}) \leq P(M(U) \geq C(k_1 \ldots k_d)^{(1/2 - \varepsilon_6)(\alpha+1)})$$

$$= \mathsf{P}\left(M(U) \geq C(k_1^\alpha k_2^{\alpha+1} \ldots k_d^{\alpha+1})^{1/2} \frac{(k_1 \ldots k_d)^{(1/2-\varepsilon_6)(\alpha+1)}}{(k_1^\alpha k_2^{\alpha+1} \ldots k_d^{\alpha+1})^{1/2}}\right)$$

$$= \mathsf{P}\left(M(U) \geq C(k_1^\alpha k_2^{\alpha+1} \ldots k_d^{\alpha+1})^{1/2} k_1^{-\alpha/2+(\alpha+1)/2-\varepsilon_6(\alpha+1)}(k_2 \ldots k_d)^{-\varepsilon_6(\alpha+1)}\right).$$

As $N_k \in G_\rho$, one has

$$k_1 \geq C\langle k\rangle^{\rho/2} \Rightarrow k_1^{1/2}(k_2 \ldots k_d)^{-\varepsilon_6(\alpha+1)} \geq C\langle k\rangle^{\rho/4-\varepsilon_6(\alpha+1)}.$$

Let $\varepsilon_6 < \rho/4(\alpha+1)$, then, for all k large enough,

$$(\log|U|)^{d+1} \leq (\log\langle N_k\rangle)^{d+1} \leq \langle k\rangle^{\rho/4-\varepsilon_6(\alpha+1)}.$$

Thus, by Lemma 2.18,

$$\mathsf{P}(M(U) \geq \langle N_k\rangle^{1/2-\varepsilon_6}) \leq \mathsf{P}(M(U) \geq C(k_1^\alpha k_2^{\alpha+1} \ldots k_d^{\alpha+1})^{1/2} \log|U|)^{d+1})$$

$$\leq C|U|^{-\gamma_1} \leq Ck_1^{-\alpha\gamma_1/2}(k_2 \ldots k_d)^{-(\alpha+1)\gamma_1/2}.$$

Since $\alpha\gamma_1 > 2$, by the Borel–Cantelli lemma $M(U) = O(\langle N_k\rangle^{1/2-\varepsilon_6})$ a.s. $\square$

Now, if we take γ_0, α, β to satisfy the conditions

$$\gamma_0 > \left(1 + \frac{1}{\rho}\right)\left(1 - \frac{1}{d}\right), \quad \frac{\alpha}{\beta}\left(1 - \frac{\mu\delta}{8(1+\delta)}\right) < 1, \quad \beta > \frac{6}{\rho},$$

$$\alpha - \beta > \frac{6}{\rho}, \quad \beta > \frac{2\gamma_0}{\rho}, \quad \alpha > \frac{8}{3\tau} - 1, \quad \alpha\gamma_1 > 2,$$

then Theorem 2.1 follows from (2.7), Lemmas 2.11, 2.12, 2.13, 2.16, 2.19 and 2.20 with $\varepsilon = \min_{i=1,\ldots,6} \varepsilon_i$. To satisfy the mentioned conditions it suffices to pick α and β large enough (so that $\alpha - \beta$ is large but $\alpha/\beta - 1$ is small enough). The proof of Theorem 2.1 is completed. $\square$

8°. Comments. If the random field X in the above Theorem is **PA** or **NA**, then the standard Gaussian random variables $\{\eta_k, \ k \in \mathbb{N}^d\}$ introduced after (2.11) are **PA** (resp. **NA**) as the function $x \mapsto \Phi^{-1}(F_k(x))$ is nondecreasing. Nevertheless, it does not follow that the whole field $\{\eta_k, \ k \in \mathbb{N}^d\}$ is Gaussian. Note that if one-dimensional distributions of some random vector are independent and Gaussian, then the distribution of this vector is Gaussian. Such a conclusion, however, can not be made if we replace independence with positive (negative) association as the following example shows.

Example 2.21. Let (X, Y) be a random vector with density

$$p(x,y) = \varphi(x)\varphi(y)(1 + \varepsilon(2\Phi(x) - 1)(2\Phi(y) - 1)), \quad x, y \in \mathbb{R},$$

where, as before, φ and Φ stand for the density and distribution function of a random variable $Z \sim N(0,1)$, and $\varepsilon \in (-1,1)$. Such (X, Y) belongs to well-known examples of a non-Gaussian random vector with Gaussian marginals. Indeed, since $\mathsf{E}(2\Phi(Z) - 1) = 0$, one has $Law(X) = Law(Y) = N(0,1)$. Consequently $\mathsf{E}X = \mathsf{E}Y = 0$. Note that the density $p(x,y)$ does not have a local maximum at point

$(0,0)$. Thus (X,Y) is not Gaussian (if a Gaussian random vector has a density then the expectation of this vector is the unique extremal point of the density). Now let $f,g : \mathbb{R} \to \mathbb{R}$ be two bounded twice differentiable nondecreasing functions such that $f(0) = g(0) = 0$. Then we have

$$cov(f(X), g(Y)) = \iint_{\mathbb{R}^2} f(x)g(y)\varphi(x)\varphi(y)dxdy$$

$$+ \varepsilon \int_{\mathbb{R}} (2\Phi(x) - 1)f(x)\varphi(x)dx \int_{\mathbb{R}} (2\Phi(y) - 1)g(y)\varphi(y)dy$$

$$-\mathsf{E}f(X)\mathsf{E}g(Y) = \varepsilon\mathsf{E}f(Z)(2\Phi(Z) - 1)\mathsf{E}g(Z)(2\Phi(Z) - 1).$$

The random variable $U = 2\Phi(Z) - 1$ has uniform distribution on $[-1,1]$. Thus $f(Z)(2\Phi(Z) - 1) \geq 0$ a.s., since $f(z) \leq 0$ for $z < 0$ and $f(z) \geq 0$ for $z \geq 0$. That is, $\mathsf{E}f(Z)(2\Phi(Z) - 1)\mathsf{E}g(Z)(2\Phi(Z) - 1) \geq 0$ and the sign of covariance is determined by the sign of ε. Consequently, (X,Y) is **PA** if $\varepsilon > 0$ and **NA** if $\varepsilon < 0$.

In the paper [123] by Dedecker and Prieur the strong invariance principle for τ-dependent random sequences was proved through coupling technique. **NA** random fields and their linear transformations were considered by Lu in [280]. In recent works by Kulik [244] and Wang [411] the authors consider the strong invariance principle for associated random processes ($d = 1$) to the situation when the covariance function decreases in a power way. However in Theorem 2.1 we investigate larger class of random systems than associated ones.

Chapter 6

Law of the Iterated Logarithm

In Chapter 6 we deal with results on the LIL, functional LIL and the law of a single logarithm for associated and related random fields. Note that one can view the statements of this research direction as the refinements of the SLLN describing the almost sure behavior for normalized partial sums. The early refinements of this kind were initiated at the beginning of the last century by Hausdorff, Hardy and Littlewood. It was Khintchine who gave in 1922 the first version of the LIL for a sequence of i.i.d. binary random variables. Now one can discover a lot of famous results (we refer to some of them in the text) concerning various generalizations of LIL. In the first Section of this Chapter we provide LIL for above mentioned random fields, here one can find the new results by Shashkin. In the next Section we establish the functional LIL for associated random fields under simple moment conditions for summands and power-type decrease rate of the Cox–Grimmett coefficient. This turns possible on account of the maximal inequality obtained in the second Chapter and estimates of the convergence rates in CLT proved in Chapter 3. In 1994 Qi introduced the law of a single logarithm for arrays of i.i.d. random variables. The last Section of the Chapter is devoted to some extensions of this law to random fields established recently by Hu, Liang and Su. The inequality by Lewis, employed here, is of independent interest.

1   Extensions of the classical LIL

The previous Chapter 5 is devoted to the study of a.s. fluctuations of partial sums processes constructed by means of weakly dependent, in a sense, random fields (in particular, stochastic sequences). Now, continuing the investigation on these lines, we will be mostly concerned with various refinements of SLLN. Recall that for i.i.d. random variables $X_1, X_2, \ldots$, taking values 0 and 1 with equal probability, the first estimates of the growth of partial sums were obtained as long ago as the beginning of the 20th century (see, e.g., [248, Ch. 2, §11]). In fact the first results were achieved for X_n arising in the dyadic expansion of a number $x \in [0, 1]$, i.e. $X_n = b_n(x)$ where $x = \sum_{n=1}^{\infty} b_n(x) 2^{-n}$ and $b_n(x)$ is 0 or 1. To avoid centering we set $X_n = 2b_n(x) - 1$

283

for $x \in [0,1]$ and $S_n = \sum_{k=1}^n X_k$ where $n \in \mathbb{N}$.

In 1913 Hausdorff proved that $|S_n| = O(n^{1/2+\varepsilon})$ a.s. for any $\varepsilon > 0$. In 1914 Hardy and Littlewood provided an estimate $|S_n| = O(\sqrt{n \log n})$ a.s. In 1922 Khintchine gave an estimate $|S_n| = O(\sqrt{n \log\log n})$ a.s. and in 1924 he established the following *law of the iterated logarithm* (LIL):

$$\limsup_{n\to\infty} \frac{|S_n|}{\sqrt{2n \log\log n}} = 1 \text{ a.s.}$$

After that there were a number of beautiful results devoted to various generalizations of this famous law. In 1927 Kolmogorov proved LIL for independent, bounded (not necessarily identically distributed) summands (see, [326, Ch. VII, §1]). Due to Marcienkiewicz and Zygmund and later due to Egorov we know that the conditions of this theorem are optimal in a sense (see the bibliography in [326, Ch. VII]). It is worthwhile to note the classical result of Hartman and Wintner published in 1940. Let $X_1, X_2, \ldots$ be i.i.d. random variables such that $\mathsf{E}X_1 = 0$ and $\mathsf{E}X_1^2 = 1$. Then

$$\limsup_{n\to\infty} \frac{S_n}{\sqrt{2n \log\log n}} = 1 \text{ a.s. and } \liminf_{n\to\infty} \frac{S_n}{\sqrt{2n \log\log n}} = -1 \text{ a.s.} \qquad (1.1)$$

Moreover, Strassen [392] and Martikainen [283] proved the statement converse to this theorem. Namely, the validity of (1.1) (or even of only one of the relations (1.1)) for i.i.d. random variables implies that $\mathsf{E}X_1 = 0$ and $\mathsf{E}X_1^2 = 1$.

We mention in passing that there are interesting ways of other generalizations. Lévy proposed the concept of lower and upper functions. One can use in (1.1) the normalizations which are different from $\sqrt{2n \log\log n}$ ([60], [71]). There are deep links between the behavior of partial sums processes and the Brownian motion (such problems were discussed in Chapter 5). One can refer to the integral test by Kolmogorov–Petrovski–Erdös–Feller (see, e.g., [215]). The early results on LIL for multiparameter processes can be found in papers by Zimmerman [433], Park [315], Wichura [415]. The case of mixing sequences was studied by Iosifescu [214].

The new important step was made in 1964 by Strassen [390] who proved the famous functional law of the iterated logarithm (FLIL). There also appeared appropriate versions of the LIL (and CLT) for random variables with values in Hilbert and Banach spaces (see, e.g., [253]). In 1990s the problems of the convergence rate in FLIL were studied by Grill, Talagrand, whereas the relation between the using of nonclassical normalizations and the rates in FLIL was investigated by Bulinski and Lifshits (see [75] and references therein). Various generalizations concerning the LIL for random processes indexed by sets were obtained by Bass and Pyke [26], Alexander [5], Shashkin [380].

After the paper by Qi [336] the *law of a single logarithm* appeared in 1994 for arrays of i.i.d. random variables.

Thus there are quite a number of publications devoted to LIL and its extensions. In Chapter 6 our purpose is to study versions of the LIL for associated random fields and their modifications, paying attention to FLIL as well.

1°. LIL for associated random fields. We start with an associated random field. In this Chapter we write $\mathrm{Log}\, x := \log(x \vee e)$, $LLx := \mathrm{Log}\,(\mathrm{Log}\, x)$, $x \geq 0$. As before, $S_n = S((0, n])$, $n \in \mathbb{N}^d$, see the beginning of Section 2.1; $n \to \infty$ means that $n_i \to \infty$, $i = 1, \ldots, d$.

Theorem 1.1. ([381]) *Let* $X = \{X_j, j \in \mathbb{Z}^d\}$ *be a wide-sense stationary, centered, associated random field such that*

1) $\mathsf{E}|X_0|^{2+\delta} < \infty$ *for some* $\delta > 0$;
2) *the Cox–Grimmett coefficient* $u_r = O(r^{-\lambda})$ *as* $r \to \infty$, *for some* $\lambda > 0$.
Then with probability one

$$\limsup_{n \to \infty} \frac{S_n}{\sqrt{2\langle n \rangle LL\langle n \rangle}} = \sigma\sqrt{d}, \quad \liminf_{n \to \infty} \frac{S_n}{\sqrt{2\langle n \rangle LL\langle n \rangle}} = -\sigma\sqrt{d},$$

where $\sigma \geq 0$ *is the same as in* (1.5.3).

Proof is typical for LIL. We need only be concerned with the upper limit, since to handle the other, one can consider the field $\{-X_j, j \in \mathbb{Z}^d\}$.

If $\sigma = 0$, then $X_j = 0$ a.s. for any $j \in \mathbb{Z}^d$, and the Theorem is true. Thus further we consider the case $\sigma > 0$. At first we show that the upper limit is no greater than $\sigma\sqrt{d}$. In the proofs we often employ the customary multiindex notation, in particular, $[c^k] = ([c^{k_1}], \ldots, [c^{k_d}]) \in \mathbb{Z}^d$ for $c > 1$ and $k \in \mathbb{Z}^d$; whereas $m \leq n$, for $m, n \in \mathbb{Z}^d$, means $m_i \leq n_i$, $i = 1, \ldots, d$.

Let $\varepsilon > 0$ be arbitrary and take $c > 1$ to be specified later. For a multiindex $k \in \mathbb{N}^d$ such that $\langle [c^k] \rangle$ is large enough and any $\tau \in (0, 1)$, in view of Theorem 2.4.6, one may write

$$\mathsf{P}\left(\max_{[c^{k-1}] \leq n \leq [c^k]} \frac{S_n}{\sqrt{2\langle n \rangle LL\langle n \rangle}} > \sqrt{d}(\sigma + \varepsilon)\right)$$

$$\leq \mathsf{P}\left(\max_{[c^{k-1}] \leq n \leq [c^k]} S_n > \sqrt{d}(\sigma + \varepsilon)\sqrt{2\langle [c^{k-1}] \rangle LL\langle [c^{k-1}] \rangle}\right)$$

$$\leq \mathsf{P}\left(\max_{1 \leq n \leq [c^k]} S_n > \sqrt{d}(\sigma + \varepsilon)\sqrt{\langle [c^k] \rangle}\,\frac{\sqrt{2\langle [c^{k-1}] \rangle LL\langle [c^{k-1}] \rangle}}{\sqrt{\langle [c^k] \rangle}}\right)$$

$$\leq 4\mathsf{P}\left(\frac{S_{[c^k]}}{\sqrt{\mathsf{Var}S_{[c^k]}}} > \tau\sqrt{d}(\sigma + \varepsilon)\frac{\sqrt{2\langle [c^{k-1}] \rangle LL\langle [c^{k-1}] \rangle}}{\sqrt{\mathsf{Var}S_{[c^k]}}}\right). \tag{1.2}$$

From Theorem 3.1.21 one sees that the last probability in (1.2) does not exceed

$$C\langle [c^k] \rangle^{-\mu} + \mathsf{P}\left(|Z| > \tau\sqrt{d}(\sigma + \varepsilon)\frac{\sqrt{2\langle [c^{k-1}] \rangle LL\langle [c^{k-1}] \rangle}}{\sqrt{\mathsf{Var}S_{[c^k]}}}\right) \tag{1.3}$$

where positive C, μ are independent of k, and $Z \sim N(0, 1)$. A standard estimate for the tail of normal distribution yields that, for all k large enough, the probability appearing in (1.3) admits the upper bound

$$\exp\left\{-\tau^2 d(\sigma + \varepsilon)^2 \frac{\langle [c^{k-1}] \rangle LL\langle [c^{k-1}] \rangle}{\mathsf{Var}S_{[c^k]}}\right\} \leq \exp\left\{-\tau^2 d\frac{(\sigma + \varepsilon)^2}{\sigma^2}\frac{\langle [c^{k-1}] \rangle LL\langle [c^{k-1}] \rangle}{\langle [c^k] \rangle}\right\}.$$

The last inequality is true because, due to association, $\mathrm{Var}S_{[c^k]} \leq \sigma^2\langle[c^k]\rangle$. An elementary calculation shows that, for any $c > 1$ and all $l \in \mathbb{N}$ large enough, one has $[c^{l-1}]/[c^l] \geq c^{-2}$. Thus $\langle[c^{k-1}]\rangle/\langle[c^k]\rangle \geq c^{-2d}$. Therefore, for all k except for a finite number, we can continue the estimation of the term from (1.3) to obtain the following upper bound

$$\exp\left\{-\tau^2 d\frac{(\sigma+\varepsilon)^2}{\sigma^2}c^{-2d}LL\langle[c^{k-1}]\rangle\right\} = (\mathrm{Log}\,\langle[c^{k-1}]\rangle)^{-\tau^2 d(1+\varepsilon/\sigma)^2 c^{-2d}}. \tag{1.4}$$

Recall that $\sum_{k\in\mathbb{N}^d}(k_1+\ldots+k_d)^{-\nu} < \infty$ if $\nu > d$. Now, when $\tau = \tau(\varepsilon)$ and $c = c(\varepsilon)$ are taken so close to 1 that $c^{-2d}\tau^2(1+\varepsilon/\sigma)^2 > 1$, from (1.2)—(1.4) it follows that

$$\sum_{k\in\mathbb{N}^d}\mathsf{P}\left(\max_{[c^{k-1}]\leq n\leq[c^k]}\frac{S_n}{\sqrt{2\langle n\rangle LL\langle n\rangle}} > \sqrt{d}(\sigma+\varepsilon)\right) < \infty,$$

which by virtue of Borel–Cantelli lemma means that the events in brackets occur only finitely often, with probability one. Consequently, the upper limit of the sums $S_n/\sqrt{2\langle n\rangle LL\langle n\rangle}$ (as $n \to \infty$) does not exceed $(\sigma+\varepsilon)\sqrt{d}$, a.s. Thus it is no greater than $\sigma\sqrt{d}$.

Next we are going to obtain the lower estimate for the same upper limit of normalized partial sums. Take again an arbitrary $\varepsilon \in (0,1)$ and some integer $c > 1$ (large enough) to be specified later. For some fixed $\rho \in (0,1/d)$, let

$$J_\rho = \{k \in \mathbb{N}^d : k_i \geq \rho\|k\|_1 \text{ for } i = 1,\ldots,d\}$$

where $\|k\|_1 = k_1 + \ldots + k_d$, $k = (k_1,\ldots,k_d)$. Introduce events

$$E_k = \left\{S(B_k) \geq (1-\varepsilon)\sqrt{2d\mathrm{Var}S(B_k)LL\langle c^k\rangle}\right\} \tag{1.5}$$

where the blocks $B_k = (c^{k-1}, c^k]$, $k \in J_\rho$.

Our goal is to show that the events E_k occur with probability one for an infinite sequence $k^{(r)} = k^{(r)}(\omega) \in J_\rho$, $k^{(r)} \to \infty$ as $r \to \infty$. We have divided the proof into a series of lemmas. Set

$$R_{d,m}(\rho) = \sharp\{k = (k_1,\ldots,k_d) \in J_\rho : \|k\|_1 = m\}, \quad m \in \mathbb{N}. \tag{1.6}$$

Lemma 1.2. *Given $d \in \mathbb{N}$, for $\rho \in (0,1/d)$ one has*

$$R_{d,m}(\rho) \sim \frac{(1-d\rho)^{d-1}}{(d-1)!}m^{d-1} \quad \text{as} \quad m \to \infty.$$

Proof. First of all we show that

$$M_{d,r} := \sharp\{k \in \mathbb{Z}_+^d : \|k\|_1 = r\} \sim ((d-1)!)^{-1}r^{d-1}, \quad r \to \infty. \tag{1.7}$$

To this end note that $M_{d,r}$ is the number of possible ways to pack r undistinguishable balls into d boxes, so

$$M_{d,r} = \binom{d+r-1}{r}.$$

Indeed, to any packing we can assign a unique set of $d + r - 1$ elements (r balls and $d - 1$ walls of boxes), and we need to mark the places for balls (or the walls). Consequently,

$$M_{d,r} = \frac{(d+r-1)!}{(d-1)!r!} = \frac{1}{(d-1)!} \prod_{j=1}^{d-1} (r+j) \sim \frac{r^{d-1}}{(d-1)!}, \quad r \to \infty,$$

for any fixed $d \in \mathbb{N}$.

Let $m \in \mathbb{N}$ be large enough, and denote by $\lceil \rho m \rceil$ the smallest integer exceeding ρm. Clearly, for any $k \in \mathbb{N}^d$ with $\|k\|_1 = m$, one has $k \in J_\rho$ if and only if $k_i - \lceil \rho m \rceil \geq 0$. Therefore, R_m is equal to the number of points $l \in \mathbb{Z}_+^d$ such that $l_1 + \cdots + l_d = m - d\lceil \rho m \rceil$. Thus by (1.7)

$$R_m \sim \frac{1}{(d-1)!}(m - d\lceil \rho m \rceil)^{d-1} \sim \frac{(1 - d\rho)^{d-1}}{(d-1)!} m^{d-1}, \quad m \to \infty,$$

since $\lceil \rho m \rceil \sim \rho m$ as $m \to \infty$ and $\rho < 1/d$. The established relation is the statement of the Lemma. $\square$

Lemma 1.3. *Let U_1, U_2 be two disjoint blocks in $\mathbb{N}^d$. Then*

$$\frac{1}{|U_1| \vee |U_2|} cov(S(U_1), S(U_2)) \leq 6(c_0 + 2d\sigma^2)l(U_1, U_2)^{-\lambda/(1+\lambda)} \tag{1.8}$$

where $l(U_1, U_2)$ is the minimal length of edge of either U_1 or U_2.

Proof. Obviously,

$$2\,cov(S(U_1), S(U_2)) = \mathsf{Var}S(U_1 \cup U_2) - \mathsf{Var}S(U_1) - \mathsf{Var}S(U_2)$$

$$= \mathsf{Var}S(U_1 \cup U_2) - \sigma^2(|U_1| + |U_2|) - \mathsf{Var}S(U_1) + \sigma^2|U_1| - \mathsf{Var}S(U_2) + \sigma^2|U_2|.$$

By association and Lemma 5.2.2 (its second assertion and Remark 5.2.3) the left-hand side of (1.8) is majorized by

$$\frac{1}{2}\left(\left|\sigma^2 - \frac{\mathsf{Var}S(U_1 \cup U_2)}{|U_1| + |U_2|}\right| + \left|\sigma^2 - \frac{\mathsf{Var}S(U_1)}{|U_1|}\right| + \left|\sigma^2 - \frac{\mathsf{Var}S(U_2)}{|U_2|}\right|\right),$$

which can be estimated as $6(c_0 + 2d\sigma^2)l(U_1, U_2)^{-\lambda/(1+\lambda)}$. Here we have used that $\sigma^2|U| \geq \mathsf{Var}S(U)$ for any block $U \subset \mathbb{Z}^d$ and $|U_1| + |U_2| \leq 2(|U_1| \vee |U_2|)$.

Lemma 1.4. *If the conditions of Theorem 1.1 hold, then the double sum*

$$\sum_{\substack{k,l \in J_\rho, \\ k \neq l}} \frac{cov(S(B_k), S(B_l))}{\sqrt{|B_k||B_l|}} =: I < \infty.$$

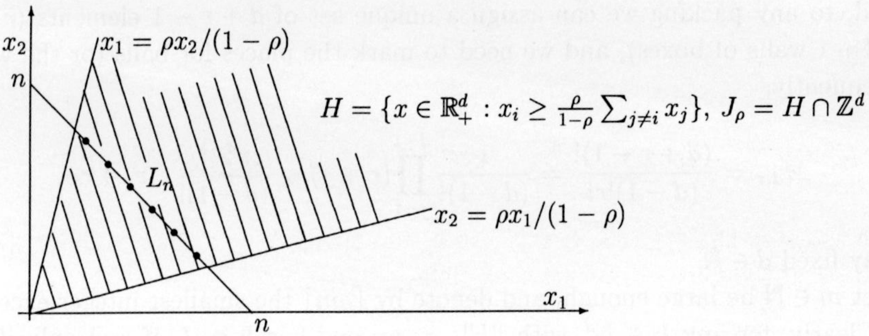

Fig. 6.1 $(d = 2)$

Proof. By association, all summands in I are nonnegative. Since $B_k = (c^{k-1}, c^k]$, with $c \in \mathbb{N}$, $k \in \mathbb{N}^d$, one has

$$|B_k| = \prod_{i=1}^{d}(c^{k_i} - c^{k_i-1}) = c^{\|k\|_1-d}(c-1)^d.$$

For $n \in \mathbb{N}$, under *n-th layer* we will mean the set $L_n(\rho) = \{k \in J_\rho : \|k\|_1 = n\}$.

We represent $I = I_1 + I_2$ where I_1 is the sum over pairs $k, l \in J_\rho$, $k \neq l$, such that $|\|k\|_1 - \|l\|_1| \leq d$, and I_2 is the sum over the remaining pairs of multiindices.

First of all we estimate I_1. If $k \in L_m$ and $l \in L_n$, then $|B_k| = c^{m-d}(c-1)^d$, $|B_l| = c^{n-d}(c-1)^d$. Consequently,

$$I_1 \leq 2c^d(c-1)^{-d}\sum_{n=1}^{\infty}\sum_{m=n}^{n+d} c^{-m/2-n/2}\sum_{\substack{k\in L_m(\rho), l\in L_n(\rho), \\ k\neq l}} cov(S(B_k), S(B_l))$$

$$\leq 2c^d(c-1)^{-d}\sum_{n=1}^{\infty}\sum_{m=n}^{n+d} c^{d/2-m}R_{d,n}R_{d,m}\max_{\substack{k\in L_m(\rho), l\in L_n(\rho), \\ k\neq l}} cov(S(B_k), S(B_l)). \quad (1.9)$$

Besides (1.6) we used here that $c^{-n/2} \leq c^{d/2-m/2}$ if $m \leq n + d$. If $k \in L_m \cap J_\rho$, then the minimal edge of B_k has length not less than $(c-1)c^{\rho m}$. Thus, by Lemma 1.3 we have

$$cov(S(B_k), S(B_l)) \leq 6(c_0 + 2^d\sigma^2)\left((c-1)c^{\rho n}\right)^{-\lambda/(1+\lambda)} c^{m-d}(c-1)^d$$

for all $k \neq l$, $k \in L_m(\rho)$, $l \in L_n(\rho)$. From this observation and Lemma 1.2 it follows that the right-hand side of (1.9) is no greater than

$$12(c_0 + 2^d\sigma^2)(c-1)^{-\lambda/(\lambda+1)} c^{d/2}\sum_{n=1}^{\infty} R_{d,n}(\rho)\sum_{m=n}^{n+d} R_{d,m}(\rho)c^{-\lambda\rho n/(\lambda+1)}$$

$$\leq 12(c_0 + 2^d\sigma^2)(c-1)^{-\lambda/(\lambda+1)} c^{d/2}d\sum_{n=1}^{\infty}(n+d)^{2(d-1)}c^{-\lambda\rho n/(\lambda+1)} < \infty$$

where we applied an elementary estimate $R_{d,n}(\rho) \le n^{d-1}$ (here $d, n \in \mathbb{N}$), proved by induction.

To find an estimate for I_2, rewrite

$$I_2 = 2c^d(c-1)^{-d} \sum_{n=1}^{\infty} \sum_{m=n+d+1}^{\infty} c^{-(n+m)/2} cov(S(U_m(\rho)), S(U_n(\rho))) \tag{1.10}$$

where $U_n(\rho) = \cup_{k \in L_n(\rho)} B_k$, $n \in \mathbb{N}$. Note that $\sharp(U_n(\rho)) = c^{n-d}(c-1)^d R_{d,n}(\rho)$. By definition of the Cox–Grimmett coefficient,

$$cov(S(U_m), S(U_n)) \le c_0 |U_n| \, (dist(U_n, U_m))^{-\lambda}. \tag{1.11}$$

Thus we will be able to bound I_2 if we manage to measure the distance between U_m and U_n, $m > n$. Let $x \in U_m$ and $y \in U_n$ be arbitrary points. Then there are some $k \in L_m$, $l \in L_n$ such that $x \in B_k$ and $y \in B_l$. Let $i \in \{1, \ldots, d\}$ be any index such that $k_i - l_i \ge (m-n)/d$. It clearly exists since $\sum_{i=1}^{d}(k_i - l_i) = m - n$. One has

$$x_i - y_i \ge c^{k_i-1} + 1 - c^{l_i} = c^{l_i}(c^{k_i - l_i - 1} - 1) + 1 \ge c^{\rho n}(c^{(m-n)/d-1} - 1) + 1.$$

Combining this estimate with (1.10) and (1.11) gives

$$I_2 \le 2c_0 c^d (c-1)^{-d} \sum_{n=1}^{\infty} \sum_{m=n+d+1}^{\infty} c^{-(m+n)/2} c^{n-d}(c-1)^d n^{d-1} \left(c^{\rho n}(c^{(m-n)/d-1} - 1) \right)^{-\lambda}$$

$$= 2c_0 \sum_{n=1}^{\infty} \sum_{m=n+d+1}^{\infty} c^{n/2-m/2} n^{d-1} c^{-\lambda \rho n} \left(c^{(m-n)/d-1} - 1 \right)^{-\lambda}$$

$$= 2c_0 \sum_{n=1}^{\infty} \sum_{q=d+1}^{\infty} c^{-q/2-\lambda \rho n} n^{d-1} \left(c^{q/d-1} - 1 \right)^{-\lambda} < \infty.$$

The last expression is a product of two convergent series, hence is finite. The Lemma is proved. $\square$

2°. Employment of the Borel–Cantelli lemma generalization. The version of the Borel–Cantelli lemma which we use to prove that $P(E_k \text{ i. o.}) = 1$ is the classical result by Erdös and Rényi (see, e.g., [326, Ch. VI]), "i.o." means "infinitely often".

Lemma 1.5. *Let $A_1, A_2, \ldots$ be a sequence of events such that the following two conditions hold:*

$$\sum_{n=1}^{\infty} P(A_n) = \infty, \tag{1.12}$$

$$\liminf_{n \to \infty} \frac{\sum_{1 \le i,j \le n}(P(A_i A_j) - P(A_i)P(A_j))}{\left(\sum_{i=1}^{n} P(A_i)\right)^2} = 0. \tag{1.13}$$

Then $P(A_n \text{ i. o.}) = 1$.

Proof. The Markov inequality yields

$$\mathsf{P}\left(\left|\sum_{i=1}^{n}\mathbb{I}\{A_i\} - \sum_{i=1}^{n}\mathsf{P}(A_i)\right| \geq \frac{1}{2}\sum_{i=1}^{n}\mathsf{P}(A_i)\right) \leq 4\frac{\mathsf{Var}\sum_{i=1}^{n}\mathbb{I}\{A_i\}}{\left(\sum_{i=1}^{n}\mathsf{P}(A_i)\right)^2}. \tag{1.14}$$

By (1.13), the lower limit of the right-hand side of (1.14) is 0, as $n \to \infty$ (hence the same is true of the left-hand side). Consequently, there exists an increasing sequence of positive integers $r_n \to \infty$ such that

$$\sum_{n=1}^{\infty}\mathsf{P}\left(\sum_{i=1}^{r_n}\mathbb{I}\{A_i\} \leq \frac{1}{2}P(r_n)\right) \leq \sum_{n=1}^{\infty}\mathsf{P}\left(\left|\sum_{i=1}^{r_n}\mathbb{I}\{A_i\} - P(r_n)\right| \geq \frac{1}{2}P(r_n)\right) < \infty$$

where $P(r_n) := \sum_{i=1}^{r_n}\mathsf{P}(A_i)$. By the Borel–Cantelli lemma we have, almost surely,

$$\sum_{i=1}^{r_n}\mathbb{I}\{A_i\} > \frac{1}{2}\sum_{i=1}^{r_n}\mathsf{P}(A_i)$$

for all n large enough. This inequality together with (1.12) provide that A_i occur infinitely often with probability one. $\square$

Let us enumerate the events $\{E_k, k \in J_\rho\}$ introduced in (1.5) by increase of $\|k\|_1$ (for $k \in L_n = J_\rho \cap \{k : \|k\|_1 = n\}$ one can use the lexicographic order) and denote the members of resulting sequence as $A_1, A_2, \ldots$. That is, we pick at first all k in L_1, then in L_2 and so on. Let $q_n = R_1 + R_2 + \ldots + R_n$, $n \in \mathbb{N}$, i.e. q_n is the total number of points in the first n layers. We will use Lemma 1.5 and show that the lower limit in (1.13) taken over some subsequence A_{q_n} is zero. As usual, $C_1, C_2, \ldots$ are positive and independent of n.

Lemma 1.6. *For any $\varepsilon \in (0,1)$ one has*

$$\sum_{i=1}^{q_n}\mathsf{P}(A_i) \sim C_1(\log n)^{-1/2}n^{d(2\varepsilon-\varepsilon^2)} \quad \text{as} \ \ n \to \infty.$$

Proof. Evidently,

$$\sum_{i=1}^{q_n}\mathsf{P}(A_i) = \sum_{m=1}^{n}\sum_{k \in L_m}\mathsf{P}(E_k), \tag{1.15}$$

and due to Theorem 3.1.21

$$\mathsf{P}(E_k) = \mathsf{P}\left(Z \geq (1-\varepsilon)\sqrt{2dLLc^m}\right) + a_k =: p_m(\varepsilon) + a_k,$$

whenever $k \in L_m$, here $|a_k| \leq Cc^{-m\mu}$ and C does not depend on m. We took into account that, for $k \in L_m$, $|B_k| = c^{m-d}(c-1)^d$. A standard estimate for the tail of a normal distribution shows that one has, as $n \to \infty$,

$$p_m(\varepsilon) \sim \frac{\exp\{-(1-\varepsilon)^2dLLc^m\}}{2(1-\varepsilon)\sqrt{\pi dLLc^m}} = \frac{(m\log c)^{-d(1-\varepsilon)^2}}{2(1-\varepsilon)\sqrt{\pi d\log(m\log c)}}. \tag{1.16}$$

Consequently,

$$\sum_{i=1}^{q_n} \mathsf{P}(A_i) = \sum_{m=1}^{n} R_{d,m}(\rho)\mathsf{P}(Z \geq (1-\varepsilon)\sqrt{2dLLc^m}) + \sum_{m=1}^{n}\sum_{k\in L_m} a_k.$$

Clearly, the last sum is finite as

$$\sum_{m=1}^{n}\sum_{k\in L_m} a_k \leq C\sum_{m=1}^{n} R_{d,m}(\rho)c^{-m\mu} \leq C\sum_{m=1}^{\infty} m^{d-1}c^{-m\mu} < \infty, \ n \in \mathbb{N}.$$

Due to Lemma 1.2 and to (1.16), for any $\gamma \in (0,1)$ there exists $N = N(\gamma)$ such that for all $m \geq N$ one has

$$(1-\gamma)\frac{(1-\rho d)^{d-1}}{(d-1)!}m^{d-1} \leq R_{d,m}(\rho) \leq (1+\gamma)\frac{(1-\rho d)^{d-1}}{(d-1)!}m^{d-1},$$

$$(1-\gamma)\frac{(m\log c)^{-d(1-\varepsilon)^2}}{2(1-\varepsilon)\sqrt{\pi d\log m}} \leq \mathsf{P}(Z \geq (1-\varepsilon)\sqrt{2dLLc^m}) \leq (1+\gamma)\frac{(m\log c)^{-d(1-\varepsilon)^2}}{2(1-\varepsilon)\sqrt{\pi d\log m}}.$$

Therefore, if $n \to \infty$ then

$$\sum_{i=1}^{q_n} \mathsf{P}(A_i) \sim \frac{(1-d\rho)^{d-1}(\log c)^{-d(1-\varepsilon)^2}}{2(1-\varepsilon)\sqrt{\pi d}} \sum_{m=1}^{n} \frac{m^{d-1-d(1-\varepsilon)^2}}{(\log m)^{1/2}} \sim Cn^{d(2\varepsilon-\varepsilon^2)}(\log n)^{-1/2}.$$

Now we estimate the numerator in (1.13). Note that for any $k,l \in \mathbb{N}^d$, in view of association property of the field X,

$$C(k,l) := \mathsf{P}(E_k E_l) - \mathsf{P}(E_k)\mathsf{P}(E_l) = cov(\mathbb{I}\{E_k\},\mathbb{I}\{E_l\}) \geq 0,$$

since $\mathbb{I}\{S(B_k) \geq x\}$ is a nondecreasing function in $S(B_k)$.

To obtain the upper estimate for $cov(\mathbb{I}\{E_k\},\mathbb{I}\{E_l\})$ when $k \in L_m(\rho)$, $l \in L_n(\rho)$ $(m,n \in \mathbb{N})$, introduce a "smoothed version" of the indicator function. Set, for $x \in \mathbb{R}$ and $\gamma > 0$,

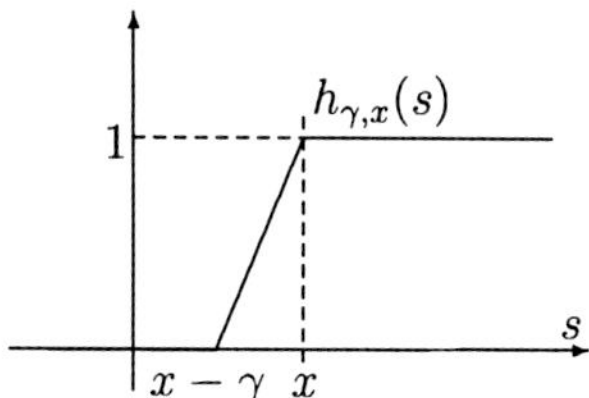

Fig. 6.2

$$h_{\gamma,x}(s) = \begin{cases} 0, & s \leq x - \gamma, \\ \gamma^{-1}(s-x), & x - \gamma < s \leq x, \\ 1, & s > x. \end{cases} \tag{1.17}$$

For all $s \in \mathbb{R}$ and any $\gamma > 0$,

$$\mathbb{I}\{s \geq x\} \leq h_{\gamma,x}(s), \quad \mathbb{I}\{s \geq x\} \geq h_{\gamma,x+\gamma}(s). \tag{1.18}$$

For $n \in \mathbb{N}$, take $\gamma_n = n^{-\alpha}$, $\alpha > 0$ being a positive parameter, and write $x_n = (1-\varepsilon)\sqrt{2dLLc^n}$, here $\varepsilon \in (0,1)$ and $n \in \mathbb{N}$. Let $Z_k = S(B_k)/\sqrt{\mathsf{Var}S(B_k)}$, $k \in \mathbb{N}^d$. For arbitrary $k \in L_m(\rho)$ and $l \in L_n(\rho)$ we have, in view of (1.18),

$$C(k,l) \leq \mathbb{EI}\{Z(B_k) \geq x_m\}\mathbb{I}\{Z(B_l) \geq x_n\} - \mathbb{EI}\{Z(B_k) \geq x_m\}\mathbb{EI}\{Z(B_l) \geq x_n\}$$

$$\leq \mathsf{E}h_{\gamma_m,x_m}(Z_k)h_{\gamma_n,x_n}(Z_l) - \mathsf{E}h_{\gamma_m,x_m+\gamma_m}(Z_k)\mathsf{E}h_{\gamma_n,x_n+\gamma_n}(Z_l)$$

$$= cov\left(h_{\gamma_m,x_m}(Z_k), h_{\gamma_n,x_n}(Z_l)\right)$$

$$+ \mathsf{E}h_{\gamma_m,x_m}(Z_k)\mathsf{E}h_{\gamma_n,x_n}(Z_l) - \mathsf{E}h_{\gamma_m,x_m+\gamma_m}(Z_k)\mathsf{E}h_{\gamma_n,x_n+\gamma_n}(Z_l). \tag{1.19}$$

Lemma 1.7. *Let $\eta \in (\varepsilon, 1)$ and $\alpha > 0$ be such that $\alpha \neq d\eta(2-\eta)$. Then*

$$\sum_{i,j=1}^{q_n} (\mathsf{P}(A_i A_j) - \mathsf{P}(A_i)\mathsf{P}(A_j)) = O\left((\log n)^{-1/2}n^{d(2\eta-\eta^2)} + n^{2\alpha} + n^{-\alpha+2d(2\eta-\eta^2)}\right)$$

as $n \to \infty$.

Proof. By (1.19),

$$\sum_{i,j=1}^{q_n} (\mathsf{P}(A_i A_j) - \mathsf{P}(A_i)\mathsf{P}(A_j)) = \sum_{m,v=1}^{n} \sum_{k \in L_m, l \in L_v} (\mathsf{P}(E_k E_l) - \mathsf{P}(E_k)\mathsf{P}(E_l))$$

$$\leq \sum_{m,v=1}^{n} \sum_{k \in L_m, l \in L_v} cov\left(h_{\gamma_m,x_m}(Z_k), h_{\gamma_v,x_v}(Z_l)\right) + \left(\sum_{m=1}^{n} \sum_{k \in L_m} \mathsf{E}h_{\gamma_m,x_m}(Z_k)\right)^2$$

$$-\left(\sum_{m=1}^{n} \sum_{k \in L_m} \mathsf{E}h_{\gamma_m,x_m+\gamma_m}(Z_k)\right)^2 =: J_1 + J_2 - J_3.$$

Moreover, J_1 admits the following upper bound

$$\sum_{m=1}^{n} \sum_{k \in L_m} \mathsf{E}h^2_{\gamma_m,x_m}(Z_k) + \sum_{m,v=1}^{n} \sum_{\substack{k \in L_m, l \in L_v, \\ k \neq l}} cov(h_{\gamma_m,x_m}(Z_k), h_{\gamma_n,x_n}(Z_l)) =: J_{11} + J_{12}.$$

To estimate J_{11} note that, for all m large enough,

$$x_m - \gamma_m = (1-\varepsilon)\sqrt{2dLLc^m} - m^{-\alpha} \geq (1-\eta)\sqrt{2dLLc^m}.$$

Observe also that

$$|h^2_{\gamma_m,x_m}(t)| \leq \mathbb{I}\{t \geq x_m - \gamma_m\} \leq \mathbb{I}\left\{t \geq (1-\eta)\sqrt{2dLLc^m}\right\}, \quad t \in \mathbb{R}.$$

Consequently, Lemma 1.6 (applied with η instead of ε) shows that

$$J_{11} = O\left((\log n)^{-1/2}n^{d(2\eta-\eta^2)}\right) \quad \text{as } n \to \infty. \tag{1.20}$$

By association, $\mathsf{Var}S(U) \geq |U|\mathsf{E}X_0^2$ for any finite $U \subset \mathbb{Z}^d$. Clearly $Lip(h_{\gamma,x}) = \gamma^{-1}$. Therefore, Theorem 1.5.3 and Lemma 1.4 ensure that

$$J_{12} \leq \sum_{m,v=1}^{n} \sum_{k \in L_m, l \in L_v, k \neq l} Lip(h_{\gamma_m,x_m}) Lip(h_{\gamma_v,x_v}) cov\,(Z_k, Z_l)$$

$$= \sum_{m,v=1}^{n} \sum_{k \in L_m, l \in L_v, k \neq l} \gamma_m^{-1} \gamma_v^{-1} cov\left(\frac{S(B_k)}{\sqrt{\mathsf{Var}S(B_k)}}, \frac{S(B_l)}{\sqrt{\mathsf{Var}S(B_l)}}\right)$$

$$\leq (\mathsf{E}X_0^2)^{-1} \sum_{m,v=1}^{n} \sum_{k \in L_m, l \in L_v, k \neq l} (mv)^\alpha cov\left(\frac{S(B_k)}{\sqrt{|B_k|}}, \frac{S(B_l)}{\sqrt{|B_l|}}\right)$$

$$\leq (\mathsf{E}X_0^2)^{-1} n^{2\alpha} \sum_{m,v=1}^{n} \sum_{k \in L_m, l \in L_v, k \neq l} cov\left(\frac{S(B_k)}{\sqrt{|B_k|}}, \frac{S(B_l)}{\sqrt{|B_l|}}\right) = O(n^{2\alpha}). \qquad (1.21)$$

The next step is to estimate $J_2 - J_3$. We have

$$\left(\sum{}' \mathsf{E}h_{\gamma_m,x_m}(Z_k)\right)^2 - \left(\sum{}' \mathsf{E}h_{\gamma_m,x_m+\gamma_m}(Z_k)\right)^2$$

$$= \left(\sum{}' \mathsf{E}(h_{\gamma_m,x_m}(Z_k) - h_{\gamma_m,x_m+\gamma_m}(Z_k))\right)\left(\sum{}' \mathsf{E}(h_{\gamma_m,x_m}(Z_k) + h_{\gamma_m,x_m+\gamma_m}(Z_k))\right)$$

$$\leq \left(\sum{}' \mathsf{P}(x_m - \gamma_m \leq Z_k \leq x_m + \gamma_m)\right)\left(2\sum{}' \mathsf{P}(Z_k \geq x_m - \gamma_m)\right) =: Q_1 Q_2,$$

here $\sum'$ means the summation over $m \in \{1, \ldots, n\}$ and $k \in L_m$; the last inequality is due to (1.18). Analogously to (1.20), on account of Lemma 1.6 we see that

$$Q_2 = O\left((\log n)^{-1/2} n^{d(2\eta - \eta^2)}\right) \quad \text{as } n \to \infty. \qquad (1.22)$$

If $Z \sim N(0,1)$ then, for any $a > 0$ and $b > a$, one has the inequality

$$\mathsf{P}(b - a \leq Z \leq b + a) \leq \frac{2}{\sqrt{2\pi}} a \exp\{-(b-a)^2/2\} < a \exp\{-(b-a)^2/2\}.$$

Therefore, for all m large enough

$$\mathsf{P}(x_m - \gamma_m \leq Z \leq x_m + \gamma_m) \leq \gamma_m \exp\{-(x_m - \gamma_m)^2/2\}$$

$$\leq m^{-\alpha} \exp\{-((1-\eta)\sqrt{2dLLc^m})^2/2\} = m^{-\alpha}(m \log c)^{-d(1-\eta)^2}. \qquad (1.23)$$

From (1.23) and the Berry–Esseen type estimate (Theorem 3.1.21) one infers that, for $k \in L_m$,

$$\mathsf{P}(x_m - \gamma_m \leq Z_k \leq x_m + \gamma_m) \leq Cc^{-m\mu} + m^{-\alpha}(m \log c)^{-d(1-\eta)^2}$$

where C does not depend on m. Note that $|L_m| = R_m \leq m^{d-1}$, $m \in \mathbb{N}$. Thus summation in m yields

$$Q_1 \leq C \sum_{m=1}^{n} m^{d-1} c^{-m\mu} + 2 \sum_{m=1}^{n} m^{d-1-\alpha}(m \log c)^{-d(1-\eta)^2}$$

$$\leq C + (\log c)^{-d(1-\eta)^2} \sum_{m=1}^{n} m^{d-1-\alpha-d(1-\eta)^2} = O(1 + n^{-\alpha+d(2\eta-\eta^2)}), \qquad (1.24)$$

if $\alpha \neq d\eta(2 - \eta)$. Gathering (1.20)—(1.22) and (1.24), we come to the result. $\square$

Lemmas 1.6 and 1.7 guarantee that

$$\frac{\sum_{i,j=1}^{q_n} (\mathsf{P}(A_i A_j) - \mathsf{P}(A_i)\mathsf{P}(A_j))}{\left(\sum_{i=1}^{n} \mathsf{P}(A_i)\right)^2} \leq C_5 \frac{(\log n)^{-1/2} n^{d(2\eta-\eta^2)} + n^{2\alpha} + n^{-\alpha+2d(2\eta-\eta^2)}}{(\log n)^{-1} n^{2d(2\varepsilon-\varepsilon^2)}}.$$

$$(1.25)$$

Now, if we choose $\eta \in (\varepsilon, 1)$ in such a way that $2\eta - \eta^2 < 3(2\varepsilon - \varepsilon^2)/2$ and then take α so that

$$2d(2\eta - \eta^2 - 2\varepsilon + \varepsilon^2) < \alpha < d(2\varepsilon - \varepsilon^2),$$

then the right-hand side of (1.25) clearly tends to zero as $n \to \infty$. Hence (1.13) is verified. By Lemma 1.5, the events A_i occur infinitely often with probability one. In other words, there exists a sequence of multiindices $\{k_n(\omega)\}_{n\in\mathbb{N}} \subset J_\rho$ such that $k_n(\omega) \to \infty$ and

$$S(B_{k_n}) \geq (1 - \varepsilon)\sqrt{2d \operatorname{Var} S(B_{k_n})LL\langle c^{k_n}\rangle} \quad \text{a.s.} \qquad (1.26)$$

Since the lengths of all edges of B_k grow to infinity as $k \to \infty$, by Theorem 3.1.8 from (1.26) it follows that

$$\frac{S(B_{k_n})}{\sqrt{2d\sigma^2|B_{k_n}|LL\langle c^{k_n}\rangle}} \geq (1 - 2\varepsilon) \quad \text{a.s.} \qquad (1.27)$$

Recall that $|B_k| = \langle c^k\rangle(1 - c^{-1})^d$, so (1.27) amounts to

$$\frac{S(B_{k_n})}{\sqrt{2d\langle c^{k_n}\rangle LL\langle c^{k_n}\rangle}} \geq \sigma(1 - c^{-1})^{d/2}(1 - 2\varepsilon) \quad \text{a.s.}$$

Write $S(B_k)$ as a signed sum of the variables $S((0, v_z])$, that is

$$S(B_k) = S_{c^k} + \sum_z t_z S_{c^{k-z}}$$

where v_z are the vertices of B_k and the sum is over $2^d - 1$ vectors $z = (z_1, \ldots, z_d)$ such that $z_i = 0$ or $z_i = 1$ for $i = 1, \ldots, d$ but z is not identically zero, and $t_z \in \{-1, 1\}$. Note that $\langle c^{k-z}\rangle \leq \langle c^k\rangle/c$ for any such z. Having the already proved upper bound in the LIL, we deduce from (1.26) and (1.27) that a.s.

$$\frac{S_{c^{k_n}(\omega)}}{\sqrt{2d\langle c^{k_n}(\omega)\rangle LL\langle c^{k_n}(\omega)\rangle}} \geq \sigma(1 - c^{-1})^{d/2}(1 - 2\varepsilon) - \sigma(2^d - 1)c^{-1/2}.$$

Since c could be chosen arbitrarily large while ε made arbitrarily small, the lower bound is proved because

$$\limsup_{k\to\infty} \frac{S_k}{\sqrt{2d\langle c^k\rangle LL\langle c^k\rangle}} \geq \sigma.$$

The proof of Theorem 1.1 is complete. $\square$

3°. The case of associated random sequences. If $d = 1$, one can weaken the conditions of Theorem above, by demanding only the logarithmic rate of decrease for the covariance function. This is due to the fact that the maximal inequality for associated sequences (Theorem 2.4.5) is insensitive to the rate of decrease of the Cox–Grimmett coefficient. The next theorem is a refinement of the result by Yu.

Theorem 1.8. ([422]) *Let $X = \{X_j, j \in \mathbb{Z}\}$ be a wide-sense stationary, centered, associated random sequence such that*
1) $\mathsf{E}|X_0|^{2+\delta} < \infty$ *for some $\delta > 0$;*
2) *the Cox–Grimmett coefficient $u_r = O((\operatorname{Log} r)^{-\lambda})$ as $r \to \infty$, for some $\lambda > 2$.*
Then, with probability one,

$$\limsup_{n \to \infty} \frac{S_n}{\sqrt{2nLLn}} = \sigma, \quad \liminf_{n \to \infty} \frac{S_n}{\sqrt{2nLLn}} = -\sigma,$$

where $\sigma \geq 0$ is defined according to (1.5.3).

Proof. Again the case $\sigma = 0$ is trivial and will be omitted. The upper estimate is performed as in the proof of Theorem 1.1, with the only exception that instead of the maximal inequality provided by Theorem 2.4.6 one can use the second assertion of Theorem 2.4.5. Namely, for any $c > 1$ and arbitrary $\varepsilon > 0$, due to the mentioned Theorem 2.4.5,

$$\mathsf{P}\left(\max_{[c^{k-1}] \leq n \leq [c^k]} S_n > (\sigma + \varepsilon)\sqrt{2[c^{k-1}]LL[c^{k-1}]}\right)$$

$$\leq \mathsf{P}\left(|S_{[c^k]}| > \left((\sigma + \varepsilon)\frac{\sqrt{2[c^{k-1}]LL[c^{k-1}]}}{\sqrt{\mathsf{Var}S_{[c^k]}}} - \sqrt{2}\right)\sqrt{\mathsf{Var}S_{[c^k]}}\right)$$

$$\leq \mathsf{P}\left(\frac{|S_{[c^k]}|}{\sqrt{\mathsf{Var}S_{[c^k]}}} > c^{-1/2}\left(1 + \frac{\varepsilon}{2\sigma}\right)\sqrt{2LL[c^{k-1}]}\right) \qquad (1.28)$$

for all k large enough (since $\mathsf{Var}S_n \sim \sigma^2 n$ as $n \to \infty$). From Theorem 3.1.21 one infers that the last probability in (1.28) does not exceed

$$C(\operatorname{Log}[c^k])^{-\lambda/2} + \mathsf{P}\left(|Z| > c^{-1/2}\left(1 + \frac{\varepsilon}{2\sigma}\right)\sqrt{2LL[c^{k-1}]}\right).$$

If c was taken in such a way that $c^{-1/2}(1 + \varepsilon/(2\sigma)) > 1$, the Borel–Cantelli lemma implies that $|S_n| \leq (\sigma + \varepsilon)\sqrt{2nLLn}$ for all n large enough, almost surely.

For the lower estimate take arbitrary $\varepsilon \in (0, 1)$ and integer $c > 1$ to be specified later. Again we consider blocks $B_k = (c^{k-1}, c^k]$ and events

$$E_k = \{S(B_k) \geq (1 - \varepsilon)\sqrt{2\mathsf{Var}S(B_k)LLc^k}\}, \quad k \in \mathbb{N},$$

and prove that E_k take place infinitely often, with probability one.

The assertion of Lemma 1.4 remains true. Indeed, for $k, l \in \mathbb{N}$ and $k > l$, by Theorem 1.5.3,

$$\frac{cov(S(B_k), S(B_l))}{\sqrt{|B_k||B_l|}} \leq C(\text{Log}\,(c^{k-1} - c^l))^{-\lambda}.$$

Therefore

$$\sum_{l=1}^{\infty} \sum_{k=l+1}^{\infty} \frac{cov(S(B_k), S(B_l))}{\sqrt{|B_k||B_l|}} \leq C \sum_{l=1}^{\infty} \sum_{m=1}^{\infty} (\text{Log}\,(c^{l+m-1} - c^l))^{-\lambda}$$

$$\leq C \sum_{l=1}^{\infty} \sum_{m=1}^{\infty} (\text{Log}\,(c^l) + \text{Log}\,(c^{m-1} - 1))^{-\lambda},$$

and the last double sum converges since $\sum_{k_1, k_2 = 1}^{\infty} (k_1 + k_2)^{-\lambda} < \infty$ when $\lambda > 2$.

From this point the proof goes exactly as for Theorem 1.1. The only distinction is that the Berry–Esseen type estimate (appearing after (1.15)) is of order $a_m = O(m^{-\lambda/2})$, $m \to \infty$ (by Theorem 3.1.21), thus the corresponding series $\sum_{m=1}^{\infty} a_m$ converges. $\square$

Remark 1.9. In [422] the proof of the "lower" part of the LIL was based on the following interesting observation.

Lemma 1.10. *If (X, Y) is an associated random vector such that $cov(X, Y)$ exists, then, for any $a > 0$, one has*

$$\text{mes}\{(x, y) \in \mathbb{R}^2 : \mathrm{P}(X \geq x, Y \geq y) - \mathrm{P}(X \geq x)\mathrm{P}(Y \geq y) \geq a\} \leq a^{-1} cov(X, Y).$$

This bound is proved by application of the Chebyshev inequality to the integral appearing in the Hoeffding formula (see Appendix A.1). However, our proof of the lower part of the LIL, given in Theorem 1.8, is applicable to a more general case (it is true for (BL, θ)-dependent random sequences or fields).

4°. LIL for (BL, θ)-dependent random fields. If the Cox–Grimmett coefficient decreases fast enough, the LIL may be easily deduced from the strong invariance principle.

Theorem 1.11. ([80]) *Suppose that $X = \{X_j,\ j \in \mathbb{Z}^d\}$ is a wide-sense stationary, (BL, θ)-dependent random field satisfying conditions (5.2.1) and (5.2.2). Then for any $\tau \in (0, 1/(d-1))$ one has, almost surely,*

$$\limsup \frac{S_n}{\sqrt{2d\langle n \rangle LL\langle n \rangle}} = \sigma \quad \text{and} \quad \liminf \frac{S_n}{\sqrt{2d\langle n \rangle LL\langle n \rangle}} = -\sigma,$$

as $n \to \infty$, $N \in G_\tau$. Here G_τ is the set defined in (5.2.5).

Proof. Fix $\tau \in (0, 1/(d-1))$. By symmetry we need only to establish the claim involving the upper limit. If $\sigma^2 = \sum_{j \in \mathbb{Z}^d} cov(X_0, X_j) > 0$ then X satisfies all the conditions of Theorem 5.2.1. Consequently, by the assertion of that Theorem there exists some probability space with a field $X^* = \{X_j^*, j \in \mathbb{Z}^d\}$ distributed as X and a d-parameter Brownian motion $\{W_t, t \in \mathbb{R}_+^d\}$ such that $S_n^* - \sigma W_n = O(\langle n \rangle^{1/2 - \varepsilon})$ a.s., for some $\varepsilon > 0$ and $n \in G_\tau$. Here $S_n^* = \sum_{j \in (0,n]} X_j^*$, $n \in \mathbb{N}^d$. Therefore, one observes that

$$\frac{S_n^* - \sigma W_n}{\sqrt{2d \langle n \rangle LL \langle n \rangle}} \to 0 \text{ a.s.}, \quad n \to \infty, \quad n \in G_\tau. \tag{1.29}$$

Clearly, $W_n = \sum_{j \in (0,n]} \zeta_j$ where $\{\zeta_j, j \in \mathbb{Z}^d\}$ are i.i.d. $N(0,1)$ random variables. Thus, by Theorem 1.1 we have

$$\limsup_{n \to \infty} \frac{W_n}{\sqrt{2d \langle n \rangle LL \langle n \rangle}} \leq 1. \tag{1.30}$$

Alternatively one may refer to the classical LIL for d-parameter Wiener process (see, e.g., [231, 433]). As closer inspection of the proof of Theorem 1.1 shows, actually we have proved that

$$\limsup_{n \to \infty, n \in G_\tau} \frac{W_n}{\sqrt{2d \langle n \rangle LL \langle n \rangle}} \geq 1 \tag{1.31}$$

(remember that one considered indices k from the set J_ρ). We mention in passing that (1.31) can be deduced from the LIL for Brownian motion restricted to an arbitrary set proved in [380]. By (1.29)—(1.31) one has

$$\limsup_{n \to \infty, n \in G_\tau} \frac{S_n^*}{\sqrt{2d \langle n \rangle LL \langle n \rangle}} = \sigma \text{ a.s.}$$

The distributions of the random fields $\{S_n, n \in G_\tau\}$ and $\{S_n^*, n \in G_\tau\}$ coincide, so the same relation holds when S_n^* is replaced with S_n. Whence Theorem is proved under the additional assumption $\sigma^2 > 0$.

Suppose now that $\sigma^2 = 0$. Let $\{Z_j, j \in \mathbb{Z}^d\}$ be a field consisting of i.i.d. $N(0,1)$ random variables independent from X. For $k \in \mathbb{N}$, consider random fields $X^{(k)} = \{X_j + k^{-1} Z_j, j \in \mathbb{Z}^d\}$. Then, for any $k \in \mathbb{N}$,

$$\sum_{j \in \mathbb{Z}^d} cov(X_0^{(k)}, X_j^{(k)}) = k^{-2} > 0.$$

Besides that, by Lemma 5.2.4 all the fields $X^{(k)}$ are (BL, θ)-dependent with the same dependence coefficients $(\theta_r)_{r \in \mathbb{N}}$ as for X. Thus, from what has already been proved, we have

$$\limsup_{n \to \infty, \, n \in G_\tau} \frac{|S_n^{(k)}|}{\sqrt{2d \langle n \rangle LL \langle n \rangle}} = k^{-1} \text{ and } \limsup_{n \to \infty, \, n \in G_\tau} \frac{|T_n|}{\sqrt{2d \langle n \rangle LL \langle n \rangle}} = 1 \text{ a.s.} \tag{1.32}$$

where $S_n^{(k)} = \sum_{j \in (0,n]} X_j^{(k)}$ and $T_n = \sum_{j \in (0,n]} Z_j$. From the relations (1.32) one easily deduces that

$$\limsup_{n \to \infty, \, n \in G_\tau} \frac{|S_n|}{\sqrt{2d \langle n \rangle LL \langle n \rangle}} \leq 2k^{-1}.$$

As k could be taken arbitrarily large, the last upper limit is a.s. zero. $\square$

2   Functional LIL

This Section is devoted to proving the law of iterated logarithm in Strassen's form for the partial sums process constructed by means of an associated random field.

For any bounded set $R \subset \mathbb{R}^d$ write $Int(R) = |R \cap \mathbb{Z}^d|$. If $R = (a, b]$ is a rectangle and $b_i - a_i \geq 2$, then

$$2^{-d}\mathrm{mes}(R) \leq Int(R) \leq 2^d \, \mathrm{mes}(R). \qquad (2.1)$$

Define a sequence of random elements H_n in the Skorokhod space $D([0, 1]^d)$ by

$$H_n(t) = \frac{S(n[0, t])}{\sqrt{2n^d LLn}}, \quad t = (t_1, \ldots, t_d) \in [0, 1]^d, \quad n \in \mathbb{N}.$$

Theorem 2.1. ([65]) *Let* $X = \{X_j, j \in \mathbb{Z}^d\}$ *be a centered, wide-sense stationary, associated random field. Suppose that, for some* $s \in (2, 3]$,

$$D_s = \sup_{j \in \mathbb{Z}^d} \mathsf{E}|X_j|^s < \infty$$

and the Cox–Grimmett coefficient $u_n \leq c_0 n^{-\lambda}$ *for some* $c_0 > 0$, $\lambda > d/(s - 1)$ *and all* $n \in \mathbb{N}$. *Then the limit points set of the family* $\{H_n, n \in \mathbb{N}\}$ *in the Skorokhod space* $D([0, 1]^d)$ *endowed with uniform topology is almost surely the set*[1]

$$\mathbb{K}_\sigma = \left\{ x(t), t \in [0, 1]^d : x(t) = \int_0^{t_1} \cdots \int_0^{t_d} h(z)dz, \int_0^1 \cdots \int_0^1 h^2(z)dz \leq \sigma^2 \right\}.$$

Proof. At first we note that if $\sigma = 0$, the Theorem is trivial. Hence in what follows we assume that $\sigma > 0$.

1°. The continuity modulus of $(H_n)_{n \in \mathbb{N}}$ **subsequence.**

Lemma 2.2. *For each* $c > 1$ *and any* $\varepsilon > 0$ *there exists some* $\delta = \delta(\varepsilon) > 0$ *such that*

$$\limsup_{l \to \infty} \sup_{\substack{t, v \in [0, 1]^d \\ |t-v| \leq \delta}} |H_{[c^l]}(t) - H_{[c^l]}(v)| \leq \varepsilon \quad \text{a.s.}$$

Proof. For a block $Q \subset \mathbb{R}^d$ let $H_n(Q) := S(nQ)/(2n^d LLn)^{1/2}$. Then $H_n(t) = H_n([0, t])$ for $t \in [0, 1]^d$. By the Borel–Cantelli lemma it suffices to prove that, for some $N \in \mathbb{N}$,

$$\sum_{l=1}^{\infty} \mathsf{P}\left(\sup_{Q \in \mathcal{A}_N} |H_{[c^l]}(Q)| \geq \varepsilon 2^{-d} \right) < \infty \qquad (2.2)$$

where $\mathcal{A}_N$ is the set of all rectangles $V \subset [0, 1]^d$ whose minimal edge $\Delta_0(V)$ is no longer than $1/N$. Let P_l be an element of the series in (2.2). Then, for $l \in \mathbb{N}$,

$$\mathsf{P}_l \leq \mathsf{P}(\max |S(V)| \geq x_l) \leq 2 \sum_{i=1}^{d} \sum_{k=1}^{N} \mathsf{P}(M(T_l(i, k)) \geq x_l 2^{-d-1})$$

---

[1]It is called the Strassen ball of radius σ.

where the maximum is taken over rectangles $V \subset [0, [c^l]]^d$ with $\Delta_0(V) \leq [c^l]/N$ and we use the notation

$$x_l = \varepsilon 2^{-d}(2[c^l]^d LL[c^l])^{1/2}, \quad T_l(i,k) := [c^l]\Big\{[0,1]^d \cap \{x \in \mathbb{R}^d : (k-1) < Nx_i \leq k\}\Big\},$$

$i = 1, \ldots, d$, $k = 1, \ldots, N$, and $M(Q) := M(Q \cap \mathbb{Z}^d)$, see (2.1.1). Note that

$$\frac{x_l}{Int(T_l(i,k))} \sim \varepsilon N^{1/2} 2^{-d+1/2} \sqrt{\log l}, \quad l \to \infty. \tag{2.3}$$

By Theorem 2.4.6 and (2.3), for all l large enough,

$$\mathsf{P}_l \leq 4 \sum_{i=1}^{d} \sum_{k=1}^{N} \mathsf{P}(|S(T_l(i,k))| \geq x_l 2^{-d-2}). \tag{2.4}$$

Here we have replaced an inequality for maxima with that for sums, hence we can use the normal approximation. By Theorems 3.3.1 and 1.1.8, taking into account that $\lambda(s-1) > d$, we deduce the bound

$$\left| \mathsf{P}\left(|S(T_l(i,k))| \geq x_l 2^{-d-2}\right) - \mathsf{P}\left(|\xi| \geq \frac{x_l 2^{-d-2}}{\sqrt{\mathsf{Var}S(T_l(i,k))}}\right) \right|$$

$$\leq C_0 \, \mathrm{mes}(T_l(i,k))^{-\tau_1} \leq C_1([c^l]^d/N)^{-\tau_1} \tag{2.5}$$

where $\xi \sim N(0,1)$, $\tau > 0$ and $C_0, C_1, \ldots$ are some positive factors independent of l. Now (2.3) ensures that if $N \geq \alpha \sigma^2 2^{4d+3} \varepsilon^{-2}$ with $\alpha > 2$, then, for any $i = 1, \ldots, d$, $k = 1, \ldots, N$, and all sufficiently large l,

$$y_l := \frac{x_l 2^{-d-2}}{\sqrt{\mathsf{Var}S(T_l(i,k))}} \geq (\alpha \log l)^{1/2}$$

provided that $N = N(\varepsilon, \sigma, d)$ is taken large enough. Since $\mathsf{P}(|\xi| \geq x) \leq x^{-1}(2/\pi)^{1/2} e^{-x^2/2}$ for any $x > 0$, one has

$$\mathsf{P}(|\xi| \geq y_l) \leq C_2 l^{-\alpha/2}. \tag{2.6}$$

From (2.4), (2.5) and (2.6) the relation (2.2) follows. $\square$

2°. **The normal approximation for random vectors constructed by partial sums.** Suppose that $r > 1$ is a positive integer and $V_1, \ldots, V_r \subset \mathbb{R}^d$ are the sets such that $\mathsf{Var}S(V_i) > 0$, $i = 1, \ldots, r$. Let $f(t)$ be the characteristic function of the random vector

$$Y = \Big(S(V_1)/\sqrt{\mathsf{Var}S(V_1)}, \ldots, S(V_r)/\sqrt{\mathsf{Var}S(V_r)}\Big)$$

and $f_0(t)$ be the characteristic function of the standard normal distribution in $\mathbb{R}^r$.

Lemma 2.3. *Suppose that $V_1, \ldots, V_r$ are rectangles in $\mathbb{R}^d$ such that*

$$\min_{i,p:1 \leq i < p \leq r} dist(V_i, V_p) \geq c_1 v^{\mu_1}$$

for some $c_1, \mu_1 > 0$ and $v = \min_{i=1,\ldots,r} |V_i|$. Then under conditions of Theorem 2.1 it is possible to find $\delta, \gamma, c_2, c_3 > 0$, depending only on $r, c_1, D_s, \mathsf{E}X_0^2$, such that

$$|f(t) - f_0(t)| \leq c_2 v^{-\delta} \quad \text{when} \quad |t| \leq c_3 v^\gamma.$$

Proof. If $v \leq 1$ one can take $c_2 = 2$ and arbitrary positive δ and γ. In what follows we assume that $v > 1$. If $f_i(t)$ is the characteristic function of $S(V_i)/\sqrt{\mathrm{Var}S(V_i)}$, then for any $t \in \mathbb{R}^r$

$$|f(t) - f_0(t)| \leq |f(t) - \prod_{i=1}^{r} f_i(t_i)| + \sum_{i=1}^{r} |f_i(t_i) - e^{-t_i^2/2}|. \qquad (2.7)$$

The first summand in the right-hand side of (2.7) can be bounded by Newman's inequality or by Theorem 1.5.3. Namely, for t such that $|t| \leq c_3 v^\gamma$

$$|f(t) - f_0(t)| \leq 2 \sum_{i \neq p} |t_i t_p| \frac{cov(S(V_i), S(V_p))}{\sqrt{\mathrm{Var}S(V_i)\mathrm{Var}S(V_p)}}$$

$$\leq r^2 c_1^{-\lambda} c_3^2 (\mathsf{E}X_0^2)^{-1} v^{2\gamma - \lambda\mu_1} < C_3 v^{-\lambda\mu_1/2} \qquad (2.8)$$

if one takes here $\gamma \leq \lambda\mu_1/4$. We also have used that $\mathrm{Var}S(U) \geq \mathsf{E}X_0^2 |U|$ for any finite $U \subset \mathbb{Z}^d$.

By the second assertion of Theorem 3.1.21, there exist $C_4, C_5, \delta_1, \gamma_2$ such that for any $i = 1, \ldots, r$

$$|f_i(t_i) - e^{-t_i^2/2}| \leq C_4 |V_i|^{-\delta_1} \leq C_4 v^{-\delta_1} \quad \text{if} \quad |t_i| \leq C_5 v^{\gamma_1}.$$

From this relation and (2.8) the Lemma follows if we set

$$\gamma = \gamma_1 \wedge \frac{\lambda\mu_1}{4}, \quad \delta = \delta_1 \wedge \frac{\lambda\mu_1}{2}. \quad \square$$

Lemma 2.4. *Let the conditions of Lemma 2.3 be satisfied. Then, for any bounded convex set $B \subset \mathbb{R}^r$,*

$$|\mathsf{P}(Y \in B) - \mathsf{P}(Z \in B)| \leq C_6 \left(1 + \mathrm{mes}(B^{(1)}) \right) v^{-\beta}$$

where $Z \sim N(0, I_r)$ and $\beta > 0$ does not depend on B; as usual, $B^{(1)}$ denotes the neighborhood of B of radius 1.

Proof. By the von Bahr multidimensional analogue of the Esseen inequality ([12]) there exist positive numbers E_1, E_2, E_3, depending only on r, such that

$$|\mathsf{P}(Y \in B) - \mathsf{P}(Z \in B)| \leq \frac{E_1}{T} + E_2 \mathrm{mes}\left(B^{(E_3/T)} \right) \int_{t:|t| \leq T} |f(t) - f_0(t)| dt$$

for any $T > 0$. Lemma 2.3 and optimization in T (i.e. letting $T = E_3(1 \vee v^\nu)$ under appropriate choice of $\nu > 0$) yield the desired result. $\square$

$3°$. **All limit points of $(H_n)_{n \in \mathbb{N}}$ belong to the Strassen ball.** Now we may start proving the Theorem. Let $\varepsilon > 0$ be arbitrary and $Lim\{H_n\}$ be the (random) set of limit points of H_n in the space $D([0,1]^d)$ endowed with uniform metric. At first we are going to show that $Lim\{H_n\} \subset \mathbb{K}_\sigma^{(\varepsilon)}$ with probability one, where $\mathbb{K}_\sigma^{(\varepsilon)}$ is the ε-neighborhood of $\mathbb{K}_\sigma$ in the same metric.

For $m \in \mathbb{N}$ and $j \in \mathbb{N}^d$ write $B_{jm} = m^{-1}(j - \mathbf{1}, j] \subset \mathbb{R}^d$. Then for $t \in [0, 1]^d$ and $m \in \mathbb{N}$ introduce the functions $\Pi_m H_n$ by

$$\Pi_m H_n(t) := \sum_{j \in J} \int_{[0,t] \cap B_{jm}} m^d H_n(B_{jm}) dz_1 \dots dz_d$$

where $J = J_m = \{j \in \mathbb{N}^d : |j| \leq m\}$ and $H_n(B) := S(B)/\sqrt{2n^d LLn}$ for a block $B \in \mathcal{U}$, $n \in \mathbb{N}$.

Choose $r = r(\varepsilon) \in (1, 1 + \varepsilon/4\sigma)$. We will show that

$$P(H_n \notin \mathbb{K}_\sigma^{(\varepsilon)}) \leq P(r^{-1} \Pi_m H_n \notin \mathbb{K}_\sigma) + P\left(\sup_{t \in [0,1]^d} |H_n(t) - \Pi_m H_n(t)| > \frac{\varepsilon}{2} \right). \quad (2.9)$$

Set $A = \left\{ \sup_{t \in [0,1]^d} |H_n(t) - \Pi_m H_n(t)| \leq \varepsilon/2 \right\}$. It is clear that

$$P(H_n \notin \mathbb{K}_\sigma^{(\varepsilon)}) \leq P(\{H_n \notin \mathbb{K}_\sigma^{(\varepsilon)}\} \cap A) + P(A^c).$$

Let $\omega \in \{H_n \notin \mathbb{K}_\sigma^{(\varepsilon)}\} \cap A$, then, for all functions $x \in \mathbb{K}_\sigma$ and some $t \in [0, 1]^d$ (depending on ω and x), one has $|H_n(t) - x(t)| \geq \varepsilon$. Consequently, at this point t

$$|\Pi_m H_n(t) - x(t)| \geq \varepsilon/2. \quad (2.10)$$

The Cauchy–Bunyakowski–Schwarz inequality implies the bound

$$|x(t)| = \left| \int_0^{t_1} \dots \int_0^{t_d} \dot{x}(t) dt \right| \leq \left(\int_0^{t_1} \dots \int_0^{t_d} \dot{x}^2(t) dt \right)^{1/2} \leq \sigma$$

where $\dot{x}$ is the Radon–Nikodym derivative of x. Due to our choice of r and (2.10),

$$|r^{-1} \Pi_m H_n(t) - x(t)| \geq r^{-1} |\Pi_m H_n(t) - x(t)| - (1 - r^{-1})|x(t)| \geq \frac{\varepsilon}{2r} - \frac{r-1}{r} \sigma > 0.$$

Since this argument is true for any $x \in \mathbb{K}_\sigma$, it follows that $r^{-1} \Pi_m H_n(\cdot, \omega) \notin \mathbb{K}_\sigma$ and thus (2.9) is checked.

Next define the cubes $F_{jm} = F_{jm}(n) = (a_{jm}(n), b_{jm}(n)] \subset B_{jm}$ $(j \in \mathbb{N}^d, m \in \mathbb{N})$ by the following rules:

a) for any $m \in \mathbb{N}$ and $j \in \mathbb{N}^d$, the centers of F_{jm} and B_{jm} coincide;

b) for some $\nu > 0$ and for any $j \in J_m$, $n \in \mathbb{N}$,

$$dist(\partial F_{jm}(n), \partial B_{jm}) = n^{-\nu} m^{-1}. \quad (2.11)$$

More precisely, F_{jm} is homothetic to B_{jm}, being obtained from the latter by a linear contraction of $\mathbb{R}^d$ centered at the center point of B_{jm} and such that (2.11) holds. Let $R_{jm} = B_{jm} \setminus F_{jm}$ $(R_{jm} = R_{jm}(n))$. Note that the Radon–Nikodym derivative $d/dt(\Pi_m H_n(t))$ is a step function with values $m^d H_n(B_{jm})$ for $t \in B_{jm}$ (where $m \in \mathbb{N}, j \in J_m$). Then we can write:

$$P\left(r^{-1} \Pi_m H_n \notin \mathbb{K}_\sigma\right) = P\left(\sum_{j \in J} H_n(B_{jm})^2 > r^2 \sigma^2 \right)$$

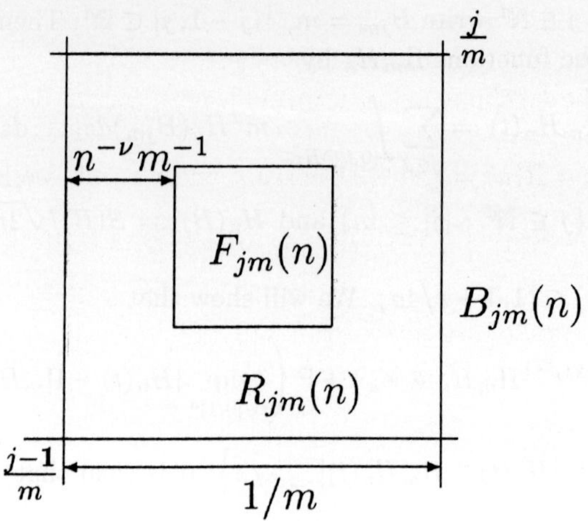

Fig. 6.3

$$\leq \mathsf{P}\left(|\eta| > n^d \kappa \sigma^2 LLn\right) + \mathsf{P}\left(\zeta > n^d(1+\kappa/2)2\sigma^2 LLn\right) \tag{2.12}$$

with

$$\kappa = r^2 - 1, \ \eta = \sum_{j\in J}(S^2(nR_{jm}) + 2S(nR_{jm})S(nF_{jm})), \ \zeta = \sum_{j\in J} S^2(nF_{jm}).$$

Clearly, $Int(nF_{jm}) \sim |nF_{jm}|$ as $n \to \infty$ ($m \in J_m$). Therefore, for all n large enough and any $j \in J_m$,

$$\frac{1}{2}\left(\frac{n}{m}\right)^d \leq |nF_{jm}| \leq \left(\frac{n}{m}\right)^d. \tag{2.13}$$

It is easily seen that

$$|nR_{jm}(n)| \leq n^d d 2^{d+1} m^{-d} n^{-\nu}, \ n \in \mathbb{N}, \ m \in J_m.$$

By (2.1), the Markov and the Cauchy–Bunyakowski–Schwarz inequalities, we get

$$\mathsf{P}\left(|\eta| > n^d \kappa \sigma^2 LLn\right) \leq \mathsf{E}|\eta|(n^d \kappa \sigma^2 LLn)^{-1} \leq 5d2^d(\kappa LLn)^{-1}n^{-\nu/2}. \tag{2.14}$$

Consequently, in view of (2.13) we derive the bound

$$\mathsf{P}\left(\zeta > n^d(1+\kappa/2)2\sigma^2 LLn\right) \leq \mathsf{P}\left(\sum_{j\in J}Y_j^2 > m^d(1+\kappa/2)2LLn\right)$$

where $Y_j = Y_j(n,m) = S(nF_{jm})/\sqrt{\mathsf{Var}S(nF_{jm})}$. For any $j,q \in J, j \neq q$, one has

$$dist(nF_{jm}, nF_{qm}) \geq n^{1-\nu}m^{-1}. \tag{2.15}$$

Hence, using Theorem 3.1.8, (2.13) and Lemma 2.4, we come to an estimate

$$|\mathsf{P}(Y \in B_R(0)) - \mathsf{P}(Z \in B_R(0))| \leq C_6 2^\beta \left(1 + \mathrm{mes}(B_R(0)^{(1)})\right)\left(\frac{n}{m}\right)^{-d\beta}.$$

Here Y is the m^d-dimensional random vector having components $Y_j(n,m)$, $j \in J$, and $B_R(0)$ is the m^d-dimensional Euclidean ball of radius R centered at 0. Note that here and in the formulas below the coefficients C_i can depend on the (fixed) number m.

Thus, taking $R = R(n) = \sqrt{m^d(2+\kappa)LLn}$ and $r = m^d$, we have

$$\left| P\left(\sum_{j \in J} Y_j^2 > R(n)^2 \right) - P(Z \notin B_{R(n)}(0)) \right| \leq C_7 (LLn)^{r/2} n^{-d\beta}. \tag{2.16}$$

If χ_p^2 $(p > 1)$ is a χ-square distributed random variable[2] with p degrees of freedom, then changing variables and integrating by parts yields

$$P(\chi_p^2 > x^2) = \frac{1}{(2\pi)^{p/2}} \int_{t \in \mathbb{R}^p : \|t\| \geq x} e^{-\|t\|^2/2} dt$$

$$= \frac{S_p}{(2\pi)^{p/2}} \int_x^\infty r^{p-1} e^{-r^2/2} dr \leq \frac{S_p}{(2\pi)^{p/2}} x^{p-2} e^{-x^2/2} = \frac{x^{p-2} e^{-x^2/2}}{\Gamma(p/2) 2^{p/2-1}} \tag{2.17}$$

where $S_p = (2\pi)^{p/2} 2^{1-p/2}/\Gamma(p/2)$ is the surface area of the p-dimensional unit sphere. Moreover, by the same argument one sees that, for p fixed, the first and the last expressions in (2.17) are equivalent as $x \to \infty$.

Consequently,

$$P(Z \notin B_{R(n)}(0)) \leq C_8 (\mathrm{Log}\, n)^{-1-\kappa/4}, \quad n \in \mathbb{N}.$$

From (2.12), (2.14) and (2.16) it follows that

$$P(r^{-1} \Pi_m H_n \notin \mathbb{K}_\sigma^\varepsilon) \leq C_9 (\mathrm{Log}\, n)^{-1-\kappa/4}, \quad n \in \mathbb{N}. \tag{2.18}$$

Set $t_{jm}^* = m^{-1}(j_1 - 1, \ldots, j_d - 1)$, then

$$P\left(\sup_{t \in [0,1]^d} |H_n(t) - \Pi_m H_n(t)| > \frac{\varepsilon}{2} \right) \leq P\left(\max_{j \in J} \sup_{t \in B_{jm}} |H_n(t) - H_n(t_{jm}^*)| > \frac{\varepsilon}{4} \right)$$

$$+ P\left(\max_{j \in J} \sup_{t \in B_{jm}} |\Pi_m H_n(t) - H_n(t_{jm}^*)| > \frac{\varepsilon}{4} \right) =: p_1 + p_2.$$

Obviously, for the first summand we have

$$p_1 \leq P\left(\sup_{\substack{t,v \in [0,1]^d \\ |t-v| \leq 1/m}} |H_n(t) - H_n(v)| > \frac{\varepsilon}{4} \right). \tag{2.19}$$

For any $j \in J$, one has $H_n(t_{jm}^*) = \Pi_m H_n(t_{jm}^*)$. Hence, letting $V(t, t_{jm}^*) = [0,t] \triangle [0, t_{jm}^*]$, we have, for any $j \in J$ and all $t \in B_{jm}$,

$$|\Pi_m H_n(t) - H_n(t_{jm}^*)| \leq \sum_{k \in J} \int_{V(t,t_j^*) \cap B_{km}} \cdots \int m^d |H_n(B_{km})| dz_1 \ldots dz_d$$

---

[2] That is, $Law(\chi_p^2) = Law(\xi_1^2 + \ldots + \xi_p^2)$ where $\xi_1, \ldots, \xi_p$ are i.i.d. $N(0,1)$ random variables.

$$\leq m^d \max_{k \in J} \operatorname{mes}(V(t,t_j^*) \cap B_{km})^{1/2} \sum_{k \in J} \operatorname{mes}(V(t,t_j^*) \cap B_{km})^{1/2} |H_n(B_{km})|$$

$$\leq m^d \max_{k \in J} \operatorname{mes}(V(t,t_j^*) \cap B_{km})^{1/2} \operatorname{mes}(V(t,t_j^*))^{1/2} \left(\sum_{k \in J} H_n(B_{km})^2 \right)^{1/2}$$

$$\leq \left(\frac{d}{m} \sum_{j \in J} m^d H_n(B_{jm})^2 \right)^{1/2}.$$

Therefore

$$p_2 \leq \mathsf{P}\left(\sum_{j \in J} m^d H_n(B_{jm})^2 > \frac{m\varepsilon^2}{16d} \right).$$

Choose m large enough to ensure that $m\varepsilon^2 > 16dr^2\sigma^2$. Then, on account of (2.12), p_2 can be estimated similarly to (2.18). Using also (2.19) and (2.2), we have, for any $\varepsilon > 0$ and $m \in \mathbb{N}$, the relation

$$\sum_{l=1}^{\infty} \mathsf{P}\left(\sup_{t \in [0,1]^d} |H_{[c^l]}(t) - \Pi_m H_{[c^l]}(t)| > \frac{\varepsilon}{2} \right) < \infty. \tag{2.20}$$

By (2.9), (2.18), and (2.20) it follows that, for any $\omega \in \Omega_1$ with $\mathsf{P}(\Omega_1) = 1$,

$$H_{[c^l]}(\cdot, \omega) \in \mathbb{K}_\sigma^\varepsilon \quad \text{for} \quad l > N_1(\varepsilon, c, \omega). \tag{2.21}$$

Therefore, for the same ω and l,

$$\sup_{t \in [0,1]^d} |H_{[c^l]}(t, \omega)| \leq \sigma + \varepsilon. \tag{2.22}$$

The final step is to approximate H_n with $H_{[c^l]}$. For $l \in \mathbb{N}$ introduce the set $I_l = \{k \in \mathbb{N} : [c^{l-1}] < k \leq [c^l]\}$. Then we can write

$$\max_{k \in I_l} \sup_{t \in [0,1]^d} |H_{[c^l]}(t) - H_k(t)| \leq \max_{k \in I_l} \sup_{t \in [0,1]^d} \left| H_{[c^l]}(t) - H_{[c^l]}\left(\frac{k}{[c^l]} t \right) \right|$$

$$+ \max_{k \in I_l} \left| \left(\frac{[c^l]^d LL[c^l]}{k^d LLk} \right)^{1/2} - 1 \right| \sup_{t \in [0,1]^d} \left| H_{[c^l]}\left(\frac{k}{[c^l]} t \right) \right| =: \delta_1^{(1)} + \delta_2^{(2)}.$$

For any $t \in [0,1]^d$ and $k \in I_l$,

$$\limsup_{l \to \infty} \left| t - \left(\frac{k}{[c^l]} t \right) \right| \leq \frac{c-1}{c},$$

and

$$\limsup_{l \to \infty} \left| \left(\frac{[c^l]^d LL[c^l]}{k^d LLk} \right)^{1/2} - 1 \right| = c^{d/2} - 1.$$

Thus, taking $c > 1$ close enough to 1, from (2.21) and (2.22) one sees that $Lim\{H_n\} \subset \mathbb{K}_\sigma^{2\varepsilon}$.

4°. The Berkes version of the Borel–Cantelli lemma.

Lemma 2.5. ([34]) *Let* $(A_k)_{k\in\mathbb{N}}$ *be a sequence of events such that, for some numbers* $\alpha_1,\alpha_2,\rho_1,\rho_2>0$ *and* $\nu_0<1$, *the following inequalities hold* :

$$P(A_k)\geq\alpha_1 k^{-\nu_0}\text{ for all }k,\qquad(2.23)$$

$$|P(A_k A_l)-P(A_k)P(A_l)|\leq\alpha_2 l^{\rho_1}e^{-\rho_2 k}\text{ for all }k,l:1\leq k<l<\infty.\qquad(2.24)$$

Then $P(A_k\text{ i. o.})=1$.

Proof. The Erdös–Renyi theorem (Lemma 1.5) claims that the desired conclusion holds if the relations (1.12) and (1.13) are satisfied for a sequence of events $(A_k)_{k\in\mathbb{N}}$.

The first relation trivially follows from (2.23), so we will prove (1.13). Write

$$\sum_{k,l=1}^{n}(P(A_k A_l)-P(A_k)P(A_l))=S_0+S_1+S_2+S_3+S_4,$$

here $S_0=\sum_{k=1}^{n}(P(A_k)-P(A_k)^2)\leq\sum_{k=1}^{n}P(A_k)$ and in the following sums S_1,S_2,S_3,S_4 the summation is implemented for $k\neq l$:

$$S_1=\sum_{k=[\log^2 n]+1}^{n}\sum_{l=[\log^2 n]+1}^{n}d_{kl},\quad S_2=\sum_{k=1}^{[\log^2 n]}\sum_{l=[\log^2 n]+1}^{n}d_{kl},$$

$$S_3=\sum_{k=[\log^2 n]+1}^{n}\sum_{l=1}^{[\log^2 n]}d_{kl},\quad S_4=\sum_{k=1}^{[\log^2 n]}\sum_{l=1}^{[\log^2 n]}d_{kl}$$

where $d_{kl}=P(A_k A_l)-P(A_k)P(A_l)$. By Lemma's assumptions, one has for all n large enough

$$|S_1|\leq\alpha_2 n^{2+\rho_1}e^{-\rho_2[\log^2 n]}\leq C_{10},$$

$$|S_2+S_3|\leq 2\sum_{k=1}^{n}\sum_{l=1}^{[\log^2 n]}P(A_k)+2\sum_{k=1}^{[\log^2 n]}\sum_{l=1}^{n}P(A_l)\leq 4\log^2 n\sum_{j=1}^{n}P(A_j),$$

$$|S_4|\leq 2\log^4 n,\quad\sum_{k=1}^{n}P(A_k)\geq C_{11}n^{1-\nu_0}.$$

Thus,

$$\frac{\sum_{k,l=1}^{n}(P(A_k A_l)-P(A_k)P(A_l))}{(\sum_{k=1}^{n}P(A_k))^2}$$

$$\leq\frac{\sum_{k=1}^{n}P(A_k)+2\log^4 n+4\log^2 n\sum_{k=1}^{n}P(A_k)}{(\sum_{k=1}^{n}P(A_k))^2}\to 0,\ n\to\infty.\quad\square$$

5°. Every point of $\mathbb{K}_\sigma$ is a limiting one for $(H_n)_{n\in\mathbb{N}}$. Now we prove the "lower part" of the LIL, that is we show that any point in $\mathbb{K}_\sigma$ is a limiting one for H_n, almost surely. Clearly, it suffices to prove this only for points forming a dense subset in $\mathbb{K}_\sigma$, in uniform metric.

Consider in K the class of functions $g(t) = \int_{[0,t]} h(z)dz$, $t \in [0,1]^d$ where

$$\int_{[0,1]^d} h^2(t)dt = \sigma^2(1-\delta), \quad \delta > 0, \quad h \in C([0,1]^d). \tag{2.25}$$

This class is a dense subset of $\mathbb{K}_\sigma$ (in the sense mentioned above). Indeed, if $f \in \mathbb{K}_\sigma$ then there exists $u : [0,1]^d \to \mathbb{R}$ such that

$$f(t) = \int_{[0,t]} u(z)dz, \ t \in [0,1]^d, \quad \int_{[0,1]^d} u^2(t)dt \leq \sigma^2.$$

For an arbitrary $\varepsilon > 0$ let $w \in C([0,1]^d)$ be a function verifying $\|w-u\|_{L^2([0,1]^d)} < \varepsilon$. Set $h_\varepsilon := \sigma w/(\sigma+2\varepsilon)$ and $g_\varepsilon(t) = \int_{[0,t]} h_\varepsilon(z)dz$, $t \in [0,1]^d$. Then, for any $t \in [0,1]^d$, due to the Cauchy–Bunyakowski–Schwarz inequality one has

$$|f(t) - g_\varepsilon(t)| = \left|\int_{[0,t]} (u(z) - h_\varepsilon(z))dz\right| \leq \|h_\varepsilon - u\|_{L^2([0,1]^d)}$$

$$\leq \|h_\varepsilon - w\|_{L^2([0,1]^d)} + \|w - u\|_{L^2([0,1]^d)} < \left(1 - \frac{\sigma}{\sigma + 2\varepsilon}\right)\|w\|_{L^2([0,1]^d)} + \varepsilon$$

$$\leq \frac{2\varepsilon}{\sigma + 2\varepsilon}(\|u\|_{L^2([0,1]^d)} + \varepsilon) + \varepsilon \leq \frac{2\varepsilon}{\sigma + 2\varepsilon}(\sigma + \varepsilon) + \varepsilon < 3\varepsilon. \tag{2.26}$$

In addition,

$$\left(\int_{[0,1]^d} h_\varepsilon^2(t)dt\right)^{1/2} = \|h_\varepsilon\|_{L^2([0,1]^d)} = \frac{\sigma}{\sigma + 2\varepsilon}\|w\|_{L^2([0,1]^d)} \leq \frac{\sigma(\sigma + \varepsilon)}{\sigma + 2\varepsilon} < \sigma. \tag{2.27}$$

Thus (2.26) and (2.27) imply that our class of functions is everywhere dense in $\mathbb{K}_\sigma$.

Now fix $\delta > 0$ and a function $g \in \mathbb{K}_\sigma$ satisfying (2.25). We will show that, for any $\varepsilon > 0$, there exist, almost surely, $m \in \mathbb{N}$ and a subsequence $(n_k)_{k\in\mathbb{N}}$ of the sequence $(m^n)_{n\in\mathbb{N}}$ such that

$$\sup_{t\in[0,1]^d} |H_{n_k}(t) - g(t)| < \varepsilon, \quad k \in \mathbb{N}.$$

Consider a rectangle $B \subset [0,1]^d$. Let $g(B)$ be the mixed difference (see (5.1.5)) of g at vertices of B and introduce events

$$E_n = \{|H_n(B_{jm}) - g(B_{jm})| < \varepsilon m^{-d}, \ j \in J \setminus \mathbf{1}\}$$

where $\mathbf{1} = (1,\ldots,1) \in \mathbb{Z}^d$. It suffices to prove that, for all m large enough, the event $A_k := E_{m^k}$ takes place infinitely often with probability one. Then, according to Lemma 2.2, for any $\varepsilon > 0$, every m large enough and an infinite number of indices $k > N(\varepsilon, m, g, \omega)$ one has

$$|H_{m^k}(t) - g(t)| \leq |H_{m^k}(t) - H_{m^k}(0)| + \left|\int_{[0,t]^d} h(z)dz\right| \leq \varepsilon + \sigma m^{-d/2}$$

whenever $t \in [0, 1/m]^d$. Furthermore, for $j \in J \setminus \{1\}$ and $t \in B_{jm}$,

$$|H_{m^k}(t) - g(t)| \leq |H_{m^k}(t) - H_{m^k}(t_{jm}^*)| + |H_{m^k}(t_{jm}^*) - g(t_{jm}^*)| + |g(t_{jm}^*) - g(t)|$$

$$\leq |H_{m^k}(t_{jm}^*) - g(t_{jm}^*)| + \varepsilon + \int_{V(t, t_{jm}^*)} |h(z)| \sigma \, dz$$

$$\leq \sum_{j \in J \setminus \{1\}} |H_{m^k}(B_{jm}) - g(B_{jm})| + 2\varepsilon + \sigma m^{-d/2} + (d/m)^{1/2} \sigma.$$

We prove the relation $\mathsf{P}(A_k \text{ i.o.}) = 1$ with the help of Lemma 2.5. To check that the conditions of that Lemma are met we need one more auxiliary result.

Lemma 2.6. *Let ξ, η be random vectors with values in $\mathbb{R}^r$, $r \in \mathbb{N}$. Then, for each rectangle $B = (a, b) \subset \mathbb{R}^r$ and any nonrandom vector τ with components such that $0 < \tau_i < (b_i - a_i)/2$, $i = 1, \ldots, r$, one has*

$$\mathsf{P}(\xi \in B_\tau) - \sum_{i=1}^{r} \mathsf{P}(|\eta_i| > \tau_i) \leq \mathsf{P}(\xi + \eta \in B) \leq \mathsf{P}(\xi \in B^\tau) + \sum_{i=1}^{r} \mathsf{P}(|\eta_i| > \tau_i).$$

Here we use the notation

$$B_\tau = \prod_{i=1}^{r} (a_i + \tau_i, b_i - \tau_i), \quad B^\tau = \prod_{i=1}^{r} (a_i - \tau_i, b_i + \tau_i).$$

Proof. For the first inequality note that

$$\{\xi \in B_r\} = \left(\{\xi \in B_\tau\} \cap \left(\bigcap_{i=1}^{r} \{|\eta_i| \leq \tau_i\}\right)\right) \cup \left(\{\xi \in B_\tau\} \cap \left(\bigcup_{i=1}^{r} \{|\eta_i| > \tau_i\}\right)\right)$$

$$\subset \{\xi + \eta \in B\} \cup \left(\bigcup_{i=1}^{r} \{|\eta_i| > \tau_i\}\right).$$

The second inequality is handled in the same way. $\square$

Next we notice that

$$\mathsf{P}(E_n) = \mathsf{P}(Y \in B) \tag{2.28}$$

where $Y = Y_n(m)$ is the $(m^d - 1)$-dimensional random vector with components

$$S(nB_{jm}) \Big/ \sqrt{\mathrm{Var}S(nB_{jm})}, \quad j \in J \setminus \{1\},$$

and

$$B = B(n, m) = \{x : a_j < x_j < b_j, \, j \in J \setminus \{1\}\},$$

$$a_j := \delta_j (LLn)^{1/2}(g(B_{jm}) - \varepsilon m^{-d}), \, b_j := \delta_j (LLn)^{1/2}(g(B_{jm}) + \varepsilon m^{-d}),$$

$$\delta_j = \delta_j(n, m) = (2n^d)^{1/2} \Big/ \sqrt{\mathrm{Var}S(nB_{jm})}.$$

Let ξ and η be $(m^d - 1)$-dimensional random vectors having components

$$S(nF_{jm})/\sqrt{\mathrm{Var}S(nB_{jm})} \quad \text{and} \quad S(nR_{jm})/\sqrt{\mathrm{Var}S(nB_{jm})}$$

where $j \in J \setminus \{1\}$ and the sets F_{jm} and R_{jm} were defined after (2.11). If one takes $\tau_j = n^{-\nu/4}$, then by the Chebyshev inequality

$$P(|\eta_j| > \tau_j) \leq \tau_j^{-2} \frac{\mathrm{Var}S(nR_{jm})}{\mathrm{Var}S(nB_{jm})} \leq C_{12} n^{-\nu/2} \tag{2.29}$$

for all n large enough. Set

$$G = G(\tau, n, m) = \{x : \varkappa_j(a_j - \tau_j) < x_j < \varkappa_j(b_j + \tau_j),\ j \in J \setminus \{1\}\}, \tag{2.30}$$

where

$$\varkappa_j = \varkappa_j(n, m) = \frac{\mathrm{Var}S(nB_{jm})}{\mathrm{Var}S(nF_{jm})}.$$

Note that for any $j \in J$ and every $m \in \mathbb{N}$,

$$\delta_j(n, m) \to \frac{m^{d/2}}{\sigma} \quad \text{as } n \to \infty, \tag{2.31}$$

$$\varkappa_j(n, m) = 1 + O(n^{-\lambda_1}) \quad \text{for some } \lambda_1 > 0 \tag{2.32}$$

(the second formula is easily proved by virtue of Theorem 3.1.8). Applying Lemma 2.4 with $r = m^d - 1$ and the set B from (2.28), and afterwards using Theorem 3.1.8, (2.13), (2.15), (2.31) and (2.32), for all n large enough we obtain the estimate

$$|P(\xi \in B^\tau) - P(Z \in G)|$$

$$\leq C_{13}\left(1 + \prod_{j \in J \setminus \{1\}} (2 + \varkappa_j(b_j - a_j + 2\tau_j))\right)\left(\frac{n}{m}\right)^{-\beta} \leq C_{14}(LLn)^{r/2} n^{-\beta}, \tag{2.33}$$

here $Z \sim N(0, I_r)$. Due to (2.32),

$$|P(Z \in G) - P(Z \in B)| \leq C_{15} n^{-\lambda_1}. \tag{2.34}$$

An analogous estimate is possible for $|P(\xi \in B_\tau) - P(Z \in B)|$. From Lemma 2.6, (2.29), (2.33) and (2.34), summing up, we have, for any $m \in \mathbb{N}$,

$$P(E_n) = P(Z \in B(n, m)) + O(n^{-\lambda_2}) \tag{2.35}$$

with some $\lambda_2 > 0$ not depending on n.

Lemma 2.7. ([390]) *Let Z be a standard normal random variable. Then*
   1) *for any $\alpha \in \mathbb{R}$ and $\gamma \geq 0$ one has*

$$P(\alpha - \gamma < Z < \alpha + \gamma) \geq P(|\alpha| < Z < |\alpha| + \gamma);$$

   2) *for any $a \in \mathbb{R}$ and $b > (a \vee 0)$ one has*

$$P(a < Z < b) \geq \frac{1}{b\sqrt{2\pi}} e^{-a^2/2}(1 - e^{-(b^2 - a^2)/2}).$$

Proof. Due to symmetry of Z it suffices to prove the first relation for nonnegative α, and for such α one has $[\alpha - \gamma, \alpha + \gamma] \supset [\alpha, \alpha + \gamma]$.

Further, let us rewrite the second relation as

$$e^{-b^2/2} + b \int_a^b e^{-x^2/2} dx \geq e^{-a^2/2}.$$

However, the left-hand side is no less than

$$e^{-b^2/2} + \int_a^b x e^{-x^2/2} dx = e^{-a^2/2}. \quad \square$$

The function g was taken in such a way that, in particular,

$$0 \leq |g(B_{jm})| \leq \sigma m^{-d/2}$$

for $j \in J$ and $m \in \mathbb{N}$. Applying Lemma 2.7, we see that $\mathsf{P}(Z \in B)$ admits the following lower bound

$$\frac{\exp\left\{-LLn \sum_{j \in J} g(B_{jm})^2 \delta_j^2\right\}}{\left((4\pi LLn)^{1/2}(\sigma m^{-d/2} + \varepsilon m^{-d})\right)} \prod_{j \in J \setminus \{1\}} \delta_j^{-1} \left(1 - \exp\left\{-2\delta_j \varepsilon^2 m^{-2d} LLn\right\}\right).$$

Note that

$$\sum_{j \in J} m^d g(B_{jm})^2 \to \int_{[0,1]^d} h(t)^2 dt, \quad m \to \infty,$$

because the Riemann sums tend to the integral since the integrated function is continuous. Taking m large enough and using (2.31) and (2.25) we have

$$\mathsf{P}(Z \in B) \geq C_{15}(Log\, n)^{-1-\delta/2}. \tag{2.36}$$

Now (2.35) and (2.36) imply (2.23) with $A_k = E_{m^k}$.

To check (2.24) we estimate $\mathsf{P}(A_k A_l)$, i.e. $\mathsf{P}(Y_{m^k} \in B(m^k, m), Y_{m^l} \in B(m^l, m))$ where Y_n and $B(n, m)$ appear in (2.28). We apply Lemma 2.6 having chosen the $2(m^d - 1)$-dimensional random vector Y with components

$$\frac{S(m^k F_{jm}(m^k))}{\sqrt{\mathsf{Var}S(m^k F_{jm}(m^k))}}, \quad \frac{S(m^k F_{qm}(m^k))}{\sqrt{\mathsf{Var}S(m^k F_{qm}(m^k))}}, \quad j, q \in J \setminus \{1\}.$$

We also take $\tau_j = m^{-\nu k/4}, j \in J \setminus \{1\}$, and $\varrho_q = m^{-\nu l/4}, q \in J \setminus \{1\}$, where ν is from (2.11). If m is large enough, then

$$\mathsf{Var}S(m^k F_{jm}(m^k)) \geq \frac{\sigma^2}{2} m^{d(k-1)} \text{ for } k \in \mathbb{N},$$

$$\min_{j,q \in J \setminus \{1\}} \{|m^k F_{jm}(m^k)| \wedge |m^l F_{jm}(m^l)|\} \geq \frac{1}{2} m^{d(k-1)},$$

$$dist(m^k F_{jm}(m^k), m^l F_{jm}(m^l)) \geq m^{k(1-\nu)-1} \text{ for } 1 \leq k < l \text{ and } j, q \in J \setminus \{1\}.$$

Now apply Lemma 2.4 with $B = B(m^k, m) \times B(m^l, m)$ and $r = 2(m^d - 1)$. If $\tau = \{\tau_j, j \in J \setminus \{1\}, \varrho_q, q \in J \setminus \{1\}\}$, an estimate similar to (2.33) shows that, for $1 \leq k < l$ and m large enough,

$$|\mathsf{P}(Y \in B^\tau) - \mathsf{P}(Z \in G_1 \times G_2)| \leq C_{16}(LLm^l)^r m^{-\lambda_3 k}. \tag{2.37}$$

Here $\lambda_3 > 0$ does not depend on k and l, $Z \sim N(0, I_r)$ and the set G_1 (resp. G_2) is defined in (2.30) with the choice $n = m^k$, $\tau_j = m^{-\nu k/4}$, $j \in J \setminus \{1\}$ (resp. $n = m^l$, $\varrho_q = m^{-\nu l/4}$, $q \in J \setminus \{1\}$). An analogous estimate of $\mathsf{P}(Y \in B_\tau)$ and invoking the relations (2.31)—(2.35), (2.37) allow us to conclude that, for $k < l$,

$$\mathsf{P}(A_k A_l) = \mathsf{P}\left(Z \in B(m^k, m) \times B(m^l, m)\right) + f_m(k, l) \tag{2.38}$$

if m is large enough. Note that here $|f_m(k, l)| \leq C_{17}(\log l)^r m^{-\lambda_4 k}$, $\lambda_4 > 0$. Relations (2.35), (2.37) and (2.38) imply (2.24) with $\rho_1 = 2(m^d - 1)$, $\rho = \lambda_4 \log m$. Hence Lemma 2.6 applies and the Theorem follows. $\square$

3 Law of a single logarithm

In 1994 Qi [336] proved the result, called the *law of a single logarithm* (or *logarithmic law*) for a triangular array $X = \{X_{nk}, k = 1, \ldots, n, n \in \mathbb{N}\}$ consisting of i.i.d. centered random variables. To be exact, for the corresponding partial sums given by $S_n = \sum_{k=1}^{n} X_{nk}$, $n \in \mathbb{N}$, both

$$\limsup_{n \to \infty} \frac{S_n}{\sqrt{2n \mathrm{Log}\, n}} = 1 \quad \text{and} \quad \liminf_{n \to \infty} \frac{S_n}{\sqrt{2n \mathrm{Log}\, n}} = -1$$

hold almost surely if and only if the conditions $\mathsf{E}X_{11}^2 = 1$ and

$$\mathsf{E}\frac{X_{11}^4}{(\mathrm{Log}\, |X_{11}|)^2} < \infty$$

are true. To underline the beauty of this statement recall that for a sequence of i.i.d. random variables $\xi_1, \xi_2, \ldots$, by the Kolmogorov theorem, the SLLN is equivalent to the hypothesis that $\mathsf{E}|\xi_1| < \infty$, while theorems of Hartman–Wintner, Strassen and Martikainen imply that there is a criterion of the LIL described in terms of the first two moments of summands.

Note that one can consider the following natural generalization of the above mentioned scheme. Namely, let $\{X_t, t \in \mathbb{Z}^d\}$ be a random field consisting of i.i.d. random variables and U_n, $n \in \mathbb{N}$, be finite disjoint subsets of $\mathbb{Z}^d$. Set $S(U_n) = \sum_{t \in U_n} X_t$, $n \in \mathbb{N}$. Then we obtain an equivalent formulation of the Qi law taking (disjoint) U_n with $|U_n| = n$ and $S_n = S(U_n)$, $n \in \mathbb{N}$.

Moreover, if we consider more general situation when $|U_n| = k_n$ for $k_n \in \mathbb{N}$, $k_n \to \infty$ ($n \to \infty$) then we come to study of subsequences S_{k_n} in the scheme of array. In this more general case studied by Mikusheva ([298]) it was shown that there is a nontrivial connection between the growth rate of a sequence $(k_n)_{n \in \mathbb{N}}$ and the moments of the summands.

1°. The extension of the Qi theorem. Following Hu, Su and Liang we consider the extension of Qi's theorem for an array $X = \{X_{nk}, k = 1\ldots, n, n \in \mathbb{N}\}$ possessing the **NA** property and such that, for some strictly stationary sequence $\xi = (\xi_n)_{n\in\mathbb{Z}} \in \mathbf{NA}$,

$$Law(X_{n1}, \ldots, X_{nn}) = Law(\xi_1, \ldots, \xi_n), \quad n \in \mathbb{N}. \tag{3.1}$$

Theorem 3.1. ([204]) *Suppose that X and ξ are the array and the sequence respectively introduced above which satisfy (3.1). Assume that*

$$\mathsf{E}\xi_0 = 0 \quad \text{and} \quad \mathsf{E}\frac{\xi_0^4}{(\mathrm{Log}\,|\xi_0|)^2} < \infty, \tag{3.2}$$

the Cox–Grimmett coefficient of the sequence ξ (see (1.5.4))

$$u_r = O(r^{-\lambda}) \quad \text{as } r \to \infty \tag{3.3}$$

for some $\lambda > 2$, and also

$$\sigma^2 = \sum_{j\in\mathbb{Z}} cov(\xi_0, \xi_j) > 0.$$

Then, with notation $S_n = \sum_{k=1}^n X_{nk}$, $n \in \mathbb{N}$, one almost surely has

$$\limsup_{n\to\infty} \frac{S_n}{\sqrt{2n\mathrm{Log}\,n}} = \sigma \quad \text{and} \quad \liminf_{n\to\infty} \frac{S_n}{\sqrt{2n\mathrm{Log}\,n}} = -\sigma.$$

Remark 3.2. This theorem was proved in [204] with restriction $\lambda > 4$ in (3.3).

Proof. As usual we only prove the assertion concerning the upper limit. Without loss of generality we can assume that $\mathsf{Var}\xi_0 = 1$.

2°. Truncation. Let $(m_n)_{n\in\mathbb{N}}$ be a sequence of positive integers such that $m_n = O(n^{1-\delta})$ as $n \to \infty$, for some $\delta \in (0, 1)$. Set $S(m_n) = \sum_{j=1}^{m_n} \xi_j$, $n \in \mathbb{N}$.

Lemma 3.3. *Let the conditions of Theorem 3.1 hold. Then, for any $\eta \in (0, 1/4)$,*

$$\sum_{n=1}^{\infty} \mathsf{P}\left(|S(m_n)| > \eta\sqrt{n\mathrm{Log}\,n}\right) < \infty.$$

Proof. Let $H_M(\cdot)$ be the truncation function introduced in (2.1.5). Introduce

$$S^{(1)}(m_n) = \sum_{k=1}^{m_n}(H_M(\xi_k) - \mathsf{E}H_M(\xi_k)), \quad S^{(2)}(m_n) = S(m_n) - S^{(1)}(m_n), \quad n \in \mathbb{N},$$

with $M = M_n = \eta^2\sqrt{n/\mathrm{Log}\,n}$. Take some $\tau \in (0, 1/(4\eta) - 1)$ and set $z_n = \eta\sqrt{n\mathrm{Log}\,n}$, $\gamma_n = m_n Var H_{M_n}(\xi_0)$. Note that, for all n large enough, one has

$$|\mathsf{E}H_{M_n}(\xi_0)| < \tau$$

as $\mathsf{E}\xi_1 = 0$. Then for such n, by Corollary 2.2.10 (with $R = (1+\tau)M_n$ and $x = z_n$),

$$\mathsf{P}(|S^{(1)}(m_n)| > \eta\sqrt{n\mathrm{Log}\,n}) \leq \begin{cases} 2\exp\left\{-z_n^2/(4\gamma_n)\right\}, & \text{if } (1+\tau)\eta^3 n \leq \gamma_n, \\ 2n^{-1/(4\eta+4\eta\tau)}, & \text{otherwise.} \end{cases} \tag{3.4}$$

Since $m_n/n \to 0$ and $\operatorname{Var} H_{M_n}(\xi_0) \to 1$ as $n \to \infty$, the second case in (3.4) takes place, for all n large enough. Therefore,

$$\sum_{n=1}^{\infty} \mathsf{P}(|S(m_n)^{(1)}| > z_n) < \infty. \tag{3.5}$$

To estimate $S^{(2)}(m_n)$ notice that

$$\mathsf{P}(S(m_n)^{(2)} \neq 0) \leq 2m_n \mathsf{P}(|\xi_0| > \eta^2 \sqrt{n/\operatorname{Log} n}) \leq C_1 n^{1-\delta} \mathsf{P}(\xi_0^2 > \eta^4 n/\operatorname{Log} n)$$

for some $C_1 > 0$ not depending on $n \in \mathbb{N}$. It is easy to see that if the function $\varphi : \mathbb{R} \to \mathbb{R}$ is inverse to the function $x \mapsto x/\operatorname{Log} x$ for all x large enough, then $\varphi(x) \sim x \log x$ as $x \to \infty$. Indeed, for any $a > 1$ and all $x > x_0(a)$ with some $x_0(a) > 0$ one has

$$x \log x \leq \varphi(x) \leq ax \log x.$$

Hence,

$$\sum_{n=1}^{\infty} n^{1-\delta} \mathsf{P}\left(\varphi\left(\frac{\xi_0^2}{\eta^4}\right) > n\right) < \infty \tag{3.6}$$

as $\mathsf{E}(\xi_0^2 \operatorname{Log}|\xi_0|)^{2-\delta} < \infty$. The Lemma follows from (3.5) and (3.6). $\square$

3°. The upper bound for limsup of normalized sums. Let us start proving the upper estimate in the law of a single logarithm. We intend to show that, for any fixed $\varepsilon > 0$,

$$\limsup_{n \to \infty} \frac{|S_n|}{\sqrt{2\sigma^2 n \operatorname{Log} n}} \leq 1 + 15\varepsilon.$$

Clearly we may only consider $\varepsilon \in (0,1)$ so small that

$$1 + 2\varepsilon \leq \frac{1-\varepsilon}{2^{3/2}\varepsilon}. \tag{3.7}$$

Theorem 3.1.9 shows that for any $\varepsilon > 0$ one can find $m \in \mathbb{N}$ such that

$$\sigma^2 \leq \sigma_m^2 := \frac{1}{m}\mathsf{E}\left(\sum_{k=1}^{m}\xi_k\right)^2 < \sigma^2(1+\varepsilon). \tag{3.8}$$

Let $k_n = m[n/m]$ and $m_n = n - k_n$. Then $m_n = O(n^{1-\delta})$ as $n \to \infty$, for any $\delta \in (0,1)$ (in fact m_n is a bounded sequence). Consequently, the Borel–Cantelli lemma and Lemma 3.3 together imply that it suffices to prove the relation

$$\sum_{n=m}^{\infty} \mathsf{P}(|S(k_n)| > (1+14\varepsilon)\sqrt{2\sigma^2 n \operatorname{Log} n}) < \infty$$

where $S(k_n) = \sum_{j=1}^{k_n} X_{nj}$, $n \in \mathbb{N}$. Introduce the sequences

$$y_n = \frac{\varepsilon\sigma}{m}\sqrt{\frac{n}{\operatorname{Log} n}}, \quad z_n = \varepsilon\sqrt{2\sigma^2 n \operatorname{Log} n}, \ n \geq m,$$

and random variables

$$X_{nj}^{(1)} = H_{y_n}(X_{nj}), \quad X_{nj}^{(2)} = H_{z_n}\left(X_{nj} - X_{nj}^{(1)}\right), \quad X_{nj}^{(3)} = X_{nj} - X_{nj}^{(2)}$$

where $j = 1, \ldots, k_n$, $n \geq m$ and the functions H_{y_n} and H_{z_n} determine the same truncation as above. In other words, we use two successive truncations of the initial random variables. The corresponding partial sums are

$$S^{(q)}(k_n) := \sum_{j=1}^{k_n} X_{nj}^{(q)}, \quad q = 1, 2, 3.$$

The easiest part of the forthcoming calculations is to bound $S^{(3)}(k_n)$. Namely,

$$\sum_{n=m}^{\infty} \mathsf{P}(S^{(3)}(k_n) \neq 0) \leq \sum_{n=1}^{\infty} n\mathsf{P}(|\xi_0| > z_n)$$

$$= \sum_{n=1}^{\infty} n\mathsf{P}(\xi_0^2 > 2\varepsilon^2\sigma^2 n\mathrm{Log}\, n) = \sum_{n=1}^{\infty} n\mathsf{P}(\psi(\xi_0^2/2\varepsilon^2\sigma^2) > n) < \infty$$

where ψ is the inverse function to $x \mapsto x\log x$ for x large enough. Indeed, since $\psi(x) \sim x/\log x$ as $x \to \infty$, one has $\mathsf{E}\psi(\alpha\xi_0^2) < \infty$ by (3.2), for any $\alpha > 0$. By the definition of random variables $X_{nj}^{(2)}$,

$$\mathsf{P}(|S^{(2)}(k_n)| > 6\varepsilon\sqrt{2\sigma^2 n\mathrm{Log}\, n}) \leq \mathsf{P}\left(\sum_{j=1}^{k_n}\left(X_{nj}^{(2)}\right)^+ > 3\varepsilon\sqrt{2\sigma^2 n\mathrm{Log}\, n}\right)$$

$$+ \mathsf{P}\left(\sum_{j=1}^{k_n}\left(X_{nj}^{(2)}\right)^- > 3\varepsilon\sqrt{2\sigma^2 n\mathrm{Log}\, n}\right) =: J_{n,1} + J_{n,2}.$$

Due to the **NA** property (Corollary 1.1.10),

$$J_{n,1} \leq \sum_{1 \leq i < j < v \leq k_n} \mathsf{P}(X_{ni} > y_n, X_{nj} > y_n, X_{nv} > y_n) \leq k_n^3 (\mathsf{P}(\xi_0 > y_n))^3$$

$$\leq n^3 \mathsf{P}\left(\xi_0 > \frac{\varepsilon\sigma}{m}\sqrt{\frac{n}{\mathrm{Log}\, n}}\right)^3 = O(n^{-3/2}(\mathrm{Log}\, n)^{9/2}), \quad n \to \infty,$$

by the Markov inequality. An analogous estimate holds for $J_{n,2}$. Consequently,

$$\sum_{n=m}^{\infty} \mathsf{P}(|S^{(2)}(k_n)| > 6\varepsilon\sqrt{2\sigma^2 n\mathrm{Log}\, n}) < \infty.$$

Note that by Markov's inequality

$$|\mathsf{E}S^{(1)}(k_n)| \leq k_n|\mathsf{E}X_{n1}^{(1)}| \leq n\left(y_n\mathsf{P}(|\xi_0| > y_n) + \mathsf{E}|\xi_0|\mathbb{I}\{|\xi_0| > y_n\}\right) = o(\sqrt{n\mathrm{Log}\, n})$$

as $n \to \infty$. Thus the upper estimate in the law of a single logarithm will be established if one proves that

$$\sum_{n=1}^{\infty} \mathsf{P}\left(\left|S^{(1)}(k_n) - \mathsf{E}S^{(1)}(k_n)\right| > (1 + 7\varepsilon)\sqrt{2\sigma^2 n\mathrm{Log}\, n}\right) < \infty. \qquad (3.9)$$

Set $l_n = [n/m]$ and $Y_{nj} = X_{nj}^{(1)} - \mathsf{E}X_{nj}^{(1)}$, $Z_{nj} = \sum_{k=(j-1)m+1}^{jm} Y_{nk}$, $j = 1, \ldots, l_n$,

$$T_n = S^{(1)}(k_n) - \mathsf{E}S^{(1)}(k_n) = \sum_{j=1}^{l_n} Z_{nj}, \quad n \geq m.$$

By Theorem 1.1.8, (d), the random variables $Z_{n1}, Z_{n2}, \ldots, Z_{n,l_n}$ are **NA** and identically distributed (for each $n \geq m$). Also

$$\frac{1}{m}\mathsf{E}Z_{n1}^2 \to \sigma_m^2, \quad \frac{1}{n}\sum_{j=1}^{l_n}\mathsf{E}Z_{nj}^2 = \frac{l_n}{n}\mathsf{E}Z_{n1}^2 = \frac{k_n}{nm}\mathsf{E}Z_{n1}^2 \to \sigma_m^2$$

as $n \to \infty$, where σ_m was introduced in (3.8). Therefore, for all n large enough,

$$n\sigma_m^2(1 - \varepsilon) < \sum_{j=1}^{l_n}\mathsf{E}Z_{nj}^2 = l_n\mathsf{E}Z_{n1}^2 < n\sigma^2(1 + 2\varepsilon). \tag{3.10}$$

By the definition of Z_{nj},

$$|Z_{nj}| \leq 2my_n = 2\varepsilon\sigma\sqrt{\frac{n}{\mathrm{Log}\, n}}, \quad j = 1, \ldots, l_n, \quad n \geq m.$$

Then we apply Corollary 2.2.10 with

$$x = (1 + 2\varepsilon)\sqrt{2\sigma^2 n\mathrm{Log}\, n}, \quad R = 2\varepsilon\sigma\sqrt{\frac{n}{\mathrm{Log}\, n}}, \quad B_n = \sum_{j=1}^{l_n}\mathsf{E}Z_{nj}^2 \geq n\sigma_m^2(1 - \varepsilon).$$

An easy calculation taking into account the requirement (3.7) implies the inequality $x \leq B_n/H$. Therefore, by the assertion mentioned above and relation (3.10), we have an estimate, for all n large enough,

$$\mathsf{P}\left(|T(k_n)| > (1 + 2\varepsilon)\sqrt{2\sigma^2 n\mathrm{Log}\, n}\right) \leq 2\exp\left\{-\frac{(1 + 2\varepsilon)^2 2\sigma^2 n\mathrm{Log}\, n}{2B_n}\right\}$$

$$\leq 2\exp\left\{-\frac{(1 + 2\varepsilon)^2 2\sigma^2 n\mathrm{Log}\, n}{2n\sigma^2(1 + 2\varepsilon)}\right\} = 2\exp\left\{-(1 + 2\varepsilon)\mathrm{Log}\, n\right\} \tag{3.11}$$

from which (3.9) follows.

4°. **The Lewis inequality.** We use a generalization of the Newman inequality due to Lewis.

Lemma 3.4. ([259]) *Let* $X = (X_1, \ldots, X_n)$ *be* **PA** *or* **NA** *random vector with values in* $\mathbb{R}^n$ *such that* $\mathsf{E}\|X\|^2 < \infty$ *and let* $(Y_1, \ldots, Y_n)$ *be its decoupled version[3]. Suppose that* $f : \mathbb{R} \to \mathbb{R}$ *is a function having second derivative which is bounded and continuous. Then*

$$|\mathsf{E}f(X_1 + \ldots + X_n) - \mathsf{E}f(Y_1 + \ldots + Y_n)| \leq \|f''\|_\infty \sum_{\substack{1 \leq i,j \leq n \\ i \neq j}} |cov(X_i, X_j)|. \tag{3.12}$$

---

[3]See Section 2.2.

Proof. First of all note that the moment condition on X ensures that the expectations in (3.12) exist (by application of the Taylor formula). The Lemma is proved by induction on n. For $n = 1$ the assertion is obvious as $\mathsf{E}f(X_1) = \mathsf{E}f(Y_1)$ and $\sum_\varnothing := 0$; suppose that $n > 1$ and the Lemma has been verified for $(n-1)$-dimensional random vectors. The left-hand side of (3.12) admits the upper bound

$$|\mathsf{E}f(X_1 + \ldots + X_{n-1} + X_n) - \mathsf{E}f(X_1 + \ldots + X_{n-1} + Y_n)|$$

$$+ |\mathsf{E}f(X_1 + \ldots + X_{n-1} + Y_n) - \mathsf{E}f(Y_1 + \ldots + Y_n)| =: Q_1 + Q_2. \qquad (3.13)$$

The function $y \mapsto \mathsf{E}f(y + Y_n)$ is twice differentiable with bounded continuous second derivative, as is seen by the Lagrange theorem and the dominated convergence theorem. Therefore by the induction hypothesis

$$Q_2 \leq \|f''\|_\infty \sum_{\substack{1 \leq i,j < n \\ i \neq j}} |cov(X_i, X_j)|. \qquad (3.14)$$

Let $S = X_1 + \ldots + X_{n-1}$ and $T = Y_1 + \ldots + Y_{n-1}$, then (S, X_n) is either **PA** or **NA** random vector with decoupled version (T, Y_n). By the generalized Hoeffding identity (Appendix A.1)

$$Q_1 = |\mathsf{E}f(S + Y_n) - \mathsf{E}f(T + Y_n)|$$

$$= \left| \int\!\!\int_{\mathbb{R}^2} f''(t + w)(\mathsf{P}(S \geq t, Y_n \geq w) - \mathsf{P}(S \geq t)\mathsf{P}(Y_n \geq w))dtdw \right|$$

$$\leq \|f''\|_\infty \, |cov(S, X_n)| \leq \|f''\|_\infty \sum_{j=1}^{n-1} |cov(X_j, X_n)|. \qquad (3.15)$$

The Lemma follows from (3.13)—(3.15). $\square$

5°. The lower bound for limsup of normalized sums. Now let us prove the lower estimate, namely that

$$\limsup_{n \to \infty} \frac{S_n}{\sqrt{2\sigma^2 n \mathrm{Log}\, n}} \geq 1 \text{ a.s.}$$

Fix some $\varepsilon \in (0, 1/5)$ and define events

$$A_n = \{S_n > (1 - 5\varepsilon)\sqrt{2\sigma^2 n \mathrm{Log}\, n}\}, \quad n \in \mathbb{N}.$$

Since all X_{nj} are **NA**, for any $m, n \in \mathbb{N}$ ($m \neq n$) we have

$$\mathsf{P}(A_n A_m) \leq \mathsf{P}(A_n)\mathsf{P}(A_m).$$

Thus, the sequence of events $(A_n)_{n \in \mathbb{N}}$ satisfies condition (1.12) of Lemma 1.5 if

$$\sum_{n=1}^\infty \mathsf{P}(A_n) = \infty, \qquad (3.16)$$

so our goal is to establish (3.16).

Take $\delta \in (0,1)$ to be specified later and define

$$d_n = [n^{1-\delta}], \quad \tau_n = n^2 + (n-1)d_n, \quad n \in \mathbb{N},$$

$$a_0 = b_0 = 0, \quad a_1 = n, \quad b_1 = a_1 + d_n, \dots,$$

$$a_{j+1} = b_j + n, \quad b_{j+1} = a_{j+1} + d_n, \quad j \in \mathbb{N}, \quad J_n = \{k \in \mathbb{N} : b_{j-1} < k \le a_j, \ j = 1,\dots,n\}.$$

Moreover, for any positive integer $l \in [\tau_n, \tau_{n+1}]$, we set

$$L = L_l = \{1,\dots,l\} \setminus J_n, \quad S_l^{(1)} = \sum_{k \in J_n} X_{lk}, \quad S_l^{(2)} = S_l - S_l^{(1)}.$$

That is the sum S_n is split into a sum over "big" blocks J_n and other points. Clearly $|J_n| = n^2$, $n \in \mathbb{N}$, $m_l := |L| \le \tau_{n+1} - n^2 = (n+1)^2 - n^2 + nd_{n+1} \le 2n^{2-\delta} = O(l^{1-\delta/2})$. Since $S_l^{(2)}$ is the sum of m_l identically distributed **NA** random variables, Lemma 3.3 shows that

$$\sum_{n=1}^{\infty} \mathsf{P}\left(|S_l^{(2)}| > \varepsilon\sqrt{2\sigma^2 l \mathrm{Log}\, l}\right) < \infty.$$

As $S_l = S_l^{(1)} + S_l^{(2)}$, relation (3.16) will be a consequence of

$$\sum_{l=1}^{\infty} \mathsf{P}\left(S_l^{(1)} > (1-4\varepsilon)\sqrt{2\sigma^2 l \mathrm{Log}\, l}\right) = \infty. \tag{3.17}$$

Pick $\eta > 0$ so small that $(1+\eta)(1-4\varepsilon) < 1 - 3\varepsilon$. For all n large enough one has

$$\max_{\tau_n \le l \le \tau_{n+1}} l\mathrm{Log}\, l < (1+\eta)^2 n^2 \mathrm{Log}\,(n^2),$$

since $1 < \tau_n/n^2 \le l/n^2 < \tau_{n+1}/n^2 \to 1$ as $n \to \infty$, and Log is a slowly varying function. Thus to establish (3.17) we may prove that

$$\sum_{n=1}^{\infty} \sum_{\tau_n \le l < \tau_{n+1}} \mathsf{P}\left(S_l^{(1)} > (1-3\varepsilon)\sqrt{2\sigma^2 n^2 \mathrm{Log}\,(n^2)}\right) = \infty. \tag{3.18}$$

Introduce random variables

$$Y_{nj} = n^{-1/2} \sum_{k=b_{j-1}+1}^{b_{j-1}+n} X_{\tau_n, k}, \quad j = 1,\dots,n.$$

For any $l = \tau_n, \dots, \tau_{n+1} - 1$, one has $Law(Y_{n1}, \dots, Y_{nn}) = Law(\zeta_{n1}, \dots, \zeta_{nn})$ where

$$\zeta_{nj} = n^{-1/2} \sum_{k=b_{j-1}+1}^{a_j} \xi_k, \quad j = 1,\dots,n.$$

Set $V_n = \sum_{j=1}^{n} Y_{nj}$, then $Law(S_l^{(1)}) = Law(\sqrt{n} V_n)$, $\tau_n \le l < \tau_{n+1}$. Since $\sigma_n^2 \ge \sigma^2$ for all $n \in \mathbb{N}$ and $\tau_{n+1} - \tau_n \sim 2n$ as $n \to \infty$, instead of (3.18) one can prove that

$$\sum_{n=1}^{\infty} n\mathsf{P}\left(V_n > (1-2\varepsilon)\sqrt{2\sigma_n^2 n \mathrm{Log}\, n^2}\right) = \infty.$$

For any $n \in \mathbb{N}$, let $(Z_{n1}, \ldots, Z_{nn})$ be independent random variables distributed as ζ_{n1}. We need the result on large deviations from [337].

Lemma 3.5. ([337]) *Let* $(Z_{nk})_{k=1,\ldots,n,\, n\in\mathbb{N}}$ *be an array of centered random variables such that*

$$D_3 := \sup_{k=1,\ldots,n,\, n\in\mathbb{N}} \mathsf{E}|Z_{nk}|^3 < \infty.$$

Suppose that, for each n*, the variables* $Z_{n1}, \ldots, Z_{nn}$ *are i.i.d. with* $\mathsf{E}Z_{n1}^2 = 1$*. Set* $T_n = \sum_{j=1}^{n} Z_{nj}$*. Then, for any* $c > 0$ *and* $\delta > 0$*,*

$$\mathsf{P}(T_n \geq c\sqrt{n\mathrm{Log}\,n}) \geq n^{-\delta-c^2/2}$$

for all n *large enough.*

Theorem 2.3.3 implies that

$$\sup_{n\in\mathbb{N}} \mathsf{E}|Z_{n1}|^3 = \sup_{n\in\mathbb{N}} n^{-3/2}\mathsf{E}\left|\sum_{k=1}^{n}\xi_k\right|^3 \leq 2^9 \sup_{n\in\mathbb{N}} n^{-3/2}\left(n\mathsf{E}|\xi_0|^3 + (n\mathsf{E}\xi_0^2)^{3/2}\right) < \infty.$$

Therefore, Lemma 3.5 with $c = 2(1-2\varepsilon)$, $\delta = \varepsilon/2$ shows that, for some $n_0 \in \mathbb{N}$,

$$\sum_{n=n_0}^{\infty} n\mathsf{P}\left(V_n > (1-2\varepsilon)\sqrt{2\sigma_n^2 n \log n^2}\right)$$

$$= \sum_{n=n_0}^{\infty} n\mathsf{P}\left(\frac{V_n}{\sigma_n} > (1-2\varepsilon)\sqrt{4n\log n}\right) \geq \sum_{n=n_0}^{\infty} n^{1-2(1-2\varepsilon)^2-\varepsilon/2} = \infty \qquad (3.19)$$

(note that $\sigma_n^2 \in [\sigma^2, \mathsf{E}\xi_0^2]$, so we can divide all the variables Z_{n1} by σ_n to fit the formulation of Lemma 3.5).

Introduce $h_n = 2\sigma_n\sqrt{\mathrm{Log}\,n}, n \in \mathbb{N}$. Then $0 < h_n/\sqrt{\mathrm{Log}\,n} \to 2\sigma$, $n \to \infty$. Let $(f_n)_{n\in\mathbb{N}}$ be a sequence of nondecreasing infinitely differentiable functions $f_n : \mathbb{R} \to \mathbb{R}$ such that

$$Lip(f_n') \leq Kh_n^{-3}, \ \ 0 \leq f_n(x) \leq 1, \ \ x \in \mathbb{R},$$

$$f_n(x) = 0 \text{ if } x \leq (1-2\varepsilon)h_n, \quad f_n(x) = 1 \text{ if } x \geq (1-\varepsilon)h_n,$$

where $K > 0$ is some fixed number and $n \in \mathbb{N}$. Then one has

$$\mathsf{P}\left(V_n > (1-2\varepsilon)\sqrt{2\sigma_n^2 n\mathrm{Log}\,n}\right) \geq \mathsf{E}f_n\left(\frac{V_n}{\sqrt{n}}\right)$$

$$\geq \mathsf{E}f_n\left(\frac{T_n}{\sqrt{n}}\right) - I(n) \geq \mathsf{P}\left(T_n > (1-\varepsilon)\sqrt{2\sigma_n^2 n\mathrm{Log}\,n}\right) - I(n)$$

where

$$I(n) = \left|\mathsf{E}f_n\left(\frac{V_n}{\sqrt{n}}\right) - \mathsf{E}f_n\left(\frac{T_n}{\sqrt{n}}\right)\right|.$$

So, in view of (3.19), the Theorem will be proved if we show that

$$\sum_{n=1}^{\infty} nI(n) < \infty. \tag{3.20}$$

By Lemma 3.4 we have, for $n \in \mathbb{N}$,

$$nI(n) \le -\frac{Kn}{h_n^3} \sum_{1 \le j < k \le n} \mathsf{E}\zeta_{nj}\zeta_{nk} \le -\frac{Kn}{h_n^3} u_{d_n} \le -C_2 n (\mathrm{Log}\, n)^{-3/2} n^{-\lambda(1-\delta)}, \tag{3.21}$$

according to the definition of random variables ζ_{nj} and condition (3.3). Now the last estimate implies (3.20) if one uses (3.3) to select such $\delta \in (0,1)$ that $\lambda(1 - \delta) > 2$.
$\square$

To finish this Chapter we note that the first work on the LIL in the association setup was by Dabrowski [116], and in the **NA** case by Shao and Su [372]. The quasi-associated random sequences were studied in [230], the linear transforms of **NA** random fields in Huang [208].

Functional laws of the iterated logarithm for **PA/NA** vector-valued fields are established in [117, 428]. The LIL for associated random measures was proved by Evans [164].

There are various interesting problems to further investigations, for example, to obtain for (BL, θ)-dependent random fields the analogs of results established in [75] concerning the estimate of the convergence rate in the FLIL.

Chapter 7

Statistical Applications

Chapter 7 provides some statistical applications of limit theorems. It opens with the study of self-normalized partial sums of multiindexed summands. One can recall the famous studentization, used for a sequence of independent observations, permitting to construct the approximate confidence intervals for unknown mean. For independent observations essential progress was also achieved for vector-valued case. In this regard we refer to the papers by Sepański, Giné and Götze. However for dependent observations it is more convenient to use statistics of other kind involving "local averaging". For mixing sequences of random variables two important statistics of this type were introduced by Peligrad and Shao. Here we employ the family of statistics proposed by Bulinski and Vronski for strictly stationary associated random fields and regularly growing finite subsets of $\mathbb{Z}^d$. Moreover, we discuss further extensions, specifically, for (BL, θ)-dependent random fields with values in $\mathbb{R}^k$ and for random matrix normalization. Next we concentrate on the kernel (or the Parzen–Rosenblatt) density estimates. This is a very popular object of investigation in non-parametric Mathematical Statistics. The conditions are given to guarantee the CLT (with estimate of the convergence rate) for mentioned statistics and their a.s. behavior is studied as well. In particular we come to the recent result by Doukhan and Louhichi. We also refer to the papers by Roussas for other important statistical applications.

1 Statistics involving random normalization

It is well-known (see, e.g., [111]), that even in the case of i.i.d. observations one has to involve[1] statistical estimates of the unknown variance of the mentioned data to construct the asymptotic confidence interval for the unknown expectation. To perform this one can use the classical studentization procedure based on so-called self-normalizations; see, e.g., [74, 112] and references therein. In this Section we study the asymptotic behavior of self-normalized partial sums constructed by means of dependent random variables.

---

[1]Under assumption that the second moment is finite.

1°. The case of summation over rectangles. Let $X = \{X_j, j \in \mathbb{Z}^d\}$ be a wide-sense stationary random field. For a finite subset U of $\mathbb{Z}^d$, we assume known the observations of $\{X_j, j \in U\}$. If the CLT holds and the set U is "large enough" in the sense of regular growth (see the Definition 3.1.4), then the distribution of the random variable

$$\frac{S(U) - |U|\mathsf{E}X_0}{\sqrt{\mathsf{Var}S(U)}}, \quad \text{where} \quad S(U) = \sum_{j \in U} X_j,$$

is close to the standard Gaussian law (under conditions considered in Chapter 3). In particular, by Theorem 3.1.8 we have $\mathsf{Var}S(U) \sim \sigma^2 |U|$, for regularly growing sets U, where σ^2 is defined via the covariance function of the field in (1.5.3). Thus if $U_n \to \infty$ in a regular manner ($U_n \subset \mathbb{Z}^d$, $n \in \mathbb{N}$), then, for example by Theorem 3.1.12 (for strictly stationary random field X), we have

$$\frac{S(U_n) - |U|\mathsf{E}X_0}{\sqrt{|U_n|}} \to N(0, \sigma^2), \quad n \to \infty. \tag{1.1}$$

If $\sigma \neq 0$ and we know some sequence of consistent estimates[2] σ (i.e. $\hat{\sigma}_n \to \sigma$ in probability as $n \to \infty$), then instead of (1.1) we can apply

$$\frac{S(U_n) - |U|\mathsf{E}X_0}{\hat{\sigma}_n \sqrt{|U_n|}} \to N(0, 1), \quad n \to \infty, \tag{1.2}$$

which gives a method to construct approximate confidence intervals for $\mathsf{E}X_0$. The problem of using such intervals, e.g., in Radiobiological stochastic models is discussed in [73].

We describe two families of statistics allowing to estimate the asymptotic variance of partial sums of a random field, taken over growing sets. For a finite set $U \subset \mathbb{Z}^d$ introduce the statistics

$$B(U) = \sum_{j \in U} f_j^U \left(\frac{S(Q_j)}{|Q_j|} - \frac{S(U)}{|U|} \right) \tag{1.3}$$

where $f_j^U : \mathbb{R} \to \mathbb{R}_+$, $Q_j = Q_j(U) \subset U$, $Q_j \neq \varnothing$ for $j \in U$.

At first, let U be a block $(a, b] \in \mathcal{U}$, $a, b \in \mathbb{Z}^d$, such that $l_k(U) = b_k - a_k > 1$, $k = 1, \ldots, d$, and $R_j = \{q \in \mathbb{Z}^d : q \leq j\}$. For $j \in U$ and $s \in [0, 2]$ set

$$Q_j(U) = U \cap R_j, \quad f_j^U(x) = G_d(U)|Q_j|^{s/2-1}|x|^s$$

where

$$G_d(U) = \left(\prod_{k=1}^{d} \log l_k(U) \right)^{-1}.$$

The following theorems are due to Bulinski and Vronski. Similar methods of estimating variance of a random sequence (in mixing setup, in the cases $s = 1$ and $s = 2$) were employed in [322].

---

[2] The random expression (1.2) must be modified when $\hat{\sigma}_n(\omega) = 0$.

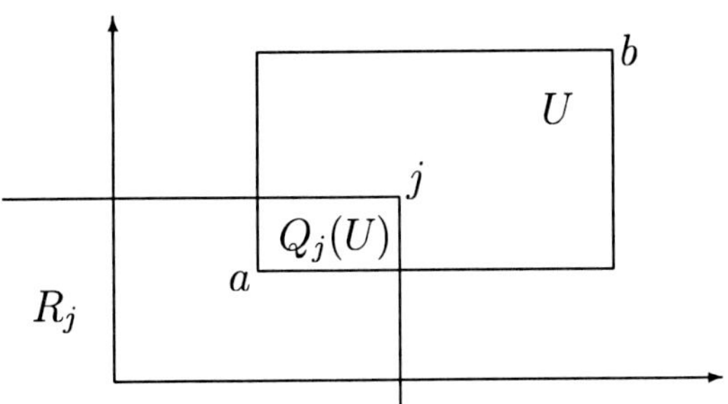

Fig. 7.1

Theorem 1.1. ([83]) *Let $X = \{X_j, j \in \mathbb{Z}^d\}$ be a strictly stationary, associated random field satisfying the finite susceptibility condition (1.5.4). Then, for any sequence of blocks $U_n \in \mathcal{U}$ ($n \in \mathbb{N}$) tending to infinity in a regular manner[3] and every $s \in (0, 2]$, one has*

$$B(U_n) \to \sigma^s \mathsf{E}|Z|^s \quad \text{in} \quad L^{(2/s)\wedge 2}(\Omega, \mathcal{F}, \mathsf{P}) \quad \text{as} \quad n \to \infty \tag{1.4}$$

where σ^2 is defined by (1.5.3) and $Z \sim N(0,1)$. Moreover,

$$\frac{S(U_n) - |U_n|\mathsf{E}X_0}{B(U_n)^{1/s}|U_n|^{1/2}} C_s^{1/s} \to N(0,1) \tag{1.5}$$

in law as $k \to \infty$, here

$$C_s = (2^s/\pi)^{1/2}\Gamma((s+1)/2) \quad \text{and} \quad \Gamma(t) = \int_0^\infty x^{t-1}e^{-x}dx, \quad t > 0.$$

Proof. The second assertion of the Theorem follows from the first one, Theorem 3.1.12, Lemma 3.1.27 and the fact that

$$\mathsf{E}|Z|^s = \sqrt{\frac{2}{\pi}} \int_0^\infty x^s e^{-x^2/2}dx = C_s.$$

Thus we have to prove (1.4). Without loss of generality we can assume that $\mathsf{E}X_0 = 0$. It is enough to prove the validity (as $n \to \infty$) of the following three relations:

$$\triangle_n^{(1)} := G_d(U_n)\left\|\sum_{j \in U_n} |Q_j|^{s/2-1}\left(\left|\frac{S(Q_j)}{|Q_j|} - \frac{S(U_n)}{|U_n|}\right|^s - \left|\frac{S(Q_j)}{|Q_j|}\right|^s\right)\right\|_{(2/s)\wedge 2} \to 0, \tag{1.6}$$

$$\triangle_n^{(2)} := G_d(U_n)\left\|\sum_{j \in U_n} |Q_j|^{s/2-1}\left(\frac{|S(Q_j)|^s}{|Q_j|} - \mathsf{E}\frac{|S(Q_j)|^s}{|Q_j|}\right)\right\|_{(2/s)\wedge 2} \to 0, \tag{1.7}$$

$$\triangle_n^{(3)} := G_d(U_n)\sum_{j \in U_n} |Q_j|^{-1}\frac{\mathsf{E}|S(Q_j)|^s}{|Q_j|^{s/2}} - \sigma^s\mathsf{E}|Z|^s \to 0. \tag{1.8}$$

---

[3]See Section 3.1.

By stationarity of X we can consider the sequence of blocks $U_n = (0, l_1] \times \ldots \times (0, l_d]$ where $l_i = l_i(n)$, $i = 1, \ldots, d$, $n \in \mathbb{N}$, and

$$Q_j = \{0 < q_i \leq j_i, \ i = 1, \ldots, d\}. \tag{1.9}$$

As usual, for an array $\{\beta_j, j \in \mathbb{N}^d\}$ of real numbers the relation $\beta_j \to \beta$ as $j \to \infty$ means that $\beta \in \mathbb{R}$ and, for any $\varepsilon > 0$, there exists $N = N(\varepsilon) \in \mathbb{N}$ such that

$$|\beta_j - \beta| < \varepsilon \quad \text{for all} \quad j \geq (N, \ldots, N), \tag{1.10}$$

i.e. for all $j \in \mathbb{N}^d : j_k \geq N$, $k = 1, \ldots, d$. The following fact is a multidimensional analog of the well-known Toeplitz lemma.

Lemma 1.2. *Let $\{\alpha_j \geq 0, j \in \mathbb{N}^d\}$ and $\{\beta_j, j \in \mathbb{N}^d\}$ be d-dimensional arrays of real numbers. Suppose that β_j are bounded and $\beta_j \to \beta$ as $j \to \infty$. Assume also that, for any fixed $q \in \mathbb{N}^d$,*

$$\frac{\sum_{1 \leq j \leq k} \alpha_j}{\sum_{1 \leq j \leq k, \, j \geq q} \alpha_j} \to 1 \quad \text{as} \quad k \to \infty, \ k \in \mathbb{N}^d. \tag{1.11}$$

Then

$$\frac{\sum_{1 \leq j \leq k} \alpha_j \beta_j}{\sum_{1 \leq j \leq k} \alpha_j} \to \beta \quad \text{as} \quad k \to \infty.$$

Proof of the Lemma is standard. Namely, let $\varepsilon > 0$ be arbitrary and let $N \in \mathbb{N}$ be such that $|\beta_j - \beta| < \varepsilon$ whenever $j \geq q = (N, \ldots, N)$. Then

$$\frac{\sum_{1 \leq j \leq k} \alpha_j \beta_j}{\sum_{1 \leq j \leq k} \alpha_j} - \beta = \frac{\sum_{1 \leq j \leq k} \alpha_j (\beta_j - \beta)}{\sum_{1 \leq j \leq k} \alpha_j}$$

$$= \frac{\sum_{1 \leq j \leq k, \, j \geq q} \alpha_j (\beta_j - \beta)}{\sum_{1 \leq j \leq k} \alpha_j} + \frac{\sum_{1 \leq j \leq k, \, j \not\geq q} \alpha_j (\beta_j - \beta)}{\sum_{1 \leq j \leq k} \alpha_j} := I_{k,1} + I_{k,2}$$

where $\{j \in \mathbb{N}^d : j \not\geq q\} = \mathbb{N}^d \setminus \{j : j \geq q\}$.

In view of (1.10) and (1.11) the upper limit of $|I_{k,1}|$ as $k \to \infty$ does not exceed ε. Now we can use the assumption that $|\beta_j| \leq c_0$ for some $c_0 > 0$ and all $j \in \mathbb{N}^d$. So, due to (1.11) one has

$$|I_{k,2}| \leq 2c_0 \frac{\sum_{1 \leq j \leq k} \alpha_j - \sum_{1 \leq j \leq k, \, j \geq q} \alpha_j}{\sum_{1 \leq j \leq k} \alpha_j} \to 0 \text{ as } k \to \infty. \ \square$$

Remark 1.3. If $d = 1$ then (1.11) is equivalent to the relation

$$\sum_{1 \leq j \leq k} \alpha_j \to \infty, \ k \to \infty. \tag{1.12}$$

For $d = 1$ the assumption $\beta_j \to \beta$ implies that the sequence $(\beta_j)_{j \in \mathbb{N}}$ is bounded. However if $d > 1$ then (1.12) in general is not equivalent to relation (1.11), and the assumption $\beta_j \to \beta$ does not imply the boundedness of $(\beta_j)_{j \in \mathbb{N}}$.

Now we prove (1.8). By Theorem 3.1.8,

$$\frac{\mathsf{E}S^2(Q_j)}{|Q_j|} \to \sigma^2, \quad j \to \infty. \tag{1.13}$$

Let $s \in (0,2)$. Theorem 3.1.12 yields that

$$\frac{S(Q_j)}{\sqrt{|Q_j|}} \to \sigma Z \quad \text{in law}, \quad j \to \infty. \tag{1.14}$$

Therefore, for any $s > 0$,

$$\frac{|S(Q_j)|^s}{|Q_j|^{s/2}} \to \sigma^s |Z|^s \quad \text{in law as} \quad j \to \infty. \tag{1.15}$$

Taking into account that $2/s > 1$ for $s \in (0,2)$ and

$$(|S(Q_j)|^s/|Q_j|^{s/2})^{2/s} = S^2(Q_j)/|Q_j|, \tag{1.16}$$

using (1.13) we conclude that a family $\{|S(Q_j)|^s/|Q_j|^{s/2}, j \in \mathbb{N}^d\}$ is uniformly integrable. Now (1.15) makes it obvious that

$$\mathsf{E}|S(Q_j)|^s/|Q_j|^{s/2} \to \sigma^s \mathsf{E}|Z|^s, \quad j \to \infty. \tag{1.17}$$

If $s = 2$ then $\mathsf{E}Z^2 = 1$. Thus in view of (1.13) we can claim that (1.17) holds for $s \in (0,2]$.

Set $\alpha_j = |Q_j|^{-1}$, $\beta_j = \mathsf{E}|S(Q_j)|^s/|Q_j|^{s/2}$, $j \in \mathbb{N}^d$. Then (1.13) implies that $\beta_j \to \beta = \sigma^s \mathsf{E}|Z|^s$ as $j \to \infty$ and $\beta_j \le \sigma^s \mathsf{E}|Z|^s$ for all $j \in \mathbb{N}^d$, due to (1.16). In view of (1.9) we obtain that (1.11) holds. Finally it remains to observe that

$$G_d(U_n) \sum_{j \in U_n} |Q_j|^{-1} \to 1, \quad n \to \infty, \tag{1.18}$$

and apply Lemma 1.2. The relation (1.8) is proved.

Now we are going to establish (1.7). Let us consider two cases.

Case 1: $s \in [1,2]$. For positive M define a function $H_M(\cdot)$ by (2.1.5) and set

$$T_j = \left| H_M\left(\frac{S(Q_j)}{\sqrt{|Q_j|}}\right) \right|^s - \mathsf{E}\left| H_M\left(\frac{S(Q_j)}{\sqrt{|Q_j|}}\right) \right|^s, \quad Y_j = \left| \frac{S(Q_j)}{\sqrt{|Q_j|}} \right|^s - \mathsf{E}\left| \frac{S(Q_j)}{\sqrt{|Q_j|}} \right|^s - T_j \tag{1.19}$$

where $j \in U_n$. Then we obtain a bound

$$\triangle_n^{(2)} = G_d(U_n) \left\| \sum_{j \in U_n} |Q_j|^{-1} \left(\left| \frac{S(Q_j)}{\sqrt{|Q_j|}} \right|^s - \mathsf{E}\left| \frac{S(Q_j)}{\sqrt{|Q_j|}} \right|^s \right) \right\|_{2/s}$$

$$\le G_d(U_n) \left\| \sum_{j \in U_n} |Q_j|^{-1} T_j \right\|_{2/s} + G_d(U_n) \left\| \sum_{j \in U_n} |Q_j|^{-1} Y_j \right\|_{2/s} =: I_{n,1} + I_{n,2}.$$

Note that $Lip(|H_M(x)|^s) = sM^{s-1}$. Thus, by Theorem 1.5.3,

$$I_{n,1}^2 \le G_d(U_n)^2 \mathsf{E}\left(\sum_{j \in U_n} |Q_j|^{-1} T_j \right)^2 = G_d(U_n)^2 \sum_{j \in U_n} \sum_{q \in U_n} |Q_j|^{-1} |Q_q|^{-1} cov(T_j, T_q)$$

$$\leq s^2 M^{2(s-1)} G_d(U_n)^2 \sum_{j\in U_n}\sum_{q\in U_n} |Q_j|^{-3/2}|Q_q|^{-3/2} cov(S(Q_j),S(Q_q))$$

$$= s^2 M^{2(s-1)} G_d(U_n)^2 \sum_{j\in U_n}\sum_{q\in U_n} |Q_j|^{-3/2}|Q_q|^{-3/2} \sum_{t,p\in U_n;\, t\leq j, p\leq q} cov(X_t,X_p). \tag{1.20}$$

Obviously $|Q_j| = j_1\ldots j_d$ and, for any $m,n\in\mathbb{N}$,

$$\sum_{k=m}^{m+n} k^{-3/2} \leq \frac{3}{\sqrt{m}}.$$

Thus, changing the order of summation in the last expression of (1.20) we see that

$$I_{n,1}^2 \leq 3^{2d} s^2 M^{2(s-1)} G_d(U_n)^2 \sum_{t\in U_n}\sum_{p\in U_n} cov(X_t,X_p)(\langle t\rangle\langle p\rangle)^{-1/2} \tag{1.21}$$

where $\langle t\rangle = t_1\ldots t_d$ for $t\in\mathbb{R}^d$. Note that $0 \leq cov(X_t,X_p) \leq \sigma^2$ for all $t,p\in\mathbb{Z}^d$ and

$$\{t,p\in U_n\} = \{t,p\in U_n, \langle t\rangle \leq \langle p\rangle\} \cup \{t,p\in U_n, \langle t\rangle > \langle p\rangle\}.$$

Hence

$$\sum_{t\in U_n}\sum_{p\in U_n} cov(X_t,X_p)(\langle t\rangle\langle p\rangle)^{-1/2} \leq 2\sigma^2 \sum_{t\in U_n} \langle t\rangle^{-1}. \tag{1.22}$$

Then relations (1.18), (1.21) and (1.22) imply that $I_{n,1} \to 0$ as $n\to\infty$.

Now we prove that $I_{n,2} \to 0$ as $n\to\infty$. For $Y\in L^p(\Omega,\mathcal{F},\mathsf{P})$, $p\geq 1$, one has $\|Y - \mathsf{E}Y\|_p \leq 2\|Y\|_p$. Therefore, using an obvious estimate

$$\left| |x|^s - |H_M(x)|^s \right| \leq |x|^s \mathbb{I}\{|x|\geq M\}, \quad x\in\mathbb{R}, \quad s>0,$$

we obtain

$$I_{n,2} \leq G_d(U_n) \sum_{j\in U_n} |Q_j|^{-1} \|Y_j\|_{2/s}$$

$$\leq 2G_d(U_n) \sum_{j\in U_n} |Q_j|^{-1} \left(\mathsf{E}\left(\frac{|S(Q_j)|^s}{|Q_j|^{s/2}} \mathbb{I}\left\{ \frac{|S(Q_j)|}{|Q_j|} \geq M \right\} \right)^{2/s} \right)^{s/2}. \tag{1.23}$$

We have proved that, for each $s\in(0,2)$, the family of random variables $\{|S(Q_j)|^s/|Q_j|^{s/2}, j\in\mathbb{N}^d\}$ is uniformly integrable. Note that (1.13) and (1.14) imply that a family of nonnegative random variables $\{S^2(U_n)/|U_n|, n\in\mathbb{N}\}$ is also uniformly integrable (by Lemma 3.2.6). Thus, taking M large enough and using (1.18) we see that, for any $s\in[1,2]$, $\limsup_{n\to\infty} I_{n,2}$ can be made less than any given positive number, whence it equals zero.

Finally we have $I_{n,2} \to 0$, $n\to\infty$ for $s\in[1,2]$. Therefore $\triangle_n^{(2)} \to 0$ as $n\to\infty$ for $s\in[1,2]$.

Case 2: $s\in(0,1)$. For $M>0$ and $j\in U_n$ define

$$L_j = M^s \vee \left| \frac{S(Q_j)}{\sqrt{|Q_j|}} \right|^s - \mathsf{E}\left(M^s \vee \left| \frac{S(Q_j)}{\sqrt{|Q_j|}} \right|^s \right), \quad W_j = \left| \frac{S(Q_j)}{\sqrt{|Q_j|}} \right|^s - \mathsf{E}\left| \frac{S(Q_j)}{\sqrt{|Q_j|}} \right|^s - L_j.$$

$$\tag{1.24}$$

One may write

$$G_d(U_n)\left\|\sum_{j\in U_n}|Q_j|^{s/2-1}\left(\left|\frac{S(Q_j)}{|Q_j|}\right|^s-\mathsf{E}\left|\frac{S(Q_j)}{|Q_j|}\right|^s\right)\right\|_2$$

$$\leq G_d(U_n)\left\|\sum_{j\in U_n}|Q_j|^{-1}L_j\right\|_2+G_d(U_n)\left\|\sum_{j\in U_n}|Q_j|^{-1}W_j\right\|_2=:J_{n,1}+J_{n,2}.$$

Since $h(x)=|x|^s\vee M^s$ is a Lipschitz function with $Lip(h)=sM^{s-1}$ and $cov(L_j,L_q)$ exists for all $j,q\in U_n$, we can estimate $J_{n,1}$ in the same way as $I_{n,1}$ in the case 1. Therefore, for every $M>0$, $J_{n,1}\to 0$ as $n\to\infty$.

Since $|h(x)-|x|^s|\leq M^s$, $x\in\mathbb{R}$, we have $|W_j|\leq 2M^s$ and thus

$$J_{n,2}\leq G_d(U_n)\Big(\sum_{j\in U_n}|Q_j|^{-1}2M^s\Big).$$

Hence, by (1.18), $\limsup(J_{n,1}+J_{n,2})\leq 2M^s$. We can take $M>0$ arbitrary small, so $\triangle_n^{(2)}\to 0$ as $n\to\infty$ for all $s\in(0,1)$. We are done with Case 2.

Our next goal is to prove (1.6). For any $x,y\in\mathbb{R}$ we have

$$\big||x+y|^s-|y|^s\big|\leq\begin{cases}|x|^s, & s\in(0,1],\\ s|x|\left(|x|^{s-1}+|y|^{s-1}\right), & s\in(1,2],\\ s2^{s-1}|x|\left(|x|^{s-1}+|y|^{s-1}\right), & s>2.\end{cases}\tag{1.25}$$

Set

$$J_{n,3}=\left\|\sum_{j\in U_n}|Q_j|^{s/2-1}\left(\left|\frac{S(Q_j)}{|Q_j|}-\frac{S(U_n)}{|U_n|}\right|^s-\left|\frac{S(Q_j)}{|Q_j|}\right|^s\right)\right\|_{(2/s)\wedge 2}.$$

If $s\in(0,2]$ then for any finite $U\subset\mathbb{Z}^d$

$$\mathsf{E}|S(U)|^s\leq(\mathsf{E}S^2(U))^{s/2}\leq\sigma^s|U|^{s/2},\tag{1.26}$$

as $\mathsf{E}S^2(U)\leq\sigma^2|U|$, $U\subset\mathbb{Z}^d$. Then, for $s\in(0,1]$, (1.25) yields

$$J_{n,3}\leq\frac{\mathsf{E}|S(U_n)|^s}{|U_n|^s}\sum_{j\in U_n}|Q_j|^{s/2-1}\leq\sigma^s|U_n|^{-s/2}\sum_{j\in U_n}|Q_j|^{s/2-1}.\tag{1.27}$$

Thus $J_{n,3}$ is uniformly bounded, because $k^{-s/2}\sum_{m=1}^k m^{s/2-1}<C(s)$ for some $C(s)>0$ and all $k\in\mathbb{N}$. Consequently, $\triangle_n^{(1)}=G_d(U_n)I_{n,3}\to 0$ as $n\to\infty$ and (1.6) is proved for $s\in(0,1]$.

If $s\in(1,2]$ then by (1.25)

$$J_{n,3}\leq s\left\|\sum_{j\in U_n}|Q_j|^{s/2-1}\left(\frac{|S(U_n)|}{|U_n|}\frac{|S(Q_j)|^{s-1}}{|Q_j|^{s-1}}+\frac{|S(U_n)|^s}{|U_n|^s}\right)\right\|_{2/s}$$

$$\leq s\left\|\frac{|S(U_n)|}{|U_n|}\sum_{j\in U}|Q_j|^{-1/2}\frac{|S(Q_j)|^{s-1}}{|Q_j|^{(s-1)/2}}\right\|_{2/s}+s\sum_{j\in U_n}\frac{(\mathsf{E}S^2(U_n))^{s/2}}{|U_n|^s}|Q_j|^{s/2-1}.\tag{1.28}$$

The Hölder inequality for sums implies that

$$\sum_{j\in U_n} |Q_j|^{-1/2}\frac{|S(Q_j)|^{s-1}}{|Q_j|^{(s-1)/2}} \leq \Big(\sum_{j\in U_n} |Q_j|^{-1/(3-s)}\Big)^{(3-s)/2}\Big(\sum_{j\in U_n}\frac{S^2(Q_j)}{|Q_j|}\Big)^{(s-1)/2}.$$

$$(1.29)$$

By the Hölder inequality for integrals and (1.29)

$$\Big(\mathsf{E}\frac{|S(U_n)|^{2/s}}{|U_n|^{2/s}}\Big(\sum_{j\in U_n}\frac{S^2(Q_j)}{|Q_j|}\Big)^{(s-1)/s}\Big)^{s/2}$$

$$\leq \frac{(\mathsf{E}S^2(U_n))^{1/2}}{|U_n|}\Big(\mathsf{E}\sum_{j\in U_n}\frac{S^2(Q_j)}{|Q_j|}\Big)^{(s-1)/2} \leq \sigma^s|U_n|^{s/2-1}\qquad(1.30)$$

where we have also used (1.26) with $s = 2$. Note in addition that

$$\sum_{j\in U_n} |Q_j|^{-1/(3-s)} \leq \begin{cases} C_d^{(s)}|U_n|^{(2-s)/(3-s)}, & s\in(1,2),\\ C_d^{(2)}\log|U_n|, & s=2. \end{cases}\qquad(1.31)$$

Here $C_d^{(s)}$ are positive factors depending only on s and d.

In view of (1.29)—(1.31) the first summand in the right-hand side of (1.28) can be bounded by $s\sigma^s C_d^{(s)}$ if $s\in(1,2)$ and $2\sigma^2 C_d^{(2)}(\log|U_n|)^{1/2}$ if $s=2$. The second summand in the right-hand side of (1.28) can be bounded analogously to (1.27). Finally,

$$\triangle_n^{(1)} = G_d(U_n)J_{n,3} \to 0 \quad\text{as}\quad n\to\infty$$

which proves (1.6) for $s\in(1,2]$. The proof of the Theorem is complete. $\square$

2°. The case of regularly growing sets of summation. In this Theorem the sets Q_j, used in estimating, were blocks $(0,j]$ contained in the "sample index set" $U_n = (0,n]$. If we take instead of R_j (see (1.2)) the set K_j defined as a neighborhood of a point $j \in U$, it becomes possible to extend the estimation procedure and include growing sets more general than blocks U_n. In case of simple U_n (cubes) this latter statistics proves to converge faster that the former with R_j, as we will show later.

For a point $j\in\mathbb{Z}^d$ and any $r > 0$ define

$$K_j(r) = \{q\in\mathbb{Z}^d : |q-j|\leq r\}.$$

For a finite $U\subset\mathbb{Z}^d$, some $j\in U, s\in(0,2]$ and $r = r(U) > 0$ put

$$Q_j(U) = U\cap K_j(r), \quad f_j^U(x) = |U|^{-1}|Q_j|^{s/2}|x|^s, \quad x\in\mathbb{R}.\qquad(1.32)$$

Note that these f_j^U will be used below (in the same capacity but) instead of the f_j^U that appeared in the previous Theorem. As before, for finite $U\subset\mathbb{Z}^d$ and $m\in\mathbb{N}$ we denote the closed m-neighborhood of U by $U^{(m)}$, i.e.

$$U^{(m)} = \bigcup_{j\in U} K_j(m).$$

Let δU be defined by (3.1.3). Then we have

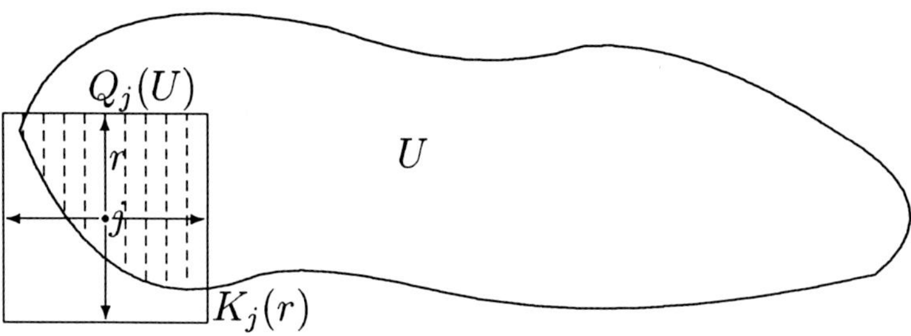

Fig. 7.2 $(Q_j(U)$ is shaded)

Theorem 1.4. ([83]) *Let the conditions of Theorem 1.1 hold and let $(U_n)_{n \in \mathbb{N}}$ be an arbitrary sequence of regularly growing finite sets of $\mathbb{Z}^d$. Suppose that $(r_n)_{n \in \mathbb{N}}$ is a sequence of positive integers $r_n = r(U_n)$ such that $r_n \to \infty$ as $n \to \infty$, $r_n^{2d}/|U_n| \leq R$ for some positive R and*

$$\frac{|(\delta U_n)^{(2r_n)}|}{|U_n|} \to 0, \quad n \to \infty. \tag{1.33}$$

Then, for any $s \in (0, 2]$, the asymptotic relations (1.4) and (1.5) are valid, where the statistics $B(U_n)$ are defined by (1.3) with $f_j^U(\cdot)$ appearing in (1.32).

Proof. Again it suffices to establish (1.4). Note that

$$\|B(U_n) - \sigma^s \mathsf{E}|Z|^s\|_{(2/s) \wedge 2} \leq J_{1,n} + J_{2,n} + J_{3,n}$$

where[4]

$$J_{1,n} := |U_n|^{-1} \left\| \sum_{j \in U_n} |Q_j|^{s/2} \left(\left| \frac{S(Q_j)}{|Q_j|} - \frac{S(U)}{|U|} \right|^s - \left| \frac{S(Q_j)}{|Q_j|} \right|^s \right) \right\|_{(2/s) \wedge 2},$$

$$J_{2,n} := |U_n|^{-1} \left\| \sum_{j \in U_n} |Q_j|^{-s/2} \left(|S(Q_j)|^s - \mathsf{E}|S(Q_j)|^s \right) \right\|_{(2/s) \wedge 2},$$

$$J_{3,n} := \left| |U_n|^{-1} \sum_{j \in U_n} |Q_j|^{-s/2} \mathsf{E}|S(Q_j)|^s - \sigma^s \mathsf{E}|Z|^s \right|.$$

As in the proof of Theorem 1.1, the simplest part is to check that $J_{3,n} \to 0$ as $n \to \infty$. Namely, for $n \in \mathbb{N}$ and $r \in \mathbb{N}$, let $U_n(r) = U_n \setminus ((\mathbb{Z}^d \setminus U_n)^{(r)})$.
Note that

$$U_n = U_n(2r_n) \cup (U_n \setminus U_n(2r_n)), \quad n \in \mathbb{N}.$$

---

[4]We write $J_{i,n}$ because in Theorem 1.1 we used $J_{n,i}$ $(i = 1, 2, 3)$.

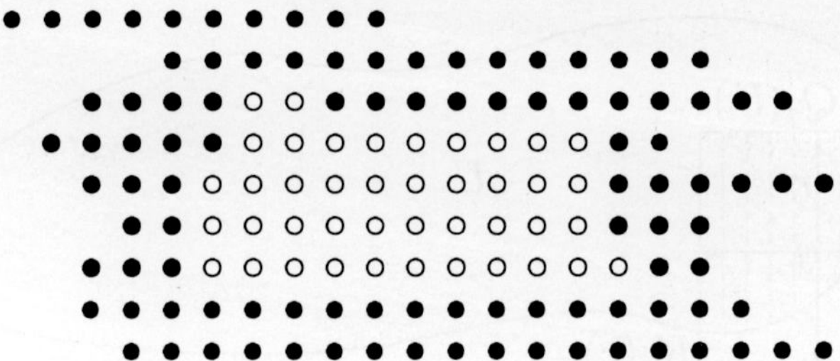

Fig. 7.3   (white circles denote elements of $U_n(2)$, black ones constitute $U_n \setminus U_n(2)$)

Due to stationarity of X, one has $\mathsf{E}|S(Q_j(r_n))|^s = \mathsf{E}|S(K_0(r_n))|^s$ for $j \in U_n(2r_n)$. Therefore,

$$|U_n|^{-1} \sum_{j \in U_n} |Q_j|^{-s/2} \mathsf{E}|S(Q_j)|^s = |U_n|^{-1}|U_n(2r_n)|\frac{\mathsf{E}|S(K_0(r_n))|^s}{|K_0(r_n)|^{s/2}}$$

$$+ |U_n|^{-1} \sum_{j \in U_n \setminus U_n(2r_n)} |Q_j|^{-s/2}\mathsf{E}|S(Q_j)|^s. \tag{1.34}$$

By (1.17),

$$\frac{\mathsf{E}|S(K_0(r_n))|^s}{|K_0(r_n)|^{s/2}} \to \sigma^s \mathsf{E}|Z|^s$$

as $r_n \to \infty$ $(n \to \infty)$. Obviously $U_n \setminus U_n(2r_n) \subset (\delta\, U_n)^{(2r_n)}$, $n \in \mathbb{N}$. Thus by virtue of (1.33) the first summand at the right-hand side of (1.34) tends to zero. In view of (1.26) and taking into account that, for $U \subset \mathbb{Z}^d$ and $r \in \mathbb{N}$,

$$U \setminus U(r) \subset \bigcup_{j \in \delta\, U} K_j(r),$$

we conclude that

$$\frac{1}{|U_n|} \sum_{j \in U_n \setminus U_n(2r_n)} |Q_j|^{-s/2}\mathsf{E}|S(Q_j)|^s \le \frac{\sigma^s |U_n \setminus U_n(2r_n)|}{|U_n|} \le \frac{\sigma^s |(\delta\, U_n)^{(2r_n)}|}{|U_n|} \to 0$$

$$\tag{1.35}$$

as $n \to \infty$, due to (1.33). Thus $J_{3,n} \to 0$, $n \to \infty$.

The term $J_{2,n}$ is estimated analogously to $\Delta_n^{(2)}$ in the proof of Theorem 1.1, with the following amendments. Introducing T_j and Y_j as it was made in (1.19), we see that

$$J_{2,n} \le |U_n|^{-1}\Big\| \sum_{j \in U_n} T_j \Big\|_{(2/s)\wedge 2} + |U_n|^{-1}\Big\| \sum_{j \in U_n} Y_j \Big\|_{(2/s)\wedge 2} =: J_{2,n}^{(1)} + J_{2,n}^{(2)}. \tag{1.36}$$

Note that

$$\Big\| \sum_{j \in U_n} T_j \Big\|_{(2/s)\wedge 2} \le \Big\| \sum_{j \in U_n} T_j \Big\|_2.$$

Set $C(t,p) := cov(X_t, X_p)$ and $A(t) := \sum_{j \in U_n : Q_j \ni t} |Q_j|^{-s/2}$ $(A(t) = A(t,s,n))$, $t \in U_n$. Consequently,

$$(J_{2,n}^{(1)})^2 \leq s^2 M^{2(s-1)} |U_n|^{-2} \sum_{j,q \in U_n} |Q_j|^{-s/2} |Q_q|^{-s/2} cov(S(Q_j), S(Q_q))$$

$$\leq s^2 M^{2(s-1)} |U_n|^{-2} \sum_{t,p \in U_n} C(t,p) A(t) A(p)$$

$$\leq s^2 M^{2(s-1)} |U_n|^{-2} \Big(\sum_{t,p \in D_n} C(t,p) A(t)^2 + \sum_{t,p \in U_n \setminus D_n} C(t,p) A(p)^2 \Big)$$

where $D_n = \{(t,p) \in U_n : A(t) \geq A(p)\}$. Due to stationarity of X, for any $t \in \mathbb{Z}^d$, one has $\sum_{p \in \mathbb{Z}^d} cov(X_t, X_p) = \sigma^2$ and

$$(J_{2,n}^{(1)})^2 \leq 2s^2 M^{2(s-1)} |U_n|^{-2} \sum_{t \in U_n} A(t)^2 \sum_{p \in \mathbb{Z}^d} C(t,p)$$

$$\leq 2s^2 M^{2(s-1)} \sigma^2 |U_n|^{-2} \sum_{t \in U_n} A(t)^2. \tag{1.37}$$

If $t \in U_n(2r_n)$ and $Q_j \ni t$ then $|Q_j| = |K_0(r_n)|$. Since also $|Q_j| \geq 1$, we have

$$(J_2^{(1)})^2 \leq 2s^2 M^{2(s-1)} \sigma^2 |U_n|^{-1} |K_0(r_n)|^2 (|K_0(r_n)|^{-s} + |U_n|^{-1} |U_n \setminus U_n(2r_n)|) \to 0$$

as $n \to \infty$ by the choice of r_n satisfying the conditions of Theorem.

Now we use (1.36) to estimate $J_{2,n}^{(2)}$. For any $M > 0$, $n \in \mathbb{N}$ and $s \in [1,2]$, write

$$J_{2,n}^{(2)} \leq |U_n|^{-1} \sum_{j \in U_n(2r_n)} \|Y_j\|_{2/s} + |U_n|^{-1} \sum_{j \in U_n \setminus U_n(2r_n)} \|Y_j\|_{2/s}$$

$$\leq 2 \left(\mathsf{E} \frac{S^2(K_0(r_n))}{|K_0(r_n)|} \mathbb{I} \left\{ \frac{S^2(K_0(r_n))}{|K_0(r_n)|} \geq M^2 \right\} \right)^{s/2} + 4\sigma^s \frac{|U_n \setminus U_n(2r_n)|}{|U_n|},$$

where the last estimate is due to (1.26). The family $\{S^2(K_0(r_n))/|K_0(r_n)|, n \in \mathbb{N}\}$ is uniformly integrable (see the part of the proof of Theorem 1.1 after (1.23)). From this and (1.35) it follows that the upper limit of $J_{2,n}^{(2)}$ (as $n \to \infty$) can be made arbitrary small by appropriate choice of M.

For $s \in (0,1]$ the estimation of $J_{2,n}^{(2)}$ is performed as in the case 2 in the proof of Theorem 1.1. Namely, we use auxiliary random variables L_j and W_j defined in (1.24), to verify that

$$J_{2,n}^{(2)} \leq |U_n|^{-1} \Big\| \sum_{j \in U_n} L_j \Big\|_2 + |U_n|^{-1} \Big\| \sum_{j \in U_n} W_j \Big\|_2 =: R_{2,n}^{(1)} + R_{2,n}^{(2)}.$$

Analogously to (1.20) one has

$$(R_{2,n}^{(1)})^2 = |U_n|^{-2} \sum_{j \in U_n} \sum_{q \in U_n} cov(L_j, L_q) \leq (sM^{s-1})^2 |U_n|^{-2} \sum_{t,p \in U_n} C(t,p) A(t) A(p).$$

In the same way as while proving (1.37) we have

$$(R_{2,n}^{(1)})^2 \leq 2 \left(sM^{s-1}\right)^2 |U_n|^{-2}\sigma^2 \sum_{t \in U_n} A(t)^2$$

$$\leq 2 \left(sM^{s-1}\right)^2 |U_n|^{-1}\sigma^2 |K_0(r_n)|^2 \left(|K_0(r_n)|^{-1/2} + |U_n|^{-1}|U_n \setminus U_n(2r_n)|\right) \to 0$$

if only $r_n^{3d/2}/|U_n| \to 0$ as $n \to \infty$, which is true by assumption. Note that

$$R_{2,n}^{(2)} \leq |U_n|^{-1} \sum_{j \in U_n} \|W_j\|_2 \leq 2M^s|U_n|^{-1}|U_n| = 2M^s.$$

Thus $\limsup_{n \to \infty} J_{2,n} \leq 2M^s$ for any $M > 0$. Therefore $J_{2,n} \to 0$, $n \to \infty$, in the case $s \in (0, 1]$.

It remains to study the behavior of $J_{1,n}$. If $s \in (0, 1]$ then by (1.25) and (1.26)

$$J_{1,n} \leq |U_n|^{-1-s} \sum_{j \in U_n} |Q_j|^{s/2}(\mathsf{E}|S(U_n)|^{2s})^{1/2} \leq |U_n|^{-1-s} \sum_{j \in U_n} |Q_j|^{s/2}\sigma^s|U_n|^{s/2}$$

$$\leq \sigma^s \frac{|K_0(r_n)|^{s/2}}{|U_n|^{s/2}} = \sigma^s \left(\frac{|K_0(r_n)|^2}{|U_n|}\right)^{s/2} |K_0(r_n)|^{-s/2} \to 0 \text{ as } n \to \infty, \qquad (1.38)$$

since the fraction in the last expression is bounded.

If $s \in (1, 2]$ then, by (1.25), $J_{1,n}$ is majorized by

$$s|U_n|^{-2}\left\| |S(U_n)| \sum_{j \in U_n} |Q_j|^{1-s/2}|S(Q_j)|^{s-1}\right\|_{2/s} + s\sigma^s|U_n|^{-1-s/2} \sum_{j \in U_n} |Q_j|^{s/2}. \quad (1.39)$$

The second summand in (1.39) tends to zero as $n \to \infty$ in view of (1.38) (this conclusion is valid for $s \in (1, 2]$ as well). The first one, by the Hölder inequality, has the estimate

$$s|U_n|^{-1}|K_0(r_n)|^{1-s/2} \max_{j \in U_n}\left\| |S(U_n)||S(Q_j)|^{s-1}\right\|_{2/s}$$

$$\leq s|U_n|^{-1}|K_0(r_n)|^{1-s/2} \max_{j \in U_n}(\mathsf{E}S^2(U_n))^{1/2}(\mathsf{E}S^2(Q_j))^{\frac{s-1}{2}}$$

$$\leq \sigma^s s|U_n|^{-1}|K_0(r_n)|^{1-s/2}|U_n|^{1/2}|K_0(r_n)|^{\frac{s-1}{2}} = s\sigma^s \frac{|K_0(r_n)|^{1/2}}{|U_n|^{1/2}}$$

$$= s\sigma^s \left(\frac{|K_0(r_n)|^2}{|U_n|}\right)^{1/2} |K_0(r_n)|^{-1/2} \to 0, \ \ n \to \infty,$$

vanishing is due to the choice of r_n. The proof is complete. $\square$

Remark 1.5. Note that $|Q^{(r)}| \leq (2r+1)^d|Q|$ for a finite $Q \subset \mathbb{Z}^d$. Thus we can provide the following version of Theorem 1.4.

Corollary 1.6. *Let the conditions of Theorem 1.1 hold and let $(U_n)_{n\in\mathbb{N}}$ be an arbitrary regularly growing sequence of finite subsets of $\mathbb{Z}^d$. Then for any sequence $(r_n)_{n\in\mathbb{N}}$ of positive integers such that $r_n = r_n(U_n) \to \infty$, $r_n = O(|U_n|^{1/(2d)})$ and*

$$r_n^d \frac{|\delta U_n|}{|U_n|} \to 0, \quad n \to \infty,$$

the statements of Theorem 1.4 are true. In particular, for the cubes $U_n = (0,n]^d$ one can take $r_n = o(n^{1/d})$ if $d > 1$ and $r_n = O(n^{1/2})$ if $d = 1$ $(n \to \infty)$.

3°. The comparison of the introduced statistics. Now we will compare the introduced families of statistics in a "good" case, when the additional moments exist and correlations decrease fast. We will confine ourselves to the case $d = 1$ when the index sets U_n are "integer segments", i.e. $U_n = (0,n] \cap \mathbb{Z}^d$, $n \in \mathbb{N}$. Let us introduce

$$\Delta_1(n) = B_1(U_n) - \sigma^s \mathsf{E}|Z|^s, \quad \Delta_2(n) = B_2(U_n) - \sigma^s \mathsf{E}|Z|^s$$

where B_1 and B_2 are the statistics defined by (1.3) with specified choice of f_j, used in Theorems 1.1 and 1.4 respectively. Clearly $B_1 = B_1(s, U_n)$ and $B_2 = B_2(s, r_n, U_n)$, $s \in (0,2]$, $n \in \mathbb{N}$ and $(r_n)_{n\in\mathbb{N}}$ is a sequence of positive integers satisfying the conditions of Theorem 1.4.

Theorem 1.7. ([405]) *Let $X = (X_j)_{j\in\mathbb{Z}}$ be a strictly stationary centered associated random sequence such that*
   1) $\mathsf{E}|X_0|^{2+\delta} < \infty$ *for some $\delta > 0$;*
   2) *for some $\lambda > 0$, the Cox–Grimmett coefficient $u_r = O(\exp\{-\lambda r\})$ as $r \to \infty$;*
   3) $\sigma^2 = \sum_{j\in\mathbb{Z}} cov(X_0, X_j) > 0$.
Then, for any $s \in (0,2]$, one has

$$\sqrt{\log n}\,\Delta_1(n) \to N(0, A_s^2) \quad \text{in law}$$

where $A_s > 0$. If $r_n = [n^\beta]$ with positive β small enough, then

$$\mathsf{P}(|\Delta_2(n)| \geq n^{-\gamma}) = O(n^{-\delta_0})$$

for some positive γ and δ_0 (which may depend on β). Thus the second estimate converges faster.

Proof. Without loss of generality we may assume that $\sigma = 1$. By the strong invariance principle (Theorem 5.2.1) the sequence X can be redefined on another probability space together with a Wiener process $W = \{W_t, t \geq 0\}$ such that $S_n - W_n = O(n^{1/2-\varepsilon})$ a.s. where $\varepsilon > 0$. Our claim is that

$$(\log n)B_1(n) - \sum_{j=1}^n j^{-1-s/2}|W_j|^s = o(\sqrt{\log n}) \quad \text{a.s.,} \quad n \to \infty. \tag{1.40}$$

If $s \in [1,2]$ then by (1.25) the absolute value of the left-hand side is no greater than

$$\sum_{j=1}^n \frac{1}{j}\left(\left|\left|\frac{S_j}{\sqrt{j}} - \sqrt{j}\frac{S_n}{n}\right|^s - \left|\frac{S_j}{\sqrt{j}}\right|^s\right| + \left|\left|\frac{S_j}{\sqrt{j}}\right|^s - \left|\frac{W_j}{\sqrt{j}}\right|^s\right|\right)$$

$$\leq \sum_{j=1}^{n} \frac{s}{j}\left(|S_n|^s\frac{j^{s/2}}{n^s} + |S_n|\frac{j^{1-s/2}}{n}|S_j|^{s-1} + \frac{|S_j - W_j|^s}{j^{s/2}} + \frac{|S_j - W_j||W_j|^{s-1}}{j^{s/2}}\right)$$

$$=: s(I_1 + I_2 + I_3 + I_4). \tag{1.41}$$

For $s \in (0,1)$, by (1.25) one has $I_1 + I_3$ instead of $\sum_{q=1}^{4} I_q$ in the right-hand side of (1.41). The law of the iterated logarithm (Theorem 6.1.1) implies that

$$I_1 \vee I_2 = O((LLn)^{s/2}) = o(\sqrt{\log n}), \quad n \to \infty.$$

By the strong invariance principle $I_3 \vee I_4$ is a.s. bounded. This proves (1.40).

Now we are going to show that

$$\sum_{j=1}^{n} j^{-1-s/2}|W_j|^s - \int_1^n |W_u|^s u^{-1-s/2}du = o(\sqrt{\log n}) \text{ in probability}, \tag{1.42}$$

which is equivalent, by the LIL, to

$$\sum_{j=1}^{n} j^{-1-s/2}|W_j|^s - \int_1^{n+1} |W_u|^s u^{-1-s/2}du = o(\sqrt{\log n}) \text{ in probability} \tag{1.43}$$

as $n \to \infty$. The absolute value of the left-hand side of (1.43) does not exceed

$$\sum_{j=1}^{n} |W_j|^s \int_j^{j+1} (j^{-1-s/2} - u^{-1-s/2})du + \sum_{j=1}^{n} \int_j^{j+1} \frac{||W_j|^s - |W_u|^s|}{u^{1+s/2}}du =: J_1 + J_2.$$

We have, with $Z \sim N(0,1)$,

$$\mathsf{E}J_1 = \sum_{j=1}^{n} \mathsf{E}|Z|^s j^{s/2} \int_j^{j+1} (j^{-1-s/2} - u^{-1-s/2})du$$

$$\leq \mathsf{E}|Z|^s \sum_{j=1}^{n} j^{s/2}(j^{-1-s/2} - (j+1)^{-1-s/2}) \leq \mathsf{E}|Z|^s \sum_{j=1}^{n} j^{s/2}(1 + s/2)j^{-2-s/2} = O(1)$$

as $n \to \infty$. If $s \in (0,1]$ then, by the Fubini theorem,

$$\mathsf{E}J_2 = \sum_{j=1}^{n} \mathsf{E}\int_j^{j+1} \frac{||W_j + W_u - W_j|^s - |W_j|^s|}{u^{1+s/2}}du \leq \sum_{j=1}^{n}\int_j^{j+1} \frac{\mathsf{E}|W_u - W_j|^s}{u^{1+s/2}}du$$

$$= \sum_{j=1}^{n} \int_j^{j+1} \frac{\mathsf{E}|Z|^s(u-j)^{s/2}}{u^{1+s/2}}du \leq \mathsf{E}|Z|^s \sum_{j=1}^{n} j^{-1-s/2}\int_j^{j+1}(u-j)^{s/2}du = O(1)$$

as $n \to \infty$. If $s > 1$ then, by independence of increments of W and (1.25),

$$\mathsf{E}J_2 \leq s\sum_{j=1}^{n}\int_j^{j+1} \frac{\mathsf{E}|W_u - W_j|(|W_j|^{s-1} + |W_u - W_j|^{s-1})}{u^{1+s/2}}du$$

$$\leq s\sum_{j=1}^{n} j^{-1-s/2}\int_j^{j+1} \left((u-j)^{1/2}\mathsf{E}|Z|j^{\frac{s-1}{2}}\mathsf{E}|Z|^{s-1} + (u-j)^{s/2}\mathsf{E}|Z|^s\right)du = O(1)$$

as $n \to \infty$. Consequently, for every $s \in (0, 2]$ and any $\varepsilon > 0$,

$$\mathsf{P}(J_1 + J_2 \geq \varepsilon\sqrt{\log n}) \to 0 \quad \text{as} \quad n \to \infty,$$

which gives (1.42).

From (1.40) and (1.42) one sees that to prove the first assertion of Theorem it suffices to show that

$$(\mathrm{Log}\, n)^{-1/2} \left(\int_1^n |W_u|^s u^{-1-s/2} du - (\log n)\mathsf{E}|Z|^s \right) \to N(0, A_{1,s}^2) \quad \text{in law as } n \to \infty.$$

Here the notation $A_{1,s}^2 > 0$ reminds that we consider $d = 1$. By the change of variable and Theorem 3.1.25 we have

$$(\mathrm{Log}\, n)^{-1/2} \left(\int_1^n |W_u|^s u^{-1-s/2} du - (\log n)\mathsf{E}|Z|^s \right)$$

$$= (\mathrm{Log}\, n)^{-1/2} \left(\int_0^{\log n} |W(e^t)|^s e^{-st/2} dt - (\log n)\mathsf{E}|Z|^s \right) \to N(0, A_{1,s}^2)$$

in law as $n \to \infty$.

For the second assertion we need two lemmas. The first of them is a well-known result on non-uniform bounds in CLT, see, e.g., [326, Ch. V, §5].

Lemma 1.8. *Let F be a distribution function such that*

$$\rho := \sup_{x \in \mathbb{R}} |F(x) - \Phi(x)| \leq e^{-1/2}$$

where $\Phi(x) = \mathsf{P}(Z \leq x)$, $Z \sim N(0, 1)$, $x \in \mathbb{R}$. Suppose that $b := \int_{\mathbb{R}} x^2 dF(x) < \infty$. Then there exists some constant $R > 0$ such that

$$|F(x) - \Phi(x)| \leq \frac{R\rho(-\log \rho) + D}{1 + x^2}$$

for any $x \in \mathbb{R}$, where

$$D = \left| \int_{\mathbb{R}} x^2 dF(x) - \int_{\mathbb{R}} x^2 d\Phi(x) \right| = |b - 1|.$$

Proof. Let $a \geq 1$ be such that a and $-a$ are continuity points of F. Write, integrating by parts,

$$\int_{-a}^a x^2 dF(x) = a^2(F(a) - \Phi(a)) - a^2(F(-a) - \Phi(-a)) - 2 \int_0^a x(F(x) - \Phi(x))dx$$

$$+ 2 \int_{-a}^0 |x|(F(x) - \Phi(x))dx + \int_{-a}^a x^2 d\Phi(x).$$

Thus one has

$$\int_{-a}^a x^2 dF(x) \geq -4a^2 \rho + \int_{-a}^a x^2 d\Phi(x),$$

and consequently

$$\int_{x:|x|\geq a} x^2 \, dF(x) \leq D + 4a^2\rho + \int_{x:|x|\geq a} x^2 \, d\Phi(x). \tag{1.44}$$

Likewise,

$$\int_{x:|x|\geq a} x^2 \, d\Phi(x) \leq D + 4a^2\rho + \int_{x:|x|\geq a} x^2 \, dF(x). \tag{1.45}$$

Now, if $|x| \leq a$, one obviously has an estimate

$$x^2 |F(x) - \Phi(x)| \leq a^2\rho. \tag{1.46}$$

For any $x \in (-\infty, -a]$, by (1.44) and (1.45), $x^2(\Phi(x) - F(x))$ admits the upper bounds

$$x^2 \Phi(x) \leq x^2 \int_{y:|y|\geq|x|} d\Phi(y) \leq \int_{y:|y|\geq|x|} y^2 \, d\Phi(y) \leq \int_{y:|y|\geq a} y^2 \, d\Phi(y),$$

and

$$x^2(F(x) - \Phi(x)) \leq x^2 \int_{y:|y|\geq|x|} dF(y) \leq \int_{y:|y|\geq|x|} y^2 \, dF(y)$$

$$\leq \int_{y:|y|\geq a} y^2 \, dF(y) \leq \int_{y:|y|\geq a} y^2 \, d\Phi(y) + D + 4a^2\rho.$$

That is,

$$x^2 |F(x) - \Phi(x)| \leq \int_{y:|y|\geq a} y^2 \, d\Phi(y) + D + 4a^2\rho. \tag{1.47}$$

Considering the distribution function $x \mapsto 1 - F(-x)$ instead of F and repeating all the arguments one readily sees that for $x \geq a$

$$x^2 |F(x) - \Phi(x)| \leq \int_{y:|y|\geq a} y^2 \, d\Phi(y) + D + 4a^2\rho. \tag{1.48}$$

Therefore, by (1.46)—(1.48)

$$(1 + x^2)|F(x) - \Phi(x)| \leq \int_{y:|y|\geq a} y^2 \, d\Phi(y) + D + 5a^2\rho$$

for any $a \geq 1$ and all $x \in \mathbb{R}$. Take $a = \sqrt{-2\log\rho} \geq 1$. Then the desired estimate holds with $R = 2(5 + K_2\sqrt{2/\pi})$ where

$$K_2 = \sup_{a\geq 1} a^{-1} e^{a^2/2} \int_a^{+\infty} y^2 e^{-y^2/2} \, dy \leq 2. \quad \square$$

Let

$$Y_j = S(Q_j)/\sqrt{|Q_j|}, \quad j = 1, \dots, n$$

and $\mu > 0$ be some number to be specified later. Introduce the functions

$$h_{n,\mu}(x) = \begin{cases} |x|^s \wedge n^{-\mu s}, & s \in (0, 1], \; x \in \mathbb{R}, \\ |x|^s \vee n^{\mu s}, & s \in (1, 2], \; x \in \mathbb{R}. \end{cases}$$

To simplify the notation we write $h(x)$ instead of $h_{n,\mu}(x)$. Then

$$Lip(h) = sn^{\mu|s-1|}. \tag{1.49}$$

In the sequel we use positive $\delta_1, \delta_2, \ldots$ independent of n.

Lemma 1.9. *For $Z \sim N(0,1)$ and any $\mu > 0$ there exists some $\eta > 0$ such that*

$$\left| n^{-1} \sum_{j=1}^{n} \mathsf{E}h(Y_j) - \sigma^s \mathsf{E}|Z|^s \right| \leq Cn^{-\eta}.$$

Proof. Let at first $s \in (1,2]$. One has

$$\left| n^{-1} \sum_{j=1}^{n} \mathsf{E}h(Y_j) - \sigma^s \mathsf{E}|Z|^s \right| \leq \max_{j=1,\ldots,n} |\mathsf{E}h(Y_j) - \sigma^s \mathsf{E}|Z|^s|$$

$$\leq \max_{j=1,\ldots,n} |\mathsf{E}h(Y_j) - \mathsf{E}h(\sigma Z)| + \sigma^s \mathsf{E}|Z|^s \mathbb{I}\{|\sigma Z| \geq n^\mu\} \tag{1.50}$$

It is obvious that

$$\mathsf{E}|Z|^s \mathbb{I}\{|Z| \geq \sigma^{-1}n^\mu\} \leq \frac{\mathsf{E}|Z|^3}{(\sigma^{-1}n^\mu)^{3-s}}.$$

To estimate the first summand in (1.50) let us recall that by Lemma 5.2.2 one has

$$D_n := \left| 1 - (\sigma^2|Q_j|)^{-1}\mathsf{Var}S(Q_j) \right| \leq C|Q_j|^{-\delta_1} \tag{1.51}$$

for any Q_j, $j \in \{1,\ldots,n\}$. Introduce a function $g(y) = h(\sigma y)$, $y \in \mathbb{R}$. Integrating by parts we can write

$$I_1 := |\mathsf{E}h(Y_j) - \mathsf{E}h(\sigma Z)| = |\mathsf{E}g(Y_j/\sigma) - \mathsf{E}g(Z)| \leq \int_{\mathbb{R}} |g'(x)||F_j^{(n)}(x) - \Phi(x)|dx$$

where $F_j^{(n)}$ and Φ are the distribution functions of Y_j/σ and Z respectively. By Theorem 3.1.21

$$\rho_n := \sup_{x \in \mathbb{R}} |F_j^{(n)}(x) - \Phi(x)| \leq Cn^{-\delta_2} \tag{1.52}$$

for some $\delta_2 > 0$. Thus applying (1.51), (1.52) and Lemma 1.8 we have, for all n large enough,

$$I_1 \leq C \int_{\mathbb{R}} |g'(x)| \frac{|\rho_n \log \rho_n| + D_n}{1 + x^2} dx \leq C \left(\rho_n|\log \rho_n| + D_n\right) \int_{\mathbb{R}} \frac{\sigma h'(\sigma x)}{1 + x^2} dx$$

$$\leq C \left(\rho_n|\log \rho_n| + D_n\right) \int_{-n^\mu/\sigma}^{n^\mu/\sigma} \frac{|x|^{s-1}}{1 + x^2} dx$$

where, as usual, $0 \cdot \log 0 := 0$. Therefore

$$I_1 = \begin{cases} O(\rho_n|\log \rho_n| + D_n), & \text{if } s \in [1,2), \\ O(\rho_n|\log \rho_n|\log n + D_n \log n), & \text{if } s = 2 \end{cases}$$

(as $n \to \infty$). Hence $I_1 = O(n^{-\delta_3})$ for small enough $\delta_3 > 0$. If $s < 1$ the proof remains the same except that instead of the second summand in (1.50) we have $\sigma^s E|Z|^s \mathbb{I}\{|\sigma Z| \le n^{-\mu}\}$ which is estimated trivially. Note also that in this case $\int_{\mathbb{R}} |x|^{s-1}/(1+x^2)dx < \infty$.

Thus, in both cases, the left-hand side of (1.50) is $O(n^{-\delta_4})$. $\square$

In order to prove the Theorem, take $\gamma > 0$ to be specified later and write

$$P(|\Delta_2(n)| \ge n^{-\gamma}) \le P\left(\left|B_2(s, r_n, U_n) - n^{-1}\sum_{j=1}^{n}|Y_j|^s\right| \ge \frac{1}{2}n^{-\gamma}\right)$$

$$+ P\left(\left|n^{-1}\sum_{j=1}^{n}|Y_j|^s - \sigma^s E|Z|^s\right| \ge \frac{1}{2}n^{-\gamma}\right) =: J_1 + J_2.$$

Case 1: $s \in (0, 1]$. Then, by (1.25) and the Markov inequality,

$$J_1 \le P\left(\frac{|S_n|^s}{n^s}(2r_n + 1)^{s/2} \ge \frac{1}{2}n^{-\gamma}\right) \le C\frac{(ES_n^2/n)^{s/2}}{n^{(s/2)(1-\beta)-\gamma}} = O(n^{-\delta_5}),$$

provided that $\gamma < s(1-\beta)/2$. Write $T_j = h(Y_j) - Eh(Y_j)$, $j = 1, \ldots, n$. Then

$$J_2 \le P\left(\left|\frac{1}{n}\sum_{j=1}^{n}T_j\right| + \frac{1}{n}\sum_{j=1}^{n}|E|Y_j|^s - Eh(Y_j)| + \left|\frac{1}{n}\sum_{j=1}^{n}Eh(Y_j) - \sigma^s E|Z|^s\right| \ge \frac{1}{2}n^{-\gamma}\right)$$

$$\le P\left(\left|\frac{1}{n}\sum_{j=1}^{n}T_j\right| \ge \frac{1}{6}n^{-\gamma}\right) + P\left(\left|\frac{1}{n}\sum_{j=1}^{n}Eh(Y_j) - \sigma^s E|Z|^s\right| \ge \frac{1}{6}n^{-\gamma}\right) \qquad (1.53)$$

if $\gamma < s\mu$ and n is large enough. We took into account that $P(n^{-s\mu} \ge \frac{1}{6}n^{-\gamma}) = 0$ for such γ and n. The last term decreases as a power of n by Lemma 1.9 if γ is small enough. To estimate the first term note that, by the Markov inequality and Theorem 1.5.3,

$$P\left(\left|n^{-1}\sum_{j=1}^{n}T_j\right| \ge \frac{1}{6}n^{-\gamma}\right) \le n^{2\gamma-2}\text{Var}\sum_{j=1}^{n}T_j = n^{2\gamma-2}\sum_{j,q=1}^{n}cov(T_j, T_q)$$

$$\le n^{2\gamma-2}s^2 n^{2(1-s)\mu}(2r_n + 1)n = O(n^{2\gamma-1+2(1-s)\mu+\beta}) = O(n^{-\delta_6})$$

if $2\gamma + 2(1-s)\mu + \beta < 1$.

Case 2: $s \in (1, 2]$. Set $R_n = \sqrt{2r_n + 1}$. Then, again by (1.25),

$$J_1 \le P\left(sn^{-2}|S_n|R_n\sum_{j=1}^{n}|Y_j|^{s-1} \ge \frac{1}{4}n^{-\gamma}\right) + P\left(sn^{1-s}|S_n|^s R_n^s \ge \frac{1}{4}n^{-\gamma}\right)$$

$$\le P\left(n^{-1}\sum_{j=1}^{n}|Y_j|^{s-1} - \sigma^{s-1}E|Z|^{s-1} \ge \frac{1}{2}\sigma^{s-1}E|Z|^{s-1}\right)$$

$$+ P(R_n|S_n| \ge x_1(n)) + P(R_n|S_n| \ge x_2(n)) \qquad (1.54)$$

where

$$x_1(n) = \frac{n}{3n^\gamma s\sigma^{s-1}\mathsf{E}|Z|^{s-1}}, \quad x_2 = \frac{1}{(4sn^{\gamma-1})^{1/s}}.$$

The first summand in the right-hand side of (1.54) decreases in a power way by Lemma 1.9. The second and third ones are easily estimated via the Markov inequality. That is,

$$\mathsf{P}(R_n|S_n| \geq x_1(n)) = O(n^{2\gamma+\beta-1}),$$

$$\mathsf{P}(R_n|S_n| \geq x_2(n)) = O(n^{1+\beta-2s(1-\gamma)}), \quad n \to \infty.$$

For handling J_2 note that, again by Lemma 1.9, we only need to prove that

$$n \max_{j=1,\ldots,n} \mathsf{P}(|Y_j| > n^\mu) + \mathsf{P}\left(\left|n^{-1}\sum_{j=1}^{n} T_j\right| \geq \frac{1}{4}n^{-\gamma}\right) = O(n^{-\delta_7}) \tag{1.55}$$

where T_j is the same as in (1.53). On account of Theorem 2.1.4 there exists some $p > 2$ such that $\mathsf{E}|Y_j|^p \leq C$, with C independent of n and j. Thus we have

$$n \max_{j=1,\ldots,n} \mathsf{P}(|Y_j| > n^\mu) \leq Cn^{1-p\mu}.$$

The second term in (1.55) is estimated via the Theorem 1.5.3:

$$\mathsf{P}\left(\left|n^{-1}\sum_{j=1}^{n} T_j\right| \geq \frac{1}{4}n^{-\gamma}\right) \leq Cn^{2\gamma-2} \sum_{1\leq j,q\leq n} cov(T_j, T_q) \leq Cn^{2\gamma-2+\beta+2\mu(s-1)+1}.$$

Now we gather all the necessary inequalities containing $\beta \in (0,1)$, $\gamma > 0$, $\mu > 0$. If $s \in (0,1)$ then one requires that

$$\gamma < \min\left\{s\mu, s(1-\beta)/2\right\}, \quad 2\gamma + 2(1-s)\mu + \beta < 1,$$

but such numbers can be easily chosen (by selecting at first β and μ, and then γ). For $s \in [1,2]$ these numbers must satisfy the inequalities

$$\gamma < (1-\beta)/2, \quad p\mu > 1, \quad 1+\beta < 2s(1-\gamma), \quad 2\gamma + 2(s-1)\mu + \beta < 1.$$

Since $p > 2$, we can pick $\mu \in (1/p, 1/(2s-2))$. Then it remains to pick $\beta \in (0, 1-2(s-1)\mu)$ so that $\beta < 2s-1$ and, finally, take positive γ small enough. $\square$

Remark 1.10. In [69] a statement analogous to Theorems 1.1 and (1.4) was established for (BL, θ)-dependent random fields. In [74] the rates of convergence for vector-valued statistics involving matrix self-normalization were investigated, see also the forthcoming paper [242].

2　Kernel density estimation

Let $X = \{X_j, j \in \mathbb{Z}^d\}$ be a strictly stationary (real-valued) random field. Suppose that X_0 has some unknown density $f(x)$. A classical way to estimate $f(x)$ by inspecting the outcomes of X with indices in a finite set U is to construct the *kernel density estimate*, or the Rosenblatt–Parzen estimate [316, 348]. A *kernel* is a measurable function K such that

$$K(x) \geq 0, \quad x \in \mathbb{R}, \quad \text{and} \quad \int_{\mathbb{R}} K(x)dx = 1.$$

One may construct the estimate

$$\widehat{f}_{U_n}(x) = \widehat{f}_{U_n, h_n}(x) = \frac{1}{|U_n|h_n} \sum_{j \in U_n} K\left(\frac{x - X_j}{h_n}\right), \quad x \in \mathbb{R}, \quad n \in \mathbb{N}, \tag{2.1}$$

where $(h_n)_{n \in \mathbb{N}}$ is a positive sequence such that $\lim_{n \to \infty} h_n = 0$, and U_n is a finite subset of $\mathbb{Z}^d$. Note that

$$\widehat{f}_{U_n, h_n}(x) = \widehat{f}_{U_n, h_n}(x, K, \{X_j, j \in U_n\}).$$

Clearly, each $\widehat{f}_{U_n}(x)$ is a probability density. One can be interested in the limit behavior of $\widehat{f}_{U_n, h_n}$ (consistency, asymptotic normality, etc.) as U_n grows to infinity. These properties can be studied under different conditions imposed on X, f, K, $(h_n)_{n \in \mathbb{N}}$ and $(U_n)_{n \in \mathbb{N}}$. Note that while the properties of f belong to intrinsic ones of a field, the parameters K, h_n and the sets U_n can be chosen in the most convenient way possible. There is a vast literature devoted to the kernel estimates, see, e.g., [130] and references therein.

All the integrals throughout the Section are over $\mathbb{R}$, unless other is stated.

1°. Asymptotic unbiasedness and normality of the estimates. The behavior of the expectation of $\widehat{f}_{U_n}$ is determined not by the type of dependence of a field, but only by the regularity of f and K. In fact, one has the following simple

Lemma 2.1. *Suppose that $U_n \subset \mathbb{Z}^d$ $(n \in \mathbb{N})$ are finite sets and $h_n \searrow 0$ as $n \to \infty$. Assume that f is bounded and continuous on $\mathbb{R}$. Then, for any $x \in \mathbb{R}$,*

$$\mathsf{E}\widehat{f}_{U_n, h_n}(x) \to f(x) \text{ as } n \to \infty.$$

Proof. By the change of variable one has

$$\mathsf{E}\widehat{f}_{U_n, h_n}(x) - f(x) \int K(z)\left(f(x - zh_n) - f(x)\right)dz \to 0$$

as $n \to \infty$, by the dominated convergence theorem. $\square$

Below we will often use the following group of assumptions:

$$Lip(f) < \infty, \tag{2.2}$$

for any $j \in \mathbb{Z}^d$; the vector (X_0, X_j) has a density $f_{0,j}$ such that

$$M := \sup_{j \in \mathbb{Z}^d} \|f_{0,j}\|_\infty < \infty. \tag{2.3}$$

Clearly, (2.2) implies that f is bounded as well. Indeed, if $x \in \mathbb{R}$ is such that $f(x) > M_0$ for some $M_0 > 0$, then

$$1 \geq \int_{x-1}^{x+1} f(y)dy = \int_{x-1}^{x+1} (f(y) - f(x))dy + 2f(x)$$

$$\geq 2M_0 - \int_{x-1}^{x+1} Lip(f)|x - y|dy = 2M_0 - Lip(f),$$

hence $M_0 \leq (1 + Lip(f))/2$.

Note that if a random field X is associated (**PA**, **NA** or (BL,θ)-dependent) and K is a Lipschitz function, then $Y = \{Y_j = h^{-1}K(h^{-1}(x - X_j)), j \in \mathbb{Z}^d\}$ is a (BL,θ)-dependent random field by Lemma 1.5.16. Indeed, the function

$$y \mapsto \frac{1}{h}K\left(\frac{x - y}{h}\right)$$

has the Lipschitz constant $h^{-2}Lip(K)$. Note that for a field Y the coefficients $\theta_r^{(Y)}$ depend heavily on h, so there is no opportunity to apply directly previous limit theorems to obtain the convergence of $\widehat{f}_{U_n,h_n}(x)$ to $f(x)$. Thus it is necessary to modify the mentioned results.

Theorem 2.2. *Let $X = \{X_j, j \in \mathbb{Z}^d\}$ be a (BL,θ)-dependent, strictly stationary random field such that $\theta_r = O(r^{-\lambda})$ for some $\lambda > 3d$. Assume that (2.2) and (2.3) hold. Suppose that $K : \mathbb{R} \to \mathbb{R}_+$ is a Lipschitz kernel and $(U_n)_{n \in \mathbb{N}}$ is a sequence of blocks in $\mathbb{Z}^d$ (belonging to $\mathcal{U}$) regularly growing to infinity[5]. Then there exists a sequence $(h_n)_{n \in \mathbb{N}}$ such that $h_n \searrow 0$ as $n \to \infty$ and, for any $x \in \mathbb{R}$ with $f(x) > 0$, one has*

$$L_n(x) := \sqrt{h_n|U_n|}\left(\widehat{f}_{U_n,h_n}(x) - \mathsf{E}\widehat{f}_{U_n,h_n}(x)\right) \to N\left(0, \sigma^2(x)\right) \qquad (2.4)$$

in law as $n \to \infty$, here

$$\sigma^2(x) := f(x) \int K^2(z)dz.$$

Proof. For any fixed $h > 0$ and $x \in \mathbb{R}$ we have, by Theorem 3.1.12,

$$\sqrt{h|U_n|}\left(\widehat{f}_{U_n,h}(x) - \mathsf{E}\widehat{f}_{U_n,h}(x)\right)$$

$$= \frac{1}{\sqrt{|U_n|}}\left(\sum_{j \in U_n} \frac{1}{\sqrt{h}}K\left(\frac{x - X_j}{h}\right) - \frac{1}{\sqrt{h}}\mathsf{E}K\left(\frac{x - X_0}{h}\right)\right) \to N(0, \sigma^2(x, h)) \qquad (2.5)$$

in law as $n \to \infty$, $\widehat{f}_{U_n,h}$ being defined according to (2.1) with h instead of h_n,

$$\sigma^2(x, h) = \frac{1}{h}\sum_{j \in \mathbb{Z}^d} cov\left(K\left(\frac{x - X_0}{h}\right), K\left(\frac{x - X_j}{h}\right)\right),$$

and the series converges absolutely for any fixed $h > 0$.

Lemma 2.3. *Suppose that Theorem's conditions are satisfied. Then*

$$\sigma^2(x, h) \to \sigma^2(x) \quad as \quad h \to 0+.$$

---

[5]See Section 3.1, Definition 3.1.4.

Proof. First of all, to estimate the summand with $j = 0$, we can write

$$\frac{1}{h}\mathsf{E}K^2\left(\frac{x - X_0}{h}\right) = \frac{1}{h}\int K^2\left(\frac{x - y}{h}\right)f(y)dy$$

$$= \int K^2(z)f(x - zh)dz \to \int K^2(z)f(x)dz = \sigma^2(x), \qquad (2.6)$$

by the dominated convergence theorem. For any $j \in \mathbb{Z}^d$, $j \neq 0$, one has

$$\frac{1}{h}cov\left(K\left(\frac{x - X_0}{h}\right), K\left(\frac{x - X_j}{h}\right)\right)$$

$$= \frac{1}{h}\iint_{\mathbb{R}^2} K\left(\frac{x - y_1}{h}\right)K\left(\frac{x - y_2}{h}\right)f_{0,j}(y_1, y_2)dy_1 dy_2 - \frac{1}{h}\left(\int_{\mathbb{R}} K\left(\frac{x - y}{h}\right)f(y)dy\right)^2$$

$$= h\iint_{\mathbb{R}^2} K(z_1) K(z_2) f_{0,j}(x - hz_1, x - hz_2)dz_1 dz_2 - h\left(\int_{\mathbb{R}} K(z) f(x - hz)dz\right)^2.$$

Having employed (2.2) and (2.3) we see that

$$\left|\frac{1}{h}cov\left(K\left(\frac{x - X_0}{h}\right), K\left(\frac{x - X_j}{h}\right)\right)\right| \leq Ch, \qquad (2.7)$$

with C depending only on $Lip(f)$ and M.

Hence the left-hand side of (2.7) is $O(h)$ as $h \to 0+$. Clearly, by Theorem 1.5.3 for any finite $U \not\ni \{0\}$, $U \subset \mathbb{Z}^d$, one has

$$\left|cov\left(K\left(\frac{x - X_0}{h}\right), \sum_{j \in U} K\left(\frac{x - X_j}{h}\right)\right)\right| \leq \frac{Lip^2(K)}{h^2}\theta_l \qquad (2.8)$$

where $l = dist(\{0\}, U)$.

Therefore, for any $r \in \mathbb{N}$, treating separately j with $|j| \leq r$ and $|j| > r$ and using (2.7) and (2.8) we obtain

$$\frac{1}{h}\left|\sum_{j \in \mathbb{Z}^d, j \neq 0} cov\left(K\left(\frac{x - X_0}{h}\right), K\left(\frac{x - X_j}{h}\right)\right)\right| \leq C(2r + 1)^d h + \frac{Lip^2(K)}{h^3}\theta_r \quad (2.9)$$

where C depends on K and M only. By the conditions of the Lemma

$$\theta_r = O(r^{-\lambda}) \text{ as } r \to \infty, \text{ with} \lambda > 3d.$$

Take $r = r(h) = [h^{-\gamma}]$ where $\gamma > 0$. Selecting $\gamma = 4/(d + \lambda)$ we note that the left-hand side of inequality (2.9) admits an estimate $O(h^{(\lambda - 3d)/(\lambda + d)})$. This fact and (2.6) yield the Lemma. $\square$

Recall that if $(\xi_n)_{n \in \mathbb{N}}$ is a sequence of random variables such that $\xi_n \to \xi$ in law as $n \to \infty$ for some random variable ξ, then

$$\sup_{t \in \mathbb{R}} |\mathsf{P}(\xi_n \leq t) - \mathsf{P}(\xi \leq t)| \to 0 \text{ as } n \to \infty,$$

provided that the distribution function of ξ is continuous. Let $F_{n,h,x}(\cdot)$ be the distribution function of the left-hand side of (2.5) (for $n \in \mathbb{N}$, $h > 0$, $x \in \mathbb{R}$). Let

$$a_{n,h,x} = \sup_{t \in \mathbb{R}} \left| F_{n,h,x}(t) - \Phi_{\sigma^2(x,h)}(t) \right|$$

where Φ_c is the distribution function of $N(0,c)$, $c \geq 0$. Then $a_{n,h,x} \to 0$ for any fixed $h > 0$ and $x \in \mathbb{R}$, by (2.5). Therefore, there exists a sequence $(h_n)_{n \in \mathbb{N}}$ such that $h_n \searrow 0$ and $a_{n,h_n,x} \to 0$, $n \to \infty$. By Lemma 2.3 we have

$$\Phi_{\sigma^2(x,h)}(t) \to \Phi_{\sigma^2(x)}(t), \quad h \to 0+,$$

for any $t \in \mathbb{R}$. Consequently, $F_{n,h_n,x} \to \Phi_{\sigma^2(x)}$ as $n \to \infty$ for any $x \in \mathbb{R}$ which is the desired conclusion. $\square$

2°. Convergence rate estimate. Imposing more restrictive conditions on blocks U_n and the decrease of a sequence $\theta = (\theta)_{n \in \mathbb{N}}$ we can prove the same result for a given bandwidth h_n. It is also possible to establish a bound for the rate of convergence to the normal law.

Theorem 2.4. *Let the conditions of Theorem 2.2 be satisfied, $U_n = (0,n] \cap \mathbb{Z}^d$ where $n \in \mathbb{N}$. Assume also that*

$$\int |z| K(z) dz < \infty. \tag{2.10}$$

Then, for $L_n(x)$ defined in (2.4), at any point $x \in \mathbb{R}$ such that $f(x) > 0$ one has

$$\sup_{t \in \mathbb{R}} \left| \mathsf{P}\left(L_n(x) \leq t \right) - \Phi_{\sigma^2(x)}(t) \right| \leq C(h_n^{(\lambda - 3d)/(\lambda + d)} + h_n^{-3} n^{-\mu})$$

where $\mu = \lambda d / 4(\lambda + d(2 + 3\lambda))$ and C does not depend on n and x.

Proof. We apply Theorem 3.1.21 and Remark 3.1.23. The field

$$Y(x,h) = \{Y_j(x,h), j \in \mathbb{Z}^d\} = \left\{ \frac{1}{\sqrt{h}} K\left(\frac{x - X_j}{h} \right), j \in \mathbb{Z}^d \right\}$$

is (BL, θ)-dependent with dependence coefficient $\theta_r(Y(x,h)) \leq h^{-3} Lip^2(K)\theta_r(X)$, $r \in \mathbb{N}$. Moreover, it is worthwhile to mention a simple

Lemma 2.5. *For any $s > 0$ and $n \in \mathbb{N}$ one has*

$$\mathsf{E}|Y(x,h)|^s \leq h^{-s/2+1} \|f\|_\infty \int |K(z)|^s dz$$

Proof. Note that since K is integrable and Lipschitz, it is bounded (see the argument after (2.3)). Thus by the change of variable

$$\mathsf{E}|Y(x,h)|^s = h^{-s/2} \int \left| K\left(\frac{x - y}{h} \right) \right|^s f(y) dy$$

$$= h^{-s/2+1} \int |K(z)|^s f(x - hz) dz \leq h^{-s/2+1} \|f\|_\infty \int |K(z)|^s dz. \quad \square$$

Applying Lemma 2.3 one observes that, for all n large enough,

$$\mathsf{Var} L_n(x) \geq \frac{1}{2} f(x) \int K^2(z) dz > 0.$$

This property and Lemma 2.5 enable us to employ Remark 3.1.23 (with $\delta = 1$). That remark yields a bound

$$\sup_{t\in\mathbb{R}} \left| \mathsf{P}\left(\frac{\sum_{j\in U_n}(Y_j(x,h) - \mathsf{E}Y_j(x,h))}{\sqrt{\mathsf{Var}\sum_{j\in U_n} Y_j(x,h)}} \leq t \right) - \mathsf{P}(Z \leq t) \right|$$

$$= \sup_{t\in\mathbb{R}} |\mathsf{P}\left(L_n(x) \leq tv_n \right) - \mathsf{P}(Z \leq t)| \leq C(1 + h_n^{-3})n^{-\mu} \tag{2.11}$$

where $\mu = \lambda d/(\lambda + d(2 + 3\lambda))$ and $v_n = (\mathsf{Var}\sum_{j\in U_n} Y_j(x,h)/|U_n|)^{1/2}$.
  Furthermore, by (2.11)

$$\sup_{t\in\mathbb{R}} \left| \mathsf{P}\left(L_n(x) \leq t \right) - \Phi_{\sigma^2(x,h)}(t) \right| = \sup_{t\in\mathbb{R}} \left| \mathsf{P}\left(L_n(x) \leq tv_n \right) - \Phi_{\sigma^2(x,h)}(tv_n) \right|$$

$$\leq C(1 + h_n^{-3})n^{-\mu} + \sup_{t\in\mathbb{R}} \left| \mathsf{P}(Z \leq t) - \Phi_{\sigma^2(x,h)} tv_n \right|, \tag{2.12}$$

so it only remains to estimate the last term in (2.12).

Lemma 2.6. *(see, e.g., [326]) If* $Z \sim N(0,1)$, *then for any* $a > 0$ *one has*

$$\sup_{t\in\mathbb{R}} |\mathsf{P}(Z \leq at) - \mathsf{P}(Z \leq t)| \leq (2\pi e)^{-1/2}(a \vee a^{-1} - 1).$$

  Proof. Clearly it suffices to consider the case $a \geq 1$ and take the supremum only over $t \geq 0$. Then

$$\mathsf{P}(Z \leq at) - \mathsf{P}(Z \leq t) = (2\pi)^{-1/2} \int_t^{at} e^{-x^2/2} dx \leq (2\pi)^{-1/2}(a - 1)te^{-t^2/2}.$$

Now note that $\sup_{t\geq 0} te^{-t^2/2} = e^{-1/2}$. $\square$
  By virtue of Lemma 2.6 we may write that

$$\left| \mathsf{P}(Z \leq t) - \Phi_{\sigma^2(x,h_n)}(tv_n) \right|$$

$$= \left| \mathsf{P}(Z \leq t) - \mathsf{P}\left(Z \leq tR_n \right) \right| \leq (2\pi e)^{-1/2}|1 - R_n \vee R_n^{-1}|$$

where

$$R_n = R_n(x, h_n) = \frac{1}{\sigma} \left(\frac{1}{|U_n|} \mathsf{Var} \sum_{j\in U_n} Y_j(x, h_n) \right)^{1/2}.$$

So we are reduced to proving that $R_n - 1 = O(n^{-\nu})$ as $n \to \infty$, or equivalently that

$$\sigma^2 - \frac{1}{|U_n|} \mathsf{Var} \sum_{j\in U_n} Y_j(x, h_n) = O(n^{-\nu}), \; n \to \infty.$$

The optimization of (2.9) in r shows that

$$\left| \sum_{j \in \mathbb{Z}^d,\, j \neq 0} cov\Big(Y_0(x, h_n), Y_j(x, h_n)\Big) \right| = O(h_n^{(\lambda - 3d)/(d + \lambda)}) \qquad (2.13)$$

as $n \to \infty$. An argument similar to (2.6) yields the estimate

$$\left| \mathsf{E}(Y_0(x, h_n))^2 - \sigma^2(x) \right| \leq Lip(f) h_n \int K(z)|z|dz = O(h_n), \quad n \to \infty. \qquad (2.14)$$

Now the Theorem follows from (2.11), (2.13) and (2.14). $\square$

3°. Almost sure convergence of kernel density estimates. Similar ideas allow us to deduce analogs of other limit theorems for kernel density estimates. We give the result on almost sure convergence. To this end we start with

Theorem 2.7. *Suppose that the conditions of Theorem 2.4 are satisfied. Then*

$$\limsup_{n \in \mathbb{N}} \sup_{x \in \mathbb{R}} (|U_n|h_n)^2 \mathsf{E}(\widehat{f}_{U_n, h_n}(x) - \mathsf{E}\widehat{f}_{U_n, h_n}(x))^4 < \infty$$

if the sequence $h_n \to 0+$ is taken in such a way that $|U_n|h_n^{1+\delta+3d/\lambda} \geq 1$ for some $\delta > 0$ and all $n \in \mathbb{N}$.

Proof. Let $x \in \mathbb{R}$ and $n \in \mathbb{N}$. Introduce

$$Z_j = Z_j(x, h_n) = K\left(\frac{x - X_j}{h_n}\right) - \mathsf{E}K\left(\frac{x - X_j}{h_n}\right), \quad j \in U_n.$$

Take $p > 4$ such that $\lambda(p - 4)/d(p - 2) > 3$. Theorem 2.1.26 and Remark 2.1.27 ensure the existence of some $C_1 > 0$, independent of n and x, such that

$$\mathsf{E}\left(\sum_{j \in U_n} Z_j\right)^4 \leq C_1\left(|U_n|\mathsf{E}Z_0^4 + |U_n|^2(\mathsf{E}Z_0^2 + \vartheta_1)^2 + |U_n|\Big(\mathsf{E}|Z_0|^p\Big)^{2/(p-2)}\right). \qquad (2.15)$$

Integration by parts shows that for $Z_0 = Z_0(x, h_n)$

$$\mathsf{E}Z_0^4 \leq 16 \int K^4\left(\frac{x - y}{h_n}\right) f(y)dy$$

$$= 16 h_n \int K^4(z) f(x - z h_n)dz \leq 16\|f\|_\infty \|K\|_\infty^3 h_n, \qquad (2.16)$$

$$\mathsf{E}Z_0^2 \leq 4\|f\|_\infty \|K\|_\infty h_n, \qquad (2.17)$$

$$(\mathsf{E}|Z_0|^p)^{2/(p-2)} \leq \Big(2^p \|f\|_\infty \|K\|_\infty^{p-1} h_n\Big)^{2/(p-2)}. \qquad (2.18)$$

To estimate ϑ_1 (see Remark 2.1.27 for its definition) we use exactly the same argument as in (2.9) establishing that $\vartheta_1 = O(h_n)$, $n \to \infty$. Employing this fact and (2.16)—(2.18), we infer from (2.15) that, for some $C_2 > 0$,

$$\limsup_{n \in \mathbb{N}} \sup_{x \in \mathbb{R}} (|U_n|h_n)^2 \mathsf{E}(\widehat{f}_{U_n, h_n}(x) - \mathsf{E}\widehat{f}_{U_n, h_n}(x))^4$$

$$= \limsup_{n \in \mathbb{N}} \sup_{x \in \mathbb{R}} (|U_n| h_n)^{-2} \mathsf{E} \Big(\sum_{j \in U_n} Z_j(x, h_n) \Big)^4$$

$$\leq C_2 \limsup_{n \in \mathbb{N}} \Big(\frac{1}{|U_n| h_n} + 1 + \frac{1}{|U_n| h_n^{2 - 2/(p-2)}} \Big) < \infty$$

provided that p is chosen close enough to $2(2\lambda - 3d)/(\lambda - 3d)$. $\square$

As a consequence we obtain the result by Doukhan and Louhichi.

Theorem 2.8. ([142]) *Let* $X = \{X_j, j \in \mathbb{Z}^d\}$ *be a* (BL, θ)-*dependent random field with* $\theta_r = O(r^{-\lambda})$ *for some* $\lambda > 3d$. *Assume that* K *is a Lipschitz function and conditions* (2.3) *and* (2.10) *hold. Let blocks* U_n *and numbers* h_n *be chosen in such a way that*

$$|U_n| h_n^3 \to \infty \ \text{ as } n \to \infty, \tag{2.19}$$

$$\sum_{n=1}^{\infty} \frac{1}{h_n^{5/2} |U_n|^{3/2}} < \infty. \tag{2.20}$$

Then, for any $M > 0$, *one has*

$$\sup_{x \in [-M, M]} |\widehat{f}_{U_n, h_n}(x) - f(x)| \to 0 \ \text{ a.s., } \ n \to \infty.$$

Proof. Note that the conditions on K and f imply that $\mathsf{E} \widehat{f}_{U_n, h_n}(x) \to f(x)$ as $n \to \infty$ uniformly in $x \in \mathbb{R}$. Indeed,

$$|\mathsf{E} \widehat{f}_{U_n, h_n}(x) - f(x)| = \frac{1}{h_n} \left| \int K \left(\frac{x - y}{h_n} \right) (f(y) - f(x)) dy \right|$$

$$\leq \int K(z) |f(x - z h_n) - f(x)| dz \leq h_n Lip(f) \int |z| K(z) dz \to 0$$

uniformly in x. Thus we only need to check that, for any $M > 0$,

$$\sup_{x \in [-M, M]} |\widehat{f}_{U_n, h_n}(x) - \mathsf{E} \widehat{f}_{U_n, h_n}(x)| \to 0 \ \text{ a.s., when } n \to \infty.$$

For $n \in \mathbb{N}$ let $q_n = [\sqrt{|U_n|/h_n}]$, then $h_n q_n \to \infty$ as $n \to \infty$. Divide the segment $[-M, M]$ into q_n equal segments (more precisely, one segment and $q_n - 1$ half-intervals). Let I_k be the k-th segment and x_k be its center, $k = 1, \ldots, q_n$.

Fix $\varepsilon > 0$. By Theorem 2.7 and the Chebyshev inequality there exists some $C_1 > 0$ such that

$$\sup_{x \in \mathbb{R}} \mathsf{P} \left(|\widehat{f}_{U_n, h_n}(x) - \mathsf{E} \widehat{f}_{U_n, h_n}(x)| \geq \varepsilon \right) \leq \frac{C_1}{\varepsilon^4 |U_n|^2 h_n^2} \tag{2.21}$$

for any $n \in \mathbb{N}$ (the conditions imposed on h_n by the mentioned Theorem are satisfied due to (2.19), as $1 + 3d/\lambda < 3$). We have, by (2.21),

$$\mathsf{P} \left(\sup_{k=1,\ldots,q_n} |\widehat{f}_{U_n, h_n}(x_k) - \mathsf{E} \widehat{f}_{U_n, h_n}(x_k)| \geq \varepsilon \right)$$

$$\leq q_n \sup_{k=1,\ldots,q_n} \mathsf{P}\left(|f_n(x_k) - \mathsf{E}f_n(x_k)| \geq \varepsilon\right) \leq C_1 \frac{q_n}{\varepsilon^4(|U_n|h_n)^2} \leq \frac{C_1}{\varepsilon^4 h_n^{5/2}|U_n|^{3/2}}.$$

As $\varepsilon > 0$ was arbitrary, (2.20) and the Borel–Cantelli lemma yield that $\mathsf{P}(A) = 1$ where

$$A = \left\{\omega : \sup_{k=1,\ldots,q_n} |f_n(x_k) - \mathsf{E}f_n(x_k)| \to 0 \text{ as } n \to \infty\right\}.$$

For all $\omega \in \Omega$ the function

$$x \mapsto \widehat{f}_{U_n,h_n}(x) - \mathsf{E}\widehat{f}_{U_n,h_n}(x) = \frac{1}{|U_n|h_n} \sum_{j \in U_n} \left(K\left(\frac{x - X_j}{h_n}\right) - \mathsf{E}K\left(\frac{x - X_j}{h_n}\right)\right)$$

has Lipschitz constant not exceeding $2Lip(K)h_n^{-2}$. Therefore, for any elementary outcome $\omega \in A$,

$$\sup_{x \in [-M,M]} |f_n(x) - \mathsf{E}f_n(x)| \leq \sup_{k=1,\ldots,q_n} |f_n(x_k) - \mathsf{E}f_n(x_k)|$$

$$+ \sup_{x \in [-M,M]} \inf_{k=1,\ldots,q_n} |x - x_k|(2Lip(K)h_n^{-2}). \tag{2.22}$$

Clearly $\inf_{k=1,\ldots,q_n} |x - x_k| \leq 2M/q_n$. Thus the second term on the right-hand side of (2.22) has the upper bound

$$4MLip(K)q_n^{-1}h_n^{-2} = 4MLip(K)q_n^{-1}h_n^{-2} \sim 4MLip(K)h_n^{-3/2}|U_n|^{-1/2} \to 0$$

as $n \to \infty$ by (2.19). Consequently, the right-hand side of (2.22) tends to zero as $n \to \infty$, which is the desired statement. $\square$

Diverse further generalizations of the problems considered above are possible ([132, 193–195, 264, 350, 351, 356, 357]). E.g., one can estimate the density of random vectors or construct estimates at several points simultaneously ([76], [79], [284]). There are also results on regression estimate ([59, 217, 285, 355]), empirical likelihood [426]. We refer to a review by Roussas [354].

4°. **The regression function estimation.** As an example of a typical result we present a simplified proof of a theorem by Roussas.

Theorem 2.9. *Let $Z = (Z_n)_{n \in \mathbb{Z}}$ be a strictly stationary, square-integrable, associated random sequence, let $\varphi : \mathbb{R} \to \mathbb{R}$ be a Lipschitz function, and let $d \in \mathbb{N}$ be a fixed number. Set $Y_i = Z_{i+d}$ and $X_i = (Z_i, \ldots, Z_{i+d-1})$, $i \in \mathbb{Z}$. Suppose that the function $m : \mathbb{R}^d \to \mathbb{R}$, defined by the relation*

$$m(x) = \mathsf{E}(\varphi(Y_0)|X_0 = x), \quad x \in \mathbb{R}^d,$$

is bounded and Lipschitz one. Assume also that the random vector (X_0, Y_0) has a bounded continuous density p, and there exist densities of random vectors (Y_0, X_0, Y_j, X_j) $(j > d)$, these densities being bounded uniformly in j. Suppose that the Cox–Grimmett coefficients $(u_r)_{r \in \mathbb{N}}$ (see (1.5.4)) of Z obey the bound

$$u_r = O(r^{-\lambda}) \text{ as } r \to \infty, \text{ where } \lambda > (d + 2)/d. \tag{2.23}$$

Finally, take a kernel $K : \mathbb{R}^d \to \mathbb{R}$ which is a Lipschitz function such that $\int_{\mathbb{R}^d} |z| K(z) dz < \infty$. Then, for any $x \in \mathbb{R}^d$ with $f(x) > 0$, f being the density of X_0, one has

$$\sqrt{nh_n^d}(m_n(x) - m(x)) \to N(0, \sigma^2(x)) \quad \text{in law,} \quad n \to \infty, \tag{2.24}$$

where $(h_n)_{n \in \mathbb{N}}$ is a sequence verifying $h_n \searrow 0$, $nh_n^d \to \infty$ and $nh_n^{d+2} \to 0$, $n \to \infty$,

$$m_n(x) = \frac{1}{nh^d \widehat{f}_{n,h_n}(x)} \sum_{j=1}^{n} \varphi(Y_j) K\left(\frac{x - X_j}{h_n}\right), \quad \widehat{f}_{n,h_n}(x) = \frac{1}{nh^d} \sum_{j=1}^{n} K\left(\frac{x - X_j}{h_n}\right),$$

$$\gamma^2(x) = \frac{\int_{\mathbb{R}^d} K^2(t) dt}{f(x)} \mathsf{E}\left((\varphi(Y_0) - m(x))^2 | X_0 = x\right).$$

Proof. Fix $x \in \mathbb{R}^d$ with $f(x) > 0$. By Theorem 1.1.8 d), the vector-valued random sequence $(X_n, Y_n)_{n \in \mathbb{N}}$ is **PA** with Cox–Grimmett coefficients $(\theta_r)_{r \in \mathbb{N}}$ admitting the same estimate as in (2.23).

A slight modification of Lemma 2.1 (adaptation to multivariate case) shows that $\mathsf{E}\widehat{f}_{n,h_n}(x) \to f(x)$ as $n \to \infty$. Take some $\eta \in (d/(2 + d)\lambda, 1)$ and let $(r_n)_{n \in \mathbb{N}}$ be a sequence of positive integers such that $r_n \sim n^\eta$ as $n \to \infty$. Then using the same argument as in (2.7)—(2.9) and the assumed existence of density, one has

$$\mathsf{Var}\widehat{f}_{n,h_n}(x) = \frac{1}{n^2 h_n^{2d}} \mathsf{Var} \sum_{j=1}^{n} K\left(\frac{x - X_j}{h_n}\right)$$

$$= O(n^{-1}h_n^{-d} + n^{\eta-1} + n^{-1}\theta_{r_n} h_n^{-2d-2}) \to 0, \, n \to \infty$$

(indeed, $r_n^{-\lambda} h_n^{-d-2} \to 0, \, n \to \infty$), provided that $nh_n^d \to \infty$ as $n \to \infty$. Consequently, the last stipulation yields

$$\widehat{f}_{n,h_n}(x) \to f(x) \quad \text{in probability as} \quad n \to \infty. \tag{2.25}$$

Set

$$V_j(x, h_n) = K\left(\frac{x - X_j}{h_n}\right)$$

and write

$$\sqrt{nh_n^d}(m_n(x) - m(x)) = \frac{1}{f_n(x)\sqrt{nh_n^d}} \sum_{j=1}^{n} (\varphi(Y_j) - m(X_j))V_j(x, h_n)$$

$$+ \frac{1}{f_n(x)\sqrt{nh_n^d}} \sum_{j=1}^{n} (m(X_j) - m(x))V_j(x, h_n).$$

It is easily seen that the following inequality holds:

$$\sqrt{nh_n^{-d}}\mathsf{E}\,|m(X_0) - m(x)|\,V_0(x, h_n) \le Lip(m)\sqrt{nh_n^{-d}}\mathsf{E}\|X_0 - x\|_1 V_0(x, h_n)$$

$$= \int_{\mathbb{R}^d} Lip(m) \sqrt{nh_n^{-d}} \|x - t\|_1 K\left(\frac{x-t}{h_n}\right) f(t)dt$$

$$= \sqrt{nh_n^d} \int_{\mathbb{R}^d} Lip(m)\|h_n z\|_1 K(z) f(x - h_n z)dz \to 0, \qquad (2.26)$$

provided that $nh_n^{d+2} \to 0, \ n \to \infty$.

Thus by Slutsky Lemma 3.1.27, (2.25) and (2.26), instead of (2.24) one can prove that

$$\frac{1}{\sqrt{nh_n^d}} \sum_{j=1}^{n} \left(\varphi(Y_j) - m(X_j)\right) V_j(x, h_n) \to N(0, \gamma^2(x)f^2(x)) \quad \text{in law} \qquad (2.27)$$

as $n \to \infty$. The random variables standing inside the sum in (2.27) are centered, since by the standard properties of conditional expectations,

$$\mathsf{E}\left(\varphi(Y_j) - m(X_j)\right) V_j(x, h_n) = \mathsf{E}\left(\varphi(Y_j) - \mathsf{E}(\varphi(Y_j)|X_j)\right) V_j(x, h_n)$$

$$\mathsf{E}\mathsf{E}\left(\left(\varphi(Y_j) - \mathsf{E}(\varphi(Y_j)|X_j)\right) V_j(x, h_n) \Big| X_j\right)$$

$$= \mathsf{E}V_j(x, h_n)\mathsf{E}\left(\left(\varphi(Y_j) - \mathsf{E}(\varphi(Y_j)|X_j)\right) \Big| X_j\right) = 0.$$

The proof of (2.27) is similar to one of Theorem 2.4. Namely, in the estimate analogous to (2.11) at the right-hand side we have $C(1 + h_n^{-d-2})n^{-\mu}$, with $\mu = \lambda/(2 + 4\lambda)$. The counterpart of (2.9) and (2.13) is the expression

$$\frac{1}{h_n^d} \left| \sum_{j \in \mathbb{Z}^d, j \neq 0} cov\left((\varphi(Y_0) - m(X_0)) V_0(x, h_n), (\varphi(Y_j) - m(X_j)) V_j(x, h_n)\right) \right|$$

$$\leq C(h_n^d r + h_n^{-d-2}\theta_r),$$

which tends to zero as $n \to \infty$, if one selects $r = r_n \sim h_n^{-\tau}$ as $n \to \infty$, with some $\tau \in (\lambda^{-1}(d+2), d)$. Finally, similarly to (2.14) one may write (note that we are not interested below in the rate of convergence)

$$h_n^{-d}\mathsf{E}\left(\varphi(Y_j) - m(X_j)\right)^2 V_j(x, h_n)^2 = \int_{\mathbb{R}^d} \int_{\mathbb{R}} (\varphi(y) - m(t))^2 K^2\left(\frac{x-t}{h_n}\right) p(t, y)dy\, dt$$

$$= \int_{\mathbb{R}^d} \int_{\mathbb{R}} (\varphi(y) - m(x - h_n z))^2 K^2(z) p(x - h_n z, y)dy\, dz$$

$$\to \int_{\mathbb{R}^d} \int_{\mathbb{R}} (\varphi(y) - m(x))^2 K^2(z) p(x, y)dy\, dz$$

as $n \to \infty$, by the dominated convergence theorem. The last integral equals

$$\int_{\mathbb{R}^d} K^2(z)dz \int_{\mathbb{R}} (\varphi(y) - m(x))^2 p(x, y)dy = \gamma^2(x)f^2(x). \qquad \square$$

3   Empirical processes

We devote this short Section to the estimation of the distribution function by a sample taken from a random sequence satisfying one of the covariance dependence conditions. Recall that for random variables $X_1, X_2, \ldots$ and $n \in \mathbb{N}$ the *empirical distribution function* is the random function

$$F_n^*(x, \omega) = \frac{1}{n} \sum_{j=1}^{n} \mathbb{I}\{X_j(\omega) \leq x\}, \quad x \in \mathbb{R}, \quad \omega \in \Omega.$$

For convenience we will use the agreement that $G(+\infty) = 1$ and $G(-\infty) = 0$ for any distribution function G (in particular F_n^*).

    $1°$. **The generalization of the classical Glivenko–Cantelli theorem.** The following theorem is due to Yu.

Theorem 3.1. ([423]) *Suppose that $X = (X_n)_{n\in\mathbb{N}}$ is a* **PA** *or* **NA** *sequence of random variables such that all X_n have common distribution function $F = F(x)$. In the* **PA** *case assume in addition that (4.1.2) holds. Then, if $F \in C(\mathbb{R})$, one has almost surely*

$$\sup_{x \in \mathbb{R}} |F_n^*(x) - F(x)| \to 0 \text{ as } n \to \infty. \tag{3.1}$$

    Proof is analogous to one in the independent case, that is the main step is the application of the strong law of large numbers. We begin with studying $X \in$ **PA**. Let $x \in \mathbb{R}$. Take arbitrary $\varepsilon > 0$ and choose $\delta > 0$ so that $F(x+\delta) - F(x-\delta) < \varepsilon$. Let $h_{\delta,x}$ be the function defined in (6.1.17). For fixed $\delta > 0$ and $x \in \mathbb{R}$ let $h_+(s) = h_{\delta,x}(s)$ and $h_-(s) = h_{\delta,x}(s + \delta)$, $s \in \mathbb{R}$. Then

$$h_+(X_1) \leq \mathbb{I}\{X_1 > x\} \leq h_-(X_1) \tag{3.2}$$

and

$$\mathsf{E}(h_-(X_1) - h_+(X_1)) \leq \mathsf{P}(x - \delta < X_1 \leq x + \delta) \leq \varepsilon. \tag{3.3}$$

Clearly, $\{h_+(X_n), n \in \mathbb{N}\}$ and $\{h_-(X_n), n \in \mathbb{N}\}$ are positively associated random systems according to Theorem 1.1.8, (d). By Theorem 1.5.3 we also have

$$cov(h_\pm(X_j), h_\pm(X_k)) \leq \delta^{-2} cov(X_j, X_k) \tag{3.4}$$

for any $j, k \in \mathbb{N}$, $j \neq k$. Here the signs at the index are simultaneously $+$ or $-$. Thus by Theorem 4.1.1

$$n^{-1} \sum_{i=1}^{n} h_-(X_i) \to \mathsf{E}h_-(X_1) \text{ a.s.}, \quad n^{-1} \sum_{i=1}^{n} h_+(X_i) \to \mathsf{E}h_+(X_1) \text{ a.s } (n \to \infty).$$

Therefore, due to (3.2) and (3.3), one almost surely has

$$1 - F(x) - \varepsilon \leq \mathsf{E}h_+(X_1) \leq \liminf_{n\to\infty}(1 - F_n^*(x))$$

$$\leq \limsup_{n\to\infty}(1 - F_n^*(x)) \leq \mathsf{E}h_-(X_1) \leq 1 - F(x) + \varepsilon.$$

Hence, as ε can be arbitrarily small, it follows that $F_n^*(x) \to F(x)$ a.s., $n \to \infty$.

Now let $\mu \in (0,1)$ be arbitrary and let $x_1, \ldots, x_N$ be real numbers such that

$$F(x_1) < \mu, \quad 1 - F(x_N) < \mu, \quad F(x_{i+1}) - F(x_i) < \mu \text{ for } i = 1, \ldots, N-1.$$

Also we stipulate that $x_0 = -\infty$, $x_{N+1} = +\infty$. For $x \in \mathbb{R}$ let $i(x) \in \{0, \ldots, N\}$ be the integer such that $x \in [x_i, x_{i+1}]$. Then, for any $x \in \mathbb{R}$ and $\omega \in \Omega$,

$$|F_n^*(x) - F(x)| \le F_n^*(x_{i(x)+1}) \vee F(x_{i(x)+1}) - F_n^*(x_{i(x)}) \wedge F(x_{i(x)}).$$

Consequently,

$$\limsup_{n \to \infty} \sup_{x \in \mathbb{R}} |F_n^*(x) - F(x)|$$

$$\le \limsup_{n \to \infty} \sup_{i=0,1,\ldots,N} |F_n^*(x_{i+1}) \vee F(x_{i+1}) - F_n^*(x_i) \wedge F(x_i)| = \mu.$$

As μ could be taken arbitrarily small, the limit of the left-hand side of (3.1) exists and equals zero.

For the case of **NA** random variables the proof is analogous, with the step (3.4) omitted. $\square$

Remark 3.2. If $(X_n)_{n \in \mathbb{N}}$ is a strictly stationary **PA** random sequence, then in Theorem 3.1 one can impose the (weaker) condition (4.1.9) instead of (4.1.2). The proof remains the same.

2°. The functional limit theorem. The above application of LLN for associated (**PA**, **NA**) random variables has resulted in the analog of Glivenko–Cantelli theorem. Employment of other limit theorems (CLT and maximal inequalities) permits to establish the functional limit theorem for an empirical process (the Kolmogorov criterion), to the proof of which we are passing.

For a nondecreasing function $F : \mathbb{R} \to \mathbb{R}$ we denote by F^{inv} the *generalized inverse* to F, that is, $F^{\mathrm{inv}}(t) = \inf\{z \in \mathbb{R} : F(z) > t\}$.

Theorem 3.3. *Let $X = (X_n)_{n \in \mathbb{Z}}$ be a strictly stationary **PA** or **NA** random sequence such that the following two conditions hold:*

1) the distribution function F of X_0 is continuous on $\mathbb{R}$ and has a bounded derivative f everywhere on $\mathbb{R}$, except for a finite number of points;

2) $\mathsf{E}X_0^2 < \infty$ and $|cov(X_0, X_r)| = O(r^{-\lambda})$ as $r \to \infty$, for some $\lambda > 6$.

Then the random processes $(B_n^*(x), x \in [0,1])_{n \in \mathbb{N}}$, defined by

$$B_n^*(x) = \sqrt{n}(F_n(F^{\mathrm{inv}}(x)) - F_n(F(x))),$$

converge in law in the Skorokhod space $D([0,1])$ to a centered Gaussian process with covariance function

$$\mathsf{E}B(s)B(t) = s \wedge t - st + \sum_{k \in \mathbb{Z}} cov(\mathbb{I}\{F(X_0) \le s\}, \mathbb{I}\{F(X_k) \le t\}).$$

Remark 3.4. In [423] this theorem was proved for associated random sequences under assumption that $\sum_{k=1}^{\infty} k^{13/2+\nu} cov(X_0, X_k) < \infty$. In [273] this requirement was relaxed to the condition $cov(X_0, X_k) = O(k^{-\lambda})$ as $k \to \infty$, $\lambda > 4$.

Proof. Again we study the **PA** case only. At first let us consider the case when X_0 has uniform distribution on the segment $[0, 1]$. As usual we intend to establish that the sequence $(B_n^*(x), x \in [0, 1])_{n \in \mathbb{N}}$ is dense in $D([0, 1])$ and any of its partial limits in law coincides with B.

As for the first condition, we show that (see (5.1.2))

$$\lim_{\delta \to 0+} \limsup_{n \to \infty} \mathsf{P}\left(\sup_{t,s \in [0,1]:|t-s|<\delta} |B_n^*(t) - B_n^*(s)| \geq \varepsilon \right) = 0$$

for any $\varepsilon > 0$. A well-known criterion of tightness of a family of random processes ([39, §15, Th. 15.6]) states that a sufficient condition is the following:

$$\limsup_{n \in \mathbb{N}} \mathsf{E}|B_n^*(t) - B_n^*(s)|^4 \leq R|t - s|^{2+\nu} \tag{3.5}$$

for any $t, s \in [0, 1]$, where $\nu > 0$ and $R > 0$ do not depend on t and s. We also use two auxiliary results.

Lemma 3.5. ([423]) *Let (ξ, η) be a* **PA** *or* **NA** *random vector with* $\mathsf{E}(\xi^2 + \eta^2) < \infty$. *Assume that each of its components has a density bounded by some constant* $a > 0$. *Then, for any* $x, y \in \mathbb{R}$,

$$|\mathsf{P}(\xi \geq x, \eta \geq y) - \mathsf{P}(\xi \geq x)\mathsf{P}(\eta \geq y)| \leq 3 \cdot 2^{2/3} a^{2/3} |cov(\xi, \eta)|^{1/3}.$$

Proof. If $cov(\xi, \eta) = 0$ then ξ and η are independent in view of Corollary 1.5.5, so the Lemma is obviously true. Thus we assume that $|cov(\xi, \eta)| > 0$. Let $\delta > 0$ be a number to be specified later. Take the functions $h_{\delta,x}$, $h_{\delta,y}$ defined in (1.17). Then

$$|\mathsf{P}(\xi \geq x, \eta \geq y) - \mathsf{P}(\xi \geq x)\mathsf{P}(\eta \geq y)| = |cov(\mathbb{I}\{\xi \geq x\}, \mathbb{I}\{\eta \geq y\})|$$

$$\leq |cov(h_{\delta,x}(\xi), h_{\delta,y}(\eta))| + |cov(\mathbb{I}\{\xi \geq x\} - h_{\delta,x}(\xi), h_{\delta,y}(\eta))|$$

$$+ |cov(\mathbb{I}\{\xi \geq x\}, \mathbb{I}\{\eta \geq y\} - h_{\delta,y}(\eta))| \leq \delta^{-2}|cov(\xi, \eta)| + 4a\delta$$

where the last inequality is due to Theorem 1.5.3 and the bound for densities of ξ and η. It remains to optimize the estimate in $\delta > 0$. $\square$

Lemma 3.6. *There exist some* $R > 0$ *and* $\nu > 0$ *such that, for any* $s, t \in [0, 1]$ *and all* $n \in \mathbb{N}$, *one has*

$$\mathsf{E}|B_n^*(t) - B_n^*(s)|^4 \leq R|t - s|^{2+\nu}.$$

Proof. Without loss of generality one may assume that $t > s$. Obviously

$$B_n^*(t) - B_n^*(s) = n^{-1/2} \sum_{i=1}^{n} (\mathbb{I}\{X_i \leq t\} - F(t) - \mathbb{I}\{X_i \leq s\} + F(s)) = n^{-1/2} \sum_{i=1}^{n} \xi_i$$

where $(\xi_i)_{i\in\mathbb{N}}$ is a strictly stationary, centered sequence of bounded random variables. We claim that this sequence is (BL,θ)-dependent with dependence coefficient $\theta_r = O(r^{-\lambda/3})$ as $r \to \infty$. Indeed, the sequence of 2-dimensional random vectors

$$(Y_n, Z_n) := (\mathbb{I}\{X_n \le t\} - F(t), \mathbb{I}\{X_n \le s\} - F(s))_{n\in\mathbb{N}}$$

is positively associated with the Cox–Grimmett coefficient

$$\theta_r^{(2)} = 2\sum_{k=r}^{\infty}\Big(|cov(\mathbb{I}\{X_0 \le t\},\mathbb{I}\{X_k \le t\})| + |cov(\mathbb{I}\{X_0 \le t\},\mathbb{I}\{X_k \le s\})|$$

$$+ |cov(\mathbb{I}\{X_0 \le s\},\mathbb{I}\{X_k \le t\})| + |cov(\mathbb{I}\{X_0 \le s\},\mathbb{I}\{X_k \le s\})|\Big),\ r \in \mathbb{N}.$$

Note that $\xi_n = Y_n - Z_n$, $n \in \mathbb{N}$. Therefore, by Lemma 1.5.16 and Lemma 3.5, the sequence $(\xi_i)_{i\in\mathbb{N}}$ is (BL,θ)-dependent. More precisely, let $I, J \subset \mathbb{N}$ be disjoint sets with $|I| + |J| \le 4$ and suppose that $f : \mathbb{R}^{|I|} \to \mathbb{R}$, $g : \mathbb{R}^{|J|} \to \mathbb{R}$ are bounded Lipschitz functions. Then the same Lemma 3.5 together with Theorem 1.5.3 show that

$$|cov(f(\xi_I), g(\xi_J))| \le Lip(f)Lip(g)\kappa_r,$$

with $\kappa_r = O(r^{-\lambda/3})$. Consequently, the dependence coefficient θ_r of the sequence $(\xi_i)_{i\in\mathbb{N}}$ is $O(r^{-\lambda/3+1})$ as $r \to \infty$.

Let $p > 4$ be so large that $3 - \lambda(p - 4)/3(p - 2) < 1$. By Theorem 2.1.26 and Remark 2.1.27, for any $n \in \mathbb{N}$,

$$\mathsf{E}(B_n^*(t) - B_n^*(s))^4 = n^{-2}\mathsf{E}\Big(\sum_{j=1}^{n}\xi_j\Big)^4$$

$$\le Cn^{-2}\Big(\sum_{j=1}^{n}\mathsf{E}(\xi_j - \mathsf{E}\xi_j)^4 + \Big(\sum_{j=1}^{n}\mathsf{Var}\xi_j + \vartheta_1 n\Big)^2 + n\, b_n(\lambda, p)\Big(\mathsf{E}\big|\xi - \mathsf{E}\xi\big|^p\Big)^{2/(p-2)}\Big) \quad (3.6)$$

with $C > 0$ independent of n, t, s and

$$b_n(\lambda, p) = B(n, 3 - \lambda(p - 4)/3(p - 2))$$

(see Remark 2.1.27 for the definition of ϑ_1).

The second term in the right-hand side of (3.6) is the main one as $n \to \infty$ (recall that $B(n, 3 - \lambda(p - 4)/3(p - 2)) = o(n)$, $n \to \infty$). Clearly

$$\mathsf{Var}\xi_j \le 4\mathsf{E}\xi_j^2 = 4(t - s),\ \ j = 1,\ldots,n. \quad (3.7)$$

Further, for any $r \in \mathbb{N}$ one can write

$$\vartheta_1 = \sup_{j\in\mathbb{N}}\ \sup_{V\subset\mathbb{N},\, j\notin V}\ \sup_{\alpha}\ \Big|\sum_{k\in V}\alpha_k cov(\xi_j, \xi_k)\Big|$$

$$\le 16r(t - s) + \sup_{j\in\mathbb{N}}\ \sup_{V\subset\mathbb{N},\, dist(\{j\},V)\ge r}\ \sup_{\alpha}\ \Big|\sum_{k\in V}\alpha_k cov(\xi_j, \xi_k)\Big| \le 16r(t - s) + \theta_r. \quad (3.8)$$

Using the condition $\theta_r = O(r^{-\nu})$ as $r \to \infty$, with $\nu = \lambda/3 - 1 > 1$, and taking $r = [(t-s)^{-1/(1+\nu)}]$ we prove that there exists some $C_1 > 0$ (not depending on t and s) such that

$$\vartheta_1 \le C_1(t-s)^{\nu/(1+\nu)}. \tag{3.9}$$

From (3.6)—(3.9) it follows that

$$\limsup_{n\to\infty} \mathsf{E}(B_n^*(t) - B_n^*(s))^4 \le C(4 + C_1)(t-s)^{2\nu/(1+\nu)}.$$

As $\nu/(1+\nu) > 1/2$, the Lemma is proved. $\square$

Lemma 3.6 ensures that (3.5) is fulfilled, hence the family of processes under consideration is tight.

To prove the convergence of finite-dimensional distributions, take some $k \in \mathbb{N}$ and arbitrary points $s_1, \ldots, s_k \in [0,1]$. Then

$$(B_n^*(s_1), \ldots, B_n^*(s_k)) = n^{-1/2} \sum_{i=1}^{n} \eta_i$$

where $(\eta_i)_{i\in\mathbb{Z}}$ is a strictly stationary (BL,θ)-dependent (in fact, even **PA** or **NA**, in accordance with the alternative condition of Theorem) sequence of centered bounded random vectors. The covariance matrix Σ of η_0 has entries

$$\Sigma_{lm} = s_l \wedge s_m - s_l s_m + \sum_{j\in\mathbb{Z}, j\neq 0} cov(\mathbb{I}\{X_0 \le s_l\}, \mathbb{I}\{X_0 \le s_m\}), \quad l,m = 1, \ldots, k.$$

Therefore, by Corollary 3.1.13 one has

$$n^{-1/2} \sum_{i=1}^{n} \eta_i \to N(0, \Sigma) \quad \text{as } n \to \infty,$$

i.e. the finite-dimensional distributions of B_n^* converge to those of B, which proves the Theorem in case when X_i are uniformly distributed on $[0,1]$.

In general case note that the random sequence $\{F(X_i)\}_{i\in\mathbb{N}}$ consists of uniformly distributed on $[0,1]$ random variables and satisfies the assumptions of Theorem (because $Lip(F) < \infty$). Therefore, the Theorem follows from the already proved case. $\square$

To conclude the Section we mention that various problems concerning estimation of distribution function or related objects in the association setup can be found in papers [11], [91], [90], [92], [94], [93], [156, 157], [353]. A Glivenko–Cantelli result for **NA** random variables is proved in [290]. The functional limit theorems for empirical processes were obtained in [140, 141, 318, 373]. An interesting subject studied in [339] is a statistical test checking that the law of observed stable random vector is associated.

Chapter 8

Integral Functionals

In Chapter 8 we use the previous material to study the integral functionals arising in analysis of the Cauchy problem for multidimensional Burgers equation with random data. This famous model equation was applied to various physical phenomena. One can refer, e.g., to the papers by Zeldovich and his school, see also the book by Gurbatov, Malahov and Saichev (and references therein) to find the results concerning the problems of acoustic waves as well as the surprising effects of formation of pancakes and planets after the Big Bang. Non-uniform distribution of the media in the Universe and appearance of the Voronoi tesselation can be explained by means of 3-dimensional Burgers equation. We consider the transformed solution of the Burgers equation with random data. CLT for finite dimensional distributions of such solution was proved in 1990 by Bulinski and Molchanov (under certain conditions) with a shot-noise field as initial data. After that a number of deep results were published by Sinai, Khanin, Leonenko, Barndorff-Nielsen, Woyczynski and other researchers. Of the main interest in the present Chapter is the result by Bakhtin (2001) providing the FCLT for transformed solution in appropriate functional space when the data is the stationary associated (in particular, independently scattered) random measure. As a tool we apply the properties of stationary associated random measures (mostly studied by Evans and Bakhtin) and limit theorems for integrals over such measures, which are of own independent interest.

1 Stationary associated measures

Let $(\Omega, \mathcal{F}, \mathrm{P})$ be a probability space. Let M be an associated random measure on $\mathcal{B}_0(\mathbb{R}^d)$, which is the ring of bounded Borel subsets of $\mathbb{R}^d$. This measure is called *stationary* if, for any collection of sets $B_1, \ldots, B_m$ belonging to $\mathcal{B}_0(\mathbb{R}^d)$ and any vector $t \in \mathbb{R}^d$, the distributions of random vectors $(M(B_1), \ldots, M(B_m))$ and $(M(B_1 + t), \ldots, M(B_m + t))$ coincide. In this Section, we study the properties of integrals over stationary associated random measures. We rely on Section 1.3 for basic properties of M and on Section 3.1 for the results on normal approximation.

1°. Finite susceptibility condition for random measures. Let[1]

$$I = (-1/2, 1/2]^d \subset \mathbb{R}^d \quad \text{and} \quad a_M = \mathsf{E}M(I). \tag{1.1}$$

If $a_M < \infty$, then $\mu(B) := \mathsf{E}M(B) < \infty$, for $B \in \mathcal{B}_0(\mathbb{R}^d)$, and μ can be extended to a σ-finite (non-random) measure on $\mathcal{B}(\mathbb{R}^d)$ due to the Caratheodori theorem. This measure μ is invariant with respect to translations in $\mathbb{R}^d$. Therefore, it is the Lebesgue measure *mes* multiplied by a_M (see, e.g., [239, Ch. V, §3, 1–3,5], [47]).

We write $L^p(\mathbb{R}^d) = L^p(\mathbb{R}^d, \mathcal{B}(\mathbb{R}^d), mes)$, $p \in [1, \infty]$, and $L = L^1(\mathbb{R}^d) \cap L^\infty(\mathbb{R}^d)$. The *support* of a real-valued function f, denoted by $supp(f)$, is, as usual, the closure (in Euclidean metric) of the set $\{x \in \mathbb{R}^d : f(x) \neq 0\}$. The sup-norm of a function $f : \mathbb{R}^n \to \mathbb{R}$, denoted by $\|f\|_\infty$, will be understood as the *essential supremum*, i.e.

$$\|f\|_\infty = \inf_{B \in \mathcal{B}(\mathbb{R}^n),\, mes(B)=0} \ \sup_{x \in \mathbb{R}^n \setminus B} |f(x)|.$$

Throughout the Section we often use the notation

$$xB = \{(xy_1, \ldots, xy_d) : y \in B\} \quad \text{for} \quad x \in \mathbb{R} \quad \text{and} \quad B \subset \mathbb{R}^d.$$

All integrals in this Section are taken over $\mathbb{R}^d$ unless other is mentioned, dx means $mes(dx)$.

Lemma 1.1. *Let M be a stationary random measure such that $a_M < \infty$. Then any function $f \in L$ is integrable with respect to M and*

$$\mathsf{E} \int f(x)M(dx) = a_M \int f(x)dx.$$

Proof. If $f \in L$ is nonnegative, then the assertion follows from Lemma 1.3.25, with $\mu(\cdot) = a_M mes(\cdot)$. For a general function $f \in L$ write $f = f^+ - f^-$ and apply the previous argument. $\square$

Lemma 1.2. *Let $(\xi_1, \xi_2, \xi_3, \xi_4)$ be an associated random vector with values in $\mathbb{R}^4$. Then $cov(\xi_1, \xi_3) \leq cov(\xi_1 + \xi_2, \xi_3 + \xi_4)$, provided that both covariances exist. Given an associated random measure M and nonnegative functions $h_1, h_2, h_3, h_4 \in L$, one has*

$$cov\left(\int h_1 M(dx), \int h_3 M(dx)\right) \leq cov\left(\int (h_1 + h_2)M(dx), \int (h_3 + h_4)M(dx)\right)$$

if the covariances exist.

Proof. Employ the association property and refer to Theorem 1.3.27. $\square$

From now on we assume that, for any $B \in \mathcal{B}_0(\mathbb{R}^d)$, one has $\mathsf{E}M^2(B) < \infty$. Let

$$K(U) = cov(M(I), M(U)) \quad \text{for} \quad U \in \mathcal{B}_0(\mathbb{R}^d).$$

Definition 1.3. A random measure M satisfies the *finite susceptibility* (FS) condition if

$$\Gamma_M := \sup_{t>0} K(tI) < \infty. \tag{1.2}$$

---

[1] We do not use the cube $(0, 1]^d$ since we want to obtain the whole space $\mathbb{R}^d$ by dilation, multiplying that cube's points by growing t.

Lemma 1.4. *Let M be an associated stationary random measure which satisfies the* (FS) *condition. Then, for any $B \in \mathcal{B}_0(\mathbb{R}^d)$, one has*

$$\sup_{t>0} cov(M(B), M(tI)) = \lim_{t\to\infty} cov(M(B), M(tI)) = \Gamma_M \mathrm{mes}(B). \qquad (1.3)$$

Proof. The expression inside the limit is nondecreasing in t by Lemma 1.2, thus the limit and supremum coincide (a priori they might be infinite).

Let us prove that the supremum in (1.3) does not change when B is translated by a vector $v \in \mathbb{R}^d$. From stationarity of M and Lemma 1.2, for any $t > 0$, one has

$$cov(M(B + v), M(tI)) = cov(M(B), M(tI - v)) \le cov(M(B), M(sI))$$

if $s = s(t) > 0$ is chosen in such a way that $tI - v \subset sI$ (clearly, the last inclusion holds for s large enough). Therefore,

$$cov(M(B + v), M(tI)) \le \sup_{s>0} cov(M(B), M(sI)),$$

and taking supremum in t we obtain that

$$\sup_{t>0} cov(M(B + v), M(tI)) \le \sup_{s>0} cov(M(B), M(sI)). \qquad (1.4)$$

Now, applying (1.4) to the set $B + v$ instead of B and the vector $(-v)$ instead of v, we come to (1.4) with reversed sign. Thus in fact one has equality in (1.4).

If $B = I$, the last assertion in (1.3) is true in view of (1.2). Suppose now that $B = (a, b] \subset \mathbb{R}^d$ is a cube with edge's length equal to p^{-1} where $p \in \mathbb{N}$. Construct pairwise disjoint cubes $B_1 = B, B_2, \ldots, B_{p^d}$ congruent to B and such that $\cup_{m=1}^{p^d} B_m =: V$ where V is congruent to I. Then, by (1.4) (with equality), the additivity of M and linearity of covariance function in each argument, one has

$$\Gamma_M = \lim_{t\to\infty} cov(M(V), M(tI))$$

$$= \sum_{m=1}^{p^d} \lim_{t\to\infty} cov(M(B_m), M(tI)) = p^d \lim_{t\to\infty} cov(M(B), M(tI)),$$

which implies the result for B since $\mathrm{mes}(B) = p^{-d}$.

Let now $B = (a, b] \subset \mathbb{R}^d$ be a rectangle with rational $b_i - a_i$, $i = 1, \ldots, d$. Then $B = \cup_{m=1}^{N} B_j$, and B_m $(m = 1, \ldots, N)$ are pairwise disjoint cubes having edge length p^{-1}. The already proved part of Lemma implies that

$$\lim_{t\to\infty} cov(M(B), M(tI)) = \sum_{m=1}^{N} \lim_{t\to\infty} cov(M(B_m), M(tI)) = \Gamma_M N p^{-d} = \Gamma_M \mathrm{mes}(B).$$

Clearly this relation holds also for arbitrary $B = (a, b] \subset \mathbb{R}^d$ as we can find approximating rectangles $B', B'' \subset \mathbb{R}^d$ with rational edge's length and such that $B' \subset B \subset B''$. Consequently, the last relation in (1.3) is satisfied for a ring $\mathcal{A}$ consisting of finite unions of pair-wise disjoint rectangles of the form $(a, b] \subset \mathbb{R}^d$.

For any $n \in \mathbb{N}$ define a set function ν_n by the equality

$$\nu_n(B) = cov(M(B), M(nI)), \quad B \in \mathcal{B}_0(\mathbb{R}^d).$$

This function is clearly additive, and for a sequence of bounded Borel sets $(B_k)_{k\in\mathbb{N}}$ decreasing to an empty set[2] one has

$$\nu_n(B_k) \leq 2\left(\mathsf{E}M(nI)^2\mathsf{E}M^2(B_k)\right)^{1/2} \to 0, \quad k \to \infty,$$

by the Cauchy-Bunyakowski-Schwarz inequality and the monotone convergence theorem (because $M(\omega, B_k) \searrow 0$ for any $\omega \in \Omega$). Consequently, by the Caratheodori theorem, ν_n can be extended to a σ-finite measure on $\mathcal{B}(\mathbb{R}^d)$ (one can consider an algebra $\mathcal{A}_j = \mathcal{A}\cap(j, j+1]$ for $1, j \in \mathbb{N}^d$ and the restriction of ν_n on $\mathcal{A}_j$). Moreover, due to Lemma 1.2, for any $B \in \mathcal{B}_0(\mathbb{R}^d)$ and all $n \in \mathbb{N}$, one has $\nu_n(B) \leq \nu_{n+1}(B)$. If $m \in \mathbb{N}$ is so large that $B \subset mI$, then by the same Lemma, stationarity of M and (1.2) we have

$$\nu_n(B) \leq cov(M(mI), M(nI)) \leq m^d\Gamma_M < \infty.$$

Thus the sequence $\nu_n(B)$ is nondecreasing and bounded, hence converges to some $\nu(B)$ where $\nu(\cdot)$ is an additive set function. We claim that ν is σ-additive on $\mathcal{A}_j$ $(j \in \mathbb{N})$, which is equivalent to continuity at "zero" (i.e. the empty set). Suppose that $(B_k)_{k\in\mathbb{N}}$ is a sequence of finite unions of pair-wise disjoint rectangles belonging to $(j, j + 1]$ and such that $B_{k+1} \subset B_k$, $k \in \mathbb{N}$, and $\cap_k B_k = \varnothing$. Taking arbitrary $\varepsilon > 0$ and k so large that $\mathrm{mes}(B_k) < \varepsilon$, we infer that $\nu(B_k) \leq \Gamma_M\varepsilon$. In other words, ν is continuous at "zero". This ensures that ν is σ-additive, hence it is a measure on $\mathcal{A}_j$ which can be extended to $\mathcal{B}_0(\mathbb{R}^d)$. As this measure assigns the value $\Gamma_M\mathrm{mes}(B)$ to any $B = (a, b] \subset \mathbb{R}^d$, it is in fact the Lebesque measure multiplied by Γ_M.

Lemma 1.5. *Let M be an associated stationary random measure which satisfies (1.2). Suppose that h is a nonnegative bounded Borel function having compact support. Then*

$$J_n(h) := cov\left(\int h(x)M(dx), M(nI)\right) \nearrow \Gamma_M \int h(x)dx \quad \text{as } n \to \infty. \qquad (1.5)$$

Proof. The covariance clearly exists, since $\int h(x)M(dx) \leq \sup(h)M(U)$ a.s. where $U = supp(h)$. Hence the random variable $\int h(x)M(dx)$ is square-integrable.

Lemma 1.4 implies that (1.5) holds for a function $h = \mathbb{I}_B$ where $B \in \mathcal{B}_0(\mathbb{R}^d)$. By linearity the Lemma is also valid for simple functions of the type

$$h = \sum_{k=1}^{N} c_k\mathbb{I}\{B_k\}, \quad c_k \in \mathbb{R}, \quad B_k \in \mathcal{B}_0(\mathbb{R}^d), \quad k = 1,\ldots,N.$$

Clearly, multiplying h by a positive constant we may confine ourselves to the case $0 \leq h < 1$. Then there exist sequences $(f_m)_{m\in\mathbb{N}}$ and $(g_m)_{m\in\mathbb{N}}$ converging uniformly to h and such that $f_m(x) \leq h(x) \leq g_m(x)$, $x \in \mathbb{R}^d$, $m \in \mathbb{N}$. Namely, we can take

$$f_m(x) = \sum_{k=1}^{m} \frac{k-1}{m}\mathbb{I}\left\{\frac{k-1}{m} \leq h(x) < \frac{k}{m}\right\},$$

$$g_m(x) = \sum_{k=1}^{m} \frac{k}{m}\mathbb{I}\left\{\frac{k-1}{m} \leq h(x) < \frac{k}{m}\right\}\mathbb{I}\{x \in supp(h)\}. \qquad (1.6)$$

---

[2] I.e. $B_k \subset B_{k+1}$, $k \in \mathbb{N}$, and $\cap_{k\in\mathbb{N}}B_k = \varnothing$.

By Lemma 1.2 one has

$$J_n(f_m) \leq J_n(h) \leq J_n(g_m), \quad m, n \in \mathbb{N}.$$

Therefore, for each $m \in \mathbb{N}$, by the assertion already proved for simple functions,

$$\liminf_{n \to \infty} J_n(h) \geq \lim_{n \to \infty} J_n(f_m) = \Gamma_M \int f_m(x) dx,$$

$$\limsup_{n \to \infty} J_n(h) \leq \lim_{n \to \infty} J_n(g_m) = \Gamma_M \int g_m(x) dx.$$

Letting $m \to \infty$ we come to the statement. $\square$

2°. Bound for variance of integral with respect to random measure.
For $L^p = L^p(\mathbb{R}^d)$ with $p \geq 1$ we denote by $\|f\|_{L^p}$ the usual L^p-norm of a function $f : \mathbb{R}^d \to \mathbb{R}$, i.e. $\|f\|_{L^p}^p = \int |f(x)|^p dx$.

Lemma 1.6. *Let M be an associated stationary random measure which satisfies the condition (1.2). Then, for any $f \in L$, the random variable $\int f(x)M(dx)$ is square-integrable and*

$$\mathrm{Var} \int f(x)M(dx) \leq \Gamma_M \|f\|_{L^1} \|f\|_\infty.$$

Proof. We can assume that f is bounded by $\|f\|_\infty$ (redefining it, if necessary, on a set of zero Lebesgue measure). At first suppose that f is nonnegative and has compact support. Then, by Lemmas 1.2 and 1.5,

$$\mathrm{Var} \int f(x)M(dx) = cov\left(\int f(x)M(dx), \int f(x)M(dx) \right)$$

$$\leq cov\left(\int f(x)M(dx), \int \|f\|_\infty \mathbb{I}\{x \in supp(f)\}M(dx) \right)$$

$$= \|f\|_\infty cov\left(\int f(x)M(dx), M(supp(f)) \right)$$

$$\leq \|f\|_\infty cov\left(\int f(x)M(dx), M(nI) \right) \leq \|f\|_\infty \Gamma_M \int f(x)dx = \Gamma_M \|f\|_{L^1} \|f\|_\infty,$$

here $n \in \mathbb{N}$ is taken so large that $supp(h) \subset nI$.

Now consider any f with compact support. Write $f = f^+ - f^-$, then both f^+ and f^- are nonnegative and have compact support. Again we may and will assume that, for all $x \in \mathbb{R}^d$,

$$f^+(x) \leq \|f^+\|_\infty \leq \|f\|_\infty, \quad f^-(x) \leq \|f^-\|_\infty \leq \|f\|_\infty. \tag{1.7}$$

Therefore, in view of (1.7),

$$\mathrm{Var} \int f(x)M(dx) \leq \mathrm{Var} \int f^+(x)M(dx) + \mathrm{Var} \int f^-(x)M(dx)$$

$$\leq \Gamma_M \|f\|_\infty \left(\|f^+\|_{L^1} + \|f^-\|_{L^1} \right) \leq \Gamma_M \|f\|_\infty \|f\|_{L^1}$$

where we used that

$$cov\left(\int f^+(x)M(dx), \int f^-(x)M(dx) \right) \geq 0,$$

due to Theorem 1.3.27. Thus the Lemma is proved for f having compact support. Let $f \in L$. Again we assume that f is bounded by $\|f\|_\infty$. For any $n \in \mathbb{N}$, let

$$f_n = f\mathbb{I}\{x \in \mathbb{R}^d : \|x\| \le n\}.$$

By the already proved part of the Lemma and Lemma 1.1, for all $n \in \mathbb{N}$,

$$\mathsf{E}\Big(\int f_n(x)M(dx)\Big)^2 = \mathsf{Var}\int f_n(x)M(dx) + \Big(\mathsf{E}\int f_n(x)M(dx)\Big)^2$$

$$\le \Gamma_M\|f_n\|_{L^1}\|f_n\|_\infty + \Big(a_M\int f_n(x)dx\Big)^2. \tag{1.8}$$

By the dominated convergence theorem and Lemma 1.1,

$$\int f_n(x)M(dx) \to \int f(x)M(dx) \ \text{a.s.,} \ n \to \infty.$$

Combining (1.8) and the Fatou lemma shows that $\int f_n(x)M(dx)$ is square-integrable and

$$\mathsf{E}\Big(\int f(x)M(dx)\Big)^2 \le \liminf_{n\to\infty}\Big(\Gamma_M\|f_n\|_{L^1}\|f_n\|_\infty + \Big(a_M\int f_n(x)dx\Big)^2\Big)$$

$$= \Gamma_M\|f\|_{L^1}\|f\|_\infty + \Big(a_M\int f(x)dx\Big)^2$$

where the limit is evaluated by the dominated convergence theorem. The last inequality is equivalent to the Lemma's assertion. $\square$

For positive t and $f \in L$ set

$$M_t(f) = \int f\Big(t^{-1}x\Big)M(dx), \quad Z_t(f) = t^{-d/2}\Big(M_t(f) - \mathsf{E}M_t(f)\Big).$$

3°. CLT by Evans. Now we are able to prove the CLT for integrals over associated random measures.

Theorem 1.7. ([164]) *Let M be an associated stationary random measure which satisfies the condition (1.2). Then, for any $f \in L$,*

$$Z_t(f) \to N\Big(0, \|f\|_{L^2}^2\Gamma_M\Big) \ \text{in law} \ \text{as} \ t \to \infty.$$

Proof. For $q \in \mathbb{N}$ and $k \in \mathbb{Z}^d$ we consider the set $B(q,k) = (q^{-1}k, q^{-1}(k+1)] \subset \mathbb{R}^d$, i.e. $B(q,k)$ is the image of the unit cube $(0,1]I \subset \mathbb{R}^d$ after the space $\mathbb{R}^d$ was shifted by the vector k and uniformly q-fold contracted. We will check that, for any $m \in \mathbb{N}$ and arbitrary pairwise disjoint points $k^1, \ldots, k^m \in \mathbb{Z}^d$, one has

$$Y_t = (Z_t(\mathbb{I}\{B(q,k^v)\}))_{v=1,\ldots,m} \to N(0, \Gamma_M q^{-d}I_m) \ \text{in law as} \ t \to \infty. \tag{1.9}$$

Here $t \in \mathbb{N}$ only and I_m is the identity matrix of order m. Introduce

$$X_j = M(B(q,j)), \ j \in \mathbb{Z}^d.$$

Then the field $X = \{X_j, j \in \mathbb{Z}^d\}$ is square-integrable (i.e. $\mathsf{E}X_j^2 < \infty, j \in \mathbb{Z}^d$), strictly stationary and associated. Moreover, by stationarity and Lemma 1.5

$$\sum_{j \in \mathbb{Z}^d} cov(X_0, X_j) = \lim_{n \to \infty} cov\left(M(B(q,0)), \sum_{j: |j| \leq n} M(B(q,j))\right)$$

$$= \lim_{n \to \infty} cov\left(M\left(B(q,0) - \frac{q}{2}\mathbf{1}\right), \sum_{j: |j| \leq n} M\left(B(q,j) - \frac{q}{2}\mathbf{1}\right)\right)$$

$$= \Gamma_M \mathrm{mes}(B(q,0)) = \left(\Gamma_M\right)q^{-d}. \tag{1.10}$$

Calculating the limit we used, along with Lemma 1.5, the fact that all the summands are nonnegative. Consequently, the field X satisfies condition (1.5.3).

Now for any $v = 1, \ldots, m$ we have

$$Z_t(\mathbb{I}\{B(q, k^v)\}) = t^{-d/2}\left(M((q^{-1}tk^v, q^{-1}t(k^v + 1)]) - \mathsf{E}M((q^{-1}tk^v, q^{-1}t(k^v + 1)])\right)$$

$$= t^{-d/2} \sum_{l \in \mathbb{Z}^d: \, k^v t \leq l < (k^v + 1)t} \left(X_l - \mathsf{E}X_l\right).$$

The relation (1.9) follows from Theorem 3.1.16.

The next step of the proof is to remove the restriction that t is a positive integer. As usual, $[t]$ denotes the lower integer part of a number $t \in \mathbb{R}$. For $f \in L$ and $t > 0$ set $\widetilde{M}_t(f) := M_t(f) - \mathsf{E}M_t(f)$ (thus $\widetilde{M}_t(f) = t^{d/2}Z_t(f)$). By the multidimensional version of Slutsky's lemma (Lemma 3.1.27), to establish (1.9) for $t \in \mathbb{R}$, $t \to \infty$, it suffices to show that

$$\mathsf{E}(Y_{t,v} - Y_{[t],v})^2 = \mathsf{E}(t^{-d/2}\widetilde{M}_t(\mathbb{I}\{B(q, k^v)\}) - [t]^{-d/2}\widetilde{M}_{[t]}(\mathbb{I}\{B(q, k^v)\}))^2 \to 0 \tag{1.11}$$

as $t \to \infty$, for $v = 1, \ldots, m$. For simplicity of notation we prove (1.11) for an arbitrary $B = (a, b] \subset \mathbb{R}^d$ instead of $B(n, k^v)$ and write $\widetilde{M}(C) := M(C) - \mathsf{E}M(C)$. For $t > 1$, one has

$$\mathsf{E}\left(t^{-d/2}\widetilde{M}(tB) - [t]^{-d/2}\widetilde{M}([t]B)\right)^2$$

$$\leq 2[t]^{-d}\mathsf{E}\left(\widetilde{M}(tB) - (\widetilde{M}([t]B)\right)^2 + 2\mathsf{E}\left((t^{-d/2} - [t]^{-d/2})\widetilde{M}(tB)\right)^2. \tag{1.12}$$

The estimation on the right-hand side of (1.12) is based on Lemma 1.6. Namely,

$$[t]^{-d}\mathsf{E}\left(\widetilde{M}(tB) - \widetilde{M}([t]B)\right)^2 = [t]^{-d}\mathsf{Var}\int(\mathbb{I}_B(t^{-1}x) - \mathbb{I}_B([t]^{-1}x))M(dx)$$

$$= [t]^{-d}\mathsf{Var}\int(\mathbb{I}\{tB\}(x) - \mathbb{I}\{[t]B\}(x))M(dx) \leq \Gamma_M[t]^{-d}\mathrm{mes}\,((tB)\triangle([t]B)). \tag{1.13}$$

For $B = (a, b]$ where $a, b \in \mathbb{R}^d$, one infers that

$$[t]^{-d}\mathrm{mes}\,((tB)\triangle([t]B)) = \mathrm{mes}\,((t[t]^{-1}B)\triangle B) \to 0 \quad \text{as } t \to \infty. \tag{1.14}$$

Indeed, $((1+\alpha)B)\triangle B \subset (\partial B)^{(2\alpha)}$ for any $\alpha > 0$; $\mathrm{mes}(\partial B) = 0$ and $t[t]^{-1} \to 1$ as $t \to \infty$.

For any $t > 1$ we have $[t]^{-d/2} - t^{-d/2} \le d[t]^{-d/2-1}$. So, by Lemma 1.6,

$$\mathsf{E}\left(\left(t^{-d/2} - [t]^{-d/2}\right)\widetilde{M}(tB)\right)^2 = (t^{-d/2} - [t]^{-d/2})^2\,\mathsf{Var}\int \mathbb{I}_B(t^{-1}x)M(dx)$$

$$\le d^2[t]^{-d-2}\Gamma_M\mathrm{mes}(tB) = d^2[t]^{-d-2}t^d\Gamma_M\mathrm{mes}(B) \to 0, \quad t \to \infty. \tag{1.15}$$

From (1.13)—(1.15) it is easily seen that the left-hand side of (1.12) tends to zero as $t \to \infty$. Consequently, (1.9) is true for $t \to \infty$, $t \in \mathbb{R}$. In the rest of the proof we let t take all positive real values.

Now let $f \in L$ be a "rationally simple function" i.e.

$$f(x) = \sum_{v=1}^{m} a_v\mathbb{I}\{B(q,k^v)\}(x), \quad x \in \mathbb{R}^d,$$

where $q, m \in \mathbb{N}$, $k^1,\ldots,k^m \in \mathbb{Z}^d$ and $a_1,\ldots,a_m \in \mathbb{R}$.

Since convergence in law is preserved under continuous mapping and the inner product is a continuous function on $\mathbb{R}^m$, from (1.9) one easily deduces for such f that

$$Z_t(f) \to N\left(0, \Gamma_M \int f^2(x)dx\right) \quad \text{in law} \quad \text{as } t \to \infty. \tag{1.16}$$

To make the next step of the proof we turn to a general function $f \in L$. At first we assume that f has compact support. We can approximate f by rationally simple functions. Note that the standard approximation procedure (1.6) does not satisfy this requirement.

Lemma 1.8. *Let $f \in L$ be a function having compact support. Then there exists a sequence of rationally simple functions f_n such that $f_n \to f$ mes-a.e. and in $L^1(\mathbb{R}^d)$ as $n \to \infty$; this sequence can be chosen in such a way that $\|f_n\|_\infty \le \|f\|_\infty, n \in \mathbb{N}$.*

Proof. Without loss of generality we assume that f is nonnegative (otherwise we take $f = f^+ - f^-$ and notice that in the forthcoming proof for a nonnegative function the approximating functions are nonnegative). As previously, we suppose that f is bounded by $\|f\|_\infty$. By linearity we can also assume that $\|f\| < 1$. Note that the Borel functions

$$g_n(x) = \sum_{k=1}^{n} \frac{k-1}{n}\mathbb{I}\left\{\frac{k-1}{n} \le f(x) < \frac{k}{n}\right\} =: \sum_{k=1}^{n} c_{k,n}\mathbb{I}\{B_{k,n}\}(x)$$

converge uniformly to f as $n \to \infty$, and $g_n(x) \le f(x)$, $x \in \mathbb{R}^d$. Now, for any $n \in \mathbb{N}$ and any $k \in \{1,\ldots,n\}$, take $U_{k,n}$ to be a finite union of rational rectangles such that

$$\mathrm{mes}(U_{k,n}\triangle B_{k,n}) < n^{-3}. \tag{1.17}$$

It can be done, since such finite unions form a ring $\mathcal{A}$, and $\sigma(\mathcal{A}) = \mathcal{B}(\mathbb{R}^d)$. For $n \in \mathbb{N}$ introduce $V_{1,n} = U_{1,n}$, $V_{2,n} = U_{2,n} \setminus U_{1,n}$, $V_{k,n} = U_{k,n} \setminus (U_{1,n} \cup \ldots \cup U_{k-1,n})$ for $k = 2, \ldots, n$, and also let $V_{0,n} = \mathbb{R}^d \setminus \bigcup_{k=1}^{n} U_{k,n}$. Furthermore, define functions

$$f_n := \sum_{k=0}^{n} c_{k,n} \mathbb{I}\{V_{k,n}\}$$

where $c_{0,n} = 0$, $n \in \mathbb{N}$. Evidently, for any $n \in \mathbb{N}$ we have $V_{k,n} \cap V_{m,n} = \varnothing$, if $k \neq m$ ($k, m \in \{1, \ldots, n\}$). Hence,

$$\|f_n\|_\infty \leq \max_{0 \leq k \leq n} c_{k,n} \leq \|f\|_\infty, \; n \in \mathbb{N}. \tag{1.18}$$

For arbitrary sets $D_1, D_2, D_3 \in \mathcal{B}(\mathbb{R}^d)$ one can easily check the inclusion

$$(D_1 \triangle D_3) \subset (D_1 \triangle D_2) \cup (D_2 \triangle D_3). \tag{1.19}$$

Fix $n \in \mathbb{N}$. Note that the sets $\{B_{k,n}, k = 1, \ldots, n\}$ are pairwise disjoint. Therefore for all $k, m \in \{1, \ldots, n\}$ such that $k \neq m$ one has an estimate

$$\mathrm{mes}(U_{k,n} \cap U_{m,n}) \leq \mathrm{mes}((U_{k,n} \setminus B_{k,n}) \cap U_{m,n})$$

$$+ \mathrm{mes}(U_{k,n} \cap (U_{m,n} \setminus B_{m,n})) + \mathrm{mes}((U_{k,n} \setminus B_{k,n}) \cap (U_{m,n} \setminus B_{m,n})) < 3n^{-3}, \tag{1.20}$$

due to (1.17). From (1.19) we infer that

$$\int |f_n(x) - g_n(x)| dx \leq \|f\|_\infty \int \sum_{k=1}^{n} \mathbb{I}\{V_{k,n} \triangle B_{k,n}\}(x) dx = \|f\|_\infty \sum_{k=1}^{n} \mathrm{mes}(V_{k,n} \triangle B_{k,n})$$

$$\leq \|f\|_\infty \sum_{k=1}^{n} \mathrm{mes}(V_{k,n} \triangle U_{k,n}) + \|f\|_\infty \sum_{k=1}^{n} \mathrm{mes}(U_{k,n} \triangle B_{k,n}) =: \|f\|_\infty (R_1 + R_2).$$

$$\tag{1.21}$$

Clearly,

$$\sum_{k=1}^{n} \mathrm{mes}(U_{k,n} \triangle B_{k,n}) \leq n \cdot n^{-3} = n^{-2}, \tag{1.22}$$

while the definition of sets $V_{k,n}$ and the estimate (1.20) imply

$$\sum_{k=1}^{n} \mathrm{mes}(V_{k,n} \triangle U_{k,n}) = \sum_{k=1}^{n} \mathrm{mes}\left(U_{k,n} \cap \left(\bigcup_{m=1}^{k-1} U_{m,n}\right)\right)$$

$$\leq \sum_{k=1}^{n} \sum_{m=1}^{k-1} \mathrm{mes}(U_{k,n} \cap U_{m,n}) \leq n^2 \cdot 3n^{-3} = 3n^{-1}. \tag{1.23}$$

By the relations (1.21)—(1.23) we have now

$$\int |f_n(x) - g_n(x)| dx \to 0 \; \text{ as } n \to \infty.$$

We conclude that $f_n \to f$ in $L^1(\mathbb{R}^d)$ since $g_n \to f$ in $L^1(\mathbb{R}^d)$, as $n \to \infty$. Hence there exists a sequence of positive integers $(n_k)_{k \in \mathbb{N}}$ such that $f_{n_k} \to f$ a.s., $k \to \infty$. Recalling also (1.18) we get the Lemma. $\square$

Continue the proof of the Theorem. Lemma 1.8 allows to take as $(f_r)_{r \in \mathbb{N}}$ a uniformly bounded by $\|f\|_\infty$ sequence of rationally simple functions converging to f in $L^1(\mathbb{R}^d)$. Obviously $\mathsf{E}Z_t(f) = 0$ for any $f \in L$ and all $t > 1$. Consequently, for any $t > 0$ and every $r \in \mathbb{N}$, by Lemma 1.6 one has

$$\mathsf{E}\left(Z_t(f) - Z_t(f_r)\right)^2 = t^{-d}\mathsf{Var}\int \left(f(t^{-1}x) - f_r(t^{-1}x)\right) M(dx)$$

$$\leq t^{-d}\Gamma_M(\|f\|_\infty + \|f_r\|_\infty)\int \left|f(t^{-1}x) - f_r(t^{-1}x)\right| dx$$

$$\leq 2\Gamma_M\|f\|_\infty \int |f(y) - f_r(y)|dy, \tag{1.24}$$

here we used the change of variable $y = t^{-1}x$ in the integral. Since the right-hand side of the last inequality in (1.24) tends to 0 as $r \to \infty$, we have

$$Z_t(f_r) \to Z_t(f) \quad \text{in quadratic mean} \quad \text{as } r \to \infty \tag{1.25}$$

uniformly in $t \in [1, +\infty)$.

Note that since $(f_r)_{r \in \mathbb{N}}$ is a uniformly bounded sequence, the convergence $f_r \to f$ holds also in $L^2(\mathbb{R}^d)$, $r \to \infty$, and therefore

$$\|f_r\|_{L^2} \to \|f\|_{L^2}, \quad r \to \infty. \tag{1.26}$$

Now let us take any sequence $(t_k)_{k \in \mathbb{N}}$ such that $t_k \nearrow \infty$, $k \to \infty$. Introduce

$$\xi_{k,r} = Z_{t_k}(f_r), \quad \xi_{0,r} \sim N(0, \Gamma_M\|f_r\|_{L^2}^2),$$

$$\xi_{k,0} = Z_{t_k}(f), \quad \xi_{0,0} \sim N(0, \Gamma_M\|f\|_{L^2}^2), \quad k, r \in \mathbb{N}.$$

From (1.16), established for rationally simple functions $f_r \in L$, (1.25) and (1.26) we conclude that the conditions of Lemma 5.1.13 are satisfied. Thus $\xi_{k,0} \to \xi_{0,0}$ in law as $k \to \infty$, i.e. (1.16) holds for any $f \in L$ having bounded support.

The final step is to consider $f \in L$ with unbounded support. Note that the functions $f_n = f\mathbb{I}\{x \in \mathbb{R}^d : \|x\| \leq n\}$, $n \in \mathbb{N}$, are uniformly bounded and converge to f in L^1 and L^2, as $n \to \infty$. These functions have compact supports. So it remains to apply Lemma 5.1.13 once again. The Theorem is proved. $\square$

4°. Some auxiliary results. The following two theorems play, for stationary associated random measures, the roles of the LLN and Theorem 3.1.8.

Theorem 1.9. *Let M be an associated stationary random measure which satisfies the condition (1.2). Then, for any $f \in L$ and a_M introduced in (1.1),*

$$t^{-d}M_t(f) \to a_M \int f(x)dx \tag{1.27}$$

in probability as $t \to \infty$.

Proof. The convergence in (1.27) actually holds in the quadratic mean sense. Indeed, by Lemmas 1.1 and 1.6, using the change of variable in the integral, we have

$$\mathsf{E}\left(t^{-d}M_t(f) - a_M\int f(x)dx\right)^2 = \mathsf{E}\left(t^{-d}M_t(f) - t^{-d}a_M\int f(t^{-1}x)dx\right)^2$$

$$= \mathsf{E}\left(t^{-d}\int f(t^{-1}x)M(dx) - t^{-d}\mathsf{E}\int f(t^{-1}x)M(dx)\right)^2$$

$$= t^{-2d}\mathsf{Var}\int f(t^{-1}x)M(dx) \le \Gamma_M\|f\|_\infty t^{-d}\int f(x)dx \to 0, \ \ t \to \infty. \ \Box$$

Theorem 1.10. ([15]) *Let M be a random measure which satisfies the conditions of the previous Theorem. Then, for any $f \in L$,*

$$t^{-d}\mathsf{Var}M_t(f) \to \|f\|_{L^2}^2\Gamma_M \ \ as \ \ t \to \infty.$$

Proof. We begin with the case when f is the indicator of a rational cube $B = B(q,k) = (q^{-1}k, q^{-1}(k+1)]$, here $q \in \mathbb{N}$ and $k \in \mathbb{Z}^d$. Also at first we consider only $t \in \mathbb{N}$. Then, using the notation introduced before (1.9) and taking into account Theorem 3.1.8 and relation (1.10), we may write

$$t^{-d}\mathsf{Var}M_t(f) = t^{-d}\mathsf{Var}\int \mathbb{I}_{B(q,k)}(t^{-1}x)M(dx) = t^{-d}\mathsf{Var}M(tB(q,k))$$

$$= t^{-d}\mathsf{Var}\sum_{l\in\mathbb{Z}^d:\, kt\le l<(k+1)t} X_l \to \sum_{j\in\mathbb{Z}^d} cov(X_0, X_j) = \Gamma_M \mathrm{mes}(B(q,k)) \tag{1.28}$$

as $t \to \infty$. Now note that $\mathrm{mes}(B(q,k)) = \|f\|_{L^2}^2$. We will establish the above asymptotic relation without the restriction that t be an integer. Clearly it suffices to prove that

$$t^{-d}\mathsf{Var}M_t(f) - [t]^{-d}\mathsf{Var}M_{[t]}(f) \to 0, \ \ t \to \infty. \tag{1.29}$$

It is obvious that (1.28) and the Lagrange formula imply, for $t > 1$,

$$([t]^{-d} - t^{-d})\mathsf{Var}M_{[t]}(f) \le 2d[t]^{-d-1}\mathsf{Var}M_{[t]}(f) \to 0, \ \ t \to \infty. \tag{1.30}$$

If X and Y are square-integrable random variables, then

$$\mathsf{Var}X - \mathsf{Var}Y = \mathsf{Var}(X - Y) + 2cov(Y, X - Y)$$

and accordingly,

$$|\mathsf{Var}X - \mathsf{Var}Y| \le \mathsf{Var}(X - Y) + 2\sqrt{\mathsf{Var}Y\mathsf{Var}(X - Y)}. \tag{1.31}$$

Applying this estimate with $f = \mathbb{I}_B$, we see that

$$t^{-d}\left|\mathsf{Var}M_t(f) - \mathsf{Var}M_{[t]}(f)\right| \le t^{-d}\mathsf{Var}\left(M(tB) - M([t]B)\right)$$

$$+ 2t^{-d}\left(\mathsf{Var}M([t]B)\mathsf{Var}\left(M(tB) - M([t]B)\right)\right)^{1/2} \le 2\Gamma_M t^{-d}\mathrm{mes}((tB)\triangle([t]B))$$

$$+ 2 \left(t^{-2d} \mathrm{Var} M([t]B) 2\Gamma_M \mathrm{mes}((tB)\triangle([t]B)) \right)^{1/2} \to 0 \quad \text{as} \ t \to \infty,$$

due to (1.28), Lemma 1.6 and (1.14). From this observation and (1.30) the required relation (1.29) follows.

Before we proceed further, let us consider a set $B = \cup_{k \in V} B(q, k)$ where $q \in \mathbb{N}$ and finite $V \subset \mathbb{Z}^d$. By the argument leading to (1.28) (with appropriate change of sets $U(t, k)$) we see that

$$t^{-d} \mathrm{Var} M_t(\mathbb{I}\{B\}) = t^{-d} \mathrm{Var} M(tB) \to \Gamma_M \mathrm{mes}(B), \quad t \to \infty. \tag{1.32}$$

On the other hand, writing B as the union of small cubes $B(q, k), k \in V$, we have

$$t^{-d} \mathrm{Var} M(tB) - t^{-d} \sum_{k \in V} \mathrm{Var} M(tB(q, k))$$

$$= t^{-d} \sum_{j, k \in V, \, j \neq k} cov(M(tB(q, j)), M(tB(q, k))). \tag{1.33}$$

By virtue of (1.28) and (1.32) the left-hand side of (1.33) tends to zero as $t \to \infty$. Therefore, the right-hand side of this formula also tends to zero.

Now we can prove the Theorem for rationally simple function

$$f = \sum_{j \in V} a_j \mathbb{I}\{B(q, j)\}$$

where $q \in \mathbb{N}$ and V is a finite subset of $\mathbb{Z}^d$. In view of (1.33) and the association property,

$$\left| \sum_{j, k \in V, \, j \neq k} a_j a_k cov(M(tB(q, j)), M(tB(q, k))) \right|$$

$$\leq \max_{j=1,\dots,N} a_j^2 \sum_{j, k \in V, \, j \neq k} cov(M(tB(q, j)), M(tB(q, k))) \to 0, \quad t \to \infty.$$

Then $t^{-d} \mathrm{Var} M_t(f)$ is equal to the following expression:

$$t^{-d} \sum_{j \in V} a_j^2 \mathrm{Var} M(tB(q, j)) + t^{-d} \sum_{j, k \in V, \, j \neq k} a_j a_k cov(M(tB(q, j)), M(tB(q, k)))$$

$$\to \Gamma_M \sum_{j \in V} a_j^2 \mathrm{mes}(B(q, j)) = \Gamma_M \int f^2(x) dx = \Gamma_M \|f\|_{L^2}^2, \quad t \to \infty, \tag{1.34}$$

by (1.28) and (1.33).

Let $f \in L$. Clearly, one can assume that $|f(x)| \leq \|f\|_\infty, x \in \mathbb{R}^d$. If f has a compact support, then Lemma 1.8 ensures that there exists a sequence of rationally simple functions $(f_n)_{n \in \mathbb{N}}$ such that $f_n \to f$ $(n \to \infty)$ mes-a.e. and in $L^1(\mathbb{R}^d)$. By the same Lemma, this sequence can be chosen uniformly bounded by $\|f\|_\infty$. Thus $f_n \to f$ in $L^2(\mathbb{R}^d)$ by the dominated convergence theorem.

Write an obvious estimate

$$\delta_t := \left| t^{-d} \mathsf{Var} \int f(t^{-1}x) M(dx) - \Gamma_M \|f\|_{L^2}^2 \right|$$

$$\leq \left| t^{-d} \mathsf{Var} \int f(t^{-1}x) M(dx) - t^{-d} \mathsf{Var} \int f_n(t^{-1}x) M(dx) \right|$$

$$+ \left| t^{-d} \mathsf{Var} \int f_n(t^{-1}x) M(dx) - \Gamma_M \|f_n\|_{L^2}^2 \right| + \Gamma_M \left| \|f_n\|_{L^2}^2 - \|f\|_{L^2}^2 \right| . \tag{1.35}$$

Relation (1.31) for $X = \int f(t^{-1}x) M(dx)$ and $Y = \int f_n(t^{-1}x) M(dx)$ and Lemma 1.6 yield

$$\left| t^{-d} \mathsf{Var} \int f(t^{-1}x) M(dx) - t^{-d} \mathsf{Var} \int f_n(t^{-1}x) M(dx) \right|$$

$$\leq 2\|f\|_\infty \Gamma_M t^{-d} \int \left| f(t^{-1}x) - f_n(t^{-1}x) \right| dx$$

$$+ 2\sqrt{2}\|f\|_\infty \Gamma_M t^{-d} \left(\int \left| f(t^{-1}x) - f_n(t^{-1}x) \right| dx \int |f_n(t^{-1}x)| dx \right)^{1/2}$$

$$= 2\|f\|_\infty \Gamma_M \left(\|f - f_n\|_{L^1} + (2\|f_n\|_{L^1} \|f - f_n\|_{L^1})^{1/2} \right) . \tag{1.36}$$

On account of (1.34) and the above mentioned properties of the sequence $(f_n)_{n \in \mathbb{N}}$ from (1.35) and (1.36) it follows that $\limsup_{t \to \infty} \delta_t$ can be made arbitrarily small by an appropriate choice of n. This proves the Theorem for $f \in L$ with compact support.

If $f \in L$ does not have compact support, take the sequence $(f_n)_{n \in \mathbb{N}}$ where $f_n(x) = f(x) \mathbb{I}\{\|x\| \leq n\}$, $x \in \mathbb{R}^d$. Then $f_n \to f$ mes-a.e. and in $L^1(\mathbb{R}^d)$, as $n \to \infty$. Since all f_n are bounded by $\|f\|_\infty$, the convergence holds also in $L^2(\mathbb{R}^d)$. It remains to apply the same argument once again. $\square$

To finish the Section we present one more auxiliary result due to Bakhtin.

Lemma 1.11. ([15]) *Let M be an associated, square-integrable random measure on $(\mathbb{R}^d, \mathcal{B}_0(\mathbb{R}^d))$ and let $G_1, G_2 \subset \mathbb{R}^d$ be Borel sets such that $G_1 \cap G_2 = \varnothing$ and $G_1 \cup G_2 = \mathbb{R}^d$. Assume that $\mathrm{cov}(M(B_1), M(B_2)) = 0$ for any bounded Borel sets $B_i \subset G_i$, $i = 1, 2$. Let $f_1, \ldots, f_k : \mathbb{R}^d \to \mathbb{R}$ be Borel functions such that $f_i(x) \geq 0$ for $x \in G_1$ and $f_i(x) \leq 0$ for $x \in G_2$, $i = 1, \ldots, k$. Suppose also that all these functions are integrable with respect to the measure $\mu(\cdot) = \mathsf{E}M(\cdot)$. Then the random vector*

$$X = \left(\int_{\mathbb{R}^d} f_1(x) M(dx), \ldots, \int_{\mathbb{R}^d} f_k(x) M(dx) \right) \in \mathbf{A}.$$

Proof. Let $X(+) = (\int_{G_1} f_i(x) M(dx))_{i=1}^k$ and $X(-) = (\int_{G_2} f_i(x) M(dx))_{i=1}^k$. The random vector $X(+)$ is associated on account of Theorem 1.3.27. By the same result the random vector $(-X(-))$ is associated, hence $X(-)$ is also associated. The restrictions of M onto G_1 and G_2 (more precisely, onto the classes of bounded Borel subsets of G_1 and G_2) are two random systems which are uncorrelated and jointly associated. Therefore, they are independent. Consequently, $X(+)$ and $X(-)$ are independent associated random vectors and their sum X is associated by Theorem 1.1.8, (d). $\square$

2   PDE with random initial data

The important source of functionals in random fields are the partial differential equations with random initial data, random perturbations, or random coefficients etc. In this regard we refer, e.g., to works [77], [228], [257], [148]. Here we restrict ourselves to the study of asymptotical behavior of the transformed Burgers equation. This equation is widely used in various domains of Modern Physics and Astronomy (see, e.g., [369], [420]).

1°. The multidimensional Burgers equation. A prominent role in mathematical physics (hydrodynamics, astronomy etc.) is played by the *Burgers equation* and the corresponding Cauchy problem

$$\begin{cases} \frac{\partial v}{\partial t} + (v, \nabla_x)v = \varkappa \Delta v, \\ v(0, x) = v_0(x) = -\nabla_x \xi(x), \\ (t, x) \in \mathbb{R}_+ \times \mathbb{R}^d, \, v \in \mathbb{R}^d. \end{cases}$$

It is convenient to take the viscosity coefficient of a system $\varkappa = 1/2$. The so-called *Hopf–Cole substitution* $v(t, x) = -\nabla_x \log u(t, x)$ allows to rewrite the above system:

$$\begin{cases} \frac{\partial u}{\partial t} = \frac{1}{2}\Delta u, \\ u(0, x) = u_0(x) = e^{-\xi(x)}, \\ (t, x) \in \mathbb{R}_+ \times \mathbb{R}^d, \, v \in \mathbb{R}^d. \end{cases}$$

The unique solution is given by the fraction

$$v(t, x) = \frac{\int_{\mathbb{R}^d} \nabla_x g(t, x, y) \exp\{\xi(y)\} dy}{\int_{\mathbb{R}^d} g(t, x, y) \exp\{\xi(y)\} dy}$$

$$= \frac{\int_{\mathbb{R}^d} t^{-1}(x - y) \exp\left\{\xi(y) - \|x - y\|^2/2t\right\} dy}{\int_{\mathbb{R}^d} \exp\left\{\xi(y) - \|x - y\|^2/2t\right\} dy}, \tag{2.1}$$

with

$$g(t, x, y) = \frac{1}{(2\pi t)^{d/2}} \exp\left\{-\|x - y\|^2/2t\right\},$$

if the integral in the denominator of (2.1) converges for any $(t, x) \in \mathbb{R}_+ \times \mathbb{R}^d$, $t > 0$.

Consider the situation when the initial potential ξ is some random field. More precisely, for a random measure M we write (formally) $e^{\xi(y)} dy = M(dy)$ and treat the corresponding solution as

$$v(t, x) = \frac{A(t, x)}{tB(t, x)} = \frac{\int_{\mathbb{R}^d} t^{-1}(x - y) \exp\left\{-\|x - y\|^2/2t\right\} M(dy)}{t \int_{\mathbb{R}^d} \exp\left\{-\|x - y\|^2/2t\right\} M(dy)} \tag{2.2}$$

where $t > 0$, $x \in \mathbb{R}^d$. Note that by our choice of M the denominator in (2.2) is positive. Introduce, for $T > 0$, the *parabolic rescaled* solution of the Burgers equation by the formula

$$V_T(t, x) = T^{d/4+1/2} v(tT, x\sqrt{T})$$

and set

$$A_T(t, x) = (A_{T,i}(t, x))_{i=1,\ldots,d} = T^{-d/4-1/2} A\left(tT, x\sqrt{T}\right), \qquad (2.3)$$

$$B_T(t, x) = T^{-d/4-1/2} B\left(tT, x\sqrt{T}\right), \quad T > 0, \ t > 0, \ x \in \mathbb{R}^d.$$

2°. Auxiliary results for the transformed velocity. At first we study separately the asymptotical behavior of the numerator and denominator in (2.2).

Lemma 2.1. ([13]) *Let M be a stationary associated random measure on $(\mathbb{R}^d, \mathcal{B}_0(\mathbb{R}^d))$ satisfying the (FS) condition (see Definition 1.3). Then the finite-dimensional distributions of the random fields $(A_T(\cdot, \cdot))_{T>0}$ converge, as $T \to \infty$, to ones of a Gaussian centered random field A_∞ having the covariance function*

$$\mathsf{E}A_{\infty,i}(t, a)A_{\infty,j}(s, b) = H_{ij}(t, a, s, b) = (2\pi)^{d/2}\Gamma_M \exp\left\{-\frac{\|a - b\|^2}{2(t + s)}\right\} G_{ij}(t, a, s, b)$$

where

$$G_{ij}(t, a, s, b) = \begin{cases} \left(ts(t + s)^{-1}\right)^{d/2+1} \left(1 - (t + s)^{-1}(a_i - b_i)^2\right), & i = j, \\ -\left(ts(t + s)^{-1}\right)^{d/2+1} (t + s)^{-1}(a_i - b_i)(a_j - b_j), & i \neq j, \end{cases}$$

$i, j = 1, \ldots, d.$

Proof. Apply the Cramér-Wald device. For $N \in \mathbb{N}$ take $b \in \mathbb{R}^N$, $t(1), \ldots, t(N) > 0$, $x(1), \ldots, x(N) \in \mathbb{R}^d$ and $i(1), \ldots, i(N) \in \{1, \ldots, d\}$. Write

$$\overline{A}_T := \sum_{k=1}^{N} b_k A_{T,i(k)}(t(k), x(k))$$

$$= T^{-d/4-1/2} \sum_{k=1}^{N} b_k \int_{\mathbb{R}^d} \left(x_{i(k)}(k)\sqrt{T} - y_{i(k)}\right) \exp\left\{-\frac{\|x(k)\sqrt{T} - y\|^2}{2t(k)T}\right\} M(dy)$$

$$= \sum_{k=1}^{N} b_k T^{-d/4} \int_{\mathbb{R}^d} f_k(yT^{-1/2}) M(dy) = \sum_{k=1}^{N} b_k T^{-d/4} M_T(f_k)$$

where M_T denotes the last integral and

$$f_k(y) = (x_{i(k)}(k) - y_{i(k)}) \exp\left\{-\frac{\|x(k) - y\|^2}{2t(k)}\right\}.$$

Clearly, $f_k \in L^1(\mathbb{R}^d) \cap L^\infty(\mathbb{R}^d)$ and $\int_{\mathbb{R}^d} f_k(y)dy = 0$. Therefore, according to Lemma 1.1 and Theorem 1.7, we conclude that

$$\overline{A}_T \to N\left(0, \sum_{1 \leq k,l \leq N} b_k b_l H_{i(k),i(l)}(t(k), x(k), t(l), x(l))\right) \quad \text{in law,} \quad \text{as } T \to \infty. \ \square$$

Lemma 2.2. ([13]) *Let the conditions of Lemma 2.1 be satisfied. Then, for any $t > 0$ and $x \in \mathbb{R}^d$, one has $B_T(t, x) \to a_M(2\pi t)^{d/2}$ in probability as $T \to \infty$.*

Proof. By Theorem 1.9 the following convergence in probability holds

$$B_T(t,x) = T^{-d/2} \int_{\mathbb{R}^d} \exp\left\{-\frac{\|x - yT^{-1/2}\|^2}{2t}\right\} M(dy)$$

$$\to a_M \int_{\mathbb{R}^d} \exp\left\{-\frac{\|x - y\|^2}{2t}\right\} dy = a_M(2\pi t)^{d/2}. \quad \Box$$

For any $T > 0$ define

$$V_T(t,x) = T^{d/4+1/2} v\left(tT, x\sqrt{T}\right). \tag{2.4}$$

Theorem 2.3. ([13]) *Let the conditions of Lemma 2.1 be fulfilled. Then the finite-dimensional distributions of the random fields $(V_T(\cdot,\cdot))_{T>0}$ converge, as $T \to \infty$, to ones of a Gaussian centered field $V_\infty = (V_{\infty,i})_{i=1,\ldots,d}$ with covariance function*

$$\mathsf{E}V_{\infty,i}(t,a)V_{\infty,j}(s,b)$$

$$= \begin{cases} R(M,t,s)\exp\left\{-\|a-b\|^2/(2t+2s)\right\}\left(1 - (a_i - b_i)^2/(t+s)\right), & i = j, \\ -R(M,t,s)\exp\left\{-\|a-b\|^2/(2t+2s)\right\}(a_i - b_i)(a_j - b_j)(t+s)^{-1}, & i \neq j, \end{cases}$$

$R(M,t,s) = \Gamma_M/(a_M^2(2\pi)^{d/2}(t+s)^{d/2+1})$, $i,j = 1,\ldots,d$; Γ_M *was defined in* (1.2).

Proof. This Theorem follows from Lemmas 2.1, 2.2, relation (2.2) and the easy multidimensional modification of Lemma 3.1.27. $\Box$

Let M be an associated measure satisfying the finite susceptibility condition and such that $\mathsf{P}(E_0) = 0$ where $E_0 = \{\omega : M(\mathbb{R}^d) = 0\}$. Consider

$$v(t) = v(t,0) = -\frac{\int_{\mathbb{R}^d} y \exp\left\{-\|y\|^2/2t\right\} M(dy)}{t \int_{\mathbb{R}^d} \exp\left\{-\|y\|^2/2t\right\} M(dy)} =: -\frac{A(t)}{tB(t)}, \quad t > 0.$$

We are going to study the behavior of $v(t)$ (being properly rescaled and normalized) as a random process. To this end, let us at first exclude from the probability space the events E_0 and

$$E_\infty = \{\omega : \int_{\mathbb{R}^d} e^{-\|y\|} M(dy) = +\infty\}.$$

The latter event is of zero probability (e.g., by Lemma 1.3.25). Thus the trajectories of A and B are continuous for $\omega \in \Omega \setminus (E_0 \cup E_\infty)$.

Introduce

$$A_T(t) = \frac{A(tT)}{T^{d/4+1/2}}, \quad T > 0, \ t > 0. \tag{2.5}$$

As usual, for metric spaces S and M the symbol $C(S,M)$ will denote the space of continuous functions on S taking values in M.

Lemma 2.4. *For any $r \in (0,1)$ the processes $(A_T)_{T>0}$ converge in law in the space $C([r,r^{-1}], \mathbb{R}^d)$, endowed with uniform topology, to a Gaussian centered process A_∞ with covariance function*

$$\mathsf{E}A_{\infty,i}(t)A_{\infty,j}(s) = \delta_{ij}(2\pi)^{d/2}\Gamma_M\left(\frac{ts}{t+s}\right)^{d/2+1}.$$

Here δ_{ij} is the Kronecker symbol.

Proof. The convergence of corresponding finite-dimensional distributions has been already proved (Lemma 2.1), thus we concentrate on the tightness. More exactly, we prove that, for any real sequence $(T_n)_{n\in\mathbb{N}}$ such that $T_n \nearrow \infty$ as $n \to \infty$, the sequence of processes $(A_{T_n,i})_{n\in\mathbb{N}}$ is tight.

Without loss of generality we may consider a single (spatial) coordinate in $\mathbb{R}^d$, say, $i = 1$. For $i = 1, \ldots, d$ set

$$A_{T,i}^{\pm}(t) = T^{-d/4-1/2} \int_{\mathbb{R}^d} y_i^{\pm} \exp\left\{ -\frac{\|y\|^2}{2tT} \right\} M(dy),$$

here and further on it is agreed to choose the same upper sign ($+$ or $-$) at both sides of the relation. These processes have continuous trajectories. In contrast to A these processes are not centered, so we introduce the processes $(X_n^{\pm}(\cdot))_{n\in\mathbb{N}}$ by

$$X_n^{\pm}(t) = A_{T_n,1}^{\pm}(t) - \mathsf{E} A_{T_n,1}^{\pm}(t).$$

We proceed with X_n^+ only, the study of X_n^- being analogous. On account of the monotone convergence theorem $\lim_{t\searrow 0} X_n^+(t) = 0$ a.s. Due to (5.1.2) for the tightness[3] we check that, for any $\eta, \varepsilon > 0$, there exist $\delta \in (0,1)$ and $n_0 \in \mathbb{N}$ such that, for any $n > n_0$ and any $t \in [r, r^{-1}]$,

$$\delta^{-1}\mathsf{P}\left(\sup |X_n^+(s) - X_n^+(t)| > \varepsilon \right) \leq \eta \tag{2.6}$$

where the supremum is taken over $s \in [r, r^{-1}]$, $t \leq s \leq t+\delta$. This relation is implied by the following condition: for any $\eta, \varepsilon > 0$ there exist $\delta \in (0,1)$ and $n_0 \in \mathbb{N}$ such that, for all $m \in \mathbb{N}$, any $n > n_0$ and every $t \in [r, r^{-1}]$,

$$\delta^{-1}\mathsf{P}\left(\sup_{k\in\mathbb{N}\cap[0,2^m\delta]} |X_n^+(t + 2^{-m}k) - X_n^+(t)| > \varepsilon \right) \leq \eta. \tag{2.7}$$

Indeed, since the trajectories of X_n^+ are continuous, the events which stand inside the probability sign in (2.7) increase, as $m \to \infty$, to the event which stands in (2.6). By Lemma 1.6, for any n such that $T_n > 1$, we have

$$\mathsf{Var}(X_n^+(t) - X_n^+(s))$$

$$\leq T_n^{-d/2-1}\Gamma_M \sup_{y\in\mathbb{R}^d, y_1>0} y_1 F(y, s, t, T_n) \int_{\mathbb{R}^d} y_1^+ F(y, s, t, T_n)) dy$$

$$\leq (2\pi)^{\frac{d}{2}-\frac{1}{2}}\Gamma_M |t^{\frac{d}{2}+\frac{1}{2}} - s^{\frac{d}{2}+\frac{1}{2}}| C_1$$

where

$$F(y, s, t, T) = \left| \exp\left\{ -\frac{\|y\|^2}{2tT} \right\} - \exp\left\{ -\frac{\|y\|^2}{2sT} \right\} \right|, \quad C_1 \leq \sup_{x\in\mathbb{R}^d,\, x_1>0} x_1 e^{-\frac{r}{2}\|x\|^2} < \infty.$$

Consequently,

$$\mathsf{Var}(X_n^+(t) - X_n^+(s)) \leq C_2(d, \Gamma_M, r)|t - s|.$$

---

[3]The condition given there for a space of functions on $[0,1]$ is easily adapted to functions defined on the segment $[r, r^{-1}]$.

Let $\delta \in (0,1)$ be specified later. Denote by p_δ the left-hand side of inequality (2.7). Hence p_δ admits an upper bound

$$\delta^{-1}\mathsf{P}\left(\sup_{k\in\mathbb{N}\cap[0,2^m\delta]}|X_n^+(t+2^{-m}k)-X_n^+(t)|>\frac{\varepsilon}{\sqrt{C_2\delta}}\sqrt{\mathsf{Var}(X_n^+(t+\delta)-X_n^+(t))}\right).$$

The vector $(X_n^+(t+2^{-m}(k+1))-X_n^+(t+2^{-m}k))_{k=0,\dots,[2^m\delta]-1}$ is associated. Thus by Theorem 2.4.5 there exist constants $C_3>0$ and $\delta_0>0$ such that, for any $\delta\in(0,\delta_0)$,

$$p_\delta \leq \frac{3}{\delta}\mathsf{P}\left(|X_n^+(t+\delta)-X_n^+(t)|\geq\frac{C_3\varepsilon}{\sqrt{\delta}}\sqrt{\mathsf{Var}(X_n^+(t+\delta)-X_n^+(t))}\right).$$

The probability in the last expression tends to $\mathsf{P}(|Z|\geq C_3\varepsilon\delta^{-1/2})$ for $Z\sim N(0,1)$, by virtue of Theorem 1.7. Hence there exists $n_0\in\mathbb{N}$ such that the inequality

$$p_\delta \leq \frac{3}{\delta}\sqrt{\frac{2}{\pi}}\int_{C_3\delta^{-1/2}}e^{-u^2/2}du \to 0 \quad\text{as}\quad \delta\to 0$$

holds for $n>n_0$, which completes the proof. $\square$

Analogously to (2.5) introduce $B_T(t)=T^{-d/2}B(tT)$ and

$$B_T^0(t)=T^{-d/2}\left(B(tT)-\mathsf{E}B(tT)\right)=T^{-d/2}B(tT)-a_M(2\pi t)^{d/2},\quad T,t>0.$$

Lemma 2.5. *For any* $r\in(0,1)$ *the processes* $(B_T^0)_{T>0}$ *converge in law in the space* $C([r,r^{-1}],\mathbb{R})$, *endowed with uniform topology, to the identically zero process.*

Proof. The convergence of finite-dimensional distributions was already proved in Lemma 2.2. The proof of tightness is analogous to that of Lemma 2.4. Namely, due to Lemma 1.6, for $t>s$, $t,s\in[r,r^{-1}]$, we obtain the bound

$$\mathsf{Var}(B_T^0(s)-B_T^0(t))\leq T^{-d}\mathsf{Var}\int_{\mathbb{R}^d}F(y,s,t,T)M(dy)$$

$$\leq T^{-d}\Gamma_M\sup_{y\in\mathbb{R}^d}F(y,s,t,T)\int_{\mathbb{R}^d}F(y,s,t,T)dy\leq C_4(d,r,\Gamma_M)T^{-d/2}|t-s|.$$

Therefore one has

$$p_\delta \leq \frac{2}{\delta}\mathsf{P}\left(|B_T^0(t+\delta)-B_T^0(t)|>\frac{C_5T^{d/4}}{\sqrt{\delta}}\sqrt{\mathsf{Var}(B_T^0(t+\delta)-B_T^0(t))}\right)\leq\frac{C_6}{T^{d/2}}.\square$$

3°.The Polish space $C((0,+\infty),\mathbb{R}^d)$. At first we introduce a metric in the space $C_\infty=C((0,+\infty),\mathbb{R}^d)$. For functions $f,g\in C_\infty$ set

$$\rho_n(f,g)=\sup_{x\in[(n+1)^{-1},n+1]}|f(x)-g(x)|,\quad n\in\mathbb{N},$$

$$\rho(f,g)=\sum_{n=1}^{\infty}2^{-n}\frac{\rho_n}{1+\rho_n}.$$

The topology induced by metric ρ will be called the *LU-topology*. The restriction of a function $f\in C_\infty$ onto $\mathbf{I}_n:=[(n+1)^{-1},n+1]$, $n\in\mathbb{N}$, is denoted by $f|_{\mathbf{I}_n}$.

Lemma 2.6. *The following statements are true:*

(a) *The space (C_∞, ρ) is Polish.*

(b) *Suppose that the set K_n is a compact in $(C([(n+1)^{-1}, n+1], \mathbb{R}^d), \rho_n)$ for $n \in \mathbb{N}$. Then the closure of the set*

$$K = \{f \in C_\infty : f|_{\mathbf{I}_n} \in K_n \text{ for any } n \in \mathbb{N}\} \tag{2.8}$$

is a compact in (C_∞, ρ).

(c) *Let $(X_k)_{k \in \mathbb{N}}$ be a sequence of random elements taking values in (C_∞, ρ). Assume that there exists a random element X with values in (C_∞, ρ) such that for any $n \in \mathbb{N}$ the restrictions $X_k\big|_{\mathbf{I}_n}$ converge (as $k \to \infty$) in law in the space $(C(\mathbf{I}_n, \mathbb{R}^d), \rho_n)$ to $X\big|_{\mathbf{I}_n}$. Then $X_k \to X$ in law, as $k \to \infty$.*

Proof. (a) The fact that $\rho(\cdot, \cdot)$ is a metric is easy (as ρ_n is a metric on $C(\mathbf{I}_n, \mathbb{R}^d)$, for any $n \in \mathbb{N}$). If $(f_k)_{k \in \mathbb{N}}$ is a Cauchy sequence in C_∞ then, for any $n \in \mathbb{N}$, the restrictions of f_k onto $\mathbf{I}_n$ form a Cauchy sequence in $C(\mathbf{I}_n, \mathbb{R}^d)$. Thus, for any $n \in \mathbb{N}$, there exists a corresponding limit $f^{(n)} : \mathbf{I}_n \to \mathbb{R}^d$. If $m > n$, then $f^{(n)}$ and $f^{(m)}$ coincide on $\mathbf{I}_n$. Thus there exists a continuous function $f : (0, +\infty) \to \mathbb{R}^d$ which is a pointwise limit of f_k as $k \to \infty$. Moreover, for any $n \in \mathbb{N}$, one has $\rho_n(f_k, f) \to 0$ as $k \to \infty$. Thus the relation $\rho(f_k, f) \to 0$ follows by the dominated convergence theorem. This proves that C_∞ is complete.

Finally, for any $n \in \mathbb{N}$, let $\mathcal{D}_n$ be a countable dense subset of $C(\mathbf{I}_n, \mathbb{R}^d)$. Let $\mathcal{S}_n \subset C_\infty$ be the set of functions $h \in \mathcal{D}_n$ extended onto $(0, +\infty)$ as follows: $h(x) = 0$ for $x \notin [(n+2)^{-1}, n+2]$ and h is linear on the segments $[(n+2)^{-1}, (n+1)^{-1}]$ and $[n+1, n+2]$. Fix arbitrarily some $f \in C_\infty$ and some $\varepsilon \in (0, 1)$. Let $N \in \mathbb{N}$ ($N > 1$) be a number such that $\sum_{n \geq N} 2^{-n} < \varepsilon/2$. Take $g \in C_\infty$ to be the function which coincides with f on the segment $\mathbf{I}_{N-1}$ and vanishes outside the set $\mathbf{I}_N$. Then $\rho_m(f, g) = 0$ whenever $m < N$. There exists some $h \in \mathcal{S}_N$ such that $\rho_N(h, g) < \varepsilon/2$. Then, obviously, $\rho_m(h, g) < \varepsilon/2$ for any $m \leq N$. Furthermore, for $m > N$ one has

$$\rho_m(h, g) = \rho_N(h, g) \vee \sup_{x \notin \mathbf{I}_N} |h(x)| \leq \varepsilon/2$$

as $|h(N+1)| \vee |h(1/(N+1))| < \varepsilon/2$. So,

$$\rho(f, h) \leq \rho(f, g) + \rho(g, h) = \sum_{n=1}^{\infty} 2^{-n} \frac{\rho_n(f, g)}{1 + \rho_n(f, g)} + \sum_{n=1}^{\infty} 2^{-n} \frac{\rho_n(g, h)}{1 + \rho_n(g, h)}$$

$$\leq \sum_{n=N}^{\infty} 2^{-n} \frac{\rho_n(f, g)}{1 + \rho_n(f, g)} + \frac{\varepsilon}{2} \sum_{n=1}^{\infty} 2^{-n} \leq \varepsilon$$

by the choice of N. We have proved that the countable set $\cup_{n \in \mathbb{N}} \mathcal{S}_n$ is dense in C_∞, hence this space is separable.

(b) Take arbitrary $\varepsilon > 0$. Let $N \in \mathbb{N}$ ($N > 1$) be a number such that $\sum_{n \geq N} 2^{-n} < \varepsilon/2$. By compactness, it is possible to find a finite set $R \subset C(\mathbf{I}_N, \mathbb{R}^d)$ such that[4] for any $f \in K_N$ there is some $g \in R$ with $\rho_N(f, g) < \varepsilon/2$.

---

[4]Such set is usually called an $\varepsilon/2$-net in the compact K_N, see, e.g., [239, Ch. II, §7, 2].

For an arbitrary function $f \in K$ we let $h \in R$ be a function verifying $\rho_{N-1}(h, f|_{\mathbf{I}_{N-1}}) < \varepsilon/2$. Extend this function h onto the whole real line $\mathbb{R}$ in such a way that $h \in C_\infty$. Then

$$\rho(f, h) = \sum_{n=1}^\infty 2^{-n} \frac{\rho_n(f, h)}{1 + \rho_n(f, h)} \le \frac{\varepsilon}{2} + \sum_{n=1}^{N-1} 2^{-n} \rho_n(f, h) \le \frac{\varepsilon}{2} + \sum_{n=1}^{N-1} 2^{-n} \rho_{N-1}(f, h) \le \varepsilon.$$

We have proved that R is a finite ε-net in K (under the proviso that any function in R is continued onto $(0, +\infty)$). Thus K is a subset of compact, as required.

(c) Take arbitrary $\varepsilon > 0$. For any $n \in \mathbb{N}$ let K_n be a compact subset of $C(\mathbf{I}_n, \mathbb{R}^d)$ such that

$$P\left(\omega : X_k(\omega)|_{\mathbf{I}_n} \notin K_n\right) < \frac{\varepsilon}{2^n} \quad \text{for any } k \in \mathbb{N}.$$

Such a compact exists by the Prohorov theorem (see, e.g., [39, §6, Th. 6.1,6.2]; its "necessity" part is applicable because C_∞ is a Polish space) since the restrictions of X_k onto $\mathbf{I}_n$ converge in law when $k \to \infty$, hence their distributions are tight in C_∞.

Let K be the subset of C_∞ defined by formula (2.8). Then obviously

$$P\left(\omega : X_k(\omega) \notin K\right) < \varepsilon \quad \text{for any } k \in \mathbb{N}.$$

Since K is compact (by Lemma 2.6, (b) which is already proved), by the Prohorov theorem the set of distributions of $(X_k)_{k \in \mathbb{N}}$ is tight in C_∞. Therefore, each of its subsequences contains in its turn a subsequence converging in law to some random function Y, and it only remains to show that the distributions of X and Y coincide. But this property is true as, for any $N \in \mathbb{N}$ and all $t_1, \ldots, t_N > 0$, the laws of random vectors $(X_{t_1}, \ldots, X_{t_N})$ and $(Y_{t_1}, \ldots, Y_{t_N})$ coincide. The last assertion follows easily from the Lemma's condition by taking n so large that $t_1, \ldots, t_N \in \mathbf{I}_n$. $\square$

4°. Functional limit theorems for the transformed solution of the Burgers equation with random data and fixed spatial argument. The last two Lemmas provide us with a possibility to derive a functional limit theorem. For the processes

$$Z_T(x, t) = T^{d/4+1/2} v(x, Tt), \quad T, t > 0, \ x \in \mathbb{R}^d,$$

we prove the weak convergence of their distributions in the space C_∞.

Theorem 2.7. ([13]) *Let M be a stationary, associated random measure satisfying the (FS) condition. Assume that $a_M > 0$. Then the distributions of rescaled solutions $(Z_T)_{T>0}$ in C_∞, endowed with LU-topology, converge, as $T \to \infty$, to a Gaussian centered process Z_∞ with covariance function*

$$\mathsf{E} Z_{\infty,i}(t) Z_{\infty,j}(s) = \delta_{ij} \frac{\Gamma_M}{(2\pi)^{d/2} a_M^2 (t + s)^{d/2+1}}. \tag{2.9}$$

Proof. By Lemma 2.6, (c), it suffices to prove that, for any $n \in \mathbb{N}$, the restrictions of processes $Z_T(\cdot)$ onto the segment $\mathbf{I}_n = [(n+1)^{-1}, n+1]$ converge in law to a Gaussian centered process with covariance function determined by (2.9), as $T \to \infty$.

In the further proof $n \in \mathbb{N}$ is fixed, and all the random processes are considered on the segment $\mathbf{I}_n$. The convergence of finite-dimensional distributions of Z_T to ones of Z_∞ is implied by Lemmas 2.4 and 2.5. Thus it remains to prove the tightness.

Fix any $\varepsilon > 0$ and take a compact $K_A \subset C(\mathbf{I}_n, \mathbb{R}^d)$ such that, for any $T > 0$, $\mathsf{P}(A_T(\cdot) \in K_A) > 1 - \varepsilon/2$. This compact exists because Lemma 2.4 implies that the family $(A_T)_{T>0}$ is tight. Analogously, Lemma 2.5 implies the existence of a compact $K_B^0 \subset C(\mathbf{I}_n, \mathbb{R}^d)$ such that $\mathsf{P}(B_T \in K_B^0) > 1 - \varepsilon/4$ for any $T > 0$. Note that $\mathsf{P}(\inf_{t \in \mathbf{I}_n} B_T(t) > 0) = 1$. Consequently, by continuity of trajectories we can find $\mu > 0$ so that

$$\mathsf{P}(\inf_{t \in \mathbf{I}_n} B_T(t) \geq \mu) > 1 - \varepsilon/4.$$

The set

$$D_\mu := \left\{ f \in C(\mathbf{I}_n, \mathbb{R}^d) : \inf_{t \in \mathbf{I}_n} f(t) \geq \mu \right\}$$

is closed in the metric ρ_n. Thus the set $K_B := K_B^0 \cap D_\mu$ is a compact and, moreover, $\mathsf{P}(B_T \in K_B) > 1 - \varepsilon/2$.

Define a set of functions

$$K_Z := \left\{ t \mapsto -\frac{f(t)}{tg(t)}, f \in K_A, g \in K_B \right\} \subset C(\mathbf{I}_n, \mathbb{R}^d).$$

Then $\mathsf{P}(Z_T \notin K_Z) < \varepsilon$ for any $T > 0$. So by the Prohorov theorem the assertion will be proved if we show that K_Z is a subset of a compact[5].

Let $\varepsilon > 0$ be arbitrary and let $f_1, \ldots, f_N$ and $g_1, \ldots, g_L$ be ε-nets in K_A and K_B respectively. One may think that these nets are subsets of K_A and K_B respectively. For any function $h \in K_Z$ defined by way of $h(t) = -tf(t)/g(t)$, $f \in K_A$, $g \in K_B$, let $i \in \{1, \ldots, N\}$, $j \in \{1, \ldots, L\}$ be integers such that

$$\|f_i - f\|_\infty < \varepsilon, \quad \|g_j - g\|_\infty < \varepsilon,$$

here $\| \cdot \|_\infty$ is, as usual, the sup-norm of a continuous function (i.e. the norm corresponding to the metric ρ_n). Then one can write, for $t \in \mathbf{I}_n$,

$$\left| \frac{f(t)}{tg(t)} - \frac{f_i(t)}{tg_j(t)} \right| \leq (n+1) \left(|f(t)| \left| \frac{1}{g(t)} - \frac{1}{g_j(t)} \right| + \frac{|f(t) - f_i(t)|}{g_j(t)} \right)$$

$$\leq (n+1) \left(\sup_{\varphi \in K_A} \|\varphi\|_\infty \frac{|g(t) - g_j(t)|}{\mu^2} + \frac{|f(t) - f_i(t)|}{\mu} \right)$$

$$\leq (n+1) \left(\sup_{\varphi \in K_A} \|\varphi\|_\infty \frac{\varepsilon}{\mu^2} + \frac{\varepsilon}{\mu} \right).$$

As $t \in \mathbf{I}_n$ and $\varepsilon > 0$ were arbitrary, we see that K_Z is completely bounded, hence a subset of a compact. $\square$

---

[5] In fact K_Z is a compact, since it is completely bounded (see the current proof) and closed.

3   Asymptotical behavior of transformed solutions of the Burgers equation

In the previous Section we established the functional central limit theorem for the solution of Burgers equation with fixed spatial argument (we took $x = 0$ but could take any other point). In this Section we are going to consider the parabolic-rescaled solution as a (continuous-parameter) random field on $\mathbb{R}_+ \times \mathbb{R}^d$ and prove the corresponding FCLT.

A crucial difference between these cases is that in the first one we applied the Newman-Wright maximal inequality and thus needed only the finite susceptibility condition. The rate at which the correlations between values of measure M on separated sets decrease was not essential. Now we are going to treat the multi-dimensional situation, hence this rate becomes important (recall from Ch. 2 that the Bulinski-Keane maximal inequality and other inequalities for associated and related random fields require some restrictions on the Cox-Grimmett coefficient). We introduce at first an analog of the Cox-Grimmett coefficient.

Let M be a stationary associated random measure on $\mathbb{R}^d$. Define a sequence $(u_r)_{r\in\mathbb{N}} = (u_r(M))_{r\in\mathbb{N}}$ by

$$u_r = \sum_{j\in\mathbb{Z}^d,|j|\geq r} cov(M(I), M(I+j)), \quad j \in \mathbb{Z}^d, \tag{3.1}$$

where $I = (-1/2, 1/2]^d$ (one can say that we use an auxiliary random field $X = \{M(I+j), j \in \mathbb{Z}^d\}$). If M meets the finite susceptibility condition, then $u_r \to 0$ as $r \to \infty$.

1°. The application of cumulant techniques. We study separately a simple case when M is independently scattered (e.g., defined by a Poisson point field). Suppose that $\mathsf{E}M(I)^k < \infty$ for some $k \in \mathbb{N}$.

Recall that if ξ is a random variable such that $\mathsf{E}|\xi|^k < \infty$, then the *k-th cumulant* of ξ is the number

$$\gamma_k(\xi) = i^{-k} \frac{d^k}{dt^k}\bigg|_{t=0} \log \varphi_\xi(t) \tag{3.2}$$

where φ_ξ is the characteristic function of a random variable ξ, $i^2 = -1$ and log is understood as the main branch of the complex logarithmic function, i.e. $\log 1 = 0$. If $\xi_1, \ldots, \xi_N$ are independent random variables from $L^k(\Omega, \mathcal{F}, \mathsf{P})$, and $a_1, \ldots, a_N \in \mathbb{R}$, then the definition (3.2) implies that

$$\gamma_k(a_1\xi_1 + \ldots + a_N\xi_N) = \sum_{j=1}^{N} a_j^k \gamma_k(\xi_j). \tag{3.3}$$

Therefore the real-valued set function $\gamma_k(M(B))$, $B \in \mathcal{B}_0(\mathbb{R}^d)$, is additive and invariant with respect to shifts in $\mathbb{R}^d$. Moreover, this function is σ-additive as it is continuous at "zero" $(\varnothing)$. To see that, let $(B_n)_{n\in\mathbb{N}}$ be a sequence of bounded Borel sets decreasing to $\varnothing$ as $n \to \infty$; then $\mathsf{E}M(B_n)^m \to 0$ as $n \to \infty$, $m = 1, \ldots, k$, by the monotone convergence theorem (see the beginning of proof of Lemma 1.5 for similar argument). Note that, for any random variable ξ, there exists a polynomial in k arguments such that $\gamma_k(\xi)$ can be written as its value at the point $(\mathsf{E}\xi, \ldots, \mathsf{E}\xi^k)$. Namely (see, e.g., [326, Ch. 1, §3]),

$$\gamma_k(\xi) = k! \sum_{\substack{m \in \mathbb{Z}_+^k: \\ m_1 + 2m_2 + \ldots + km_k = k}} (-1)^{r_k - 1}(r_k - 1)! \prod_{v=1}^{k} \frac{1}{m_v!} \left(\frac{\mathsf{E}\xi^v}{v!} \right)^{m_v}. \tag{3.4}$$

Therefore, $\gamma_k(M(B_n)) \to 0$ as $n \to \infty$. Consequently, $\gamma_k(M(B)) = G_{k,M}\mathrm{mes}(B)$ for any $B \in \mathcal{B}_0(\mathbb{R}^d)$ where

$$G_{k,M} = \gamma_k(M(I)), \quad k \in \mathbb{N}.$$

Lemma 3.1. ([14]) *Let M be a stationary, independently scattered random measure such that $\mathsf{E}M(I)^k < \infty$. Then, for any $f \in L = L^1(\mathbb{R}^d) \cap L^\infty(\mathbb{R}^d)$,*

$$\gamma_k \left(\int f(x)M(dx) \right) = G_{k,M} \int f^k(x)dx.$$

Here and further on all integrals are taken over $\mathbb{R}^d$.

Proof. If f is an indicator of a bounded Borel set B then our assertion is $\gamma_k(M(B)) = G_{k,M}(B)$, hence it is true. If f is a simple function, i.e. $f(x) = \sum_{j=1}^{N} a_j \mathbb{I}\{B_j\}$ where $B_j \in \mathcal{B}_0(\mathbb{R}^d)$ are pairwise disjoint, then the Lemma follows from (3.3).

For nonnegative $f \in L$ let $(f_n)_{n \in \mathbb{N}}$ be a sequence of simple functions increasing to f as $n \to \infty$ (see (1.6)). Then, for any $m = 1, \ldots, k$,

$$\int f_n^m(x)dx \to \int f^m(x)dx \text{ and } \mathsf{E}\left(\int f_n(x)M(dx) \right)^m \to \mathsf{E}\left(\int f(x)M(dx) \right)^m < \infty$$

as $n \to \infty$, by the monotone convergence theorem. From the polynomial formula (3.4) we conclude that

$$\gamma_k \left(\int f_n(x)M(dx) \right) \to \gamma_k \left(\int f(x)M(dx) \right), \quad n \to \infty.$$

For any $f \in L$ write $f = f^+ - f^-$. Note that the sets $\{x : f^+(x) \neq 0\}$ and $\{x : f^-(x) \neq 0\}$ are disjoint, hence the random variables $\int f^+(x)M(dx)$ and $\int f^-(x)M(dx)$ are independent. By (3.3) and the already proved part of the Lemma,

$$\gamma_k \left(\int f(x)M(dx) \right) = \gamma_k \left(\int f^k(x)M(dx) \right) + (-1)^k \gamma_k \left(\int f^-(x)M(dx) \right)$$

$$= G_{k,M} \int (f^+)^k(x)dx + G_{k,M} \int (f^-)^k(x)dx = G_{k,M} \int f^k(x)dx. \;\square$$

If $p \in \mathbb{N}$ then M_p will stand for p-th central moment of a random variable, i.e. $\mathsf{M}_p Z = \mathsf{E}(Z - \mathsf{E}Z)^p$ whenever $\mathsf{E}|Z|^p < \infty$. Set

$$J(f) = \int f(x)M(dx), \quad f \in L.$$

Theorem 3.2. ([15]) *Let k be an even positive integer and M be a stationary, independently scattered random measure such that $\mathsf{E}M(I)^k < \infty$. Then there exist homogeneous polynomials $P_{k,M}$ and $Q_{k,M}$ such that $\deg P_{k,M} = k$, $\deg Q_{k,M} = k/2$ and, for any $f \in L$, one has*

$$\mathsf{M}_k J(f) \leq P_{k,M}(\|f\|_{L^2}, \ldots, \|f\|_{L^k}) \leq \|f\|_{\infty}^{k/2} Q_{k,M}(\|f\|_{L^1}, \|f\|_{\infty}). \tag{3.5}$$

Proof. The formula expressing moments in terms of cumulants (converse to (3.4)) is the following (see, e.g., [383, Ch. II, §12.8]):

$$\mathsf{E}\xi^k = \sum_{q=1}^{k} \sum_{\substack{j \in \mathbb{N}^q: \\ j_1 + \ldots + j_q = k}} c_k(q,j) \prod_{v=1}^{q} \gamma_{j_v}(\xi), \quad c_k(q,j) := \frac{k!}{q! j_1! \ldots j_q!}. \tag{3.6}$$

Obviously $\gamma_1(\xi) = \mathsf{E}\xi$ and

$$\gamma_r(\xi) = \gamma_r(\xi + a) \quad \text{for all } r > 1, \ r \in \mathbb{N}, \ a \in \mathbb{R}$$

and any random variable $\xi \in L^r(\Omega, \mathcal{F}, \mathsf{P})$. By (3.6) and Lemma 3.1, for integer $k \geq 2$, we have

$$\mathsf{M}_k J(f) = \sum_{q=1}^{k} \sum_{\substack{j \in \mathbb{N}^q: \\ j_1 + \ldots + j_q = k}} c_k(q,j) \prod_{v=1}^{q} \gamma_{j_v}\left(J(f) - \mathsf{E}J(f)\right)$$

$$= \sum {}'c_k(q,j) \prod_{v=1}^{q} \gamma_{j_v}(J(f)) \leq \sum {}'c_k(q,j) \prod_{v=1}^{q} G_{j_v,M} \int |f^{j_v}(y)| dy$$

$$\leq \sum {}'c_k(q,j) \prod_{v=1}^{q} G_{j_v,M} \|f\|_{L^{j_v}}^{j_v} \tag{3.7}$$

where $q \in \mathbb{N}$ and the sum $\sum'$ is taken over all q-tuples of nonnegative integers $(j_1, \ldots, j_q)$ such that $j_1 + \ldots + j_q = k$, $j_v \geq 2$, $v = 1, \ldots, q$. The last expression in (3.7) is a homogeneous polynomial (of degree k) in $\|f\|_{L^2}, \ldots, \|f\|_{L^k}$. This proves the first inequality in (3.5). To prove the second one use the obvious bound

$$\|f\|_{L^m}^m \leq \|f\|_{L^1} \|f\|_{\infty}^{m-1} \quad \text{for } f \in L. \quad \square$$

For a nonnegative function $f : \mathbb{R}^d \to \mathbb{R}$ and $x \in \mathbb{R}^d$ set

$$R(f) = \sum_{j \in \mathbb{Z}^d} \sup_{y \in I+j} f(y).$$

Theorem 3.3. ([14]) *Let M be a stationary, associated random measure such that*

$$D := \mathsf{E} M(I)^{k+\delta} < \infty,$$

for some even integer $k \geq 4$ and some $\delta > 0$. Assume that M obeys the (FS) *condition, and the coefficients $(u_r)_{r \in \mathbb{N}}$ defined in* (3.1) *are such that*

$$\sum_{r=1}^{\infty} u_r^{\delta/(k+\delta-2)} r^{d(k-1)-1} < \infty.$$

Then, for any $f \in L$ such that $R(f) < \infty$,

$$\mathsf{M}_k J(f) \leq \|f\|_\infty^{k/2} P_{k,\delta,u,D}(R(f), \|f\|_\infty)$$

where $P_{k,\delta,u,D}$ is a homogeneous polynomial of degree $k/2$.

Proof. Set

$$\widetilde{f}(x) = \sum_{j \in \mathbb{Z}^d} \sup_{y \in I+j} f(y) \mathbb{I}\{I+j\}(x), \quad x \in \mathbb{R}^d.$$

Then $\widetilde{f} \geq f \geq 0$. Consequently, according to Theorem 1.3.27 the random vector

$$(X, Y) = \left(J(f) - \mathsf{E} J(f), J(\widetilde{f} - f) - \mathsf{E} J(\widetilde{f} - f) \right)$$

is associated. As $\mathsf{E} X = \mathsf{E} Y = 0$ and the function $x \to x^k$ is convex, Lemma 2.1.18 entails the estimate

$$\mathsf{M}_k J(f) = \mathsf{E} X^k \leq \mathsf{E}(X+Y)^k = \mathsf{M}_k J(\widetilde{f})$$

$$= \mathsf{E}\Big(\sum_{j \in \mathbb{Z}^d} \sup_{y \in I+j} f(y) \Big(M(I+j) - \mathsf{E} M(I+j) \Big) \Big)^k.$$

The desired statement follows now from Theorem 2.3.5. $\square$

2°. The functional limit theorem for rescaled solutions. Let $C_\infty(1, d; m)$ be the space of continuous functions $f : (0, +\infty) \times \mathbb{R}^d \to \mathbb{R}^m$, $m \in \mathbb{N}$. Endow $C_\infty(1, d; m)$ with LU-topology corresponding to the metric ρ, given by

$$\rho(f, g) = \sum_{n=2}^{\infty} 2^{-n} \frac{\rho_n(f, g)}{1 + \rho_n(f, g)},$$

for $f, g \in C_\infty(1, d; m)$, where

$$\rho_n(f, g) = \sup_{(t,x) \in J_n} \|f(t, x) - g(t, x)\|, \quad J_n = [n^{-1}, n] \times ([-n, n]^d). \tag{3.8}$$

The notation $\| \cdot \|$ stands, as usual, for the Euclidean norm. By the same reasoning as in the proof of Lemma 2.6, $(C_\infty(1, d; m), \rho)$ is a Polish space. Thus to check the convergence in law of a sequence $(X_k)_{k \in \mathbb{N}}$ to X (as $k \to \infty$) it is enough to prove such convergence in any space $C(J_n, \mathbb{R}^m)$ for the restrictions to J_n, $n \in \mathbb{N}$. Here X_k, X are random elements with values in $C_\infty(1, d; m)$.

Theorem 3.4. ([14]) *Let M be a stationary associated random measure on $\mathbb{R}^d$ having the finite susceptibility property. Let any of the following three groups of conditions hold.*

$\mathbf{1^0}.$ $d \leq 2$;

$\mathbf{2^0}.$ $d \geq 3$, $\displaystyle\sum_{r=1}^{\infty} u_r^{\delta/(k+\delta-2)} r^{d(k-1)-1} < \infty$ and $\mathsf{E}M(I)^{k+\delta} < \infty$

     for some $\delta > 0$ and even integer $k > \dfrac{d+1}{2}$;

$\mathbf{3^0}.$ $d \in \mathbb{N}$, M is independently scattered and $\mathsf{E}M(I)^k < \infty$

     for some even integer $k > \dfrac{d+1}{2}$.

Then the distributions of rescaled solutions $(V_T)_{T>0}$ (defined in (2.4)) in $C_\infty(1, d, d)$ endowed with LU-topology converge, as $T \to \infty$, to a Gaussian centered process V_∞ with covariance function given in Theorem 2.3.

Proof. We fix a positive integer N and prove the weak convergence $V_T \to V_\infty$ $(T \to \infty)$ for time and spatial arguments $(t, x) \in J_N$. The convergence of finite-dimensional distributions follows from Theorem 2.3, so we need to establish the tightness only. To do this we employ two lemmas. Beforehands it will be convenient to define k and δ for all three conditions. Namely, in cases of conditions $\mathbf{1^0}$ or $\mathbf{3^0}$ set $\delta = 0$; in case $\mathbf{1^0}$ set $k = 2$.

Lemma 3.5. *Let any of conditions $\mathbf{1^0}$—$\mathbf{3^0}$ be satisfied, and define the function $f(y) = \exp\{-\|y\|\}$, $y \in \mathbb{R}^d$. Then there exist $H_{M,k,\delta} > 0$ and $s_0 > 0$ such that*

$$\mathsf{M}_k \int f(s^{-1}x)M(dx) \leq H_{M,k,\delta} s^{dk/2}$$

for any $s > s_0$.

Proof. Note that $\int f(s^{-1}x)dx = s^d \int f(x)dx$. Thus, if either $\mathbf{1^0}$ or $\mathbf{3^0}$ is true, the assertion follows from Lemma 1.6 or Theorem 3.2 respectively. For the condition $\mathbf{2^0}$ we apply Theorem 3.3. Namely, if f_s is such function that $f_s(x) = f(x/s)$, $s > 0$, then

$$R(f_s) = \sum_{j \in \mathbb{Z}^d} e^{-\|j\|/s} = s^d \sum_{j \in \mathbb{Z}^d} e^{-\|j\|/s} s^{-d},$$

but $\sum_{j \in \mathbb{Z}^d} e^{-\|j\|/s} s^{-d}$ is the Riemann sum approximation for the integral $\int f(x)dx$. Consequently, for s large enough we have

$$\sum_{j \in \mathbb{Z}^d} e^{-\|j\|/s} s^{-d} \leq 2 \int f(x)dx.$$

Now Theorem 3.3 implies the required estimate. $\square$

Lemma 3.6. *Let the conditions of Theorem 3.4 hold. Then, for any $i = 1, \ldots, d$ and any sequence $(T_n)_{n \in \mathbb{N}}$ such that $T_n \nearrow \infty$ as $n \to \infty$, the family of random processes $\{A_{T_n,i}\}_{n \in \mathbb{N}}$ is tight in $C(J_N, \mathbb{R}^d)$, here J_N is defined in (3.8).*

Proof. Without loss of generality we take $i = 1$. It suffices to prove that, for any $\varepsilon > 0$,

$$\lim_{\delta \to 0} \limsup_{n \to \infty} \mathsf{P}(M_n^\delta > \varepsilon) = 0$$

where

$$M_n^\delta = \sup_{\substack{\|(t,x)-(s,z)\| < \delta, \\ (t,x),(s,z) \in J_N}} |A_{T_n,1}(t,x) - A_{T_n,1}(s,z)|.$$

To this end, we show that

$$\lim_{l \to \infty} \sum_{j \in \mathbb{Z}^d : 0 \leq j < (l,\ldots,l)} \limsup_{n \to \infty} \mathsf{P} \left(\sup_{(t,x) \in Q_l(j)} |A_{T_n,1}(t,x) - A_{T_n,1}(q_l(j))| > \varepsilon \right) = 0,$$

here

$$q_l(j) = (N^{-1} + j_0(N - N^{-1})l^{-1}, -N + 2j_1 N l^{-1}, \ldots, -N + 2j_d N l^{-1})$$

$$Q_l(j) = Q_l + q_l(j), \quad Q_l = [0, (N - N^{-1})l^{-1}] \times \prod_{i=1}^{d} [0, 2N l^{-1}].$$

Introduce the random processes

$$Z_n(t,x) := A_{T_n,1}(t,x) - A_{T_n,1}(q_l(j)), \quad (t,x) \in Q_l(j), \ n \in \mathbb{N}.$$

We also use the functions

$$h_n(t,x,y) = T_n^{-d/4} \left(x_1 - \frac{y_1}{T_n^{1/2}} \right) \exp \left\{ -\frac{\|x - T_n^{-1/2} y\|^2}{2t} \right\},$$

$$f_n(t,x,y) = h_n(t,x,y) - h_n(q_l(j), y).$$

Then one can write

$$Z_n(t,x) = \int f_n(t,x,y) M(dy).$$

Note that any first-order partial derivative of f_n can be expressed as

$$T_n^{-d/4} P(t^{-1}, x, y T_n^{-1/2}) \exp \left\{ -\frac{\|x - T_n^{-1/2} y\|^2}{2t} \right\}$$

where P does not depend on n and is a polynomial in $(2d + 1)$ real arguments. Therefore, there exists $C_1 = C_1(N) > 0$ such that, for any $(t,x) \in J_N$, one has

$$\left| \frac{\partial f_n(t,x,y)}{\partial t} \right| \vee \max_{i=1,\ldots,d} \left| \frac{\partial f_n(t,x,y)}{\partial x_i} \right| \leq g_n(y) = \frac{C_1}{T_n^{d/4}} \exp \left\{ -\frac{\|y\|^2}{T_n^{1/2}} \right\}.$$

Define, for $n \in \mathbb{N}$, $(t, x) \in J_N$ and $y \in \mathbb{R}^d$, the functions

$$f_{n,+}(t, x, y) = g_n(y)\|(t, x) - q_l(j)\|_1, \quad f_{n,-} = f_{n,+} - f_n \qquad (3.9)$$

where, as usual, $\|v\|_1 = |v_1| + \ldots + |v_{d+1}|$ for $v \in \mathbb{R}^{d+1}$. Let

$$Z_{n,\pm}(t, x) = \int f_{n,\pm}(t, x, y) M(dy) - \mathsf{E} \int f_{n,\pm}(t, x, y) M(dy),$$

here the sign in the subscript is either $+$ or $-$ at both sides of the formula.

From (3.9) one immediately infers that $f_{n,+}$ is a coordinatewise nondecreasing (in (t, x)) function on $Q_l(j)$, for any $y \in \mathbb{R}^d$. We claim that the same fact is true for $f_{n,-}$. Indeed, if $(s, z) \geq (t, x) \geq q_l(j)$ then

$$f_{n,+}(s, z, y) - f_n(s, z, y) - (f_{n,+}(t, x, y) - f_n(t, x, y))$$

$$= f_{n,+}(s, z, y) - f_{n,+}(t, x, y) - (f_n(s, z, y) - f_n(t, x, y))$$

$$\geq g_k(y) \left(\|(s, z) - q_l(j)\|_1 - \|(t, x) - q_l(j)\|_1 - \|(s, z) - (t, x)\|_1\right) = 0.$$

Thus, if $(t, x) \leq (s, z) \leq (w, u)$ are points in $Q_l(j)$, Lemma 1.11 implies that 2-dimensional random vectors

$$(Z_{n,\pm}(w, u) - Z_{n,\pm}(s, z), Z_{n,\pm}(s, z) - Z_{n,\pm}(t, x))$$

(one with sign "+", another with "−") are associated.

By the definition of functions $f, f_{n,\pm}$,

$$Z_{n,+}(t, x) - Z_{n,-}(t, x) = \int \left(f_{n,+}(t, x, y) - f_{n,-}(t, x, y)\right) M(dy)$$

$$- \mathsf{E} \int \left(f_{n,+}(t, x, y) - f_{n,-}(t, x, y)\right) M(dy) = \int f_n(t, x, y) M(dy)$$

$$- \mathsf{E} \int f_n(t, x, y) M(dy) = \int f_n(t, x, y) M(dy) = Z_n(t, x), \quad (t, x) \in Q_l(j).$$

Whence we have

$$\mathsf{P}\left(\sup_{(t,x)\in Q_l(j)} |A_{T_n,1}(t, x) - A_{T_n,1}(q_l(j))| > \varepsilon\right)$$

$$\leq \mathsf{P}\left(\sup_{(t,x)\in Q_l(j)} |Z_{n,+}(t, x)| > \frac{\varepsilon}{2}\right) + \mathsf{P}\left(\sup_{(t,x)\in Q_l(j)} |Z_{n,-}(t, x)| > \frac{\varepsilon}{2}\right). \qquad (3.10)$$

By association, if $(t, x) \leq (s, z) \leq (w, u)$ are points belonging to $Q_l(j)$, then

$$\mathsf{P}(Z_{n,+}(s, z) - Z_{n,+}(t, x) > a, \; Z_{n,+}(s, z) - Z_{n,+}(w, u) > a)$$

$$\leq \mathsf{P}(Z_{n,+}(s, z) - Z_{n,+}(t, x) > a)\mathsf{P}(Z_{n,+}(s, z) - Z_{n,+}(w, u) > a)$$

$$\leq a^{-2k} \mathsf{E}(Z_{n,+}(s, z) - Z_{n,+}(t, x))^k \mathsf{E}(Z_{n,+}(w, u) - Z_{n,+}(s, z))^k. \qquad (3.11)$$

Similarly,

$$P(Z_{n,+}(s,z) - Z_{n,+}(t,x) < -a, \; Z_{n,+}(s,z) - Z_{n,+}(w,u) < -a)$$

$$\leq P(Z_{n,+}(s,z) - Z_{n,+}(t,x) < -a)P(Z_{n,+}(s,z) - Z_{n,+}(w,u) < -a)$$

$$\leq a^{-2k}\mathsf{E}(Z_{n,+}(s,z) - Z_{n,+}(t,x))^k \mathsf{E}(Z_{n,+}(w,u) - Z_{n,+}(s,z))^k. \tag{3.12}$$

By Lemma 3.5, for any pair of points $(t,x), (s,z) \in Q_l(j)$ such that $(t,x) \leq (s,z)$, we have

$$\mathsf{E}(Z_{n,+}(s,z) - Z_{n,+}(t,x))^k = \mathsf{M}_k \int g_n(y)\|(s,z) - (t,x)\|_1 M(dy)$$

$$= C_1\|(s,z) - (t,x)\|_1^k T_n^{-dk/4} \mathsf{M}_k \int \exp\left\{-\frac{\|y\|}{T_n^{1/2}}\right\} M(dy)$$

$$\leq C_1\|(s,z) - (t,x)\|_1^k T_n^{-dk/4} H_{M,k,\delta} T_n^{dk/4} = C_2(M,N,k,\delta)\|(s,z) - (t,x)\|_1^k \tag{3.13}$$

with some $C_2(M,N,k,\delta) > 0$. On inserting (3.13) into (3.11) and (3.12) we see that the conditions of Theorem 2.4.9 are satisfied uniformly in n and j with

$$\alpha = k, \; K = C_2^2, \; \gamma = 2k, \; Z(y) = Z_{n,+}(y + q_l(j)) \quad \text{for} \quad y \in \Pi = Q_l \subset \mathbb{R}^{d+1}$$

and $d + 1$ instead of d. So, for some $\tau > 1$, by virtue of Theorem 2.4.9,

$$P\left(\sup_{(t,x)\in Q_l(j)} Z_{n,\pm}(t,x) > \frac{\varepsilon}{2}\right) \leq P\left(Z_{n,+}(q_l(j+1)) > \frac{\varepsilon}{2\tau}\right) + C_3(M,N,\varepsilon,k)\left(\frac{2N}{l}\right)^{2k}.$$

The same estimate can be obtained for the processes $-Z_{n,+}$, $Z_{n,-}$ and $-Z_{n,-}$. Thus from these estimates and (3.10) we deduce that

$$P\left(\sup_{(t,x)\in Q_l(j)} |A_{T_n,1}(t,x) - A_{T_n,1}(q_l(j))| > \varepsilon\right)$$

$$\leq P\left(|Z_{n,+}(q_l(j+1))| > \frac{\varepsilon}{2\tau}\right) + P\left(|Z_{n,-}(q_l(j+1))| > \frac{\varepsilon}{2\tau}\right) + C_4(M,N,\varepsilon,k)l^{-2k}.$$

Recall that by construction (see (3.9))

$$Z_{n,+}(q_l(j+1)) = C_1\|q_l(j+1) - q_l(j)\|_1 \left(\int h(y,T_n)M(dy) - \mathsf{E}\int h(y,T_n)M(dy)\right)$$

where $h(y,T_n) = \exp\{-T_n^{-1/2}\|y\|\}$. Therefore, by Theorem 1.7,

$$Z_{n,+}(q_l(j+1)) \to N\left(0, \Gamma_M C_1^2 \|q_l(j+1) - q_l(j)\|_1^2 \int e^{-2\|y\|} dy\right) \quad \text{in law as } n \to \infty.$$

Here the variance of the limit distribution is clearly bounded by $C_5(M,N)l^{-2}$.

An analogous argument is valid for $Z_{n,-}$. Consequently,

$$\limsup_{n\to\infty} P\left(\sup_{(t,x)\in Q_l(j)} |A_{T_n,1}(t,x) - A_{T_n,1}(q_l(j))| > \varepsilon\right) \leq C_6 l^{-1}\exp\{-C_7 l^2\} + C_4 l^{-2k}$$

where $C_i = C_i(M, N, \varepsilon, \tau) > 0$, $i = 6, 7$. Hence

$$\lim_{l \to \infty} \sum_{j \in \mathbb{Z}^d : 0 \leq j < (l, \ldots, l)} \limsup_{n \to \infty} \mathsf{P}\left(\sup_{(t,x) \in Q_l(j)} |A_{T_n,1}(t, x) - A_{T_n,1}(q_l(j))| > \varepsilon \right)$$

$$\leq \lim_{l \to \infty} C_6 l^{d-1} \exp\{-C_7 l^2\} + C_4 l^{d-2k} = 0$$

because $d + 1 < 2k$ under each of conditions $\mathbf{1^0}$—$\mathbf{3^0}$. This proves the Lemma. $\square$

To prove the Theorem let $\varepsilon > 0$ be arbitrary and take an arbitrary sequence $(T_n)_{n \in \mathbb{N}}$ such that $T_n \nearrow \infty$. Lemma 3.6 implies that, for any $i = 1, \ldots, d$, there exists a compact set $K_i \subset C(J_N, \mathbb{R})$ such that $\mathsf{P}(A_{T_n,i} \notin K_i) \leq \varepsilon/d$ for all $n \in \mathbb{N}$. Then $K = K_1 \times \ldots \times K_d$ is a compact in $C(J_N, \mathbb{R}^d)$.

Indeed, take arbitrary $\delta > 0$ and for each $i = 1, \ldots, d$ construct a δ/d-net N_i in K_i. For a function $f = (f_1, \ldots, f_d) \in K$ choose functions $g_i \in N_i$ so that $\rho(f_i, g_i) < \delta/d$. Then considering a function $g = (g_1, \ldots, g_d) \in K$ one has

$$\rho(f, g) \leq \sum_{n=2}^{\infty} 2^{-n} \frac{\sup \sum_{i=1}^{d} |f_i(t, x) - g_i(t, x)|}{1 + \sup \sum_{i=1}^{d} |f_i(t, x) - g_i(t, x)|}$$

$$\leq \sum_{i=1}^{d} \sum_{n=2}^{\infty} 2^{-n} \frac{\sup |f_i(t, x) - g_i(t, x)|}{1 + \sup |f_i(t, x) - g_i(t, x)|} < \delta.$$

Here the suprema are over $(t, x) \in J_N$ and we took into account that the function $x \mapsto x/(1 + x)$ is increasing when $x \geq 0$, and used the estimate $\|x\| \leq \|x\|_1$, $x \in \mathbb{R}^d$, as well as the inequality

$$\frac{\sum_{i=1}^{d} x_i}{1 + \sum_{i=1}^{d} x_i} \leq \sum_{i=1}^{d} \frac{x_i}{1 + x_i}, \quad x_i \geq 0, \quad i = 1, \ldots, d.$$

Besides, $\mathsf{P}(A_{T_n} \notin K) \leq \varepsilon$ for all $n \in \mathbb{N}$. Thus we have proved that $\{A_{T_n}, n \in \mathbb{N}\}$ is a tight family of vector-valued random processes. The rest of the proof is completed in the same fashion as that of Theorem 2.7. $\square$

To finish the Chapter we also note that there are important stochastic models based on the classical Burgers equation and more general equations with initial data given by various stochastic processes and fields (white noise, Lévy processes, stationary Markov processes etc.). The perturbations of the equation with external force are investigated as well. An interesting problem is to analyze not only the velocity but also the density of media. The stochastic differential Burgers system and its analogs containing fractional derivatives are also considered. Here we have to restrict ourselves only by several references, e.g., [4, 178, 303, 417].

Appendix A

Auxiliary Statements

A.1 Extensions of the Hoeffding lemma

We establish the classical formula (1.1.8) by means of the result which is due to Khoshnevisan and Lewis.

Theorem A.1. ([230]) *Let $h : \mathbb{R}^2 \to \mathbb{R}$ be a function having a continuous second-order derivative $\partial^2 h/\partial x \partial y$. Suppose that X, Y, Z are random variables such that $Law(Y) = Law(Z)$ and Z is independent of X. Then*

$$\mathsf{E}h(X,Y) - \mathsf{E}h(X,Z) = \iint_{\mathbb{R}^2} \frac{\partial^2 h(x,y)}{\partial x \partial y} C(x,y) dx dy \qquad (A.1.1)$$

where $C(x,y) = \mathsf{P}(X \geq x, Y \geq y) - \mathsf{P}(X \geq x)\mathsf{P}(Y \geq y)$, provided that the expectations and the double integral in (A.1.1) exist.

Proof. Clearly it suffices to consider the case when Z is independent not only of X but of Y also. Let U be a random variable (possibly, on an enlargement of the initial probability space) such that the random vector (U, Z) is independent of (X, Y) and has the same distribution as (X, Y). After rewriting the integrand in the right hand side of (A.1.1) in terms of covariance and using the Fubini theorem, we obtain

$$\iint_{\mathbb{R}^2} \frac{\partial^2 h(x,y)}{\partial x \partial y} C(x,y) dx dy = \iint_{\mathbb{R}^2} \frac{\partial^2 h(t,w)}{\partial t \partial w} cov\left(\mathbb{I}\{X \geq t\}, \mathbb{I}\{Y \geq w\}\right) dt dw$$

$$= \frac{1}{2} \iint_{\mathbb{R}^2} \frac{\partial^2 h(t,w)}{\partial t \partial w} \mathsf{E}(\mathbb{I}\{X \geq t\} - \mathbb{I}\{U \geq t\})(\mathbb{I}\{Y \geq w\} - \mathbb{I}\{Z \geq w\}) dt dw$$

$$= \frac{1}{2}\mathsf{E} \iint_{\mathbb{R}^2} \frac{\partial^2 h(t,w)}{\partial t \partial w} (\mathbb{I}\{X \geq t\} - \mathbb{I}\{U \geq t\})(\mathbb{I}\{Y \geq w\} - \mathbb{I}\{Z \geq w\}) dt dw$$

$$= \frac{1}{2}\mathsf{E} \int_U^X \int_Z^Y \frac{\partial^2 h(t,w)}{\partial t \partial w} dt dw = \frac{1}{2}\mathsf{E} \int_U^X \left(\frac{\partial h(t,Y)}{\partial t} - \frac{\partial h(t,Z)}{\partial t}\right) dt$$

$$= \frac{1}{2}\mathsf{E}(h(X,Y) - h(U,Y) - h(X,Z) + h(U,Z)) = \mathsf{E}h(X,Y) - \mathsf{E}h(X,Z),$$

383

since $cov\left(\mathbb{I}\{X \geq x\}, \mathbb{I}\{Y \geq y\}\right)$ is a bounded Borel function[6] defined on $\mathbb{R}^2$. $\square$

As a corollary of this result one gets

Theorem A.2. ([200]) *Suppose that $f, g : \mathbb{R} \to \mathbb{R}$ have continuous derivatives. Let X, Y be random variables such that $\mathsf{E}|f(X)|$, $\mathsf{E}|g(Y)|$, $\mathsf{E}|f(X)g(Y)| < \infty$. Then*

(a) *The following formula is true:*

$$cov(f(X), g(Y)) = \iint_{\mathbb{R}^2} f'(x)g'(y)C(x,y)dxdy, \qquad (A.1.2)$$

provided that the integral on the right-hand side exists.

(b) *If, in addition, the random vector (X, Y) is **PA** or **NA** and the functions f and g are nondecreasing, then the integral on the right-hand side of (A.1.2) exists.*

Proof. The assertion (a) follows from Theorem A.1 with $h(x, y) = f(x)g(y)$, $x, y \in \mathbb{R}$. To prove (b) one can use the appropriate approximation. Namely, let $\chi : \mathbb{R} \to \mathbb{R}$ be an even differentiable function such that $\chi(x) = 1$ if $|x| \leq 1/2$ while $\chi(x) = 0$ if $|x| \geq 1$ and χ is nonincreasing on $\mathbb{R}_+$. For $n \in \mathbb{N}$ consider functions f_n and g_n where

$$f_n(x) = f(x)\chi(x/n), \quad g_n(x) = g(x)\chi(x/n), \quad x \in \mathbb{R}.$$

Clearly f_n and g_n are differentiable and $f_n'(x) = g_n'(x) = 0$ whenever $|x| \geq n$, $n \in \mathbb{N}$. Thus, by the already proved part of assertion,

$$cov(f_n(X), g_n(Y)) = \int_{-n}^{n} \int_{-n}^{n} (f_n'(x)g_n'(y)C(x,y)dxdy. \qquad (A.1.3)$$

By the dominated convergence theorem,

$$\lim_{n \to \infty} cov(f_n(X), g_n(Y)) = cov(f(X), g(Y))$$

$$= \lim_{n \to \infty} \int_{-n}^{n} \int_{-n}^{n} f'(x)g'(y)C(x,y)dxdy.$$

Note that if (X, Y) is **PA** (resp. **NA**), then the expression inside the integral in (A.1.3) is nonnegative (resp. nonpositive) provided that $f', g' \geq 0$. Therefore by the monotone convergence theorem the integral in (A.1.2) exists. $\square$

Remark A.3. If $f(x) = g(x) = x$ for $x \in \mathbb{R}$ then formula (A.1.2) coincides with the classical Hoeffding formula (1.8.8).

---

[6] As it is upper-continuous, see Definition 5.1.2.

A.2 Markov processes. Background

We briefly review some basic properties of Markov processes. Let S be a finite state space endowed with a collection $\mathcal{B}$ of all its subsets. Consider an S-valued *Markov process* (or *Markov chain*) $\mathbf{X} = \{X_t, t \geq 0\}$ with continuous time. The Markov property of $\mathbf{X}$ amounts to

$$P(X_t = x | X_{t_n} = x_n, \ldots, X_{t_1} = x_1) = P(X_t = x | X_{t_n} = x_n) \qquad (A.2.1)$$

for any $0 \leq t_1 \leq \cdots \leq t_n \leq t$ and every $x, x_1, \ldots, x_n \in S$, $n \in \mathbb{N}$.

A function $p(s, t; x, y) := P(X_t = y | X_s = x)$ where $0 \leq s \leq t$ and $x, y \in S$ is called the *transition probability*. If $p(s, t; x, y) = p(s + h, t + h; x, y)$ for all $0 \leq s \leq t$, $h \geq 0$ and $x, y \in S$, the Markov process $\mathbf{X}$ is called *homogeneous*. In this case one writes

$$p(t, x, y) := p(0, t; x, y) = P(X_t = y | X_0 = x), \quad x, y \in S, \ t \geq 0. \qquad (A.2.2)$$

So, for all such x, y, t, evidently,

$$p(t, x, y) \geq 0, \qquad (A.2.3)$$

$$p(0, x, y) = \delta_{xy} \qquad (A.2.4)$$

where δ_{xy} is the Kronecker symbol. Moreover,

$$\sum_{y \in S} p(t, x, y) = 1 \qquad (A.2.5)$$

and the following *Chapman–Kolmogorov equation* holds

$$p(u + v, x, y) = \sum_{z \in S} p(u, x, z) p(v, z, y), \quad u, v \geq 0. \qquad (A.2.6)$$

One often enumerates the points of S as $x_1, \ldots, x_N$, $N = |S|$, and instead of $p(t, x_i, x_j)$ writes simply $p_{ij}(t)$.

It is well-known (see, e.g., [272, p. 570]) that, for a given family of matrices $(p(t, x, y))_{x, y \in S}$, $t \geq 0$, having the properties (A.2.3)—(A.2.6) and for any probability measure μ_0 on S (i.e. on $(S, \mathcal{B})$ where $\mathcal{B} = 2^S$), there exists a Markov chain $\mathbf{X}$ such that (A.2.2) holds and $Law(X_0) = \mu_0$, that is *initial distribution* of $\mathbf{X}$ is μ_0.

Thus we have a description of homogeneous Markov chains in terms of transition probabilities and initial distribution (in fact the same description is valid for countable S and, moreover, an analogue of this statement holds for non-homogeneous chains).

As usual (see, e.g., [151, Ch. 2 § 1]), define *transition operators* ($|S| < \infty$) by way of

$$(T_t f)(x) = \sum_{y \in S} p(t, x, y) f(y), \quad (\mu T_t)(y) = \sum_{x \in S} \mu(x) p(t, x, y) \qquad (A.2.7)$$

where $t \geq 0$, $f : S \to \mathbb{R}$ and μ is a probability measure on $(S, \mathcal{B})$.

Clearly, the family $(T_t)_{t\geq 0}$ in view of (A.2.6) forms a semigroup which we consider on a (Banach) space of functions $f : S \to \mathbb{R}$ endowed with a norm $|f| := \sup_{x\in S}|f(x)|$. Thus for all $u,v \geq 0$

$$(T_{u+v}f)(x) = (T_u(T_vf))(x), \quad x \in S,$$

and according to (A.2.4) the identity operator $I = T_0$ is the unit of semigroup. Note also that μT_t is the distribution of X_t if μ is the initial distribution of $\mathbf{X}$.

One says that $(p(t,x,y))_{x,y\in S}$ is *standard* if

$$\lim_{t\to 0+} p(t,x,y) = \delta_{xy} \text{ for any } x,y \in S. \tag{A.2.8}$$

It is easily seen that (A.2.8) is equivalent to the statement that trajectories $X(t)$ are continuous in probability at $t = 0$, i.e.

$$X(t) \to X(0) \text{ in probability as } t \to 0+. \tag{A.2.9}$$

Furthermore, (A.2.9) is equivalent to the condition that trajectories $X(t)$ are continuous in probability on $\mathbb{R}_+$. Note also that (A.2.8) holds if and only if, for any $f : S \to \mathbb{R}$, one has

$$|T_tf - f| \to 0 \text{ as } t \to 0+,$$

that is the semigroup $(T_t)_{t\geq 0}$ is continuous at $t = 0$.

It is also well-known (see, e.g., [272, pp. 590, 610]) that (A.2.8) implies the existence of *infinitesimal matrix* A with finite elements (as $|S| < \infty$) defined by

$$A(x,y) := \lim_{t\to 0+} t^{-1}(p(t,x,y) - \delta_{xy}), \quad x,y \in S. \tag{A.2.10}$$

It is immediately seen that

$$A(x,y) \geq 0 \text{ if } x \neq y \text{ and } \sum_{y\in S} A(x,y) = 0 \tag{A.2.11}$$

due to (A.2.3) and (A.2.5). In particular $A(x,x) \leq 0$ for all $x \in S$. If the matrix defined in (A.2.10) exists, then, for any $f : S \to \mathbb{R}$,

$$\lim_{t\to 0+} \frac{T_tf(x) - f(x)}{t} = \sum_{y\in S} A(x,y)f(y) =: (Af)(x), \quad x \in S. \tag{A.2.12}$$

Therefore A is the *generator* of the semigroup $(T_t)_{t\geq 0}$. Moreover, if (A.2.8) is satisfied, then

$$T_t = e^{tA}, \quad t \geq 0, \tag{A.2.13}$$

that is for any $f : S \to \mathbb{R}$ one has

$$(T_tf)(x) = \sum_{k=0}^{\infty} \frac{t^k(A^kf)(x)}{k!}, \quad x \in S, \tag{A.2.14}$$

where A^k is the k-th power of matrix A ($A^0 = I = (\delta_{xy})_{x,y\in S}$). It is easy to verify that for every $n \in \mathbb{N}$

$$\Delta_n := \sup_{x\in S}\left|T_tf(x) - \sum_{k=0}^{n} \frac{t^k(A^kf)(x)}{k!}\right| \leq \frac{(|t|\||A\||)^{n+1}e^{\||A\||}|f|}{(n+1)!} \tag{A.2.15}$$

where $\| \cdot \|$ is the operator norm, i.e. $\|A\| = \sup_{f:|f|=1} |Af|$.

Note in passing the following important result (see e.g., [247]). If $|S| < \infty$ and $A = (A(x,y))_{x,y \in S}$ is a matrix such that (A.2.11) holds then formula (A.2.13) provides a semigroup $(T_t)_{t \geq 0}$ (acting on a space of functions $f : S \to \mathbb{R}$) which is standard and A is the generator of that semigroup. Moreover, if $p(t,x,y) := (T_t f)(x)$ where $f = \mathbb{I}_{\{y\}}$ then this function p possesses properties (A.2.3)—(A.2.6).

The study of Markov processes with infinite phase space S is much more involved. We recall (see, e.g., [151, Ch. 2,§1]) only a few basic definitions and results which will be used.

Let $P(t,x,B)$ be a *Markov transition function* defined for $t \geq 0$, $x \in S$ and $B \in \mathcal{B}$ (where $(S,\mathcal{B})$ is some measurable space) and taking values in $[0,1]$. This function satisfies the following conditions.

(1^0) $P(t,x,\cdot)$ is a probability measure on $(S,\mathcal{B})$ for any $t \geq 0$ and $x \in S$.

(2^0) $P(0,x,B) = \delta_x(B)$ where δ_x is the Dirac measure.

(3^0) $P(t,\cdot,B)$ is $\mathcal{B}|\mathcal{B}(\mathbb{R})$-measurable for any $t \geq 0$ and $B \in \mathcal{B}$.

(4^0) For all $u,v \geq 0$, $x \in S$ and $B \in \mathcal{B}$ one has

$$P(u+v,x,B) = \int_S P(u,x,dy)P(v,y,B).$$

A stochastic process $\mathbf{X} = \{X_t, t \geq 0\}$ with values in $(S,\mathcal{B})$ is called a *homogeneous Markov process* corresponding to Markov transition function $P(t,x,B)$ if

$$\mathsf{P}(X_t \in B | X_0 = x) = P(t,x,B)$$

for all $t \geq 0, x \in S$ and $B \in \mathcal{B}$.

Note that in case of finite S (and $\mathcal{B} = 2^S$) one can set $P(t,x,B) = \sum_{y \in B} p(t,x,y)$ where $p(t,x,y)$ is a function with properties (A.2.3)—(A.2.6).

Analogously to (A.2.12) we introduce operators

$$(T_t f)(x) := \int_S f(y)P(t,x,dy), \quad (\mu T_t)(B) := \int_S P(t,x,B)\mu(dx) \qquad \text{(A.2.16)}$$

where f belongs to a Banach space

$$\mathbf{B} = \{f : S \to \mathbb{R}, \quad f \in \mathcal{B}|\mathcal{B}(\mathbb{R}), \quad |f| := \sup_{x \in S} |f(x)| < \infty\}$$

and μ is a probability measure on $(S,\mathcal{B})$.

One can verify that $(T_t)_{t \geq 0}$ is a semigroup on the space $\mathbf{B}$. In view of condition (2^0) the identity operator $I = T_0$ is the unit of semigroup. An *infinitesimal operator* (or *generator*) G of a semigroup $(T_t)_{t \geq 0}$ is defined by the formula

$$Gf := \lim_{t \to 0+} \frac{T_t f - f}{t}$$

where the limit is taken in the sup-norm introduced above. In contrast to the case of finite S the generator G is in general unbounded and defined on a domain D_G which is a proper subset of $\mathbf{B_0} = \{f \in \mathbf{B} : |T_t f - f| \to 0, t \to 0+\}$. Thus, unfortunately, a simple relation (A.2.14) between $(T_t)_{t\geq 0}$ and G holds only when G is bounded (and $D_G = \mathbf{B}$).

Note that if $\mu = Law(X_0)$ then

$$\mu T_t = Law(X_t), \quad t \geq 0.$$

Moreover, for any $f \in \mathbf{B}$ and $t \geq 0$, one has $T_t f \in \mathbf{B}$ and

$$\int_S T_t f d\mu = \int_S f d(\mu T_t). \tag{A.2.17}$$

In fact, for $f = \mathbb{I}_B$, $B \in \mathcal{B}$, it is clear that

$$(T_t f)(x) = \int_S \mathbb{I}_B P(t, x, dy) = P(t, x, B).$$

Therefore in view of (A.2.16) formula (A.2.17) is established for step functions. Consequently (A.2.17) holds for any $f \in \mathbf{B}$ by the standard approximation technique.

We need the following result (see, e.g., [151, Ch.1, §2]).

Theorem A.4. *For any $f \in D_G$ and $t \geq 0$ one has $T_t f \in \mathcal{D}_G$, $Gf \in \mathbf{B_0}$ and*

$$G(T_t f) = T_t(Gf) = \frac{dT_t}{dt} f.$$

Here

$$\frac{dT_t}{dt} f := \lim_{h \to 0} h^{-1}(T_{t+h} f - T_t f)$$

for $t > 0$ and such f that this limit exists; it is the right derivative at $t = 0$ (in this case $h \to 0+$).

A.3 Poisson spatial process

Consider a partition of $\mathbb{R}^n$ into bounded Borel sets $K_m, m \in \mathbb{N}$. Let $(\tau_m, X_{mj})_{m,j\in\mathbb{N}}$ be an array of independent random variables[7] defined on a probability space $(\Omega, \mathcal{F}, \mathsf{P})$ such that for all $m, j \in \mathbb{N}$

(a) $\tau_m \sim Pois(\Lambda(K_m))$;

(b) for any $C \in \mathcal{B}(\mathbb{R}^n)$

$$\mathsf{P}(X_{mj} \in C) = \begin{cases} \Lambda(C \cap K_m)/\Lambda(K_m), & \Lambda(K_m) \neq 0, \\ 0, & \text{otherwise.} \end{cases}$$

Clearly $\mathsf{P}(X_{mj} \in C) = \mathsf{P}(X_{mj} \in C \cap K_m)$.

Set for $B \in \mathcal{B}(\mathbb{R}^n)$ and $m \in \mathbb{N}$

$$Z_m(B) = \sum_{j=1}^{\tau_m} \mathbb{I}\{X_{mj} \in B\}, \quad Z(B) = \sum_{m=1}^{\infty} Z_m(B), \tag{A.3.1}$$

if $\tau_m = 0$ then the sum over empty set is equal to 0 as usual.

Obviously $Z_m(B)$ is a random variable for every $m \in \mathbb{N}$ and any $B \in \mathcal{B}(\mathbb{R}^n)$. Moreover, $0 \leq Z_m(B) \leq \tau_m < \infty$ for any $B \in \mathcal{B}(\mathbb{R}^n), m \in \mathbb{N}$ and every $\omega \in \Omega$. However, in general $Z(B)$, for $B \in \mathcal{B}(\mathbb{R}^n)$, take values in $\overline{\mathbb{R}}_+$ (more exactly, in $\overline{\mathbb{Z}}_+$).

Theorem A.5. *For any partition of $\mathbb{R}^n$ into bounded Borel sets $K_m, m \in \mathbb{N}$, a process $Z = \{Z(B), B \in \mathcal{B}_0(\mathbb{R}^n)\}$ introduced above is a Poisson spatial stochastic process with intensity measure Λ.*

Proof. At first, for arbitrary fixed $m \in \mathbb{N}$, consider any pairwise disjoint Borel sets $B_1, \ldots, B_k \subset K_m$ $(k \in \mathbb{N})$. We shall show that $Z(B_1), \ldots, Z(B_k)$ are independent and $Z(B_r) \sim Pois(\Lambda(B_r)), r = 1, \ldots, k$. Clearly $Z(B_r) = Z_m(B_r), r = 1, \ldots, k$. In this case, due to Remark 3.3, it is sufficient to verify that the characteristic function of $(Z_m(B_1), \ldots, Z_m(B_k))$ is equal to the right-hand side of (1.3.4). One has

$$\mathsf{E} \exp\{i(t_1 Z_m(B_1) + \ldots + t_k Z_m(B_k))\}$$

$$= \sum_{N=0}^{\infty} \mathsf{E}\left(\exp\left\{ i \sum_{j=1}^{N} (t_1 \mathbb{I}\{X_{mj} \in B_1\} + \ldots + t_k \mathbb{I}\{X_{mj} \in B_k\}) \right\} \mathbb{I}\{\tau_m = N\} \right)$$

$$= \sum_{N=0}^{\infty} \varphi_{\eta_N}(t_1, \ldots, t_k) \frac{\Lambda(K_m)^N}{N!} e^{-\Lambda(K_m)}, \tag{A.3.2}$$

here $t_1, \ldots, t_k \in \mathbb{R}$, $i^2 = -1$ and φ_{η_N} is the characteristic function of a vector $\eta_N = (\eta_{N1}, \ldots, \eta_{Nk})$ with components $\eta_{Nr} = \sum_{j=1}^{N} \mathbb{I}\{X_{mj} \in B_r\}, r = 1, \ldots, k$.

---

[7] τ_m and X_{mj} take values in $\mathbb{Z}_+$ and $\mathbb{R}^n$ respectively.

Let $\eta_{N0} = \sum_{j=1}^{N} \mathbb{I}\{X_{mj} \in B_0\}$ where $B_0 = K_m \setminus \cup_{r=1}^{N} B_r$. Then $(\eta_{N0}, \ldots, \eta_{NK})$ has a polynomial distribution with parameters N and $p_0, \ldots, p_k$ where $p_r = \mathrm{P}(X_{mj} \in B_r) = \Lambda(B_r)/\Lambda(K_m)$, $r = 0, \ldots, k$, that is

$$\mathrm{P}(\eta_{N0} = l_0, \ldots, \eta_{Nk} = l_k) = \binom{N}{l_0, \ldots, l_k} p_0^{l_0} \ldots p_k^{l_k}$$

with $l_0, \ldots, l_k \in \mathbb{Z}_+$, $\sum_{r=0}^{k} l_r = N$. Therefore

$$\varphi_{(\eta_{N0}, \ldots, \eta_{Nk})}(t_0, \ldots, t_k) = \left(p_0 e^{it_0} + \ldots + p_k e^{it_k}\right)^N, \quad t_0, \ldots, t_k \in \mathbb{R}.$$

Consequently,

$$\varphi_{(\eta_{N1}, \ldots, \eta_{Nk})}(t_1, \ldots, t_k) = \varphi_{(\eta_{N0}, \ldots, \eta_{Nk})}(0, t_1, \ldots, t_k) = \left(p_0 + \ldots + p_k e^{it_k}\right)^N.$$
$$(A.3.3)$$

Note that $p_0 = 1 - p_1 - \ldots - p_k$. Thus on account of (A.3.2) and (A.3.3) one has

$$\varphi_{(Z_m(B_1), \ldots, Z_m(B_k))}(t_1, \ldots, t_k) = e^{-\Lambda(K_m)} \sum_{N=0}^{\infty} \left(\Lambda(K_m)(p_0 + \ldots + p_k e^{it_k})\right)^N / N!$$

$$= \exp\{\Lambda(K_m)(p_1(e^{it_1} - 1) + \ldots + p_k(e^{it_k} - 1))\} = \prod_{r=1}^{k} \exp\{\Lambda(B_r)(e^{it_r} - 1)\}.$$

We have obtained the same expression as the right-hand side of (1.3.4), as required. For the general case we need the following elementary statement.

Lemma A.6. *Let $(\xi_{1m}, \ldots, \xi_{km})_{m \in \mathbb{N}}$ be a sequence of independent random vectors in $\mathbb{R}^k$ with independent components such that $\xi_{rm} \sim Pois(a_{rm})$, $r = 1, \ldots, k$ and $m \in \mathbb{N}$. Assume that $\sum_{m=1}^{\infty} a_{rm} = a_r < \infty$, $r = 1, \ldots, k$. Then $\xi_r = \sum_{m=1}^{\infty} \xi_{rm} < \infty$ a.s., $\xi_1, \ldots, \xi_k$ are independent and $\xi_r \sim Pois(a_r), r = 1, \ldots, k$.*

Proof. For any $N \in \mathbb{N}$, one has, in view of (3.3),

$$\sum_{m=1}^{N} \xi_{rm} \sim Pois\left(\sum_{m=1}^{N} a_{rm}\right), \quad r = 1, \ldots, k.$$

Therefore the sums $\sum_{m=1}^{N} \xi_{rm}$ converge in law as $N \to \infty$ to a Poisson random variable with parameter a_r $(r = 1, \ldots, k)$. For a series of independent summands convergence in law is equivalent (see, e.g., [272, p.237]) to a.s. convergence. Thus $\sum_{m=1}^{\infty} \xi_{rm} = \xi_r$ a.s. and $\xi_r \sim Pois(a_r), r = 1, \ldots, k$. The independence of $\xi_1, \ldots, \xi_k$ follows immediately, due to the properties of characteristic functions, as $\sum_{m=1}^{N} \xi_{1m}, \ldots, \sum_{m=1}^{N} \xi_{km}$ are independent for each $N \in \mathbb{N}$ and

$$\left(\sum_{m=1}^{N} \xi_{1m}, \ldots, \sum_{m=1}^{N} \xi_{km}\right) \to (\xi_1, \ldots, \xi_k) \text{ a.s., } N \to \infty. \quad \square$$

Now consider any pairwise disjoint $B_1, \ldots, B_k \in \mathcal{B}_0(\mathbb{R}^n)$. Set $\xi_{rm} = Z_m(B_r \cap K_m)$, $a_{rm} = \Lambda(B_r \cap K_m)$, $r = 1, \ldots, k$ and $m \in \mathbb{N}$. Using the already established part of the Theorem (for pairwise disjoint Borel $B_1, \ldots, B_k \subset K_m$) and Lemma A.6 we complete the proof. $\square$

Remark A.7. This Theorem shows that the law of a field Z introduced in (A.3.1) does not depend on the choice of the sets $K_1, K_2, \ldots$ forming a partition of $\mathbb{R}^n$. Lemma A.6 implies that $Z(B) \sim Pois(\Lambda(B))$ for any set $B \in \mathcal{B}(\mathbb{R}^n)$ such that $\Lambda(B) < \infty$. At the same time, for $B \in \mathcal{B}(\mathbb{R}^n)$, one has $Z(B) = \infty$ a.s. if $\Lambda(B) = \infty$. This follows by simple reasoning. In fact, let $\xi_1, \xi_2, \ldots$ be independent random variables such that $\xi_m \sim Pois(a_m)$, $m \in \mathbb{N}$, and $\sum_{m=1}^\infty a_m = \infty$, then $\sum_{m=1}^\infty \xi_m = \infty$ a.s. (one can take $\xi_m = Z(B \cap K_m)$). Indeed, $S_N(\omega) = \sum_{m=1}^N \xi_m(\omega)$ is a nondecreasing function in $N \in \mathbb{N}$ for each $\omega \in \Omega$. So there exists $S(\omega) = \sum_{m=1}^\infty \xi_m(\omega) \leq \infty$. The Kolmogorov $0-1$ law (see, e.g., [167]) implies that a series $\sum_{m=1}^\infty \xi_m$ converges or diverges with probability 1. Suppose that $S(\omega) < \infty$ a.s. Then $S_N \to \xi$ in law as $N \to \infty$ and therefore, for all $t \in \mathbb{R}$,

$$\varphi_{S_N}(t) \to \varphi_S(t), \quad N \to \infty,$$

where φ_{S_N} and φ_S are characteristic functions of S_N and S respectively. In view of (3.3) the condition $\sum_{m=1}^\infty a_m \to \infty$ $(N \to \infty)$ yields $\varphi_{S_N}(t) \to 1$ if $t = 2\pi l$, $l \in \mathbb{Z}$ and $\varphi_{S_N}(t) \to 0$ otherwise. This limiting function is not continuous and cannot be a characteristic function. The contradiction shows that $\mathsf{P}(\xi = \infty) > 0$. Thus we conclude that $\xi = \infty$ a.s.

Hence $Z(\cdot, B)$, for each $B \in \mathcal{B}(\mathbb{R}^n)$, is a random variable with values in $\overline{\mathbb{Z}}_+$.

Remark A.8. For any $\omega \in \Omega$, we have a set

$$Q(\omega) = \{X_{11}(\omega), \ldots, X_{1\tau_1(\omega)}(\omega), X_{21}(\omega), \ldots, X_{2\tau_2(\omega)}(\omega), \ldots\}$$

in $\mathbb{R}^n$ (if $\tau_m = 0$ then $\{X_{m1}(\omega), \ldots, X_{m\tau_m(\omega)}(\omega)\} = \varnothing$) and one can think that $Q(\omega)$ is a configuration of "random points" $x_i = x_i(\omega)$ used in the notation $X = \{x_i\}$. Moreover, formula (A.3.1) shows that

$$Z(B, \omega) = \sum_{i=1}^\infty \mathbb{I}\{x_i(\omega) \in B\}, \quad B \in \mathcal{B}(\mathbb{R}^n).$$

Therefore, for every $\omega \in \Omega$, a function $Z(\omega, \cdot)$ is a measure on $(\mathbb{R}^n, \mathcal{B}(\mathbb{R}^n))$ and we conclude that $Z = Z(\omega, B)$ is a random measure. It is worth to emphasize another fact. For every $\omega \in \Omega$ we can enumerate the points of $Q(\omega)$ to get $x_1(\omega), x_2(\omega), \ldots$ (e.g., using the lexicographical order). However we can not claim in this case that $x_1(\omega), x_2(\omega), \ldots$ are random vectors because $x_i(\omega) = X_{m(\omega), j(\omega)}(\omega)$ and we can not guarantee $\mathcal{F}|\mathcal{B}(\mathbb{R}^n)$-measurability of such function. Thus $Q(\omega)$ represents the support of the measure $Z(\cdot, \omega)$ where $x_i(\omega)$ are taken in general with multiplicities (i.e. some of $x_i(\omega)$ can coincide).

A.4 Electric currents

The aim of this short Appendix is to prove some auxiliary results used to study negative dependence in Chapter 1.

Lemma A.9. *Let $\{e^1, \ldots, e^n\}$ be the standard orthonormal basis in the Euclidean space $\mathbb{R}^n$ and let M and L be linear subspaces of $\mathbb{R}^n$ such that $M \subset L$. Then, for any $i = 1, \ldots, n$, one has $0 \leq a_i^i \leq b_i^i$, where $a^i = (a_1^i, \ldots, a_n^i)$ and $b^i = (b_1^i, \ldots, b_n^i)$ are the respective orthogonal projections of e^i onto M and L.*

Proof. Clearly, it suffices to consider $i = 1$. To simplify the notation we write $a^1 = a = (a_1, \ldots, a_n)$. Since the orthogonal projection does not increase the Euclidean norm, we have $a_1^2 + \ldots + a_n^2 \leq 1$. It is well-known that

$$\inf_{x \in M} \|x - e^1\| = \|a - e^1\|.$$

Assume that $a_1 < 0$. Clearly, the vector $-a \in M$ and thus $\| - a - e^1\| \geq \|a - e^1\|$. However,

$$\|-a-e^1\|^2 = (1+a_1)^2+(a_2)^2+\ldots+(a_n)^2 < (1-a_1)^2+(a_2)^2+\ldots+(a_n)^2 = \|a-e^1\|^2.$$

Therefore a is not the projection of e^1 onto M. Consequently, $a_1 \geq 0$.

Let K be the orthogonal complement of M in L. Then $b^1 = a+c^1$, where $c^1 \in K$ is the projection of e^1 onto K. By the previous argument $c_1^1 \geq 0$, consequently $b_1^1 \geq a_1$. $\square$

For the notation used in the next Lemma see Section 1.4, subsection 5.

Lemma A.10. *Let G be a finite directed graph. Let $e = (x, y) \in E$ and P be the linear operator projecting $H(G)$ onto the orthogonal complement of $\Diamond(G)$. Then*

$$(P\chi^e)(e) = \frac{\tau_e(G)}{\tau(G)}.$$

Remark A.11. The function $P\chi^e$ is called the *unit electric current* from x to y.

Proof. For convenience we may assume that $e = (x \rightarrowtail y)$.

For a spanning tree[8] T of G, let $h_1(T), \ldots, h_{k(T)}(T) \in E$ be the unique path in T from x to y (the path is defined as in subsection 1.4.4). That is, there are vertices $v_1 = x, v_2, \ldots, v_{k(T)+1} = y$ such that $h_i(T) = (v_i, v_{i+1})$, $i = 1, \ldots, k(T)$. Define a function $\Phi_T : E \to \mathbb{R}$ as follows:

$$\Phi_T(h) = \begin{cases} 1, & h = h_i(T) \text{ for some } i = 1, \ldots, k(T) \text{ and } h_i(T) = (v_i \rightarrowtail v_{i+1}), \\ -1, & h = h_i(T) \text{ for some } i = 1, \ldots, k(T) \text{ and } h_i(T) = (v_{i+1} \rightarrowtail v_i), \\ 0, & \text{otherwise.} \end{cases}$$

Set

$$\Phi(h) = \frac{1}{\tau(G)} \sum_T \Phi_T(h), \quad h \in E,$$

---

[8]See Definition 1.4.29.

where the last sum is over all spanning trees T of G. Then, obviously, $\Phi(e) = \tau_e(G)/\tau(G)$ (since $\Phi_T(e) = 1$ if T contains e and $\Phi_T(e) = 0$ otherwise). Thus our goal is to show that

$$(P\chi^e)(e) = \Phi(e).$$

Let $(\cdot, \cdot)$ be the usual inner product in $H(G)$, i.e. $(\phi, \psi) = \sum_{h \in E} \phi(h)\psi(h)$. Then the desired relations are

$$\chi^e - \Phi \in \Diamond(G), \tag{A.4.1}$$

$$(\Phi, \psi) = (\chi^e, \psi) \text{ for any } \psi \perp \Diamond(G). \tag{A.4.2}$$

Checking relation (A.4.1) is simple. We have

$$\chi^e - \Phi = \sum_T \frac{1}{\tau(G)}(\chi^e - \Phi_T),$$

but each summand is either zero or a cyclic function obtained by adding the edge e to the path from x to y in T.

To prove (A.4.2) we examine a more delicate structure of the space $H(G)$. Let T^0 be some non-random spanning tree of G. Deleting one edge e^0 from it we obtain a non-connected graph $T_1 \cup T_2$ with two connected components $T_1 = (V_1, E_1)$ and $T_2 = (V_2, E_2)$. We call a *cut set* the system of edges

$$\mathcal{C} = \{e \in E : \text{ the addition of } e \text{ to } T_1 \cup T_2 \text{ gives a connected graph}\}.$$

A *pseudoflow* over (T^0, e^0) is a function F_{e^0} defined by the relations

$$F_{e^0}(e) = \begin{cases} 1, & e \in \mathcal{C} \text{ and } e = (a \mapsto b) \text{ for some } a \in V_1,\, b \in V_2, \\ -1, & e \in \mathcal{C} \text{ and } e = (a \mapsto b) \text{ for some } a \in V_2,\, b \in V_1, \\ 0, & e \notin \mathcal{C}. \end{cases}$$

Denote by $\bigstar$ the linear span of all pseudoflows in $H(G)$ (over all possible T^0, e^0). Then $\bigstar = \Diamond^\perp$. Indeed, the fact that these subspaces are orthogonal is easily verified directly (because each cycle starting in T_1 leaves T_1 the same number of times as it enters it). Thus we only need to check that the direct sum of $\bigstar$ and $\Diamond$ is $H(G)$, i.e. that

$$\dim(\bigstar \oplus \Diamond) = \dim H(G) = |E|. \tag{A.4.3}$$

Let $T^0 = (V, E^0)$ be again some spanning tree of G. For an edge $h \in E^0$ define a pseudoflow F_h to be a pseudoflow over (T_0, h) obtained as above. Thus all pseudoflows $\{F_h, h \in E^0\}$ form a linearly independent system of vectors, as $F_h(w) = 0$ for any edge $w \in E^0$, $w \neq h$, while $F_h(h) \neq 0$. Therefore, $\dim \bigstar \geq |E^0|$. For an edge $h \notin E^0$ let F^h be a cyclic function over a cycle $\mathcal{E}_h$ obtained by adding of h to T_0. Then $F^h(w) = 0$ for any edge $w \notin E^0$, $w \neq h$, while $F^h(h) \neq 0$. So these cyclic functions are linearly independent and their total number is $|E| - |E^0|$. In other words, $\dim \Diamond \geq |E| - |E^0|$. This proves (A.4.3).

Thus we have to verify relation (A.4.2) for any function $\psi \in \bigstar$. It suffices to consider ψ which is a pseudoflow over some tree $T^0 = (V, E^0)$ and edge $h \in E^0$. Indeed, such functions generate the linear space $\bigstar$.

Let $T_1 = (V_1, E_1)$, $T_2 = (V_2, E_2)$ be the components remaining after deleting h from T^0 and let $\mathcal{C}_h$ be the corresponding cut set of edges.

Case 1: $e \in \mathcal{C}_h$. In this case $(\chi^e, \psi) = 1$ if $x \in V_1, y \in V_2$ and $(\chi^e, \psi) = -1$ if $y \in V_1, x \in V_2$. Let now T be (another) spanning tree and let $h_1(T), \ldots, h_{k(T)}(T)$ be the unique path in T from x to y. The number of crossings of the cut set $\mathcal{C}_h$ by this path is odd. Each crossing of $\mathcal{C}_h$ from T_1 to T_2 adds to (Φ_T, ψ) a value 1, whereas the return crossing subtracts it. Thus $(\Phi_T, \psi) = (\chi^e, \psi)$, which leads to (A.4.2).

Case 2: $e \notin \mathcal{C}_h$. Then $(\chi^e, \psi) = 0$. In this case x and y belong to the same connected component of $T_1 \cup T_2$, and each path from x to y crosses the cut set even number of times. Likewise as in the previous case one deduces that $(\Phi_T, \psi) = 0$ for any spanning tree T.

We have proved (A.4.2), hence the Lemma. $\square$

Remark A.12. The model above can be further generalized to comprise, e.g., the situation when the edges are not identical (in terms of the electric current theory, they might have different conductivities). See, e.g., [27].

A.5 The Móricz theorem

In this Appendix $X = \{X_j, j \in \mathbb{Z}^d\}$ is a real-valued random field and, as usual,

$$S(U) = \sum_{j \in U} X_j, \quad M(U) = \max_{V \triangleleft U} |S(V)|, \quad U \in \mathcal{U}.$$

Throughout this Section $\log x = \log_2 x$, $x > 0$, and $\mathrm{Log}\, x = \log_2(2 \vee x)$. We refer to Section 2.1 for other basic notation. Móricz established the following result.

Theorem A.13. ([304]) *Let $d \in \mathbb{N}$ and $\gamma \geq 1$. Suppose that there exist functions $\varphi : \mathcal{U} \to \mathbb{R}_+$ and $\psi : \mathbb{R}_+ \times \mathbb{Z}^d \to \mathbb{R}_+$ such that φ is superadditive, ψ is coordinatewise nondecreasing, and for any block $U = (a, b] \in \mathcal{U}$ with $m_i = b_i - a_i \geq 1$, $i = 1, \ldots, d$, one has*

$$\mathsf{E}|S(U)|^\gamma \leq \varphi(U)\psi^\gamma(\varphi(U), m_1, \ldots, m_d). \tag{A.5.1}$$

Then $\mathsf{E}M(U)^\gamma$, for any $U = (a, b] \in \mathcal{U}$, admits the following upper bound

$$\left(\frac{5}{2}\right)^d \varphi(U) \left(\sum_{k_1=0}^{[\log m_1]} \cdots \sum_{k_d=0}^{[\log m_d]} \psi\left(2^{-k_1 - \cdots - k_d}\varphi(U), [2^{-k_1}m_1], \ldots, [2^{-k_d}m_d] \right) \right)^\gamma.$$

Proof uses the induction on d. Let us consider the case $d = 1$. It is convenient to modify the notation slightly. Set for $a \in \mathbb{Z}$, $m \in \mathbb{Z}_+$ and $t > 0$

$$S(a, m) = S((a, a+m]) = \sum_{j=a+1}^{a+m} X_j \quad \text{and} \quad M(a, m) = \max_{j=1,\ldots,m} |S(a, j)|,$$

$$\varphi(a, m) = \varphi((a, a+m]), \quad \Psi(t, 1) = \psi(t, 1), \quad \Psi(t, m) = \sum_{k=0}^{[\log m]} \psi\left(2^{-k}t, [2^{-k}m] \right).$$

Then one easily checks that

$$\Psi(t, m) = \psi(t, m) + \Psi(t/2, [m/2]), \quad m > 1, \tag{A.5.2}$$

and we have to prove that for any $a \in \mathbb{Z}$, $m \in \mathbb{N}$

$$\mathsf{E}M(a, m)^\gamma \leq \frac{5}{2}\varphi(a, m)\Psi(\varphi(a, m), m)^\gamma. \tag{A.5.3}$$

Inequality (A.5.3), for each $a \in \mathbb{Z}$, is established by induction on m. Clearly, if $m = 1$, then (A.5.3) is a consequence of (A.5.1).

Now suppose that $n > 1$ and (A.5.3) has been verified for all $a \in \mathbb{Z}$ and any $m < n$. Fix $a \in \mathbb{Z}$. If $\varphi(a, n) = 0$, then $S(a, n) = 0$ a.s. by (A.5.1) and the assertion is evident. By superadditivity, the (finite) sequence $\varphi(a, k)$, $k = 0, \ldots, n$, is nondecreasing if we agree that $\varphi(\varnothing) = 0$. Otherwise it is possible to find such $p \in \mathbb{N}$, $p \leq n$, that

$$\varphi(a, p-1) \leq \frac{\varphi(a, n)}{2} < \varphi(a, p).$$

Then, again by superadditivity, we have

$$\varphi(a+p, n-p) \le \varphi((a,n]) - \varphi((a,p]) < \frac{\varphi(a,n)}{2}.$$

Set

$$p_1 = \left[\frac{p-1}{2}\right], \quad p_2 = \left[\frac{n-p}{2}\right], \quad q_1 = p_1 + \mathbb{I}\{p \text{ is even}\}, \quad q_2 = p_2 + \mathbb{I}\{n-p \text{ is odd}\}.$$

Note that in this case $p_1 + q_1 = p - 1$ and $p + q_2 + p_2 = n$. Then for any $k = 1, \ldots, n$ one has

$$|S(a,k)| \le \begin{cases} M(a, p_1), & k \le p_1, \\ |S(a, q_1)| + M(a + q_1, p_1), & q_1 \le k \le p - 1, \\ |S(a, p)| + M(a + p, p_2), & p \le k \le p + p_2, \\ |S(a, p + q_2)| + M(a + p + q_2, n - p - q_2), & p + q_2 \le k \le n. \end{cases} \qquad (A.5.4)$$

Therefore a trivial estimate yields

$$M(a,n) \le \left(|S(a, q_1)|^\gamma + |S(a,p)|^\gamma + |S(a, p+q_2)|^\gamma\right)^{1/\gamma}$$

$$+ \left(M(a, p_1)^\gamma + M(a+q_1, p_1)^\gamma + M(a+p, p_2)^\gamma + M(a+p+q_2, n-p-q_2)^\gamma\right)^{1/\gamma}.$$

Consequently, by the Minkowski inequality

$$\mathsf{E}M(a,n)^\gamma \le \left(A^{1/\gamma} + B^{1/\gamma}\right)^\gamma \qquad (A.5.5)$$

where $A = \mathsf{E}|S(a,q_1)|^\gamma + \mathsf{E}|S(a,p)|^\gamma + \mathsf{E}|S(a,p+q_2)|^\gamma$ and

$$B = \mathsf{E}M(a, p_1)^\gamma + \mathsf{E}M(a+q_1, p_1)^\gamma + \mathsf{E}M(a+p, p_2)^\gamma + \mathsf{E}M((a+p+q_2, n])^\gamma =: \sum_{j=1}^{4} B_j.$$

According to (A.5.1) and the choice of p we have

$$A \le \varphi((a,q_1])\psi(\varphi((a,q_1]), q_1)^\gamma + \varphi((a,p])\psi(\varphi((a,p]), p)^\gamma$$

$$+ \varphi((a, p+q_2])\psi(\varphi((a, p+q_2]), p+q_2)^\gamma$$

$$\le \psi(\varphi((a,n]), n)^\gamma \Big(\varphi((a,q_1]) + \varphi((a,p]) + \varphi((a,p+q_2])\Big)$$

$$\le \psi(\varphi((a,n]), n)^\gamma \left(\frac{\varphi((a,n])}{2} + \varphi((a,n]) + \varphi((a,n])\right) = \frac{5}{2}\psi(\varphi((a,n]), n)^\gamma \varphi((a,n]).$$
$$(A.5.6)$$

The quantities B_1, B_2, B_3, B_4 are estimated via the induction hypothesis; namely,

$$B_1 \le \frac{5}{2}\varphi((a, p_1])\Psi(\varphi((a, p-1]), p_1)^\gamma \le \frac{5}{2}\varphi((a, p_1])\Psi\left(\frac{\varphi((a,n])}{2}, [n/2]\right)^\gamma$$

where the choice of p and the monotonicity of Ψ were used. Furthermore,

$$B_2 \le \frac{5}{2}\varphi((a+q_1, p_1])\Psi(\varphi((a+q_1, p_1]), p_1)^\gamma \le \frac{5}{2}\varphi((a+q_1, p_1])\Psi\left(\frac{\varphi((a,n])}{2}, [n/2]\right)^\gamma.$$

We estimate B_3 and B_4 in the same way. Finally B^γ admits an upper bound

$$\frac{5}{2}\Big(\varphi((a,p_1])+\varphi((a+q_1,p_1])+\varphi((a+p,p_2])+\varphi((a+p+q_2,n])\Big)\Psi\left(\frac{\varphi((a,n])}{2},\left[\frac{n}{2}\right]\right)^\gamma$$

$$\leq \frac{5}{2}\varphi((a,n])\Psi\left(\frac{\varphi((a,n])}{2},[n/2]\right)^\gamma \tag{A.5.7}$$

due to superadditivity of φ.

From (A.5.5), (A.5.6) and (A.5.7) one infers that

$$\|M(a,n)\|_\gamma \leq \left(\frac{5}{2}\right)^{1/\gamma}\varphi((a,n])^{1/\gamma}\left(\Psi\left(\frac{\varphi((a,n])}{2},[n/2]\right)+\psi(\varphi((a,n]),n)\right),$$

which is the desired conclusion (A.5.3), in view of (A.5.2).

Thus Theorem is proved for $d=1$, which gives the induction base.

Suppose that $d>1$ and the Theorem is true for the random fields on lattices $\mathbb{Z}^k$, for all $k<d$. Fix a nonempty block $U=(a,b]\in\mathcal{U}$, and let $m=b-a\in\mathbb{N}^d$. Now, for any block $W\subset\mathbb{Z}^{d-1}$, $t>0$ and $k\in\mathbb{Z}^{d-1}$ introduce functions

$$\varphi_0(W):=\varphi(W\times(a_d,b_d]),\quad \psi_0(t,k):=\psi(t,k_1,\ldots,k_{d-1},m_d).$$

Then φ_0 is superadditive and ψ_0 is coordinatewise nondecreasing. For $k\in\mathbb{Z}^{d-1}$ denote

$$Y_k=\sum_{j=1}^{m_d}X_{(k_1,\ldots,k_{d-1},a_d+j)}.$$

By assumption (A.5.1), for any block $W\subset\mathbb{Z}^{d-1}$ we have

$$\mathsf{E}\left|\sum_{k\in W}Y_k\right|^\gamma \leq \varphi_0(W)\psi_0^\gamma(\varphi_0(W),l_1,\ldots,l_{d-1})$$

where l_i equals the length of i-th edge of W. Consequently, if one introduces the "partial maxima" for a block by way of

$$M_{d-1}(U)=\max_{1\leq j_1\leq m_1}\cdots\max_{1\leq j_{d-1}\leq m_{d-1}}\left|\sum_{1\leq k_1\leq j_1}\cdots\sum_{1\leq k_{d-1}\leq j_{d-1}}\sum_{1\leq k_d\leq m_d}X_k\right|,$$

then from the induction hypothesis (applied to the random field $\{Y_j, j\in\mathbb{Z}^{d-1}\}$) it follows that

$$\mathsf{E}M_{d-1}(U)^\gamma \leq \left(\frac{5}{2}\right)^{d-1}\varphi(U)\Psi_{d-1}(\varphi(U),m_1,\ldots,m_d)^\gamma \tag{A.5.8}$$

where, for $t>0$ and $m\in\mathbb{N}^d$,

$$\Psi_{d-1}(t,m):=\sum_{k_1=0}^{[\log m_1]}\cdots\sum_{k_{d-1}=0}^{[\log m_{d-1}]}\psi\left(\frac{t}{2^{k_1+\ldots+k_{d-1}}},\left[\frac{m_1}{2^{k_1}}\right],\ldots,\left[\frac{m_{d-1}}{2^{k_{d-1}}}\right],m_d\right).$$

Set

$$\Psi_d(t, m_1, \ldots, m_d) = \sum_{k_1=0}^{[\log m_1]} \cdots \sum_{k_d=0}^{[\log m_d]} \psi\left(\frac{\varphi(U)}{2^{k_1+\ldots+k_d}}, \left[\frac{m_1}{2^{k_1}}\right], \ldots, \left[\frac{m_d}{2^{k_d}}\right]\right).$$

Then we have to prove that

$$\mathsf{E}M(U)^\gamma \le \left(\frac{5}{2}\right)^d \varphi(U)\Psi_d(\varphi(U), m_1, \ldots, m_d)^\gamma. \tag{A.5.9}$$

Analogously to (A.5.2) one can verify directly the relations

$$\Psi_d(t, m_1, \ldots, m_{d-1}, 1) = \Psi_{d-1}(t, m_1, \ldots, m_{d-1}, 1), \quad t > 0,$$

$$\Psi_d(t, m_1, \ldots, m_d) = \Psi_{d-1}\left(t, m_1, \ldots, m_{d-1}, \left[\frac{m_d}{2}\right]\right) + \Psi_d\left(\frac{t}{2}, m_1, \ldots, m_{d-1}, \left[\frac{m_d}{2}\right]\right)$$

for $m_d > 1$ and $t > 0$. The desired estimate (A.5.9) is proved by induction on m_d (the numbers $m_1, \ldots, m_{d-1}$ being fixed). The induction base is (A.5.8). For the transition we let $m_d = n$, and for $p, q \in \mathbb{Z}$ such that $a_d \le p < q \le a_d + n$ define the block

$$U(p, q) = \{k \in U : p < k_d \le p + q\}.$$

From this point the proof runs exactly as for the case $d = 1$, if we replace a with a_d, b with b_d, $\varphi((p, q])$ with $\varphi(U(p, q))$, $S(p, q)$ with $M_{d-1}(U(p, q))$ and $M(p, q)$ with $M(U(p, q))$. Indeed, we note that $M(U) = \max_{1 \le s \le n} M_{d-1}(U(a_d, a_d + s))$ and that the analogue of (A.5.4) is true, which is the only place where we used substantially the relation between sums and maxima. $\square$

Corollary A.14. ([304]) *Let $d \in \mathbb{N}$ and $\gamma \ge 1$, $\alpha \ge 1$. Suppose that there exists a nonnegative and superadditive function $f : \mathcal{U} \to \mathbb{R}_+$ such that, for any block $U = (a, b]$, one has*

$$\mathsf{E}|S(U)|^\gamma \le f^\alpha(U).$$

Then, for any $U = (a, b] \in \mathcal{U}$,

$$\mathsf{E}M(U)^\gamma \le \begin{cases} (5/2)^d \left(1 - 2^{(1-\alpha)/\gamma}\right)^{-d\gamma} f^\alpha(U), & \text{if } \alpha > 1, \\ 5^d 2^{d(\gamma-1)} f(U) \left([\mathrm{Log}\, m_1] \ldots [\mathrm{Log}\, m_d]\right)^\gamma, & \text{if } \alpha = 1; \end{cases} \tag{A.5.10}$$

here $m_i = b_i - a_i$, $i = 1, \ldots, d$. In particular, these estimates are true for the function $f(U) = c|U|$, where $U \in \mathcal{U}$, $c \ge 0$.

Proof. Let $\alpha > 1$. Apply Theorem A.13 with $\varphi(U) = f(U)$ and

$$\psi(t, m_1, \ldots, m_d) = t^{(\alpha-1)/\gamma}, \quad t > 0, \quad m_1, \ldots, m_d \in \mathbb{N}.$$

By the mentioned Theorem, to establish (A.5.10) one can prove that

$$\sum_{k_1=0}^{[\log m_1]} \cdots \sum_{k_d=0}^{[\log m_d]} \psi\left(2^{-k_1-\ldots-k_d}t, [2^{-k_1}m_1], \ldots, [2^{-k_d}m_d]\right)$$

$$\leq t^{(\alpha-1)/\gamma}\left(1 - 2^{(1-\alpha)/\gamma}\right)^{-d}, \quad t > 0.$$

We have

$$\sum_{k_1=0}^{[\log m_1]} \cdots \sum_{k_d=0}^{[\log m_d]} \psi\left(2^{-k_1-\ldots-k_d}t, \left[\frac{m_1}{2^{k_1}}\right], \ldots, \left[\frac{m_d}{2^{k_d}}\right]\right)$$

$$= \sum_{k_1=0}^{[\log m_1]} \cdots \sum_{k_d=0}^{[\log m_d]} t^{(\alpha-1)/\gamma} 2^{-(k_1+\ldots+k_d)(\alpha-1)/\gamma} = t^{(\alpha-1)/\gamma} \prod_{j=1}^{d} \sum_{k_j=0}^{[\log m_j]} 2^{-k_j(\alpha-1)/\gamma}$$

$$\leq t^{(\alpha-1)/\gamma}\left(\sum_{k=0}^{\infty} 2^{-k(\alpha-1)/\gamma}\right)^{d} = t^{(\alpha-1)/\gamma}\left(1 - 2^{(1-\alpha)/\gamma}\right)^{-d},$$

which is the desired statement.

If $\alpha = 1$ then application of Theorem A.13 with $\varphi(U) = f(U)$ and $\psi \equiv 1$ yields the result. $\square$

A.6   Gaussian approximation

Here we provide some variants of the coupling techniques results. The next theorem is close to one proved by Berkes and Philipp in [37].

Theorem A.15. *Let $Y_k \sim N(0,1)$, $k \in \mathbb{N}$. Suppose that there exist sequences $(\varkappa_m)_{m\in\mathbb{N}}$ and $(z_m)_{m\in\mathbb{N}}$ of (strictly) positive numbers such that for $m \in \mathbb{N}, m > 1$, and $t = (t_1, \ldots, t_m) \in \mathbb{R}^m$ with $\|t\|^2 \leq z_m^2$ one has*

$$\left| \mathsf{E}\exp\left\{i\sum_{l=1}^{m} t_l Y_l\right\} - \mathsf{E}\exp\left\{i\sum_{l=1}^{m-1} t_l Y_l\right\}\mathsf{E}\exp\left\{it_m Y_m\right\}\right| \leq \varkappa_m, \qquad \text{(A.6.1)}$$

$$mz_m^{-2} + m^3\exp\{-m^{-6}z_m^2/2\}z_m^{2m-1} + \varkappa_m z_m^{2m} = O(m^{-2}), \ m \to \infty. \qquad \text{(A.6.2)}$$

Then one can redefine the sequence $Y = (Y_k)_{k\in\mathbb{N}}$ without changing its distribution on another probability space together with a sequence $Z = (Z_k)_{k\in\mathbb{N}}$ of independent standard Gaussian random variables such that

$$\mathsf{P}\left(|Y_k - Z_k| \geq k^{-2} \ \text{i. o.}\right) = 0.$$

Proof. First of all we establish a series of lemmas. The first of them is the classical measure-theoretic result ascending to Ornstein, Strassen and Dobrushin. Recall that if μ is a signed measure on a space $(S, \mathcal{B})$ then its *total variation* is

$$\|\mu\|_{TV} := \sup_{B\in\mathcal{B}} |\mu(B) - \mu(S \setminus B)|,$$

If μ_1 and μ_2 are two probability measures on $(S, \mathcal{B})$, the *total variation distance* between μ_1 and μ_2 is the quantity

$$\|\mu_1 - \mu_2\|_{TV} := 2\sup_{B\in\mathcal{B}} |\mu_1(B) - \mu_2(B)|$$

which introduce a metric on the space of probability measures on $(S, \mathcal{B})$.

Further on we use $S = \mathbb{R}^n$ and $\mathcal{B} = \mathcal{B}(\mathbb{R}^n)$. Denote by D the diagonal in $\mathbb{R}^n \times \mathbb{R}^n$, i.e. the set

$$D = \{(x_1, \ldots, x_n, x_1, \ldots, x_n) : x = (x_1, \ldots, x_n) \in \mathbb{R}^n\}.$$

Lemma A.16. ([365]) *Let μ_1 and μ_2 be two probability measures on $(\mathbb{R}^n, \mathcal{B}(\mathbb{R}^n))$. Then there exists a probability measure ν on $(\mathbb{R}^n \times \mathbb{R}^n, \mathcal{B}(\mathbb{R}^n) \otimes \mathcal{B}(\mathbb{R}^n))$ such that the projection of ν onto the first n coordinates (resp. onto the last n ones) is μ_1 (resp. μ_2) and*

$$\nu\left((\mathbb{R}^n \times \mathbb{R}^n) \setminus D\right) = \frac{1}{2}\|\mu_1 - \mu_2\|_{TV}. \qquad \text{(A.6.3)}$$

Proof. Let $A, B \in \mathcal{B}(\mathbb{R}^n)$ be the sets forming the Hahn-Jordan decomposition of the signed measure $\mu_1 - \mu_2$, i.e. $A \cup B = \mathbb{R}^n$, $A \cap B = \varnothing$,

$$\mu_1(F) - \mu_2(F) \geq 0 \text{ if } F \subset A, F \in \mathcal{B}(\mathbb{R}^n), \ \ \mu_1(G) - \mu_2(G) \leq 0 \text{ if } G \subset B, G \in \mathcal{B}(\mathbb{R}^n)$$

(see, e.g., [239, Ch. VI, §5, 1]). Define, for $C \in \mathcal{B}(\mathbb{R}^n) \otimes \mathcal{B}(\mathbb{R}^n)$, a measure ν by the following four relations:

$$\nu(C) = \nu(C \cap D) := \mu_2\left(\{(x_1, \ldots, x_n) : (x_1, \ldots, x_n, x_1, \ldots, x_n) \in C\}\right), \quad C \subset A \times A;$$

$$\nu(C) = \nu(C \cap D) := \mu_1\left(\{(x_1, \ldots, x_n) : (x_1, \ldots, x_n, x_1, \ldots, x_n) \in C\}\right), \quad C \subset B \times B;$$

$$\nu(C) = \frac{1}{\mu_2(B) - \mu_1(B)}(\mu_1 - \mu_2) \otimes (\mu_2 - \mu_1)(C), \quad C \subset A \times B;$$

$$\nu(C) = 0, \quad C \subset B \times A.$$

Here $(1/0)0 := 0$ and the product measure is understood in a usual way. Now, if $C = K \times \mathbb{R}^n$ where $K \in \mathcal{B}(\mathbb{R}^n)$, one has

$$\nu(C) = \nu((K \cap A) \times A) + \nu((K \cap A) \times B) + \nu((K \cap B) \times A) + \nu((K \cap B) \times B)$$

$$= \mu_2(K \cap A) + \mu_1(K \cap B) + \frac{1}{\mu_2(B) - \mu_1(B)}(\mu_1(K \cap A) - \mu_2(K \cap A))(\mu_2(B) - \mu_1(B))$$

$$= \mu_2(K \cap A) + \mu_1(K \cap B) + \mu_1(K \cap A) - \mu_2(K \cap A) = \mu_1(K).$$

If $C = \mathbb{R}^n \times K$ with $K \in \mathcal{B}(\mathbb{R}^n)$, then

$$\nu(C) = \nu(A \times (K \cap A)) + \nu(B \times (K \cap A)) + \nu(A \times (K \cap B)) + \nu(B \times (K \cap B))$$

$$= \mu_2(K \cap A) + \mu_1(K \cap B) + \frac{1}{\mu_2(B) - \mu_1(B)}(\mu_1(A) - \mu_2(A))(\mu_2(K \cap B) - \mu_1(K \cap B))$$

$$= \mu_2(K \cap A) + \mu_1(K \cap B) + \mu_2(K \cap B) - \mu_1(K \cap B) = \mu_2(K)$$

where we have used that $\mu_2(B) - \mu_1(B) = \mu_1(A) - \mu_2(A)$. Let us verify (A.6.3).

$$\nu\left((\mathbb{R}^n \times \mathbb{R}^n) \setminus D\right) = \nu(A \times B) = \frac{1}{\mu_2(B) - \mu_1(B)}(\mu_1 - \mu_2) \otimes (\mu_2 - \mu_1)(A \times B)$$

$$= \mu_1(A) - \mu_2(A) = \sup_{K \in \mathcal{B}(\mathbb{R}^n)} |\mu_1(K) - \mu_2(K)| = \frac{1}{2}\|\mu_1 - \mu_2\|_{TV}. \quad \square$$

For a random element X, we denote its distribution by P_X. For random elements X and Y with values in Polish spaces, their common distribution will be denoted by $P_{X,Y}$ and the regular conditional distribution of X, given $Y = y$, by $P_{X|Y=y}$.

Recall that for σ-algebras $\mathcal{V}$ and $\mathcal{W}$ the β-mixing coefficient (or the absolute regularity coefficient) is defined as

$$\beta(\mathcal{V}, \mathcal{W}) = \frac{1}{2}\sup \sum_{i,j} |\mathsf{P}(A_i B_j) - \mathsf{P}(A_i)\mathsf{P}(B_j)|,$$

the supremum being taken over all finite partitions of $\mathcal{V}$ and $\mathcal{W}$ into pairwise disjoint events. The β-mixing coefficient X and Y by definition equals $\beta(\sigma(X), \sigma(Y))$.

A well-known property of β-mixing, due to Volkonskii and Rozanov [402], is

Lemma A.17. *Let $(S_1, \mathcal{B}_1)$ and $(S_2, \mathcal{B}_2)$ be Polish spaces, ξ and η be random elements on a common probability space, taking values in S_1 and S_2 respectively. Then the function $\beta : S_2 \to \mathbb{R}$, where*

$$\beta(y) = \frac{1}{2}\|P_{\xi|\eta=y} - P_\xi\|_{TV}, \quad y \in S_2,$$

is measurable in y. Moreover,

$$\int_{S_2} \beta(y) P_\eta(dy) = \frac{1}{2}\|P_{\xi,\eta} - P_\xi \otimes P_\eta\|_{TV}. \tag{A.6.4}$$

Proof. Let $\mathcal{A}$ be a countable algebra of sets of $\mathcal{B}_1$ such that $\sigma(\mathcal{A}) = \mathcal{B}_1$. Such an algebra exists since S_1 is a Polish space. For example, one can take a countable dense subset $\mathcal{D}$ of S_1, then consider all balls with rational radii centered at points of $\mathcal{D}$ and introduce $\mathcal{A}$ as the class of their finite unions, intersections and complements to such sets in S_1.

It is easy to verify that for an arbitrary signed measure Q on a space $(S_1, \mathcal{B}_1)$, any $D \in \mathcal{B}_1$ and each $\varepsilon > 0$ one can find such $A \in \mathcal{A}$ that $|Q(A \triangle D)| < \varepsilon$. Therefore, for any $y \in S_2$ we have

$$\beta(y) = \sup_{A \in \mathcal{B}_1} (P_{\xi|\eta=y}(A) - P_\xi(A)) = \sup_{A \in \mathcal{A}} (P_{\xi|\eta=y}(A) - P_\xi(A)). \tag{A.6.5}$$

Thus the left-hand side of (A.6.5) is measurable in y (being a supremum over a countable set of a family of functions, each of them measurable). Fix arbitrary $\varepsilon > 0$. Enumerate the elements of $\mathcal{A}$ to obtain a sequence of sets $(A_j)_{j \in \mathbb{N}}$. According to (A.6.5), for any $y \in S_2$, there exists some $j = j(y) \in \mathbb{N}$ (in fact $j = j(\varepsilon, y)$) such that

$$(P_{\xi|\eta=y}(A_{j(y)}) - P_\xi(A_{j(y)})) > \sup_{A \in \mathcal{A}} (P_{\xi|\eta=y}(A) - P_\xi(A)) - \varepsilon. \tag{A.6.6}$$

One can think that $j(y)$ is uniquely determined by y (e.g., taking the minimal $j(y)$ such that (A.6.6) holds). Next introduce the sets

$$B_k = \{y \in S_2 : j(y) = k\}, \quad k \in \mathbb{N}.$$

Note that the sets $B_1, B_2, \ldots$ are Borel ones as, for $k \in \mathbb{N}$,

$$B_k = \{y \in S_2 : f_v(y) \le g_v(y) \text{ for all } v < k \text{ and } f_k(y) > g_k(y)\}$$

where the functions

$$f_r(y) = P_{\xi|\eta=y}(A_{j(y)}) - P_\xi(A_{j(y)}), \quad g_r(y) = \sup_{A \in \mathcal{A}} (P_{\xi|\eta=y}(A) - P_\xi(A)) - \varepsilon$$

are measurable in y for any $r \in \mathbb{N}$.

Now we define a set in $\mathcal{B}_1 \otimes \mathcal{B}_2$ as follows

$$C = \{(x, y) : x \in A_{j(y)}, y \in S_2\} = \bigcup_{k=1}^{\infty} \{(x, y) : x \in A_k, j(y) = k\} = \bigcup_{k=1}^{\infty} (A_k \times B_k)$$

and observe that

$$P_{\xi,\eta}(C) - (P_\xi \otimes P_\eta)(C) = \int_{S_2} \left(P_{\xi|\eta=y}(A_{j(y)}) - P_\xi(A_{j(y)}) \right) P_\eta(dy)$$

$$\ge \int_{S_2} \left(\sup_{A \in \mathcal{A}} (P_{\xi|\eta=y}(A) - P_\xi(A)) - \varepsilon \right) P_\eta(dy) = 2 \int_{S_2} \beta(y) P_\eta(dy) - \varepsilon.$$

As ε can be taken arbitrarily small, this can happen only if

$$\frac{1}{2} \|P_{\xi,\eta} - P_\xi \otimes P_\eta\|_{TV} \ge \int_{S_2} \beta(y) P_\eta(dy).$$

The converse inequality is easy to handle, since for any $C \in \mathcal{B}_1 \otimes \mathcal{B}_2$ and $y \in S_2$, letting $C_y = \{x \in S_1 : (x, y) \in C\}$, one has

$$|P_{\xi,\eta}(C) - (P_\xi \otimes P_\eta)(C)| = \left| \int_{S_2} \left(P_{\xi|\eta=y}(C_y) - P_\xi(C_y) \right) P_\eta(dy) \right|$$

$$\leq \int_{S_2} \sup_{A \in \mathcal{B}_1} |P_{\xi|\eta=y}(A) - P_\xi(A)| P_\eta(dy) = 2 \int_{S_2} \beta(y) P_\eta(dy).$$

Taking the supremum over all $C \in \mathcal{B}_1 \otimes \mathcal{B}_2$ we come to (A.6.4). $\square$

We need also

Lemma A.18. ([37]) *Let S_i, $i = 1, 2, 3$, be Polish spaces. Suppose that (X, Y) and (R, T) are random elements with values respectively in*

$$(S_1 \times S_2, \mathcal{B}(S_1) \otimes \mathcal{B}(S_2)) \text{ and } (S_2 \times S_3, \mathcal{B}(S_2) \otimes \mathcal{B}(S_3))$$

and such that $Law(Y) = Law(R)$. Then there exist a probability space and random elements ζ^1, ζ^2 and ζ^3 defined on it such that ζ^i takes values in $(S_i, \mathcal{B}(S_i))$ (for $i = 1, 2, 3$), $Law(\zeta^1, \zeta^2) = Law(X, Y)$ and $Law(\zeta^2, \zeta^3) = Law(R, T)$.

Proof. Assume at first that X, Y and T take only finite number of values, i.e. there exist $n \in \mathbb{N}$ and sets

$$L_1 = \{x_1, \ldots, x_n\} \subset S_1, \ L_2 = \{y_1, \ldots, y_n\} \subset S_2, \ L_3 = \{t_1, \ldots, t_n\} \subset S_3$$

such that $\mathsf{P}(X \in L_1, Y \in L_2) = \mathsf{P}(R \in L_2, T \in L_3) = 1$. We may require that there are exactly n points in L_i, $i = 1, 2, 3$ (if some of the spaces S_i consist of less than n points, then we can add to it the required number of points agreeing that the corresponding random element takes values in the set of these added points with probability zero). Define a measure μ on $(S_1 \times S_2 \times S_3, \mathcal{B}(S_1) \otimes \mathcal{B}(S_2) \otimes \mathcal{B}(S_3))$ by the formula

$$\mu((x_i, y_j, t_k)) = \mathsf{P}(X = x_i | Y = y_j) \mathsf{P}(T = t_k | R = y_j) \mathsf{P}(Y = y_j), \ i, j, k = 1, \ldots, n,$$

and $\mu((S_1 \times S_2 \times S_3) \setminus (L_1 \times L_2 \times L_3)) = 0$. The conditional probabilities are understood in the classical sense, i.e. $\mathsf{P}(A_1 | A_2) = \mathsf{P}(A_1 A_2) / \mathsf{P}(A_2)$ for events A_1, A_2 with agreement that $0/0 = 0$. We have defined a probability measure, since all n^3 points of the type (x_i, y_j, t_k) are different, and

$$\sum_{i,j,k=1}^{n} \mu((x_i, y_j, t_k)) = \sum_{i,j=1}^{n} \mathsf{P}(X = x_i | Y = y_j) \sum_{k=1}^{n} \mathsf{P}(T = t_k | R = y_j) \mathsf{P}(Y = y_j) = 1.$$

Take a probability space on which there is a random element $(\zeta^1, \zeta^2, \zeta^3)$ having distribution μ. For any sets $B_1 \in \mathcal{B}(S_1)$, $B_2 \in \mathcal{B}(S_2)$ let

$$J = \{1 \leq i, j, k \leq n : (x_i, y_j, t_k) \in B_1 \times B_2 \times S_3\}.$$

Then one has, by definition of μ,

$$\mathsf{P}(\zeta^1 \in B_1, \zeta^2 \in B_2) = \sum_{i,j,k \in J} \mathsf{P}(X = x_i | Y = y_j) \mathsf{P}(T = t_k | R = y_j) \mathsf{P}(Y = y_j)$$

$$= \sum{}' \mathsf{P}(X = x_i, Y = y_j) = \mathsf{P}(X \in B_1, Y \in B_2)$$

where $\sum'$ denotes the double sum over $i, j \in \{1, \dots, n\}$ such that $(x_i, y_j) \in B_1 \times B_2$. In the same way one checks that $Law(\zeta^2, \zeta^3) = Law(R, T)$.

Now suppose that the random elements under consideration take infinite number of values. As the spaces S_1, S_2, S_3 are separable, there exists a sequence of random elements $(X_n, Y_n)_{n \in \mathbb{N}}$ (defined on the same probability space as (X, Y)) such that, for any $n \in \mathbb{N}$, X_n and Y_n take finite number of values, and $(X_n, Y_n) \to (X, Y)$ almost surely as $n \to \infty$. Likewise, there exist a sequence $(R_n, T_n)_{n \in \mathbb{N}}$ of discrete random elements such that $(R_n, T_n) \to (R, T)$ a.s., $n \to \infty$. Clearly, these sequences can be chosen in such a way that $Law(Y_n) = Law(R_n)$, $n \in \mathbb{N}$.

By the part of Lemma already proved, for any $n \in \mathbb{N}$ there exists a probability measure μ_n on $(S_1 \times S_2 \times S_3, \mathcal{B}(S_1) \otimes \mathcal{B}(S_2) \otimes \mathcal{B}(S_3))$ for which $\mu_n(B_1 \times B_2 \times S_3) = (Law(X_n, Y_n))(B_1 \times B_2)$ and $\mu_n(S_1 \times B_2 \times B_3) = (Law(R_n, T_n))(B_2 \times B_3)$.

Let $\varepsilon > 0$. By the necessity part of the Prohorov theorem ([39, §6, Th. 6.2]) there exist compact sets $K_i \subset S_i$ $(i = 1, 2, 3)$ such that, for any $n \in \mathbb{N}$, one has

$$Law(X_n)(S_1 \setminus K_1) < \varepsilon, \; Law(Y_n)(S_2 \setminus K_2) < \varepsilon, \; Law(T_n)(S_3 \setminus K_3) < \varepsilon.$$

Let $K = K_1 \times K_2 \times K_3$. Then K is a compact set in $(S_1 \times S_2 \times S_3, \mathcal{B}(S_1) \otimes \mathcal{B}(S_2) \otimes \mathcal{B}(S_3))$ (see the argument finishing the proof of Theorem 8.3.4). Moreover,

$$\mu_n((S_1 \times S_2 \times S_3) \setminus (K_1 \times K_2 \times K_3)) \le 3\varepsilon.$$

Consequently, the sequence of measures $(\mu_n)_{n \in \mathbb{N}}$ is tight. By the Prohorov theorem it contains a subsequence $(\mu_v)_{v \in \mathbb{N}}$ weakly converging to some limit measure μ. Thus on some probability space there exist random elements $(\zeta^1, \zeta^2, \zeta^3)$ with common distribution μ. For any bounded continuous function $f : S_1 \times S_2 \to \mathbb{R}$ one has

$$\mathsf{E}f(X_v, Y_v) = \int_{S_1 \times S_2 \times S_3} f(x, y) \mu_v(dx\, dy\, dt) \to \mathsf{E}f(\zeta^1, \zeta^2), \;\; v \to \infty.$$

Consequently $Law(\zeta^1, \zeta^2) = Law(X, Y)$. In the same way one checks that $Law(\zeta^2, \zeta^3) = Law(R, T)$. $\square$

The following result is a modification of a coupling lemma of [29].

Lemma A.19. *Let $U = (U_1, \dots, U_n)$ be a random vector with values in $\mathbb{R}^n$, $n \ge 2$. Suppose that*

$$\frac{1}{2} \| P_{U_1, \dots, U_k} - P_{U_1, \dots, U_{k-1}} \otimes P_{U_k} \|_{TV} \le \beta_k \qquad (A.6.7)$$

for any $k = 2, \dots, n$ and some nonnegative numbers $\beta_2, \dots, \beta_n$. Then one can redefine U on another probability space which also supports a decoupled version[9] $Z = (Z_1, \dots, Z_n)$ of U and such that

$$\mathsf{P}(U_j \neq Z_j) \le \beta_j, \; 2 \le j \le n. \qquad (A.6.8)$$

---

[9]See Section 2.2.

Proof. We apply the induction on n. Set $V = (U_1, \ldots, U_{n-1})$ and let $v = (v_1, \ldots, v_{n-1}) \in \mathbb{R}^{n-1}$ be an arbitrary point. By Lemma A.16, for any $v \in \mathbb{R}^{n-1}$ there exists a probability measure $\nu_n^{(v)}$ on $\mathbb{R}^2$ such that

$$\nu_n^{(v)}\left((x,y) \in \mathbb{R}^2 : x \neq y\right) = \frac{1}{2}\|P_{U_n|V=v} - P_{U_n}\|_{TV} =: \beta_n(v) \qquad \text{(A.6.9)}$$

and the first and second marginals of $\nu_n^{(v)}$ are respectively $P_{U_n|V=v}$ and P_{U_n}. From the argument proving the Lemma A.16 (the construction of measure ν on the product space) one sees that $\nu_n^{(v)}(C)$ is a measurable function in v, for any $B \in \mathcal{B}(\mathbb{R}^2)$. Define a measure μ on $\mathcal{B}(\mathbb{R}^{n+1})$ by the following relation:

$$\mu(B \times C) = \int_{B_1} \nu_n^{(v)}(C) P_V(dv), \quad B \in \mathcal{B}(\mathbb{R}^{n-1}), \quad C \in \mathcal{B}(\mathbb{R}^2). \qquad \text{(A.6.10)}$$

Let us take on some probability space a random vector $(\xi_1, \ldots, \xi_{n+1})$ such that $Law(\xi_1, \ldots, \xi_{n+1}) = \mu$, then in view of (A.6.10) one has, for $B \in \mathcal{B}(\mathbb{R}^{n-1})$ and $D \in \mathcal{B}(\mathbb{R})$,

$$\mathsf{P}((\xi_1, \ldots, \xi_{n-1}) \in B, \xi_n \in D) = \mu(B \times D \times \mathbb{R})$$

$$= \int_B \mathsf{P}(U_n \in D | V = v) P_V(dv) = \mathsf{P}(V \in B, U_n \in D).$$

Therefore, $Law(\xi_1, \ldots, \xi_n) = Law(U)$. Moreover, due to (A.6.9), (A.6.10) and Lemma A.17 we obtain

$$\mathsf{P}(\xi_n \neq \xi_{n+1}) = \mu\left(\mathbb{R}^{n-1} \times \{(x,y) \in \mathbb{R}^2 : x \neq y\}\right)$$

$$= \int_{\mathbb{R}^{n-1}} \nu_n^{(v)}\left((x,y) \in \mathbb{R}^2 : x \neq y\right) P_V(dv) \leq \int_{\mathbb{R}^{n-1}} \beta_n(v) P_V(dv) \leq \beta_n.$$

One easily sees from (A.6.10) that $Law(\xi_{n+1}) = Law(\xi_n)$ and ξ_{n+1} is independent of $\xi_1, \ldots, \xi_{n-1}$. If $n = 2$, the Lemma is proved, which gives the base of induction.

Suppose that $n > 2$. By the induction hypothesis, there exists some probability space which supports a random vector $(\eta_1, \ldots, \eta_{n-1})$ and its decoupled version $\tau = (\tau_1, \ldots, \tau_{n-1})$ with the following properties:

$$Law(\xi) = Law(V), \quad \mathsf{P}(\eta_j \neq \tau_j) \leq \beta_j, \quad j = 2, \ldots, n-1.$$

Obviously, one can assume (enlarging the probability space if necessary) that there exists a random variable ϑ which is independent of (η, τ) and $Law(\vartheta) = Law(U_n)$. Applying now Lemma A.18 with $S_1 = \mathbb{R}$, $S_2 = \mathbb{R}^n$, $S_3 = \mathbb{R}^{n-1}$ and

$$X = \xi_n, \quad Y = (\xi_1, \ldots, \xi_{n-1}, \xi_{n+1}), \quad L = (\eta, \vartheta), \quad T = \tau,$$

we can define random vectors $\zeta^1, \zeta^2, \zeta^3$ such that

$$Law(\zeta^1, \zeta^2) = Law(X, Y), \quad Law(\zeta^2, \zeta^3) = Law(L, T).$$

On the last constructed probability space (on which ζ^i have been defined) let us introduce

$$U_1' = \zeta_1^2, \ldots, U_{n-1}' = \zeta_{n-1}^2, \; U_n' = \zeta^1, \; Z_1 = \zeta_1^3, \ldots, Z_{n-1} = \zeta_{n-1}^3, \; Z_n = \zeta_n^2.$$

Then we can write

$$Law(U_1', \ldots, U_n') = Law(\xi_1, \ldots, \xi_n) = Law(U_1, \ldots, U_n),$$

$$Law(Z_1, \ldots, Z_n) = Law(\tau_1, \ldots, \tau_{n-1}, \vartheta)$$

and

$$P(U_n' \neq Z_n) \leq P(\xi_n \neq \xi_{n+1}) \leq \beta_n. \tag{A.6.11}$$

Now it remains to notice that $Z = (Z_1, \ldots, Z_n)$ is a decoupled version of $U = (U_1, \ldots, U_n)$ and, for any $j = 2, \ldots, n$, by the induction hypothesis and (A.6.11) we come to (A.6.8) with U_j identified with U_j'. $\square$

Lemma A.20. *Let $U = (U_n)_{n \in \mathbb{N}}$ be a sequence of Gaussian random variables such that $\mathsf{E}U_n = 0$ and $\mathsf{Var}U_n = 1 + \rho_n$ with $0 \leq \rho_n \leq 1$, $n \in \mathbb{N}$. Suppose that*

$$\frac{1}{2}\|P_{U_1,\ldots,U_k} - P_{U_1,\ldots,U_{k-1}} \otimes P_{U_k}\|_{TV} \leq \beta_k$$

for any $k \geq 2$ and some nonnegative sequence $(\beta_k)_{k \in \mathbb{N}}$. Then one can redefine U (preserving its distribution) on another probability space such that it supports a sequence $\zeta = (\zeta_k)_{k \in \mathbb{N}}$ of mutually independent random variables with $Law(\zeta_k) = Law(U_k)$, $k \in \mathbb{N}$, and

$$\mathsf{P}(U_j \neq \zeta_j) \leq \beta_j, \quad j \in \mathbb{N}, \quad j > 1. \tag{A.6.12}$$

Proof. Lemma A.19 implies that, for each $k \in \mathbb{N}$, there exist random vectors $U^{(k)} = (U_1^{(k)}, \ldots, U_1^{(k)})$, $\zeta^{(k)} = (\zeta_1^{(k)}, \ldots, \zeta_1^{(k)})$ with values in $\mathbb{R}^k$ defined (for all k) on the same probability space $(\Omega', \mathcal{F}', \mathsf{P}')$ such that

$$Law(U^{(k)}) = P_{U_1,\ldots,U_k}, \quad Law(\zeta^{(k)}) = P_{U_1} \otimes \ldots \otimes P_{U_k} \tag{A.6.13}$$

and

$$\mathsf{P}'\left(U_j^{(k)} \neq \zeta_j^{(k)}\right) \leq \beta_j, \quad j = 2, \ldots, k. \tag{A.6.14}$$

For any $k, n \in \mathbb{N}$, $n > k$, set $U_n^{(k)} = \zeta_n^{(k)} = 0$. Fix any $n \in \mathbb{N}$ and consider random vectors $T_n^{(k)} = (U_1^{(k)}, \ldots, U_n^{(k)}, \zeta_1^{(k)}, \ldots, \zeta_n^{(k)})$ in $\mathbb{R}^{2n}$, $k \in \mathbb{N}$. If $T = (T_1, \ldots, T_{2n})$ is a random vector in $\mathbb{R}^{2n}$ (defined on $(\Omega', \mathcal{F}', \mathsf{P}')$) such that $T_i \sim N(0, 1 + \rho_i)$, $i = 1, \ldots, 2n$, then, for any $\varepsilon > 0$, there exists $q = q(\varepsilon, n)$ such that

$$\mathsf{P}'\left(\bigcup_{i=1}^{2n}\{|T_i| > q\}\right) \leq \frac{n}{\sqrt{\pi}}\int_{|u|>q} e^{-u^2/8}du < \varepsilon.$$

In view of (A.6.13) all the components of $T_k^{(n)}$ are Gaussian centered random variables with variance belonging to $[1, 2]$, if $k \geq n$. Therefore the family of distributions of $T_n^{(k)}$, $k \in \mathbb{N}$, on $(\mathbb{R}^{2n}, \mathcal{B}(\mathbb{R}^{2n}))$ is tight. Thus by the Prohorov theorem each sequence of these distributions contains a weakly convergent subsequence. Consequently, there exists a sequence $(k_v)_{v \in \mathbb{N}}$ of positive integers such that $k_v \to \infty$ as $v \to \infty$ and

$$Law(T_1^{(k_v)}) \to Q_1$$

where Q_1 is a probability measure on $(\mathbb{R}^2, \mathcal{B}(\mathbb{R}^2))$. By the same reason one can extract a subsequence $(k_l)_{l \in \mathbb{N}}$ of $(k_v)_{v \in \mathbb{N}}$ such that $k_l \to \infty$ as $l \to \infty$ and
$$Law(T_2^{(k_l)}) \to Q_2$$
where Q_2 is some probability measure on $(\mathbb{R}^4, \mathcal{B}(\mathbb{R}^4))$, etc.

Using the Cantor diagonal device one can choose a sequence $(k_m)_{m \in \mathbb{N}}$ such that $k_m \to \infty$ as $m \to \infty$ and $Law(T_n^{(k_m)}) \to Q_n$ in $(\mathbb{R}^{2n}, \mathcal{B}(\mathbb{R}^{2n}))$ for each $n \in \mathbb{N}$. Clearly, for $n \in \mathbb{N}$, there exists a random vector $L_n = (\eta_1^{(n)}, \ldots, \eta_n^{(n)}, \gamma_1^{(n)}, \ldots, \gamma_n^{(n)})$ defined[10] on $(\Omega', \mathcal{F}', \mathsf{P}')$ such that $Q_n = Law(L_n)$. Thus, for any $n \in \mathbb{N}$,
$$T_n^{(k_m)} \to L_n \quad \text{in law}, \quad m \to \infty.$$
Obviously, if $(Y_1^{(k)}, \ldots, Y_N^{(k)}) \to (Y_1, \ldots, Y_N)$ in law as $k \to \infty$, then for every $1 \le n \le N$ and any $i_1, \ldots, i_n \in \{1, \ldots, N\}$ one has
$$(Y_{i_1}^{(k)}, \ldots, Y_{i_n}^{(k)}) \to (Y_{i_1}, \ldots, Y_{i_n}) \quad \text{in law}, \quad k \to \infty. \tag{A.6.15}$$
This fact and the definition of $T_n^{(k)}$ yield that, for any $1 \le n \le N$,
$$Q_n = Law(\eta_1^{(N)}, \ldots, \eta_n^{(N)}, \gamma_1^{(N)}, \ldots, \gamma_n^{(N)}).$$
Consequently, the Kolmogorov consistency theorem (e.g., [81, Ch. 1]) guarantees that on $(\Omega', \mathcal{F}', \mathsf{P}')$ there exists[11] a sequence of vectors $(\eta_n, \gamma_n)_{n \in \mathbb{N}}$ with values in $\mathbb{R}^2$ such that
$$Q_n = Law(\eta_1, \ldots, \eta_n, \gamma_1, \ldots, \gamma_n), \ n \in \mathbb{N}. \tag{A.6.16}$$
Moreover, due to our construction γ_k, $k \in \mathbb{N}$, are independent $N(0, 1 + \rho_k)$ random variables. Indeed, for each $n \in \mathbb{N}$, $\gamma_1^{(n)}, \ldots, \gamma_n^{(n)}$ are independent centered Gaussian random variables with $\mathsf{Var}\, \gamma_i^{(n)} = 1 + \rho_i$, $i = 1, \ldots, n$ (because $\zeta_1^{(n)}, \ldots, \zeta_n^{(n)}$ have the same property). In view of (A.6.16) we conclude that, for any $n \in \mathbb{N}$,
$$(U_1^{(k_m)}, \ldots, U_n^{(k_m)}, \zeta_1^{(k_m)}, \ldots, \zeta_n^{(k_m)}) \to (\eta_1, \ldots, \eta_n, \gamma_1, \ldots, \gamma_n) \quad \text{in law}, m \to \infty.$$
On account of (A.6.15) we infer that
$$(U_j^{(k_m)}, \zeta_j^{(k_m)}) \to (\eta_j, \gamma_j) \quad \text{in law}, m \to \infty. \tag{A.6.17}$$
Recall (see, e.g., [39, §1]) that if $X_k \to X$ in law, as $k \to \infty$, where X, X_k ($k \in \mathbb{N}$) take values in a metric space (S, ρ) with Borel σ-algebra $\mathcal{B}(S)$, then, for any open set $G \in \mathcal{B}(S)$, one has
$$\liminf_{k \to \infty} P_{X_k}(G) \ge P_X(G), \tag{A.6.18}$$
here P_X, P_{X_k} stand for the laws of X, X_k respectively. Note that the set $G = \{(x, y) \in \mathbb{R}^2 : x \ne y\}$ is open in $\mathbb{R}^2$. Thus, using (A.6.14), (A.6.17) and (A.6.18) we obtain that
$$\beta_j \ge \liminf_{m \to \infty} \mathsf{P}'\left(U_j^{(k_m)} \ne \zeta_j^{(k_m)}\right) \ge \mathsf{P}'(\eta_j \ne \zeta_j), \ j \ge 2.$$
To complete the proof we redefine $(U_n)_{n \in \mathbb{N}}$ on $(\Omega', \mathcal{F}', \mathsf{P}')$ as a sequence $(\eta_n)_{n \in \mathbb{N}}$. $\square$

Remark A.21. The requirement that U_n be Gaussian is immaterial for the proof of Lemma A.20. One could assume instead that the family of probability measures $(Law(U_n))_{n \in \mathbb{N}}$ on $(\mathbb{R}, \mathcal{B}(\mathbb{R}))$ is tight.

---

[10]We can always enlarge this space keeping the same notation for a new one.
[11]See previous footnote.

Continue the proof of Theorem. Let $\rho_k = k^{-6}$, $k \in \mathbb{N}$. Suppose that $(\chi_k)_{k \in \mathbb{N}}$ is a sequence of independent random variables which is independent of Y and such that $\chi_k \sim N(0, \rho_k)$, $k \in \mathbb{N}$. Set $U_k = Y_k + \chi_k$, $k \in \mathbb{N}$. We establish (A.6.7) for the sequence $(U_k)_{k \in \mathbb{N}}$ and appropriate positive numbers $(\beta_k)_{k \in \mathbb{N}}$. Before proceeding further we introduce some notation. Let $p_k : \mathbb{R}^k \to \mathbb{R}$ denote the joint density of $(U_1, \ldots, U_k)$ and $p^{(k)} : \mathbb{R} \to \mathbb{R}$ to be the density of U_k, $k \in \mathbb{N}$. These densities clearly exist since the characteristic function of $(U_1, \ldots, U_k)$ belongs to $L^1(\mathbb{R}^k)$ (cf. [383, Ch. II, §12]). Set $\vec{u} = (u_1, \ldots, u_{k-1})$. We have

$$\sup_{B \in \mathcal{B}(\mathbb{R}^k)} \left| P_{U_1, \ldots, U_k}(B) - P_{U_1, \ldots, U_{k-1}} \otimes P_{U_k}(B) \right|$$

$$\leq \int_{\mathbb{R}^k} \left| p_k(\vec{u}, u_k) - p_{k-1}(\vec{u}) p^{(k)}(u_k) \right| d\vec{u} du_k$$

$$\leq \int_{-z_k}^{z_k} \cdots \int_{-z_k}^{z_k} \left| p_k(\vec{u}, u_k) - p_{k-1}(\vec{u}) p^{(k)}(u_k) \right| d\vec{u} du_k + 2 \sum_{j=1}^{k} \mathsf{P}(|U_j| > z_k) =: I_1 + I_2$$

where $d\vec{u} = du_1 \ldots du_{k-1}$, z_k being the same as in (A.6.1) and (A.6.2). A sequence $(\chi_k)_{k \in \mathbb{N}}$ is independent of $(Y_k)_{k \in \mathbb{N}}$, thus for any $k \in \mathbb{N}$ and all $t = (t_1, \ldots, t_k) \in \mathbb{R}^k$ one has

$$\left| \mathsf{E} \exp\left\{ i \sum_{l=1}^{k} t_l U_l \right\} - \mathsf{E} \exp\left\{ i \sum_{l=1}^{k-1} t_l U_l \right\} \mathsf{E} \exp\{ i t_k U_k \} \right|$$

$$= \left| \mathsf{E} \exp\left\{ i \sum_{l=1}^{k} t_l Y_l + i \sum_{l=1}^{k} t_l \chi_l \right\} - \mathsf{E} \exp\left\{ i \sum_{l=1}^{k-1} t_l Y_l + i \sum_{l=1}^{k-1} t_l \chi_l \right\} \mathsf{E} \exp\{ i t_k Y_k + i t_k \chi_k \} \right|$$

$$= \left| \mathsf{E} \exp\left\{ i \sum_{l=1}^{k} t_l Y_l \right\} - \mathsf{E} \exp\left\{ i \sum_{l=1}^{k-1} t_l Y_l \right\} \mathsf{E} \exp\{ i t_k Y_k \} \right| \left| \prod_{j=1}^{k} \mathsf{E} \exp\{ i t_j \chi_j \} \right|$$

$$= \left| \mathsf{E} \exp\left\{ i \sum_{l=1}^{k} t_l Y_l \right\} - \mathsf{E} \exp\left\{ i \sum_{l=1}^{k-1} t_l Y_l \right\} \mathsf{E} \exp\{ i t_k Y_k \} \right| \exp\left\{ -\frac{1}{2} \sum_{j=1}^{k} t_j^2 \rho_j \right\} =: \psi_k(t).$$

$$(\text{A.6.19})$$

Recall (see, e.g., [383, Ch. II, §12]) that if $T = (T_1, \ldots, T_k)$ is a random vector in $\mathbb{R}^k$ with the characteristic function $\varphi_T(t_1, \ldots, t_k)$, $t = (t_1, \ldots, t_k) \in \mathbb{R}^k$, belonging to $L^1(\mathbb{R}^k)$, then there exists a density $p_T(x)$, $x = (x_1, \ldots, x_k) \in \mathbb{R}^k$. Moreover,

$$p_T(x) = \frac{1}{(2\pi)^k} \int_{\mathbb{R}^k} e^{-i(t,x)} \varphi_T(t) dt, \quad x \in \mathbb{R}^k,$$

here (t, x) is the inner product in $\mathbb{R}^k$ and $dt = dt_1 \ldots dt_k$. Consequently, by (A.6.19) and the latter inversion formula,

$$I_1 \leq (2 z_k)^k \sup_{u \in [-z_k, z_k]^k} \left| p_k(\vec{u}, u_k) - p_{k-1}(\vec{u}) p^{(k)}(u_k) \right|$$

$$\leq \frac{z_k^k}{\pi^k} \int_{\mathbb{R}^k} \left| \mathsf{E}\exp\left\{i\sum_{l=1}^{k} t_l U_l\right\} - \mathsf{E}\exp\left\{i\sum_{l=1}^{k-1} t_l U_l\right\}\mathsf{E}\exp\{it_k U_k\}\right| \exp\left\{-\frac{1}{2}\sum_{j=1}^{k} t_j^2 \rho_j\right\} dt$$

$$= \frac{z_k^k}{\pi^k} \int_{B_{z_k}(0)} \psi_k(t)dt + \frac{z_k^k}{\pi^k} \int_{\mathbb{R}^k \setminus B_{z_k}(0)} \psi_k(t)dt =: I_{11} + I_{12}$$

where $B_r(0)$ is the open ball of radius r, in Euclidean metric, in $\mathbb{R}^k$ centered at 0.
According to (A.6.1)

$$I_{11} \leq \frac{z_k^k}{\pi^k}\mathrm{mes}(B_{z_k}(0))\varkappa_k. \tag{A.6.20}$$

Using the change of variables and (6.2.17) one derives the bound

$$I_{12} \leq 2\frac{z_k^k}{\pi^k} \int_{\mathbb{R}^k \setminus B_{z_k}(0)} \exp\left\{-\frac{1}{2}\sum_{j=1}^{k} t_j^2 \rho_j\right\} dt$$

$$\leq \frac{2z_k^k}{\pi^k} \int_{\mathbb{R}^k \setminus B_{z_k}(0)} \exp\left\{-\frac{\rho_k}{2}\sum_{j=1}^{k} t_j^2\right\} dt = \frac{2z_k^k}{\pi^k \rho_k^{k/2}} \int_{\mathbb{R}^k \setminus B_{z_k \sqrt{\rho_k}}(0)} \exp\left\{-\sum_{j=1}^{k} \frac{t_j^2}{2}\right\} dt$$

$$\leq 2\frac{z_k^k}{\pi^k \rho_k^{k/2}} \frac{z_k^{k-2}\rho_k^{(k-2)/2}}{2^{k/2-1}\Gamma(k/2)} \exp\{-z_k^2 \rho_k/2\} \leq \rho_k^{-1/2} z_k^{2k-2} \exp\{-z_k^2 \rho_k/2\}. \tag{A.6.21}$$

Note that $\mathsf{E}U_j^2 = \mathsf{E}Y_j^2 + \mathsf{E}\chi_j^2 = 1 + \rho_j \leq 2$, $j \in \mathbb{N}$. Thus by the Markov inequality

$$I_2 \leq 4kz_k^{-2}. \tag{A.6.22}$$

From (A.6.2) and (A.6.20)—(A.6.22) one sees that the conditions of Lemma A.20 are
satisfied with $\beta_k = O(k^{-2})$, $k \to \infty$. Thus by the assertion of the mentioned Lemma,
one can enlarge the initial probability space to define a sequence of independent
random variables $(\zeta_k)_{k\in\mathbb{N}}$ such that $\zeta_k \sim N(0, 1+\rho_k)$, $k \in \mathbb{N}$, and (A.6.12) be true.

Equip the linear space $\mathbb{R}^\infty$ (the space of all real-valued functions on $\mathbb{N}$) with the
metric

$$\rho_\infty(f,g) = \sum_{j=1}^{\infty} 2^{-j} \frac{|f(j) - g(j)|}{1 + |f(j) - g(j)|}, \quad f,g : \mathbb{N} \to \mathbb{R}.$$

Then $(\mathbb{R}^\infty, \rho_\infty)$ is a Polish space and each random sequence $X = (X_j)_{j\in\mathbb{N}}$ is a
random element taking values in $(\mathbb{R}^\infty, \rho_\infty)$. Indeed, for any $N \in \mathbb{N}$, all $t_1, \ldots, t_N \in \mathbb{N}$
and all sets $B_1, \ldots, B_n \in \mathcal{B}(\mathbb{R})$ one has $\{\omega : X(t_j) \in B_j, j = 1, \ldots, N\} \in \mathcal{F}$, but
the finite-dimensional sets in $\mathbb{R}^\infty$ generate the Borel σ-algebra $\mathcal{B}(\mathbb{R}^\infty)$. Let

$$S_1 = S_2 = S_3 = (\mathbb{R}^\infty, \rho_\infty).$$

Applying Lemma A.18 one sees that it is possible to redefine sequences Y, U and ζ
on a common probability space with all their properties preserved.

Let $Z_k := \zeta_k/\sqrt{1+\rho_k}$, $k \in \mathbb{N}$. . Then Z_k are i.i.d. random variables such that $Z_1 \sim N(0,1)$ and

$$\sum_{k=1}^{\infty} \mathsf{P}(|Y_k - Z_k| \geq k^{-2}) \leq \mathsf{P}(|Y_k - U_k| \geq k^{-2}/3) + \mathsf{P}(U_k \neq \zeta_k) + \mathsf{P}(|\zeta_k - Z_k| \geq k^{-2}/3)$$

$$\leq 9\sum_{k=1}^{\infty} k^4 \mathsf{E}\chi_k^2 + \sum_{k=1}^{\infty} \mathsf{P}(U_k \neq \zeta_k) + 9\sum_{k=1}^{\infty} k^4 \mathsf{E}\zeta_k^2 \left(\sqrt{1+\rho_k} - 1\right)^2 < \infty,$$

by the definition of ζ_k, Markov inequality and the choice of $(\rho_k)_{k\in\mathbb{N}}$. The Theorem follows by the application of the Borel-Cantelli lemma. $\square$

Theorem A.22. *Let $X = \{X_j, j \in \mathbb{N}^d\}$ be a random field and $Z = (Z_l)_{l\in\mathbb{N}}$ be a sequence of i.i.d. $N(0,1)$ random variables defined on the common probability space with X. Suppose that $(B_l)_{l\in\mathbb{N}}$ is a sequence of blocks in $\mathbb{R}^d$ such that $\mathrm{mes}(B_i\cap B_j) = 0$ whenever $i \neq j$. Then one can redefine X and Z on some probability space together with a d-parameter Brownian motion $W = \{W_t, t \in \mathbb{R}_+^d\}$ such that*

$$\frac{W(B_l)}{\sqrt{\mathrm{mes}(B_l)}} = Z_l \quad \text{for all} \quad l \in \mathbb{N}, \quad \text{a.s.}$$

Here $W(B)$ is the increment of W over B, see (5.1.5).

Remark A.23. The assertion is trivial without the requirement that X and Z are defined on a common probability space (i.e. their common distribution is fixed). In the presence of such requirement one, in contrast, has to apply a special argument to preserve their common distribution while the processes are redefined.

Proof. Let $(C_\infty(\mathbb{R}_+^d), \rho)$ be the space of continuous real-valued functions on $\mathbb{R}_+^d$ with a metric

$$\rho(f,g) = \sum_{n=2}^{\infty} 2^{-n} \frac{\rho_n(f,g)}{1 + \rho_n(f,g)}, \quad f,g \in C_\infty(\mathbb{R}_+^d),$$

where $\rho_n(f,g) = \sup_{x\in[n^{-1},n]^d} |f(x) - g(x)|$, $n \in \mathbb{N}$, $n \geq 2$. Set

$$S_1 = S_2 = (\mathbb{R}^\infty, \rho_\infty), \quad S_3 = (C_\infty(\mathbb{R}_+^d), \rho).$$

Let W be a d-parameter Wiener process and $Y_l = W(B_l)/\sqrt{\mathrm{mes}(B_l)}$, $l \in \mathbb{N}$. Then $Y = (Y_l)_{l\in\mathbb{N}}$ is a sequence of i.i.d. $N(0,1)$ random variables. By Lemma A.18 there exists a probability space $(\Omega_0, \mathcal{F}_0, \mathsf{P}_0)$ supporting a random element (ξ^1, ξ^2, ξ^3) with values in $S_1 \times S_2 \times S_3$ such that $Law(\xi^1, \xi^2) = Law(X, Z)$ and $Law(\xi^2, \xi^3) = Law(Y, W)$. The proof is complete. $\square$

Bibliography

[1] Ahlswede, R. and Daykin, D. E. An inequality for the weights of two families of sets, their unions and intersections, *Z. Wahrsch. verw. Geb.*, **43**, 3, pp. 183–185.

[2] Ahmed, A. N., Leon, R., Proschan, F. (1978). Two concepts of positive dependence with applications in multivariate analysis, *Tech. Report 78-6, Department of statistics, Florida State University.*

[3] Alam, K. and Saxena, K. M. L. (1981). Positive dependence in multivariate distributions, *Commun. Stat., Theory Methods*, **A10**, pp. 1183–1196.

[4] Albeverio, S. and Belopolskaya, Ya. (2006). Probabilistic approach to systems of nonlinear PDEs and vanishing viscosity method, *Markov Proc. Rel. Fields*, **12**, 1, pp. 59–94.

[5] Alexander, K. S. (1984). Probability inequalities for empirical processes and a law of the iterated logarithm, *Ann. Probab.*, **12**, 4, pp. 1041–1067.

[6] Alexander, K. S. and Uzun, H. B. (2003). Lower bounds for boundary roughness for droplets in Bernoulli percolation, *Probab. Theory Rel. Fields*, **27**, 1, pp. 62–88.

[7] Amini, M. and Bozorgnia, A. (2000). Negatively dependent bounded random variable probability inequalities and the strong law of large numbers, *J. Appl. Math. Stochastic Anal.*, **13**, 3, pp. 261–267.

[8] Baddeley, A. J. (2002). Time-invariance estimating equations, *Bernoulli*, **6**, 5, pp. 783–808.

[9] Baek, J.-I (1997). A weakly dependence structure of multivariate processes, *Statist. Probab. Lett.*, **34**, 4, pp. 355–363.

[10] Baek, J.-I. and Kim, T.-S. (1995). A functional central limit theorem for positively dependent random vectors, *Comm. Korean Math. Soc.*, **10**, 3, pp. 707–714.

[11] Bagai, I. and Prakasa Rao, B. L. S. (1991). Estimation of the survival function for stationary associated processes, *Statist. Probab. Lett.*, **12**, 5, pp. 385–391.

[12] von Bahr, B. (1967). Multi-dimensional integral limit theorem, *Ark. Math.*, **7**, 1, pp. 71–88.

[13] Bakhtin, Yu. Yu. (2000). Functional central limit theorem for the solution of a many-dimensional Burgers equation with initial data given by an associated random measure, *Mosc. Univ. Math. Bull.*, **55**, 6, pp. 7–14.

[14] Bakhtin, Yu. Yu. (2001). A functional central limit theorem for transformed solutions of the multidimensional Burgers equation with random initial data, *Theory Probab. Appl.*, **46**, 3, pp. 387–405.

[15] Bakhtin, Yu. Yu. (2001). Limit theorems for the solutions of the stochastic Burgers equation. PhD thesis, Moscow.

[16] Bakhtin, Yu. Yu., Bulinski, A. V. (1997). Moment inequalities for the sums of de-

pendent multiindexed random variables, *Fundam. Appl. Math.*, **3**, 4, pp. 1101–1108 (in Russian).

[17] Balan, R. (2005). A strong invariance principle for associated random fields, *Ann. Probab.*, **33**, 2, pp. 823–840.

[18] Barbato, D. (2005). FKG inequality for Brownian motion and stochastic differential equations, *Electron. Commun. Probab.*, **10**, 1, pp. 7–16.

[19] Barbour, A. D. and Chen, L. H. Y. (eds.) (2005) *Stein's method and applications. A program in honor of Charles Stein, Proceedings of a workshop* (Singapore, World Scientific).

[20] Barbour, A. D. and Chen, L. H. Y. (eds.) (2005). *An introduction to Stein's method* (Singapore, World Scientific).

[21] Barbour, A. D. and Eagleson, G. K. (1985). Multiple comparisons and sums of dissociated random variables, *Adv. Appl. Probab.*, **17**, 1, pp. 147–162.

[22] Barbour, A. D. and Hall, P. (1984). Stein's method and the Berry-Esseen theorem, *Austral. J. Statist.*, **26**, 1, pp. 8–15.

[23] Barbour, A. D., Holst, L., Janson, S. (1992). *Poisson approximation* (Oxford, Clarendon Press).

[24] Barlow, R. E., Proschan, F. (1975), *Statistical Theory of Reliability and Life Testing* (New York, Holt, Rinehart and Winston, Inc. XIII).

[25] Barndorff-Nielsen, O. E. and Leonenko, N. N. (2005). Burgers' turbulence problem with linear or quadratic external potential, *J. Appl. Probab.*, **42**, 2, pp. 550–565.

[26] Bass, R. F. and Pyke, R. (1984). Functional law of the iterated logarithm and uniform central limit theorem for partial-sum processes indexed by sets, *Ann. Probab.*, **12**, 1, pp. 13–34.

[27] Benjamini, I., Lyons, R., Peres, Y., Schramm, O. (2001). Uniform spanning forests, *Ann. Probab.*, **29**, 1, pp. 1–65.

[28] Bentkus, V. (2004). A Lyapunov-type bound in $\mathbb{R}^d$, *Theory Probab. Appl.*, **49**, 2, pp. 311–323.

[29] Berbee, H. (1987). Convergence rates in the strong law for bounded mixing sequences, *Probab. Th. Rel. Fields*, **74**, 2, pp. 255–270.

[30] van den Berg, J. (1996). A note on disjoint-occurrence inequalities for marked Poisson point processes, *J. Appl. Probab.*, **33**, 2, pp. 420–426.

[31] van den Berg, J. and Fiebeg, U. (1987). On a combinatorial conjecture concerning disjoint occurrences of events, *Ann. Probab.*, **15**, 1, pp. 354–374.

[32] van den Berg, J., Häggström, O., Kahn, J. (2005). Some conditional correlation inequalities for percolation and related processes, *Rand. Struct. Algor.*, **29**, 4, pp. 417–435.

[33] van den Berg, J. and Kesten, H. (1985). Inequalities with applications to percolation and reliability, *J. Appl. Probab.*, **22**, 3, pp. 556–569.

[34] Berkes, I. (1973). The functional law of the iterated logarithm for dependent random variables, *Z. Wahrsch. verw. Geb.*, **26**, 3, pp. 245–258.

[35] Berkes, I., Csáki, E. (2001). A universal result in almost sure central limit theory, *Stoch. Proc. Appl.*, **94**, 1, pp. 105–134.

[36] Berkes, I. and Morrow, G. J. (1981). Strong invariance principle for mixing random fields, *Z. Wahrsch. verw. Geb.*, **57**, 1, pp. 15–37.

[37] Berkes, I. and Philipp, W. (1979). Approximation theorems for independent and weakly dependent random vectors, *Ann. Probab.*, **7**, 1, pp. 29–54.

[38] Bernstein, S. N. (1939). Generalization of limit theorems of probability theory on sums of dependent random variables. Uspehi Mat. Nauk, **10**, pp. 65–114 (in Russian).

[39] Billingsley, P. (1968). *Convergence of probability measures* (New York–London–Sydney–Toronto, Wiley, Inc. XII).

[40] Bingham, N. H. and Nili Sani, H. R. (2004). Summability methods and negatively associated random variables, *J. Appl. Probab.*, **41A**, pp. 231–238.

[41] Birkel, T. (1987). The invariance principle for associated processes, *Stoch. Proc. Appl.*, **27**, 1, pp. 57–71.

[42] Birkel, T. (1988). A note on the strong law of large numbers for positively dependent random variables, *Statist. Probab. Lett.*, **7**, 1, pp. 17–20.

[43] Birkel, T. (1988). Moment bounds for associated sequences, *Ann. Probab.*, **16**, 3, pp. 1184–1193.

[44] Birkel, T. (1988). On the convergence rate in central limit theorem for associated processes, *Ann. Probab.*, **16**, 4, pp. 1685–1698.

[45] Birkhoff, G. (1967). *Lattice Theory,* (Providence, Amer. Math. Society, **25**).

[46] Block, H. W., Savitz, T. H., Shaked, M. (1982). Some concepts of negative dependence, *Ann. Probab.*, **10**, 3, pp. 765–772.

[47] Bogachev, V. I. (2003). *Measure Theory* (Berlin, Springer).

[48] Bollobas, B. (1990) *Graph Theory. An Introductory Course, Graduate Texts in Mathematics,* **63** (New York, Springer).

[49] Bolthausen, E. (1982). On the central limit theorem for stationary mixing random fields, *Ann. Probab.*, **10**, 4, pp. 1047–1050.

[50] Borovkov, A. A. (1973). On the rate of convergence for the invariance principle, *Theory Probab. Appl.*, **18**, 2, pp. 207–225.

[51] Boutsikas, M. V. and Koutras, M. V. (2000). A bound for the distribution of the sum of discrete associated or negatively associated random variables, *Ann. Appl. Probab.*, **10**, 4, pp. 1137–1150.

[52] Boutsikas, M. V. and Vaggelatou, E. (2002). On the distance between the convex-ordered random variables, with applications, *Adv. Appl. Probab.*, **34**, 2, pp. 349–374.

[53] Bradley, R. C. (1989). A caution on mixing conditions for random fields, *Statist. Probab. Lett.*, **8**, 5, pp. 489–491.

[54] Bradley, R. C. (2002), On positive spectral density functions, *Bernoulli*, **8**, 2, pp. 175–193.

[55] Bradley, R. C and Bryc, W. (1985). Multilinear forms and measures of dependence between random variables, *J. Multivar. Anal.*, **16**, 3, pp. 335–367.

[56] Bryc, W. (1982). On the approximation theorem of I. Berkes and W. Philipp, *Demonstr. Math.*, **15**, 3, pp. 807–816.

[57] Broadbent, S. B. and Hammersley, J. M. (1957). Percolation processes. I: Crystals and mazes, *Proc. Cambr. Phil. Soc.*, **53**, 3, pp. 629–641.

[58] Brosamler, G. A. (1988). An almost everywhere central limit theorem, *Proc. Cambr. Phil. Soc.*, **104**, 3, pp. 561–574.

[59] Bühlmann, P., Doukhan, P., Ango Nze, P. (2002). Weak dependence beyond mixing and asymptotics for nonparametric regression, *Ann. Statist.*, **30**, 2, pp. 397–430.

[60] Bulinski, A. V. (1977). On normalization in the law of the iterated logarithm, *Theory Probab. Appl.*, **22**, 2, pp. 398–399.

[61] Bulinski, A. V. (1989). *Limit Theorems under Weak Dependence Conditions* (Moscow, MSU) (in Russian).

[62] Bulinski, A. V. (1992). Central limit theorem for shot noise fields, *J. Math. Sci.*, **61**, 1, pp. 1840–1845.

[63] Bulinski, A. V. (1992). Some problems of asymptotical analysis of nonlinear diffusion. In: Shiryaev A. N. et al. (eds.), *Probability theory and mathematical statistics. Proceedings of the 6th USSR-Japan symposium, Kiev, USSR, 5-10 August, 1991,*

Singapore, World Scientific, pp. 32–46.

[64] Bulinski, A. V. (1993). Inequalities for the moments of sums of associated multi-indexed variables, *Theory Probab. Appl.*, **38**, 2, pp. 342–349.

[65] Bulinski, A. V. (1995). The functional law of the iterated logarithm for associated random fields, *Fundam. Prikl. Mat.*, **1**, 3, pp. 623–639 (in Russian).

[66] Bulinski, A. V. (1995). Rate of convergence in the central limit theorem for fields of associated random variables, *Theory Probab. Appl.*, **40**, 1, pp. 136-144.

[67] Bulinski, A. V. (1996). On the convergence rates in the central limit theorem for positively and negatively dependent random fields. In: bragimov, I. A. and Zaitsev, A. Yu. (eds.), *Probab. Theory and Math. Statist.*, Gordon and Breach, pp. 3–14.

[68] Bulinski, A. V. (2000), Asymptotical Gaussian behavior of quasi-associated vector-valued random fields, *Obozr. Prikl. Prom. Mat.*, **7**, 2, pp. 482–483 (in Russian).

[69] Bulinski, A. V. (2004). A statistical version of the central limit theorem for vector-valued random fields, *Math. Notes*, **76**, 4, pp. 455–464.

[70] Bulinski, A. (2007). CLT for random fields with applications. In:*Transactions of ASMDA 2007*, Chania (Greece), pp. 1–10.

[71] Bulinski, A. V. and Dilman, S. V. (2002). Universal normalization and the law of the iterated logarithm, *Russ. Math. Surv.*, **57**, 2, pp. 418–419.

[72] Bulinski, A. V., Keane, M. S. (1996). Invariance principle for associated random fields, *J. Math. Sci.*, **81**, 5, pp. 2905–2911.

[73] Bulinski, A. and Khrennikov, A. (2005). Generalization of the Critical Volume NTCP Model in the Radiobiology, `http://arxiv.org/abs/math/0504225`

[74] Bulinski, A. and Kryzhanovskaya, N. (2006). Convergence rate in CLT for vector-valued random fields with self-normalization, *Probab. Math. Stat.*, **26**, 2, pp. 261–281.

[75] Bulinski, A. V. and Lifshits, M. A. (1995). Rate of convergence in the functional law of the iterated logarithm with non-standard normalizing factors, *Russ. Math. Surv.*, **50**, 5, pp. 925–944.

[76] Bulinski, A. V. and Millionshchikov, N. V. (2003). The normal approximation rate for kernel estimates of the density of a quasi-associated random field, *Theory Probab. Math. Stat.*, **66**, pp. 37–48.

[77] Bulinski, A. V. and Molchanov, S. A. (1991). Asymptotical normality of a solution of Burgers' equation with random initial data, *Theory Probab. Appl.*, **36**, 2, pp. 217–236.

[78] Bulinski, A. V. and Shabanovich, E. (1998). Asymptotical behaviour for some functionals of positively and negatively dependent random fields, *Fundam. Prikl. Mat.*, **4**, 2, pp. 479–492 (in Russian).

[79] Bulinski, A.V. and Shashkin, A.P. (2004). Rates in the central limit theorem for dependent multiindexed random vectors. J. Math. Sci. **122**, 4, pp. 3343–3358.

[80] Bulinski, A. and Shashkin, A. (2006). Strong Invariance Principle for Dependent Random Fields, *IMS Lect. Notes – Monograph Series Dynamics and Stochastics*, **48**, pp. 128–143.

[81] Bulinski, A. V. and Shiryaev, A. N. (2003). *Theory of Stochastic Processes*, Moscow, Fizmatlit (in Russian).

[82] Bulinski, A. and Suquet, Ch. (2001). Normal approximation for quasi associated random fields, *Statist. Probab. Lett.*, **54**, 2, pp. 215–226.

[83] Bulinski, A. V. and Vronski, M. A. (1996). Statistical variant of the central limit theorem for associated random fields, *Fundam. Prikl. Mat.*, **2**, 4, pp. 999–1018 (in Russian).

[84] Bulinski, A. V. and Zhurbenko, I. G. (1976). A central limit theorem for additive

random functions, *Theory Probab. Appl.*, **21**, 4, pp. 687–697.

[85] Burton, R., Dabrowski, A. R., Dehling, H. (1986). An invariance principle for weakly associated random vectors, *Stoch. Proc. Appl.*, **23**, 2, pp. 301–306.

[86] Burton, R. and Kim, T.-S. (1988). An invariance principle for associated random fields, *Pacific J. Math.*, **132**, 1, pp. 11–19.

[87] Burton, R. and Waymire, E. (1985). Scaling limits for associated random measures, *Ann. Probab.*, **13**, 4, pp. 1267–1278.

[88] Burton, R. M. and Waymire, E. (1986). The central limit problem for infinitely divisible random measures. In: *Dependence in probability and statistics, Conf. Oberwolfach 1985*, Prog. Probab. Stat. **11**, pp. 383–395,

[89] Caffarelli, L. A. (2000). Monotonicity properties of optimal transportation and the FKG and related inequalities, *Comm. Math. Phys.*, **214**, 3, pp. 547–563.

[90] Cai, Z. and Roussas, G. G. (1997). Smooth estimate of quantiles under association, *Statist. Probab. Lett.*, **36**, 3, pp. 275–287.

[91] Cai, Z. and Roussas, G. G. (1998). Efficient estimation of a distribution function under quadrant dependence, *Scand. J. Statist.*, **25**, 1, pp. 211–224.

[92] Cai, Z. and Roussas, G. G. (1998). Kaplan-Meier estimator under association, *J. Multivar. Anal.*, **67**, 2, pp. 318–348.

[93] Cai, Z. and Roussas, G. G. (1999). Weak convergence for smooth estimator of a distribution function under negative association, *Stoch. Anal. Appl.*, **17**, 2, pp. 245–268.

[94] Cai, Z. and Roussas, G.G. (1999). Berry-Esseen bounds for smooth estimator of a distribution function under association, *J. Nonparametric Statist.*, **11**, 1, pp. 79–106.

[95] Calvó-Armengol, A. and Jackson, M. O. (2005). Networks in Labor Markets: Wage and Employment Dynamics and Inequality. Mimeo, Caltech.

[96] Chayes, L. and Lei, H. K. (2006). Random cluster models on the triangular lattice, *J. Stat. Phys.*, **122**, 4, pp. 647–670.

[97] Chayes, J. Y., Puha, A. L., Sweet, T. (1999). Independent and dependent percolation. In: Hsu, E. P. and Varadhan, S. R. S. (eds), *Probability Theory and Applications*, IAS/PARK CITY Math. Series, **6**, AMS, Institute for Advance Study, pp. 50–116.

[98] Chen, L. H. Y. (1975). Poisson approximation for dependent trials, *Ann. Probab.*, **3**, 3, pp. 534–545.

[99] Chen, L. H. Y. and Shao, Q.-M. (2001). A non-uniform Berry-Esseen bound via Stein's method, *Probab. Theory Rel. Fields*, **120**, 2, pp. 236–254.

[100] Cherny, A. S. and Engelbert, H.-P. (2004). *Singular Stochastic Differential Equations* (Springer).

[101] Chow, Y. S. and Teicher, H. (2003). *Probability Theory: Independence, Interchangeability, Martingales* (New York, Springer).

[102] Christofides, T. C. (2000). Maximal inequalities for demimartingales and a strong law of large numbers, *Statist. Probab. Lett.*, **50**, 4, pp. 357–363.

[103] Christofides, T. C. and Vaggelatou, E. (2004). A connection between supermodular ordering and positive/negative association, *J. Multivar. Anal.*, **88**, 1, pp. 138–151.

[104] Colangelo, A., Müller, A., Scarsini, M. (2006). Positive dependence and weak convergence, *J. Appl. Probab.*, **43**, 1, pp. 48–59.

[105] Colangelo, A., Scarsini, M., Shaked, M. (2006). Some positive dependence stochastic orders, *J. Multivar. Anal.*, **97**, 1, pp. 46–78.

[106] Coulon-Prieur, C. and Doukhan, P. (2000). A triangular central limit theorem under a new weak dependence condition, *Statist. Probab. Lett.*, **47**, 1, pp. 61–68.

[107] Coupier, D., Doukhan, P., Ycart, B. (2006). Zero-one laws for binary random fields. arXiv:math.PR/0605502 v1 18 May2006.

[108] Courant, R. (1988), *Differential and Integral Calculus,* **I** (New York, Wiley).

[109] Cox, J. T. (1984). An alternative proof of a correlation inequality of Harris, *Ann. Probab.*, **12**, 1, pp. 272–273.

[110] Cox, J. T. and Grimmett, G. (1984). Central limit theorems for associated random variables and the percolation model, *Ann. Probab.*, **12**, pp. 514–528.

[111] Cramér, H. (1999). *Mathematical Methods of Statistics* (Princeton University Press).

[112] Csörgő, M., Horvath, L., Szyszkowicz, B. (eds.) (2004). *Asymptotic Methods In Stochastics : Festschrift For Miklos Csorgo*, AMS.

[113] Csörgő and M., Révész, P. (1975). A new method to prove Strassen type laws of invariance principle I, *Z.Wahrsch. verw. Geb.*, **31**, 4, pp. 255–259.

[114] Csörgő and M., Révész, P. (1975). A new method to prove Strassen type laws of invariance principle II, *Z. Wahrsch. verw. Geb.*, **31**, 4, pp. 261–269.

[115] Dabrowski, A. R. (1985). A functional law of the iterated logarithm for associated sequences, *Stat. Probab. Lett.*, **3**, pp. 209–212.

[116] Dabrowski, A. R. (1985). Joint characteristic functions and associated sequences. In: *Dependence in Probability and Statistics, Conf. Oberwolfach, Prog. Probab. Stat.* **11**, pp. 349–360.

[117] Dabrowski, A. R. and Dehling, H. (1988). A Berry-Esseen theorem and a functional law of the iterated logarithm for weakly associated random vectors, *Stoch. Proc. Appl.*, **30**, 2, pp. 277–289.

[118] Dabrowski, A. R. and Jakubowski, A. (1994). Stable limits for associated random variables, *Ann. Probab.*, **22**, 1, pp. 1–16.

[119] Daley, D. J. and Vere-Jones, D. (1988). *An Introduction to the Theory of Point Processes* (New York, Springer).

[120] Dedecker, J. (2004). Inégalités de covariance. *C.R.Acad. Sci. Paris. Math.*, **339**, 7, pp. 503–506.

[121] Dedecker, J. and Doukhan, P. (2003). A new covariance inequality and its applications, *Stoch. Proc. Appl.*, **106**, 1, pp. 63–80.

[122] Dedecker, J. and Louhichi, S. (2005). Convergence to infinitely divisible distributions with finite variance for some weakly dependent sequences, *ESAIM Probab. Stat.*, **9**, 1, pp. 38–73.

[123] Dedecker, J. and Prieur, C. (2004). Coupling for τ-dependent sequences and applications, *J. Theor. Probab.*, **17**, 4, pp. 861–895.

[124] Dedecker, J. and Prieur, C. (2005). New dependence coefficients. Examples and applications to statistics, *Probab. Th. Rel. Fields*, **132**, 2, pp. 203–236.

[125] Dedecker, J., De Fitte, P. R., Prieur, C. (2006). Parametrized Kantorovich-Rubinstein theorem and application to the coupling of randon variables. In:*Dependence in Probability and Statistics, Lecture Notes in Statistics,* **187** (Berlin, Springer).

[126] Dehling, H., Denker, M., Philipp, W. (1986). Central limit theorems for mixing sequences of random variables under minimal conditions, *Ann. Probab.*, **14**, 4, pp. 1359-1370.

[127] Denuit, M., Dhaene, J., Goovaerts, M. (2005). *Actuarial Theory for Dependent Risks: Measures, Orders and Models* (Wiley).

[128] Denuit, M., Lefèvre, C., Utev, S. (1999). Stochastic orderings of convex/concave-type on an arbitrary grid, *Mathematics of Operations Research*, **24**, 4, pp. 835–846.

[129] Denuit, M., Lefèvre, C., Utev, S. (2002). Measuring the impact of dependence between claims occurrences, *Insurance: Mathematics and Economics*, **30**, 1, pp. 1–19.

[130] Devroye, L. and Györfi, L. (1985). *Nonparametric density estimation. The L_1 view* (New York, Wiley).

[131] Dewan, I. and Dharamadhikaria, A. D. (2006). Association in time of a vector valued process, *Statist. Probab. Lett.*, **76**, 11, pp. 1147–1155.

[132] Dewan, I. and Prakasa Rao, B. L. S. (1999). A general method of density estimation for associated random variables, *J. Nonparametric Statist.*, **10**, 4, pp. 405–420.

[133] Dewan, I. and Prakasa Rao, B. L. S. (2001). Asymptotic normality for U-statistics of associated random variables, *J. Stat. Planning Inf.*, **97**, 2, pp. 201–225.

[134] Dewan, I. and Prakasa Rao, B. L. S. (2005). Wilcoxon-signed rank test for associated sequences, *Statist. Probab. Lett.*, **71**, 2, pp. 131–142.

[135] Dil'man, S. V. (2006). The asymptotics in the Baum-Katz formula for random fields, *Math. Notes*, **79**, 5.

[136] Dobrushin, R. L. (1968). The description of the random field by its conditional distributions and its regularity conditions (in Russian), *Teor. Veroyatn. Primen.*, **13**, 2, pp. 201–229.

[137] Dobrushin, R. L. (1970). Prescribing a system of random variables by conditional distributions, *Theory Probab. Appl.*, **15**, 3, pp. 458–486.

[138] Doukhan, P. (1994). *Mixing: Properties and Examples, Lecture Notes in Statistics*, **85** (New York, Springer).

[139] Doukhan, P. and Lang, G. (2002). Rates in the empirical central limit theorem for stationary weakly dependent random fields, *Statist. Inf. for Stoch. Proc.*, **5**, 2, pp. 199–228.

[140] Doukhan, P., Lang, G., Surgailis, D., Viano, M.-C. (2005). Functional limit theorem for the empirical process of a class of Bernoulli shifts with long memory, *J. Theor. Probab.*, **18**, 1, pp. 161–186.

[141] Doukhan, P. and Louhichi, S. (1999). A new weak dependence condition and application to moment inequalities, *Stoch. Proc. Appl.*, **84**, 2, pp. 313–342.

[142] Doukhan, P. and Louhichi, S. (2001). Functional estimation of a density under a new weak dependence condition, *Scand. J. Stat.*, **28**, 2, pp. 325–341.

[143] Doukhan, P., Oppenheim, G., Taqqu, M. (eds.) (2003). *Theory and Applications of Long-range Dependence* (Boston, Birkhäuser).

[144] Doukhan, P. and Wintenberger, O. (2006). An invariance principle for new weakly dependent stationary models using sharp moment assumptions. `arXiv:math.PR/0603321 v1 9 Mar2006`.

[145] Doyle, P. G. and Snell, J. L. (1984). *Random Walks and Electric Networks* (New York, Wiley).

[146] Dronov, S. V. and Sakhanenko, A. I. (1987). Rate of convergence in a multidimensional invariance principle for functionals of integral form, *Sib. Math. J.*, **28**, 3, pp. 415–423.

[147] Dubhashi, D., Jonasson, J., Ranjan, D. (2007). Positive Influence and Negative Dependence, *Combinatorics, Probability and Computing*, **16**, 1, pp. 29–41.

[148] Dudnikova, T. V. and Komech, A. I. (1996). Ergodic properties of hyperbolic equations with mixing, *Theory Probab. Appl.*, **41**, 3, pp. 436–448.

[149] Dudziński, M. (2003). A note on the almost sure central limit theorem for some dependent random variables, *Statist. Probab. Lett.*, **61**, 1, pp. 31–40.

[150] Dykstra, R. L., Henett, J. E., Thompson, W. A. (1973). Events which are almost independent, *Ann. Statist.*, **1**, 4, pp. 674–681.

[151] Dynkin, E. B. (1965). *Markov Processes* (Berlin-Göttingen-Heidelberg, Springer).

[152] Ebrahimi, N. (2002). On the dependence structure of certain multi-dimensional Itô processes and corresponding hitting times, *J. Multivar. Anal.*, **81**, 1, pp. 128–137.

[153] Ebrahimi, N. and Ghosh, M. (1981). Multivariate negative dependence, *Commun. Statist. Theory Methods*, **A10**, pp. 307–337.

[154] Echeverria, P. E. (1982). A criterion for invariant measures of Markov processes, *Z. Wahrsch. verw. Geb.*, **61**, pp. 1–16.

[155] Efron, B. (1965). Increasing properties of Pólya frequency functions, *Ann. Math. Statist.*, **36**, 1, pp. 272–279.

[156] Ekisheva, S. V. (1999). The Bahadur representation of a sample quantile for an associated random sequence, *Vestn. Syktyvkar. Univ., Ser. 1, Mat. Mekh. Inform.*, **3**, pp. 23–38 (in Russian).

[157] Ekisheva, S. V. (2001). Limit theorems for sample quantiles of associated random sequences, *Fundam. Prikl. Mat.*, **7**, 3, pp. 721–734 (in Russian).

[158] Erdös, P. and Kac, M. (1946). On certain limit theorems in the theory of probability, *Bull. Amer. Math. Soc.*, **52**, 4, pp. 292–302.

[159] Erickson, R. V. (1974). L_1 bounds for asymptotic normality of m-dependent sums using Stein's techniques, *Ann. Probab.*, **2**, 3, pp. 522–529.

[160] Esary, J., Proschan, F., Walkup, D. (1967). Association of random variables, with applications, *Ann. Math. Statist.*, **38**, 5, pp. 1466–1474.

[161] Esary, J. and Proschan, F. (1972). Relationships among some concepts of bivariate dependence, *Ann. Math. Statist.*, **43**, 2, pp. 651–655.

[162] Etemadi, N. (1983). On the laws of large numbers for nonnegative random variables, *J. Multivar. Anal.*, **13**, 1, pp. 187–193.

[163] Etemadi, N. (1983). Stability of sums of weighted nonnegative random variables, *J. Multivar. Anal.*, **13**, 2, pp. 361–365.

[164] Evans, S. N. (1990). Association and random measures, *Probab. Theory Rel. Fields*, **86**, 1, pp. 1–19.

[165] Fazekas, I., Klesov, O. (2000). A general approach to the strong law of large numbers, *Theory Probab. Appl.*, **45**, 3, pp. 436–449.

[166] Feder, T. and Mihail, M. (1992). Balanced matroids. In: *Proceedings of the 24th Annual ACM symposium on the theory of computing*, pp. 26–38.

[167] Feller, W. (1971). *An Introduction to Probability and its Applications* (New York, Wiley).

[168] Ferreira, H. (2003). Extremes of associated variables, *Statist. Probab. Lett.*, **63**, 4, pp. 333–338.

[169] Fishburn, P. C. and Shepp, L. A. (1991). On the FKB conjecture for disjoint intersections, *Discrete Math.*, **98**, 2, pp. 105–122.

[170] Fishburn, P. C., Doyle, P. G., Shepp, L. A. (1988). The match set of a random permutation has the FKG property, *Ann. Probab.*, **16**, 3, pp. 1194–1214.

[171] Fontes, L. R. G., Isopi, M., Newman, C. M., Ravishankar, K. (2004). The Brownian web: characterization and convergence, *Ann. Probab.*, **32**, 4, pp. 2857–2883.

[172] Fortuin, C., Kasteleyn, P., Ginibre, J. (1971). Correlation inequalities on some partially ordered sets, *Comm. Math. Phys.*, **22**, 2, pp. 89–103.

[173] Gaposhkin, V.F. (1997). On almost everywhere convergence of the Riesz averages of homogeneous random fields, *Theory Probab. Appl.*, **42**, 3, pp. 405–415.

[174] Garet, O. (2001). Limit theorems for the painting of graphs by clusters, *ESAIM, Probab. Stat.*, **5**, pp. 105–118.

[175] Garet, O. (2005). Central limit theorems for the Potts model, *Math. Phys. Electron. J.*, **11**, Paper 4.

[176] Garet, O. and Marchand, R. (2005). Coexistence in two-type first-passage percolation models, *Ann. Appl. Probab.*, **15**, 1A, pp. 298–330.

[177] Georgii, H.-O. (1988). *Gibbs Measures and Phase Transitions* (Berlin, Walter de

Gruyter).

[178] Giraud, C. (2003). On a shock front in Burgers turbulence, *J. Stat. Phys.*, **111**, 1-2, pp. 387–402.

[179] Gnedenko, B. V. and Korolev, V. Yu. (1996). *Random summation. Limit theorems and applications* (Boca Raton, CRC Press).

[180] Ginibre, J. (1970). General formulation of Griffiths' inequalities, *Comm. Math. Phys.*, **16**, 4, pp. 310–328.

[181] Götze, F. (1991). On the rate of convergence in the multivariate CLT, *Ann. Probab.*, **19**, 2, pp. 724–739.

[182] Goldstein, L. (2005). Berry-Esseen bounds for combinatorial central limit theorems and pattern occurrences, using zero and size biasing, *J. Appl. Probab.*, **42**, 3, pp. 661–683.

[183] Griffiths, R. B. (1967). Correlations in Ising ferromagnets. II. External magnetic fields, *J. Math. Phys.*, **8**, 3, pp. 484–489.

[184] Grimmett, G. (1999). *Percolation* (Berlin, Springer).

[185] Grimmett, G. and Winkler, S. N. (2004). Negative association in uniform forests and connected graphs, *Random Struct. Algor.*, **24**, 4, pp. 444–460.

[186] Griniv, O. O. (1991). A central limit theorem for Burgers equation, *Theor. Math. Phys.*, **88**, 1, pp. 678–682.

[187] Gurbatov, S. N., Malakhov, A. N., Saichev, A. I. (1991). *Nonlinear random waves and turbulence in nondispersive media: waves, rays, particles*, Manchester, Manchester University Press,

[188] Guyon, X. and Richardson, S. (1984). Vitesse de convergence du théorème de la limite centrale pour des champs faiblement dépendants, *Z. Wahrsch. verw. Geb.*, **66**, 2, pp. 297–314.

[189] Hardy, G. H., Littlewood, J. E., Pólya, G. (1988). *Inequalities* (Cambridge University Press).

[190] Hargé, G. (2004). A convex/log-concave correlation inequality for Gaussian measure and an application to abstract Wiener spaces. *Probab. Theory Relat. Fields*, **130**, 3, pp. 415–440.

[191] Harris, T. E. (1960). A lower bound for the critical probability in a certain percolation process, *Proc. Cambr. Phil. Soc.*, **56**, 1, pp. 13–20.

[192] Harris, T. E. (1977). A correlation inequality for Markov processes in partialy ordered state spaces, *Ann. Probab.*, **5**, 1, pp. 451–454.

[193] Henriques, C. and Oliveira, P. E. (2002). Almost optimal convergence rates for kernel density estimation under association, *Pré-Publicações do Departamento de Matemática Universidade de Coimbra*, Preprint 04-06.

[194] Henriques, C. and Oliveira, P. E. (2003). Estimation of a two-dimensional distribution function under association, *J. Statist. Plann. Inf.*, **113**, 1, pp. 137–150.

[195] Henriques, C. and Oliveira, P. E. (2005). Exponential rates for kernel density estimation under association, *Statistica Neerlandica*, **59**, 4, pp. 448–466.

[196] Herbst, I. and Pitt, L. D. (1991). Diffusion equations technique in stochastic monotonicity and positive correlations, *Probab. Theory Rel. Fields*, **87**, 3, pp. 275–312.

[197] Herrndorf, N. (1984). An example on the central limit theorem for associated sequences, *Ann. Probab.*, **12**, 3, pp. 912–917.

[198] Hitczenko, P. (1994). Sharp inequality for randomly stopped sums of independent non-negative random variables, *Stoch. Proc. Appl.*, **51**, 1, pp. 63–73.

[199] Hjort, N., Natvig, B., Funnemark, R. (1985). The association in time of a Markov process with application to multistate reliability theory, *J. Appl. Probab.*, **22**, 2, pp. 473–479.

[200] Hoeffding, W. (1940). Masstabinvariante korrelationstheorie, *Schriften Math. Inst. Univ. Berlin*, **5**, pp. 179–233.

[201] den Hollander, W. Th. F. and Keane, M. (1986). Inequalities of FKG type, *Physica A*, **138**, 1-2, pp. 167–182.

[202] Holley, R. (1974). Remarks on the FKG inequalities, *Comm. Math. Phys.*, **36**, 3, pp. 227–231.

[203] Hu, T., Müller, A., Scarsini, M. (2004). Some counterexamples in positive dependence, *J. Statist. Plann. Inf.*, **124**, 1, pp. 153–158.

[204] Hu, T., Su, Ch., Liang, H. (2001). On the logarithm law for strictly stationary and negatively associated arrays, *Theory Probab. Appl.*, **46**, 2, pp. 369–379.

[205] Hu, T., Su, C., Yuan, M. (2003). A central limit theorem for random fields of negatively associated processes, *J. Theor. Probab.*, **16**, 2, pp. 309–323.

[206] Hu, T., Xie, C., Ruan, L. (2005). Dependence structures of multivariate Bernoulli random vectors. *J. Multivar. Anal.*, **94**, 1, pp. 172–195.

[207] Hu, T. and Yang, J. (2004). Further developments on sufficient conditions for negative dependence of random variables, *Statist. Probab. Lett.*, **66**, 3, pp. 369–381.

[208] Huang, W. (2003). A law of the iterated logarithm for geometrically weighted series of negatively associated random variables, *Statist. Probab. Lett.*, **63**, 2, pp. 133–143.

[209] Huang, W. and Zhang, L-X. (2006). Asymptotic normality for U-statistics of negatively associated random variables, *Statist. Probab. Lett.*, **76**, 11, pp. 1125–1131.

[210] Ibragimov, I. A. and Linnik, Yu. V. (1971). *Independent and Stationary Sequences of Random Variables* (Groningen, Wolters-Noordhoff).

[211] Ibragimov, I. A. (1996). On almost-everywhere variants of limit theorems, *Dokl. Math.*, **54**, 2, pp. 703–705.

[212] Ibragimov, I.A. and Lifshits, M.A. (1999). On almost sure limit theorems, *Theory Probab. Appl.*, **44**, 2, pp. 254–272.

[213] Ioannides, D. and Roussas, G. (1999). Exponential inequalities for associated random variables, *Statist. Probab. Lett.*, **42**, pp. 423–431.

[214] Iosifescu, M. (1968). The law of the iterated logarithm for a class of dependent random variables, *Theory Probab. Appl.*, **13**, 2, pp. 304–313.

[215] Itô and K., McKean, H. (1996). *Diffusion Processes and their Sample Paths* (Berlin, Springer-Verlag).

[216] Jacod, J. and Shiryaev, A. N. (2003). *Limit Theorems for Stochastic Processes, Grundlehren der Mathematischen Wissenschaften*, **288** (Berlin, Springer).

[217] Jing, B. and Liang, H. (2005). Asymptotic properties for estimates of nonparametric regression models based on negatively associated sequences, *J. Multivar. Anal.*, **95**, 2, pp. 227–245.

[218] Joag-Dev, K., Perlman, M. D., Pitt, L. D. (1983). Association of normal random variables and Slepian's inequality, *Ann. Probab.*, **11**, 2, pp. 451–455.

[219] Joag-Dev, K. and Proschan, F. (1983). Negative association of random variables, with applications, *Ann. Statist.*, **11**, 1, pp. 286–295.

[220] Johnson, O. (2006). An information-theoretic central limit theorem for finitely susceptible FKG systems, *Theory Probab. Appl.*, **50**, 2, pp. 214–224.

[221] Johnson, O. (2006). A central limit theorem for non-overlapping return times, *J. Appl. Probab.*, **43**, 1, pp. 32–47.

[222] Johnson, O. and Suhov, Yu. (2001). Entropy and random vectors, *J. Stat. Phys.*, **104**, 1-2, pp. 145–195.

[223] Kamae, T., Krengel, U., O'Brien, G. L. (1977). Stochastic inequalities on partially ordered spaces, *Ann. Probab.*, **5**, 6, pp. 899–912.

[224] Kallenberg, O. (2002). *Foundations of Modern Probability* (New York, Springer).

[225] Karatzas, I. and Shreeve, S. (1997). *Brownian Motion and Stochastic Calculus*, New York, Springer-Verlag,

[226] Karlin, S. and Rinott, Y. (1980). Classes of orderings of measures and related correlation inequalities. I, *J. Multivar. Anal.*, **10**, 4, pp. 467–498.

[227] Karlin, S. and Rinott, Y. (1980). Classes of orderings of measures and related correlation inequalities. II. Multivariate reverse rule distributions, *J. Multivar. Anal.*, **10**, 4, pp. 499–516.

[228] Khanin, K. M. and Sinai, Ya. G. (1997). Hyperbolicity of minimizing trajectories for two-dimensional Hamiltonian systems with random forcing, *Proc. Steklov Inst. Math.*, **216**, pp. 169–173.

[229] Khasminskii, R. Z. (1960). Ergodic properties of recurrent diffusion processes and stabilization of the solution of the Cauchy problem for parabolic equation, *Theory Probab. Appl.*, **5**, pp. 179–195.

[230] Khoshnevisan, D. and Lewis, T. (1998). A law of the iterated logarithm for stable processes in random scenery, *Stoch. Proc. Appl.*, **74**, 1, pp. 89–121.

[231] Khoshnevisan, D. (2002). *Multiparameter Processes. An Introduction to Random Fields* (New York, Springer).

[232] Kim, T.-S. and Ko, M.-H. (2002). A functional central limit theorem for associated random fields, *Honam Math. J.*, **24**, 1, pp. 121–130.

[233] Kim, T.-S. and Ko, M.-H. (2003). On a functional central limit theorem for stationary linear processes generated by associated processes, *Bull. Korean Math. Soc.*, **40**, 4, pp. 715–772.

[234] Kim, T.-S. and Ko, M.-H. (2005). Almost sure convergence for weighted sums of negatively orthant dependent random variables, *J. Korean Math. Soc.*, **42**, 5, pp. 949–957.

[235] Kimball, A. W. (1951). On dependent tests of significance in the analysis of variance, *Ann. Math. Statist.*, **22**, 4, pp. 600–602.

[236] Klesov, O. I. (1998). A new method for the strong law of large numbers for random fields, *Theory Stoch. Proc.*, **4(20)**, 1-2, pp. 122–128.

[237] Klesov, O. I. (2001). The existence of moments of the suprema of multiple sums and the strong law of large numbers. Theory *Probab. Math. Stat.*, **62**, 1, pp. 27–37.

[238] Klesov, O. I, Rosalsky, A., Volodin, A. I. (2005). On the almost sure growth rate of sums of lower negatively dependent nonnegative random variables, *Stat. Probab. Lett.*, **71**, 2, pp. 193–202.

[239] Kolmogorov, A. N. and Fomin, S. V. (1970). *Introductory Real Analysis*, (Englewood Cliffs, N. J., Prentice-Hall, Inc.).

[240] Kómlos, J., Major, P., Tusnady, G. (1975). An approximation of partial sums of independent random variables and the sample distribution function I, *Z. Wahrsch. verw. Geb.*, **32**, 1-2, pp. 111–131.

[241] Kómlos, J., Major, P., Tusnady, G. (1975). An approximation of partial sums of independent random variables and the sample distribution function II, *Z. Wahrsch. verw. Geb.*, **34**, 1, pp. 33–58.

[242] Kryzhanovskaya, N. Yu. (2007). Moment inequalities for sums of dependent random variables over finite sets. In preparation.

[243] Kuczmaszewska, A. (2005). The strong law of large numbers for dependent random variables, *Statist. Probab. Lett.*, **73**, 3, pp. 305–314.

[244] Kulik, R. (2007). Rate of convergence in invariance principle for associated sequences, *Ann. Probab.*, to appeear.

[245] Kulik, R. and Szekli, R. (2001). Sufficient conditions for long-range count dependence of stationary point processes on the real line, *J. Appl. Probab.*, **38**, pp. 2,

570–581.

[246] Kwieciński, A. and Szekli, R. (1996). Some monotonicity and dependence properties of self-exciting point processes, *Ann. Appl. Probab.*, **6**, 4, pp. 1211–1231.

[247] Lamperti, J. W. (1977). *Stochastic processes. A survey of the mathematical theory*, (New York–Heidelberg–Berlin, Springer).

[248] Lamperti, J. W. (1996). *Probability. A Survey of the Mathematical Theory* (New York, Wiley).

[249] Lebedev, A.V. (2002). Extremes of subexponential shot noise, *Math. Notes*, **71**, 2, pp. 206–210.

[250] Lebowitz, J. L. (1971). Griffiths inequalities for anti-ferromagnets, *Physics Lett.* **36A**, 2, pp. 99–100.

[251] Lebowitz, J. L. (1972). Bounds on the correlations and analiticity properties of ferromagnetic Ising spin systems, *Comm. Math. Phys.*, **28**, pp. 313–321.

[252] Lebowitz, J. L. and Penrose, O. (1968). Analytic and clustering properties of thermodynamic functions and distribution functions for classical lattice and continuum systems, *Commun. Math. Phys.*, **11**, 2, pp. 99–124.

[253] Ledoux, M. and Talagrand, M. (1991). *Probability in Banach Spaces. Isoperimetry and Processes* (Berlin, Springer).

[254] Lee, M.-L. T., Rachev, S. T., Samorodnitsky, G. (1990). Association of stable random variables, *Ann. Probab.*, **18**, 4, pp. 1759–1764.

[255] Lefèvre, C. and Utev, S. (2003). Exact norms of a Stein-type operator and associated stochastic orderings, *Probab. Theory Rel. Fields*, **127**, 3, pp. 353–366.

[256] Lehmann, E. L. (1996). Some concepts of dependence, *Ann. Math. Statist.*, **37**, 5, pp. 1137–1153.

[257] Leonenko, N. N. (1999). *Limit Theorems for Random Fields with Singular Spectrum, Mathematics and its Applications (Dordrecht)*, **465** (Dordrecht, Kluwer Academic Publishers).

[258] Leonenko, N. N. and Woyczynski, W. A. (1998). Exact parabolic asymptotics for singular n-D Burgers' random fields: Gaussian approximation, *Stoch. Proc. Appl.*, **76**, 2, pp. 141–165.

[259] Lewis, T. (1998). Limit theorems for partial sums of quasi-associated random variables. In: B.Szysszkowicz (ed.), *Asymptotic methods in probability and statistics*, Elsevier, pp. 31–48.

[260] Lewis, T. and Pritchard, G. (1999). Tail properties of correlation measures. *J. Theor. Probab.*, **16**, 3, pp. 771–788.

[261] Li, H. (2003). Association of multivariate phase-type distributions, with applications to shock models, *Statist. Probab. Lett.*, **64**, 4, pp. 381–392.

[262] Li, W. V. and Shao, Q.-M. (2002). A normal comparison inequality and its applications, *Probab. Theory Rel. Fields*, **122**, 4, pp. 494–508.

[263] Li, Y.-X. and Zhang, L.-X. (2004). Complete moment convergence of moving-average processes under dependence assumptions, *Statist. Probab. Lett.*, **70**, 3, pp. 191–197.

[264] Liang, H.-Y. and Baek, J.-I. (2004). Asymptotic normality of recursive density estimates under some dependence assumptions, *Metrika*, **60**, 2, pp. 155–166.

[265] Liggett, T. M. (2005). *Interacting Particle Systems* (New York, Springer).

[266] Liggett, T. M. (2006). Conditional association and spin systems, *Lat. Am. J. Probab. Math. Stat.*, **1**, 1, pp. 1–19.

[267] Liggett, T. M. and Steif, J. E. (2007). Stochastic domination: The contact process, Ising models and FKG measures, *Ann. Inst. H.Poincaré, Probab. Stat.*, **42**, 2, pp. 223–243.

[268] Lindqvist, B. H. (1987). Monotone and associated Markov chains, with applications

to reliability theory, *J. Appl. Probab.*, **24**, 3, pp. 679–695.

[269] Lindqvist, B. H. (1988). Association of probability measures on partially ordered spaces, *J. Multivar. Anal.*, **26**, 2, pp. 111–132.

[270] Linnik, Yu. V. (1960). An information-theoretic proof of the central limit theorem under the Lindeberg conditions, *Theory Probab. Appl.*, **4**, 3, pp. 288–299.

[271] Lipšic, I. M., Gredeskul, S. A., Pastur, L. A. (1982). *An Introduction to the Disordered System Theory* (Moscow, Nauka) (in Russian).

[272] Loéve, M. (1963). *Probability Theory* (Princeton–Toronto–New York–London, Van Nostrand Company, Inc.).

[273] Louhichi, S. (2000). Weak convergence for empirical processes of associated sequences, *Ann. Inst. H. Poincaré, Probabilites et Statistiques*, **36**, 5, pp. 547–567.

[274] Louhichi, S. (1999). Rosenthal's inequality for LPQD sequences, *Statist. Probab. Lett.*, **42**, 2, pp. 139–144.

[275] Louhichi, S. (2000). Convergence rates in the strong law of large numbers for associated random variables, *Probab. Math. Stat.*, **20**, 1, pp. 203–214.

[276] Louhichi, S. (2001). Rates of convergence in the CLT theorem for some weakly dependent random variables, *Theory Probab. Appl.*, **46**, 2, pp. 297–315.

[277] Louhichi, S. (2002). Moment inequalities for sums of certain dependent random variables, *Theory Probab. Appl.*, **47**, pp. 747–763.

[278] Louhichi, S. and Louhichi, S. (2005). Some multivariate inequalities with applications, *Periodica Math. Hung.*, **51**, 2, pp. 37–58.

[279] Louhichi, S. and Soulier, P. (2002). The central limit theorem for stationary associated sequences, *Acta Math. Hung.*, **97**, 1-2, pp. 15–36.

[280] Lu, C. (2003). The invariance principle for linear processes generated by a negatively associated sequence and its applications, *Acta Math. Appl. Sinica.*, **19**, 4, pp. 641–646.

[281] Lyons, R. (2003). Determinantal probability measures, *Publ. Math., Inst. Hautes Étud. Sci.*, **98**, 1, pp. 167–212.

[282] Marshall, A. W. and Olkin, I. (1983). Domains of attraction of multivariate extreme value distributions, *Ann. Probab.*, **11**, 1, pp. 168–177.

[283] Martikajnen, A. I. (1980). A converse to the law of the iterated logarithm for a random walk, *Theory Probab. Appl.*, **25**, 2, pp. 361–362.

[284] Masry, E. (2002). Multivariate probability density estimation for associated processes: strong consistency and rates, *Statist. Probab. Lett.*, **58**, 2, pp. 205–219.

[285] Masry, E. (2003). Local polynomial fitting under association, *J. Multivar. Anal.*, **86**, 2, pp. 330–359.

[286] Matula, P. (1992). A note on almost sure convergence of sums of negatively dependent random variables, *Statist. Probab. Lett.*, **15**, 3, pp. 209–213.

[287] Matula, P. (1996). A remark on the weak convergence of sums of associated random variables, *Ann. Univ. Mariae Curie-Sklodowska, Sect. A*, **50**, 1, pp. 115–123.

[288] Matula, P. (1996). Convergence of weighted averages of associated random variables, *Probab. Math. Stat.*, **16**, 2, pp. 337–343.

[289] Matula, P. (1996). Probability and moment bounds for sums of negatively associated random variables, *Theory Probab. Math. Stat.*, **55**, pp. 135–141.

[290] Matula, P. (1997). The Glivenko-Cantelli lemma for a class of discrete associated random variables, *Ann. Univ. Mariae Curie-Sklodowska, Sect. A*, **51**, 1, pp. 129–132.

[291] Matula, P. (1998). On the almost sure central limit theorem for associated random variables, *Probab. Math. Stat.*, **18**, 2, pp. 411–416.

[292] Matula, P. (1999). A note on mixing properties of certain associated processes, *Acta*

Math. Hung., **82**, 1-2, pp. 107–112.

[293] Matula, P. (2001). Limit theorems for sums of nonmonotonic functions of associated random variables, *J. Math. Sci.*, **105**, 6, pp. 2590–2593.

[294] Matula, P. (2005). On almost sure limit theorems for positively dependent random variables, *Statist. Probab. Lett.*, **74**, 1, pp. 59–66.

[295] Matula, P. and Rychlik, Z. (1990). Distribution and moment convergence of sums of associated random variables, *Math. Pannonica*, **1**, 2, pp. 117–123.

[296] Matula, P. and Rychlik, Z. (1990). The invariance principle for nonstationary sequences of associated random variables, *Ann. Inst. H. Poincaré, Probab. Statist.*, **26**, 3, pp. 387–397.

[297] Mikusheva, A. E. (2000). On the complete convergence of sums of negatively associated random variables, *Math. Notes*, **68**, 3, pp. 355–362.

[298] Mikusheva, A. E. (2000). The law of large numbers and the logarithmic law for arrays, *Fundam. Prikl. Mat.*, **6**, 1, pp. 195–206 (in Russian).

[299] Mikusheva, A. E. (2001). An analog of the Baum-Katz theorem for negatively associated random fields, *Mosc. Univ. Math. Bull.*, **56**, 3, pp. 30–35.

[300] Mikusheva, A. E. (2001). The complete convergence and limit theorems in the series scheme. PhD thesis, Moscow,

[301] Millionshchikov, N. V. (2006). The almost sure convergence of the kernel density estimates for weakly dependent random fields, *Russ. Math. Surv.*, **61**, 1, pp. 176–178.

[302] Mishura, Yu. (2000). Skorokhod space and Skorokhod topology in probabilistic considerations during 1956–1999. In: V.Korolyuk et al (eds), *Skorokhod's ideas in probability theory*, Kyiv, Proceedings of Institute of Mathematics of National Academy of Science of Ukraine, Math. Appl. **32**, pp. 281–297.

[303] Molchanov, S. A., Surgailis, D., Woyczynski, W. A. (1997). The large-scale structure of the Universe and quasi-Voronoi tesselation of shock fronts in forced Burgers turbulence in $\mathbb{R}^d$. *Ann. Appl. Probab.*, **7**, 1, pp. 200–228.

[304] Móricz, F. (1983). A general moment inequality for the maximum of the rectangular partial sums of multiple series, *Acta Math. Hung.*, **41**, 3-4, pp. 337–346.

[305] Müller and A., Scarsini, M. (2000). Some remarks on the supermodular order, *J. Multivar. Anal.*, **73**, 1, pp. 107–119.

[306] Nahapetyan, B. S. and Pogosyan, S. K. (1995). Decay of correlations in classical lattice spin systems with vacuum, *J. Contemp. Math. Anal., Armen. Acad. Sci.*, **30**, 6, pp. 29–49.

[307] Newman, C. M. (1980). Normal fluctuations and the FKG inequalities, *Comm. Math. Phys.*, **74**, 2, pp. 119–128.

[308] Newman, C. M. (1984). Asymptotic independence and limit theorems for positively and negatively dependent random variables. In: *Inequalities in Statist. and Probab.* (Y. L.Tong ed.), Hayward, pp. 127–140.

[309] Newman, C. M and Wright, A. L. (1981). An invariance principle for certain dependent sequences, *Ann. Probab.*, **9**, 4, pp. 671–675.

[310] Newman, C. M and Wright, A. L. (1982). Associated random variables and martingale inequalities, *Prob. Th. Rel. Fields*, **59**, 3, pp. 361–171.

[311] Øksendal, B. (2003). *Stochastic Differential Equations. An Introduction with Applications* (Berlin, Springer).

[312] Oliveira, P. E. (2005). An exponential inequality for associated variables, *Statist. Probab. Lett.*, **73**, 2, pp. 189–197.

[313] Oliveira, P. E. and Suquet, Ch. (1996). An $L^2[0,1]$ invariance principle for LPQD random variables, *Portugaliae mathematica*, **53**, 2, pp. 367–379.

[314] Pan, J. M. (1997). On the convergence rate in the central limit theorem for negatively associated sequences. Chinese *J. Appl. Probab. Stat.*, **13**, 2, pp. 183–192.

[315] Park, W. J. (1974). On Strassen's version of the law of the iterated logarithm for the two-parameter Gaussian process, *J. Multivar. Anal.*, **4**, 4, pp. 479–485.

[316] Parzen, E. (1962). On estimation of a probability density function and mode, *Ann. Math. Statist.*, **33**, 3, pp. 1065–1076.

[317] Peligrad, M. (1998). Maximum of partial sums and an invariance principle for a class of weak dependent random variables, *Proc. Amer. Math. Soc.*, **126**, 4, pp. 1181–1189.

[318] Peligrad, M. (1998). On the blockwise bootstrap for empirical processes for stationary sequences, *Ann. Probab.*, **26**, 2, pp. 877–901.

[319] Peligrad, M. (2002). Some remarks on coupling of dependent random variables, *Stat. Probab. Lett.*, **60**, 2, pp. 201–209.

[320] Peligrad, M. and Shao, Q.-M. (1994). Self-normalized central limit theorem for sums of weakly dependent random variables, *J. Theor. Probab.*, **7**, 2, pp. 309–338.

[321] Peligrad, M. and Shao, Q.-M. (1995). A note on the almost sure central limit theorem for weakly dependent random variables, *Stat. Probab. Lett.*, **22**, 2, pp. 131–136.

[322] Peligrad, M. and Shao, Q.-M. (1996). A note on estimation of variance for ρ-mixing sequences, *Stat. Probab. Lett.*, **26**, 2, pp. 141–145.

[323] Perera, G. (2002). Irregular sets and central limit theorems, emphBernoulli, **8**, 5, pp. 627–642.

[324] Peligrad, M. and Utev, S. (1997). Central limit theorem for linear processes, *Ann. Probab.*, **25**, 1, pp. 443–456.

[325] Pemantle, R. (2000). Towards a theory of negative dependence, *J. Math. Phys.*, **41**, 3, pp. 1371–1390.

[326] Petrov, V. V. (1995). *Limit Theorems of Probability Theory. Sequences of Independent Random Variables* (Oxford, Clarendon Press).

[327] de la Peña and V. H., Giné, E. (1999). *Decoupling: From Dependence to Independence* (New York, Springer).

[328] Philipp, W. and Stout, W. (1975). Almost Sure Invariance Principles for Partial Sums of Weakly Dependent Random Variables, *Mem. Amer. Math. Soc.*, **161**, pp. 1–140.

[329] Pitt, L. D. (1977). A Gaussian correlation inequality for symmetric convex sets, *Ann. Probab.*, **5**, 3, pp. 470–474.

[330] Pitt, L. D. (1982). Positively correlated normal variables are associated, *Ann. Probab.*, **10**, 2, pp. 496–499.

[331] Plackett, R. L. (1954). A reduction formula for normal multivariate integrals, *Biometrika*, **41**, 3-4, pp. 351–360.

[332] Pollard D. (1984). *Convergence of Stochastic Processes* (New York, Springer).

[333] Prakasa Rao, B. L. S. Bernstein-type inequality for associated sequences. In: Ghosh, J. K. et al (eds.), *Statistics and Probability: A Raghu Raj Bahadur Festschrift*, Wiley, pp. 499–509.

[334] Preston, C. J. (1974). A generalization of FKG inequalities, *Comm. Math. Phys.*, **36**, 3, pp. 232–241.

[335] Preston, C. J. (1974). *Gibbs States on Countable Sets* (Cambridge University Press).

[336] Qi, Y.-C. (1994). On strong convergence of arrays, *Bull. Aust. Math. Soc.*, **50**, 2, pp. 219–223.

[337] Qi, Y.-C. (1999). On the one-sided logarithmic law for arrays, *Systems Sci. Math. Sci.*, **12**, 2, pp. 123–132.

[338] Rachev, S. T. (1991). *Probability metrics and the stability of stochastic models*

(Chichester, Wiley).

[339] Rachev, S. T. and Xin, H. (1993). Test on association of random variables in the domain of attraction of a multivariate stable law, *Prob. Math. Stat.*, **14**, 1, pp. 125–141.

[340] Raic, M. (2004). A multivariate CLT for decomposable random vectors with finite second moments, *J. Theor. Probab.*, **17**, 3, pp. 573–603.

[341] Resnick, S. (1988). Association and multivariate extreme value distributions. In: *Studies in statistical Modelling and statistical science* (C.C.Heyde ed.), Statistical Society of Australia.

[342] Resnick, S. (1988). Association and multivariate extreme value distributions, *Austral. J. Statist.*, **30A**, pp. 261–271.

[343] Revuz, D. and Yor, M. (1999). *Continuous martingales and Brownian motion*, Graduate Texts in Mathematics **293** (Berlin, Springer).

[344] Richards, D. S. (2004). Algebraic methods towards higher-order probability inequalities. II, *Ann. Probab.*, **32**, 2, pp. 1509–1544.

[345] Rinott, Y. and Rotar, V. (2003). On Edgeworth expansions for dependency-neighborhoods chain structures and Stein's method. *Probab. Theory Rel. Fields*, **126**, 4, pp. 528–570.

[346] Rinott, Y. and Scarsini, M. (2006). Total positivity order and the normal distribution, *J. Multivar. Anal.*, **97**, 5, pp. 1251–1261.

[347] Rogers, L. C. G. and Williams, D. (1979, 1987). *Diffusion. Markov processes and Martingales*, Chichester, Wiley.

[348] Rosenblatt, M. (1956). Remarks on some nonparametric estimates of a density function, *Ann. Math. Statist.*, **27**, 3, pp. 832–837.

[349] Rosenthal, H. P. (1970). On the subspaces of L^p ($p > 2$) spanned by sequences of independent random variables, *Israel J. Math.*, **8**, 3, pp. 273–303.

[350] Roussas, G. G. (1991). Kernel estimates under association: strong uniform consistency, *Statist. Probab. Lett.*, **12**, 5, pp. 393–403.

[351] Roussas, G. G. (1993). Curve estimation in random fields of associated processes, *J. Nonparametric Statist.*, **2**, 3, pp. 215–224.

[352] Roussas, G. G. (1994). Asymptotic normality of random fields of positively and negatively associated processes, *J. Multivar. Anal.*, **50**, 1, pp. 152–173.

[353] Roussas, G. G. (1995). Asymptotic normality of a smooth estimate of a random field distribution function under association, *Statist. Probab. Lett.*, **24**, 1, pp. 77–90.

[354] Roussas, G. G. (1999). Positive and negative dependence with some statistical applications. In: Asymptotics, nonparametrics, and time series, *A tribute to Madan Lal Puri*, (S.Ghosh ed.), New York, Marcel Dekker, Textb. Monogr. 158, pp. 757–788.

[355] Roussas, G. G. (2000). Prediction under association. In: *Recent Advances in Reliability Theory. Methodology, Practice and Inference* (Limnios et al eds.), Boston, Birkhauser, pp. 451–475.

[356] Roussas, G. G. (2000). Asymptotic normality of the kernel estimate of a probability density function under association, *Statist. Probab. Lett.*, **50**, 1, pp. 1–12.

[357] Roussas, G. G. (2001). An Esseen-type inequality for probability density functions, with an application, *Statist. Probab. Lett.*, **51**, 4, pp. 397–408 (see also ibid. **54**, p. 449).

[358] Royer G. (1999) *Une initiation aux inégalités de Sobolev logarithmiques*, Société Mathématique de France, Cours spécialisé, **5**.

[359] Ruelle, D. (1999). *Statistical Mechanics. Rigorous Results*, Singapore, World Scientific; London, Imperial College Press.

[360] Sakhanenko, A. I. (1985). Estimates in the invariance principle (in Russian), *Pre-*

del'nye Teoremy Teorii Veroyatnostej, Tr. Inst. Mat., **5**, pp. 27–44.

[361] Samorodnitsky, G. (1988). Extrema of skewed stable processes, *Stoch. Proc. Appl.*, **30**, 1, pp. 17–39.

[362] Samorodnitsky, G. (1995). Association of infinitely divisible random vectors, *Stoch. Proc. Appl.*, **55**, 1, pp. 45–55.

[363] Sarkar, T. K. (1969). Some lower boundes of reliability. Tech. Rep. 124, Dept. of Operation research and statistics, Stanford University.

[364] Schatte, P. (1988). On strong versions of the central limit theorem, *Math. Nachr.*, **137**, pp. 249–256.

[365] Schwarz, G. (1980). Finitely determined processes - an indiscrete approach, *J. Math. Anal. Appl.*, **76**, pp. 146–158.

[366] Seneta, E. (1976). *Regularly Varying Functions, Lecture Notes Math.* **508** (Berlin–Heidelberg–New York, Springer).

[367] Seymour, P. D. and Welsh, D. J. A. (1975). Combinatorial applications of an inequality from statistical mechanics. *Proc. Cambr. Phil. Soc.*, **77**, 3, pp. 485–495.

[368] Shaked, M. and Shantikumar, J. G. (1997). Supermodular stochastic orders and positive dependence of random vectors, *J. Multivar. Anal.*, **61**, 1, pp. 86–101.

[369] Shandarin, S. F. and Zeldovich, Ya. B. (1989). The large-scale structure of the universe: Turbulence, intermittency, structures in a self-gravitating medium, *Rev. Mod. Phys.*, **61**, 2, pp. 185–220.

[370] Shao, Q.-M. (1989). A Berry-Esseen inequality and an invariance principle for associated random fields. *Chinese J. Appl. Probab. Statist.*, **5**, 1, pp. 1–8.

[371] Shao, Q.-M. (2000). A comparison theorem on moment inequalities between negatively associated and independent random variables, *J. Theor. Probab.*, **13**, 2, pp. 343–356.

[372] Shao, Q.-M. and Su, C. (1999). The law of iterated logarithm for negatively associated random variables, *Stoch. Proc. Appl.*, **83**, 1, pp. 139–148.

[373] Shao, Q.-M. and Yu, H. (1996). Weak convergence for weighted empirical processes of dependent sequences, *Ann. Probab.*, **24**, 4, pp. 2098–2127.

[374] Shashkin, A. P. (2002). Quasi-associatedness of a Gaussian system of random vectors, *Russ. Math. Surv.*, **57**, 6, pp. 1243–1244.

[375] Shashkin, A. P. (2004). Maximal inequality for weakly dependent random fields, *Math. Notes*, **75**, 5, pp. 717–725.

[376] Shashkin, A. P. (2004). Invariance principle for a class of weakly dependent random fields, *Mosc. Univ. Math. Bull.*, **59**, 4, pp. 24–29.

[377] Shashkin, A. P. (2004). A dependence property of a spin system. In: *Transactions of XXIV International Seminar on Stability Problems for Stochastic Models*, Yurmala, Latvia, pp. 30–35.

[378] Shashkin, A. P. (2005). On Newman's central limit theorem, *Theory Probab. Appl.*, **50**, 2, pp. 330–337.

[379] Shashkin, A. P. (2005). Some limit theorems for weakly dependent random fields. PhD thesis, Moscow,

[380] Shashkin, A. P. (2005). A variant of the LIL for multiparameter Brownian motion. In: *Transactions of XXV International Seminar on Stability Problems for Stochastic Models*, Maiori, Italy, pp. 258–263.

[381] Shashkin, A. P. (2006). The law of the iterated logarithm for an associated random field, *Usp. Matem. Nauk*, **61**, pp. 173–174 (in Russian).

[382] Shimizu, R. (1975). On Fisher's amount of information for location family. In: Patil, G. P. et al. (eds.), *Statistical Distributions in Scientific Work*, (Reidel), **3**, pp. 305–312.

[383] Shiryaev, A. N. (1995). *Probability*, Graduate Texts in Mathematics, **95** (New York, Springer),

[384] Shiryaev, A. N. (1999). *Essentials of Stochastic Finance*, **1,2** (Singapore, World Scientific).

[385] Simon, B. (1973). Correlation inequalities and the mass gap in $P(\phi)_2$. I. Domination by the two point function, *Comm. Math. Phys.*, **31**, 2, pp. 127–136.

[386] Simon, B. (1980). Correlation inequalities and the decay of correlation in ferromagnets, *Comm. Math. Phys.*, **77**, 2, pp. 111–126.

[387] Skorokhod, A. V. (1965). *Studies in the Theory of Random Processes* (Reading Mass., Addison–Wesley).

[388] Stein, C. (1972). A bound for the error in the normal approximation to the distribution of a sum of dependent random variables, *Proc. 6th Berkeley Symposium Math. Statist. Probab.*, Univ. Calif., **2**, pp. 583–602.

[389] Stein, C. (1986). Approximate Computation of Expectations, *IMS Lecture Notes*, **7**. Inst. Math. Statist., Hayward, CA.

[390] Strassen, V. (1964). An invariance principle for the law of the iterated logarithm, *Z. Wahrsch. verw. Geb.*, **3**, 3, pp. 211–226.

[391] Strassen, V. (1965). The existence of probability measures with given marginals, *Ann. Math. Stat.*, **36**, 2, pp. 423–439.

[392] Strassen, V. (1966). A converse to the law of the iterated logarithm, *Z. Wahrsch. verw. Geb.*, **4**, 4, pp. 265–268.

[393] Strassen, V. (1967). Almost sure behavior of sums of independent random variables and martingales, *Proc. 5th Berkeley Symp. Math. Stat. Probab.*, Univ. Calif. 1965/66, 2, Part 1, pp. 315–343.

[394] Stroock, D. and Varadhan, S. R. S. (2006). *Multidimensional diffusion processes* (Berlin, Springer).

[395] Su, C., Zhao, L. C., Wang, Y. B. (1997). Moment inequalities and weak convergence for negatively associated sequences, *Sci. China Ser. A*, **40**, 2, pp. 172–182.

[396] Sunklodas, J. (1995). On the global central limit theorem for φ-mixing random variables, *Lith. Math. J.*, 2, **35**, pp. 185–196.

[397] Talagrand, M. (1994). Some remarks on the Berg-Kesten inequality, *Probab. in Banach Spaces*, **9** (Sandiberg, 1993), Prog. Probab. 35, Birkhäuser, Boston. pp. 293–297.

[398] Tang, Q. (2006). Insensitivity to negative dependence of the asymptotic behavior of precise large deviations, *Electron. J. Probab.*, **11**, 4, pp. 107–120.

[399] Tang, Q. and Wang, D. (2004). Maxima of sums and random sums for negatively associated random variables with heavy tails, *Statist. Probab. Lett.*, **68**, 3, pp. 287–295.

[400] Thomas, L. E. (1980). Stochastic coupling and thermodynamic inequalities, *Comm. Math. Phys.*, **77**, 3, pp. 211–218.

[401] Tikhomirov, A. N. (1981). On the convergence rate in the central limit theorem for weakly dependent random variables, *Theory Probab. Appl.*, **25**, 4, pp. 790-809.

[402] Volkonskij, V. A. and Rozanov, Yu. A. (1961). Some limit theorems for random functions, *Theory Probab. Appl.*, **6**, 2, pp. 186–198.

[403] Volodin, A. I. (1993). Applications of the weak ℓ_p exponential inequalities to the laws of large numbers for weighted sums, *Probab. Math. Stat.*, **14**, 1, pp. 25–32.

[404] Vronskii, M. A. (1998). Rate of convergence in the SLLN for associated sequences and fields, *Theory Probab. Appl.*, **43**, 3, pp. 449–462.

[405] Vronski, M. A. (1998). Some limit theorems for associated random fields. PhD Thesis, Moscow.

[406] Vronskii, M. A. (2000). Refinement of the almost sure central limit theorem for associated processes, *Math. Notes*, **68**, 4, pp. 444–451.

[407] Vronskii, M. A. (2000). On the approximate confidence intervals for the unknown mean of stationary associated random field. *Fundam. Prikl. Mat.*, **6**, 1, pp. 63–71 (in Russian).

[408] Wang, J. (2004). Maximal inequalities for associated random variables and demi-martingales, *Statist. Probab. Lett.*, **66**, 3, pp. 347–354.

[409] Wang, J. and Zhang, L.-X. (2006). A Berry–Esseen theorem for weakly negatively dependent random variables and its applications, *Acta Math. Hung.*, **110**, 4, pp. 293–308.

[410] Wang, J. and Zhang, L.-X. (2006). A nonclassical law of the iterated logarithm for functions of positively associated random variables, *Metrika*, **64**, 3, pp. 361–378.

[411] Wang, W. S. (2005). A strong approximation theorem for quasi-associated sequences, *Acta Math. Sinica*, **21**, 6, pp. 1269–1276.

[412] Wentzell, A. D. (1981). *A Course in the Theory of Stochastic Processes* (New York, McGraw-Hill International Book Company).

[413] Weron, A. (1984). *Stable Processes and Measures: a survey. Probability theory on vector spaces III, Lecture Notes Math.* **1080**, pp. 306–364 (Berlin, Springer).

[414] Wichura, M. (1969). Inequalities with applications to the weak convergence of random processes with multi-dimensional time parameters, *Ann. Math. Statist.*, **40**, 2, pp. 681–687.

[415] Wichura, M. (1974). On the functional form of the law of the iterated logarithm for the partial maxima of independent identically distributed random variables, *Ann. Probab.*, **2**, 2, pp. 202–230.

[416] Wilks, S. (1946). *Mathematical Statistics*, Princeton,

[417] Winkel, M. (2002). Limit clusters in the inviscid Burgers turbulence with certain random initial velocities, *J. Stat. Phys.*, **107**, 3-4, pp. 893–917.

[418] Withers, C. S. (1981). Conditions for linear processes to be strong-mixing, *Z. Wahrsch. verw. Geb.*, **57**, 4, pp. 477–480.

[419] Wood, T.E. (1985). A local limit theorem for associated sequences, *Ann. Probab.*, **13**, 2, pp. 625–629.

[420] Woyczynski, W. A. (1998). *Burgers-KPZ Turbulence. Göttingen Lectures, Lecture Notes in Mathematics* **1700** (Berlin, Springer).

[421] Yakymiv, A. L. (2005). *Probabilistic Applications of Tauberian Theorems* (Moscow, Fizmatlit) (in Russian).

[422] Yu, H. (1986). The law of the iterated logarithm for associated random variables, *Acta Math. Sinica*, **29**, pp. 507–511 (in Chinese).

[423] Yu, H. (1993). A Glivenko-Cantelli lemma and weak convergence for empirical processes of associated sequences, *Probab. Th. Rel. Fields*, **95**, 3, pp. 357–370.

[424] Yu, H. (1996). A strong invariance principle for associated sequences, *Ann. Probab.*, **24**, 4, pp. 2079–2097.

[425] Zaitsev, A. Yu. (2001). Multidimensional version of a result of Sakhanenko in the invariance principle for vectors with finite exponential moments. III, *Theory Probab. Appl.*, **46**, 4, pp. 676–698.

[426] Zhang, J. (2006). Empirical likelihood for NA series, *Statist. Probab. Lett.*, **76**, 2, pp. 153–160.

[427] Zhang, L.-X. (2000). Central limit theorems for asymptotically negatively associated random fields, *Acta Math. Sin., Engl. Ser.*, **16**, 4, pp. 691–710.

[428] Zhang, L.-X. (2001). Strassen's law of the iterated logarithm for negatively associated random vectors, *Stoch. Proc. Appl.*, **95**, 2, pp. 311–328.

[429] Zhang, L.-X. (2001). The weak convergence for functions of negatively associated random variables, *J. Multivar. Anal.*, **78**, 2, pp. 272–298.

[430] Zhang, L.-X. (2006). Maximal inequalities and a law of the iterated logarithm for negatively associated random fields, `arXiv:math.PR/0610511 v1 17 Oct 2006`

[431] Zhang, L.-X. and Wen, J. (2001). A weak convergence for negatively associated fields, *Statist. Probab. Lett.*, **53**, 3, pp. 259–267.

[432] Zhao, Y. (2007). Precise rates in log laws for NA sequences. In: *Discrete Dynamics in Nature and Society,* **2007**, article ID 89107.

[433] Zimmerman, G. J. (1972). Some sample function properties of the two-parameter Gaussian process, *Ann. Math. Statist.*, **43**, 4, pp. 1235–1246.

[434] Zolotarev, V. M. (1997). *Modern theory of summation of random variables* (Utrecht, VSP).

Notation Index

Subject Index